Basic Mathematics

9th Edition

Basic Mathematics

9th EDITION

Marvin L. Bittinger

Indiana University Purdue University Indianapolis

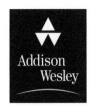

Addison
Wesley

Boston   San Francisco   New York
London   Toronto   Sydney   Singapore   Madrid
Mexico City   Munich   Paris   Cape Town   Hong Kong   Montreal

Publisher	Greg Tobin
Editor in Chief	Maureen O'Connor
Acquisitions Editor	Jennifer Crum
Project Manager	Kari Heen
Associate Editor	Lauren Morse
Editorial Assistant	Katie Nopper
Managing Editor	Ron Hampton
Production Supervisor	Kathleen A. Manley
Editorial and Production Services	Martha K. Morong/Quadrata, Inc.
Art Editor and Photo Researcher	Geri Davis/The Davis Group, Inc.
Chapter Opener Art Director	Meredith Nightingale
Marketing Manager	Dona Kenly
Marketing Coordinator	Lindsay Skay
Illustrators	Network Graphics, J. B. Woolsey Associates, Doug Hart, and Gary Torissi
Prepress Supervisor	Caroline Fell
Compositor	The Beacon Group, Inc.
Cover Designer	Dennis Schaefer
Cover Photograph	Randy Lorenteen/Index Stock Imagery
Interior Designer	Geri Davis/The Davis Group, Inc.
Print Buyer	Evelyn Beaton
Supplements Production	Sheila Spinney
Media Producers	Ruth Berry and Beth Standring
Software Development	Jozef Kubit and Gail Light

Photo credits appear on page I-8.

Library of Congress Cataloging-in-Publication Data
Bittinger, Marvin L.
    Basic mathematics.—9th ed. / Marvin L. Bittinger.
       p. cm.
    Includes index.
    ISBN 0-201-72147-3 (pbk.)—ISBN 0-201-79249-4 (AIE: pbk.)
    1. Arithmetic. I. Title.
QA107.2 .B57 2002
513'.1—dc21
                                      2001053826

5   6   7   8   9   10—WC—06 05 04

Contents

6 PERCENT NOTATION

7 DATA, GRAPHS, AND STATISTICS

8 MEASUREMENT

Preface

This text is the first in a series of texts that includes the following:

Bittinger: *Basic Mathematics*, Ninth Edition

Bittinger: *Fundamental Mathematics*, Third Edition

Bittinger: *Introductory Algebra*, Ninth Edition

Bittinger: *Intermediate Algebra*, Ninth Edition

Bittinger/Beecher: *Introductory and Intermediate Algebra: A Combined Approach*, Second Edition

Basic Mathematics, Ninth Edition, is a significant revision of the Eighth Edition, particularly with respect to design, art program, pedagogy, features, and supplements package. Its unique approach, which has been developed and refined over nine editions, continues to blend the following elements in order to bring students success:

- **Real data** Real-data applications aid in motivating students by connecting the mathematics to their everyday lives. Extensive research was conducted to find new applications that relate mathematics to the real world.
- **Art program** The art program has been expanded to improve the visualization of mathematical concepts and to enhance the real-data applications.

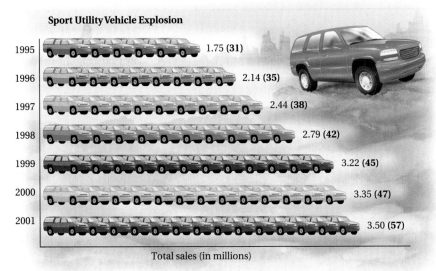

Sport Utility Vehicle Explosion

Year	Total sales (in millions)
1995	1.75 (31)
1996	2.14 (35)
1997	2.44 (38)
1998	2.79 (42)
1999	3.22 (45)
2000	3.35 (47)
2001	3.50 (57)

Total sales (in millions)

Source: Autodata

- **Writing style**   The author writes in a clear easy-to-read style that helps students progress from concepts through examples and margin exercises to section exercises.
- **Problem-solving approach**   The basis for solving problems and real-data applications is a five-step process (*Familiarize, Translate, Solve, Check,* and *State*) introduced early in the text and used consistently throughout. This problem-solving approach provides students with a consistent framework for solving applications. (See pages 68, 198, and 335.)
- **Reviewer feedback**   The author solicits feedback from reviewers and students to help fulfill student and instructor needs.
- **Accuracy**   The manuscript is subjected to an extensive accuracy-checking process to eliminate errors.
- **Supplements package**   All ancillary materials are closely tied with the text and created by members of the author team to provide a complete and consistent package for both students and instructors.

LET'S VISIT THE NINTH EDITION

The style, format, and approach of the Eighth Edition have been strengthened in this new edition in a number of ways.

Updated Applications   Extensive research has been done to make the applications in the Ninth Edition even more up to date and realistic. A large number of the applications are new to this edition, and many are drawn from the fields of business and economics, life and physical sciences, social sciences, and areas of general interest such as sports and daily life. To encourage students to understand the relevance of mathematics, many applications are enhanced by graphs and drawings similar to those found in today's newspapers and magazines. Many applications are also titled for quick and easy reference, and most real-data applications are authenticated with a source line. (See pages 8, 118, 266, 321, and 441.)

Numerous Photographs   An application becomes relevant when the connection to the real world is illustrated with a photograph. The Ninth Edition contains approximately 200 photos that immediately spark interest in examples and exercises. (See pages 78, 234, and 381.)

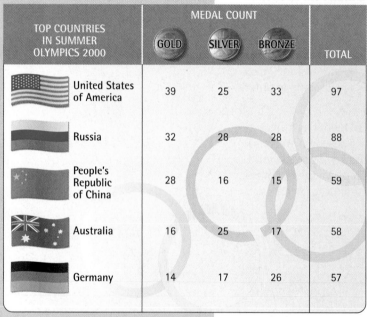

TOP COUNTRIES IN SUMMER OLYMPICS 2000	MEDAL COUNT			TOTAL
	GOLD	SILVER	BRONZE	
United States of America	39	25	33	97
Russia	32	28	28	88
People's Republic of China	28	16	15	59
Australia	16	25	17	58
Germany	14	17	26	57

Source: 2000 Olympics, Sydney, Australia

Study Tips Occurring at least twice in every chapter, these mini-lessons provide students with concrete techniques to improve studying and test-taking. These features can be covered in their entirety at the beginning of the course, encouraging good study habits early on, or they can be used as they occur in the text, allowing students to learn them gradually. These features can also be used in conjunction with Marvin L. Bittinger's "Math Study Skills for Students" Videotape, which is free to adopters. Please see your Addison-Wesley representative for details on how to obtain this videotape. (See pages 25, 189, and 315.)

Calculator Corners Designed specifically for the basic mathematics student, these optional features include scientific-calculator instruction and practice exercises (see pages 87, 208, 330, and 371). Answers to all Calculator Corner exercises appear at the back of the text.

New Art To enhance the greater emphasis on real data and applications, we have extensively increased the number of pieces of technical and situational art (see pages 32, 270, 286, and 449).

The use of color has been carried out in a methodical and precise manner so that it carries a consistent meaning, which enhances the readability of the text. For example, when perimeter is considered, figures have a red border to emphasize the perimeter. When area is considered, figures are outlined in black and screened with amber to emphasize the area. Similarly, when volume is considered, figures are three-dimensional and air-brushed blue. When fractional parts are illustrated, those parts are shown in purple.

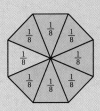

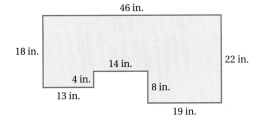

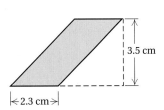

New Design The new design is more open and flexible, allowing for an expanded art and photo package and more prominent headings for the boxed definitions and rules and for the Caution boxes.

Exercises Exercises are paired, meaning that each even-numbered exercise is very much like the odd-numbered one that precedes it. This gives the instructor several options: If an instructor wants the student to have answers available, the odd-numbered exercises are assigned; if an instructor wants the student to practice (perhaps for a test), with no answers available, then the even-numbered exercises are assigned. In this way, each exercise set actually serves as two exercise sets. Answers to all odd-numbered exercises, with the exception of the Discussion and Writing exercises, and *all* Skill Maintenance exercises are provided at the back of the text. If an instructor wants the student to have access to all the answers, a complete answer book is available.

Discussion and Writing Exercises Two Discussion and Writing exercises (denoted by D_W) have been added to every exercise set and Summary and Review. Designed to develop comprehension of critical concepts, these exercises encourage students to both discuss and write about key mathematical ideas in the chapter (see pages 125, 251, and 404).

Skill Maintenance Exercises The Skill Maintenance exercises have been enhanced by the inclusion of 25% more exercises in this edition. These exercises focus on four objectives that review concepts from other sections of the text in order to prepare students for the Final Examination. Section and objective codes appear next to each Skill Maintenance exercise for easy reference. Answers to all Skill Maintenance exercises appear at the back of the book (see pages 132, 179, and 242).

Synthesis Exercises These exercises appear in every exercise set, Summary and Review, Chapter Test, and Cumulative Review. Synthesis exercises help build critical thinking skills by requiring students to synthesize or combine learning objectives from the section being studied as well as preceding sections in the book. (See pages 203, 259, and 413.)

Content We have made the following improvements to the content of *Basic Mathematics.*

- The presence of real data has been expanded and is frequently featured in a table format.

NUMBER OF E-MAILS PER DAY	PERCENT
Less than 1	28%
1–5	20%
6–10	12%
11–20	9%
21 or more	31%

Source: John J. Heldrich Center for Workforce Development

- In Chapter 6 (*Percent Notation*), a new section on interest rates on credit cards and loans has been added. To help them become financially responsible, students will learn to compare interest rates.

- Chapter 8 (*Geometry and Measures: Length and Area*) and Chapter 9 (*More Geometry and Measures*) from the Eighth Edition have been reorganized into Chapter 8 (*Measurement*) and Chapter 9 (*Geometry*).
- Coverage of medical applications has been expanded in Section 8.4 ("Weight and Mass; Medical Applications") and in Section 8.5 ("Capacity; Medical Applications").

LEARNING AIDS

Interactive Worktext Approach   The pedagogy of this text is designed to provide an interactive learning experience between the student and the exposition, annotated examples, art, margin exercises, and exercise sets. This approach provides students with a clear set of learning objectives, involves them with the development of the material, and provides immediate and continual reinforcement and assessment.

> *Section objectives* are keyed by letter not only to section subheadings, but also to exercises in the Pretest, exercise sets, and Summary and Review, as well as to answers to the Chapter Test and Cumulative Review questions. This enables students to easily find appropriate review material if they are unable to work a particular exercise.

> Throughout the text, students are directed to numerous *margin exercises,* which provide immediate reinforcement of the concepts covered in each section.

Review Material   The Ninth Edition of *Basic Mathematics* continues to provide many opportunities for students to prepare for final assessment.

> The two-column *Summary and Review* appears at the end of each chapter. The first part is a checklist of some of the Study Tips, as well as a list of important properties and formulas. The second part provides an extensive set of review exercises. Reference codes beside each exercise or direction line preceding it allow the student to easily return to the objective being reviewed (see pages 301, 352, 430, and 600).

> Also included at the end of every chapter but Chapter 1 is a *Cumulative Review,* which reviews material from all preceding chapters. At the back of the text are answers to all Cumulative Review exercises, together with section and objective references, so that students know exactly what material to study if they miss a review exercise (see pages 307, 435, 487, and 653).

> Both the Summary and Review and the Cumulative Review have been expanded to three pages, allowing for more art and a greater variety of exercises.

For Extra Help   Many valuable study aids accompany this text. At the beginning of each exercise set, references to appropriate videotape, tutorial software, and other resources make it easy for the student to find the correct support materials.

Objectives

a Add using decimal notation.

b Subtract using decimal notation.

c Solve equations of the type $x + a = b$ and $a + x = b$, where a and b may be in decimal notation.

d Balance a checkbook.

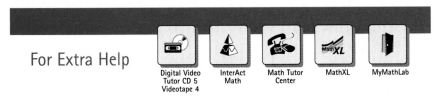

For Extra Help

Digital Video Tutor CD 5 Videotape 4   InterAct Math   Math Tutor Center   MathXL   MyMathLab

Testing   The following assessment opportunities exist in the text.

Chapter Pretests can be used to place students in a specific section of the chapter, allowing them to concentrate on topics with which they have particular difficulty (see pages 164, 312, and 546).

Chapter Tests allow students to review and test comprehension of chapter skills, as well as four objectives from earlier chapters that will be retested (see pages 96, 355, and 433).

In addition, a *Diagnostic Pretest,* found in the *Printed Test Bank/Instructor's Resource Guide* and in MyMathLab.com, can place students in the appropriate chapter for their skill level by identifying familiar material and specific trouble areas. This may be especially helpful for self-paced courses.

Answers to all Chapter Pretest and Chapter Test questions are found at the back of the book. Section and objective references for Pretest exercises are listed in blue beside each exercise or direction line preceding it. Reference codes for the Chapter Test answers are included with the answers.

SUPPLEMENTS FOR THE INSTRUCTOR

Annotated Instructor's Edition
ISBN 0-201-79249-4

The *Annotated Instructor's Edition* is a specially bound version of the student text with answers to all margin exercises and exercise sets printed in a special color near the corresponding exercises.

Instructor's Solutions Manual
ISBN 0-201-79705-4

The *Instructor's Solutions Manual* by Judith A. Penna contains brief worked-out solutions to all even-numbered exercises in the exercise sets and answers to all Discussion and Writing exercises.

Printed Test Bank/Instructor's Resource Guide
by Barbara Johnson
ISBN 0-201-79706-2

The test-bank section of this supplement contains the following:

- A diagnostic test that can place students in the appropriate chapter for their skill level
- Three alternate test forms for each chapter, with questions in the same topic order as the objectives presented in the chapter
- Five alternate test forms for each chapter, modeled after the Chapter Tests in the text
- Three alternate test forms for each chapter, designed for a 50-minute class period
- Two multiple-choice versions of each Chapter Test
- Two cumulative review tests for each chapter, with the exception of Chapter 1
- Eight final examinations: three with questions organized by chapter, three with questions scrambled as in the Cumulative Reviews, and two with multiple-choice questions
- Answers for the Diagnostic Test, Chapter Tests, and Final Examination

The resource-guide section contains the following:

- A conversion guide from the Eighth Edition to the Ninth Edition
- Extra practice exercises (with answers) for 40 of the most difficult topics in the text
- A three-column Summary and Review for each chapter, listing objectives, brief procedures, worked-out examples, multiple-choice problems similar to the example, and the answers to those problems
- Black-line masters of grids and number lines for transparency masters or test preparation
- Indexes to the videotapes and audiotapes that accompany the text

Adjunct Support Manual
ISBN 0-321-11805-7

This manual includes resources designed to help both new and adjunct faculty with course preparation and classroom management as well as offering helpful teaching tips.

Collaborative Learning Activities Manual
ISBN 0-321-11568-6

The *Collaborative Learning Activities Manual* features group activities that are tied to sections of the text. Instructions for classroom setup are also included in the manual.

Answer Book
ISBN 0-201-79708-9

The *Answer Book* contains answers to all exercises in the exercise sets in the text. Instructors can make quick reference to all answers or have quantities of these booklets made available for sale if they want students to have access to all the answers.

TestGen-EQ/QuizMaster-EQ
ISBN 0-201-79261-3

Available on a dual-platform Windows/Macintosh CD-ROM, this fully networkable software enables instructors to build, edit, print, and administer tests using a computerized test bank of questions organized according to the contents of each chapter. Tests can be printed or saved for on-line testing via a network on the Web, and the software can generate a variety of grading reports for tests and quizzes.

MathXL®: www.mathxl.com
ISBN 0-321-12986-5

The MathXL Web site provides diagnostic testing and tutorial help, all on-line using InterAct Math® tutorial software and TestGen-EQ testing software. Students can take chapter tests correlated to the text, receive individualized study plans based on those test results, work practice problems and receive tutorial instruction for areas in which they need improvement, and take further tests to gauge their progress. Instructors can customize tests and track all student test results, study plans, and practice work. An access card is required.

SUPPLEMENTS FOR THE STUDENT

Student's Solutions Manual
ISBN 0-201-79707-0

The *Student's Solutions Manual* by Judith A. Penna contains fully worked-out solutions with step-by-step annotations for all the odd-numbered exercises in the exercise sets in the text, with the exception of the Discussion and Writing exercises. It may be purchased by students from Addison-Wesley or their local college bookstore.

Videotapes
ISBN 0-201-88290-6

Digital Video
Tutor CD 5
Videotape 4

This videotape series features an engaging team of mathematics teachers who present comprehensive coverage of each section of the text in a student-interactive format. The lecturers' presentations include examples and problems from the text and support an approach that emphasizes visualization and problem solving. A video symbol at the beginning of each exercise set references the appropriate videotape or CD number (see *Digital Video Tutor*, below).

Digital Video Tutor
ISBN 0-321-11567-8, stand-alone

The videotapes for this text are also available on CD-ROM, making it easy and convenient for students to watch video segments from a computer at home or on campus. The complete digitized video set, affordable and portable for students, is ideal for distance learning or supplemental instruction.

"Math Study Skills for Students" Videotape
ISBN 0-321-11739-5

Designed to help students make better use of their math study time, this videotape helps students improve retention of concepts and procedures taught in classes from basic mathematics through intermediate algebra. Through carefully-crafted graphics and comprehensive on-camera explanation, Marvin L. Bittinger helps viewers focus on study skills that are commonly overlooked.

Audiotapes
ISBN 0-201-78618-4

The audiotapes are designed to lead students through the material in each text section. Bill Saler explains solution steps to examples, cautions students about common errors, and instructs them at certain points to stop the tape and do exercises in the margin. He then reviews the margin-exercise solutions, pointing out potential errors.

InterAct Math® Tutorial CD-ROM
ISBN 0-201-79513-2

InterAct
Math

This interactive tutorial software provides algorithmically generated practice exercises that correlate at the objective level to the odd-numbered exercises in the text. Each practice exercise is accompanied by both an example and a guided solution designed to involve students in the solution process. Selected problems also include a video clip that helps students visualize concepts. The software recognizes common student errors and provides appropriate feedback. Instructors can use InterAct Math Plus course management software to create, administer, and track on-line tests and monitor student performance during practice sessions.

MathXL®: www.mathxl.com
Stand-alone ISBN 0-201-72611-4

MathXL

The MathXL Web site provides diagnostic testing and tutorial help, all on-line, using InterAct Math® tutorial software and TestGen-EQ testing software. Students can take chapter tests correlated to the text, receive individualized study plans based on those test results, work practice problems and receive tutorial instruction for areas in which they need improvement, and take further tests to gauge their progress. An access card is required.

New! Web Site: www.MyMathLab.com

MyMathLab

Ideal for lecture-based, lab-based, and on-line courses, this state-of-the-art Web site provides students with a centralized point of access to the wide variety of on-line resources available with this text. The pages of the actual book are loaded into MyMathLab.com, and as students work through a section of the on-line text, they can link directly from the pages to supplementary resources (such as tutorial software, interactive animations, and audio and video clips) that provide instruction, exploration, and practice beyond what is offered in the printed book. MyMathLab.com generates personalized study plans for students and allows instructors to track all student work on tutorials, quizzes, and tests. Complete course-management capabilities, including a host of communication tools for course participants, are provided to create a user-friendly and interactive on-line learning environment. Contact your Addison-Wesley representative for a demonstration. An access card is required.

AW Math Tutor Center
ISBN 0-201-72170-8, stand-alone

Math Tutor
Center

The AW Math Tutor Center is staffed by qualified mathematics instructors who provide students with tutoring on examples and odd-numbered exercises from the textbook. Tutoring is available via toll-free telephone, fax, e-mail, or the Internet. White Board technology allows tutors and students to actually see problems worked while they "talk" in real time over the Internet during tutoring sessions. An access card is required.

Acknowledgments

Many of you have helped to shape the Ninth Edition by reviewing, participating in telephone surveys and focus groups, filling out questionnaires, and spending time with us on your campuses. Our deepest appreciation to all of you and in particular to the following:

Joaquin C. Armendariz, *College of Marin*
Arlene Atchison, *South Seattle Community College*
Michele Bach, *Kansas City Kansas Community College*
Roseanne Benn, *Prince George's Community College*
Maria Bennett, *West Shore Community College*
Donna Bernardy, *Lane Community College*
Martha Daniels, *Central Oregon Community College*
Drake Dennis, *Delaware Technical and Community College*
Grace Foster, *Beaufort County Community College*
Bill Graesser, *Ivy Tech State College, North*
Martha Henry, *Milwaukee Area Technical College*
Celeste Hernandez, *Richland College*
Pat Horacek, *Pensacola Junior College*
Juan Carlos Jimenez, *Springfield Technical Community College*
Michael Judge, *Houston Community College*
Rose Kaniper, *Burlington County College*
J. Barry King, *Okefenokee Technical College*
Lynette King, *Gadsden State Community College*
Thomas Lankston, *Ivy Tech State College, North Central*
Edith Lester, *Volunteer State Community College*
Pamela A. Lipka, *University of Wisconsin, Whitewater*
Debra Loeffler, *Community College of Baltimore County, Catonsville*
Marianna McClymonds, *Phoenix College*
Valerie Morgan-Krick, *Tacoma Community College*
Joyce Oster, *Johnson and Wales University*
Julie Pendleton, *Brookhaven College*
Thea Philliou, *College of Santa Fe*
Mary Rack, *Johnson County Community College*
Greg A. Rosik, *Century College*
Pat Roux, *Delgado Community College*
Nelissa Rutishauser, *Mohawk Valley Community College*
Richard Schnackenberg, *International College*
Mike Shirazi, *Germanna Community College*
Nicole Sifford, *Three Rivers Community College*
Tomesa Smith, *Wallace State Community College*

Trudy Streilein, *Northern Virginia Community College*
Sharon Testone, *Onondaga Community College*
Brad Thurmond, *Ivy Tech State College, Kokomo*
Kevin Wheeler, *Three Rivers Community College*
Jane-Marie Wright, *Suffolk County Community College*

We also wish to recognize the following people who wrote scripts, presented lessons on camera, and checked the accuracy of the videotapes:

Barbara Johnson, *Indiana University Purdue University Indianapolis*
Judith A. Penna, *Indiana University Purdue University Indianapolis*
Patricia Schwarzkopf, *University of Delaware*
Anthony Seraphin, *University of Delaware*
Clen Vance, *Houston Community College*

I wish to express my heartfelt appreciation to a number of people who have contributed in special ways to the development of this textbook. My editor, Jennifer Crum, encouraged my vision and provided marketing insight. Kari Heen, the project manager, deserves special recognition for overseeing every phase of the project and keeping it moving. The unwavering support of the Developmental Math group, including Lauren Morse, associate editor, and Kathleen Manley, production supervisor, and the endless hours of hard work by Martha Morong and Geri Davis have led to products of which I am immensely proud.

I also want to thank Judy Beecher, my co-author on many books and my developmental editor on this text. Her steadfast loyalty, vision, and encouragement have been invaluable. In addition to writing the *Student's Solutions Manual*, Judy Penna has continued to provide strong leadership in the preparation of the printed supplements, videotapes, and MyMathLab. Other strong support has come from Barbara Johnson for the *Printed Test Bank*; Bill Saler for the audiotapes; and Barbara Johnson, Judy Penna, and Vera Preston-Jaeger, for their accuracy checking.

M.L.B.

Feature Walkthrough

Chapter Openers

To engage students and prepare them for the upcoming chapter material, two-page gateway chapter openers are designed with exceptional artwork that is tied to a motivating real-world application.

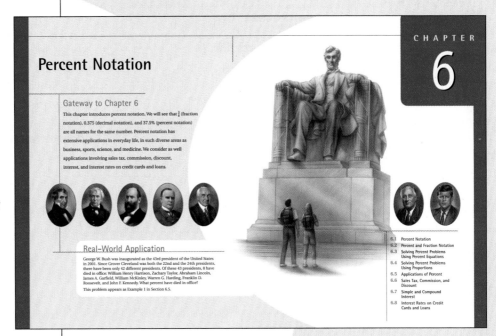

Chapter Pretests

Allowing students to test themselves before beginning each chapter, Chapter Pretests help them to identify material that may be familiar as well as targeting material that may be new or especially challenging. Instructors can use these results to assess student needs.

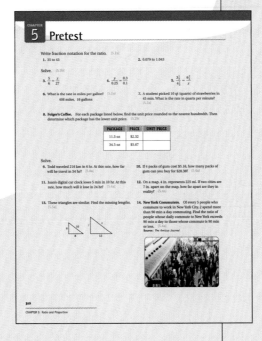

Art Program

Today's students are often visually oriented and their approach to a printed page is no exception. To appeal to students, the situational art in this edition is more dynamic and there are more photographs and art pieces overall. Where possible, mathematics is included in the art pieces to help students visualize the problem at hand.

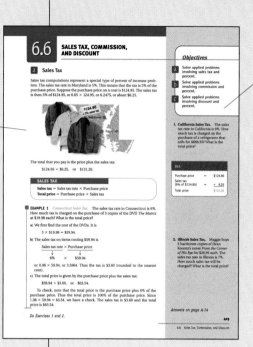

Objective Boxes

At the beginning of each section, a boxed list of objectives is keyed by letter not only to section subheadings, but also to the exercises in the Pretest, exercise sets, and Summary and Review, as well as answers to the Chapter Test and Cumulative Review questions. This correlation enables students to easily find appropriate review material if they need help with a particular exercise or skill.

Margin Exercises

Throughout the text, students are directed to numerous margin exercises that provide immediate practice and reinforcement of the concepts covered in each section.

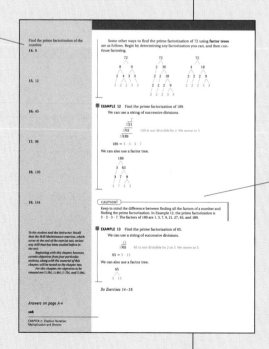

Caution Boxes

Found at relevant points throughout the text, boxes with the "Caution!" heading warn students of common misconceptions or errors made in performing a particular mathematical operation or skill.

Study Tips

Previously called "Improving Your Math Study Skills," a variety of Study Tips throughout the text give students pointers on how to develop good study habits as they progress through the course. At times short snippets and at other times more lengthy discussions, these Study Tips encourage students to input information and get involved in the learning process.

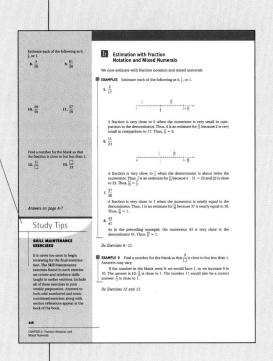

Calculator Corners

Where appropriate throughout the text, students see optional Calculator Corners. Popular in the Eighth Edition, slightly more Calculator Corners have been included in the new edition and the revised content is now more accessible to students.

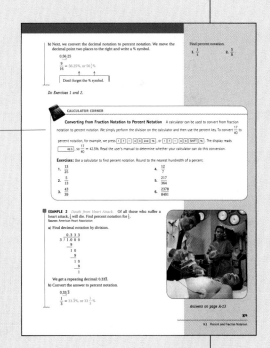

EXERCISE SETS

To give students the opportunity to practice what they have learned, each section is followed by an extensive exercise set designed to reinforce the section concepts. In addition, students also have the opportunity to synthesize the objectives from the current section as well as those from preceding sections.

Exercises

Exercises are keyed by letter to the section objectives for easy review.

For Extra Help

Many valuable study aids accompany this text. Located just before each exercise set, "For Extra Help" references list appropriate video, tutorial, and Web resources so students can easily find related support materials.

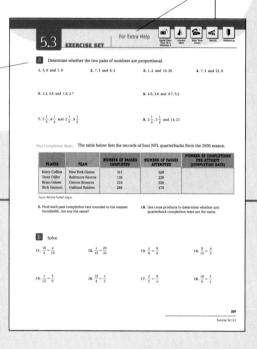

Discussion and Writing Exercises

Designed to help students develop deeper comprehension of critical concepts, Discussion and Writing exercises (indicated by the ^D_W symbol) are suitable for individual or group work. These exercises encourage students to both think and write about key mathematical ideas in the chapter.

Skill Maintenance Exercises

Found in each exercise set, these exercises review concepts from other sections in the text to prepare students for their final examination. Section and objective codes appear next to each Skill Maintenance exercise for easy reference, and in response to user feedback, the overall number of Skill Maintenance exercises has been increased.

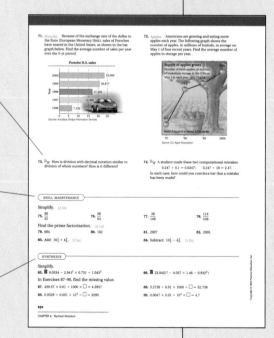

Synthesis Exercises

In most exercise sets, Synthesis exercises help build critical-thinking skills by requiring students to synthesize or combine learning objectives from the current section as well as from preceding text sections.

Real-Data Applications

This text encourages students to see and interpret the mathematics that appears every day in the world around them. Throughout the writing process, an energetic search for real-data applications was conducted, and the result is a variety of examples and exercises that connect the mathematical content with the real world. Most of these applications feature source lines and frequently include charts and graphs.

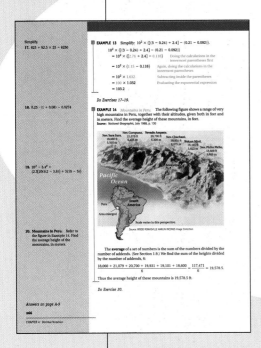

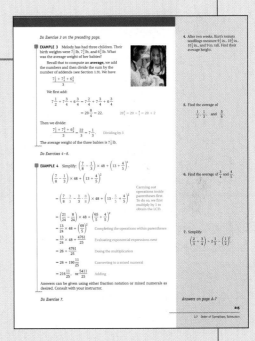

Annotated Examples

Detailed annotations and color highlights lead the student through the structured steps of the examples.

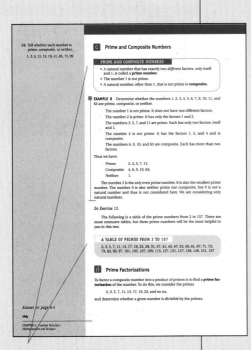

Highlighted Information

Important definitions, rules, and procedures are highlighted in titled boxes.

END-OF-CHAPTER MATERIAL

At the end of each chapter, students can practice all they have learned
as well as tie the current chapter material to material covered in earlier chapters.

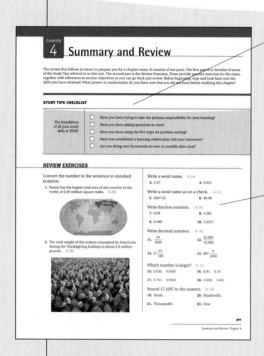

Study Tips Checklist

Each chapter review begins with a Study Tips
Checklist that reviews Study Tips introduced in
the current and previous chapters, making the
use of these Study Tips more interactive.

Review Exercises

At the end of each chapter, students are provided
with an extensive set of Review exercises.
Reference codes beside each exercise or direction
line allow students to easily review the related
objective.

Chapter Test

Following the Review exercises, a
sample Chapter Test allows students to
review and test comprehension of
chapter skills prior to taking an
instructor's exam.

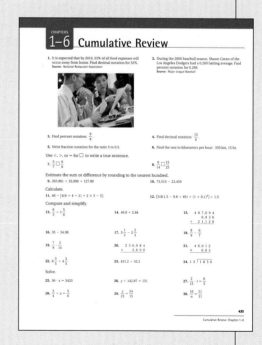

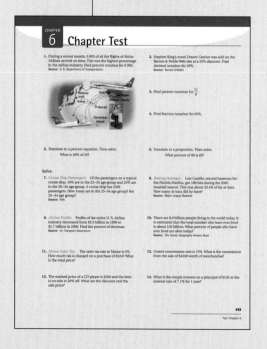

Cumulative Review

Following each chapter (beginning with
Chapter 2), students encounter a Cumulative
Review. This exercise set reviews skills and
concepts from all preceding chapters to help
students recall previously learned material and
prepare for a final exam.

Introduction to the Student

As your author, I'd like to welcome you to this study of *Basic Mathematics.* Students come to this course with all kinds of backgrounds. Many are recent graduates of high school. Some are returning to college after spending time in the job market or after raising a family.

Whatever your past experiences, I encourage you to look at this mathematics course as a fresh start. Approach your course with a positive attitude about mathematics: This will put you in the right frame of mind to learn. Mathematics is a base for life, for many majors, for personal finances, for most careers, or just for pleasure. There is power in those symbols and equations—believe me!

If you have negative thoughts about mathematics, it is probably because you have had some kind of unpleasant experience in your study of math before now. It is my belief that most people can, and will, be able to learn under the right conditions, but some changes in your approach may be in order.

Let's think about your learning as a team approach. What can "we," meaning me, your instructor, and, most of all, *you,* do to facilitate real success in your learning? Let's consider the parts of the team.

Your Author: My life has been dedicated to writing mathematics texts for over 30 years. I received my Ph.D. in Mathematics Education at Purdue University in 1968. My first and only teaching position was at Indiana University Purdue University Indianapolis. I sometimes think of myself as a "teacher on paper," or a "global professor," because of the wide use of my textbooks.

I live in Carmel, Indiana, with my wife, Elaine, who is very encouraging and supportive of my writing. We have two grown sons, Lowell and Chris, who are married to Karen and Tricia, respectively. Karen is a wonderful photographer; you will see many of her photographs in this book. Tricia is the mother of our grandchild, Margaret Grace. Needless to say, Maggie occupies much of my time when I'm away from my writing.

Apart from my family, my hobbies include hiking, baseball, golf, and bowling. I'm a terrible baseball player, but one week a year, I go to adult baseball fantasy camp and play ball like a kid. I'm a good bowler, with a 200 average, but a really poor golfer (30 handicap). Golf keeps me humble, and I enjoy learning about the game. I also have an interest in philosophy and theology, in particular, apologetics.

Your Instructor: Clearly, your instructor is at the forefront of your learning. He or she is who you learn from in the classroom. I encourage you to establish a learning relationship with your instructor early on. Feel free to ask questions in and outside of class. Do not wait too long for help or advice! Trivial though it may seem, be sure to get basic information, like his or her name, how he or she can be contacted outside of class, and the location of his or her office.

In addition, learn about your instructor's teaching style and try to adapt your learning to it. Does he or she use an overhead projector or the board? Will there be frequent in-class questions, tests, or quizzes, and so on? How is your grade ultimately determined?

Make use of what your instructor and your college has to offer in the way of help. If your campus has any kind of tutor center or learning lab, be sure to locate it and find out the hours of operation. Too often, students do not avail themselves of help that is there, free for the asking.

Yourself: You are the biggest factor in the success of your learning. This may be the first adjustment you have in college. In earlier experiences, you may have allowed yourself to sit back and let the instructor "pour in" the learning, with little or no follow-up on your part. But now you must take a more assertive and proactive stance. As soon as possible after class, you should thoroughly read the textbook and the supplements and do all you can on your own to learn. In other words, rid yourself of former habits and take responsibility for your own learning. Then, with all the help you have around you, your hard work will lead to success.

A helpful proverb comes to mind here:

"The best way to acquire a virtue is to act as if you already have it."

C. S. Lewis, English scholar and author

If you have never taken an assertive approach to learning mathematics, do so now, and you will soon realize the success of your actions.

One of the most important suggestions I can make is to allow yourself enough *time* to learn. You can have the best book, the best instructor, and the best supplements, but if you do not give yourself time to learn, how can they be of benefit? I usually ask my students the following questions:

- Are you working 40 hours or more per week?
- Are you taking 12 or more hours of classes?
- Do you dislike mathematics or have you had trouble learning in the past?

If you answered "yes" to all three of these questions, you must change one or both of the first two situations listed. You just cannot learn without proper time management!

This introduction has contained many points that fall under the category of *Study Tips*, which you will find throughout the book. The following is an example in which I summarize some of the suggestions we have considered in this introduction:

Study Tips

- Establish a learning relationship with your instructor.
- Take more of the responsibility for your learning—do not wait for someone else to provide it for you.
- Use proper time management to allow time to learn.

You may want to study all of the Study Tips before you begin the text or you may decide to wait and encounter them as you go along. Your instructor may have suggestions in this regard.

You probably sense that the purpose of this Introduction is to encourage you, and that is indeed true. This, along with the effort we have made to write the best instructional book we can, are as close as we can come to being your personal instructor.

In closing, I want to wish you well in your new start studying mathematics. I wish I could meet each of you personally, but rest assured I think of you often in the sense that most of my waking moments are spent contemplating textbooks that make your learning more effective. Best wishes!

Marv Bittinger

Whole Numbers

Gateway to Chapter 1

You are beginning a study of Basic Mathematics. In this chapter, we consider addition, subtraction, multiplication, and division of whole numbers, as well as exponential notation and order of operations. We also introduce the idea of using variables to form equations. Then we solve simple equations and use the skills of this chapter to solve applied problems.

Before starting, be sure to read the Preface and the Introduction from the Author to the Student.

Real-World Application

Boeing Corporation builds commercial aircraft. A Boeing 767 has a seating configuration with 4 rows of 6 seats across in first class and 35 rows of 7 seats across in economy class. Find the total seating capacity of the plane.

Sources: Boeing Corporation; Delta Airlines

This problem appears as Example 8 in Section 1.8.

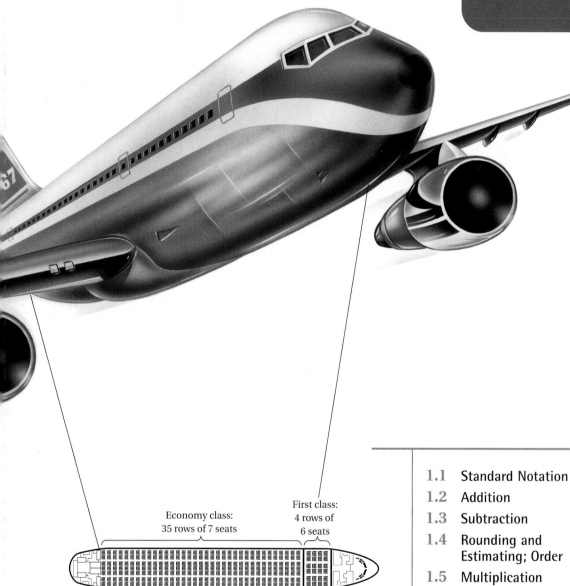

Economy class:
35 rows of 7 seats

First class:
4 rows of
6 seats

CHAPTER

1

1. Write a word name: 3,078,059. [1.1c]

2. Write expanded notation: 6987. [1.1b]

3. Write standard notation: Two billion, forty-seven million, three hundred ninety-eight thousand, five hundred eighty-nine. [1.1c]

4. What does the digit 6 mean in 2,967,342? [1.1a]

5. Round 956,449 to the nearest thousand. [1.4a]

6. Estimate the product $594 \cdot 126$ by first rounding the numbers to the nearest hundred. [1.5b]

7. Add. [1.2a]

$$\begin{array}{r} 7312 \\ + 2904 \\ \hline \end{array}$$

8. Subtract. [1.3b]

$$\begin{array}{r} 7012 \\ - 2904 \\ \hline \end{array}$$

9. Multiply: $359 \cdot 64$. [1.5a]

10. Divide: $23{,}149 \div 46$. [1.6b]

Use either $<$ or $>$ for $\square$ to write a true sentence. [1.4c]

11. $346 \ \square \ 364$

12. $54 \ \square \ 45$

Solve. [1.7b]

13. $326 \cdot 17 = m$

14. $y = 924 \div 42$

15. $19 + x = 53$

16. $34 \cdot n = 850$

Solve. [1.8a]

17. **Paper Quantity.** There are 500 sheets in a ream of paper. How many sheets are in 9 reams?

9 reams

500 sheets in each

18. **Digital Cameras.** A group of 63 language students from VaMard University is planning a year abroad to study German. They decide that each of them will buy a digital camera like the one shown in the ad below. The total cost of the purchase is $18,837. What is the cost per camera?

OPUS
Digital Camera

$???

19. **Checking Account.** You have $756 in your checking account. Using your debit card, you pay $387 for a VCR for your dorm room. How much is left in your account?

20. **College Costs.** It has been estimated that by 2012, the costs of each of the four years of college will be $7383, $7359, $7925, and $8126. Find the total cost of four years of college at that time.

Evaluate. [1.9b]

21. 5^2

22. 4^3

Simplify.

23. $8^2 \div 8 \cdot 2 - (2 + 2 \cdot 7)$ [1.9c]

24. $108 \div 9 - \{3 \cdot [18 - (5 \cdot 3)]\}$ [1.9d]

1.1 STANDARD NOTATION

Objectives

a   Give the meaning of digits in standard notation.

b   Convert between standard notation and expanded notation.

c   Convert between standard notation and word names.

We study mathematics in order to be able to solve problems. In this section, we study how numbers are named. We begin with the concept of place value.

a   Place Value

Consider the number named in the following ad.

BURGER KING® sells **1,305,716,519** WHOPPER® sandwiches each year

Source: ™ and ©2000 Burger King Brands, Inc.

A **digit** is a number 0, 1, 2, 3, 4, 5, 6, 7, 8, or 9 that names a place-value location. For large numbers, digits are separated by commas into groups of three, called **periods.** Each period has a name: *ones, thousands, millions, billions, trillions,* and so on. To understand the number in the ad, we can use a **place-value chart,** as shown below.

PLACE-VALUE CHART														
Periods →	Trillions			Billions			Millions			Thousands			Ones		
						1	3	0	5	7	1	6	5	1	9
	Hundreds	Tens	Ones	Hundreds	Tens	Ones	Hundreds	Tens	Ones	Hundreds	Tens	Ones	Hundreds	Tens	Ones

1 billion,   305 million,   716 thousand,   519 ones

EXAMPLES   What does the digit 8 mean in each number?

1. 278,342   8 thousands
2. 872,342   8 hundred thousands
3. 28,343,399,223   8 billions

Do Margin Exercises 1–4.

What does the digit 2 mean in each number?

1. 526,555

2. 265,789

3. 42,789,654

4. 24,789,654

5. Golf Balls. It is estimated that in one day Americans buy 486,575 golf balls. What does each digit name?
Source: U.S. Golf Association

Answers on page A-1

Write expanded notation.

6. 1895

7. $22,132, the average salary for a flight attendant in 1990

8. 3031 mi (miles), the diameter of Mercury

9. 4100 mi, the length of the Nile River, the longest in the world

10. 3860 mi, the length of the Missouri–Mississippi River, the longest in the United States

Answers on page A-1

EXAMPLE 4 *Pacific Ocean.* The area of the Pacific Ocean is 64,186,000 square miles. What does each digit name?

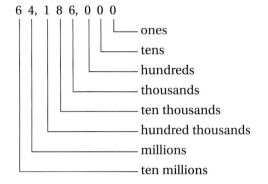

6 4, 1 8 6, 0 0 0
— ones
— tens
— hundreds
— thousands
— ten thousands
— hundred thousands
— millions
— ten millions

Do Exercise 5 on the preceding page.

b Converting Between Standard Notation and Expanded Notation

To answer questions such as "How many?", "How much?", and "How tall?", we use whole numbers. The set, or collection, of **whole numbers** is

$$0, 1, 2, 3, 4, 5, 6, 7, 8, 9, 10, 11, 12, \ldots .$$

The set goes on indefinitely. There is no largest whole number, and the smallest whole number is 0. Each whole number can be named using various notations. The set $1, 2, 3, 4, 5, \ldots$, without 0, is called the set of **natural numbers.**

Let's look at the data from the line graph shown here.

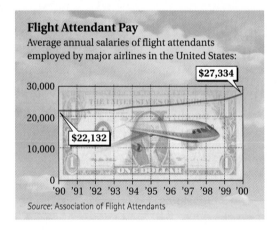

Flight Attendant Pay
Average annual salaries of flight attendants employed by major airlines in the United States:

$27,334

$22,132

30,000
20,000
10,000
0
'90 '91 '92 '93 '94 '95 '96 '97 '98 '99 '00

Source: Association of Flight Attendants

The average salary for a flight attendant in 2000 was $27,334. **Standard notation** for the salary is 27,334. We write **expanded notation** for 27,334 as follows:

$$27,334 = 2 \text{ ten thousands} + 7 \text{ thousands}$$
$$+ 3 \text{ hundreds} + 3 \text{ tens} + 4 \text{ ones}.$$

EXAMPLE 5 Write expanded notation for 4218 mi, the diameter of Mars.

$$4218 = 4 \text{ thousands} + 2 \text{ hundreds} + 1 \text{ ten} + 8 \text{ ones}$$

EXAMPLE 6 Write expanded notation for 3400.

$3400 = 3$ thousands $+ 4$ hundreds $+ 0$ tens $+ 0$ ones,   or

3 thousands $+ 4$ hundreds

Do Exercises 6–10 on the preceding page.

EXAMPLE 7 Write standard notation for 9 ten thousands + 6 thousands + 7 hundreds + 1 ten + 8 ones.

Standard notation is 96,718.

EXAMPLE 8 Write standard notation for 2 thousands + 3 tens.

Standard notation is 2030.

Do Exercises 11–13.

C Converting Between Standard Notation and Word Names

We often use **word names** for numbers. When we pronounce a number, we are speaking its word name. The People's Republic of China won 59 medals in the 2000 Summer Olympics in Sydney, Australia. A word name for 59 is "fifty-nine." Word names for some two-digit numbers like 59, 76, and 97 use hyphens. Others like 17 use only one word, "seventeen." Let's write some word names.

TOP COUNTRIES IN SUMMER OLYMPICS 2000	MEDAL COUNT			
	GOLD	SILVER	BRONZE	TOTAL
United States of America	39	25	33	97
Russia	32	28	28	88
People's Republic of China	28	16	15	59
Australia	16	25	17	58
Germany	14	17	26	57

Source: 2000 Olympics, Sydney, Australia

Write standard notation.

11. 5 thousands + 6 hundreds + 8 tens + 9 ones

12. 8 ten thousands + 7 thousands + 1 hundred + 2 tens + 8 ones

13. 9 thousands + 3 ones

Write a word name. (Refer to the figure at left.)

14. 88, the total number of medals won by Russia

15. 16, the number of silver medals won by the People's Republic of China

16. 32, the number of gold medals won by Russia

Answers on page A-1

Write a word name.

17. 204

18. $43,782, the average salary in 1998 for those who have a bachelor's degree
Source: U.S. Bureau of the Census

19. 1,879,204

20. 6,259,600,000, the world population in 2000
Source: U.S. Bureau of the Census

21. Write standard notation.

Two hundred thirteen million, one hundred five thousand, three hundred twenty-nine

EXAMPLES Write a word name.

9. 97, the total number of medals won by the United States

Ninety-seven

10. 15, the number of bronze medals won by the People's Republic of China

Fifteen

Do Exercises 14–16 on the preceding page.

For word names for larger numbers, we begin at the left with the largest period. The number named in the period is followed by the name of the period; then a comma is written and the next period is named.

EXAMPLE 11   Write a word name for 46,605,314,732.

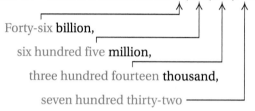

Forty-six **billion,**

six hundred five **million,**

three hundred fourteen **thousand,**

seven hundred thirty-two

The word "and" *should not* appear in word names for whole numbers. Although we commonly hear such expressions as "two hundred *and* one," the use of "and" is not, strictly speaking, correct in word names for whole numbers. For decimal notation, it is appropriate to use "and" for the decimal point. For example, 317.4 is read as "three hundred seventeen *and* four tenths."

Do Exercises 17–20.

EXAMPLE 12   Write standard notation.

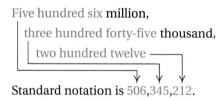

Five hundred six **million,**

three hundred forty-five **thousand,**

two hundred twelve

Standard notation is 506,345,212.

Do Exercise 21.

Study Tips

Throughout this textbook, you will find a feature called *Study Tips*. These tips are intended to help improve your math study skills. On the first day of class, you should complete this chart.

BASIC INFORMATION ON THE FIRST DAY OF CLASS

Instructor: Name _____

Office hours and location

_____

Phone number _____

Fax number _____

e-mail address _____

Find the names of two students whom you could contact for information or study questions:

1. Name _____

   Phone number _____

   e-mail address _____

2. Name _____

   Phone number _____

   e-mail address _____

Math lab on Campus:

Location _____

Hours _____

Phone _____

Tutoring:

Campus location _____

Hours _____

AW Math Tutor Center _____

To order, call _____.

(See the Preface for important information concerning this tutoring.)

Important Supplements:
(See the Preface for a complete list of available supplements.)

Supplements recommended by the instructor

_____

_____

"I know the price of success: dedication, hard work, and an unremitting devotion to the things you want to see happen."

Frank Lloyd Wright, architect

a What does the digit 5 mean in each case?

1. 235,888

2. 253,777

3. 1,488,526

4. 500,736

Skiers. In the 1999–2000 ski season, Vail, Colorado, had 1,370,000 skiers. In the number 1,370,000, what digit names the number of:
Source: *Denver Post*

5. Ones?

6. Ten thousands?

7. Millions?

8. Hundred thousands?

b Write expanded notation.

9. 5702

10. 3097

11. 93,986

12. 38,453

Step-Climbing Races. Races in which runners climb the steps inside a building are called "run-up" races. The graph below shows the number of steps in four buildings. In Exercises 13–16, write expanded notation for the number of steps in each race.

Step-Climbing Races

2058

1776

1268

1081

International Towerthon, Kuala Lumpur, Malaysia | CN Tower Run-Up, Toronto | World Financial Center, New York | Skytower Run-Up, Aukland, New Zealand

Source: New York Road Runners Club

13. 2058 steps in the International Towerthon, Kuala Lumpur, Malaysia

14. 1776 steps in the CN Tower Run-Up, Toronto, Ontario, Canada

15. 1268 steps in the World Financial Center, New York

16. 1081 steps in the Skytower Run-Up, Auckland, New Zealand

Write standard notation.

17. 2 thousands + 4 hundreds + 7 tens + 5 ones

18. 7 thousands + 9 hundreds + 8 tens + 3 ones

19. 6 ten thousands + 8 thousands + 9 hundreds + 3 tens + 9 ones

20. 1 ten thousand + 8 thousands + 4 hundreds + 6 tens + 1 one

21. 7 thousands + 3 hundreds + 0 tens + 4 ones

22. 8 thousands + 0 hundreds + 2 tens + 0 ones

23. 1 thousand + 9 ones

24. 2 thousands + 4 hundreds + 5 tens

C Write a word name.

25. 85

26. 48

27. 88,000

28. 45,987

29. 123,765

30. 111,013

31. 7,754,211,577

32. 43,550,651,808

Write standard notation.

33. Two million, two hundred thirty-three thousand, eight hundred twelve

34. Three hundred fifty-four thousand, seven hundred two

35. Eight billion

36. Seven hundred million

Write a word name for the number in each sentence.

37. *Great Pyramid.* The area of the base of the Great Pyramid in Egypt is 566,280 square feet.

38. *Population of the United States.* The population of the United States in 2000 was estimated to be 273,540,000.
Source: U.S. Bureau of the Census

39. *Monopoly.* In a recent Monopoly® game sponsored by McDonalds® restaurants, the odds of winning the grand prize were estimated to be 467,322,388 to 1.
Source: McDonald's Corporation

40. *Native American Population.* In a recent year, the population of Native Americans in Arizona was 165,385.
Source: U.S. Bureau of the Census

Write standard notation for the number in each sentence.

41. Light travels nine trillion, four hundred sixty billion kilometers in one year.

42. The distance from the sun to Pluto is three billion, six hundred sixty-four million miles.

43. *Pacific Ocean.* The area of the Pacific Ocean is sixty-four million, one hundred eighty-six thousand square miles.

44. *Gigabyte.* On a computer hard disk, one gigabyte is one billion, seventy-three million, seven hundred forty-one thousand, eight hundred twenty-four bytes of memory.

To the student and the instructor: The Discussion and Writing exercises are meant to be answered with one or more sentences. They can be discussed and answered collaboratively by the entire class or by small groups. Because of their open-ended nature, the answers to these exercises do not appear at the back of the book. They are denoted by the symbol D_W.

45. D_W Explain why we use commas when writing large numbers.

46. D_W Write an English sentence in which the number 370,000,000 is used.

SYNTHESIS

To the student and the instructor: The Synthesis exercises found at the end of every exercise set challenge students to combine concepts or skills studied in that section or in preceding parts of the text. Exercises marked with a ▦ symbol are meant to be solved using a calculator.

47. How many whole numbers between 100 and 400 contain the digit 2 in their standard notation?

48. ▦ What is the largest number that you can name on your calculator? How many digits does that number have? How many periods?

1.2 ADDITION

Objectives

a Add whole numbers.

b Use addition in finding perimeter.

a Addition of Whole Numbers

Addition of whole numbers corresponds to combining or putting things together.

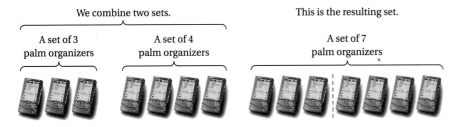

We combine two sets.   This is the resulting set.

A set of 3 palm organizers   A set of 4 palm organizers   A set of 7 palm organizers

The addition that corresponds to the figure above is

$3 + 4 = 7.$

The number of objects in a set can be found by counting. We count and find that the two sets have 3 palm organizers and 4 palm organizers, respectively. After combining, we count and find that there are 7 palm organizers. We say that the **sum** of 3 and 4 is 7. The numbers added are called **addends.**

$$\begin{array}{ccc} 3 & + \quad 4 & = \quad 7 \\ \downarrow & \downarrow & \downarrow \\ \text{Addend} & \text{Addend} & \text{Sum} \end{array}$$

Addition also corresponds to moving distances on a number line. The number line below is marked with tick marks at equal distances of 1 *unit*. The sum $3 + 4$ is shown. We first move 3 units from 0, and then 4 more units, and end up at 7. The addition that corresponds to the situation is $3 + 4 = 7$.

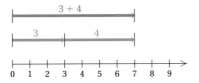

To add whole numbers, we add the ones digits first, then the tens, then the hundreds, then the thousands, and so on. Adding 0 to a number does not change the number: $a + 0 = 0 + a = a$. We say that 0 is the **additive identity.**

Add.

1. 74 + 23

2. 6 2 0 3
     + 3 5 4 2

3. 7 9 6 8
     + 5 4 9 7

4. 9 8 0 4
     + 6 3 7 8

EXAMPLE 1 Add: 7312 + 2504.

Place values are lined up in columns.

```
  7 3 1 2     Add ones.
+ 2 5 0 4
        6
```

```
  7 3 1 2     Add tens.
+ 2 5 0 4
      1 6
```

We show you this for explanation.

```
  7 3 1 2     Add hundreds.
+ 2 5 0 4
    8 1 6
```

You need write only this.

```
  7 3 1 2     Add thousands.
+ 2 5 0 4
  9 8 1 6
```

```
  7 3 1 2  ← Addends
+ 2 5 0 4  ←
  9 8 1 6  ← Sum
```

Do Exercises 1 and 2.

EXAMPLE 2 Add: 6878 + 4995.

```
      1
  6 8 7 8      Add ones. We get 13 ones, or 1 ten + 3 ones.
+ 4 9 9 5      Write 3 in the ones column and 1 above the tens.
        3      This is called *carrying*, or *regrouping*.
```

```
    1 1
  6 8 7 8      Add tens. We get 17 tens, or 1 hundred + 7 tens.
+ 4 9 9 5      Write 7 in the tens column and 1 above the
      7 3      hundreds.
```

```
  1 1 1
  6 8 7 8      Add hundreds. We get 18 hundreds, or 1
+ 4 9 9 5      thousand + 8 hundreds.
    8 7 3      Write 8 in the hundreds column and 1 above the
               thousands.
```

```
  1 1 1
  6 8 7 8      Add thousands. We get 11 thousands.
+ 4 9 9 5
1 1 8 7 3
```

Do Exercises 3 and 4.

How do we do an addition of three numbers, like $2 + 3 + 6$? We do so by adding 3 and 6, and then 2. We can show this with parentheses:

$$2 + (3 + 6) = 2 + 9 = 11. \qquad \text{Parentheses tell what to do first.}$$

We could also add 2 and 3, and then 6:

$$(2 + 3) + 6 = 5 + 6 = 11.$$

Either way we get 11. It does not matter how we group the numbers. This illustrates the **associative law of addition,** $a + (b + c) = (a + b) + c$. We can also add whole numbers in any order. That is, $2 + 3 = 3 + 2$. This illustrates the **commutative law of addition,** $a + b = b + a$. Together the commutative and associative laws tell us that to add more than two numbers, we can use any order and grouping we wish.

EXAMPLE 3 Add from the top.

$$
\begin{array}{r}
8 \\
9 \\
7 \\
+\ 6 \\
\end{array}
$$

We first add 8 and 9, getting 17; then 17 and 7, getting 24; then 24 and 6, getting 30.

$$
\begin{array}{r}
8 \\
9 \longrightarrow 17 \\
7 \qquad 7 \longrightarrow 24 \\
+\ 6 \qquad\qquad 6 \longrightarrow 30 \\
\hline
30 \longleftarrow
\end{array}
$$

You write only this.

Do Exercises 5 and 6.

EXAMPLE 4 Add from the bottom.

$$
\begin{array}{r}
8 \longrightarrow 30 \\
9 \longrightarrow 22 \\
7 \longrightarrow 13 \\
+\ 6 \\
\hline
30 \longleftarrow
\end{array}
$$

You still write the answer here.

Do Exercise 7.

5.
$$
\begin{array}{r}
9 \\
9 \\
4 \\
+\ 5 \\
\end{array}
$$

6.
$$
\begin{array}{r}
8 \\
6 \\
9 \\
7 \\
+\ 4 \\
\end{array}
$$

7. Add from the bottom.
$$
\begin{array}{r}
9 \\
9 \\
4 \\
+\ 5 \\
\end{array}
$$

Answers on page A-1

Add. Look for pairs of numbers whose sums are 10, 20, 30, and so on.

8.
```
    1 5
      7
      5
      3
  +   8
  _____
```

9. 6 + 12 + 14 + 8 + 7

10. 27 + 8 + 13 + 2 + 11

11. Add.
```
    1 9 3 2
    6 7 2 3
    9 8 7 8
  + 8 9 4 1
  _____
```

Sometimes it is easier to look for pairs of numbers whose sums are 10 or 20 or 30, and so on.

EXAMPLES Add.

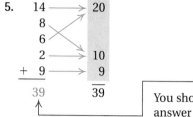

5.
```
  14  ───→  20
   8
   6
   2  ───→  10
  + 9  ───→   9
  ____      ____
  39         39
```

You should write only the answer in the position shown.

6. 23 + 19 + 7 + 21 + 4 = 74

    30 + 40 + 4
        74

Do Exercises 8–10.

EXAMPLE 7 Add: 2391 + 3276 + 8789 + 1498.

```
        2
    2 3 9 1
    3 2 7 6
    8 7 8 9
  + 1 4 9 8
  _____
          4
```
Add ones. We get 24, so we have 2 tens + 4 ones. Write 4 in the ones column and 2 above the tens.

```
      3 2
    2 3 9 1
    3 2 7 6
    8 7 8 9
  + 1 4 9 8
  _____
        5 4
```
Add tens. We get 35 tens, so we have 30 tens + 5 tens. This is also 3 hundreds + 5 tens. Write 5 in the tens column and 3 above the hundreds.

```
    1 3 2
    2 3 9 1
    3 2 7 6
    8 7 8 9
  + 1 4 9 8
  _____
      9 5 4
```
Add hundreds. We get 19 hundreds, or 1 thousand + 9 hundreds. Write 9 in the hundreds column and 1 above the thousands.

```
    1 3 2
    2 3 9 1
    3 2 7 6
    8 7 8 9
  + 1 4 9 8
  _____
  1 5 9 5 4
```
Add thousands. We get 15 thousands.

Do Exercise 11.

Answers on page A-1

b Finding Perimeter

Addition can be used when finding perimeter.

> ### PERIMETER
> The distance around an object is its **perimeter.**

EXAMPLE 8 A computer sales rep travels the following route to visit various electronics stores. How long is the route?

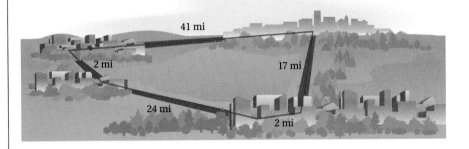

41 mi

2 mi

17 mi

24 mi

2 mi

$$2 \text{ mi} + 24 \text{ mi} + 2 \text{ mi} + 17 \text{ mi} + 41 \text{ mi} = \text{Perimeter}$$

We carry out the addition as follows.

```
    1
        2
    2   4
        2
    1   7
+   4   1
    8   6
```

The perimeter of the figure is 86 mi. The route is 86 mi long.

Do Exercises 12–14.

Solve.

12. Index Cards. Two standard sizes for index cards are 3 in. (inches) by 5 in. and 5 in. by 8 in. Find the perimeter of each card.

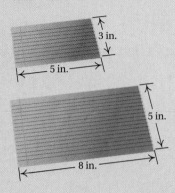

3 in.

5 in.

5 in.

8 in.

Find the perimeter of each figure.

13.

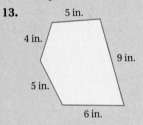

5 in.

4 in.

9 in.

5 in.

6 in.

14.

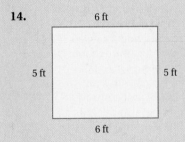

6 ft

5 ft

5 ft

6 ft

Answers on page A-1

Study Tips

USING THIS TEXTBOOK

We began our "Study Tips" in Section 1.1. You will find many of these tips throughout the book. One of the most important ways in which to improve your math study skills is to learn the proper use of the textbook. Here we highlight a few points that we consider most helpful.

■ **Be sure to note the special symbols** a , b , c , **and so on, that correspond to the objectives you are to be able to perform.** The first time you see them is in the margin at the beginning of each section; the second time is in the subheadings of each section; and the third time is in the exercise set for the section. You will also find them next to the skill maintenance exercises in each exercise set and the review exercises at the end of the chapter, as well as in the answers to the chapter tests and the cumulative reviews. These objective symbols allow you to refer to the appropriate place in the text whenever you need to review a topic.

■ **Read and study each step of each example.** The examples include important side comments that explain each step. These carefully chosen examples and notes prepare you for success in the exercise set.

■ **Stop and do the margin exercises as you study a section.** Doing the margin exercises is one of the most effective ways to enhance your ability to learn mathematics from this text. Don't deprive yourself of this benefit!

■ **Note the icons listed at the top of each exercise set.** These refer to the many distinctive multimedia study aids that accompany the book.

a Add.

1.
```
  3 6 4
+    2 3
```

2.
```
  1 5 2 1
+    3 4 8
```

3.
```
  1 7 1 6
+ 3 4 8 2
```

4.
```
  7 5 0 3
+ 2 6 8 3
```

5.
```
  8 6
+ 7 8
```

6.
```
  7 3
+ 6 9
```

7.
```
  9 9
+    1
```

8.
```
  9 9 9
+    1 1
```

9. 8113 + 390

10. 271 + 3338

11. 356 + 4910

12. 280 + 34,702

13. 3870 + 92 + 7 + 497

14. 10,120 + 12,989 + 5738

15.
```
  5 0 9 3
+ 3 2 1 7
```

16.
```
  3 6 5 4
+ 2 7 0 0
```

17.
```
  4 8 2 5
+ 1 7 8 3
```

18.
```
  6 7 7 5
+ 1 4 3 2
```

19.
```
  9 9 9 9
+ 6 7 8 5
```

20.
```
  4 5,8 7 9
+ 2 1,7 8 6
```

21.
```
  2 3,4 4 3
+ 1 0,9 8 9
```

22.
```
  6 7,6 5 4
+ 9 8,7 8 6
```

23.
```
  7 7,5 4 3
+ 2 3,7 6 7
```

24.
```
  4 4,6 5 4
+    4,7 6 5
```

25.
```
  9 9,9 9 9
+      1 1 2
```

26.
```
  1 2 7,5 5 6
+   6 8,7 6 6
```

Add from the top. Then check by adding from the bottom.

27.
```
   7
   9
   4
+  8
```

28.
```
   4
   3
   9
   1
+  8
```

29.
```
   8
   6
   2
   3
+  7
```

30.
```
   9
   4
   7
   8
+  7
```

Add. Look for pairs of numbers whose sums are 10, 20, 30, and so on.

31.
```
    7
  1 8
    3
  3 7
+   2
```

32.
```
  2 3
  1 6
  1 1
  1 8
+ 1 9
```

33.
```
  4 5
  2 5
  3 6
  4 4
+ 8 0
```

34.
```
  3 8
  2 7
  3 2
  1 4
+ 7 6
```

35.
```
  4 5 1
    3 6
+ 8 6 2
```

36.
```
    3 1
  7 5 3
+ 9 2 4
```

37.
```
  1 2,0 7 0
    2,9 5 4
+     3,4 0 0
```

38.
```
  4 2,4 8 7
  8 3,1 4 1
+ 3 6,7 1 2
```

39.
```
  3 2 7
  4 2 8
  5 6 9
  7 8 7
+ 2 0 9
```

40.
```
  9 8 9
  5 6 6
  8 3 4
  9 2 0
+ 7 0 3
```

41.
```
  4 8 3 5
    7 2 9
  9 2 0 4
  8 9 8 6
+ 7 9 3 1
```

42.
```
      5,9 4 6
        8 3 4
    1 2,9 5 6
  9 2 8,3 4 2
    3 4,9 0 1
+   5 6,0 0 0
```

CHAPTER 1: Whole Numbers

43.

44.
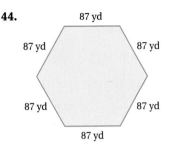

45. Find the perimeter of a standard hockey rink.

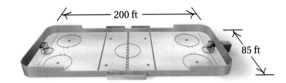

46. In major league baseball, how far does a batter travel in circling the bases when a home run has been hit?

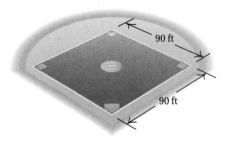

47. ^{D}W Explain in your own words what the associative law of addition means.

48. ^{D}W Describe a situation that corresponds to this mathematical expression:

80 mi + 245 mi + 336 mi.

SKILL MAINTENANCE

The exercises that follow begin an important feature called *Skill Maintenance exercises.* These exercises provide an ongoing review of any preceding objective in the book. You will see them in virtually every exercise set. It has been found that this kind of extensive review can significantly improve your performance on a final examination.

49. What does the digit 8 mean in 486,205? [1.1a]

50. Write a word name for the number in the following sentence: [1.1c]

In a recent year, the New York Yankees topped all professional baseball teams with a total payroll of $114,336,610.

Source: Major League Baseball

SYNTHESIS

51. A fast way to add all the numbers from 1 to 10 inclusive is to pair 1 with 9, 2 with 8, and so on. Use a similar approach to add all numbers from 1 to 100 inclusive.

Objectives

a Convert between addition sentences and subtraction sentences.

b Subtract whole numbers.

1.3 SUBTRACTION

a Subtraction and Related Sentences

TAKE AWAY

Subtraction of whole numbers applies to two kinds of situations. The first is called "take away." Consider the following example.

A bowler starts with 10 pins and knocks down 8 of them.

From 10 pins, the bowler "takes away" 8 pins. There are 2 pins left. The subtraction is $10 - 8 = 2$.

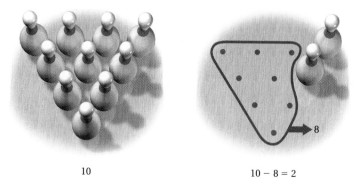

10                    $10 - 8 = 2$

We use the following terminology with subtraction:

$$10 \quad - \quad 8 \quad = \quad 2 \quad .$$

Minuend   Subtrahend   Difference

The **minuend** is the number from which another number is being subtracted. The **subtrahend** is the number being subtracted. The **difference** is the result of subtracting the subtrahend from the minuend.

Subtraction also corresponds to moving distances on a number line. The number line below is marked with tick marks at equal distances of 1 unit. The difference $10 - 8$ is shown. We first move from 0 right 10 units, and then left 8 units, and end up at 2. The subtraction that corresponds to the situation is $10 - 8 = 2$.

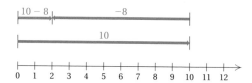

This leads us to the following definition of subtraction.

SUBTRACTION

The difference $a - b$ is that unique whole number c for which $a = c + b$.

RELATED SENTENCES

Subtraction is defined in terms of addition. For example, $5 - 2$ is that number which when added to 2 gives 5. Thus for the subtraction sentence

$5 - 2 = 3$,   Taking away 2 from 5 gives 3.

there is a *related addition sentence*

$5 = 3 + 2$.   Putting back the 2 gives 5 again.

In fact, we know that answers we find to subtractions are correct only because of the related addition, which provides a handy way to *check* a subtraction.

EXAMPLE 1   Write a related addition sentence: $8 - 5 = 3$.

$8 - 5 = 3$

↑

This number
gets added.

↓

$8 = 3 + 5$

By the commutative law of
addition, there is also another
addition sentence:

$8 = 5 + 3$.

↙

The related addition sentence is $8 = 3 + 5$.

Do Exercises 1 and 2.

EXAMPLE 2   Write two related subtraction sentences: $4 + 3 = 7$.

$4 + 3 = 7$ $4 + 3 = 7$

↑   ↑

This addend gets   This addend gets
subtracted from   subtracted from
the sum.   the sum.

↘   ↘

$4 = 7 - 3$ $3 = 7 - 4$

(7 take away 3 is 4.)   (7 take away 4 is 3.)

The related subtraction sentences are $4 = 7 - 3$ and $3 = 7 - 4$.

Do Exercises 3 and 4.

Write a related addition sentence.

1. $7 - 5 = 2$

2. $17 - 8 = 9$

Write two related subtraction
sentences.

3. $5 + 8 = 13$

4. $11 + 3 = 14$

Answers on page A-1

MISSING ADDEND

The second kind of situation to which subtraction can apply is called a "missing addend." You have 2 notebooks, but you need 7. You can think of this as "how many do I need to add to 2 to get 7?" Finding the answer can be thought of as finding a missing addend, and can be found by subtracting 2 from 7.

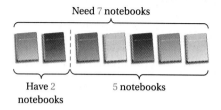

Need 7 notebooks

Have 2
notebooks

5 notebooks

What must be added to 2 to get 7? The answer is 5.

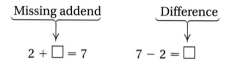

Missing addend

Difference

$2 + \square = 7$

$7 - 2 = \square$

Let's look at the following example in which a missing addend occurs: Jason wants to buy the CD player shown in this ad. He has \$30. He needs \$79. How much more does he need in order to buy the CD player?

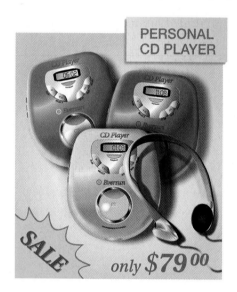

PERSONAL
CD PLAYER

SALE

only 79^{00}

Thinking of this situation in terms of a missing addend, we have:

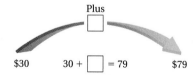

Plus
$\square$

\$30

$30 + \square = 79$

\$79

To find the answer, we think of the related subtraction sentence:

$$30 + \square = 79$$

$$\square = 79 - 30.$$

b Subtraction of Whole Numbers

To subtract numbers, we subtract the ones digits first, then the tens digits, then the hundreds, then the thousands, and so on.

EXAMPLE 3 Subtract: 9768 − 4320.

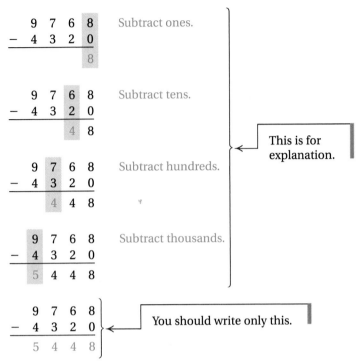

We have considered the subtraction 9768 − 4320 = □. That is, we have found the missing addend in the sentence 9768 = 4320 + □. If 5448 is indeed the missing addend, then if we add it to 4320, the answer should be 9768. The related addition sentence is the basis for adding as a *check*.

Subtraction:

```
    9  7  6  8
 −  4  3  2  0
    5  4  4  8
```

Check by Addition:

```
    5  4  4  8
 +  4  3  2  0
    9  7  6  8
```

Do Exercise 5.

5. Subtract.

```
    7  8  9  3
 −  4  0  9  2
```

Answer on page A-2

Subtract. Check by adding.

6.
```
   8 6 8 6
 − 2 3 5 8
```

7.
```
   7 1 4 5
 − 2 3 9 8
```

Subtract.

8.
```
    7 0
  − 1 4
```

9.
```
    5 0 3
  − 2 9 8
```

Subtract.

10.
```
   7 0 0 7
 − 6 3 4 9
```

11.
```
   6 0 0 0
 − 3 1 4 9
```

12.
```
   9 0 3 5
 − 7 4 8 9
```

EXAMPLE 4   Subtract: 348 − 165.

We have

$$
\begin{array}{rcl}
3 \text{ hundreds} + 4 \text{ tens} + 8 \text{ ones} &=& 2 \text{ hundreds} + 14 \text{ tens} + 8 \text{ ones} \\
-\ 1 \text{ hundred} \ -\ 6 \text{ tens} - 5 \text{ ones} &=& -\ 1 \text{ hundred} \ -\ 6 \text{ tens} - 5 \text{ ones} \\
\hline
&=& 1 \text{ hundred} \ +\ 8 \text{ tens} + 3 \text{ ones} \\
&=& 183.
\end{array}
$$

Note that in this case, although we can subtract the ones (8 − 5 = 3), we cannot do so with the tens, because 4 − 6 is *not* a whole number. To see why, consider

$$4 - 6 = \square \quad \text{and the related addition sentence} \quad 4 = \square + 6.$$

There is no whole number that when added to 6 gives 4. To complete the subtraction, we must *borrow* 1 hundred from 3 hundreds and regroup it with 4 tens. Then we can do the subtraction 14 tens − 6 tens = 8 tens. Below we consider a shortened form.

```
   3 4 8
 − 1 6 5
 ───────
       3
```
Subtract ones.

```
   2 14
   3 4̶ 8
 − 1 6 5
 ───────
       3
```
Borrow one hundred. That is, 1 hundred = 10 tens, and 10 tens + 4 tens = 14 tens. Write 2 above the hundreds column and 14 above the tens.

```
   2 14
   3 4̶ 8
 − 1 6 5
 ───────
   1 8 3
```
Subtract tens; subtract hundreds.

EXAMPLE 5   Subtract: 6246 − 1879.

```
       3 16
   6 2 4̶ 6̶
 − 1 8 7 9
 ─────────
         7
```
We cannot subtract 9 ones from 6 ones, but we can subtract 9 ones from 16 ones. We borrow 1 ten to get 16 ones.

```
     13
   1 3̶ 16
   6 2 4̶ 6̶
 − 1 8 7 9
 ─────────
       6 7
```
We cannot subtract 7 tens from 3 tens, but we can subtract 7 tens from 13 tens. We borrow 1 hundred to get 13 tens.

```
   11 13
   5 1̶ 3̶ 16
   6̶ 2 4̶ 6̶
 − 1 8 7 9
 ─────────
   4 3 6 7
```
We cannot subtract 8 hundreds from 1 hundred, but we can subtract 8 hundreds from 11 hundreds. We borrow 1 thousand to get 11 hundreds.

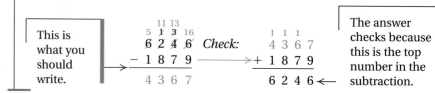

This is what you should write.

```
     11 13
   5 1̶ 3̶ 16
   6̶ 2 4̶ 6̶
 − 1 8 7 9
 ─────────
   4 3 6 7
```

Check:
```
     1 1 1
     4 3 6 7
   + 1 8 7 9
   ─────────
     6 2 4 6 ←
```

The answer checks because this is the top number in the subtraction.

Do Exercises 6 and 7.

EXAMPLE 6 Subtract: 902 − 477.

$$\begin{array}{r} \overset{8}{\cancel{9}}\,\overset{9}{\cancel{0}}\,\overset{12}{\cancel{2}} \\ -\ 4\ 7\ 7 \\ \hline 4\ 2\ 5 \end{array}$$

We cannot subtract 7 ones from 2 ones. We have 9 hundreds, or 90 tens. We borrow 1 ten to get 12 ones. We then have 89 tens.

Do Exercises 8 and 9 on the preceding page.

EXAMPLE 7 Subtract: 8003 − 3667.

$$\begin{array}{r} \overset{7}{\cancel{8}}\,\overset{9}{\cancel{0}}\,\overset{9}{\cancel{0}}\,\overset{13}{\cancel{3}} \\ -\ 3\ 6\ 6\ 7 \\ \hline 4\ 3\ 3\ 6 \end{array}$$

We have 8 thousands, or 800 tens.
We borrow 1 ten to get 13 ones. We then have 799 tens.

EXAMPLES

8. Subtract: 6000 − 3762.

$$\begin{array}{r} \overset{5}{\cancel{6}}\,\overset{9}{\cancel{0}}\,\overset{9}{\cancel{0}}\,\overset{10}{\cancel{0}} \\ -\ 3\ 7\ 6\ 2 \\ \hline 2\ 2\ 3\ 8 \end{array}$$

9. Subtract: 6024 − 2968.

$$\begin{array}{r} \overset{5}{\cancel{6}}\,\overset{9}{\cancel{0}}\,\overset{11}{\underset{1}{\cancel{2}}}\,\overset{14}{\cancel{4}} \\ -\ 2\ 9\ 6\ 8 \\ \hline 3\ 0\ 5\ 6 \end{array}$$

Do Exercises 10–12 on the preceding page.

CALCULATOR CORNER

Subtracting Whole Numbers To subtract whole numbers on a calculator, we use the $\boxed{-}$ and $\boxed{=}$ keys. For example, to find 63 − 47, we press $\boxed{6}$ $\boxed{3}$ $\boxed{-}$ $\boxed{4}$ $\boxed{7}$ $\boxed{=}$. The calculator displays $\boxed{\quad 16\ }$, so 63 − 47 = 16. We can check this result by adding the subtrahend, 47, and the difference, 16. To do this, we press $\boxed{1}$ $\boxed{6}$ $\boxed{+}$ $\boxed{4}$ $\boxed{7}$ $\boxed{=}$. The sum is the minuend, 63, so the subtraction is correct.

Exercises: Use a calculator to perform each subtraction. Check by adding.

1. 57 − 29
2. 81 − 34
3. 145 − 78
4. 612 − 493
5. $\begin{array}{r} 4\ 9\ 7\ 6 \\ -\ 2\ 8\ 4\ 8 \end{array}$
6. $\begin{array}{r} 1\ 2,4\ 0\ 6 \\ -\ \ \ 9\ 8\ 1\ 3 \end{array}$

Study Tips

HIGHLIGHTING

Reading and highlighting a section before your instructor lectures on it allows you to maximize your learning and understanding during the lecture.

■ **Try to keep one section ahead of your syllabus.** If you study ahead of your lectures, you can concentrate on what is being explained in them, rather than trying to write everything down. You can then take notes only of special points or of questions related to what is happening in class.

■ **Highlight important points.** You are probably used to highlighting key points as you study. If that works for you, continue to do so. But you will notice many design features throughout this book that already highlight important points. Thus you may not need to highlight as much as you generally do.

■ **Highlight points that you do not understand.** Use a unique mark to indicate trouble spots that can lead to questions to be asked during class, in a tutoring session, or when calling or contacting the AW Math Tutor Center.

1.3

EXERCISE SET

For Extra Help

Digital Video
Tutor CD 1
Videotape 1

InterAct
Math

Math Tutor
Center

MathXL

MyMathLab

a Write a related addition sentence.

1. $7 - 4 = 3$

2. $12 - 5 = 7$

3. $13 - 8 = 5$

4. $9 - 9 = 0$

5. $23 - 9 = 14$

6. $20 - 8 = 12$

7. $43 - 16 = 27$

8. $51 - 18 = 33$

Write two related subtraction sentences.

9. $6 + 9 = 15$

10. $7 + 9 = 16$

11. $8 + 7 = 15$

12. $8 + 0 = 8$

13. $17 + 6 = 23$

14. $11 + 8 = 19$

15. $23 + 9 = 32$

16. $42 + 10 = 52$

b Subtract.

17.
$$\begin{array}{r} 1\,6 \\ -\quad 4 \\ \hline \end{array}$$

18.
$$\begin{array}{r} 8\,6 \\ -\,1\,3 \\ \hline \end{array}$$

19.
$$\begin{array}{r} 6\,5 \\ -\,2\,1 \\ \hline \end{array}$$

20.
$$\begin{array}{r} 8\,7 \\ -\,3\,4 \\ \hline \end{array}$$

21.
$$\begin{array}{r} 8\,6\,6 \\ -\,3\,3\,3 \\ \hline \end{array}$$

22.
$$\begin{array}{r} 5\,2\,6 \\ -\,3\,2\,3 \\ \hline \end{array}$$

23.
$$\begin{array}{r} 4\,5\,4\,7 \\ -\,3\,4\,2\,1 \\ \hline \end{array}$$

24.
$$\begin{array}{r} 6\,8\,7\,5 \\ -\,2\,1\,1\,1 \\ \hline \end{array}$$

25. $86 - 47$

26. $73 - 28$

27. $625 - 327$

28. $726 - 509$

29. $835 - 609$

30. $953 - 246$

31. $981 - 747$

32. $887 - 698$

33.
$$\begin{array}{r} 7769 \\ -2387 \\ \hline \end{array}$$

34.
$$\begin{array}{r} 6431 \\ -2896 \\ \hline \end{array}$$

35.
$$\begin{array}{r} 3982 \\ -2489 \\ \hline \end{array}$$

36.
$$\begin{array}{r} 7650 \\ -1765 \\ \hline \end{array}$$

37.
$$\begin{array}{r} 5046 \\ -2859 \\ \hline \end{array}$$

38.
$$\begin{array}{r} 6308 \\ -2679 \\ \hline \end{array}$$

39.
$$\begin{array}{r} 7640 \\ -3809 \\ \hline \end{array}$$

40.
$$\begin{array}{r} 8003 \\ -599 \\ \hline \end{array}$$

41.
$$\begin{array}{r} 12{,}647 \\ -4{,}899 \\ \hline \end{array}$$

42.
$$\begin{array}{r} 16{,}222 \\ -5{,}888 \\ \hline \end{array}$$

43.
$$\begin{array}{r} 46{,}771 \\ -12{,}977 \\ \hline \end{array}$$

44.
$$\begin{array}{r} 95{,}654 \\ -48{,}985 \\ \hline \end{array}$$

45. $10{,}002 - 7834$

46. $23{,}048 - 17{,}592$

47. $90{,}237 - 47{,}209$

48. $84{,}703 - 298$

49.
$$\begin{array}{r} 80 \\ -24 \\ \hline \end{array}$$

50.
$$\begin{array}{r} 40 \\ -37 \\ \hline \end{array}$$

51.
$$\begin{array}{r} 90 \\ -54 \\ \hline \end{array}$$

52.
$$\begin{array}{r} 90 \\ -78 \\ \hline \end{array}$$

53.
$$\begin{array}{r} 140 \\ -56 \\ \hline \end{array}$$

54.
$$\begin{array}{r} 470 \\ -188 \\ \hline \end{array}$$

55.
$$\begin{array}{r} 690 \\ -236 \\ \hline \end{array}$$

56.
$$\begin{array}{r} 803 \\ -418 \\ \hline \end{array}$$

57.
$$\begin{array}{r} 903 \\ -132 \\ \hline \end{array}$$

58.
$$\begin{array}{r} 6408 \\ -258 \\ \hline \end{array}$$

59.
$$\begin{array}{r} 2300 \\ -109 \\ \hline \end{array}$$

60.
$$\begin{array}{r} 3506 \\ -1293 \\ \hline \end{array}$$

61.
$$\begin{array}{r} 6808 \\ -3059 \\ \hline \end{array}$$

62.
$$\begin{array}{r} 7840 \\ -3027 \\ \hline \end{array}$$

63.
$$\begin{array}{r} 8092 \\ -1073 \\ \hline \end{array}$$

64.
$$\begin{array}{r} 6007 \\ -1589 \\ \hline \end{array}$$

65. 5843 − 98

66. 10,002 − 398

67. 101,734 − 5760

68. 15,017 − 7809

69. 10,008 − 19

70. 21,043 − 8909

71. 83,907 − 89

72. 311,568 − 19,394

73.
```
   7 0 0 0
 − 2 7 9 4
```

74.
```
   8 0 0 1
 − 6 5 4 3
```

75.
```
   4 8,0 0 0
 − 3 7,6 9 5
```

76.
```
   1 7,0 4 3
 − 1 1,5 9 8
```

77. ^DW Describe two situations that correspond to the subtraction $20 − $17, one "take away" and one "missing addend."

78. ^DW Is subtraction commutative (is there a commutative law of subtraction)? Why or why not?

SKILL MAINTENANCE

Add. [1.2a]

79.
```
   9 4 6
 +   7 8
```

80.
```
   9 0 7 8
 + 3 6 5 4
```

81.
```
   5 7,8 7 7
 + 3 2,4 0 6
```

82.
```
   8 0 0 4
   6 7 8 9
   7 7 2 0
 + 6 8 5 1
```

83. 567 + 778

84. 901 + 23

85. 12,885 + 9807

86. 9909 + 1011

87. Write a word name for 6,375,602.   [1.1c]

88. What does the digit 7 mean in 6,375,602?   [1.1a]

SYNTHESIS

89. Fill in the missing digits to make the subtraction true:
$$9,\square 48,621 − 2,097,\square 81 = 7,251,140.$$

90. 🖩 Subtract: 3,928,124 − 1,098,947.

1.4 ROUNDING AND ESTIMATING; ORDER

a Rounding

We round numbers in various situations if we do not need an exact answer. For example, we might round to check if an answer to a problem is reasonable or to check a calculation done by hand or on a calculator. We might also round to see if we are being charged the correct amount in a store.

To understand how to round, we first look at some examples using number lines, even though this is not the way we generally do rounding.

EXAMPLE 1 Round 47 to the nearest ten.

Here is a part of a number line; 47 is between 40 and 50.

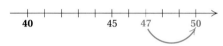

Since 47 is closer to 50, we round up to 50.

EXAMPLE 2 Round 42 to the nearest ten.

42 is between 40 and 50.

Since 42 is closer to 40, we round down to 40.

Do Exercises 1–4.

EXAMPLE 3 Round 45 to the nearest ten.

45 is halfway between 40 and 50.

We could round 45 down to 40 or up to 50. We agree to round up to 50.

> When a number is halfway between rounding numbers, round up.

Do Exercises 5–7.

Objectives

a   Round to the nearest ten, hundred, or thousand.

b   Estimate sums and differences by rounding.

c   Use < or > for ☐ to write a true sentence in a situation like 6 ☐ 10.

Round to the nearest ten.

1. 37

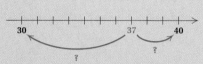

2. 52

3. 73

4. 98

Round to the nearest ten.

5. 35

6. 75

7. 85

Answers on page A-2

Round to the nearest ten.

8. 137

9. 473

10. 235

11. 285

Round to the nearest hundred.

12. 641

13. 759

14. 750

15. 9325

Round to the nearest thousand.

16. 7896

17. 8459

18. 19,343

19. 68,500

Answers on page A-2

Here is a rule for rounding.

ROUNDING WHOLE NUMBERS

To round to a certain place:

a) Locate the digit in that place.

b) Consider the next digit to the right.

c) If the digit to the right is 5 or higher, round up. If the digit to the right is 4 or lower, round down.

d) Change all digits to the right of the rounding location to zeros.

EXAMPLE 4 Round 6485 to the nearest ten.

a) Locate the digit in the tens place, 8.

    6 4 8 5
          ↑

b) Consider the next digit to the right, 5.

    6 4 8 5
            ↑

c) Since that digit is 5 or higher, round 8 tens up to 9 tens.

d) Change all digits to the right of the tens digit to zeros.

    6 4 9 0 ← This is the answer.

EXAMPLE 5 Round 6485 to the nearest hundred.

a) Locate the digit in the hundreds place, 4.

    6 4 8 5
       ↑

b) Consider the next digit to the right, 8.

    6 4 8 5
       ↑

c) Since that digit is 5 or higher, round 4 hundreds up to 5 hundreds.

d) Change all digits to the right of hundreds to zeros.

    6 5 0 0 ← This is the answer.

EXAMPLE 6 Round 6485 to the nearest thousand.

a) Locate the digit in the thousands place, 6.

    6 4 8 5
   ↑

b) Consider the next digit to the right, 4.

    6 4 8 5
   ↑

c) Since that digit is 4 or lower, round down, meaning that 6 thousands stays as 6 thousands.

d) Change all digits to the right of thousands to zeros.

    6 0 0 0 ← This is the answer.

Do Exercises 8–19.

CAUTION!

7000 is not a correct answer to Example 6. It is incorrect to round from the ones digit over, as follows:

6485, 6490, 6500, 7000.

Sometimes rounding involves changing more than one digit in a number.

EXAMPLE 7 Round 78,595 to the nearest ten.

a) Locate the digit in the tens place, 9.

    7 8,5 9 5
         ↑

b) Consider the next digit to the right, 5.

    7 8,5 9 5
           ↑

c) Since that digit is 5 or higher, round 9 tens to 10 tens. To carry this out, we think of 10 tens as 1 hundred + 0 tens, and increase the hundreds digit by 1, to get 6 hundreds + 0 tens. We then write 6 in the hundreds place and 0 in the tens place.

d) Change the digit to the right of the tens digit to zero.

    7 8,6 0 0 ← This is the answer.

Note that if we round this number to the nearest hundred, we get the same answer.

Do Exercises 20 and 21.

There are many methods of rounding. For example, in computer applications, the rounding of 8563 to the nearest hundred might be done using a different rule called **truncating,** meaning that we simply change all digits to the right of the rounding location to zeros. Thus, 8563 would round to 8500, which is not the same answer that we would get using the rule discussed in this section.

b Estimating

In the following example, we see how estimation can be used in making a purchase.

EXAMPLE 8 *Estimating the Cost of an Automobile Purchase.* Maria and Luis Vasquez are shopping for a new car. They are considering an Oldsmobile Alero. There are three basic models of this car, and each has options beyond the basic price, as shown in the chart on the following page. Maria and Luis have allowed themselves a budget of $20,000. They look at the list of options and want to make a quick estimate of the cost of model GL3 with all the options.

Refer to the chart below to answer Margin Exercises 22 and 23.

22. Suppose Maria and Luis want to buy a GL1 with all options except the sunroof and the sport package.

   a) Estimate this cost by rounding to the nearest hundred.

Estimate by rounding to the nearest hundred the cost of the GL3 with all the options and decide whether it will fit into their budget.

MODEL GL1 SEDAN (4 DOOR) 2.4-liter engine, 5 SPEED MANUAL TRANSMISSION	MODEL GL2 SEDAN (4 DOOR) 2.4-liter engine, 4 SPEED AUTOMATIC TRANSMISSION	MODEL GL3 SEDAN (4 DOOR) 3.4-liter engine, 4 SPEED AUTOMATIC TRANSMISSION	
Base price: $17,650	Base price: $18,270	Base price: $18,875	
Destination charges: $535	Destination charges: $535	Destination charges: $535	
Each of these vehicles comes with several options.			
Driver's seat with 6-way power adjustment:			$305
Sunroof:			$650
Feature package: 15" aluminum wheels, remote keyless entry, foglamps, leather-wrapped steering wheel and shift knob			$585
Sport package: 16" aluminum wheels, performance radial tires, and performance suspension			$450
Rear decklid spoiler:			$225
Radio: AM/FM cassette/CD with 6-speaker dimensional sound system			$200

Source: General Motors

   b) Can they afford this car with a budget of $20,000?

23. By eliminating options, find a way that Luis and Maria can buy the GL3 and stay within their $20,000 budget. Answers may vary.

First, we list the base price of the GL3 and then the cost of each of the options. We then round each number to the nearest hundred and add.

```
  1 8,8 7 5        1 8,9 0 0
      5 3 5            5 0 0
      3 0 5            3 0 0
      6 5 0            7 0 0
      5 8 5            6 0 0
      4 5 0            5 0 0
      2 2 5            2 0 0
  +   2 0 0        +   2 0 0
                   ─────────────
                   2 1,9 0 0  ← Estimated answer
```

The estimated total cost is $21,900. Since Maria and Luis have allowed themselves a budget of $20,000 for their car, they will need to forego some options.

Do Exercises 22 and 23.

Answers on page A-2

Estimating can be done in many ways and can have many results, even though in the problems that follow we ask you to round in a specific way.

EXAMPLE 9 Estimate this sum by first rounding to the nearest ten:

78 + 49 + 31 + 85.

We round each number to the nearest ten. Then we add.

```
  7 8        8 0
  4 9        5 0
  3 1        3 0
+ 8 5      + 9 0
           2 5 0  ← Estimated answer
```

EXAMPLE 10 Estimate this sum by first rounding to the nearest hundred:

850 + 674 + 986 + 839.

We have

```
  8 5 0        9 0 0
  6 7 4        7 0 0
  9 8 6      1 0 0 0
+ 8 3 9      +   8 0 0
             3 4 0 0
```

Do Exercises 24 and 25.

EXAMPLE 11 Estimate the difference by first rounding to the nearest thousand: 9324 − 2849.

We have

```
  9 3 2 4        9 0 0 0
− 2 8 4 9      − 3 0 0 0
               6 0 0 0
```

Do Exercises 26 and 27.

The sentence 7 − 5 = 2 says that 7 − 5 is the same as 2. When we round, the result is rarely the same as the number we started with. Thus we use the symbol ≈ when rounding. This symbol means "**is approximately equal to.**" For example, when 687 is rounded to the nearest ten, we can write

687 ≈ 690.

C Order

We know that 2 is not the same as 5. We express this by the sentence 2 ≠ 5. We also know that 2 is less than 5. We symbolize this by the expression 2 < 5. We can see this order on a number line: 2 is to the left of 5.

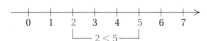

The number 0 is the smallest whole number.

---

24. Estimate the sum by first rounding to the nearest ten. Show your work.

```
  7 4
  2 3
  3 5
+ 6 6
```

25. Estimate the sum by first rounding to the nearest hundred. Show your work.

```
  6 5 0
  6 8 5
  2 3 8
+ 1 6 8
```

26. Estimate the difference by first rounding to the nearest hundred. Show your work.

```
  9 2 8 5
− 6 7 3 9
```

27. Estimate the difference by first rounding to the nearest thousand. Show your work.

```
  2 3,2 7 8
− 1 1,6 9 8
```

Answers on page A-2

Use < or > for ☐ to write a true sentence. Draw a number line if necessary.

28. 8 ☐ 12

29. 12 ☐ 8

30. 76 ☐ 64

31. 64 ☐ 76

32. 217 ☐ 345

33. 345 ☐ 217

Answers on page A-2

ORDER OF WHOLE NUMBERS

For any whole numbers *a* and *b*:

1. $a < b$ (read "*a* is less than *b*") is true when *a* is to the left of *b* on a number line.
2. $a > b$ (read "*a* is greater than *b*") is true when *a* is to the right of *b* on a number line.

We call < and > **inequality symbols.**

EXAMPLE 12 Use < or > for ☐ to write a true sentence: 7 ☐ 11.

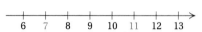

Since 7 is to the left of 11 on a number line, 7 < 11.

EXAMPLE 13 Use < or > for ☐ to write a true sentence: 92 ☐ 87.

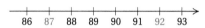

Since 92 is to the right of 87 on a number line, 92 > 87.

A sentence like 8 + 5 = 13 is called an **equation.** It is a *true* equation. The equation 4 + 8 = 11 is a *false* equation. A sentence like 7 < 11 is called an **inequality.** The sentence 7 < 11 is a *true* inequality. The sentence 23 > 69 is a *false* inequality.

Do Exercises 28–33.

Study Tips

LEARNING RESOURCES

■ The *Student's Solutions Manual* contains worked-out solutions to the odd-numbered exercises in the exercise sets.

■ An extensive set of *videotapes* supplements this text. These are available on CD-ROM by calling 1-800-282-0693.

■ *Tutorial software* called InterAct Math also accompanies this text. If it is not available in the campus learning center, you can order it by calling 1-800-282-0693.

■ The Addison-Wesley *Math Tutor Center* is available for help with the odd-numbered exercises by experienced instructors. You can order this service by calling 1-800-824-7799.

■ Extensive help is available online via MyMathLab and/or MathXL. Ask your instructor for information about these or visit MyMathLab.com and MathXL.com.

1.4

EXERCISE SET

For Extra Help

Digital Video
Tutor CD 1
Videotape 1

InterAct
Math

Math Tutor
Center

MathXL

MyMathLab

a Round to the nearest ten.

1. 48   **2.** 532   **3.** 467   **4.** 8945

5. 731   **6.** 17   **7.** 895   **8.** 798

Round to the nearest hundred.

9. 146   **10.** 874   **11.** 957   **12.** 650

13. 9079   **14.** 4645   **15.** 32,850   **16.** 198,402

Round to the nearest thousand.

17. 5876   **18.** 4500   **19.** 7500   **20.** 2001

21. 45,340   **22.** 735,562   **23.** 373,405   **24.** 6,713,855

b Estimate the sum or difference by first rounding to the nearest ten. Show your work.

25.
```
   7 8
 + 9 7
```

26.
```
   6 2
   9 7
   4 6
 + 8 8
```

27.
```
   8 0 7 4
 - 2 3 4 7
```

28.
```
   6 7 3
 -    2 8
```

Estimate the sum by first rounding to the nearest ten. Do any of the given sums seem to be incorrect when compared to the estimate? Which ones?

29.
```
   4 5
   7 7
   2 5
 + 5 6
   3 4 3
```

30.
```
   4 1
   2 1
   5 5
 + 6 0
   1 7 7
```

31.
```
   6 2 2
     7 8
     8 1
 + 1 1 1
   9 3 2
```

32.
```
   8 3 6
   3 7 4
   7 9 4
 + 9 3 8
   3 9 4 7
```

Estimate the sum or difference by first rounding to the nearest hundred. Show your work.

33.
```
   7 3 4 8
 + 9 2 4 7
```

34.
```
   5 6 8
   4 7 2
   9 3 8
 + 4 0 2
```

35.
```
   6 8 5 2
 - 1 7 4 8
```

36.
```
   9 4 3 8
 - 2 7 8 7
```

Gateway Computers. Gateway, Inc., recently sold a model of computer called the Performance 1600. The base price for this model, with the features shown in the following ad, was $1399. Options that could be included when ordering are shown in the table below.

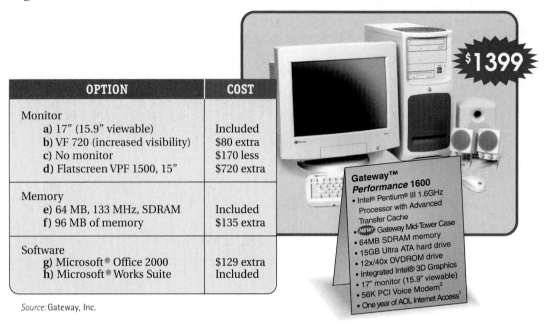

OPTION	COST
Monitor	
a) 17" (15.9" viewable)	Included
b) VF 720 (increased visibility)	$80 extra
c) No monitor	$170 less
d) Flatscreen VPF 1500, 15"	$720 extra
Memory	
e) 64 MB, 133 MHz, SDRAM	Included
f) 96 MB of memory	$135 extra
Software	
g) Microsoft® Office 2000	$129 extra
h) Microsoft® Works Suite	Included

Source: Gateway, Inc.

37. Estimate the cost, by rounding to the nearest hundred, of the computer with options (a), (f), and (h).

38. Estimate the cost, by rounding to the nearest hundred, of the computer with options (b), (f), and (h).

39. Jamaal and Natasha are shopping for a computer and have a budget of $1700. Estimate the cost, by rounding to the nearest hundred, of the computer with options (c), (e), and (g). Can they afford the computer?

40. Max is shopping for a computer and has a budget of $2100. Estimate the cost, by rounding to the nearest hundred, of the computer with options (d), (f), and (g). Can he afford the computer?

41. Suppose you need to buy a computer with a monitor, and you select this model. Decide on the options you would like and estimate the cost to the nearest ten dollars. Answers will vary.

42. Suppose you had decided on this computer and already had a monitor. Select any of the remaining options and estimate the cost to the nearest ten dollars. Answers will vary.

Estimate the sum by first rounding to the nearest hundred. Do any of the given sums seem to be incorrect when compared to the estimate? Which ones?

	43.			44.			45.			46.	
		2 1 6			4 8 1			7 5 0			3 2 6
		8 4			7 0 2			4 2 8			2 7 5
		7 4 5			6 2 3			6 3			7 5 8
	+	5 9 5		+	1 0 4 3		+	2 0 5		+	9 4 3
		1 6 4 0			1 8 4 9			1 4 4 6			2 3 0 2

Estimate the sum or difference by first rounding to the nearest thousand. Show your work.

	47.			48.			49.			50.	
		9 6 4 3			7 6 4 8			9 2,1 4 9			8 4,8 9 0
		4 8 2 1			9 3 4 8		−	2 2,5 5 5		−	1 1,1 1 0
		8 9 4 3			7 8 4 2						
	+	7 0 0 4		+	2 2 2 2						

C   Use < or > for ☐ to write a true sentence. Draw a number line if necessary.

51. 0 ☐ 17

52. 32 ☐ 0

53. 34 ☐ 12

54. 28 ☐ 18

55. 1000 ☐ 1001

56. 77 ☐ 117

57. 133 ☐ 132

58. 999 ☐ 997

59. 460 ☐ 17

60. 345 ☐ 456

61. 37 ☐ 11

62. 12 ☐ 32

Land-Speed Cars.   Two competing jet-powered cars may soon travel faster than the speed of sound. The Thrust SCC is 54 ft long and weighs 7 tons. The Spirit of America is 47 ft long and weighs 4 tons. Use this information to answer Exercises 63 and 64.

Sources: *Car & Driver,* September 1996; *Advanced Materials and Processes,* January 1998

63. Which is longer, the Thrust SCC or the Spirit of America? Express the numbers in the situation as an inequality.

64. Which is heavier, the Thrust SCC or the Spirit of America? Express the numbers in the situation as an inequality.

65. *Life Expectancy.* The life expectancy of a female in 2050 is predicted to be about 87 yr and of a male about 81 yr. Use an inequality to compare these life expectancies.

66. *Utilities.* The average yearly cost of utilities for households in the Northeast is $1644 and for households in the West is $1014. Use an inequality to compare the costs.

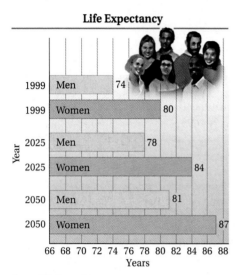

Life Expectancy

1999 Men 74
1999 Women 80
2025 Men 78
2025 Women 84
2050 Men 81
2050 Women 87

Year

66 68 70 72 74 76 78 80 82 84 86 88
Years

Source: U.S. Census Bureau

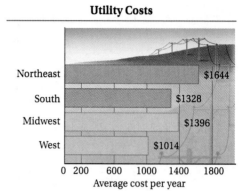

Utility Costs

Northeast $1644
South $1328
Midwest $1396
West $1014

0 200 600 1000 1400 1800
Average cost per year

Source: Energy Information Administration

67. D_W Explain how estimating and rounding can be useful when shopping for groceries.

68. D_W When rounding 748 to the nearest hundred, a student rounds to 750 and then to 800. What mistake is he making?

─────────── **SKILL MAINTENANCE** ───────────

Write standard notation.

69. 7 thousands + 9 hundreds + 9 tens + 2 ones   [1.1b]

70. Twenty-three million   [1.1c]

71. Write a word name for 246,605,004,032.   [1.1c]

72. Write expanded notation for 8017.   [1.1b]

Add.   [1.2a]

73.   6 7,7 8 9
      + 1 8,9 6 5

74.   9 0 0 2
       + 4 5 8 7

Subtract.   [1.3b]

75.   6 7,7 8 9
      − 1 8,9 6 5

76.   9 0 0 2
       − 4 5 8 7

─────────── **SYNTHESIS** ───────────

77.–80. 🖩 Use a calculator to find the sums and differences in each of Exercises 47–50. Then check your answers using estimation. Even when using a calculator it is possible to make an error if you press the wrong buttons, so it is a good idea to check by estimating.

1.5 MULTIPLICATION

Objectives

 a Multiply whole numbers.

b Estimate products by rounding.

c Use multiplication in finding area.

a   Multiplication of Whole Numbers

REPEATED ADDITION

The multiplication 3×5 corresponds to this repeated addition:

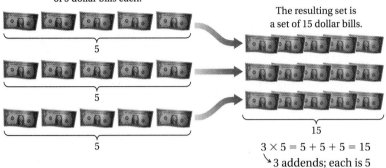

We combine 3 sets of 5 dollar bills each.

The resulting set is a set of 15 dollar bills.

$$3 \times 5 = 5 + 5 + 5 = 15$$
↘ 3 addends; each is 5

The numbers that we multiply are called **factors.** The result of the multiplication is called a **product.**

$$
\begin{array}{cccc}
3 & \times & 5 & = & 15 \\
\downarrow & & \downarrow & & \downarrow \\
\text{Factor} & & \text{Factor} & & \text{Product}
\end{array}
$$

RECTANGULAR ARRAYS

Multiplications can also be thought of as rectangular arrays. Each of the following corresponds to the multiplication 3×5.

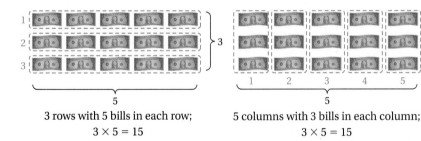

3 rows with 5 bills in each row;
$3 \times 5 = 15$

5 columns with 3 bills in each column;
$3 \times 5 = 15$

When you write a multiplication sentence corresponding to a real-world situation, you should think of either a rectangular array or repeated addition. In some cases, it may help to think both ways.

We have used an "$\times$" to denote multiplication. A dot "$\cdot$" is also commonly used. (Use of the dot is attributed to the German mathematician Gottfried Wilhelm von Leibniz in 1698.) Parentheses are also used to denote multiplication. For example,

$$3 \times 5 = 3 \cdot 5 = (3)(5) = 3(5) = 15.$$

Multiply.

1.   5 8
     × 2

2.   3 7
     × 4

3.   8 2 3
     × 6

4.   1 3 4 8
     ×   5

Answers on page A-2

The product of 0 and any whole number is 0: $0 \cdot a = a \cdot 0 = 0$. Multiplying a number by 1 does not change the number: $1 \cdot a = a \cdot 1 = a$. We say that 1 is the **multiplicative identity.**

EXAMPLE 1   Multiply: 5×734.

We have

```
        7 3 4
×           5
        ─────────
            2 0  ← Multiply the 4 ones by 5: 5 × 4 = 20.
          1 5 0  ← Multiply the 3 tens by 5: 5 × 30 = 150.
        3 5 0 0  ← Multiply the 7 hundreds by 5: 5 × 700 = 3500.
        ─────────
        3 6 7 0  ← Add.
```

Instead of writing each product on a separate line, we can use a shorter form.

```
          2
      7 3 4
×         5
      ─────────
            0
```
Multiply the ones by 5: $5 \cdot (4 \text{ ones}) = 20 \text{ ones} = 2 \text{ tens} + 0 \text{ ones}$. Write 0 in the ones column and 2 above the tens.

```
    1   2
      7 3 4
×         5
      ─────────
          7 0
```
Multiply the 3 tens by 5 and add 2 tens: $5 \cdot (3 \text{ tens}) = 15 \text{ tens}$, $15 \text{ tens} + 2 \text{ tens} = 17 \text{ tens} = 1 \text{ hundred} + 7 \text{ tens}$. Write 7 in the tens column and 1 above the hundreds.

```
    1   2
      7 3 4
×         5
      ─────────
      3 6 7 0
```
Multiply the 7 hundreds by 5 and add 1 hundred: $5 \cdot (7 \text{ hundreds}) = 35 \text{ hundreds}$, $35 \text{ hundreds} + 1 \text{ hundred} = 36 \text{ hundreds}$.

```
    1   2
      7 3 4 ⎤
×         5 ⎬  You should write only this.
      3 6 7 0 ⎦
```

Do Exercises 1–4.

Let's find the product

```
      5 4
× 3 2
```

To do this, we multiply 54 by 2, then 54 by 30, and then add.

```
      5 4              1
                       5 4
×     2            ×   3 0
────────           ────────
  1 0 8            1 6 2 0
```

Since we are going to add the results, let's write the work this way.

```
      5 4
×   3 2
──────────
  1 0 8      Multiplying by 2
1 6 2 0      Multiplying by 30
──────────
1 7 2 8      Adding to obtain the product
```

The fact that we can do this is based on a property called the **distributive law.** It says that to multiply a number by a sum, $a \cdot (b + c)$, we can multiply each part by a and then add like this: $(a \cdot b) + (a \cdot c)$. Thus, $a \cdot (b + c) = (a \cdot b) + (a \cdot c)$. Applied to the example above, the distributive law gives us

$$54 \cdot 32 = 54 \cdot (30 + 2) = (54 \cdot 30) + (54 \cdot 2).$$

EXAMPLE 2 Multiply: 43×57.

```
        2
      5 7
  ×   4 3
  ─────────
    1 7 1      Multiplying by 3
```

```
        2
        2
      5 7
  ×   4 3
  ─────────
    1 7 1
  2 2 8 0      Multiplying by 40. (We write a 0 and then multiply 57
               by 4).
```

You may have learned that such a 0 does not have to be written. You may omit it if you wish. If you do omit it, remember, when multiplying by tens, to put the answer in the tens place.

```
        2
        2
      5 7
  ×   4 3
  ─────────
    1 7 1
  2 2 8 0
  ─────────
  2 4 5 1      Adding to obtain the product
```

Do Exercises 5 and 6.

EXAMPLE 3 Multiply: 457×683.

```
      5 2
    6 8 3
  × 4 5 7
  ─────────
  4 7 8 1      Multiplying 683 by 7
```

```
    4 1
    5 2
    6 8 3
  × 4 5 7
  ─────────
  4 7 8 1
3 4 1 5 0      Multiplying 683 by 50
```

```
    3 1
    4 1
    5 2
    6 8 3
  × 4 5 7
  ─────────
      4 7 8 1
    3 4 1 5 0
  2 7 3 2 0 0      Multiplying 683 by 400
  ─────────────
  3 1 2 , 1 3 1    Adding
```

Do Exercises 7 and 8.

Multiply.

5.
```
    4 5
  × 2 3
```

6. 48×63

7.
```
    7 4 6
  ×   6 2
```

8. 245×837

Multiply.

9.
```
    4 7 2
  × 3 0 6
```

10. 408×704

11.
```
    2 3 4 4
  × 6 0 0 5
```

Answers on page A-2

Multiply.

12.
```
  4 7 2
× 8 3 0
```

13.
```
  2 3 4 4
× 7 4 0 0
```

14. 100×562

15. 1000×562

16. a) Find $23 \cdot 47$.

b) Find $47 \cdot 23$.

c) Compare your answers to parts (a) and (b).

Multiply.

17. $5 \cdot 2 \cdot 4$

18. $5 \cdot 1 \cdot 3$

CALCULATOR CORNER

Multiplying Whole Numbers To multiply whole numbers on a calculator, we use the $\boxed{\times}$ and $\boxed{=}$ keys. For example, to find 13×47, we press $\boxed{1}$ $\boxed{3}$ $\boxed{\times}$ $\boxed{4}$ $\boxed{7}$ $\boxed{=}$. The calculator displays 611, so $13 \times 47 = 611$.

Exercises: Use a calculator to find each product.

1. 56×8

2. 845×26

3. $5 \cdot 1276$

4. $126(314)$

5.
```
  3 7 6 0
×     4 8
```

6.
```
  5 2 1 8
×   4 5 3
```

Answers on page A-2

EXAMPLE 4 Multiply: 306×274.

Note that $306 = 3$ hundreds $+ 6$ ones.

```
        2 7 4
      × 3 0 6
      1 6 4 4    Multiplying by 6
    8 2 2 0 0    Multiplying by 3 hundreds. (We write 00
                 and then multiply 274 by 3.)
    8 3,8 4 4    Adding
```

Do Exercises 9–11 on the preceding page.

EXAMPLE 5 Multiply: 360×274.

Note that $360 = 3$ hundreds $+ 6$ tens.

```
        2 7 4    ┌─Multiplying by 6 tens. (We write 0 and
      ×   3 6 0  │  then multiply 274 by 6.)
      1 6 4 4 0 ←┤─Multiplying by 3 hundreds. (We write 00
    8 2 2 0 0  ←─┘  and then multiply 274 by 3.)
    9 8,6 4 0    Adding
```

Do Exercises 12–15.

Check on your own that $17 \cdot 37 = 629$ and that $37 \cdot 17 = 629$. This illustrates the **commutative law of multiplication.** It says that we can multiply two numbers in any order, $a \cdot b = b \cdot a$, and get the same answer.

Do Exercise 16.

To multiply three or more numbers, we generally group them so that we multiply two at a time. Consider $2 \cdot (3 \cdot 4)$ and $(2 \cdot 3) \cdot 4$. The parentheses tell what to do first:

$$2 \cdot (3 \cdot 4) = 2 \cdot (12) = 24. \qquad \text{We multiply 3 and 4, then 2.}$$

We can also multiply 2 and 3, then 4:

$$(2 \cdot 3) \cdot 4 = (6) \cdot 4 = 24.$$

Either way we get 24. It does not matter how we group the numbers. This illustrates that **multiplication is associative:** $a \cdot (b \cdot c) = (a \cdot b) \cdot c$.

Do Exercises 17 and 18.

b Estimating Products by Rounding

EXAMPLE 6 *Computer Memory.* Shaquille is buying a Gateway Performance 800 computer. He wants to add memory and finds that each additional block of 32 MB memory costs $44. By rounding to the nearest ten, estimate the cost if he purchases 16 additional blocks of memory.
Source: 2000 Gateway, Inc.

We want to estimate the product of 16 and 44. To do so, we round each factor to the nearest ten and multiply the rounded numbers:

Exact	Nearest ten
4 4	4 0
× 1 6	× 2 0
7 0 4	8 0 0

The additional memory will cost about $800.

Do Exercise 19.

19. Computer Memory. By rounding to the nearest ten, estimate the cost to Shaquille of 8 extra blocks of memory.

EXAMPLE 7 Estimate the following product by first rounding to the nearest ten and to the nearest hundred: 683×457.

Nearest ten	Nearest hundred	Exact
6 8 0	7 0 0	6 8 3
× 4 6 0	× 5 0 0	× 4 5 7
4 0 8 0 0	3 5 0 0 0 0	4 7 8 1
2 7 2 0 0 0		3 4 1 5 0
3 1 2 8 0 0		2 7 3 2 0 0
		3 1 2 1 3 1

Do Exercise 20.

20. Estimate the product by first rounding to the nearest ten and to the nearest hundred. Show your work.

8 3 7
× 2 4 5

c Finding Area

The area of a rectangular region is often considered to be the number of square units needed to fill it. Here is a rectangle 4 cm (centimeters) long and 3 cm wide. It takes 12 square centimeters (sq cm) to fill it.

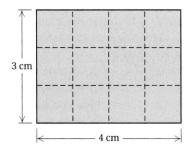

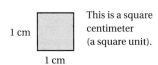
1 cm   This is a square centimeter (a square unit).
1 cm

In this case, we have a rectangular array of 3 rows, each of which contains 4 squares. The number of square units is given by $3 \cdot 4$, or 12.

Answers on page A-2

21. Table Tennis. Find the area of a standard table tennis table that has dimensions of 9 ft by 5 ft.

Professional pool player Jeanette Lee (also known as the "Black Widow")

Answer on page A-2

■ **EXAMPLE 8** *Professional Pool Table.* The playing area of a standard pool table has dimensions of 50 in. by 100 in. (There are rails 6 in. wide on the outside not included in the playing area.) Find the playing area.

50 in.

|← —————— 100 in. —————— →|

If we think of filling the rectangle with square inches, we have a rectangular array. The length $l = 100$ in. and the width $w = 50$ in. Thus the area A is given by the formula

$$A = l \cdot w = 100 \cdot 50 = 5000 \text{ sq in.}$$

Do Exercise 21.

Study Tips

TIME MANAGEMENT (PART 1)

Time is the most critical factor in your success in learning mathematics. Have reasonable expectations about the time you need to study math.

■ **Juggling time.** Working 40 hours per week and taking 12 credit hours is equivalent to working two full-time jobs. Can you handle such a load? Your ratio of number of work hours to number of credit hours should be about 40/3, 30/6, 20/9, 10/12, or 5/14.

■ **A rule of thumb on study time.** Budget about 2–3 hours for homework and study per week for every hour of class time.

■ **Scheduling your time.** Make an hour-by-hour schedule of your typical week. Include work, school, home, sleep, study, and leisure times. Try to schedule time for study when you are most alert. Choose a setting that will enable you to maximize your concentration. Plan for success and it will happen!

"You cannot increase the quality or quantity of your achievement or performance except to the degree in which you increase your ability to use time effectively."

Brian Tracy, motivational/inspirational speaker

EXERCISE SET

For Extra Help

Digital Video Tutor CD 1 Videotape 1   InterAct Math   Math Tutor Center   MathXL   MyMathLab

a Multiply.

1. 87
      × 1 0

2.   1 0 0
      ×   9 6

3.   2 3 4 0
      × 1 0 0 0

4.   8 0 0
      ×   7 0

5.   6 5
      ×   8

6.   8 7
      ×   4

7.   9 4
      ×   6

8.   7 6
      ×   9

9. 3 · 509

10. 7 · 806

11. 7(9229)

12. 4(7867)

13. 90(53)

14. 60(78)

15. (47)(85)

16. (34)(87)

17.   6 4 0
      ×   7 2

18.   7 7 7
      ×   7 7

19.   4 4 4
      ×   3 3

20.   5 0 9
      ×   8 8

21.   5 0 9
      × 4 0 8

22.   4 3 2
      × 3 7 5

23.   8 5 3
      × 9 3 6

24.   3 4 6
      × 6 5 0

25.   6 4 2 8
      × 3 2 2 4

26.   8 9 2 8
      × 3 1 7 2

27.   3 4 8 2
      ×   1 0 4

28.   6 4 0 8
      × 6 0 6 4

29.
$$\begin{array}{r} 5\ 0\ 0\ 6 \\ \times\ 4\ 0\ 0\ 8 \\ \hline \end{array}$$

30.
$$\begin{array}{r} 6\ 7\ 8\ 9 \\ \times\ 2\ 3\ 3\ 0 \\ \hline \end{array}$$

31.
$$\begin{array}{r} 5\ 6\ 0\ 8 \\ \times\ 4\ 5\ 0\ 0 \\ \hline \end{array}$$

32.
$$\begin{array}{r} 4\ 5\ 6\ 0 \\ \times\ 7\ 8\ 9\ 0 \\ \hline \end{array}$$

33.
$$\begin{array}{r} 8\ 7\ 6 \\ \times\ 3\ 4\ 5 \\ \hline \end{array}$$

34.
$$\begin{array}{r} 3\ 5\ 5 \\ \times\ 2\ 9\ 9 \\ \hline \end{array}$$

35.
$$\begin{array}{r} 7\ 8\ 8\ 9 \\ \times\ 6\ 2\ 2\ 4 \\ \hline \end{array}$$

36.
$$\begin{array}{r} 6\ 5\ 0\ 1 \\ \times\ 3\ 4\ 4\ 9 \\ \hline \end{array}$$

b Estimate the product by first rounding to the nearest ten. Show your work.

37.
$$\begin{array}{r} 4\ 5 \\ \times\ 6\ 7 \\ \hline \end{array}$$

38.
$$\begin{array}{r} 5\ 1 \\ \times\ 7\ 8 \\ \hline \end{array}$$

39.
$$\begin{array}{r} 3\ 4 \\ \times\ 2\ 9 \\ \hline \end{array}$$

40.
$$\begin{array}{r} 6\ 3 \\ \times\ 5\ 4 \\ \hline \end{array}$$

Estimate the product by first rounding to the nearest hundred. Show your work.

41.
$$\begin{array}{r} 8\ 7\ 6 \\ \times\ 3\ 4\ 5 \\ \hline \end{array}$$

42.
$$\begin{array}{r} 3\ 5\ 5 \\ \times\ 2\ 9\ 9 \\ \hline \end{array}$$

43.
$$\begin{array}{r} 4\ 3\ 2 \\ \times\ 1\ 9\ 9 \\ \hline \end{array}$$

44.
$$\begin{array}{r} 7\ 8\ 9 \\ \times\ 4\ 3\ 4 \\ \hline \end{array}$$

45. *Oldsmobile Intrigue.* Pure-Health Medical Supplies, Inc., buys an Oldsmobile Intrigue, Model GLS, for each of its 185 sales representatives. Each car costs $25,720 plus an additional $560 per car in destination charges.

a) Estimate the total cost of the purchase by rounding the cost of each car and the number of sales representatives to the nearest hundred.

b) Estimate the total cost of the purchase by rounding the cost of each car to the nearest thousand and the number of reps to the nearest hundred.

Source: General Motors

46. A travel club of 248 people decides to fly from New York to Paris. The cost of a round-trip ticket is $376.

a) Estimate the total cost of the trip by rounding the cost of the airfare and the number of travelers to the nearest ten.

b) Estimate the total cost of the trip by rounding the cost of the airfare to the nearest hundred and the number of travelers to the nearest ten.

C   What is the area of the region?

47.

728 mi

728 mi

48.

129 yd

65 yd

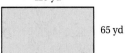

49. Find the area of the region formed by the base lines on a Major League baseball diamond.

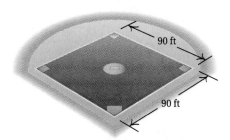

90 ft

90 ft

50. Find the area of a standard-sized hockey rink.

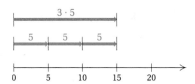

200 ft

85 ft

51. ᴰᴡ Describe a situation that corresponds to each multiplication: 4 · $150; $4 · 150.

52. ᴰᴡ Explain the multiplication illustrated in the diagram below.

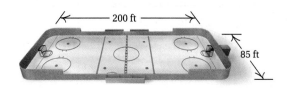

3 · 5

5   5   5

0   5   10   15   20

SKILL MAINTENANCE

Add. [1.2a]

53.
```
  4 9 0 8
  5 6 6 7
+ 2 1 1 0
```

54.
```
  9 8 7 6
    8 7 6
      7 6
+      6
```

55.
```
  3 4 0,7 9 8
+   8 6,6 7 9
```

56.
```
  8 8,7 7 7
+ 2 2,3 3 3
```

Subtract. [1.3b]

57.
```
  4 9 0 8
- 3 6 6 7
```

58.
```
  9 8 7 6
-   9 8 7
```

59.
```
  3 4 0,7 9 8
-   8 6,6 7 9
```

60.
```
  8 8,7 7 7
- 2 2,3 3 3
```

61. Round 6,375,602 to the nearest thousand. [1.4a]

62. Round 6,375,602 to the nearest ten. [1.4a]

SYNTHESIS

63. ▦ An 18-story office building is box-shaped. Each floor measures 172 ft by 84 ft with a 20-ft by 35-ft rectangular area lost to an elevator and a stairwell. How much area is available as office space?

1. Consider $54 \div 6 = 9$. Express this division in two other ways.

a Division and Related Sentences

REPEATED SUBTRACTION

Division of whole numbers applies to two kinds of situations. The first is repeated subtraction. Suppose we have 20 notebooks in a pile, and we want to find out how many sets of 5 there are. One way to do this is to repeatedly subtract sets of 5 as follows.

20 notebooks

How many sets of 5 notebooks each?

Since there are 4 sets of 5 notebooks each, we have

$$20 \div 5 = 4.$$

The division $20 \div 5$, read "20 divided by 5," corresponds to the figure above. We say that the **dividend** is 20, the **divisor** is 5, and the **quotient** is 4.

$$\underset{\text{Dividend}}{20} \quad \underset{\text{Divisor}}{\div \; 5} \quad \underset{\text{Quotient}}{= \; 4}$$

We divide the *dividend* by the *divisor* to get the *quotient*.

We can also express the division $20 \div 5 = 4$ as

$$\frac{20}{5} = 4 \quad \text{or} \quad 5\overline{)20}^{\,4}$$

Do Exercise 1.

RECTANGULAR ARRAYS AND MISSING FACTORS

We can also think of division in terms of rectangular arrays. Consider again the pile of 20 notebooks and division by 5. We can arrange the notebooks in a rectangular array with 5 rows and ask, "How many are in each row?"

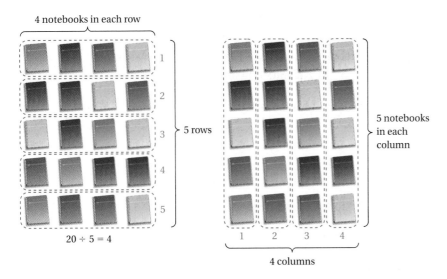

We can also consider a rectangular array with 5 notebooks in each column and ask, "How many columns are there?" The answer is still 4.

In each case, we are asking, "What do we multiply 5 by in order to get 20?"

Missing factor

$$5 \cdot \square = 20$$

Quotient

$$20 \div 5 = \square$$

This leads us to the following definition of division.

> ### DIVISION
>
> The quotient $a \div b$, where $b \neq 0$, is that unique whole number c for which $a = b \cdot c$.

RELATED SENTENCES

By looking at rectangular arrays, we can see how multiplication and division are related. The following array shows that $4 \cdot 5 = 20$.

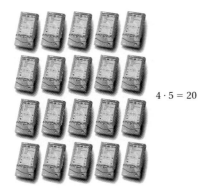

$4 \cdot 5 = 20$

The array also shows the following:

$$20 \div 5 = 4 \quad \text{and} \quad 20 \div 4 = 5.$$

Write a related multiplication sentence.

2. $15 \div 3 = 5$

3. $72 \div 8 = 9$

Write two related division sentences.

4. $6 \cdot 2 = 12$

5. $7 \cdot 6 = 42$

The division $20 \div 5$ is defined to be the number that when multiplied by 5 gives 20. Thus, for every division sentence, there is a related multiplication sentence.

$$20 \div 5 = 4 \qquad \text{Division sentence}$$

$$20 = 4 \cdot 5 \qquad \text{Related multiplication sentence}$$

To get the related multiplication sentence, we use
Dividend = Quotient · Divisor.

EXAMPLE 1 Write a related multiplication sentence: $12 \div 6 = 2$.

We have

$$12 \div 6 = 2 \qquad \text{Division sentence}$$

$$12 = 2 \cdot 6. \qquad \text{Related multiplication sentence}$$

The related multiplication sentence is $12 = 2 \cdot 6$.

By the commutative law of multiplication, there is also another multiplication sentence: $12 = 6 \cdot 2$.

Do Exercises 2 and 3.

For every multiplication sentence, we can write related divisions, as we can see from the preceding array.

EXAMPLE 2 Write two related division sentences: $7 \cdot 8 = 56$.

We have

$$7 \cdot 8 = 56 \qquad\qquad 7 \cdot 8 = 56$$

This factor becomes a divisor.　　　This factor becomes a divisor.

$$7 = 56 \div 8. \qquad\qquad 8 = 56 \div 7.$$

The related division sentences are $7 = 56 \div 8$ and $8 = 56 \div 7$.

Do Exercises 4 and 5.

b　Division of Whole Numbers

Before we consider division with remainders, let's recall four basic facts about division.

DIVIDING BY 1

Any number divided by 1 is that same number:

$$a \div 1 = \frac{a}{1} = a.$$

Answers on page A-2

DIVIDING A NUMBER BY ITSELF

Any nonzero number divided by itself is 1:

$$\frac{a}{a} = 1, \quad a \neq 0.$$

DIVIDENDS OF 0

Zero divided by any nonzero number is 0:

$$\frac{0}{a} = 0, \quad a > 0.$$

EXCLUDING DIVISION BY 0

Division by 0 is not defined. (We agree not to divide by 0.)

$$\frac{a}{0} \text{ is } \textbf{not defined.}$$

Why can't we divide by 0? Suppose the number 4 could be divided by 0. Then if $\square$ were the answer,

$$4 \div 0 = \square$$

and since 0 times any number is 0, we would have

$$4 = \square \cdot 0 = 0. \qquad \text{False!}$$

Thus, $a \div 0$ would be some number $\square$ such that $a = \square \cdot 0 = 0$. So the only possible number that could be divided by 0 would be 0 itself.

But such a division would give us any number we wish, for

$$
\left.
\begin{array}{lll}
0 \div 0 = 8 & \text{because} & 0 = 8 \cdot 0; \\
0 \div 0 = 3 & \text{because} & 0 = 3 \cdot 0; \\
0 \div 0 = 7 & \text{because} & 0 = 7 \cdot 0.
\end{array}
\right\} \quad \text{All true!}
$$

We avoid the preceding difficulties by agreeing to exclude division by 0.

Suppose we have 18 cans of soda and want to pack them in cartons of 6 cans each. How many cartons will we fill? We can determine this by repeated subtraction. We keep track of the number of times we subtract. We stop when the number of objects remaining, the **remainder,** is smaller than the divisor.

Divide by repeated subtraction. Then check.

6. $54 \div 9$

7. $61 \div 9$

8. $53 \div 12$

9. $157 \div 24$

EXAMPLE 3 Divide by repeated subtraction: $18 \div 6$.

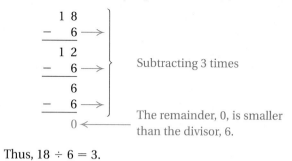

$$
\begin{array}{r}
1\ 8 \\
-\quad 6 \longrightarrow \\
\hline
1\ 2 \\
-\quad 6 \longrightarrow \\
\hline
6 \\
-\quad 6 \longrightarrow \\
\hline
0 \longleftarrow
\end{array}
$$

Subtracting 3 times

The remainder, 0, is smaller than the divisor, 6.

Thus, $18 \div 6 = 3$.

Suppose we have 22 cans of soda and want to pack them in cartons of 6 cans each. We end up with 3 cartons with 4 cans left over.

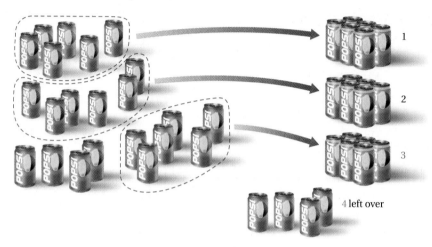

1

2

3

4 left over

EXAMPLE 4 Divide by repeated subtraction: $22 \div 6$.

$$
\begin{array}{r}
2\ 2 \\
-\quad 6 \longrightarrow \\
\hline
1\ 6 \\
-\quad 6 \longrightarrow \\
\hline
1\ 0 \\
-\quad 6 \longrightarrow \\
\hline
4 \longleftarrow
\end{array}
$$

Subtracting 3 times

Remainder

CHECK: $3 \cdot 6 = 18$,
$18 + 4 = 22$.

Note that

Quotient · Divisor + Remainder = Dividend.

We write answers to a division sentence as follows:

$$22 \div 6 = 3 \text{ R } 4$$

Dividend   Divisor   Quotient   Remainder

Do Exercises 6–9.

We can summarize our division procedure as follows.

To do division of whole numbers:
a) Estimate.
b) Multiply.
c) Subtract.

EXAMPLE 5 Divide and check: $3642 \div 5$.

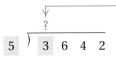
```
      ?
5 ) 3 6 4 2
```
1. Find the number of thousands in the quotient. Consider 3 thousands ÷ 5 and think 3 ÷ 5. Since 3 ÷ 5 is not a whole number, move to hundreds.

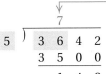
```
      7
5 ) 3 6 4 2
    3 5 0 0
      1 4 2
```
← The remainder is larger than the divisor.

2. Find the number of hundreds in the quotient. Consider 36 hundreds ÷ 5 and think 36 ÷ 5. The estimate is about 7 hundreds. Multiply 700 by 5 and subtract.

```
      7 2
5 ) 3 6 4 2
    3 5 0 0
      1 4 2
      1 0 0
        4 2
```
← The remainder is larger than the divisor.

3. Find the number of tens in the quotient using 142, the first remainder. Consider 14 tens ÷ 5 and think 14 ÷ 5. The estimate is about 2 tens. Multiply 20 by 5 and subtract. (If our estimate had been 3 tens, we could not have subtracted 150 from 142.)

```
      7 2 8
5 ) 3 6 4 2
    3 5 0 0
      1 4 2
      1 0 0
        4 2
        4 0
          2
```
← The remainder is less than the divisor.

4. Find the number of ones in the quotient using 42, the second remainder. Consider 42 ones ÷ 5 and think 42 ÷ 5. The estimate is about 8 ones. Multiply 8 by 5 and subtract. The remainder, 2, is less than the divisor, 5, so we are finished.

> You may have learned to divide like this, not writing the extra zeros. You may omit them if desired.

```
      7 2 8
5 ) 3 6 4 2
    3 5 ↓
      1 4 ↓
      1 0 ↓
        4 2
        4 0
          2
```

CHECK: $728 \cdot 5 = 3640$,
    $3640 + 2 = 3642$.

The answer is 728 R 2.

Do Exercises 10–12.

Answers on page A-2

Divide and check.
10. 4) 2 3 9

11. 6) 8 8 5 5

12. 5) 5 0 7 5

Divide.

13. 45)‾6‾0‾3‾0‾

14. 52)‾3‾2‾8‾8‾

Divide.

15. 6)‾4‾8‾4‾6‾

16. 7)‾7‾6‾1‾6‾

Answers on page A-2

Sometimes rounding the divisor helps us find estimates.

EXAMPLE 6 Divide: 8904 ÷ 42.

We mentally round 42 to 40.

$$
\begin{array}{r}
2 \\
42 \overline{)\,8\ 9\ 0\ 4} \\
8\ 4\ 0\ 0 \\
\hline
5\ 0\ 4
\end{array}
$$
← *Think*: 89 hundreds ÷ 40.
Estimate 2 hundreds. Multiply 200 · 42 and subtract.

$$
\begin{array}{r}
2\ 1 \\
42 \overline{)\,8\ 9\ 0\ 4} \\
8\ 4\ 0\ 0 \\
\hline
5\ 0\ 4 \\
4\ 2\ 0 \\
\hline
8\ 4
\end{array}
$$
← *Think*: 50 tens ÷ 40.
Estimate 1 ten. Multiply 10 · 42 and subtract.

$$
\begin{array}{r}
2\ 1\ 2 \\
42 \overline{)\,8\ 9\ 0\ 4} \\
8\ 4\ 0\ 0 \\
\hline
5\ 0\ 4 \\
4\ 2\ 0 \\
\hline
8\ 4 \\
8\ 4 \\
\hline
0
\end{array}
$$
← *Think*: 84 ones ÷ 40.
Estimate 2 ones. Multiply 2 · 42 and subtract.

CAUTION!

Be careful to keep the digits lined up correctly.

The answer is 212. *Remember*: If after estimating and multiplying you get a number that is larger than the divisor, you cannot subtract, so lower your estimate.

Do Exercises 13 and 14.

ZEROS IN QUOTIENTS

EXAMPLE 7 Divide: 6341 ÷ 7.

$$
\begin{array}{r}
9 \\
7 \overline{)\,6\ 3\ 4\ 1} \\
6\ 3\ 0\ 0 \\
\hline
4\ 1
\end{array}
$$
← *Think*: 63 hundreds ÷ 7.
Estimate 9 hundreds. Multiply 900 · 7 and subtract.

$$
\begin{array}{r}
9\ 0 \\
7 \overline{)\,6\ 3\ 4\ 1} \\
6\ 3\ 0\ 0 \\
\hline
4\ 1
\end{array}
$$
← *Think*: 4 tens ÷ 7. There are no tens in the quotient (other than the tens in 900). We write a 0 to show this.

```
        9  0  5
   7 ) 6  3  4  1
       6  3  0  0
   ─────────────
             4  1   ← Think: 41 ones ÷ 7.
             3  5       Estimate 5 ones. Multiply 5 · 7 and subtract.
       ──────────
                6   ← The remainder, 6, is less than the divisor, 7.
```

The answer is 905 R 6.

Do Exercises 15 and 16 on the preceding page.

EXAMPLE 8 Divide: 8889 ÷ 37.

We round 37 to 40.

```
              2
   3  7 ) 8  8  8  9   ← Think: 37 ≈ 40; 88 hundreds ÷ 40.
          7  4  0  0       Estimate 2 hundreds. Multiply 200 · 37
          ──────────       and subtract.
          1  4  8  9
```

```
              2  4
   3  7 ) 8  8  8  9
          7  4  0  0
          ──────────
          1  4  8  9   ← Think: 148 tens ÷ 40.
          1  4  8  0       Estimate 4 tens. Multiply 40 · 37
          ──────────       and subtract.
                   9
```

```
              2  4  0
   3  7 ) 8  8  8  9
          7  4  0  0
          ──────────
          1  4  8  9
          1  4  8  0
          ──────────
                   9   ← The remainder, 9, is less than the divisor, 37.
```

The answer is 240 R 9.

Do Exercises 17 and 18.

Divide.

17. 2 7) 9 7 2 4

18. 5 6) 4 4,8 4 7

Answers on page A-2

Study Tips

EXERCISES

- **Odd-numbered exercises.**
  Usually an instructor assigns some odd-numbered exercises. When you complete these, you can check your answers at the back of the book. If you miss any, check your work in the *Student's Solutions Manual* or ask your instructor for guidance.
- **Even-numbered exercises.**
  Whether or not your instructor assigns the even-numbered exercises, always do some on your own. Remember, there are no answers given for the chapter tests, so you need to practice doing exercises without answers. Check your answers later with a friend or your instructor.

Dividing Whole Numbers: Finding Remainders To divide whole numbers on a calculator, we use the $\boxed{\div}$ and $\boxed{=}$ keys. For example, to divide 711 by 9, we press $\boxed{7}\boxed{1}\boxed{1}\boxed{\div}\boxed{9}\boxed{=}$. The display reads $\boxed{79}$, so $711 \div 9 = 79$.

When we enter $453 \div 15$, the display reads $\boxed{30.2}$. Note that the result is not a whole number. This tells us that there is a remainder. The number 30.2 is expressed in decimal notation. The symbol "." is called a decimal point. (Decimal notation will be studied in Chapter 4.) The number to the left of the decimal point, 30, is the quotient. We can use the remaining part of the result to find the remainder. To do this, first subtract 30 from 30.2. Then multiply the difference by the divisor, 15. We get 3. This is the remainder. Thus, $453 \div 15 = 30$ R 3. The steps that we performed to find this result can be summarized as follows:

$$453 \div 15 = 30.2,$$
$$30.2 - 30 = .2,$$
$$0.2 \times 15 = 3.$$

To follow these steps on a calculator, we press $\boxed{4}\boxed{5}\boxed{3}\boxed{\div}\boxed{1}\boxed{5}\boxed{=}$ and write the number that appears to the left of the decimal point. This is the quotient. Then we continue by pressing $\boxed{-}\boxed{3}\boxed{0}\boxed{=}\boxed{\times}\boxed{1}\boxed{5}\boxed{=}$. The last number that appears is the remainder. In some cases, it will be necessary to round the remainder to the nearest one.

To check this result, we multiply the quotient by the divisor and then add the remainder.

$$30 \times 15 = 450,$$
$$450 + 3 = 453$$

Exercises: Use a calculator to perform each division. Check the results with a calculator also.

1. $92 \div 27$

2. $1\,9\,\overline{)\,5\,3\,2}$

3. $6\,\overline{)\,7\,4\,6}$

4. $3817 \div 29$

5. $1\,2\,6\,\overline{)\,3\,5{,}7\,1\,5}$

6. $3\,0\,8\,\overline{)\,2\,5\,9{,}8\,3\,1}$

1.6

EXERCISE SET

For Extra Help

Digital Video
Tutor CD 1
Videotape 2

InterAct
Math

Math Tutor
Center

MathXL

MyMathLab

a   Write a related multiplication sentence.

1. $18 \div 3 = 6$

2. $72 \div 9 = 8$

3. $22 \div 22 = 1$

4. $32 \div 1 = 32$

5. $54 \div 6 = 9$

6. $90 \div 10 = 9$

7. $37 \div 1 = 37$

8. $28 \div 28 = 1$

Write two related division sentences.

9. $9 \times 5 = 45$

10. $2 \cdot 7 = 14$

11. $37 \cdot 1 = 37$

12. $4 \cdot 12 = 48$

13. $8 \times 8 = 64$

14. $9 \cdot 7 = 63$

15. $11 \cdot 6 = 66$

16. $1 \cdot 43 = 43$

b   Divide, if possible. If not possible, write "not defined."

17. $72 \div 6$

18. $54 \div 9$

19. $\dfrac{23}{23}$

20. $\dfrac{37}{37}$

21. $22 \div 1$

22. $\dfrac{56}{1}$

23. $\dfrac{16}{0}$

24. $74 \div 0$

Divide.

25. $277 \div 5$

26. $699 \div 3$

27. $864 \div 8$

28. $869 \div 8$

29. $4\overline{)1228}$

30. $3\overline{)2124}$

31. $6\overline{)4521}$

32. $9\overline{)9110}$

33. $297 \div 4$ **34.** $389 \div 2$ **35.** $738 \div 8$ **36.** $881 \div 6$

37. $5 \overline{)8515}$ **38.** $3 \overline{)6027}$ **39.** $9 \overline{)8888}$ **40.** $8 \overline{)4139}$

41. $127{,}000 \div 10$ **42.** $127{,}000 \div 100$ **43.** $127{,}000 \div 1000$ **44.** $4260 \div 10$

45. $70 \overline{)3692}$ **46.** $20 \overline{)5798}$ **47.** $30 \overline{)875}$ **48.** $40 \overline{)987}$

49. $852 \div 21$ **50.** $942 \div 23$ **51.** $85 \overline{)7672}$ **52.** $54 \overline{)2729}$

53. $111\overline{)3219}$ **54.** $102\overline{)5612}$ **55.** $8\overline{)843}$ **56.** $7\overline{)749}$

57. $5\overline{)8047}$ **58.** $9\overline{)7273}$ **59.** $5\overline{)5036}$ **60.** $7\overline{)7074}$

61. $1058 \div 46$ **62.** $7242 \div 24$ **63.** $3425 \div 32$ **64.** $48\overline{)4899}$

65. $24\overline{)8880}$ **66.** $36\overline{)7563}$ **67.** $28\overline{)17,067}$ **68.** $36\overline{)28,929}$

69. $80\overline{)24,320}$ **70.** $90\overline{)88,560}$ **71.** $285\overline{)999,999}$

72. $306\overline{)888,888}$ **73.** $456\overline{)3,679,920}$ **74.** $803\overline{)5,622,606}$

75. D_W Is division associative? Why or why not? Give an example.

76. D_W Suppose a student asserts that "$0 \div 0 = 0$ because nothing divided by nothing is nothing." Devise an explanation to persuade the student that the assertion is false.

SKILL MAINTENANCE

77. Write expanded notation for 7882.   [1.1b]

78. Use $<$ or $>$ for $\square$ to write a true sentence: [1.4c]
888 $\square$ 788.

Write a related addition sentence.   [1.3a]

79. $21 - 16 = 5$

80. $56 - 14 = 42$

Write two related subtraction sentences.   [1.3a]

81. $47 + 9 = 56$

82. $350 + 64 = 414$

83. Add: $284 + 75$.   [1.2a]

84. Multiply: 284×75.   [1.5a]

85. Subtract: $284 - 75$.   [1.3b]

86. Subtract: $7002 - 3468$.   [1.3b]

SYNTHESIS

87. Complete the following table.

a	b	$a \cdot b$	$a + b$
	68	3672	
84			117
		32	12
		304	35

88. Find a pair of factors whose product is 36 and:
   a) whose sum is 13.
   b) whose difference is 0.
   c) whose sum is 20.
   d) whose difference is 9.

89. A group of 1231 college students is going to take buses for a field trip. Each bus can hold only 42 students. How many buses are needed?

90. Fill in the missing digits to make the equation true:
$34{,}584{,}132 \div 76\square = 4\square{,}386.$

1.7 SOLVING EQUATIONS

Objectives

a Solve simple equations by trial.

b Solve equations like $x + 28 = 54$, $28 \cdot x = 168$, and $98 \cdot 2 = y$.

a Solutions by Trial

Let's find a number that we can put in the blank to make this sentence true:

$9 = 3 + \square$.

We are asking "9 is 3 plus what number?" The answer is 6.

$9 = 3 + \boxed{6}$

Do Exercises 1 and 2.

A sentence with $=$ is called an **equation.** A **solution** of an equation is a number that makes the sentence true. Thus, 6 is a solution of

$9 = 3 + \square$ because $9 = 3 + \boxed{6}$ is true.

However, 7 is not a solution of

$9 = 3 + \square$ because $9 = 3 + \boxed{7}$ is false.

Do Exercises 3 and 4.

We can use a letter instead of a blank. For example,

$9 = 3 + x$.

We call x a **variable** because it can represent any number. If a replacement for a variable makes an equation true, it is a **solution** of the equation.

> ### SOLUTIONS OF AN EQUATION
>
> A **solution** is a replacement for the variable that makes the equation true. When we find all the solutions, we say that we have **solved** the equation.

EXAMPLE 1   Solve $x + 12 = 27$ by trial.

We replace x with several numbers.

If we replace x with 13, we get a false equation: $13 + 12 = 27$.

If we replace x with 14, we get a false equation: $14 + 12 = 27$.

If we replace x with 15, we get a true equation: $15 + 12 = 27$.

No other replacement makes the equation true, so the solution is 15.

Find a number that makes the sentence true.

1. $8 = 1 + \square$

2. $\square + 2 = 7$

3. Determine whether 7 is a solution of $\square + 5 = 9$.

4. Determine whether 4 is a solution of $\square + 5 = 9$.

Answers on page A-3

Solve by trial.

5. $n + 3 = 8$

6. $x - 2 = 8$

7. $45 \div 9 = y$

8. $10 + t = 32$

Solve.

9. $346 \times 65 = y$

10. $x = 2347 + 6675$

11. $4560 \div 8 = t$

12. $x = 6007 - 2346$

Answers on page A-3

EXAMPLES Solve.

2. $7 + n = 22$
(7 plus what number is 22?)
The solution is 15.

3. $8 \cdot 23 = y$
(8 times 23 is what?)
The solution is 184.

Do Exercises 5–8.

b Solving Equations

We now begin to develop more efficient ways to solve certain equations. When an equation has a variable alone on one side, it is easy to see the solution or to compute it. For example, the solution of

$$x = 12$$

is 12. When a calculation is on one side and the variable is alone on the other, we can find the solution by carrying out the calculation.

EXAMPLE 4 Solve: $x = 245 \times 34$.

To solve the equation, we carry out the calculation.

$$\begin{array}{r} 2\ 4\ 5 \\ \times\quad 3\ 4 \\ \hline 9\ 8\ 0 \\ 7\ 3\ 5\ 0 \\ \hline 8\ 3\ 3\ 0 \end{array} \qquad \begin{array}{l} x = 245 \times 34 \\ x = 8330 \end{array}$$

The solution is 8330.

Do Exercises 9–12.

Look at the equation

$$x + 12 = 27.$$

We can get x alone on one side of the equation by writing a related subtraction sentence:

$x = 27 - 12$ 12 gets subtracted to find the related subtraction sentence.

$x = 15.$ Doing the subtraction

It is useful in our later study of algebra to think of this as "subtracting 12 *on both sides.*" Thus

$x + 12 - 12 = 27 - 12$ Subtracting 12 on both sides

$x + 0 = 15$ Carrying out the subtraction

$x = 15.$

SOLVING $x + a = b$

To solve $x + a = b$, subtract a on both sides.

If we can get an equation in a form with the variable alone on one side, we can "see" the solution.

EXAMPLE 5 Solve: $t + 28 = 54$.

We have

$$t + 28 = 54$$
$$t + 28 - 28 = 54 - 28 \qquad \text{Subtracting 28 on both sides}$$
$$t + 0 = 26$$
$$t = 26.$$

To check the answer, we substitute 26 for t in the original equation.

CHECK: $$\frac{t + 28 = 54}{26 + 28 \;?\; 54}$$
$$54 \;\Big|\qquad \textbf{TRUE}$$

The solution is 26.

Do Exercises 13 and 14.

EXAMPLE 6 Solve: $182 = 65 + n$.

We have

$$182 = 65 + n$$
$$182 - 65 = 65 + n - 65 \qquad \text{Subtracting 65 on both sides}$$
$$117 = 0 + n \qquad\qquad \text{65 plus } n \text{ minus 65 is } 0 + n.$$
$$117 = n.$$

CHECK: $$\frac{182 = 65 + n}{182 \;?\; 65 + 117}$$
$$\Big|\; 182 \qquad \textbf{TRUE}$$

The solution is 117.

Do Exercise 15.

EXAMPLE 7 Solve: $7381 + x = 8067$.

We have

$$7381 + x = 8067$$
$$7381 + x - 7381 = 8067 - 7381 \qquad \text{Subtracting 7381 on both sides}$$
$$x = 686.$$

The check is left to the student. The solution is 686.

Do Exercises 16 and 17.

Solve. Be sure to check.

13. $x + 9 = 17$

14. $77 = m + 32$

15. Solve: $155 = t + 78$. Be sure to check.

Solve. Be sure to check.

16. $4566 + x = 7877$

17. $8172 = h + 2058$

Answers on page A-3

Solve. Be sure to check.

18. $8 \cdot x = 64$

We now learn to solve equations like $8 \cdot n = 96$. Look at

$$8 \cdot n = 96.$$

We can get n alone by writing a related division sentence:

$$n = 96 \div 8 = \frac{96}{8} \qquad \text{96 is divided by 8.}$$

$$n = 12. \qquad \text{Doing the division}$$

Note that $n = 12$ is easier to solve than $8 \cdot n = 96$. This is because we see easily that if we replace n on the left side with 12, we get a true sentence: $12 = 12$. The solution of $n = 12$ is 12, which is also the solution of $8 \cdot n = 96$.

It is useful in our later study of algebra to think of the preceding as "dividing by 8 *on both sides*." Thus,

$$\frac{8 \cdot n}{8} = \frac{96}{8} \qquad \text{Dividing by 8 on both sides}$$

$$n = 12. \qquad \text{8 times } n \text{ divided by 8 is } n.$$

19. $144 = 9 \cdot n$

> ### SOLVING $a \cdot x = b$
>
> To solve $a \cdot x = b$, divide by a on both sides.

EXAMPLE 8 Solve: $10 \cdot x = 240$.

We have

$$10 \cdot x = 240$$

$$\frac{10 \cdot x}{10} = \frac{240}{10} \qquad \text{Dividing by 10 on both sides}$$

$$x = 24.$$

CHECK:
$$10 \cdot x = 240$$
$$\overline{10 \cdot 24 \; ? \; 240}$$
$$240 \; | \qquad \textbf{TRUE}$$

The solution is 24.

20. Solve: $5152 = 8 \cdot t$.

Do Exercises 18 and 19.

EXAMPLE 9 Solve: $5202 = 9 \cdot t$.

We have

$$5202 = 9 \cdot t$$

$$\frac{5202}{9} = \frac{9 \cdot t}{9} \qquad \text{Dividing by 9 on both sides}$$

$$578 = t.$$

The check is left to the student. The solution is 578.

Do Exercise 20.

Answers on page A-3

EXAMPLE 10 Solve: $14 \cdot y = 1092$.

We have

$$14 \cdot y = 1092$$

$$\frac{14 \cdot y}{14} = \frac{1092}{14} \qquad \text{Dividing by 14 on both sides}$$

$$y = 78.$$

The check is left to the student. The solution is 78.

Do Exercise 21.

EXAMPLE 11 Solve: $n \cdot 56 = 4648$.

We have

$$n \cdot 56 = 4648$$

$$\frac{n \cdot 56}{56} = \frac{4648}{56} \qquad \text{Dividing by 56 on both sides}$$

$$n = 83.$$

The check is left to the student. The solution is 83.

Do Exercise 22.

21. Solve: $18 \cdot y = 1728$.

22. Solve: $n \cdot 48 = 4512$.

Answers on page A-3

a   Solve by trial.

1. $x + 0 = 14$ **2.** $x - 7 = 18$ **3.** $y \cdot 17 = 0$ **4.** $56 \div m = 7$

b   Solve. Be sure to check.

5. $13 + x = 42$ **6.** $15 + t = 22$ **7.** $12 = 12 + m$ **8.** $16 = t + 16$

9. $3 \cdot x = 24$ **10.** $6 \cdot x = 42$ **11.** $112 = n \cdot 8$ **12.** $162 = 9 \cdot m$

13. $45 \times 23 = x$ **14.** $23 \times 78 = y$ **15.** $t = 125 \div 5$ **16.** $w = 256 \div 16$

17. $p = 908 - 458$ **18.** $9007 - 5667 = m$ **19.** $x = 12{,}345 + 78{,}555$ **20.** $5678 + 9034 = t$

21. $3 \cdot m = 96$ **22.** $4 \cdot y = 96$ **23.** $715 = 5 \cdot z$ **24.** $741 = 3 \cdot t$

25. $10 + x = 89$ **26.** $20 + x = 57$ **27.** $61 = 16 + y$ **28.** $53 = 17 + w$

29. $6 \cdot p = 1944$ **30.** $4 \cdot w = 3404$ **31.** $5 \cdot x = 3715$ **32.** $9 \cdot x = 1269$

33. $47 + n = 84$ **34.** $56 + p = 92$ **35.** $x + 78 = 144$ **36.** $z + 67 = 133$

37. $165 = 11 \cdot n$

38. $660 = 12 \cdot n$

39. $624 = t \cdot 13$

40. $784 = y \cdot 16$

41. $x + 214 = 389$

42. $x + 221 = 333$

43. $567 + x = 902$

44. $438 + x = 807$

45. $18 \cdot x = 1872$

46. $19 \cdot x = 6080$

47. $40 \cdot x = 1800$

48. $20 \cdot x = 1500$

49. $2344 + y = 6400$

50. $9281 = 8322 + t$

51. $8322 + 9281 = x$

52. $9281 - 8322 = y$

53. $234 \times 78 = y$

54. $10{,}534 \div 458 = q$

55. $58 \cdot m = 11{,}890$

56. $233 \cdot x = 22{,}135$

57. $^{D}\mathsf{W}$ Describe a procedure that can be used to convert any equation of the form $a \cdot b = c$ to a related division equation.

58. $^{D}\mathsf{W}$ Describe a procedure that can be used to convert any equation of the form $a + b = c$ to a related subtraction equation.

59. Write two related subtraction sentences: $7 + 8 = 15$. [1.3a]

60. Write two related division sentences: $6 \cdot 8 = 48$. [1.6a]

Use $>$ or $<$ for $\square$ to write a true sentence. [1.4c]

61. $123 \,\square\, 789$

62. $342 \,\square\, 339$

63. $688 \,\square\, 0$

64. $0 \,\square\, 11$

Divide. [1.6b]

65. $1283 \div 9$

66. $1278 \div 9$

67. $1\,7\,)\,\overline{5\,6\,7\,8}$

68. $1\,7\,)\,\overline{5\,6\,8\,9}$

Solve.

69. $\boxed{\boxplus}$ $23{,}465 \cdot x = 8{,}142{,}355$

70. $\boxed{\boxplus}$ $48{,}916 \cdot x = 14{,}332{,}388$

67

APPLICATIONS AND PROBLEM SOLVING

Objective

a Solve applied problems involving addition, subtraction, multiplication, or division of whole numbers.

a **A Problem-Solving Strategy**

Applications and problem solving are the most important uses of mathematics. To solve a problem using the operations on the whole numbers, we first look at the situation. We try to translate the problem to an equation. Then we solve the equation. We check to see if the solution of the equation is a solution of the original problem. We are using the following five-step strategy.

FIVE STEPS FOR PROBLEM SOLVING

1. *Familiarize* yourself with the situation.
   a) Carefully read and reread until you understand *what* you are being asked to find.
   b) Draw a diagram or see if there is a formula that applies to the situation.
   c) Assign a letter, or *variable*, to the unknown.
2. *Translate* the problem to an equation using the letter or variable.
3. *Solve* the equation.
4. *Check* the answer in the original wording of the problem.
5. *State* the answer to the problem clearly with appropriate units.

EXAMPLE 1 *Baseball's Power Hitters.* The top three home-run hitters in the major leagues over the years from 1996 to 2000 were Sammy Sosa, Mark McGwire, and Ken Griffey, Jr. The numbers of home runs hit per year for each player are listed in the table below. Find the total number of home runs hit by Sammy Sosa over the 5-yr period.

Mark McGwire and
Ken Griffey, Jr.

YEAR	SAMMY SOSA	MARK MCGWIRE	KEN GRIFFEY, JR.
1996	40	52	49
1997	36	58	56
1998	66	70	56
1999	63	65	48
2000	50	32	40
Total	?	?	?

Source: Major League Baseball

Sammy Sosa

1. **Familiarize.** We can make a drawing or at least visualize the situation.

$$40 \; + \; 36 \; + \; 66 \; + \; 63 \; + \; 50$$

in	in	in	in	in
1996	1997	1998	1999	2000

Since we are combining numbers of home runs, addition can be used. First, we define the unknown. We let $n =$ the total number of home runs hit by Sosa in the 5-yr period.

2. **Translate.** We translate to an equation:

$$40 + 36 + 66 + 63 + 50 = n.$$

3. **Solve.** We solve the equation by carrying out the addition.

$$
\begin{array}{r}
1 \\
4\,0 \\
3\,6 \\
6\,6 \\
6\,3 \\
+\,5\,0 \\
\hline
2\,5\,5
\end{array}
\qquad
\begin{aligned}
40 + 36 + 66 + 63 + 50 &= n \\
255 &= n
\end{aligned}
$$

4. **Check.** We check 255 in the original problem. There are many ways in which this can be done. For example, we can repeat the calculation. (We leave this to the student.) Another way is to check whether the answer is reasonable. In this case, we would expect the total to be greater than the number of home runs in any of the individual years, which it is. We can also estimate by rounding. Here we round to the nearest ten:

$$40 + 36 + 66 + 63 + 50 \approx 40 + 40 + 70 + 60 + 50$$
$$= 260.$$

Since $255 \approx 260$, we have a partial check. If we had an estimate like 340 or 400, we might be suspicious that our calculated answer is incorrect. Since our estimated answer is close to our calculation, we are further convinced that our answer checks.

5. **State.** The total number of home runs hit by Sammy Sosa from 1996 to 2000 was 255.

Do Exercises 1–3.

Refer to the table on the preceding page to answer Margin Exercises 1–3.

1. Find the total number of home runs hit by Mark McGwire from 1996 to 2000.

2. Find the total number of home runs hit by Ken Griffey, Jr., from 1996 to 2000.

3. Who hit the most home runs over the 5-yr period?

Answers on page A-3

EXAMPLE 2 *Checking Account Balance.* The balance in Tyler's checking account is $528. He uses his debit card to buy the Roto Zip Spiral Saw Combo shown in this ad. Find the new balance in his checking account.

Source: Roto Zip Tool Corporation

1. **Familiarize.** We first make a drawing or at least visualize the situation. We let $M =$ the new balance in his account. This gives us the following:

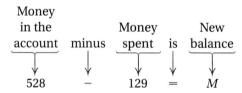

2. **Translate.** We can think of this as a "take-away" situation. We translate to an equation.

$$\underbrace{\text{Money in the account}}_{528} \underbrace{\text{minus}}_{-} \underbrace{\text{Money spent}}_{129} \underbrace{\text{is}}_{=} \underbrace{\text{New balance}}_{M}$$

3. **Solve.** This sentence tells us what to do. We subtract.

$$\begin{array}{r} \scriptstyle 11 \\ \scriptstyle 4\ \ \not{1}\ \ 18 \\ 5\ 2\ 8 \\ -\ 1\ 2\ 9 \\ \hline 3\ 9\ 9 \end{array} \qquad \begin{aligned} 528 - 129 &= M \\ 399 &= M \end{aligned}$$

4. **Check.** To check our answer of $399, we can repeat the calculation. We note that the answer should be less than the original amount, $528, which it is. We can add the difference, 399, to the subtrahend, 129: $129 + 399 = 528$. We can also estimate:

$$528 - 129 \approx 530 - 130 = 400 \approx 399.$$

5. **State.** Tyler has a new balance of $399 in his checking account.

Do Exercise 4.

Answer on page A-3

In the real world, problems may not be stated in written words. You must still become familiar with the situation before you can solve the problem.

EXAMPLE 3 *Travel Distance.* Vicki is driving from Indianapolis to Salt Lake City to work during the 2002 Winter Olympics. The distance from Indianapolis to Salt Lake City is 1634 mi. She travels 1154 mi to Denver. How much farther must she travel?

1. **Familiarize.** We first make a drawing or at least visualize the situation. We let x = the remaining distance to Salt Lake City.

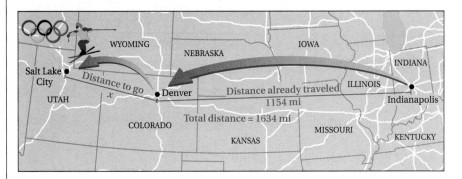

2. **Translate.** We see that this is a "missing-addend" situation. We translate to an equation.

Distance already traveled	plus	Distance to go	is	Total distance of trip
↓	↓	↓	↓	↓
1154	+	x	=	1634

3. **Solve.** To solve the equation , we subtract 1154 on both sides:

$$1154 + x = 1634$$
$$1154 + x - 1154 = 1634 - 1154$$
$$x = 480.$$

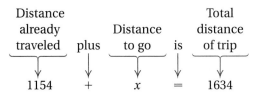

4. **Check.** We check our answer of 480 mi in the original problem. This number should be less than the total distance, 1634 mi, which it is. We can add the difference, 480, to the subtrahend, 1154: $1154 + 480 = 1634$. We can also estimate:

$$1634 - 1154 \approx 1600 - 1200$$
$$= 400 \approx 480.$$

The answer, 480 mi, checks.

5. **State.** Vicki must travel 480 mi farther to Salt Lake City.

Do Exercise 5.

---

5. Home Theatre Audio System. Bernardo has $376. He wants to purchase the Home Theatre Audio System shown in the ad below. How much more does he need?

Answer on page A-3

Answer on page A-3

6. Total Cost of Laptop Computers. What is the total cost of 12 Gateway Solo 9300 laptop computers with CD ROM drives and 1 GHz processors if each one costs $2898?

Source: Gateway Country® Stores

EXAMPLE 4 *Total Cost of DVDs.* What is the total cost of 5 DVD players if each one costs $249?

1. Familiarize. We first make a drawing or at least visualize the situation. We let $T =$ the cost of 5 DVD players. Repeated addition works well in this case.

2. Translate. We translate to an equation.

Number of DVD players	times	Cost of each player	is	Total cost
5	$\times$	$249	=	T

3. Solve. This sentence tells us what to do. We multiply.

$$
\begin{array}{r}
{\scriptstyle 2\ 4}\\
2\ 4\ 9\\
\times\quad\ 5\\
\hline
1\ 2\ 4\ 5
\end{array}
$$

$$5 \times 249 = T$$
$$1245 = T$$

4. Check. We have an answer, 1245, that is much greater than the cost of any individual DVD player, which is reasonable. We can repeat our calculation. We can also check by estimating:

$$5 \times 249 \approx 5 \times 250 = 1250 \approx 1245.$$

The answer checks.

5. State. The total cost of 5 DVD players is $1245.

Do Exercise 6.

EXAMPLE 5 *Bed Sheets.* The dimensions of a flat sheet for a king-size bed are 108 in. by 102 in. What is the area of the sheet? (The dimension labels on sheets list width $\times$ length.)

1. Familiarize. We first make a drawing. We let $A =$ the area.

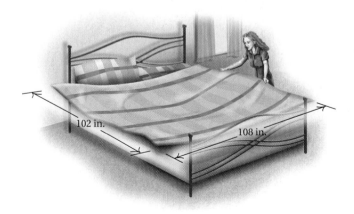

2. Translate. Using a formula for area, we have

$$A = \text{length} \cdot \text{width} = l \cdot w = 102 \cdot 108.$$

3. Solve. We carry out the multiplication.

$$
\begin{array}{r}
1\ 0\ 8 \\
\times\ \ \ 1\ 0\ 2 \\
\hline
2\ 1\ 6 \\
1\ 0\ 8\ 0\ 0 \\
\hline
1\ 1\ 0\ 1\ 6
\end{array}
\qquad
\begin{array}{l}
A = 102 \cdot 108 \\
A = 11{,}016
\end{array}
$$

4. Check. We repeat our calculation. We also note that the answer is greater than either the length or the width, which it should be. (This might not be the case if we were using fractions or decimals.) The answer checks.

5. State. The area of a king-size bed sheet is 11,016 sq in.

Do Exercise 7.

EXAMPLE 6 *Cartons of Soda.* A bottling company produces 3304 cans of soda. How many 12-can cartons can be filled? How many cans will be left over?

1. Familiarize. We first make a drawing. We let $n =$ the number of 12-can cartons that can be filled. The problem can be considered as repeated subtraction, taking successive sets of 12 cans and putting them into n cartons.

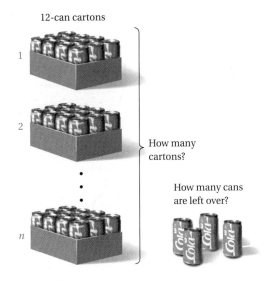

12-can cartons

How many cartons?

How many cans are left over?

2. Translate. We translate to an equation.

Number of cans	divided by	Number in each carton	is	Number of cartons
3304	÷	12	=	n

7. Bed Sheets. The dimensions of a flat sheet for a queen-size bed are 90 in. by 102 in. What is the area of the sheet?

Answer on page A-3

8. Cartons of Soda. The bottling company in Example 6 also uses 6-can cartons. How many 6-can cartons can be filled with 2269 cans of cola? How many will be left over?

3. Solve. We solve the equation by carrying out the division.

```
           2 7 5
  1 2 ) 3 3 0 4
        2 4 0 0
        ───────
          9 0 4
          8 4 0
          ─────
            6 4
            6 0
            ───
              4
```

$$3304 \div 12 = n$$
$$275 \text{ R } 4 = n$$

4. Check. We can check by multiplying the number of cartons by 12 and adding the remainder, 4:

$$12 \cdot 275 = 3300,$$
$$3300 + 4 = 3304.$$

5. State. Thus, 275 twelve-can cartons can be filled. There will be 4 cans left over.

Do Exercise 8.

EXAMPLE 7 *Automobile Mileage.* The Chrysler PT Cruiser gets 22 miles to the gallon (mpg) in city driving. How many gallons will it use in 6028 mi of city driving?
Source: DaimlerChrysler Corporation

1. Familiarize. We first make a drawing. It is often helpful to be descriptive about how we define a variable. In this case, we let g = the number of gallons ("g" comes from "gallons").

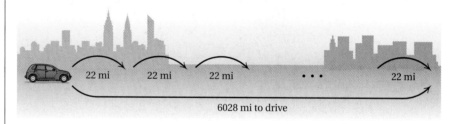

2. Translate. Repeated addition applies here. Thus the following multiplication applies to the situation.

Number of miles per gallon	times	Number of gallons needed	is	Number of miles to drive
22	·	g	=	6028

3. Solve. To solve the equation, we divide by 22 on both sides.

$$22 \cdot g = 6028$$
$$\frac{22 \cdot g}{22} = \frac{6028}{22}$$
$$g = 274$$

```
           2 7 4
  2 2 ) 6 0 2 8
        4 4 0 0
        ───────
        1 6 2 8
        1 5 4 0
        ───────
            8 8
            8 8
            ───
              0
```

Answer on page A-3

4. Check. To check, we multiply 274 by 22: $22 \cdot 274 = 6028$.

5. State. The PT Cruiser will use 274 gal.

Do Exercise 9.

Multistep Problems

Sometimes we must use more than one operation to solve a problem, as in the following example.

EXAMPLE 8 *Aircraft Seating.* Boeing Corporation builds commercial aircraft. A Boeing 767 has a seating configuration with 4 rows of 6 seats across in first class and 35 rows of 7 seats across in economy class. Find the total seating capacity of the plane.
Sources: The Boeing Company; Delta Airlines

1. Familiarize. We first make a drawing.

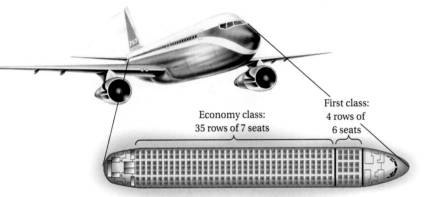

Economy class:
35 rows of 7 seats

First class:
4 rows of
6 seats

2. Translate. There are three parts to the problem. We first find the number of seats in each class. Then we add.

First-class: Repeated addition applies here. Thus the following multiplication corresponds to the situation. We let $F =$ the number of seats in first class.

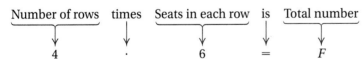

$$\text{Number of rows} \quad \text{times} \quad \text{Seats in each row} \quad \text{is} \quad \text{Total number}$$
$$4 \qquad \cdot \qquad 6 \qquad = \qquad F$$

Economy class: Repeated addition applies here. Thus the following multiplication corresponds to the situation. We let $E =$ the number of seats in economy class.

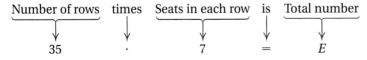

$$\text{Number of rows} \quad \text{times} \quad \text{Seats in each row} \quad \text{is} \quad \text{Total number}$$
$$35 \qquad \cdot \qquad 7 \qquad = \qquad E$$

We let $T =$ the total number of seats in both classes.

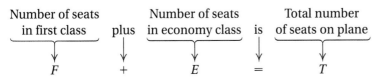

$$\text{Number of seats in first class} \quad \text{plus} \quad \text{Number of seats in economy class} \quad \text{is} \quad \text{Total number of seats on plane}$$
$$F \qquad + \qquad E \qquad = \qquad T$$

9. Automobile Mileage. The Chrysler PT Cruiser gets 26 miles to the gallon (mpg) in highway driving. How many gallons will it take to drive 884 mi of highway driving?
Source: DaimlerChrysler Corporation

Answer on page A-3

10. Aircraft Seating. A Boeing 767 used for foreign travel has three classes of seats. First class has 3 rows of 5 seats across; business class has 6 rows with 6 seats across and 1 row with 2 seats on each of the outside aisles. Economy class has 18 rows with 7 seats across. Find the total seating capacity of the plane.

Sources: The Boeing Company; Delta Airlines

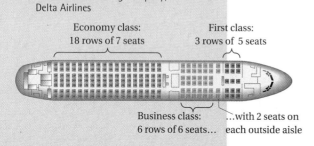

Economy class: 18 rows of 7 seats

First class: 3 rows of 5 seats

Business class: 6 rows of 6 seats...

...with 2 seats on each outside aisle

3. Solve. We solve each equation and add the solutions.

$$4 \cdot 6 = F \qquad 35 \cdot 7 = E \qquad F + E = T$$
$$24 = F \qquad 245 = E \qquad 24 + 245 = T$$
$$269 = T$$

4. Check. To check, we repeat our calculations. (We leave this to the student.) We could also check by rounding, multiplying, and adding.

5. State. There are 269 seats in a Boeing 767.

Do Exercise 10.

As you consider the following exercises, here are some words and phrases that may be helpful to look for when you are translating problems to equations.

KEY WORDS, PHRASES, AND CONCEPTS	
Addition (+)	**Subtraction (−)**
add	subtract
added to	subtracted from
sum	difference
total	minus
plus	less than
more than	decreased by
increased by	take away
	how much more
	missing addend
Multiplication (·)	**Division (÷)**
multiply	divide
multiplied by	divided by
product	quotient
times	repeated subtraction
of	missing factor
repeated addition	finding equal quantities
rectangular arrays	

Answer on page A-3

1.8

EXERCISE SET

For Extra Help

Digital Video
Tutor CD 1
Videotape 2

InterAct
Math

Math Tutor
Center

MathXL

MyMathLab

a Solve.

Top Web Properties.   The bar graph below shows the four most frequently visited Web sites, in terms of the number of visits for a recent month. Use this graph for Exercises 1–4.

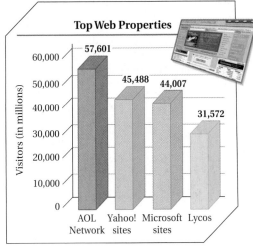

Source: Media Matrix

1. What was the total number of visits to all the sites?

2. What was the total number of visits to the three most-visited sites?

3. How many more visits were there to the AOL Network site than to the Yahoo! sites?

4. How many more visits were there to the Microsoft sites than to the Lycos site?

5. *Concorde Crash.*   The Anglo-French Concorde entered service in 1976. It had its first crash 24 yr later. In what year did it have its first crash?

6. Dwight D. Eisenhower was the 34th president of the United States. He left office in 1961 and lived another 8 yr. In what year did he die?

New England. The following table lists various data about the New England states.

NEW ENGLAND STATES	TOTAL AREA (in square miles)	TOTAL INLAND WATER AREA (in square miles)	SALARY OF THE GOVERNOR	POPULATION IN 1998
Maine	33,265	2,270	$70,000	1,244,250
New Hampshire	9,279	286	86,235	1,185,048
Vermont	9,614	341	80,725	590,883
Massachusetts	8,284	460	75,000	6,147,132
Connecticut	5,018	146	78,000	3,274,069
Rhode Island	1,212	157	69,900	988,480

Source: The New York Times Almanac

7. Find the total area of New England.

8. Find the total area of inland water in New England.

9. Find the total amount paid in salaries to the governors of the New England states.

10. Find the total population of New England in 1998.

11. *Military Downsizing.* In 2000, there were 372,000 people in the Navy. This was down from the 583,000 who were in the Navy in 1990. How many more were in the Navy in 1990 than in 2000?

12. *Baseball Salaries.* The New York Yankees led the Major Leagues in 2000 with a total payroll of $114,336,616. The Minnesota Twins had the lowest payroll at $23,499,966. How much more would the Twins have to spend on payroll to equal the Yankees?
Source: Major League Baseball

13. *Longest Rivers.* The longest river in the world is the Nile in Egypt at 4100 mi. The longest river in the United States is the Missouri–Mississippi at 3860 mi. How much longer is the Nile?

14. *Speeds on Interstates.* Recently, speed limits on interstate highways in many Western states were raised from 65 mph to 75 mph. By how many miles per hour were they raised?

15. *Automobile Mileage.* The 2000 Volkswagen New Beetle GL gets 24 miles to the gallon (mpg) in city driving. How many gallons will it use in 6144 mi of city driving?
Source: Volkswagen of America, Inc.

16. *Automobile Mileage.* The 2000 Volkswagen New Beetle GL gets 31 miles to the gallon (mpg) in highway driving. How many gallons will it use in 5859 mi of highway driving?
Source: Volkswagen of America, Inc.

17. *Pixels.* A computer screen consists of small rectangular dots called *pixels*. How many pixels are there on a screen that has 600 rows with 800 pixels in each row?

Pixel

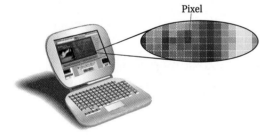

18. *Crossword.* The *USA Today* crossword puzzle is a rectangle containing 15 rows with 15 squares in each row. How many squares does the puzzle have altogether?

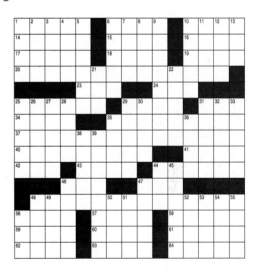

19. *Refrigerator Purchase.* Cometbucks Deli has a chain of 24 restaurants. It buys a refrigerator for each store at a cost of $499 each. Find the total cost of the purchase.

20. *Microwave Purchase.* Bridgeway College is constructing new dorms, in which each room has a small kitchen. It buys 96 microwave ovens at $88 each. Find the total cost of the purchase.

Music CD Sales. The bar graph below shows the sales of music CDs, in millions, for the years from 1995 to 1999. Use this graph for Exercises 21–24.

Music CD Sales

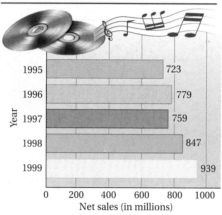

Year	Net sales (in millions)
1995	723
1996	779
1997	759
1998	847
1999	939

Net sales (in millions)
0 200 400 600 800 1000

21. How many more CDs were sold in 1999 than in 1995?

22. How many more CDs were sold in 1999 than in 1998?

23. What was the total number of CDs sold from 1997 through 1999?

24. What was the total number of CDs sold from 1995 through 1999?

25. *"Seinfeld" Episodes.* "Seinfeld" is a long-running television comedy with 177 episodes created. A local station picks up the syndicated reruns. If the station runs 5 episodes per week, how many full weeks will pass before it must start over with past episodes? How many episodes will be left for the last week?

26. A lab technician separates a vial containing 70 cubic centimeters (cc) of blood into test tubes, each of which contains 3 cc of blood. How many test tubes can be filled? How much blood is left over?

27. There are 24 hours (hr) in a day and 7 days in a week. How many hours are there in a week?

28. There are 60 min in an hour and 24 hr in a day. How many minutes are there in a day?

29. Dana borrows $5928 for a used car. The loan is to be paid off in 24 equal monthly payments. How much is each payment (excluding interest)?

30. A family borrows $4824 to build a sunroom on the back of their home. The loan is to be paid off in equal monthly payments of $134 (excluding interest). How many months will it take to pay off the loan?

31. *Atlanta Population.* The population of Atlanta was 3,857,097 in 1999. This was an increase of 897,597 from its population in 1990. What was the population of Atlanta in 1990?
Source: U.S. Bureau of the Census

32. *Orlando Population.* The population of Orlando was 1,535,004 in 1999. This was an increase of 310,160 from its population in 1990. What was the population of Orlando in 1990?
Source: U.S. Bureau of the Census

33. *Crossword.* The *Los Angeles Times* crossword puzzle is a rectangle containing 441 squares arranged in 21 rows. How many columns does the puzzle have?

34. *Sheet of Stamps.* A sheet of 100 stamps typically has 10 rows of stamps. How many stamps are in each row?

35. *Hershey Bars.* Hershey Chocolate USA makes small, fun-size chocolate bars. How many 20-bar packages can be filled with 11,267 bars? How many bars will be left over?

36. *Reese's Peanut Butter Cups.* H. B. Reese Candy Co. makes small, fun-size peanut butter cups. The company manufactures 23,579 cups and fills 1025 packages. How many cups are in a package? How many cups will be left over?

37. *High School Court.* The standard basketball court used by high school players has dimensions of 50 ft by 84 ft.
a) What is its area?
b) What is its perimeter?

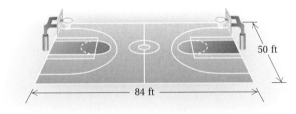

38. *NBA Court.* The standard basketball court used by college and NBA players has dimensions of 50 ft by 94 ft.
a) What is its area?
b) What is its perimeter?
c) How much greater is the area of an NBA court than a high school court? (See Exercise 37.)

39. Copies of this book are generally shipped from the Addison-Wesley warehouse in cartons containing 24 books each. How many cartons are needed to ship 840 books?

40. According to the H. J. Heinz Company, 16-oz bottles of catsup are generally shipped in cartons containing 12 bottles each. How many cartons are needed to ship 528 bottles of catsup?

41. Copies of this book are generally shipped from the warehouse in cartons containing 24 books each. How many cartons are needed to ship 1355 books?

42. Sixteen-ounce bottles of catsup are generally shipped in cartons containing 12 bottles each. How many cartons are needed to ship 1033 bottles of catsup?

43. *Map Drawing.* A map has a scale of 64 mi to the inch. How far apart *in reality* are two cities that are 6 in. apart on the map? How far apart *on the map* are two cities that, in reality, are 1728 mi apart?

44. *Map Drawing.* A map has a scale of 150 mi to the inch. How far apart *on the map* are two cities that, in reality, are 2400 mi apart? How far apart *in reality* are two cities that are 13 in. apart on the map?

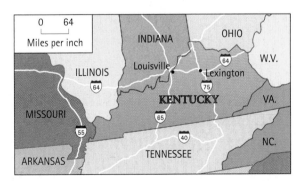

45. A carpenter drills 216 holes in a rectangular array in a pegboard. There are 12 holes in each row. How many rows are there?

46. Lou works as a CPA. He arranges 504 entries on a spreadsheet in a rectangular array that has 36 rows. How many entries are in each row?

47. Elena buys 5 video games at $64 each and pays for them with $10 bills. How many $10 bills did it take?

48. Pedro buys 5 video games at $64 each and pays for them with $20 bills. How many $20 bills did it take?

49. You have $568 in your checking account. You write checks for $46, $87, and $129. Then you deposit $94 back in the account after the return of some books. How much is left in your account?

50. The balance in your checking account is $749. You write checks for $34 and $65. Then you make a deposit of $123 from your paycheck. What is your new balance?

Weight Loss. Many Americans exercise for weight control. It is known that one must burn off about 3500 calories in order to lose one pound. The chart shown here details how much of certain types of exercise is required to burn 100 calories. Use this chart for Exercises 51–54.

To burn off 100 calories, you must:
- Run for 8 min at a brisk pace, or
- Swim for 2 min at a brisk pace, or
- Bicycle for 15 min at 9 mph, or
- Do aerobic exercises for 15 min.

51. How long must you run at a brisk pace in order to lose one pound?

52. How long must you swim in order to lose one pound?

53. How long must you do aerobic exercises in order to lose one pound?

54. How long must you bicycle at 9 mph in order to lose one pound?

55. *Bones in the Hands and Feet.* There are 27 bones in each human hand and 26 bones in each human foot. How many bones are there in all in the hands and feet?

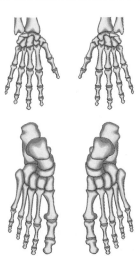

56. *Index Cards.* Index cards of dimension 3 in. by 5 in. are normally shipped in packages containing 100 cards each. How much writing area is available if one uses the front and back sides of a package of these cards?

57. Before going back to college, David buys 4 shirts at $59 each and 6 pairs of pants at $78 each. What is the total cost of this clothing?

58. An office for adjunct instructors at a community college has 6 bookshelves, each of which is 3 ft long. The office is moved to a new location that has dimensions of 16 ft by 21 ft. Is it possible for the bookshelves to be put side by side on the 16-ft wall?

59. D_W In the newspaper article, "When Girls Play, Knees Fail," the author discusses the fact that female athletes have six times the number of knee injuries that male athletes have. What information would be needed if you were to write a math problem based on the article? What might the problem be?
Source: *The Arizona Republic,* 2/9/00, p. C1

60. D_W Write a problem for a classmate to solve. Design the problem so that the solution is "The driver still has 329 mi to travel."

SKILL MAINTENANCE

Round 234,562 to the nearest: [1.4a]

61. Hundred.   **62.** Ten.   **63.** Thousand.

Estimate the computation by rounding to the nearest thousand.   [1.4b]

64. 2783 + 4602 + 5797 + 8111   **65.** 28,430 − 11,977

66. 2100 + 5800   **67.** 5800 − 2100

Estimate the product by rounding to the nearest hundred.   [1.5b]

68. 787 · 363   **69.** 887 · 799   **70.** 10,362 · 4531

SYNTHESIS

71. 🖩 *Speed of Light.* Light travels about 186,000 miles per second (mi/sec) in a vacuum as in outer space. In ice it travels about 142,000 mi/sec, and in glass it travels about 109,000 mi/sec. In 18 sec, how many more miles will light travel in a vacuum than in ice? than in glass?

72. Carney Community College has 1200 students. Each professor teaches 4 classes and each student takes 5 classes. There are 30 students and 1 teacher in each classroom. How many professors are there at Carney Community College?

Objectives

a. Write exponential notation for products such as $4 \cdot 4 \cdot 4$.

b. Evaluate exponential notation.

c. Simplify expressions using the rules for order of operations.

d. Remove parentheses within parentheses.

Write exponential notation.

1. $5 \cdot 5 \cdot 5 \cdot 5$

2. $5 \cdot 5 \cdot 5 \cdot 5 \cdot 5$

3. $10 \cdot 10$

4. $10 \cdot 10 \cdot 10 \cdot 10$

Evaluate.

5. 10^4 6. 10^2

7. 8^3 8. 2^5

Answers on page A-3

a. Writing Exponential Notation

Consider the product $3 \cdot 3 \cdot 3 \cdot 3$. Such products occur often enough that mathematicians have found it convenient to create a shorter notation, called **exponential notation,** explained as follows.

$\underbrace{3 \cdot 3 \cdot 3 \cdot 3}_{4 \text{ factors}}$ is shortened to $3^4 \leftarrow$ exponent
$\qquad\qquad\qquad\qquad\qquad\qquad\; \llcorner$ base

We read exponential notation as follows.

NOTATION	WORD DESCRIPTION
3^4	"three to the fourth power," or "the fourth power of three"
5^3	"five to the third power," or "the third power of five," or "five-cubed," or "the cube of five"
7^2	"seven to the second power," or "the second power of seven," or "seven squared," or "the square of seven"

The wording "seven squared" for 7^2 comes from the fact that a square with side s has area A given by $A = s^2$.

An expression like $3 \cdot 5^2$ is read "three times the square of five" or "three times five squared."

EXAMPLE 1   Write exponential notation for $10 \cdot 10 \cdot 10 \cdot 10 \cdot 10$.

Exponential notation is 10^5.   5 is the *exponent.*
                                       10 is the *base.*

EXAMPLE 2   Write exponential notation for $2 \cdot 2 \cdot 2$.

Exponential notation is 2^3.

Do Exercises 1–4.

b Evaluating Exponential Notation

We evaluate exponential notation by rewriting it as a product and computing the product.

EXAMPLE 3   Evaluate: 10^3.

$$10^3 = 10 \cdot 10 \cdot 10 = 1000$$

EXAMPLE 4   Evaluate: 5^4.

$$5^4 = 5 \cdot 5 \cdot 5 \cdot 5 = 625$$

CAUTION!

5^4 does not mean $5 \cdot 4$.

Do Exercises 5–8 on the preceding page.

c Simplifying Expressions

Suppose we have a calculation like the following:

$$3 + 4 \cdot 8.$$

How do we find the answer? Do we add 3 to 4 and then multiply by 8, or do we multiply 4 by 8 and then add 3? In the first case, the answer is 56. In the second, the answer is 35. We agree to compute as in the second case.

Consider the calculation

$$7 \cdot 14 - (12 + 18).$$

What do the parentheses mean? To deal with these questions, we must make some agreement regarding the order in which we perform operations. The rules are as follows.

RULES FOR ORDER OF OPERATIONS

1. Do all calculations within parentheses ( ), brackets [ ], or braces { } before operations outside.
2. Evaluate all exponential expressions.
3. Do all multiplications and divisions in order from left to right.
4. Do all additions and subtractions in order from left to right.

It is worth noting that these are the rules that computers and most scientific calculators use to do computations.

EXAMPLE 5   Simplify: $16 \div 8 \times 2$.

There are no parentheses or exponents, so we start with the third step.

$$16 \div 8 \times 2 = 2 \times 2 \quad \text{Doing all multiplications and divisions in order from left to right}$$

$$= 4$$

Simplify.

9. $93 - 14 \cdot 3$

10. $104 \div 4 + 4$

11. $25 \cdot 26 - (56 + 10)$

12. $75 \div 5 + (83 - 14)$

Answers on page A-3

Simplify and compare.

13. $64 \div (32 \div 2)$ and
$(64 \div 32) \div 2$

14. $(28 + 13) + 11$ and
$28 + (13 + 11)$

15. Simplify:

$9 \times 4 - (20 + 4) \div 8 - (6 - 2)$.

Simplify.

16. $5 \cdot 5 \cdot 5 + 26 \cdot 71$
$- (16 + 25 \cdot 3)$

17. $30 \div 5 \cdot 2 + 10 \cdot 20 + 8 \cdot 8$
$- 23$

18. $95 - 2 \cdot 2 \cdot 2 \cdot 5 \div (24 - 4)$

Answers on page A-3

EXAMPLE 6   Simplify: $7 \cdot 14 - (12 + 18)$.

$$7 \cdot 14 - (12 + 18) = 7 \cdot 14 - 30 \qquad \text{Carrying out operations inside parentheses}$$
$$= 98 - 30 \qquad \text{Doing all multiplications and divisions}$$
$$= 68 \qquad \text{Doing all additions and subtractions}$$

Do Exercises 9–12 on the preceding page.

EXAMPLE 7   Simplify and compare: $23 - (10 - 9)$ and $(23 - 10) - 9$.

We have

$$23 - (10 - 9) = 23 - 1 = 22;$$
$$(23 - 10) - 9 = 13 - 9 = 4.$$

We can see that $23 - (10 - 9)$ and $(23 - 10) - 9$ represent different numbers. Thus subtraction is not associative.

Do Exercises 13 and 14.

EXAMPLE 8   Simplify: $7 \cdot 2 - (12 + 0) \div 3 - (5 - 2)$.

$$7 \cdot 2 - (12 + 0) \div 3 - (5 - 2) = 7 \cdot 2 - 12 \div 3 - 3$$
$$\text{Carrying out operations inside parentheses}$$
$$= 14 - 4 - 3$$
$$\text{Doing all multiplications and divisions in order from left to right}$$
$$= 7 \qquad \text{Doing all additions and subtractions in order from left to right}$$

Do Exercise 15.

EXAMPLE 9   Simplify: $15 \div 3 \cdot 2 \div (10 - 8)$.

$$15 \div 3 \cdot 2 \div (10 - 8) = 15 \div 3 \cdot 2 \div 2 \qquad \text{Carrying out operations inside parentheses}$$
$$= 5 \cdot 2 \div 2 \qquad \left.\begin{array}{l} \\ \\ \end{array}\right\} \text{Doing all multiplications and divisions in order from left to right}$$
$$= 10 \div 2$$
$$= 5$$

Do Exercises 16–18.

EXAMPLE 10 Simplify: $4^2 \div (10 - 9 + 1)^3 \cdot 3 - 5$.

$$4^2 \div (10 - 9 + 1)^3 \cdot 3 - 5$$

$$= 4^2 \div (1 + 1)^3 \cdot 3 - 5 \qquad \text{Subtracting inside parentheses}$$

$$= 4^2 \div 2^3 \cdot 3 - 5 \qquad \text{Adding inside parentheses}$$

$$= 16 \div 8 \cdot 3 - 5 \qquad \text{Evaluating exponential expressions}$$

$$= 2 \cdot 3 - 5 \quad\left.\vphantom{\begin{matrix}a\\b\end{matrix}}\right\} \qquad \text{Doing all multiplications and divisions}$$
$$= 6 - 5 \qquad\qquad\quad \text{in order from left to right}$$

$$= 1 \qquad\qquad\qquad\quad \text{Subtracting}$$

Do Exercises 19–21.

EXAMPLE 11 Simplify: $2^9 \div 2^6 \cdot 2^3$.

$$2^9 \div 2^6 \cdot 2^3 = 512 \div 64 \cdot 8 \qquad \text{There are no parentheses. Evaluating exponential expressions}$$

$$= 8 \cdot 8 \quad\left.\vphantom{\begin{matrix}a\\b\end{matrix}}\right\} \qquad \text{Doing all multiplications and}$$
$$= 64 \qquad\qquad \text{divisions in order from left to right}$$

Do Exercise 22.

CALCULATOR CORNER

Order of Operations To determine whether a calculator is programmed to follow the rules for order of operations, we can enter a simple calculation that requires using those rules. For example, we enter $\boxed{3}\ \boxed{+}\ \boxed{4}\ \boxed{\times}\ \boxed{2}\ \boxed{=}$. If the result is 11, we know that the rules for order of operations have been followed. That is, the multiplication $4 \times 2 = 8$ was performed first and then 3 was added to produce a result of 11. If the result is 14, we know that the calculator performs operations as they are entered rather than following the rules for order of operations. That means, in this case, that 3 and 4 were added first to get 7 and then that sum was multiplied by 2 to produce the result of 14. For such calculators, we would have to enter the operations in the order in which we want them performed. In this case, we would press $\boxed{4}\ \boxed{\times}\ \boxed{2}\ \boxed{+}\ \boxed{3}\ \boxed{=}$.

Many calculators have parenthesis keys that can be used to enter an expression containing parentheses. To enter $5(4 + 3)$, for example, we press $\boxed{5}\ \boxed{(}\ \boxed{4}\ \boxed{+}\ \boxed{3}\ \boxed{)}\ \boxed{=}$. The result is 35.

Exercises: Simplify.

1. $84 - 5 \cdot 7$ **2.** $80 + 50 \div 10$

3. $3^2 + 9^2 \div 3$ **4.** $4^4 \div 64 - 4$

5. $15 \cdot 7 - (23 + 9)$ **6.** $(4 + 3)^2$

Simplify.

19. $5^3 + 26 \cdot 71 - (16 + 25 \cdot 3)$

20. $(1 + 3)^3 + 10 \cdot 20 + 8^2 - 23$

21. $81 - 3^2 \cdot 2 \div (12 - 9)$

22. Simplify: $2^3 \cdot 2^8 \div 2^9$.

Answers on page A-3

23. NBA Tall Men. The heights, in inches, of several of the tallest players in the NBA are given in the bar graph below. Find the average height of these players.

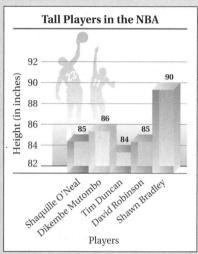

Tall Players in the NBA

Source: NBA

AVERAGES

In order to find the average of a set of numbers, we use addition and then division. For example, the average of 2, 3, 6, and 9 is found as follows.

$$\text{Average} = \frac{2 + 3 + 6 + 9}{4} = \frac{20}{4} = 5$$

The number of addends is 4.

Divide by 4.

> ### AVERAGE
>
> The **average** of a set of numbers is the sum of the numbers divided by the number of addends.

EXAMPLE 12 *Average Height of Waterfalls.* The heights of the four highest waterfalls in the world are given in the bar graph at right. Find the average height of all four.

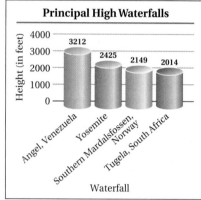

Principal High Waterfalls

Source: World Almanac

The average is given by $\dfrac{3212 + 2425 + 2149 + 2014}{4} = \dfrac{9800}{4} = 2450.$

Thus the average height of the four highest waterfalls is 2450 ft.

Do Exercise 23.

d Removing Parentheses within Parentheses

When parentheses occur within parentheses, we can make them different shapes, such as [ ] (also called "brackets") and { } (also called "braces"). All of these have the same meaning. When parentheses occur within parentheses, computations in the innermost ones are to be done first.

EXAMPLE 13 Simplify: $[25 - (4 + 3) \times 3] \div (11 - 7)$.

$[25 - (4 + 3) \times 3] \div (11 - 7)$

$= [25 - 7 \times 3] \div (11 - 7)$ Doing the calculations in the innermost parentheses first

$= [25 - 21] \div (11 - 7)$ Doing the multiplication in the brackets

$= 4 \div 4$ Subtracting

$= 1$ Dividing

Answer on page A-3

EXAMPLE 14 Simplify: $16 \div 2 + \{40 - [13 - (4 + 2)]\}$.

$16 \div 2 + \{40 - [13 - (4 + 2)]\}$

$= 16 \div 2 + \{40 - [13 - 6]\}$ — Doing the calculations in the innermost parentheses first

$= 16 \div 2 + \{40 - 7\}$ — Again, doing the calculations in the innermost parentheses

$= 16 \div 2 + 33$ — Subtracting inside the braces

$= 8 + 33$ — Doing all multiplications and divisions in order from left to right

$= 41$ — Doing all additions and subtractions in order from left to right

Do Exercises 24 and 25.

Simplify.

24. $9 \times 5 + \{6 \div [14 - (5 + 3)]\}$

25. $[18 - (2 + 7) \div 3]$
$- (31 - 10 \times 2)$

Answers on page A-3

Study Tips

TEST PREPARATION

You are probably ready to begin preparing for your first test. Here are some test-taking study tips.

- **Make up your own test questions as you study.** After you have done your homework over a particular objective, write one or two questions on your own that you think might be on a test. You will be amazed at the insight this will provide.
- **Do an overall review of the chapter, focusing on the objectives and the examples.** This should be accompanied by a study of any class notes you may have taken.
- **Do the review exercises at the end of the chapter.** Check your answers at the back of the book. If you have trouble with an exercise, use the objective symbol as a guide to go back and do further study of that objective.
- **Call the AW Math Tutor Center if you need extra help at 1-888-777-0463.**
- **Do the chapter test at the end of the chapter.** Check the answers and use the objective symbols at the back of the book as a reference for where to review.
- **Ask former students for old exams.** Working such exams can be very helpful and allows you to see what various professors think is important.
- **When taking a test, read each question carefully and try to do all the questions the first time through, but pace yourself.** Answer all the questions, and mark those to recheck if you have time at the end. Very often, your first hunch will be correct.
- **Try to write your test in a neat and orderly manner.** Very often, your instructor tries to give you partial credit when grading an exam. If your test paper is sloppy and disorderly, it is difficult to verify the partial credit. Doing your work neatly can ease such a task for the instructor.

1.9

EXERCISE SET

For Extra Help

Digital Video
Tutor CD 1
Videotape 2

InterAct
Math

Math Tutor
Center

MathXL

MyMathLab

a Write exponential notation.

1. $3 \cdot 3 \cdot 3 \cdot 3$

2. $2 \cdot 2 \cdot 2 \cdot 2 \cdot 2$

3. $5 \cdot 5$

4. $13 \cdot 13 \cdot 13$

5. $7 \cdot 7 \cdot 7 \cdot 7 \cdot 7$

6. $10 \cdot 10$

7. $10 \cdot 10 \cdot 10$

8. $1 \cdot 1 \cdot 1 \cdot 1$

b Evaluate.

9. 7^2

10. 5^3

11. 9^3

12. 10^2

13. 12^4

14. 10^5

15. 11^2

16. 6^3

c Simplify.

17. $12 + (6 + 4)$

18. $(12 + 6) + 18$

19. $52 - (40 - 8)$

20. $(52 - 40) - 8$

21. $1000 \div (100 \div 10)$

22. $(1000 \div 100) \div 10$

23. $(256 \div 64) \div 4$

24. $256 \div (64 \div 4)$

25. $(2 + 5)^2$

26. $2^2 + 5^2$

27. $(11 - 8)^2 - (18 - 16)^2$

28. $(32 - 27)^3 + (19 + 1)^3$

29. $16 \cdot 24 + 50$

30. $23 + 18 \cdot 20$

31. $83 - 7 \cdot 6$

32. $10 \cdot 7 - 4$

33. $10 \cdot 10 - 3 \cdot 4$

34. $90 - 5 \cdot 5 \cdot 2$

35. $4^3 \div 8 - 4$

36. $8^2 - 8 \cdot 2$

37. $17 \cdot 20 - (17 + 20)$

38. $1000 \div 25 - (15 + 5)$

39. $6 \cdot 10 - 4 \cdot 10$

40. $3 \cdot 8 + 5 \cdot 8$

41. $300 \div 5 + 10$

42. $144 \div 4 - 2$

43. $3 \cdot (2 + 8)^2 - 5 \cdot (4 - 3)^2$

44. $7 \cdot (10 - 3)^2 - 2 \cdot (3 + 1)^2$

45. $4^2 + 8^2 \div 2^2$

46. $6^2 - 3^4 \div 3^3$

47. $10^3 - 10 \cdot 6 - (4 + 5 \cdot 6)$

48. $7^2 + 20 \cdot 4 - (28 + 9 \cdot 2)$

49. $6 \cdot 11 - (7 + 3) \div 5 - (6 - 4)$

50. $8 \times 9 - (12 - 8) \div 4 - (10 - 7)$

51. $120 - 3^3 \cdot 4 \div (5 \cdot 6 - 6 \cdot 4)$

52. $80 - 2^4 \cdot 15 \div (7 \cdot 5 - 45 \div 3)$

53. $2^9 \cdot 2^6 \div 2^7$

54. $2^7 \div 2^5 \cdot 2^4 \div 2^2$

55. Find the average of $64, $97, and $121.

56. Find the average of four test grades of 86, 92, 80, and 78.

d Simplify.

57. $8 \times 13 + \{42 \div [18 - (6 + 5)]\}$

58. $72 \div 6 - \{2 \times [9 - (4 \times 2)]\}$

59. $[14 - (3 + 5) \div 2] - [18 \div (8 - 2)]$

60. $[92 \times (6 - 4) \div 8] + [7 \times (8 - 3)]$

61. $(82 - 14) \times [(10 + 45 \div 5) - (6 \cdot 6 - 5 \cdot 5)]$

62. $(18 \div 2) \cdot \{[(9 \cdot 9 - 1) \div 2] - [5 \cdot 20 - (7 \cdot 9 - 2)]\}$

63. $4 \times \{(200 - 50 \div 5) - [(35 \div 7) \cdot (35 \div 7) - 4 \times 3]\}$

64. $15(23 - 4 \cdot 2)^3 \div (3 \cdot 25)$

65. $\{[18 - 2 \cdot 6] - [40 \div (17 - 9)]\} + \{48 - 13 \times 3 + [(50 - 7 \cdot 5) + 2]\}$

66. $(19 - 2^4)^5 - (141 \div 47)^2$

67. ^{D}W Consider the problem in Example 8 of Section 1.8. How can you translate the problem to a single equation involving what you have learned about order of operations? How does the single equation relate to how we solved the problem?

68. ^{D}W Consider the expressions $9 - (4 \cdot 2)$ and $(3 \cdot 4)^2$. Are the parentheses necessary in each case? Explain.

SKILL MAINTENANCE

Solve. [1.7b]

69. $x + 341 = 793$

70. $4197 + x = 5032$

71. $7 \cdot x = 91$

72. $1554 = 42 \cdot y$

73. $3240 = y + 898$

74. $6000 = 1102 + t$

75. $25 \cdot t = 625$

76. $10,000 = 100 \cdot t$

Solve. [1.8a]

77. *Colorado.* The state of Colorado is roughly the shape of a rectangle that is 270 mi by 380 mi. What is its area?

78. On a long four-day trip, a family bought the following amounts of gasoline for their motor home:

23 gallons,   24 gallons,
26 gallons,   25 gallons.

How much gasoline did they buy in all?

SYNTHESIS

Each of the answers in Exercises 79–81 is incorrect. First find the correct answer. Then place as many parentheses as needed in the expression in order to make the incorrect answer correct.

79. $1 + 5 \cdot 4 + 3 = 36$

80. $12 \div 4 + 2 \cdot 3 - 2 = 2$

81. $12 \div 4 + 2 \cdot 3 - 2 = 4$

82. Use one occurrence each of 1, 2, 3, 4, 5, 6, 7, 8, and 9 and any of the symbols $+$, $-$, $\times$, $\div$, and ( ) to represent 100.

Summary and Review

The review that follows is meant to prepare you for a chapter exam. It consists of two parts. The first part is a checklist of the Study Tips referred to in this chapter. The second part is the Review Exercises. These provide practice exercises for the exam, together with references to section objectives so you can go back and review. Before beginning, stop and look back over the skills you have obtained. What skills in mathematics do you have now that you did not have before studying this chapter?

STUDY TIPS CHECKLIST

The foundation of all your study skills is TIME!	☐ Have you found adequate time to study?
	☐ Have you determined the location of the learning resource centers on your campus, such as a mathlab, tutor center, and your instructor's office?
	☐ Are you stopping to work the margin exercises when directed to do so?
	☐ Are you doing your homework as soon as possible after class?
	☐ Are you making use of any of the textbook supplements, such as the Math Tutor Center, the *Student's Solutions Manual*, and the videotapes?

REVIEW EXERCISES

The review exercises that follow are for practice. Answers are given at the back of the book. If you miss an exercise, restudy the objective indicated in blue next to the exercise or direction line that precedes it.

Write expanded notation.   [1.1b]

1. 2793   **2.** 56,078

Write standard notation.   [1.1b]

3. 8 thousands + 6 hundreds + 6 tens + 9 ones

4. 9 ten thousands + 8 hundreds + 4 tens + 4 ones

Write a word name.   [1.1c]

5. 67,819   **6.** 2,781,427

Write standard notation.   [1.1c]

7. Four hundred seventy-six thousand, five hundred eighty-eight

8. *e-books.*   The publishing industry predicts that sales of digital books will reach two billion, four hundred thousand by 2005.
Source: Andersen Consulting

9. What does the digit 8 mean in 4,678,952?   [1.1a]

10. In 13,768,940, what digit tells the number of millions? [1.1a]

Add.   [1.2a]

11. 7304 + 6968

12. 27,609 + 38,415

13. 2743 + 4125 + 6274 + 8956

14.   9 1,4 2 6
     +   7,4 9 5

15. Write a related addition sentence: [1.3a]
    10 − 6 = 4.

16. Write two related subtraction sentences: [1.3a]
    8 + 3 = 11.

Subtract.   [1.3b]

17. 8045 − 2897   **18.** 8465 − 7312

19. 6003 − 3729   **20.**   3 7,4 0 5
                                       − 1 9,6 4 8

Round 345,759 to the nearest: [1.4a]

21. Hundred.   **22.** Ten.

23. Thousand.

Estimate the sum, difference, or product by first rounding to the nearest hundred. Show your work. [1.4b], [1.5b]

24. 41,348 + 19,749   **25.** 38,652 − 24,549

26. 396 · 748

Use < or > for □ to write a true sentence.   [1.4c]

27. 67 □ 56   **28.** 1 □ 23

Multiply.   [1.5a]

29. 700 · 600   **30.** 7846 · 800

31. 726 · 698   **32.** 587 · 47

33.   8 3 0 5
        ×   6 4 2

34. Write a related multiplication sentence: [1.6a]
    56 ÷ 8 = 7.

35. Write two related division sentences: [1.6a]
    13 · 4 = 52.

Divide.   [1.6b]

36. 63 ÷ 5   **37.** 80 ÷ 16

38. 7) 6 3 9 4   **39.** 3073 ÷ 8

40. 6 0) 2 8 6   **41.** 4266 ÷ 79

42. 3 8) 1 7,1 7 6   **43.** 1 4) 7 0,1 1 2

44. 52,668 ÷ 12

Solve.   [1.7b]

45. 46 · n = 368   **46.** 47 + x = 92

47. x = 782 − 236

48. Write exponential notation: 4 · 4 · 4.   [1.9a]

Evaluate.   [1.9b]

49. 10^4 **50.** 6^2

Simplify.   [1.9c, d]

51. 8 · 6 + 17

52. 10 · 24 − (18 + 2) ÷ 4 − (9 − 7)

53. 7 + $(4 + 3)^2$

54. 7 + 4^2 + 3^2

55. (80 ÷ 16) × [(20 − 56 ÷ 8) + (8 · 8 − 5 · 5)]

56. Find the average of 157, 170, and 168.

Solve. [1.8a]

57. *Oak Desk.* Natasha has $196 and wants to buy an oak computer roll-top desk for $698. How much more does she need?
Source: Oak Express®

Desk Just...
$698

Oak Express Excalibur 48"
Computer Roll-Top Desk
Accommodates most tower or desk-top computers. Slide-out mouse pad and keyboard tray. Available in light and dark finishes.Constructed of solid oak and oak veneers.

58. Tony has $406 in her checking account. She is paid $78 for a part-time job and deposits that in her checking account. How much is then in her account?

59. *Lincoln-Head Pennies.* In 1909, the first Lincoln-head pennies were minted. Seventy-three years later, these pennies were first minted with a decreased copper content. In what year was the copper content reduced?

60. A beverage company packed 222 cans of soda into 6-can cartons. How many cartons did they fill?

61. An apple farmer keeps bees in her orchard to help pollinate the apple blossoms so more apples will be produced. The bees from an average beehive can pollinate 30 surrounding trees during one growing season. A farmer has 420 trees. How many beehives does she need to pollinate them all?
Source: Jordan Orchards, Westminster, PA

62. An apartment builder bought 3 electric ranges at $299 each and 4 dishwashers at $379 each. What was the total cost?

63. A family budgets $4950 for food and clothing and $3585 for entertainment. The yearly income of the family was $28,283. How much of this income remained after these two allotments?

64. A chemist has 2753 mL of alcohol. How many 20-mL beakers can be filled? How much will be left over?

65. *Olympic Trampoline.* Shown below is an Olympic trampoline. Find the area and the perimeter of the trampoline. [1.2b], [1.5c]
Source: International Trampoline Industry Association, Inc.

14 ft

7 ft

66. D_W Write a problem for a classmate to solve. Design the problem so that the solution is "Each of the 144 bottles will contain 8 oz of hot sauce." [1.8a]

67. D_W Is subtraction associative? Why or why not? [1.3b]

(**SYNTHESIS**)

68. ▦ Determine the missing digit d. [1.5a]

$$\begin{array}{r} 9\,d \\ \times\ \ d\,2 \\ \hline 8\,0\,3\,6 \end{array}$$

69. ▦ Determine the missing digits a and b. [1.6b]

$$2\,b\,1\,)\overline{\,2\,3\,6{,}4\,2\,1}\quad\text{with quotient}\ 9\,a\,1$$

70. A mining company estimates that a crew must tunnel 2000 ft into a mountain to reach a deposit of copper ore. Each day the crew tunnels about 500 ft. Each night about 200 ft of loose rocks roll back into the tunnel. How many days will it take the mining company to reach the copper deposit? [1.8a]

1. Write expanded notation: 8843.

2. Write a word name: 38,403,277.

3. In the number 546,789, which digit tells the number of hundred thousands?

Add.

4. 6 8 1 1
     + 3 1 7 8

5. 4 5,8 8 9
     + 1 7,9 0 2

6. 1 2
       8
       3
       7
     +   4

7. 6 2 0 3
     + 4 3 1 2

Subtract.

8. 7 9 8 3
     − 4 3 5 3

9. 2 9 7 4
     − 1 9 3 5

10. 8 9 0 7
      − 2 0 5 9

11. 2 3,0 6 7
      − 1 7,8 9 2

Multiply.

12. 4 5 6 8
      ×   9

13. 8 8 7 6
      ×   6 0 0

14. 6 5
      × 3 7

15. 6 7 8
      × 7 8 8

Divide.

16. 15 ÷ 4

17. 420 ÷ 6

18. 8 9)‾8 6 3 3

19. 4 4)‾3 5,4 2 8

Solve.

20. *Hostess Ding Dongs®.* Hostess packages its Ding Dong® snack products in 12-packs. It manufactures 22,231 cakes. How many 12-packs can it fill? How many will be left over?

21. *Largest States.* The following table lists the five largest states in terms of their area. Find the total area of these states.

STATE	AREA (in Square Miles)
Alaska	591,004
Texas	266,807
California	158,706
Montana	147,046
New Mexico	121,593

Source: The New York Times Almanac

22. *Pool Tables.* The Hartford™ pool table made by Brunswick Billiards comes in three sizes of playing area, 50 in. by 100 in., 44 in. by 88 in., and 38 in. by 76 in.

a) Find the perimeter and the area of the playing area of each table.

b) By how much area does the large table exceed the small table?

Source: Brunswick Billiards

23. *Patents Issued.* There were 169,094 patents issued in 1999. This was 70,018 more than in 1990. How many patents were issued in 1990?

Source: U.S. Patent and Trademark Office

24. A sack of oranges weighs 27 lb. A sack of apples weighs 32 lb. Find the total weight of 16 bags of oranges and 43 bags of apples.

25. A box contains 5000 staples. How many staplers can be filled from the box if each stapler holds 250 staples?

Solve.

26. $28 + x = 74$

27. $169 \div 13 = n$

28. $38 \cdot y = 532$

Round 34,578 to the nearest:

29. Thousand.

30. Ten.

31. Hundred.

Estimate the sum, difference, or product by first rounding to the nearest hundred. Show your work.

32.
$$\begin{array}{r} 2\,3{,}6\,4\,9 \\ +\,5\,4{,}7\,4\,6 \\ \hline \end{array}$$

33.
$$\begin{array}{r} 5\,4{,}7\,5\,1 \\ -\,2\,3{,}6\,4\,9 \\ \hline \end{array}$$

34.
$$\begin{array}{r} 8\,2\,4 \\ \times\,4\,8\,9 \\ \hline \end{array}$$

Use < or > for ☐ to write a true sentence.

35. $34 \,\square\, 17$

36. $117 \,\square\, 157$

37. Write exponential notation: $12 \cdot 12 \cdot 12 \cdot 12$.

Evaluate.

38. 7^3

39. 2^3

Simplify.

40. $(10 - 2)^2$

41. $10^2 - 2^2$

42. $(25 - 15) \div 5$

43. $8 \times \{(20 - 11) \cdot [(12 + 48) \div 6 - (9 - 2)]\}$

44. $2^4 + 24 \div 12$

45. Find the average of 97, 98, 87, and 86.

SYNTHESIS

46. An open cardboard shoe box is 8 in. wide, 12 in. long, and 6 in. high. How many square inches of cardboard are used?

47. Cara spends $229 a month to repay her student loan. If she has already paid $9160 on the 10-yr loan, how many payments remain?

48. Jennie scores three 90's, four 80's, and a 74 on her eight quizzes. Find her average.

49. Use trials to find the single-digit number a for which
$$359 - 46 + a \div 3 \times 25 - 7^2 = 339.$$

Fraction Notation: Multiplication and Division

Gateway to Chapter 2

We consider multiplication and division using fraction notation in this chapter. To aid our study, the chapter begins with factorizations and rules for divisibility. After multiplication and division have been discussed, those skills are used to solve equations and do problem solving.

Real-World Application

The swimming speed of a killer whale is about 30 mph. The swimming speed of a dolphin is about three-fifths that of a killer whale. Find the swimming speed of a dolphin.

Source: G. Cafiero and M. Jahoda, *Whales and Dolphins*. New York: Barnes & Noble Books, 1994

This problem appears as Exercise 47 in Section 2.6.

CHAPTER

2

1. Determine whether 59 is prime, composite, or neither. [2.1c]

2. Find the prime factorization of 420. [2.1d]

3. Determine whether 1503 is divisible by 9. [2.2a]

4. Determine whether 768 is divisible by 6. [2.2a]

Simplify. [2.3b], [2.5b]

5. $\dfrac{57}{57}$

6. $\dfrac{68}{1}$

7. $\dfrac{0}{50}$

8. $\dfrac{8}{32}$

Multiply and simplify. [2.6a]

9. $\dfrac{1}{3} \cdot \dfrac{18}{5}$

10. $\dfrac{5}{6} \cdot 24$

11. $\dfrac{2}{5} \cdot \dfrac{25}{8}$

Find the reciprocal. [2.7a]

12. $\dfrac{7}{8}$

13. 11

Divide and simplify. [2.7b]

14. $15 \div \dfrac{5}{8}$

15. $\dfrac{2}{3} \div \dfrac{8}{9}$

16. Solve: $\dfrac{7}{10} \cdot x = 21$. [2.7c]

17. Use = or ≠ for ☐ to write a true sentence: [2.5c]
$\dfrac{5}{11} \, \square \, \dfrac{1}{2}$.

Solve.

18. Julie earns $96 for working a full day. How much does she earn for working $\frac{3}{4}$ of a day? [2.6b]

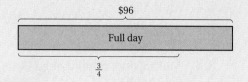

$96

Full day

$\frac{3}{4}$

19. A piece of tubing $\frac{5}{8}$ m long is to be cut into 15 pieces of the same length. What is the length of each piece? [2.7c]

2.1

FACTORIZATIONS

In this chapter, we begin our work with fractions and fraction notation. Certain skills make such work easier. For example, in order to simplify

$$\frac{12}{32},$$

it is important that we be able to *factor* the 12 and the 32, as follows:

$$\frac{12}{32} = \frac{4 \cdot 3}{4 \cdot 8}.$$

Then we "remove" a factor of 1:

$$\frac{4 \cdot 3}{4 \cdot 8} = \frac{4}{4} \cdot \frac{3}{8} = 1 \cdot \frac{3}{8} = \frac{3}{8}.$$

Thus factoring is an important skill in working with fractions.

a Factors and Factorization

In Sections 2.1 and 2.2, we consider only the **natural numbers** 1, 2, 3, and so on.

Let's look at the product $3 \cdot 4 = 12$. We say that 3 and 4 are **factors** of 12.

> **FACTOR**
>
> - In the product $a \cdot b$, a and b are called **factors.**
> - If we divide Q by d and get a remainder of 0, then the divisor d is a **factor** of the dividend Q.

EXAMPLE 1 Determine by long division whether 6 is a factor of 72.

$$
\begin{array}{r}
12 \\
6\overline{)72} \\
60 \\
\hline
12 \\
12 \\
\hline
0
\end{array}
$$

The remainder is 0, so 6 is a factor of 72. We sometimes say that 6 divides 72 "evenly" because there is a remainder of 0.

EXAMPLE 2 Determine by long division whether 15 is a factor of 7894.

$$
\begin{array}{r}
526 \\
15\overline{)7894} \\
7500 \\
\hline
394 \\
300 \\
\hline
94 \\
90 \\
\hline
4 \leftarrow \text{Not } 0
\end{array}
$$

The remainder is *not* 0, so 15 is not a factor of 7894.

Objectives

a Determine whether one number is a factor of another, and find the factors of a number.

b Find some multiples of a number, and determine whether a number is divisible by another.

c Given a number from 1 to 100, tell whether it is prime, composite, or neither.

d Find the prime factorization of a composite number.

Study Tips

STUDY TIPS REVIEW

You may or may not have studied parts of Chapter 1. In that chapter, we introduced Study Tips. If you have not studied them, go back and review the following that are in Chapter 1:

- Basic Information on the First Day of Class (Section 1.1)
- Using This Textbook (Section 1.2)
- Highlighting (Section 1.3)
- Learning Resources (Section 1.4)
- Time Management (Part 1) (Section 1.5)
- Exercises (Section 1.6)
- Test Preparation (Section 1.9)

Determine whether the second
number is a factor of the first.

1. 72; 8

2. 2384; 28

Find all the factors of the number.

3. 10

4. 45

5. 62

6. 24

7. Show that each of the numbers
5, 45, and 100 is a multiple of 5.

8. Show that each of the numbers
10, 60, and 110 is a multiple
of 10.

9. Multiply by 1, 2, 3, and so on, to
find ten multiples of 5.

Answers on page A-4

Do Exercises 1 and 2.

Consider $12 = 3 \cdot 4$. We say that $3 \cdot 4$ is a **factorization** of 12. Similarly, $6 \cdot 2$, $12 \cdot 1$, $2 \cdot 2 \cdot 3$, and $1 \cdot 3 \cdot 4$ are also factorizations of 12. Since $a = a \cdot 1$, every number has a factorization, and every number has factors. In the case of $17 = 17 \cdot 1$, the only factors of 17 are 17 and 1.

EXAMPLE 3 Find all the factors of 70.

We find as many "two-factor" factorizations as we can. We check sequentially the numbers 1, 2, 3, and so on, to see if we can form any factorizations:

70

$1 \cdot 70$
$2 \cdot 35$
$5 \cdot 14$
$7 \cdot 10$

Note that all but one of the factors of a natural number are *less* than the number.

Note that 3, 4, and 6 are not factors. If there are additional factors, they must be between 7 and 10. Since 8 and 9 are not factors, we are finished. The factors of 70 are 1, 2, 5, 7, 10, 14, 35, and 70.

Do Exercises 3–6.

b Multiples and Divisibility

A **multiple** of a natural number is a product of it and some natural number. For example, some multiples of 2 are:

2   (because $2 = 1 \cdot 2$);
4   (because $4 = 2 \cdot 2$);
6   (because $6 = 3 \cdot 2$);
8   (because $8 = 4 \cdot 2$);
10   (because $10 = 5 \cdot 2$).

Note that all but one of the multiples of a number are *larger* than the number.

We find multiples of 2 by counting by twos: 2, 4, 6, 8, and so on. We can find multiples of 3 by counting by threes: 3, 6, 9, 12, and so on.

EXAMPLE 4 Show that each of the numbers 8, 12, 20, and 36 is a multiple of 4.

$$8 = 2 \cdot 4 \qquad 12 = 3 \cdot 4 \qquad 20 = 5 \cdot 4 \qquad 36 = 9 \cdot 4$$

Do Exercises 7 and 8.

EXAMPLE 5 Multiply by 1, 2, 3, and so on, to find ten multiples of 7.

$$1 \cdot 7 = 7 \qquad 6 \cdot 7 = 42$$
$$2 \cdot 7 = 14 \qquad 7 \cdot 7 = 49$$
$$3 \cdot 7 = 21 \qquad 8 \cdot 7 = 56$$
$$4 \cdot 7 = 28 \qquad 9 \cdot 7 = 63$$
$$5 \cdot 7 = 35 \qquad 10 \cdot 7 = 70$$

Do Exercise 9.

DIVISIBILITY

The number a is **divisible** by another number b if there exists a number c such that $a = b \cdot c$. The statements "a is **divisible** by b," "a is a **multiple** of b," and "b is a **factor** of a" all have the same meaning.

Thus we have

27 is *divisible* by 3 because 27 is a *multiple* of 3 ($27 = 9 \cdot 3$);

27 is a *multiple* of 3 and 3 is a *factor* of 27.

EXAMPLE 6 Determine whether 45 is divisible by 9.

We divide 45 by 9:

$$\begin{array}{r} 5 \\ 9\overline{)45} \\ \underline{45} \\ 0 \end{array}$$

Because the remainder is 0, 45 is divisible by 9.

EXAMPLE 7 Determine whether 98 is divisible by 4.

We divide 98 by 4:

$$\begin{array}{r} 24 \\ 4\overline{)98} \\ \underline{80} \\ 18 \\ \underline{16} \\ 2 \leftarrow \text{Not 0} \end{array}$$

Since the remainder is not 0, 98 is *not* divisible by 4.

Do Exercises 10–12.

10. Determine whether 16 is divisible by 2.

11. Determine whether 125 is divisible by 5.

12. Determine whether 125 is divisible by 6.

Answers on page A-4

CALCULATOR CORNER

Divisibility and Factors   We can use a calculator to determine whether one number is divisible by another number or whether one number is a factor of another number. For example, to determine whether 387 is divisible by 18, we first press $\boxed{3}\,\boxed{8}\,\boxed{7}\,\boxed{\div}\,\boxed{1}\,\boxed{8}\,\boxed{=}$. The display reads $\boxed{21.5}$. Note that the result is not a natural number. (For a brief discussion of decimal notation, see the Calculator Corner on page 56. Decimal notation will be studied in detail in Chapter 4.) Thus we know that 387 is not a multiple of 18; that is, 387 is not divisible by 18 and 18 is not a factor of 387.

When we divide 387 by 9, the result is $\boxed{43}$. Since 43 is a natural number, we know that 387 is a multiple of 9; that is, $387 = 43 \cdot 9$. Thus, 387 is divisible by 9 and 9 is a factor of 387.

Exercises: For each pair of numbers, determine whether the first number is divisible by the second number.

1. 722; 19

2. 845; 7

3. 1047; 14

4. 5283; 9

For each pair of numbers, determine whether the second number is a factor of the first number.

5. 502; 8

6. 651; 21

7. 3875; 25

8. 8464; 12

9. 32,768; 256

10. 32,768; 864

13. Tell whether each number is prime, composite, or neither.

1, 2, 6, 12, 13, 19, 41, 65, 73, 99

C Prime and Composite Numbers

> ### PRIME AND COMPOSITE NUMBERS
>
> - A natural number that has exactly two *different* factors, only itself and 1, is called a **prime number.**
> - The number 1 is *not* prime.
> - A natural number, other than 1, that is not prime is **composite.**

EXAMPLE 8 Determine whether the numbers 1, 2, 3, 4, 5, 6, 7, 9, 10, 11, and 63 are prime, composite, or neither.

The number 1 is not prime. It does not have *two* different factors.

The number 2 is prime. It has only the factors 1 and 2.

The numbers 3, 5, 7, and 11 are prime. Each has only two factors, itself and 1.

The number 4 is not prime. It has the factors 1, 2, and 4 and is composite.

The numbers 6, 9, 10, and 63 are composite. Each has more than two factors.

Thus we have:

Prime: 2, 3, 5, 7, 11;

Composite: 4, 6, 9, 10, 63;

Neither: 1.

The number 2 is the *only* even prime number. It is also the smallest prime number. The number 0 is also neither prime nor composite, but 0 is *not* a natural number and thus is not considered here. We are considering only natural numbers.

Do Exercise 13.

The following is a table of the prime numbers from 2 to 157. There are more extensive tables, but these prime numbers will be the most helpful to you in this text.

> ### A TABLE OF PRIMES FROM 2 TO 157
>
> 2, 3, 5, 7, 11, 13, 17, 19, 23, 29, 31, 37, 41, 43, 47, 53, 59, 61, 67, 71, 73, 79, 83, 89, 97, 101, 103, 107, 109, 113, 127, 131, 137, 139, 149, 151, 157

d Prime Factorizations

To factor a composite number into a product of primes is to find a **prime factorization** of the number. To do this, we consider the primes

2, 3, 5, 7, 11, 13, 17, 19, 23, and so on,

and determine whether a given number is divisible by the primes.

Answer on page A-4

EXAMPLE 9 Find the prime factorization of 39.

a) We divide by the first prime, 2.

$$\begin{array}{r} 19 \quad R = 1 \\ 2\overline{)39} \\ \underline{38} \\ 1 \end{array}$$

Because the remainder is not 0, 2 is not a factor of 39, and 39 is not divisible by 2.

b) We divide by the next prime, 3.

$$\begin{array}{r} 13 \quad R = 0 \\ 3\overline{)39} \end{array}$$

Because 13 is a prime, we are finished. The prime factorization is

$$39 = 3 \cdot 13.$$

EXAMPLE 10 Find the prime factorization of 76.

a) We divide by the first prime, 2.

$$\begin{array}{r} 38 \quad R = 0 \\ 2\overline{)76} \end{array}$$

b) Because 38 is composite, we start with 2 again:

$$\begin{array}{r} 19 \quad R = 0 \\ 2\overline{)38} \end{array}$$

Because 19 is a prime, we are finished. The prime factorization is

$$76 = 2 \cdot 2 \cdot 19.$$

We abbreviate our procedure as follows.

$$\begin{array}{r} 19 \\ 2\overline{)38} \\ 2\overline{)76} \end{array}$$

$$76 = 2 \cdot 2 \cdot 19$$

Multiplication is commutative so a factorization such as $2 \cdot 2 \cdot 19$ could also be expressed as $2 \cdot 19 \cdot 2$ or $19 \cdot 2 \cdot 2$ (or in exponential notation, as $2^2 \cdot 19$ or $19 \cdot 2^2$), but the prime factors are still the same. For this reason, we agree that any of these is "the" prime factorization of 76.

> Every number has just one (unique) prime factorization.

EXAMPLE 11 Find the prime factorization of 72.

We can do divisions "up" as follows:

$$\begin{array}{r} 3 \leftarrow \text{Prime quotient} \\ 3\overline{)\ 9} \\ 2\overline{)18} \\ 2\overline{)36} \\ 2\overline{)72} \leftarrow \text{Begin here.} \end{array}$$

$$72 = 2 \cdot 2 \cdot 2 \cdot 3 \cdot 3$$

Or, we can also do divisions "down":

$$\begin{array}{r} 2\overline{)72} \leftarrow \text{Begin here.} \\ 2\overline{)36} \\ 2\overline{)18} \\ 3\overline{)\ 9} \\ 3 \leftarrow \text{Prime quotient} \end{array}$$

Find the prime factorization of the number.

14. 6

15. 12

16. 45

17. 98

18. 126

19. 144

To the student and the instructor: Recall that the Skill Maintenance exercises, which occur at the end of the exercise sets, review any skill that has been studied before in the text.

Beginning with this chapter, however, certain objectives from four particular sections, along with the material of this chapter, will be tested on the chapter test.

For this chapter, the objectives to be retested are [1.3b], [1.6b], [1.7b], and [1.8a].

Answers on page A-4

Some other ways to find the prime factorization of 72 using **factor trees** are as follows. Begin by determining any factorization you can, and then continue factoring.

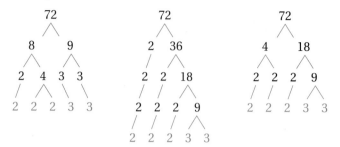

EXAMPLE 12   Find the prime factorization of 189.

We can use a string of successive divisions.

$$\begin{array}{r} 7 \\ 3\overline{)21} \\ 3\overline{)63} \\ 3\overline{)189} \end{array}$$ 189 is not divisible by 2. We move to 3.

$$189 = 3 \cdot 3 \cdot 3 \cdot 7$$

We can also use a factor tree.

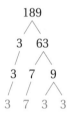

> **CAUTION!**
>
> Keep in mind the difference between finding all the factors of a number and finding the prime factorization. In Example 12, the prime factorization is $3 \cdot 3 \cdot 3 \cdot 7$. The factors of 189 are 1, 3, 7, 9, 21, 27, 63, and 189.

EXAMPLE 13   Find the prime factorization of 65.

We can use a string of successive divisions.

$$\begin{array}{r} 13 \\ 5\overline{)65} \end{array}$$ 65 is not divisible by 2 or 3. We move to 5.

$$65 = 5 \cdot 13$$

We can also use a factor tree.

65
5   13

Do Exercises 14–19.

2.1 EXERCISE SET

a Determine whether the second number is a factor of the first.

1. 52; 14   **2.** 52; 13   **3.** 625; 25   **4.** 680; 16

Find all the factors of the number.

5. 18   **6.** 16   **7.** 54   **8.** 48

9. 4   **10.** 9   **11.** 7   **12.** 11

13. 1   **14.** 3   **15.** 98   **16.** 100

b Multiply by 1, 2, 3, and so on, to find ten multiples of the number.

17. 4   **18.** 11   **19.** 20   **20.** 50

21. 3   **22.** 5   **23.** 12   **24.** 13

25. 10   **26.** 6   **27.** 9   **28.** 14

29. Determine whether 26 is divisible by 6.   **30.** Determine whether 29 is divisible by 9.

31. Determine whether 1880 is divisible by 8.   **32.** Determine whether 4227 is divisible by 3.

33. Determine whether 256 is divisible by 16.   **34.** Determine whether 102 is divisible by 4.

35. Determine whether 4227 is divisible by 9.   **36.** Determine whether 200 is divisible by 25.

37. Determine whether 8650 is divisible by 16.   **38.** Determine whether 4143 is divisible by 7.

c Determine whether the number is prime, composite, or neither.

39. 1   **40.** 2   **41.** 9   **42.** 19

43. 11   **44.** 27   **45.** 29   **46.** 49

Find the prime factorization of the number.

47. 8 **48.** 16 **49.** 14 **50.** 15

51. 42 **52.** 32 **53.** 25 **54.** 40

55. 50 **56.** 62 **57.** 169 **58.** 140

59. 100 **60.** 110 **61.** 35 **62.** 70

63. 72 **64.** 86 **65.** 77 **66.** 99

67. 2884 **68.** 484 **69.** 51 **70.** 91

71. D_W Is every natural number a multiple of 1? Explain.

72. D_W Explain a method for finding a composite number that contains exactly two factors other than itself and 1.

SKILL MAINTENANCE

Multiply. [1.5a]

73. $2 \cdot 13$ **74.** $8 \cdot 32$ **75.** $17 \cdot 25$ **76.** $25 \cdot 168$

Divide. [1.6b]

77. $0 \div 22$ **78.** $22 \div 1$ **79.** $22 \div 22$ **80.** $66 \div 22$

Solve. [1.8a]

81. Find the total cost of 7 shirts at \$48 each and 4 pairs of pants at \$69 each.

82. Sandy can type 62 words per minute. How long will it take her to type 12,462 words?

SYNTHESIS

83. *Factors and Sums.*   To *factor* a number is to express it as a product. Since $15 = 5 \cdot 3$, we say that 15 is *factored* and that 5 and 3 are *factors* of 15. In the table below, the top number in each column has been factored in such a way that the sum of the factors is the bottom number in the column. For example, in the first column, 56 has been factored as $7 \cdot 8$, and $7 + 8 = 15$, the bottom number. Such thinking will be important in understanding the meaning of a factor and in algebra.

Product	56	63	36	72	140	96		168	110			
Factor	7									9	24	3
Factor	8					8	8			10	18	
Sum	15	16	20	38	24	20	14		21			24

Find the missing numbers in the table.

2.2 DIVISIBILITY

Objective

a Determine whether a number is divisible by 2, 3, 4, 5, 6, 8, 9, or 10.

Suppose you are asked to find the simplest fraction notation for

$$\frac{117}{225}.$$

Since the numbers are quite large, you might feel that the task is difficult. However, both the numerator and the denominator have 9 as a factor. If you knew this, you could factor and simplify quickly as follows:

$$\frac{117}{225} = \frac{9 \cdot 13}{9 \cdot 25} = \frac{9}{9} \cdot \frac{13}{25} = 1 \cdot \frac{13}{25} = \frac{13}{25}.$$

How did we know that both numbers have 9 as a factor? There are fast tests for such determinations. If the sum of the digits of a number is divisible by 9, then the number is divisible by 9; that is, it has 9 as a factor. Since $1 + 1 + 7 = 9$ and $2 + 2 + 5 = 9$, both numbers have 9 as a factor.

a Rules for Divisibility

In this section, we learn fast ways of determining whether numbers are divisible by 2, 3, 4, 5, 6, 8, 9, and 10. This will make simplifying fraction notation much easier.

DIVISIBILITY BY 2

You may already know the test for divisibility by 2.

> **BY 2**
>
> A number is **divisible by 2** (is *even*) if it has a ones digit of 0, 2, 4, 6, or 8 (that is, it has an even ones digit).

Let's see why. Consider 354, which is

3 hundreds + 5 tens + 4.

Hundreds and tens are both multiples of 2. If the last digit is a multiple of 2, then the entire number is a multiple of 2.

EXAMPLES Determine whether the number is divisible by 2.

1. 355 is not a multiple of 2; 5 is *not* even.
2. 4786 is a multiple of 2; 6 is even.
3. 8990 is a multiple of 2; 0 is even.
4. 4261 is not a multiple of 2; 1 is *not* even.

Do Exercises 1–4.

Determine whether the number is divisible by 2.

1. 84

2. 59

3. 998

4. 2225

Answers on page A-4

DIVISIBILITY BY 3

> **BY 3**
>
> A number is **divisible by 3** if the sum of its digits is divisible by 3.

■ **EXAMPLES** Determine whether the number is divisible by 3.

5. 18 $1 + 8 = 9$ ⎫
6. 93 $9 + 3 = 12$ ⎬ All are divisible by 3 because the sums
7. 201 $2 + 0 + 1 = 3$ ⎭ of their digits are divisible by 3.

8. 256 $2 + 5 + 6 = 13$ The sum, 13, is not divisible by 3, so
256 is not divisible by 3.

Do Exercises 5–8.

DIVISIBILITY BY 6

A number divisible by 6 is a multiple of 6. But $6 = 2 \cdot 3$, so the number is also a multiple of 2 and 3. Thus we have the following.

> **BY 6**
>
> A number is **divisible by 6** if its ones digit is 0, 2, 4, 6, or 8 (is even) and the sum of its digits is divisible by 3.

■ **EXAMPLES** Determine whether the number is divisible by 6.

9. 720

Because 720 is even, it is divisible by 2. Also, $7 + 2 + 0 = 9$, so 720 is divisible by 3. Thus, 720 is divisible by 6.

$$720 \qquad 7 + 2 + 0 = 9$$
        ↑   ↑
       Even   Divisible by 3

10. 73

73 is *not* divisible by 6 because it is *not* even.

$$73$$
  ↑
Not even

11. 256

256 is *not* divisible by 6 because the sum of its digits is *not* divisible by 3.

$$2 + 5 + 6 = 13$$
                ↑
        Not divisible by 3

Do Exercises 9–12.

DIVISIBILITY BY 9

The test for divisibility by 9 is similar to the test for divisibility by 3.

> ### BY 9
> A number is **divisible by 9** if the sum of its digits is divisible by 9.

EXAMPLE 12 The number 6984 is divisible by 9 because

$$6 + 9 + 8 + 4 = 27$$

and 27 is divisible by 9.

EXAMPLE 13 The number 322 is *not* divisible by 9 because

$$3 + 2 + 2 = 7$$

and 7 is not divisible by 9.

Do Exercises 13–16.

DIVISIBILITY BY 10

> ### BY 10
> A number is **divisible by 10** if its ones digit is 0.

We know that this test works because the product of 10 and *any* number has a ones digit of 0.

EXAMPLES Determine whether the number is divisible by 10.

14. 3440 is divisible by 10 because the ones digit is 0.

15. 3447 is *not* divisible by 10 because the ones digit is not 0.

Do Exercises 17–20.

DIVISIBILITY BY 5

> ### BY 5
> A number is **divisible by 5** if its ones digit is 0 or 5.

EXAMPLES Determine whether the number is divisible by 5.

16. 220 is divisible by 5 because the ones digit is 0.

17. 475 is divisible by 5 because the ones digit is 5.

18. 6514 is *not* divisible by 5 because the ones digit is neither a 0 nor a 5.

Do Exercises 21–24.

Let's see why the test for 5 works. Consider 7830:

$$7830 = 10 \cdot 783 = 5 \cdot 2 \cdot 783.$$

Since 7830 is divisible by 10 and 5 is a factor of 10, 7830 is divisible by 5.

Determine whether the number is divisible by 9.

13. 16

14. 117

15. 930

16. 29,223

Determine whether the number is divisible by 10.

17. 305

18. 300

19. 847

20. 8760

Determine whether the number is divisible by 5.

21. 5780

22. 3427

23. 34,678

24. 7775

Answers on page A-4

Determine whether the number is divisible by 4.

25. 216

26. 217

27. 5865

28. 23,524

Determine whether the number is divisible by 8.

29. 7564

30. 7864

31. 17,560

32. 25,716

Answers on page A-4

Consider 6734:

$$6734 = 673 \text{ tens} + 4.$$

Tens are multiples of 5, so the only number that must be checked is the ones digit. If the last digit is a multiple of 5, the entire number is. In this case, 4 is not a multiple of 5, so 6734 is *not* divisible by 5.

DIVISIBILITY BY 4

The test for divisibility by 4 is similar to the test for divisibility by 2.

> **BY 4**
>
> A number is **divisible by 4** if the number named by its last *two* digits is divisible by 4.

EXAMPLES Determine whether the number is divisible by 4.

19. 8212   is divisible by 4 because 12 is divisible by 4.
20. 5216   is divisible by 4 because 16 is divisible by 4.
21. 8211   is *not* divisible by 4 because 11 is *not* divisible by 4.
22. 7515   is *not* divisible by 4 because 15 is *not* divisible by 4.

Do Exercises 25–28.

To see why the test for divisibility by 4 works, consider 516:

$$516 = 5 \text{ hundreds} + 16.$$

Hundreds are multiples of 4. If the number named by the last two digits is a multiple of 4, then the entire number is a multiple of 4.

DIVISIBILITY BY 8

The test for divisibility by 8 is an extension of the tests for divisibility by 2 and 4.

> **BY 8**
>
> A number is **divisible by 8** if the number named by its last *three* digits is divisible by 8.

EXAMPLES Determine whether the number is divisible by 8.

23.   5648   is divisible by 8 because 648 is divisible by 8.
24. 96,088   is divisible by 8 because 88 is divisible by 8.
25.   7324   is *not* divisible by 8 because 324 is *not* divisible by 8.
26. 13,420   is *not* divisible by 8 because 420 is *not* divisible by 8.

Do Exercises 29–32.

A NOTE ABOUT DIVISIBILITY BY 7

There are several tests for divisibility by 7, but all of them are more complicated than simply dividing by 7. So if you want to test for divisibility by 7, simply divide by 7, either by hand or using a calculator.

2.2 EXERCISE SET

a   To answer Exercises 1–8, consider the following numbers.

46	300	85	256
224	36	711	8064
19	45,270	13,251	1867
555	4444	254,765	21,568

1. Which of the above are divisible by 2?

2. Which of the above are divisible by 3?

3. Which of the above are divisible by 4?

4. Which of the above are divisible by 5?

5. Which of the above are divisible by 6?

6. Which of the above are divisible by 8?

7. Which of the above are divisible by 9?

8. Which of the above are divisible by 10?

To answer Exercises 9–16, consider the following numbers.

56	200	75	35
324	42	812	402
784	501	2345	111,111
55,555	3009	2001	1005

9. Which of the above are divisible by 3?

10. Which of the above are divisible by 2?

11. Which of the above are divisible by 5?

12. Which of the above are divisible by 4?

13. Which of the above are divisible by 9?

14. Which of the above are divisible by 6?

15. Which of the above are divisible by 10?

16. Which of the above are divisible by 8?

17. D_W How can the divisibility tests be used to find prime factorizations?

18. D_W Which of the years from 2000 to 2020, if any, also happen to be prime numbers? Explain at least two ways in which you might go about solving this problem.

SKILL MAINTENANCE

Solve. [1.7b]

19. $56 + x = 194$

20. $y + 124 = 263$

21. $3008 = x + 2134$

22. $18 \cdot t = 1008$

23. $24 \cdot m = 624$

24. $338 = a \cdot 26$

Divide. [1.6b]

25. $2106 \div 9$

26. $4\,5\,)\overline{1\,8\,0,1\,3\,5}$

Solve. [1.8a]

27. An automobile with a 5-speed transmission gets 33 mpg in city driving. How many gallons of gas will it use to travel 1485 mi?

28. There are 60 min in 1 hr. How many minutes are there in 72 hr?

SYNTHESIS

Find the prime factorization of the number. Use divisibility tests where applicable.

29. 7800

30. 2520

31. 2772

32. 1998

33. 🔢 Fill in the missing digits of the number

$$95,\square\square8$$

so that it is divisible by 99.

34. A passenger in a taxicab asks for the driver's company number. The driver says abruptly, "Sure—you can have my number. Work it out: If you divide it by 2, 3, 4, 5, or 6, you will get a remainder of 1. If you divide it by 11, the remainder will be 0 and no driver has a company number that meets these requirements and is smaller than this one." Determine the number.

CHAPTER 2: Fraction Notation:
Multiplication and Division

2.3 FRACTIONS AND FRACTION NOTATION

Objectives

a Identify the numerator and the denominator of a fraction and write fraction notation for part of an object or part of a set of objects.

b Simplify fraction notation like n/n to 1, $0/n$ to 0, and $n/1$ to n.

The study of arithmetic begins with the set of whole numbers

0, 1, 2, 3, 4, 5, 6, 7, 8, 9, 10, 11, and so on.

The need soon arises for fractional parts of numbers such as halves, thirds, fourths, and so on. Here are some examples:

- $\frac{1}{4}$ of the minimum daily requirement of calcium is provided by a cup of frozen yogurt.

- $\frac{43}{100}$ of all corporate travel money is spent on airfares.

- $\frac{1}{25}$ of the parking spaces in a commercial area in the state of Indiana are to be marked for the handicapped.

- About $\frac{1}{5}$ of the earth's surface is frozen.

Identify the numerator and the denominator.

1. $\frac{1}{6}$

2. $\frac{5}{7}$

3. $\frac{22}{3}$

a   Fractions and the Real World

The following are some additional examples of fractions:

$$\frac{1}{2}, \quad \frac{3}{4}, \quad \frac{8}{5}, \quad \frac{11}{23}.$$

This way of writing number names is called **fraction notation.** The top number is called the **numerator** and the bottom number is called the **denominator.**

EXAMPLE 1   Identify the numerator and the denominator.

$$\frac{7}{8} \begin{array}{l} \leftarrow \text{Numerator} \\ \leftarrow \text{Denominator} \end{array}$$

Do Exercises 1–3.

Answers on page A-4

What part is shaded?

4.

$1

5.

1 mile

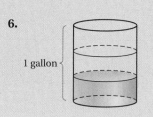

6.

1 gallon

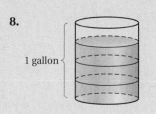

7.

1 mile

8.

1 gallon

9.

0   1   2

Inches

Let's look at various situations that involve fractions.

FRACTIONS AS A PARTITION OF AN OBJECT DIVIDED INTO EQUAL PARTS

Consider a candy bar divided into 5 equal sections. If you eat 2 sections, you have eaten $\frac{2}{5}$ of the candy bar.

The denominator 5 tells us the unit, $\frac{1}{5}$. The numerator 2 tells us the number of equal parts we are considering, 2.

EXAMPLE 2   What part is shaded?

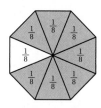

The equal parts are eighths. This tells us the unit, $\frac{1}{8}$. The *denominator* is 8. We have 7 of the units shaded. This tells us the *numerator*, 7. Thus,

$$\frac{7 \longleftarrow \text{7 units are shaded.}}{8 \longleftarrow \text{The unit is } \frac{1}{8}.}$$

is shaded.

The markings on a ruler use fractions.

EXAMPLE 3   What part of an inch is shaded?

$\frac{11}{16}$

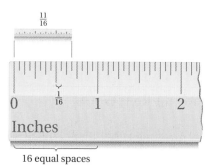

$\frac{1}{16}$

0   1   2

Inches

16 equal spaces

Each inch on the ruler shown above is divided into 16 equal parts. The shading extends to the 11th mark. Thus, $\frac{11}{16}$ is shaded.

Do Exercises 4–9.

Answers on page A-4

CHAPTER 2: Fraction Notation:
Multiplication and Division

Fractions greater than 1 correspond to situations like the following.

EXAMPLE 4 What part is shaded?

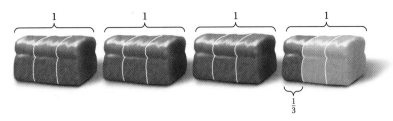

Each loaf of bread is divided into 3 equal parts. The unit is $\frac{1}{3}$. The *denominator* is 3. We have 10 of the units shaded. This tells us the *numerator* is 10. Thus, $\frac{10}{3}$ is shaded.

EXAMPLE 5 What part is shaded?

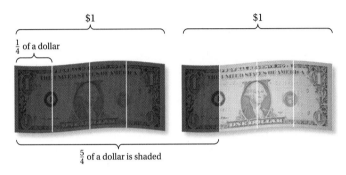

We can regard this as two objects of 4 parts each and take 5 of those parts. We have more than one whole object. Thus, $5 \cdot \frac{1}{4}$, or $\frac{5}{4}$ (also, 5 quarters) is shaded.

Do Exercises 10 and 11.

FRACTIONS AS A PART OF A SET

EXAMPLE 6 What part of this set, or collection, of people are pop stars? U.S. presidents?

| Britney Spears | Abraham Lincoln | Christina Aguilera | Ricky Martin | George W. Bush | Enrique Iglesias |

There are 6 people in the set. We know that 4 of them, Britney Spears, Christina Aguilera, Ricky Martin, and Enrique Iglesias are pop stars. Thus, 4 of 6, or $\frac{4}{6}$, are rock stars. The 2 people remaining are U.S. presidents. Thus, $\frac{2}{6}$ are U.S. presidents.

We will simplify such fraction notation in Section 2.5.

Do Exercise 12.

What part is shaded?

10.

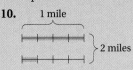

11.

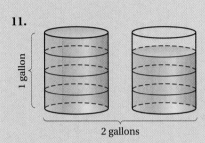

12. What part of this set, or collection, of tools are wrenches? hammers?

Answers on page A-4

13. **Baseball Standings.** Refer to the table in Example 7. The Los Angeles Dodgers finished second in the National League West in 2000. Find the ratio of Dodger wins to losses, wins to total games, and losses to total games.
Source: Major League Baseball

FRACTIONS AS RATIOS

A **ratio** is a quotient of two quantities. We can express a ratio with fraction notation. (We will consider ratios in more detail in Chapter 5.)

EXAMPLE 7 *Baseball Standings.* The following are the final standings in the National League West for 2000, when the division was won by the San Francisco Giants. Find the ratio of Giants wins to losses, wins to total games, and losses to total games.
Source: Major League Baseball

WEST	W	L	PCT.	GB	STRK	LAST 10	vs. DIV.	HOME	AWAY	vs. AL
x-San Francisco	97	65	.599	–	W-1	5-5	26-24	55-26	42-39	8-7
Los Angeles	86	76	.531	11	L-1	6-4	30-21	44-37	42-39	6-9
Arizona	85	77	.525	12	L-1	5-5	29-23	47-34	38-43	6-9
Colorado	82	80	.506	15	W-1	5-5	23-29	48-33	34-47	6-6
San Diego	76	86	.469	21	W-1	2-8	20-31	41-40	35-46	5-10

x-clinched division

The Giants won 97 games and lost 65 games. They played a total of 97 + 65, or 162 games. Thus we have the following.

The ratio of wins to losses is $\frac{97}{65}$.

The ratio of wins to total games is $\frac{97}{162}$.

The ratio of losses to total games is $\frac{65}{162}$.

Do Exercise 13.

Answer on page A-4

b Some Fraction Notation for Whole Numbers

FRACTION NOTATION FOR 1

The number 1 corresponds to situations like those shown here.

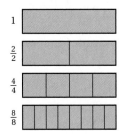

If we divide an object into n parts and take n of them, we get all of the object (1 whole object).

> **THE NUMBER 1 IN FRACTION NOTATION**
>
> $\dfrac{n}{n} = 1$, for any whole number n that is not 0.

EXAMPLES Simplify.

8. $\dfrac{5}{5} = 1$ **9.** $\dfrac{9}{9} = 1$ **10.** $\dfrac{23}{23} = 1$

Do Exercises 14–19.

FRACTION NOTATION FOR 0

Consider the fraction $\frac{0}{4}$. This corresponds to dividing an object into 4 parts and taking none of them. We get 0.

> **THE NUMBER 0 IN FRACTION NOTATION**
>
> $\dfrac{0}{n} = 0$, for any whole number n that is not 0.

EXAMPLES Simplify.

11. $\dfrac{0}{1} = 0$ **12.** $\dfrac{0}{9} = 0$ **13.** $\dfrac{0}{23} = 0$

Simplify.

14. $\dfrac{1}{1}$ **15.** $\dfrac{4}{4}$

16. $\dfrac{34}{34}$ **17.** $\dfrac{100}{100}$

18. $\dfrac{2347}{2347}$ **19.** $\dfrac{103}{103}$

Answers on page A-4

Simplify, if possible.

20. $\dfrac{0}{1}$

21. $\dfrac{0}{8}$

22. $\dfrac{0}{107}$

23. $\dfrac{4-4}{567}$

24. $\dfrac{15}{0}$

25. $\dfrac{0}{3-3}$

Simplify.

26. $\dfrac{8}{1}$

27. $\dfrac{10}{1}$

28. $\dfrac{346}{1}$

29. $\dfrac{24-1}{23-22}$

Fraction notation with a denominator of 0, such as $n/0$, is meaningless because we cannot speak of an object being divided into *zero* parts. If it is not divided at all, then we say that it is undivided and remains in one part. See also the discussion of excluding division by 0 in Section 1.6.

EXCLUDING DIVISION BY 0

$\dfrac{n}{0}$ is not defined for any whole number n.

Do Exercises 20–25.

OTHER WHOLE NUMBERS

Consider the fraction $\frac{4}{1}$. This corresponds to taking 4 objects and dividing each into 1 part. (We do not divide them.) We have 4 objects.

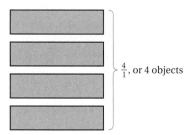

$\frac{4}{1}$, or 4 objects

DIVISION BY 1

Any whole number divided by 1 is the whole number. That is,

$$\dfrac{n}{1} = n, \quad \text{for any whole number } n.$$

EXAMPLES Simplify.

14. $\dfrac{2}{1} = 2$

15. $\dfrac{9}{1} = 9$

16. $\dfrac{34}{1} = 34$

Do Exercises 26–29.

a Identify the numerator and the denominator

1. $\dfrac{3}{4}$

2. $\dfrac{9}{10}$

3. $\dfrac{11}{20}$

4. $\dfrac{18}{5}$

What part of the object or set of objects is shaded? In Exercises 11–14, what part of an inch is shaded?

5.

6.
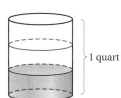

7.
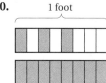

8. 1 gold bar
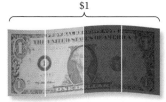

9.

10. 1 foot

11.
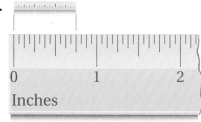

12.
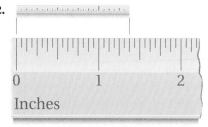

13.
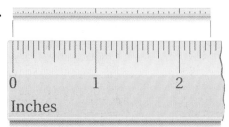

14.
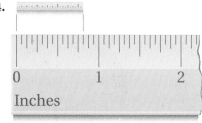

15.

16. 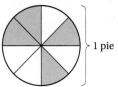 1 year

17. 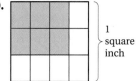 1 pie

18.

19.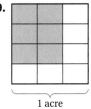
1 acre

20. 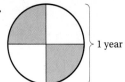 1 square inch

21.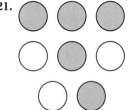

22.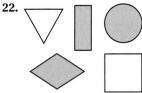

23.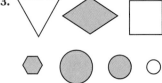

24.

For each of Exercises 25–28, give fraction notation for the amount of gas (a) in the tank and (b) used from a full tank.

25.

26.

27.

28.

CHAPTER 2: Fraction Notation:
Multiplication and Division

29. For the following set of people, what is the ratio of:

    a) women to the total number of people?
    b) women to men?
    c) men to the total number of people?
    d) men to women?

30. For the following set of nuts and bolts, what is the ratio of:

    a) nuts to bolts?
    b) bolts to nuts?
    c) nuts to the total number of elements?
    d) total number of elements to nuts?

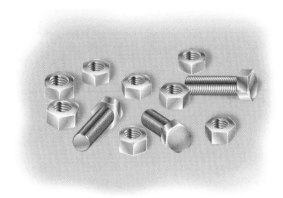

31. *Police–Resident Ratio.* Washington, D. C., has the highest ratio of police to residents in the United States, 67 police officers for every 10,000 residents. What is the ratio of police to residents?

Source: Local Police Departments 1997, the Bureau of Justice Statistics

32. *Moviegoers.* Of every 1000 people who attend movies, 340 are in the 18–24 age group. What is the ratio of moviegoers in the 18–24 age group to all moviegoers?
Source: American Demographics

33. Jake delivers car parts to auto service centers. On Thursday he had 15 deliveries. By noon he had delivered only 4 orders. What is the ratio of:

    a) orders delivered to total number of orders?
    b) orders delivered to orders not delivered?
    c) orders not delivered to total number of orders?

34. *Gas Mileage.* A 2000 Oldsmobile Intrigue will go 448 mi on 16 gal of gasoline. What is the ratio of:

    a) miles driven to gasoline burned?
    b) gasoline burned to miles driven?

b Simplify.

35. $\dfrac{0}{8}$

36. $\dfrac{8}{8}$

37. $\dfrac{8-1}{9-8}$

38. $\dfrac{16}{1}$

39. $\dfrac{20}{20}$

40. $\dfrac{20}{1}$

41. $\dfrac{45}{45}$

42. $\dfrac{11-1}{10-9}$

43. $\dfrac{0}{238}$

44. $\dfrac{238}{1}$

45. $\dfrac{238}{238}$

46. $\dfrac{0}{16}$

47. $\dfrac{3}{3}$

48. $\dfrac{56}{56}$

49. $\dfrac{87}{87}$

50. $\dfrac{98}{98}$

51. $\dfrac{18}{18}$

52. $\dfrac{0}{18}$

53. $\dfrac{18}{1}$

54. $\dfrac{8-8}{1247}$

55. $\dfrac{729}{0}$

56. $\dfrac{1317}{0}$

57. $\dfrac{5}{6-6}$

58. $\dfrac{13}{10-10}$

59. D_W Write a sentence to describe $\frac{5}{8}$ in each of the following situations.

  a) As a part of a whole
  b) As a part of a set
  c) As a ratio

60. D_W Explain in your own words why $n/n = 1$, for any natural number n.

SKILL MAINTENANCE

Round 34,562 to the nearest: [1.4a]

61. Ten.

62. Hundred.

63. Thousand.

64. Ten thousand.

Solve. [1.8a]

65. *Salaries and Education.* In 1998, the average annual salary for a person who is not a high school graduate was $16,053. The average for someone with a bachelor's degree was $43,782. How much more did the person who has a bachelor's degree earn than the person who is not a high school graduate?
Source: U.S. Bureau of the Census

66. *Gas Mileage.* The Chrysler PT Cruiser gets 22 miles to the gallon (mpg) in city driving. How many gallons will it use in 2860 mi of city driving?

Subtract. [1.3b]

67. $9001 - 6798$

68. $2037 - 1189$

69. $67,113 - 29,874$

70. $12,327 - 476$

SYNTHESIS

What part of the object is shaded?

71.

72.

73.

74.

Shade or mark the figure to show $\frac{3}{5}$.

75.

76.

77.

78.

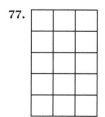

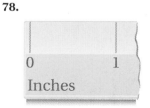

125

Objectives

a Multiply a whole number and a fraction.

b Multiply using fraction notation.

c Solve applied problems involving multiplication of fractions.

1. Find $2 \cdot \dfrac{1}{3}$.

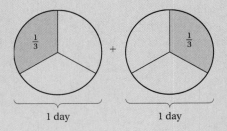

1 day   1 day

2. Find $5 \cdot \dfrac{1}{8}$.

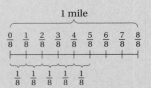

Multiply.

3. $5 \times \dfrac{2}{3}$

4. $11 \times \dfrac{3}{8}$

5. $23 \cdot \dfrac{2}{5}$

Answers on page A-5

---

a Multiplication by a Whole Number

We can find $3 \cdot \frac{1}{4}$ by thinking of repeated addition. We add three $\frac{1}{4}$'s. We see that $3 \cdot \frac{1}{4}$ is $\frac{3}{4}$.

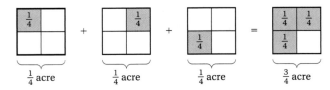

$\frac{1}{4}$ acre $\frac{1}{4}$ acre $\frac{1}{4}$ acre $\frac{3}{4}$ acre

Do Exercises 1 and 2.

To multiply a fraction by a whole number,

a) multiply the top number (the numerator) by the whole number, and

$$6 \cdot \frac{4}{5} = \frac{6 \cdot 4}{5} = \frac{24}{5}$$

b) keep the same denominator.

EXAMPLES Multiply.

1. $5 \times \dfrac{3}{8} = \dfrac{5 \times 3}{8} = \dfrac{15}{8}$

2. $\dfrac{2}{5} \cdot 13 = \dfrac{2 \cdot 13}{5} = \dfrac{26}{5}$

3. $10 \cdot \dfrac{1}{3} = \dfrac{10}{3}$

Do Exercises 3–5.

b Multiplication Using Fraction Notation

When neither factor is a whole number, multiplication using fraction notation does not correspond to repeated addition. Let's see how multiplication of fractions corresponds to situations in the real world. We consider the multiplication

$$\frac{3}{5} \cdot \frac{3}{4}.$$

We first consider some object and take $\frac{3}{4}$ of it. We divide it into 4 vertical parts, or columns of the same area, and take 3 of them. That is shown in the shading at right.

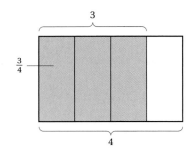

Next, we take $\frac{3}{5}$ of the result. We divide the shaded part into 5 horizontal parts, or rows of the same area, and take 3 of them. That is shown below.

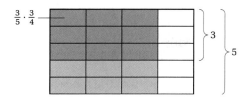

The entire object has been divided into 20 parts, and we have shaded 9 of them for a second time:

$$\frac{3}{5} \cdot \frac{3}{4} = \frac{3 \cdot 3}{5 \cdot 4} = \frac{9}{20}.$$

The figure above shows a rectangular array inside a rectangular array. The number of pieces in the entire array is $5 \cdot 4$ (the product of the denominators). The number of pieces shaded a second time is $3 \cdot 3$ (the product of the numerators). For the answer, we take 9 pieces out of a set of 20 to get $\frac{9}{20}$.

Do Exercise 6.

We find a product such as $\frac{9}{7} \cdot \frac{3}{4}$ as follows.

> To multiply a fraction by a fraction,
>
> **a)** multiply the numerators to get the new numerator, and
>
> $$\frac{9}{7} \cdot \frac{3}{4} = \frac{9 \cdot 3}{7 \cdot 4} = \frac{27}{28}$$
>
> **b)** multiply the denominators to get the new denominator.

EXAMPLES Multiply.

4. $\frac{5}{6} \times \frac{7}{4} = \frac{5 \times 7}{6 \times 4} = \frac{35}{24}$

Skip writing this step whenever you can.

5. $\frac{3}{5} \cdot \frac{7}{8} = \frac{3 \cdot 7}{5 \cdot 8} = \frac{21}{40}$

6. $\frac{3}{5} \cdot \frac{3}{4} = \frac{9}{20}$

7. $\frac{1}{4} \cdot \frac{1}{3} = \frac{1}{12}$

8. $6 \cdot \frac{4}{5} = \frac{6}{1} \cdot \frac{4}{5} = \frac{24}{5}$

Do Exercises 7–10.

6. Draw a diagram like the one at left to show the multiplication $\frac{1}{3} \cdot \frac{4}{5}$.

Multiply.

7. $\frac{3}{8} \cdot \frac{5}{7}$

8. $\frac{4}{3} \times \frac{8}{5}$

9. $\frac{3}{10} \cdot \frac{1}{10}$

10. $7 \cdot \frac{2}{3}$

Answers on page A-5

Applications and Problem Solving

Many problems that can be solved by multiplying fractions can be thought of in terms of rectangular arrays.

EXAMPLE 9 A real estate developer owns a plot of land that measures 1 square mile. He plans to use $\frac{4}{5}$ of the plot for a small strip mall and parking lot. Of this, $\frac{2}{3}$ will be needed for the parking lot. What part of the plot will be used for parking?

1. **Familiarize.** We first make a drawing to help familiarize ourselves with the problem. The land may not be rectangular. It could be in a shape like A or B below. But to think out the problem, we can think of it as a rectangle, as shown in shape C.

1 square mile   1 square mile   1 square mile

The strip mall including the parking lot uses $\frac{4}{5}$ of the plot. We shade $\frac{4}{5}$.

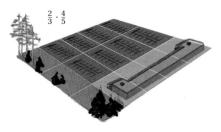

The parking lot alone takes $\frac{2}{3}$ of the preceding part. We shade that.

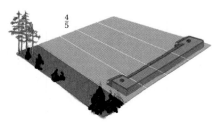

2. **Translate.** We let n = the part of the plot that is used for parking. We are taking "two-thirds of four-fifths." Recall from Section 1.8 that the word "of" corresponds to multiplication. Thus the following multiplication sentence corresponds to the situation:

$$\frac{2}{3} \cdot \frac{4}{5} = n.$$

3. Solve. The number sentence tells us what to do. We multiply:

$$\frac{2}{3} \cdot \frac{4}{5} = \frac{2 \cdot 4}{3 \cdot 5} = \frac{8}{15}.$$

4. Check. We can check partially by noting that the answer is smaller than the original area, 1, which we expect since the developer is using only part of the original plot of land. Thus, $\frac{8}{15}$ is a reasonable answer. We can also check this in the figure above, where we see that 8 of 15 parts have been shaded a second time.

5. State. The parking lot takes $\frac{8}{15}$ of the square mile of land.

Do Exercise 11.

Example 9 and the preceding discussion indicate that the area of a rectangular region can be found by multiplying length by width. That is true whether length and width are whole numbers or not. Remember, the area of a rectangular region is given by the formula

$$A = l \cdot w.$$

EXAMPLE 10 *Area of a Mosaic Tile.* The length of a tile on an inlaid mosaic table is $\frac{7}{10}$ in. The width is $\frac{3}{10}$ in. What is the area of one tile?

1. Familiarize. Recall that area is length times width. We make a drawing and let A = the area of the tile.

2. Translate. Then we translate.

Area	is	Length	times	Width
↓	↓	↓	↓	↓
A	=	$\frac{7}{10}$	×	$\frac{3}{10}$

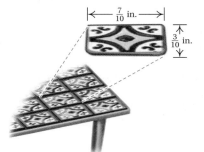

3. Solve. The sentence tells us what to do. We multiply:

$$\frac{7}{10} \cdot \frac{3}{10} = \frac{7 \cdot 3}{10 \cdot 10} = \frac{21}{100}.$$

4. Check. We check by repeating the calculation. This is left to the student.

5. State. The area is $\frac{21}{100}$ in^2.

Do Exercise 12.

---

11. A resort hotel uses $\frac{3}{4}$ of its extra land for recreational purposes. Of that, $\frac{1}{2}$ is used for swimming pools. What part of the land is used for swimming pools?

12. Area of a Fax Key. The length of a button on a fax machine is $\frac{9}{10}$ cm. The width is $\frac{7}{10}$ cm. What is its area?

Answers on page A-5

13. Of the students at Overton Junior College, $\frac{1}{8}$ participate in sports and $\frac{3}{5}$ of these play football. What fractional part of the students play football?

Answer on page A-5

■ **EXAMPLE 11** A recipe calls for $\frac{3}{4}$ cup of cornmeal. A chef is making $\frac{1}{2}$ of the recipe. How much cornmeal should the chef use?

1. **Familiarize.** We first make a drawing or at least visualize the situation. We let $n =$ the amount of cornmeal the chef should use.

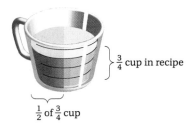

$\frac{3}{4}$ cup in recipe

$\frac{1}{2}$ of $\frac{3}{4}$ cup

2. **Translate.** The multiplication sentence $\frac{1}{2} \cdot \frac{3}{4} = n$ corresponds to the situation.

3. **Solve.** We carry out the multiplication:

$$\frac{1}{2} \cdot \frac{3}{4} = \frac{1 \cdot 3}{2 \cdot 4} = \frac{3}{8}.$$

4. **Check.** We check by repeating the calculation. This is left to the student.

5. **State.** The chef should use $\frac{3}{8}$ cup of cornmeal.

Do Exercise 13.

Study Tips

BETTER TEST TAKING

How often do you make the following statement after taking a test: "I was able to do the homework, but I froze during the test"? This can be an excuse for poor study habits. Here are two tips to help you with this difficulty. Both are intended to make test taking less stressful by getting you to practice good test-taking habits on a daily basis.

■ **Treat every homework exercise as if it were a test question.** If you had to work a problem at your job with no backup answer provided, what would you do? You would probably work it very deliberately, checking and rechecking every step. You might work it more than one time, or you might try to work it another way to check the result. Try to use this approach when doing your homework. Treat every exercise as though it were a test question with no answer at the back of the book.

■ **Be sure that you do questions without answers as part of every homework assignment whether or not the instructor has assigned them!** One reason a test may seem such a different task is that questions on a test lack answers. That is the reason for taking a test: to see if you can do the questions without assistance. As part of your test preparation, be sure you do some exercises for which you do not have the answers. Thus when you take a test, you are doing a more familiar task.

The purpose of doing your homework using these approaches is to give you more test-taking practice beforehand. Let's make a sports analogy here. At a basketball game, the players take lots of practice shots before the game. They play the first half, go to the locker room, and come out for the second half. What do they do before the second half, even though they have just played 20 minutes of basketball? They shoot baskets again! We suggest the same approach here. Create more and more situations in which you practice taking test questions by treating each homework exercise like a test question and by doing exercises for which you have no answers. Good luck!

For Extra Help

Digital Video Tutor CD 1 Videotape 3 | InterAct Math | Math Tutor Center | MathXL | MyMathLab

a Multiply.

1. $3 \cdot \dfrac{1}{5}$

2. $2 \cdot \dfrac{1}{3}$

3. $5 \times \dfrac{1}{8}$

4. $4 \times \dfrac{1}{5}$

5. $\dfrac{2}{11} \cdot 4$

6. $\dfrac{2}{5} \cdot 3$

7. $10 \cdot \dfrac{7}{9}$

8. $9 \cdot \dfrac{5}{8}$

9. $\dfrac{2}{5} \cdot 1$

10. $\dfrac{3}{8} \cdot 1$

11. $\dfrac{2}{5} \cdot 3$

12. $\dfrac{3}{5} \cdot 4$

13. $7 \cdot \dfrac{3}{4}$

14. $7 \cdot \dfrac{2}{5}$

15. $17 \times \dfrac{5}{6}$

16. $\dfrac{3}{7} \cdot 40$

b Multiply.

17. $\dfrac{1}{2} \cdot \dfrac{1}{3}$

18. $\dfrac{1}{6} \cdot \dfrac{1}{4}$

19. $\dfrac{1}{4} \times \dfrac{1}{10}$

20. $\dfrac{1}{3} \times \dfrac{1}{10}$

21. $\dfrac{2}{3} \times \dfrac{1}{5}$

22. $\dfrac{3}{5} \times \dfrac{1}{5}$

23. $\dfrac{2}{5} \cdot \dfrac{2}{3}$

24. $\dfrac{3}{4} \cdot \dfrac{3}{5}$

25. $\dfrac{3}{4} \cdot \dfrac{3}{4}$

26. $\dfrac{3}{7} \cdot \dfrac{4}{5}$

27. $\dfrac{2}{3} \cdot \dfrac{7}{13}$

28. $\dfrac{3}{11} \cdot \dfrac{4}{5}$

29. $\dfrac{1}{10} \cdot \dfrac{7}{10}$

30. $\dfrac{3}{10} \cdot \dfrac{3}{10}$

31. $\dfrac{7}{8} \cdot \dfrac{7}{8}$

32. $\dfrac{4}{5} \cdot \dfrac{4}{5}$

33. $\dfrac{1}{10} \cdot \dfrac{1}{100}$

34. $\dfrac{3}{10} \cdot \dfrac{7}{100}$

35. $\dfrac{14}{15} \cdot \dfrac{13}{19}$

36. $\dfrac{12}{13} \cdot \dfrac{12}{13}$

c Solve.

37. A rectangular table top measures $\frac{4}{5}$ m long by $\frac{3}{5}$ m wide. What is its area?

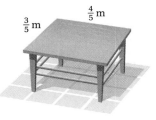

$\frac{3}{5}$ m $\frac{4}{5}$ m

38. If each piece of pie is $\frac{1}{6}$ of a pie, how much of the pie is $\frac{1}{2}$ of a piece?

39. *Football: High School to Pro.* One of 39 high school football players plays college football. One of 39 college players plays professional football. What fractional part of high school players play professional football?
Source: National Football League

40. A gasoline can holds $\frac{7}{8}$ liter (L). How much will the can hold when it is $\frac{1}{2}$ full?

41. A cereal recipe calls for $\frac{3}{4}$ cup of granola. How much is needed to make $\frac{1}{2}$ of a recipe?

42. It takes $\frac{2}{3}$ yd of ribbon to make a bow. How much ribbon is needed to make 5 bows?

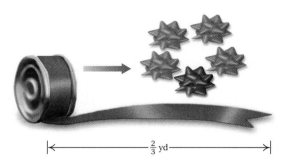

43. *Floor Tiling.* The floor of a room is being covered with tile. An area $\frac{3}{5}$ of the length and $\frac{3}{4}$ of the width is covered. What fraction of the floor has been tiled?

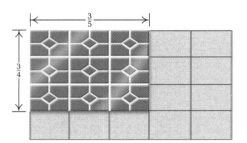

44. *Basement Carpet.* A basement floor is being covered with carpet. An area $\frac{7}{8}$ of the length and $\frac{3}{4}$ of the width is covered by lunch time. What fraction of the floor has been completed?

45. $\mathbb{D}_\mathrm{W}$ Write a problem for a classmate to solve. Design the problem so that the solution is "About $\frac{1}{30}$ of the students are left-handed women."

46. $\mathbb{D}_\mathrm{W}$ On pp. 126–127, we explained, using words and pictures, why $\frac{3}{5} \cdot \frac{3}{4}$ equals $\frac{9}{20}$. Present a similar explanation of why $\frac{2}{3} \cdot \frac{4}{7}$ equals $\frac{8}{21}$.

SKILL MAINTENANCE

Divide. [1.6b]

47. $7140 \div 35$

48. $32{,}200 \div 46$

49. $9 \overline{)\ 2\ 7{,}0\ 0\ 9}$

50. $3\ 5 \overline{)\ 7\ 1\ 4\ 8}$

What does the digit 8 mean in each number? [1.1a]

51. 4,678,952

52. 8,473,901

53. 7148

54. 23,803

Simplify. [1.9c]

55. $12 - 3^2$

56. $(12 - 3)^2$

57. $8 \cdot 12 - (63 \div 9 + 13 \cdot 3)$

58. $(10 - 3)^4 + 10^3 \cdot 4 - 10 \div 5$

SYNTHESIS

Multiply. Write the answer using fraction notation.

59. ▦ $\dfrac{341}{517} \cdot \dfrac{209}{349}$

60. ▦ $\left(\dfrac{57}{61}\right)^3$

61. ▦ $\left(\dfrac{2}{5}\right)^3 \left(\dfrac{7}{9}\right)$

62. ▦ $\left(\dfrac{1}{2}\right)^5 \left(\dfrac{3}{5}\right)$

CHAPTER 2: Fraction Notation:
Multiplication and Division

2.5 SIMPLIFYING

a Multiplying by 1

Objectives

a Use multiplying by 1 to find different fraction notation for a number.

b Simplify fraction notation.

c Use the test for equality to determine whether two fractions name the same number.

Recall the following:

$$1 = \frac{1}{1} = \frac{2}{2} = \frac{3}{3} = \frac{4}{4} = \frac{10}{10} = \frac{45}{45} = \frac{100}{100} = \frac{n}{n}.$$

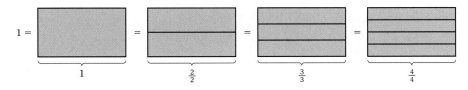

$$1 = \qquad = \frac{2}{2} \qquad = \frac{3}{3} \qquad = \frac{4}{4}$$

Any nonzero number divided by itself is 1. (See Section 1.6.)

Now recall the multiplicative identity from Section 1.5. For any whole number a, $1 \cdot a = a \cdot 1 = a$. This holds for numbers of arithmetic as well.

> ### MULTIPLICATIVE IDENTITY FOR FRACTIONS
>
> When we multiply a number by 1, we get the same number:
> $$\frac{3}{5} = \frac{3}{5} \cdot 1 = \frac{3}{5} \cdot \frac{4}{4} = \frac{12}{20}.$$

Since $\frac{3}{5} = \frac{12}{20}$, we know that $\frac{3}{5}$ and $\frac{12}{20}$ are two names for the same number. We also say that $\frac{3}{5}$ and $\frac{12}{20}$ are **equivalent.**

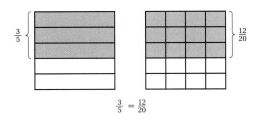

$$\frac{3}{5} = \frac{12}{20}$$

Multiply.

1. $\dfrac{1}{2} \cdot \dfrac{8}{8}$

2. $\dfrac{3}{5} \cdot \dfrac{10}{10}$

3. $\dfrac{13}{25} \cdot \dfrac{4}{4}$

4. $\dfrac{8}{3} \cdot \dfrac{25}{25}$

Do Exercises 1–4.

Suppose we want to find a name for $\frac{2}{3}$, but one that has a denominator of 9. We can multiply by 1 to find equivalent fractions:

$$\frac{2}{3} = \frac{2}{3} \cdot \frac{3}{3} = \frac{2 \cdot 3}{3 \cdot 3} = \frac{6}{9}.$$

We chose $\frac{3}{3}$ for 1 in order to get a denominator of 9.

EXAMPLE 1 Find a name for $\frac{1}{4}$ with a denominator of 24.

Since $4 \cdot 6 = 24$, we multiply by $\frac{6}{6}$:

$$\frac{1}{4} = \frac{1}{4} \cdot \frac{6}{6} = \frac{1 \cdot 6}{4 \cdot 6} = \frac{6}{24}.$$

Answers on page A-5

Find another name for the number, but with the denominator indicated. Use multiplying by 1.

5. $\dfrac{4}{3} = \dfrac{?}{9}$ **6.** $\dfrac{3}{4} = \dfrac{?}{24}$

7. $\dfrac{9}{10} = \dfrac{?}{100}$ **8.** $\dfrac{3}{15} = \dfrac{?}{45}$

9. $\dfrac{8}{7} = \dfrac{?}{49}$

Simplify.

10. $\dfrac{2}{8}$

11. $\dfrac{10}{12}$

12. $\dfrac{40}{8}$

13. $\dfrac{24}{18}$

EXAMPLE 2 Find a name for $\frac{2}{5}$ with a denominator of 35.

Since $5 \cdot 7 = 35$, we multiply by $\frac{7}{7}$:

$$\frac{2}{5} = \frac{2}{5} \cdot \frac{7}{7} = \frac{2 \cdot 7}{5 \cdot 7} = \frac{14}{35}.$$

Do Exercises 5–9.

b Simplifying Fraction Notation

All of the following are names for three-fourths:

$$\frac{3}{4}, \quad \frac{6}{8}, \quad \frac{9}{12}, \quad \frac{12}{16}, \quad \frac{15}{20}.$$

We say that $\frac{3}{4}$ is **simplest** because it has the smallest numerator and the smallest denominator. That is, the numerator and the denominator have no common factor other than 1.

To simplify, we reverse the process of multiplying by 1:

$$\frac{12}{18} = \frac{2 \cdot 6}{3 \cdot 6} \quad \begin{array}{l} \leftarrow \text{Factoring the numerator} \\ \leftarrow \text{Factoring the denominator} \end{array}$$

$$= \frac{2}{3} \cdot \frac{6}{6} \quad \text{Factoring the fraction}$$

$$= \frac{2}{3} \cdot 1 \quad \frac{6}{6} = 1$$

$$= \frac{2}{3}. \quad \text{Removing a factor of 1: } \frac{2}{3} \cdot 1 = \frac{2}{3}$$

EXAMPLES Simplify.

3. $\dfrac{8}{20} = \dfrac{2 \cdot 4}{5 \cdot 4} = \dfrac{2}{5} \cdot \dfrac{4}{4} = \dfrac{2}{5}$

4. $\dfrac{2}{6} = \dfrac{1 \cdot 2}{3 \cdot 2} = \dfrac{1}{3} \cdot \dfrac{2}{2} = \dfrac{1}{3}$
   The number 1 allows for pairing of factors in the numerator and the denominator.

5. $\dfrac{30}{6} = \dfrac{5 \cdot 6}{1 \cdot 6} = \dfrac{5}{1} \cdot \dfrac{6}{6} = \dfrac{5}{1} = 5$
   We could also simplify $\frac{30}{6}$ by doing the division $30 \div 6$. That is, $\frac{30}{6} = 30 \div 6 = 5$.

Do Exercises 10–13.

Answers on page A-5

The use of prime factorizations can be helpful for simplifying when numerators and/or denominators are larger numbers.

EXAMPLE 6 Simplify: $\dfrac{90}{84}$.

$$\frac{90}{84} = \frac{2 \cdot 3 \cdot 3 \cdot 5}{2 \cdot 2 \cdot 3 \cdot 7}$$ Factoring the numerator and the denominator into primes

$$= \frac{2 \cdot 3 \cdot 3 \cdot 5}{2 \cdot 3 \cdot 2 \cdot 7}$$ Changing the order so that like primes are above and below each other

$$= \frac{2}{2} \cdot \frac{3}{3} \cdot \frac{3 \cdot 5}{2 \cdot 7}$$ Factoring the fraction

$$= 1 \cdot 1 \cdot \frac{3 \cdot 5}{2 \cdot 7}$$

$$= \frac{3 \cdot 5}{2 \cdot 7}$$ Removing factors of 1

$$= \frac{15}{14}$$

We could have shortened the preceding example had we recalled our tests for divisibility (Section 2.2) and noted that 6 is a factor of both the numerator and the denominator. Then

$$\frac{90}{84} = \frac{6 \cdot 15}{6 \cdot 14} = \frac{6}{6} \cdot \frac{15}{14} = \frac{15}{14}.$$

The tests for divisibility are very helpful in simplifying.

EXAMPLE 7 Simplify: $\dfrac{603}{207}$.

At first glance this looks difficult. But note, using the test for divisibility by 9 (sum of digits divisible by 9), that both the numerator and the denominator are divisible by 9. Thus we can factor 9 from both numbers:

$$\frac{603}{207} = \frac{9 \cdot 67}{9 \cdot 23} = \frac{9}{9} \cdot \frac{67}{23} = \frac{67}{23}.$$

Do Exercises 14–18.

Simplify.

14. $\dfrac{35}{40}$ **15.** $\dfrac{801}{702}$

16. $\dfrac{24}{21}$ **17.** $\dfrac{75}{300}$

18. Simplify each fraction in this circle graph.

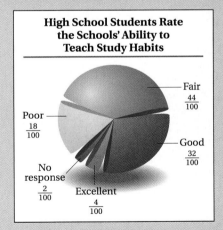

High School Students Rate the Schools' Ability to Teach Study Habits

Fair $\dfrac{44}{100}$

Poor $\dfrac{18}{100}$

Good $\dfrac{32}{100}$

No response $\dfrac{2}{100}$

Excellent $\dfrac{4}{100}$

Answers on page A-5

CANCELING

Canceling is a shortcut that you may have used for removing a factor of 1 when working with fraction notation. With *great* concern, we mention it as a possibility for speeding up your work. Canceling may be done only when removing common factors in numerators and denominators. Each common factor allows us to remove a factor of 1 in a product.

Our concern is that canceling be done with care and understanding. In effect, slashes are used to indicate factors of 1 that have been removed. For instance, Example 6 might have been done faster as follows:

$$\frac{90}{84} = \frac{2 \cdot 3 \cdot 3 \cdot 5}{2 \cdot 2 \cdot 3 \cdot 7} \qquad \text{Factoring the numerator and the denominator}$$

$$= \frac{2 \cdot 3 \cdot 3 \cdot 5}{2 \cdot 2 \cdot 3 \cdot 7} \qquad \text{When a factor of 1 is noted,}$$
$$\qquad\qquad\qquad \text{it is "canceled" as shown: } \frac{2 \cdot 3}{2 \cdot 3} = 1.$$

$$= \frac{3 \cdot 5}{2 \cdot 7} = \frac{15}{14}.$$

CAUTION!

The difficulty with canceling is that it is often applied incorrectly in situations like the following:

$$\frac{2 + 3}{2} = 3; \qquad \frac{4 + 1}{4 + 2} = \frac{1}{2}; \qquad \frac{15}{54} = \frac{1}{4}.$$

    Wrong!   Wrong!   Wrong!

The correct answers are

$$\frac{2 + 3}{2} = \frac{5}{2}; \qquad \frac{4 + 1}{4 + 2} = \frac{5}{6}; \qquad \frac{15}{54} = \frac{5}{18}.$$

In each situation, the number canceled was not a factor of 1. Factors are parts of products. For example, in $2 \cdot 3$, 2 and 3 are factors, but in $2 + 3$, 2 and 3 are *not* factors. Canceling may not be done when sums or differences are in numerators or denominators, as shown here.

> If you cannot factor, do not cancel! If in doubt, do not cancel!

C   A Test for Equality

When denominators are the same, we say that fractions have a **common denominator**. Suppose we want to compare $\frac{2}{4}$ and $\frac{3}{6}$. We find a common denominator and compare numerators. To do this, we multiply by 1 using symbols for 1 formed by looking at contrasting denominators.

The "unit" is $\frac{1}{6}$.

$$\frac{3}{6} = \frac{3}{6} \cdot \frac{4}{4} = \frac{3 \cdot 4}{6 \cdot 4} = \frac{12}{24}$$

$$\frac{2}{4} = \frac{2}{4} \cdot \frac{6}{6} = \frac{2 \cdot 6}{4 \cdot 6} = \frac{12}{24}$$

The "unit" is $\frac{1}{4}$.

Both "units" are $\frac{1}{24}$.

We see that $\frac{3}{6} = \frac{2}{4}$.

Note in the preceding that if

$$\frac{3}{6} = \frac{2}{4}, \quad \text{then} \quad 3 \cdot 4 = 6 \cdot 2.$$

We need to check only the products $3 \cdot 4$ and $6 \cdot 2$ to compare the fractions.

A TEST FOR EQUALITY

We multiply these two numbers: $3 \cdot 4$.

We multiply these two numbers: $6 \cdot 2$.

$$\frac{3}{6} \quad \frac{2}{4}$$

We call $3 \cdot 4$ and $6 \cdot 2$ **cross products.** Since the cross products are the same, that is, $3 \cdot 4 = 6 \cdot 2$, we know that

$$\frac{3}{6} = \frac{2}{4}.$$

If a sentence $a = b$ is true, it means that a and b name the same number. If a sentence $a \neq b$ is true, it means that a and b do *not* name the same number.

EXAMPLE 8 Use $=$ or $\neq$ for $\square$ to write a true sentence:

$$\frac{6}{7} \square \frac{7}{8}.$$

We multiply these two numbers: $6 \cdot 8 = 48$.

We multiply these two numbers: $7 \cdot 7 = 49$.

$$\frac{6}{7} \quad \frac{7}{8}$$

Because $48 \neq 49$ (read "48 is not equal to 49"), $\frac{6}{7} = \frac{7}{8}$ is not a true sentence. Thus,

$$\frac{6}{7} \neq \frac{7}{8}.$$

EXAMPLE 9 Use $=$ or $\neq$ for $\square$ to write a true sentence:

$$\frac{6}{10} \square \frac{3}{5}.$$

We multiply these two numbers: $6 \cdot 5 = 30$.

We multiply these two numbers: $10 \cdot 3 = 30$.

$$\frac{6}{10} \quad \frac{3}{5}.$$

Because the cross products are the same, we have

$$\frac{6}{10} = \frac{3}{5}.$$

Do Exercises 19 and 20.

Use $=$ or $\neq$ for $\square$ to write a true sentence.

19. $\dfrac{2}{6} \square \dfrac{3}{9}$

20. $\dfrac{2}{3} \square \dfrac{14}{20}$

Answers on page A-5

2.5 EXERCISE SET

a Find another name for the given number, but with the denominator indicated. Use multiplying by 1.

1. $\dfrac{1}{2} = \dfrac{?}{10}$ **2.** $\dfrac{1}{6} = \dfrac{?}{18}$ **3.** $\dfrac{5}{8} = \dfrac{?}{32}$ **4.** $\dfrac{2}{9} = \dfrac{?}{18}$

5. $\dfrac{9}{10} = \dfrac{?}{30}$ **6.** $\dfrac{5}{6} = \dfrac{?}{48}$ **7.** $\dfrac{7}{8} = \dfrac{?}{32}$ **8.** $\dfrac{2}{5} = \dfrac{?}{25}$

9. $\dfrac{5}{12} = \dfrac{?}{48}$ **10.** $\dfrac{3}{8} = \dfrac{?}{56}$ **11.** $\dfrac{17}{18} = \dfrac{?}{54}$ **12.** $\dfrac{11}{16} = \dfrac{?}{256}$

13. $\dfrac{5}{3} = \dfrac{?}{45}$ **14.** $\dfrac{11}{5} = \dfrac{?}{30}$ **15.** $\dfrac{7}{22} = \dfrac{?}{132}$ **16.** $\dfrac{10}{21} = \dfrac{?}{126}$

b Simplify.

17. $\dfrac{2}{4}$ **18.** $\dfrac{4}{8}$ **19.** $\dfrac{6}{8}$ **20.** $\dfrac{8}{12}$ **21.** $\dfrac{3}{15}$

22. $\dfrac{8}{10}$ **23.** $\dfrac{24}{8}$ **24.** $\dfrac{36}{9}$ **25.** $\dfrac{18}{24}$ **26.** $\dfrac{42}{48}$

27. $\dfrac{14}{16}$ **28.** $\dfrac{15}{25}$ **29.** $\dfrac{12}{10}$ **30.** $\dfrac{16}{14}$ **31.** $\dfrac{16}{48}$

32. $\dfrac{100}{20}$ **33.** $\dfrac{150}{25}$ **34.** $\dfrac{19}{76}$ **35.** $\dfrac{17}{51}$ **36.** $\dfrac{425}{525}$

c Use $=$ or $\neq$ for $\square$ to write a true sentence.

37. $\dfrac{3}{4}\,\square\,\dfrac{9}{12}$ **38.** $\dfrac{4}{8}\,\square\,\dfrac{3}{6}$ **39.** $\dfrac{1}{5}\,\square\,\dfrac{2}{9}$ **40.** $\dfrac{1}{4}\,\square\,\dfrac{2}{9}$

41. $\dfrac{3}{8}\,\square\,\dfrac{6}{16}$ **42.** $\dfrac{2}{6}\,\square\,\dfrac{6}{18}$ **43.** $\dfrac{2}{5}\,\square\,\dfrac{3}{7}$ **44.** $\dfrac{1}{3}\,\square\,\dfrac{1}{4}$

CHAPTER 2: Fraction Notation: Multiplication and Division

45. $\dfrac{12}{9} \;\square\; \dfrac{8}{6}$

46. $\dfrac{16}{14} \;\square\; \dfrac{8}{7}$

47. $\dfrac{5}{2} \;\square\; \dfrac{17}{7}$

48. $\dfrac{3}{10} \;\square\; \dfrac{7}{24}$

49. $\dfrac{3}{10} \;\square\; \dfrac{30}{100}$

50. $\dfrac{700}{1000} \;\square\; \dfrac{70}{100}$

51. $\dfrac{5}{10} \;\square\; \dfrac{520}{1000}$

52. $\dfrac{49}{100} \;\square\; \dfrac{50}{1000}$

53. $^\mathrm{D}\mathrm{W}$ Explain in your own words when it *is* possible to "cancel" and when it *is not* possible to "cancel."

54. $^\mathrm{D}\mathrm{W}$ Can fraction notation be simplified if its numerator and its denominator are two different prime numbers? Why or why not?

Solve.   [1.8a]

55. A playing field is 78 ft long and 64 ft wide. What is its area? its perimeter?

56. A landscaper buys 13 small maple trees and 17 small oak trees for a project. A maple costs $23 and an oak costs $37. How much is spent altogether for the trees?

Subtract.   [1.3b]

57. $34 - 23$

58. $50 - 18$

59. $803 - 617$

60. $8344 - 5607$

Solve.   [1.7b]

61. $30 \cdot x = 150$

62. $10{,}947 = 123 \cdot y$

63. $5280 = 1760 + t$

64. $x + 2368 = 11{,}369$

Simplify. Use the list of prime numbers on p. 104.

65. 🖩 $\dfrac{2603}{2831}$

66. 🖩 $\dfrac{3197}{3473}$

67. *Shy People.*   Sociologists have found that 4 of 10 people are shy. Write fraction notation for the part of the population that is shy; the part that is not shy. Simplify.

68. *Left-Handed People.*   Sociologists estimate that 3 of 20 people are left-handed. In a crowd of 460 people, how many would you expect to be left-handed?

69. 🖩 *Batting Averages.*   For the 2000 season, Todd Helton of the Colorado Rockies won the National League batting title with 216 hits in 580 times at bat. Nomar Garciaparra of the Boston Red Sox won the American League title with 197 hits in 529 times at bat. Did they have the same fraction of hits in times at bat (batting average)? Why or why not?
Source: Major League Baseball

70. 🖩 On a test of 82 questions, a student got 63 correct. On another test of 100 questions, she got 77 correct. Did she get the same portion of each test correct? Why or why not?

Objectives

a) Multiply and simplify using fraction notation.

b) Solve applied problems involving multiplication of fractions.

a   Multiplying and Simplifying Using Fraction Notation

We usually simplify after we multiply. To make such simplifying easier, it is generally best not to carry out the products in the numerator and the denominator, but to factor and simplify before multiplying. Consider the product

$$\frac{3}{8} \cdot \frac{4}{9}.$$

We proceed as follows:

$$\frac{3}{8} \cdot \frac{4}{9} = \frac{3 \cdot 4}{8 \cdot 9}$$ We write the products in the numerator and the denominator, but we do not carry them out.

$$= \frac{3 \cdot 2 \cdot 2}{2 \cdot 2 \cdot 2 \cdot 3 \cdot 3}$$ Factoring the numerator and the denominator

$$= \frac{3 \cdot 2 \cdot 2}{3 \cdot 2 \cdot 2} \cdot \frac{1}{2 \cdot 3}$$ Factoring the fraction

$$= 1 \cdot \frac{1}{2 \cdot 3}$$

$$= \frac{1}{2 \cdot 3}$$ Removing a factor of 1

$$= \frac{1}{6}.$$

The procedure could have been shortened had we noticed that 4 is a factor of the 8 in the denominator:

$$\frac{3}{8} \cdot \frac{4}{9} = \frac{3 \cdot 4}{8 \cdot 9} = \frac{3 \cdot 4}{4 \cdot 2 \cdot 3 \cdot 3} = \frac{3 \cdot 4}{3 \cdot 4} \cdot \frac{1}{2 \cdot 3} = 1 \cdot \frac{1}{2 \cdot 3} = \frac{1}{2 \cdot 3} = \frac{1}{6}.$$

To multiply and simplify:

a) Write the products in the numerator and the denominator, but do not carry out the products.

b) Factor the numerator and the denominator.

c) Factor the fraction to remove factors of 1.

d) Carry out the remaining products.

EXAMPLES   Multiply and simplify.

1. $\dfrac{2}{3} \cdot \dfrac{9}{4} = \dfrac{2 \cdot 9}{3 \cdot 4} = \dfrac{2 \cdot 3 \cdot 3}{3 \cdot 2 \cdot 2} = \dfrac{2 \cdot 3}{2 \cdot 3} \cdot \dfrac{3}{2} = 1 \cdot \dfrac{3}{2} = \dfrac{3}{2}$

2. $\dfrac{6}{7} \cdot \dfrac{5}{3} = \dfrac{6 \cdot 5}{7 \cdot 3} = \dfrac{3 \cdot 2 \cdot 5}{7 \cdot 3} = \dfrac{3}{3} \cdot \dfrac{2 \cdot 5}{7} = 1 \cdot \dfrac{2 \cdot 5}{7} = \dfrac{2 \cdot 5}{7} = \dfrac{10}{7}$

3. $40 \cdot \dfrac{7}{8} = \dfrac{40 \cdot 7}{8} = \dfrac{8 \cdot 5 \cdot 7}{8 \cdot 1} = \dfrac{8}{8} \cdot \dfrac{5 \cdot 7}{1} = 1 \cdot \dfrac{5 \cdot 7}{1} = \dfrac{5 \cdot 7}{1} = 35$

Canceling can be used as follows for these examples.

1. $\dfrac{2}{3} \cdot \dfrac{9}{4} = \dfrac{2 \cdot 9}{3 \cdot 4} = \dfrac{2 \cdot 3 \cdot 3}{3 \cdot 2 \cdot 2} = \dfrac{3}{2}$

Removing a factor of 1:
$\dfrac{2 \cdot 3}{2 \cdot 3} = 1$

2. $\dfrac{6}{7} \cdot \dfrac{5}{3} = \dfrac{6 \cdot 5}{7 \cdot 3} = \dfrac{3 \cdot 2 \cdot 5}{7 \cdot 3} = \dfrac{2 \cdot 5}{7} = \dfrac{10}{7}$

Removing a factor of 1:
$\dfrac{3}{3} = 1$

3. $40 \cdot \dfrac{7}{8} = \dfrac{40 \cdot 7}{8} = \dfrac{8 \cdot 5 \cdot 7}{8 \cdot 1} = \dfrac{5 \cdot 7}{1} = 35$

Removing a factor of 1:
$\dfrac{8}{8} = 1$

Remember, if you can't factor, you can't cancel!

Do Exercises 1–4.

b   Applications and Problem Solving

EXAMPLE 4   LeGrand Chocolate Shop is preparing Valentine boxes. How many pounds of truffles will be needed to fill 75 boxes if each box contains $\frac{2}{5}$ lb?

1. Familiarize.   We first make a drawing or at least visualize the situation. Repeated addition will work here.

75 boxes

$\frac{2}{5}$ of a pound in each box

We let $n = $ the number of pounds of truffles.

2. Translate.   The problem translates to the following equation:

$$n = 75 \cdot \dfrac{2}{5}.$$

3. Solve.   To solve the equation, we carry out the multiplication:

$$n = 75 \cdot \dfrac{2}{5} = \dfrac{75 \cdot 2}{5} \qquad \text{Multiplying}$$

$$= \dfrac{5 \cdot 5 \cdot 3 \cdot 2}{5 \cdot 1}$$

$$= \dfrac{5}{5} \cdot \dfrac{5 \cdot 3 \cdot 2}{1}$$

$$= 30. \qquad \text{Simplifying}$$

4. Check.   We check by repeating the calculation. (The check is left to the student.) We can also think about the reasonableness of the answer. We are putting less than a pound of truffles in each box, so the answer should be less than 75. Since 30 is less than 75, we have a partial check of the reasonableness of the answer. The number 30 checks.

5. State.   Thus, 30 lb of truffles will be needed.

Do Exercise 5.

Multiply and simplify.

1. $\dfrac{2}{3} \cdot \dfrac{7}{8}$

2. $\dfrac{4}{5} \cdot \dfrac{5}{12}$

3. $16 \cdot \dfrac{3}{8}$

4. $\dfrac{5}{8} \cdot 4$

5. A landscaper uses $\frac{2}{3}$ lb of peat moss for a rosebush. How much will be needed for 21 rosebushes?

Answers on page A-5

EXERCISE SET

For Extra Help

Digital Video Tutor CD 2 Videotape 4   InterAct Math   Math Tutor Center   MathXL   MyMathLab

a Multiply and simplify. | Don't forget to simplify!

1. $\dfrac{2}{3} \cdot \dfrac{1}{2}$

2. $\dfrac{3}{8} \cdot \dfrac{1}{3}$

3. $\dfrac{7}{8} \cdot \dfrac{1}{7}$

4. $\dfrac{4}{9} \cdot \dfrac{1}{4}$

5. $\dfrac{1}{8} \cdot \dfrac{4}{5}$

6. $\dfrac{2}{5} \cdot \dfrac{1}{6}$

7. $\dfrac{1}{4} \cdot \dfrac{2}{3}$

8. $\dfrac{4}{6} \cdot \dfrac{1}{6}$

9. $\dfrac{12}{5} \cdot \dfrac{9}{8}$

10. $\dfrac{16}{15} \cdot \dfrac{5}{4}$

11. $\dfrac{10}{9} \cdot \dfrac{7}{5}$

12. $\dfrac{25}{12} \cdot \dfrac{4}{3}$

13. $9 \cdot \dfrac{1}{9}$

14. $4 \cdot \dfrac{1}{4}$

15. $\dfrac{1}{3} \cdot 3$

16. $\dfrac{1}{6} \cdot 6$

17. $\dfrac{7}{10} \cdot \dfrac{10}{7}$

18. $\dfrac{8}{9} \cdot \dfrac{9}{8}$

19. $\dfrac{7}{5} \cdot \dfrac{5}{7}$

20. $\dfrac{2}{11} \cdot \dfrac{11}{2}$

21. $\dfrac{1}{4} \cdot 8$

22. $\dfrac{1}{3} \cdot 18$

23. $24 \cdot \dfrac{1}{6}$

24. $16 \cdot \dfrac{1}{2}$

25. $12 \cdot \dfrac{3}{4}$

26. $18 \cdot \dfrac{5}{6}$

27. $\dfrac{3}{8} \cdot 24$

28. $\dfrac{2}{9} \cdot 36$

29. $13 \cdot \dfrac{2}{5}$

30. $15 \cdot \dfrac{1}{6}$

31. $\dfrac{7}{10} \cdot 28$

32. $\dfrac{5}{8} \cdot 34$

33. $\dfrac{1}{6} \cdot 360$

34. $\dfrac{1}{3} \cdot 120$

35. $240 \cdot \dfrac{1}{8}$

36. $150 \cdot \dfrac{1}{5}$

37. $\dfrac{4}{10} \cdot \dfrac{5}{10}$

38. $\dfrac{7}{10} \cdot \dfrac{34}{150}$

39. $\dfrac{8}{10} \cdot \dfrac{45}{100}$

40. $\dfrac{3}{10} \cdot \dfrac{8}{10}$

41. $\dfrac{11}{24} \cdot \dfrac{3}{5}$

42. $\dfrac{15}{22} \cdot \dfrac{4}{7}$

43. $\dfrac{10}{21} \cdot \dfrac{3}{4}$

44. $\dfrac{17}{18} \cdot \dfrac{3}{5}$

b Solve.

The *pitch* of a screw is the distance between its threads. With each complete rotation, the screw goes in or out a distance equal to its pitch. Use this information to answer Exercises 45 and 46.

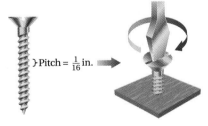

}Pitch = $\frac{1}{16}$ in.

Each rotation moves the screw in or out $\frac{1}{16}$ in.

45. The pitch of a screw is $\frac{1}{16}$ in. How far will it go into a piece of oak when it is turned 10 complete rotations clockwise?

46. The pitch of a screw is $\frac{3}{32}$ in. How far will it go out of a piece of plywood when it is turned 10 complete rotations counterclockwise?

47. *Swimming Speeds.* The swimming speed of a killer whale is about 30 mph. The swimming speed of a dolphin is about $\frac{3}{5}$ that of a killer whale. Find the swimming speed of a dolphin.

Source: G. Cafiero and M. Jahoda, *Whales and Dolphins.* New York: Barnes & Noble Books, 1994

48. After Jack completes 60 hr of teacher training at college, he can earn $75 for working a full day as a substitute teacher. How much will he receive for working $\frac{1}{5}$ of a day?

49. *Mailing-List Addresses.* Business people have determined that $\frac{1}{4}$ of the addresses on a mailing list will change in one year. A business has a mailing list of 2500 people. After one year, how many addresses on that list will be incorrect?

50. *Shy People.* Sociologists have determined that $\frac{2}{5}$ of the people in the world are shy. A sales manager is interviewing 650 people for an aggressive sales position. How many of these people might be shy?

51. A recipe for piecrust calls for $\frac{2}{3}$ cup of flour. A chef is making $\frac{1}{2}$ of the recipe. How much flour should the chef use?

52. Of the students in the freshman class, $\frac{2}{5}$ have cameras; $\frac{1}{4}$ of these students also join the college photography club. What fraction of the students in the freshman class join the photography club?

53. A house worth $154,000 is assessed for $\frac{3}{4}$ of its value. What is the assessed value of the house?

54. Roxanne's tuition was $2800. A loan was obtained for $\frac{3}{4}$ of the tuition. How much was the loan?

55. *Map Scaling.* On a map, 1 in. represents 240 mi. How much does $\frac{2}{3}$ in. represent?

56. *Map Scaling.* On a map, 1 in. represents 120 mi. How much does $\frac{3}{4}$ in. represent?

57. *Household Budgets.* A family has an annual income of $36,000. Of this, $\frac{1}{4}$ is spent for food, $\frac{1}{5}$ for housing, $\frac{1}{10}$ for clothing, $\frac{1}{9}$ for savings, $\frac{1}{4}$ for taxes, and the rest for other expenses. How much is spent for each?

58. *Household Budgets.* A family has an annual income of $29,700. Of this, $\frac{1}{4}$ is spent for food, $\frac{1}{5}$ for housing, $\frac{1}{10}$ for clothing, $\frac{1}{9}$ for savings, $\frac{1}{4}$ for taxes, and the rest for other expenses. How much is spent for each?

Family Income

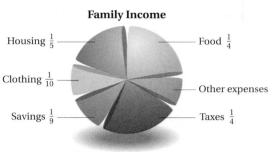

Housing $\frac{1}{5}$

Clothing $\frac{1}{10}$

Savings $\frac{1}{9}$

Food $\frac{1}{4}$

Other expenses

Taxes $\frac{1}{4}$

Source: U.S. Census Bureau

59. ^{D}W When multiplying using fraction notation, we form products in the numerator and the denominator, but do not immediately calculate the products. Why?

60. ^{D}W If a fraction's numerator and denominator have no factors (other than 1) in common, can the fraction be simplified? Why or why not?

SKILL MAINTENANCE

Solve. [1.7b]

61. $48 \cdot t = 1680$

62. $74 \cdot x = 6290$

63. $3125 = 25 \cdot t$

64. $2880 = 24 \cdot y$

65. $t + 28 = 5017$

66. $456 + x = 9002$

67. $8797 = y + 2299$

68. $10,000 = 3593 + m$

Subtract. [1.3b]

69.
$$\begin{array}{r} 9\ 0\ 6\ 0 \\ -\ 4\ 3\ 8\ 7 \\ \hline \end{array}$$

70.
$$\begin{array}{r} 7\ 8\ 0\ 0 \\ -\ 2\ 4\ 6\ 2 \\ \hline \end{array}$$

SYNTHESIS

Multiply and simplify. Use the list of prime numbers on p. 104 or a fraction calculator.

71. ▥ $\dfrac{201}{535} \cdot \dfrac{4601}{6499}$

72. ▥ $\dfrac{5767}{3763} \cdot \dfrac{159}{395}$

73. *College Profile.* Of students entering a college, $\frac{7}{8}$ have completed high school and $\frac{2}{3}$ are older than 20. If $\frac{1}{7}$ of all students are left-handed, what fraction of students entering the college are left-handed high school graduates over the age of 20?

74. *College Profile.* Refer to the information in Exercise 73. If 480 students are entering the college, how many of them are left-handed high school graduates 20 yr old or younger?

75. *College Profile.* Refer to Exercise 73. What fraction of students entering the college did not graduate high school, are 20 yr old or younger, and are left-handed?

DIVISION AND APPLICATIONS

a Find the reciprocal of a number.

b Divide and simplify using fraction notation.

c Solve equations of the type $a \cdot x = b$ and $x \cdot a = b$, where a and b may be fractions.

d Solve applied problems involving division of fractions.

Find the reciprocal.

1. $\dfrac{2}{5}$

2. $\dfrac{10}{7}$

3. 9

4. $\dfrac{1}{5}$

Answers on page A-5

a Reciprocals

Look at these products:

$$8 \cdot \frac{1}{8} = \frac{8 \cdot 1}{8} = \frac{8}{8} = 1; \qquad \frac{2}{3} \cdot \frac{3}{2} = \frac{2 \cdot 3}{3 \cdot 2} = \frac{6}{6} = 1.$$

RECIPROCALS

If the product of two numbers is 1, we say that they are **reciprocals** of each other. To find a reciprocal of a fraction, interchange the numerator and the denominator.

$$\text{Number} \longrightarrow \frac{3}{4} \to \frac{4}{3} \longrightarrow \text{Reciprocal}$$

EXAMPLES Find the reciprocal.

1. The reciprocal of $\dfrac{4}{5}$ is $\dfrac{5}{4}$. $\qquad \dfrac{4}{5} \cdot \dfrac{5}{4} = \dfrac{20}{20} = 1$

2. The reciprocal of $\dfrac{8}{7}$ is $\dfrac{7}{8}$. $\qquad \dfrac{8}{7} \cdot \dfrac{7}{8} = \dfrac{56}{56} = 1$

3. The reciprocal of 8 is $\dfrac{1}{8}$. $\qquad$ Think of 8 as $\dfrac{8}{1}$: $\dfrac{8}{1} \cdot \dfrac{1}{8} = \dfrac{8}{8} = 1.$

4. The reciprocal of $\dfrac{1}{3}$ is 3. $\qquad \dfrac{1}{3} \cdot 3 = \dfrac{3}{3} = 1$

Do Exercises 1–4.

Does 0 have a reciprocal? If it did, it would have to be a number x such that

$$0 \cdot x = 1.$$

But 0 times any number is 0. Thus we have the following.

0 HAS NO RECIPROCAL

The number 0, or $\dfrac{0}{n}$, has no reciprocal. $\left(\text{Recall that } \dfrac{n}{0} \text{ is not defined.}\right)$

b Division

Consider the division $\frac{3}{4} \div \frac{1}{8}$. We are asking how many $\frac{1}{8}$'s are in $\frac{3}{4}$. We can answer this by looking at the figure below.

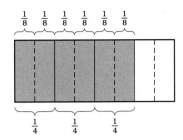

We see that there are six $\frac{1}{8}$'s in $\frac{3}{4}$. Thus,

$$\frac{3}{4} \div \frac{1}{8} = 6.$$

We can check this by multiplying:

$$6 \cdot \frac{1}{8} = \frac{6}{8} = \frac{3}{4}.$$

Here is a faster way to do this division:

$$\frac{3}{4} \div \frac{1}{8} = \frac{3}{4} \cdot \frac{8}{1} = \frac{24}{4} = 6. \qquad \text{Multiplying by the reciprocal of the divisor}$$

> To divide fractions, multiply the dividend by the reciprocal of the divisor:
>
> $$\frac{2}{5} \div \frac{3}{4} = \frac{2}{5} \cdot \frac{4}{3} = \frac{2 \cdot 4}{5 \cdot 3} = \frac{8}{15}.$$
>
> Multiply by the reciprocal of the divisor.

EXAMPLES Divide and simplify.

5. $\frac{5}{6} \div \frac{2}{3} = \frac{5}{6} \cdot \frac{3}{2} = \frac{5 \cdot 3}{6 \cdot 2} = \frac{5 \cdot 3}{3 \cdot 2 \cdot 2} = \frac{3}{3} \cdot \frac{5}{2 \cdot 2} = \frac{5}{2 \cdot 2} = \frac{5}{4}$

6. $\frac{7}{8} \div \frac{1}{16} = \frac{7}{8} \cdot 16 = \frac{7 \cdot 16}{8} = \frac{7 \cdot 2 \cdot 8}{8 \cdot 1} = \frac{8}{8} \cdot \frac{7 \cdot 2}{1} = \frac{7 \cdot 2}{1} = 14$

7. $\frac{2}{5} \div 6 = \frac{2}{5} \cdot \frac{1}{6} = \frac{2 \cdot 1}{5 \cdot 6} = \frac{2 \cdot 1}{5 \cdot 2 \cdot 3} = \frac{2}{2} \cdot \frac{1}{5 \cdot 3} = \frac{1}{5 \cdot 3} = \frac{1}{15}$

8. $\frac{3}{5} \div \frac{1}{2} = \frac{3}{5} \cdot 2 = \frac{3 \cdot 2}{5} = \frac{6}{5}$

Divide and simplify.

5. $\dfrac{6}{7} \div \dfrac{3}{4}$

6. $\dfrac{2}{3} \div \dfrac{1}{4}$

7. $\dfrac{4}{5} \div 8$

8. $60 \div \dfrac{3}{5}$

9. $\dfrac{3}{5} \div \dfrac{3}{5}$

10. Divide by multiplying by 1:

$$\dfrac{\dfrac{4}{5}}{\dfrac{6}{7}}.$$

CAUTION!

Canceling can be used as follows for Examples 5–7.

5. $\dfrac{5}{6} \div \dfrac{2}{3} = \dfrac{5}{6} \cdot \dfrac{3}{2} = \dfrac{5 \cdot 3}{6 \cdot 2} = \dfrac{5 \cdot 3}{3 \cdot 2 \cdot 2} = \dfrac{5}{2 \cdot 2} = \dfrac{5}{4}$ Removing a factor of 1: $\frac{3}{3} = 1$

6. $\dfrac{7}{8} \div \dfrac{1}{16} = \dfrac{7}{8} \cdot 16 = \dfrac{7 \cdot 16}{8} = \dfrac{7 \cdot 8 \cdot 2}{8 \cdot 1} = \dfrac{7 \cdot 2}{1} = 14$ Removing a factor of 1: $\frac{8}{8} = 1$

7. $\dfrac{2}{5} \div 6 = \dfrac{2}{5} \cdot \dfrac{1}{6} = \dfrac{2 \cdot 1}{5 \cdot 6} = \dfrac{2 \cdot 1}{5 \cdot 2 \cdot 3} = \dfrac{1}{5 \cdot 3} = \dfrac{1}{15}$ Removing a factor of 1: $\frac{2}{2} = 1$

Remember, if you can't factor, you can't cancel!

Do Exercises 5–9.

What is the explanation for multiplying by a reciprocal when dividing? Let's consider $\frac{2}{3} \div \frac{7}{5}$. We multiply by 1. The name for 1 that we will use is $(5/7)/(5/7)$; it comes from the reciprocal of $\frac{7}{5}$.

$$\dfrac{2}{3} \div \dfrac{7}{5} = \dfrac{\dfrac{2}{3}}{\dfrac{7}{5}} \qquad \text{Writing fraction notation for the division}$$

$$= \dfrac{\dfrac{2}{3}}{\dfrac{7}{5}} \cdot 1 \qquad \text{Multiplying by 1}$$

$$= \dfrac{\dfrac{2}{3}}{\dfrac{7}{5}} \cdot \dfrac{\dfrac{5}{7}}{\dfrac{5}{7}} \qquad \text{Multiplying by 1; } \tfrac{5}{7} \text{ is the reciprocal of } \tfrac{7}{5} \text{ and } \dfrac{\frac{5}{7}}{\frac{5}{7}} = 1$$

$$= \dfrac{\dfrac{2}{3} \cdot \dfrac{5}{7}}{\dfrac{7}{5} \cdot \dfrac{5}{7}} \qquad \text{Multiplying the numerators and the denominators}$$

$$= \dfrac{\dfrac{2}{3} \cdot \dfrac{5}{7}}{1} \qquad \text{After we multiplied, we got 1 for the denominator. The numerator shows the multiplication by the reciprocal.}$$

$$= \dfrac{2}{3} \cdot \dfrac{5}{7} = \dfrac{10}{21}$$

Thus,

$$\dfrac{2}{3} \div \dfrac{7}{5} = \dfrac{2}{3} \cdot \dfrac{5}{7} = \dfrac{10}{21}.$$

Do Exercise 10.

C Solving Equations

Now let's solve equations $a \cdot x = b$ and $x \cdot a = b$, where a and b may be fractions. We proceed as we did with equations involving whole numbers. We divide by a on both sides.

EXAMPLE 9   Solve: $\frac{4}{3} \cdot x = \frac{6}{7}$.

We have

$$\frac{4}{3} \cdot x = \frac{6}{7}$$

$$x = \frac{6}{7} \div \frac{4}{3} \qquad \text{Dividing by } \tfrac{4}{3} \text{ on both sides}$$

$$= \frac{6}{7} \cdot \frac{3}{4} \qquad \text{Multiplying by the reciprocal}$$

$$= \frac{2 \cdot 3 \cdot 3}{7 \cdot 2 \cdot 2} = \frac{2}{2} \cdot \frac{3 \cdot 3}{7 \cdot 2} = \frac{3 \cdot 3}{7 \cdot 2} = \frac{9}{14}.$$

The solution is $\frac{9}{14}$.

EXAMPLE 10   Solve: $t \cdot \frac{4}{5} = 80$.

Dividing by $\frac{4}{5}$ on both sides, we get

$$t = 80 \div \frac{4}{5} = 80 \cdot \frac{5}{4} = \frac{80 \cdot 5}{4} = \frac{4 \cdot 20 \cdot 5}{4 \cdot 1} = \frac{4}{4} \cdot \frac{20 \cdot 5}{1} = \frac{20 \cdot 5}{1} = 100.$$

The solution is 100.

Do Exercises 11 and 12.

d Applications and Problem Solving

EXAMPLE 11   *Test Tubes.*   How many test tubes, each containing $\frac{3}{5}$ mL, can a nursing student fill from a container of 60 mL?

1. Familiarize.   We are asking the question, "How many $\frac{3}{5}$'s are in 60?" Repeated addition will apply here. We make a drawing. We let $n =$ the number of test tubes in all.

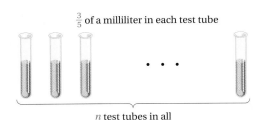

$\frac{3}{5}$ of a milliliter in each test tube

n test tubes in all

2. Translate.   The equation that corresponds to the situation is

$$n = 60 \div \frac{3}{5}.$$

Solve.

11. $\dfrac{5}{6} \cdot y = \dfrac{2}{3}$

12. $\dfrac{3}{4} \cdot n = 24$

Answers on page A-5

13. Each loop in a spring uses $\frac{3}{8}$ in. of wire. How many loops can be made from 120 in. of wire?

14. A service station tank had 175 gal of oil when it was $\frac{7}{8}$ full. How much could the tank hold altogether?

Answers on page A-5

Study Tips

THE SUPPLEMENTS

The new mathematical skills and concepts presented in the lectures will be of increased value to you if you begin the homework assignment as soon as possible after the lecture. Then if you still have difficulty with any of the exercises, you have time to access supplementary resources such as:

- *Student's Solutions Manual*
- Videotapes
- InterAct Math Tutorial CD-ROM
- AW Math Tutor Center
- MathXL

3. Solve. We solve the equation by carrying out the division:

$$n = 60 \div \frac{3}{5} = 60 \cdot \frac{5}{3} = \frac{60 \cdot 5}{3} = \frac{3 \cdot 20 \cdot 5}{3 \cdot 1}$$

$$= \frac{3}{3} \cdot \frac{20 \cdot 5}{1} = 100.$$

4. Check. We check by repeating the calculation.

5. State. The student can fill 100 test tubes.

Do Exercise 13.

EXAMPLE 12 Melissa Esplanah sells pharmaceutical supplies. After she had driven 210 mi, $\frac{5}{6}$ of her sales trip was completed. How long was the total trip?

1. Familiarize. We think: 210 mi is $\frac{5}{6}$ of the trip. We make a drawing or at least visualize the situation. We let $n =$ the length of the trip.

$\frac{5}{6}$ of the trip
210 mi

n

2. Translate. We translate to an equation.

Fraction completed	of	Total length of trip	is	Amount already traveled
$\frac{5}{6}$	$\cdot$	n	$=$	210

3. Solve. The equation that corresponds to the situation is $\frac{5}{6} \cdot n = 210$. We divide by $\frac{5}{6}$ on both sides and carry out the division:

$$n = 210 \div \frac{5}{6} = 210 \cdot \frac{6}{5} = \frac{210 \cdot 6}{5} = \frac{5 \cdot 42 \cdot 6}{5 \cdot 1} = \frac{5}{5} \cdot \frac{42 \cdot 6}{1} = 252.$$

4. Check. We check by repeating the calculation.

5. State. The total trip was 252 mi.

Do Exercise 14.

EXERCISE SET

For Extra Help

a Find the reciprocal.

1. $\dfrac{5}{6}$

2. $\dfrac{7}{8}$

3. 6

4. 4

5. $\dfrac{1}{6}$

6. $\dfrac{1}{4}$

7. $\dfrac{10}{3}$

8. $\dfrac{17}{4}$

b Divide and simplify. | Don't forget to simplify!

9. $\dfrac{3}{5} \div \dfrac{3}{4}$

10. $\dfrac{2}{3} \div \dfrac{3}{4}$

11. $\dfrac{3}{5} \div \dfrac{9}{4}$

12. $\dfrac{6}{7} \div \dfrac{3}{5}$

13. $\dfrac{4}{3} \div \dfrac{1}{3}$

14. $\dfrac{10}{9} \div \dfrac{1}{3}$

15. $\dfrac{1}{3} \div \dfrac{1}{6}$

16. $\dfrac{1}{4} \div \dfrac{1}{5}$

17. $\dfrac{3}{8} \div 3$

18. $\dfrac{5}{6} \div 5$

19. $\dfrac{12}{7} \div 4$

20. $\dfrac{18}{5} \div 2$

21. $12 \div \dfrac{3}{2}$

22. $24 \div \dfrac{3}{8}$

23. $28 \div \dfrac{4}{5}$

24. $40 \div \dfrac{2}{3}$

25. $\dfrac{5}{8} \div \dfrac{5}{8}$

26. $\dfrac{2}{5} \div \dfrac{2}{5}$

27. $\dfrac{8}{15} \div \dfrac{4}{5}$

28. $\dfrac{6}{13} \div \dfrac{3}{26}$

29. $\dfrac{9}{5} \div \dfrac{4}{5}$

30. $\dfrac{5}{12} \div \dfrac{25}{36}$

31. $120 \div \dfrac{5}{6}$

32. $360 \div \dfrac{8}{7}$

Solve.

33. $\dfrac{4}{5} \cdot x = 60$

34. $\dfrac{3}{2} \cdot t = 90$

35. $\dfrac{5}{3} \cdot y = \dfrac{10}{3}$

36. $\dfrac{4}{9} \cdot m = \dfrac{8}{3}$

37. $x \cdot \dfrac{25}{36} = \dfrac{5}{12}$

38. $p \cdot \dfrac{4}{5} = \dfrac{8}{15}$

39. $n \cdot \dfrac{8}{7} = 360$

40. $y \cdot \dfrac{5}{6} = 120$

d Solve.

41. Benny uses $\frac{2}{5}$ gram (g) of toothpaste each time he brushes his teeth. If Benny buys a 30-g tube, how many times will he be able to brush his teeth?

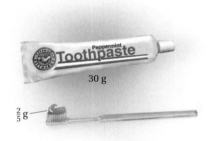

30 g

$\frac{2}{5}$ g

42. A piece of coaxial cable $\frac{4}{5}$ meter (m) long is to be cut into 8 pieces of the same length. What is the length of each piece?

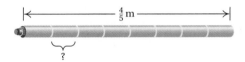

43. A pair of basketball shorts requires $\frac{3}{4}$ yd of nylon. How many pairs of shorts can be made from 24 yd of nylon?

44. A child's baseball shirt requires $\frac{5}{6}$ yd of fabric. How many shirts can be made from 25 yd of the fabric?

45. How many $\frac{2}{3}$-cup sugar bowls can be filled from 16 cups of sugar?

46. How many $\frac{2}{3}$-cup cereal bowls can be filled from 10 cups of cornflakes?

47. A bucket had 12 L of water in it when it was $\frac{3}{4}$ full. How much could it hold altogether?

48. A tank had 20 L of gasoline in it when it was $\frac{4}{5}$ full. How much could it hold altogether?

49. Yoshi Teramoto sells hardware tools. After driving 180 kilometers (km), he completes $\frac{5}{8}$ of a sales trip. How long is the total trip? How many kilometers are left to drive?

50. Alicia Simon and her road crew paint the lines in the middle and on the sides of a highway. They average about $\frac{5}{16}$ of a mile each hour. How long will it take to paint the lines on 70 mi of highway?

Pitch of a screw.   Refer to Exercises 45 and 46 in Exercise Set 2.6.

51. After a screw has been turned 8 complete rotations, it is extended $\frac{1}{2}$ in. into a piece of wallboard. What is the pitch of the screw?

52. The pitch of a screw is $\frac{3}{32}$ in. How many complete rotations are necessary to drive the screw $\frac{3}{4}$ in. into a piece of pine wood?

53. ^DW Without performing the division, explain why $5 \div \frac{1}{7}$ is a greater number than $5 \div \frac{2}{3}$.

54. ^DW A student incorrectly insists that $\frac{2}{5} \div \frac{3}{4}$ is $\frac{15}{8}$. What mistake is he probably making?

────────(**SKILL MAINTENANCE**)──

Divide.   [1.6b]

55. $268 \div 4$

56. $268 \div 8$

57. $6842 \div 24$

58. $8765 \div 85$

59. $999 \div 27$

60. $999 \div 37$

61. $289 \div 17$

62. $3136 \div 56$

Solve.   [1.7b]

63. $4 \cdot x = 268$

64. $4 + x = 268$

65. $y + 502 = 9001$

66. $56 \cdot 78 = T$

────────(**SYNTHESIS**)──

Simplify. Use the list of prime numbers on p. 104.

67. ▦ $\dfrac{711}{1957} \div \dfrac{10,033}{13,081}$

68. ▦ $\dfrac{8633}{7387} \div \dfrac{485}{581}$

69. $\left(\dfrac{9}{10} \div \dfrac{2}{5} \div \dfrac{3}{8} \right)^2$

70. $\dfrac{\left(\dfrac{3}{7} \right)^2 \div \dfrac{12}{5}}{\left(\dfrac{2}{9} \right)\left(\dfrac{9}{2} \right)}$

71. If $\frac{1}{3}$ of a number is $\frac{1}{4}$, what is $\frac{1}{2}$ of the number?

Summary and Review

The review that follows is meant to prepare you for a chapter exam. It consists of two parts. The first part is a checklist of the Study Tips referred to so far in this text. The second part is the Review Exercises. These provide practice exercises for the exam, together with references to section objectives so you can go back and review. Before beginning, stop and look back over the skills you have obtained. What skills in mathematics do you have now that you did not have before studying this chapter?

STUDY TIPS CHECKLIST

The foundation of all your study skills is TIME!	☐ Are you making use of the supplements that accompany this text?
	☐ Are you doing the exercises without answers as part of every homework assignment to better prepare you for tests?
	☐ Have you tried calling the Addison-Wesley Math Tutor Center at 1-888-777-0463?
	☐ Are you stopping to work the margin exercises when directed to do so?
	☐ Are you doing your homework as soon as possible after class?

REVIEW EXERCISES

Find the prime factorization of the number.   [2.1d]

1. 70

2. 30

3. 45

4. 150

Determine whether: [2.2a]

5. 2432 is divisible by 6.

6. 182 is divisible by 4.

7. 4344 is divisible by 9.

8. 4344 is divisible by 8.

9. Determine whether 37 is prime, composite, or neither. [2.1c]

10. Identify the numerator and the denominator of $\frac{2}{7}$. [2.3a]

11. What fractional part is shaded?   [2.3a]

12. Simplify, if possible, the fractions on this circle graph. [2.5b]

How the Business Travel Dollar is Spent

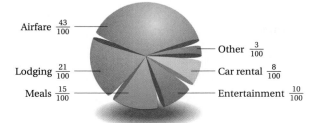

Airfare $\frac{43}{100}$

Lodging $\frac{21}{100}$

Meals $\frac{15}{100}$

Other $\frac{3}{100}$

Car rental $\frac{8}{100}$

Entertainment $\frac{10}{100}$

Simplify. [2.3b], [2.5b]

13. $\frac{0}{4}$ **14.** $\frac{23}{23}$ **15.** $\frac{48}{1}$

16. $\frac{48}{8}$ **17.** $\frac{10}{15}$ **18.** $\frac{7}{28}$

19. $\frac{21}{21}$ **20.** $\frac{0}{25}$ **21.** $\frac{12}{30}$

22. $\frac{18}{1}$ **23.** $\frac{32}{8}$ **24.** $\frac{9}{27}$

25. $\frac{18}{0}$ **26.** $\frac{5}{8-8}$

Use = or ≠ for □ to write a true sentence. [2.5c]

27. $\frac{3}{5}$ □ $\frac{4}{6}$ **28.** $\frac{4}{7}$ □ $\frac{8}{14}$

29. $\frac{4}{5}$ □ $\frac{5}{6}$ **30.** $\frac{4}{3}$ □ $\frac{28}{21}$

Multiply and simplify. [2.6a]

31. $4 \cdot \frac{3}{8}$ **32.** $\frac{7}{3} \cdot 24$

33. $9 \cdot \frac{5}{18}$ **34.** $\frac{6}{5} \cdot 20$

35. $\frac{3}{4} \cdot \frac{8}{9}$ **36.** $\frac{5}{7} \cdot \frac{1}{10}$

37. $\frac{3}{7} \cdot \frac{14}{9}$ **38.** $\frac{1}{4} \cdot \frac{2}{11}$

Find the reciprocal. [2.7a]

39. $\frac{4}{5}$ **40.** 3

41. $\frac{1}{9}$ **42.** $\frac{47}{36}$

Divide and simplify. [2.7b]

43. $6 \div \frac{4}{3}$ **44.** $\frac{5}{9} \div \frac{5}{18}$

45. $\frac{1}{6} \div \frac{1}{11}$ **46.** $\frac{3}{14} \div \frac{6}{7}$

47. $\frac{1}{4} \div \frac{1}{9}$ **48.** $180 \div \frac{3}{5}$

49. $\frac{23}{25} \div \frac{23}{25}$ **50.** $\frac{2}{3} \div \frac{3}{2}$

Solve. [2.7c]

51. $\frac{5}{4} \cdot t = \frac{3}{8}$ **52.** $x \cdot \frac{2}{3} = 160$

Solve.   [2.6b], [2.7d]

53. A road crew repaves $\frac{1}{12}$ mi of road each day. How long will it take the crew to repave a $\frac{3}{4}$-mi stretch of road?

54. After driving 60 km, the Bonewitz family has completed $\frac{3}{5}$ of their vacation. How long is the total trip?

55. Molly is making a pepper steak recipe that calls for $\frac{2}{3}$ cup of green bell peppers. How much would be needed to make $\frac{1}{2}$ recipe? 3 recipes?

56. Bernardo usually earns $105 for working a full day. How much does he receive for working $\frac{1}{7}$ of a day?

57. D_W Write, in your own words, a series of steps that can be used when simplifying fraction notation.   [2.5b]

58. D_W A student claims that "taking $\frac{1}{2}$ of a number is the same as dividing by $\frac{1}{2}$." Explain the error in this reasoning.   [2.7b]

Beginning with this chapter, certain objectives from four particular sections will be retested on the chapter test. The objectives are listed with the practice problems that follow.

Solve.   [1.7b]

59. $17 \cdot x = 408$

60. $765 + t = 1234$

Solve.   [1.8a]

61. The balance in your checking account is $789. After purchases of $78, $97, and $102 and a deposit of $400, what is your new balance?

62. A new Beetle 1.9L TDI by Volkswagen gets 43 mpg on the highway. How far can the car be driven on a full tank of 18 gal of gasoline?
Source: 2000 Volkswagen of America

63. Divide: [1.6b]
$$3\,6\,)\,\overline{1\,4,6\,9\,7}$$

64. Subtract: [1.3b]
$$\begin{array}{r} 5\,6\,0\,4 \\ -\ 1\,9\,9\,7 \\ \hline \end{array}$$

65. 🖩 In the division below, find a and b.   [2.7b]
$$\frac{19}{24} \div \frac{a}{b} = \frac{187,853}{268,224}$$

66. A prime number that becomes a prime number when its digits are reversed is called a **palindrome prime.** For example, 17 is a palindrome prime because both 17 and 71 are primes. Which of the following numbers are palindrome primes?   [2.1c]

13, 91, 16, 11, 15, 24, 29, 101, 201, 37

CHAPTER 2: Fraction Notation:
Multiplication and Division

Find the prime factorization of the number.

1. 18

2. 60

3. Determine whether 1784 is divisible by 8.

4. Determine whether 784 is divisible by 9.

5. Identify the numerator and the denominator of $\frac{4}{9}$.

6. What part is shaded?

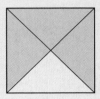

7. *Pass Completion Ratio.*   In 2000, Drew Brees completed 286 of 473 passes to become the Big Ten Player of the Year and led his Purdue University football team to a Big Ten championship and a trip to the Rose Bowl.

   a) What was the ratio of pass completions to attempts?
   b) What was the ratio of incomplete passes to attempts?
Source: Purdue University Sports Information

Simplify.

8. $\dfrac{26}{1}$

9. $\dfrac{12}{12}$

10. $\dfrac{0}{16}$

11. $\dfrac{12}{24}$

12. $\dfrac{42}{7}$

13. $\dfrac{2}{28}$

14. $\dfrac{9}{0}$

15. $\dfrac{7}{2-2}$

Use = or ≠ for ☐ to write a true sentence.

16. $\dfrac{3}{4}$ ☐ $\dfrac{6}{8}$

17. $\dfrac{5}{4}$ ☐ $\dfrac{9}{7}$

Multiply and simplify.

18. $\dfrac{4}{3} \cdot 24$

19. $5 \cdot \dfrac{3}{10}$

20. $\dfrac{2}{3} \cdot \dfrac{15}{4}$

21. $\dfrac{3}{5} \cdot \dfrac{1}{6}$

Find the reciprocal.

22. $\dfrac{5}{8}$

23. $\dfrac{1}{4}$

24. 18

Divide and simplify.

25. $\dfrac{3}{8} \div \dfrac{5}{4}$

26. $\dfrac{1}{5} \div \dfrac{1}{8}$

27. $12 \div \dfrac{2}{3}$

Solve.

28. $\dfrac{7}{8} \cdot x = 56$

29. $\dfrac{2}{5} \cdot t = \dfrac{7}{10}$

30. There are 7000 students at La Poloma College, and $\frac{5}{8}$ of them live in dorms. How many live in dorms?

31. A strip of taffy $\frac{9}{10}$ m long is cut into 12 equal pieces. What is the length of each piece?

SKILL MAINTENANCE

Solve.

32. $x + 198 = 2003$

33. $47 \cdot t = 4747$

34. It is 2060 mi from San Francisco to Winnipeg, Canada. It is 1575 mi from Winnipeg to Atlanta. What is the total length of a route from San Francisco to Winnipeg to Atlanta?

35. Divide: $2\,4\,\overline{)\,9\,1\,2\,7}$

36. Subtract:
$$\begin{array}{r} 8\,0\,0\,1 \\ -\ 3\,5\,6\,7 \\ \hline \end{array}$$

SYNTHESIS

37. A recipe for a batch of buttermilk pancakes calls for $\frac{3}{4}$ teaspoon (tsp) of salt. Jacqueline plans to cut the amount of salt in half for each of 5 batches of pancakes. How much salt will she need?

38. Grandma Jordan left $\frac{2}{3}$ of her $\frac{7}{8}$-acre apple farm to Karl. Karl gave $\frac{1}{4}$ of his share to his oldest daughter, Irene. How much land did Irene receive?

39. Simplify: $\left(\dfrac{3}{8}\right)^2 \div \dfrac{6}{7} \cdot \dfrac{2}{9} \div 5$.

40. Solve: $\dfrac{33}{38} \cdot \dfrac{34}{55} = \dfrac{17}{35} \cdot \dfrac{15}{19} x$.

CHAPTER 2: Fraction Notation:
Multiplication and Division

1. Write standard notation for the number in the following sentence:

The earth travels five hundred eighty-four million, seventeen thousand, eight hundred miles around the sun.

2. Write a word name: 5,380,621.

3. In the number 2,751,043, which digit tells the number of hundreds?

4. What part is shaded?

5. What fractional part of this set of musical groups perform pop music?

   *NSYNC

   Backstreet Boys

   The Lettermen

   Red Hot Chili Peppers

Add.

6.
$$\begin{array}{r} 1\,4,8\,6\,2 \\ +\ \ 2,9\,3\,5 \\ \hline \end{array}$$

7.
$$\begin{array}{r} 7\,9\,8\,9 \\ 7\,9\,8 \\ +\ \ \ \ 7\,9 \\ \hline \end{array}$$

Subtract.

8.
$$\begin{array}{r} 5\,3\,7\,6 \\ -\ \ \ 4\,3\,0 \\ \hline \end{array}$$

9.
$$\begin{array}{r} 2\,0\,0\,4 \\ -\ \ \ 5\,7\,9 \\ \hline \end{array}$$

Multiply and simplify.

10.
$$\begin{array}{r} 6\,2\,1 \\ \times\ \ \ 2\,7 \\ \hline \end{array}$$

11.
$$\begin{array}{r} 2\,5\,0\,5 \\ \times\,3\,3\,0\,0 \\ \hline \end{array}$$

12. $5 \times \dfrac{3}{100}$

13. $\dfrac{4}{9} \cdot \dfrac{3}{8}$

Divide and simplify.

14. $19\overline{)4\,5\,8\,0}$

15. $62\overline{)3\,8\,4\,4}$

16. $\dfrac{3}{10} \div 5$

17. $\dfrac{8}{9} \div \dfrac{15}{6}$

18. Round 427,931 to the nearest thousand.

19. Round 5309 to the nearest hundred.

Estimate the sum or product by rounding to the nearest hundred. Show your work.

20.
$$\begin{array}{r} 7\,4\,9,5\,5\,9 \\ +\,3\,0\,1,3\,6\,2 \\ \hline \end{array}$$

21.
$$\begin{array}{r} 7\,4\,9 \\ \times\ \ 5\,3\,1 \\ \hline \end{array}$$

22. Use < or > for ☐ to write a true sentence:

   26 ☐ 17.

23. Use = or ≠ for ☐ to write a true sentence:

   $\dfrac{7}{10}$ ☐ $\dfrac{5}{7}$.

Evaluate.

24. 3^4

25. 10^2

26. 2^5

27. 10^3

Simplify.

28. $35 - 25 \div 5 + 2 \times 3$

29. $\{17 - [8 - (5 - 2 \times 2)]\} \div (3 + 12 \div 6)$

30. Find all the factors of 28.

31. Find the prime factorization of 28.

32. Determine whether 39 is prime, composite, or neither.

33. Determine whether 32,712 is divisible by 3.

34. Determine whether 32,712 is divisible by 5.

Simplify.

35. $\dfrac{35}{1}$

36. $\dfrac{77}{11}$

37. $\dfrac{28}{98}$

38. $\dfrac{0}{47}$

Solve.

39. $x + 13 = 50$

40. $\dfrac{1}{5} \cdot t = \dfrac{3}{10}$

41. $13 \cdot y = 39$

42. $384 \div 16 = n$

43. *Price Comparisons.* The base price MSRP (Manufacturers Suggested Retail Price) of a 2001 Volkswagen Beetle GL was $15,900. This was $3365 more than the base price MSRP of a Saturn S Series Coupe SC1. What was the price of the Saturn?
Sources: 2000 Volkswagen of America; Saturn

44. *Pixels.* A computer screen consists of small rectangular dots called *pixels*. On a certain monitor, there are 307,200 pixels in all, with 640 pixels in each row. How many rows are there?

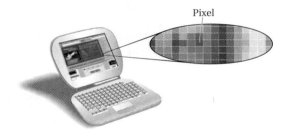

45. *Halley's Comet.* Halley's Comet passes by Earth every 76 yr. It last appeared to people on Earth in 1986.

a) When will it appear again?
b) When will it appear after that?

46. A $\frac{3}{4}$-cup serving of Kraft macaroni and cheese contains 984 calories. A box makes 2 servings.

a) How many cups of macaroni and cheese does the box make?
b) How many calories would be consumed if someone ate the macaroni and cheese from the entire box?

47. It takes Amanda 6 hr to paint the trim on a house. She can work only $\frac{3}{4}$ hr per day because she is taking 18 hr of classes. How many days will it take her to finish painting the trim?

48. *Clothing Purchase.* During a recent Christmas season, Joseph A Banks, a national clothing firm, was selling gabardine trousers for $90 a pair, wool–cashmere trousers for $115 a pair, and socks for $10 a pair. Elaine buys 5 pairs of gabardine trousers, 4 pairs of wool–cashmere trousers, and 1 pair of socks for Christmas gifts. What was the total cost of her purchase?
Source: Joseph A Banks, Inc.

49. *Matching.* Match each item in the first column with the appropriate item in the second column by drawing connecting lines.

Factors of 68	12, 54, 72, 300
Factorization of 68	2, 3, 17, 19, 23, 31, 47, 101
Prime factorization of 68	$2 \cdot 2 \cdot 17$
Numbers divisible by 6	$2 \cdot 34$
Numbers divisible by 8	8, 16, 24, 32, 40, 48, 64, 864
Numbers divisible by 5	1, 2, 4, 17, 34, 68
Prime numbers	70, 95, 215

For each of Exercises 50–53, choose the correct answer from the selections given.

50. In Arizona, people often install desert landscaping to conserve water. In a development, each home lot requires $\frac{4}{5}$ ton of gravel. How many tons of gravel are required for 40 lots?

a) $\frac{80}{100}$ **b)** 50   **c)** $\frac{80}{5}$
d) 32   **e)** None

51. Multiply and simplify: $\frac{3}{10} \cdot \frac{30}{7}$.

a) $\frac{24}{7}$ **b)** $\frac{21}{300}$ **c)** $\frac{27}{17}$
d) $\frac{9}{7}$ **e)** None

52. Divide and simplify: $\frac{3}{14} \div \frac{4}{7}$.

a) $\frac{21}{46}$ **b)** $\frac{12}{98}$ **c)** $\frac{3}{8}$
d) $\frac{21}{56}$ **e)** None

53. A gasoline tank contains 20 gal when it is $\frac{3}{4}$ full. How many gallons can it hold when full?

a) $\frac{80}{3}$ **b)** $\frac{23}{4}$ **c)** 15
d) $\frac{83}{4}$ **e)** None

SYNTHESIS

54. *Guess the Number.* Find a number having the following characteristics:

- Five digits
- No digit is a 0.
- No two digits are the same.
- The tens digit is twice the ones digit.
- The tens digit is 1 less than the hundreds digit.
- Excluding the ones digit, the sum of the digits is divisible by 9.
- The tens digit is the square of the ones digit.

55. On the popular TV show "Who Wants to Be a Millionaire?" a contestant tries to answer multiple-choice questions, each of which has 4 choices for the answer. To win $1 million, a contestant must correctly answer 15 questions in a row. The probability of guessing and getting all 15 questions correct is $1/4^{15}$. Use a calculator to evaluate 4^{15} and find equivalent fraction notation for this answer.

Fraction Notation and Mixed Numerals

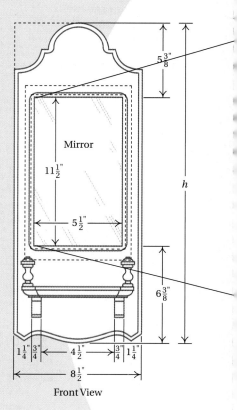

Gateway to Chapter 3

In this chapter, we consider addition and subtraction using fraction notation. Also discussed are addition, subtraction, multiplication, and division using mixed numerals. We then work with rules for order of operations, estimating, and applied problems.

Real–World Application

The mirror-backed candle shelf, shown above with a carpenter's diagram, was designed and built by Harry Cooper. Such shelves were popular in Colonial times because the mirror provided extra lighting from the candle. A rectangular walnut board is used to make the back of the shelf. Find the area of the original board and the amount left over after the space for the mirror has been cut out.

Source: Popular Science Woodworking Projects

This problem appears as Example 10 in Section 3.6.

1. Find the LCM of 15 and 24.   [3.1a]

2. Use < or > for ☐ to write a true sentence: [3.3b]
$$\frac{7}{9} \,\square\, \frac{4}{5}.$$

Add or subtract, and simplify.

3. $\frac{1}{6} + \frac{3}{4}$ [3.2a]

4. $\frac{5}{6} + \frac{7}{9}$ [3.2a]

5. $\frac{4}{5} - \frac{3}{10}$ [3.3a]

6. $\frac{2}{5} - \frac{3}{8}$ [3.3a]

7. Convert to fraction notation: $7\frac{5}{8}$.   [3.4a]

8. Convert to a mixed numeral: $\frac{11}{2}$.   [3.4a]

Perform the indicated calculation. Write a mixed numeral for the answer where appropriate.

9. $1\,2\,\overline{)\,4\,7\,8\,9}$ [3.4b]

10. $\begin{array}{r} 8\frac{11}{12} \\ +2\frac{3}{5} \\ \hline \end{array}$ [3.5a]

11. $\begin{array}{r} 14 \\ -\ 7\frac{5}{6} \\ \hline \end{array}$ [3.5b]

12. $3 \cdot 4\frac{8}{15}$ [3.6a]

13. $6\frac{2}{3} \cdot 3\frac{1}{4}$ [3.6a]

14. $35 \div 5\frac{5}{6}$ [3.6b]

15. $5\frac{5}{12} \div 3\frac{1}{4}$ [3.6b]

16. Solve: $\frac{2}{3} + x = \frac{8}{9}$.   [3.3c]

Solve.

17. At Happy Hollow Camp, the cook bought 100 lb of potatoes and used $78\frac{3}{4}$ lb. How many pounds were left over?   [3.5c]

18. **Weight of Water.**   The weight of water is $62\frac{1}{2}$ lb per cubic foot. How many cubic feet would be occupied by $265\frac{5}{8}$ lb of water?   [3.6c]

19. A courier drove $214\frac{3}{10}$ mi one day and $136\frac{9}{10}$ mi the next. How far did she travel in all?   [3.5c]

20. A cake recipe calls for $3\frac{3}{4}$ cups of flour. How much flour would be used to make 6 cakes?   [3.6c]

Estimate each of the following as 0, $\frac{1}{2}$, or 1. [3.7b]

21. $\frac{29}{30}$

22. $\frac{2}{41}$

Estimate each of the following as a whole number or as a mixed numeral where the fractional part is $\frac{1}{2}$.   [3.7b]

23. $\frac{1}{10} + \frac{7}{8} + \frac{41}{39}$

24. $33\frac{14}{15} + 28\frac{3}{4} - 4\frac{25}{28} \div \frac{75}{76}$

3.1

LEAST COMMON MULTIPLES

In this chapter, we study addition and subtraction using fraction notation. Suppose we want to add $\frac{2}{3}$ and $\frac{1}{2}$. To do so, we rewrite the numbers using the least common multiple of the denominators: $\frac{2}{3} + \frac{1}{2} = \frac{4}{6} + \frac{3}{6}$. Then we add the numerators and keep the common denominator, 6. In order to do this, we must be able to find the **least common denominator (LCD)**, or **least common multiple (LCM)** of the denominators. (A review of Section 2.1b might be helpful.)

Objective

a Find the LCM of two or more numbers.

a Finding Least Common Multiples

> **LEAST COMMON MULTIPLE, LCM**
>
> The **least common multiple,** or LCM, of two natural numbers is the smallest number that is a multiple of both.

1. By examining lists of multiples, find the LCM of 9 and 15.

EXAMPLE 1 Find the LCM of 20 and 30.

a) First list some multiples of 20 by multiplying 20 by 1, 2, 3, and so on:

20, 40, 60, 80, 100, 120, 140, 160, 180, 200, 220, 240, ...

b) Then list some multiples of 30 by multiplying 30 by 1, 2, 3, and so on:

30, 60, 90, 120, 150, 180, 210, 240, ...

c) Now list the numbers *common* to both lists, the common multiples:

60, 120, 180, 240, ...

d) These are the common multiples of 20 and 30. Which is the smallest? The LCM of 20 and 30 is 60.

Do Exercise 1.

Next we develop three methods that are more efficient for finding LCMs. You may choose to learn only one method (consult with your instructor), but if you are going to study algebra, you should definitely learn method 2.

METHOD 1: FINDING LCMS USING ONE LIST OF MULTIPLES

One method for finding LCMs uses *one* list of multiples. Let's consider finding the LCM of 9 and 12. The largest number, 12, is not a multiple of 9. The multiples of 12 are

12, 24, 36, 48, 60, ...

We check each multiple of 12 until we find a number that is also a multiple of 9.

$1 \cdot 12 = 12$, not a multiple of 9;

$2 \cdot 12 = 24$, not a multiple of 9;

$3 \cdot 12 = 36$, a multiple of 9: $4 \cdot 9 = 36$

The LCM of 9 and 12 is 36.

Answer on page A-6

2. By examining lists of multiples, find the LCM of 8 and 10.

> *Method 1.* To find the LCM of a set of numbers using a list of multiples:
>
> **a)** Determine whether the largest number is a multiple of the others. If it is, it is the LCM. That is, if the largest number has the others as factors, the LCM is that number.
>
> **b)** If not, check multiples of the largest number until you get one that is a multiple of the others.

EXAMPLE 2   Find the LCM of 12 and 15.

a) 15 is not a multiple of 12.

b) Check multiples of 15: 15, 30, 45, and so on.

$$1 \cdot 15 = 15,$$ Not a multiple of 12. When we divide 15 by 12, we get a nonzero remainder.

$$2 \cdot 15 = 30,$$ Not a multiple of 12

$$3 \cdot 15 = 45,$$ Not a multiple of 12

$$4 \cdot 15 = 60.$$ A multiple of 12

The LCM $= 60$.

Do Exercise 2.

Find the LCM.

3. 10, 15

EXAMPLE 3   Find the LCM of 4 and 14.

a) 14 is not a multiple of 4.

b) Check multiples:

$$1 \cdot 14 = 14,$$

$$2 \cdot 14 = 28.$$ A multiple of 4

The LCM $= 28$.

4. 6, 8

EXAMPLE 4   Find the LCM of 8 and 32.

a) 32 is a multiple of 8, so the LCM $= 32$.

5. 5, 10

EXAMPLE 5   Find the LCM of 10, 100, and 1000.

a) 1000 is a multiple of 10 and 100, so the LCM $= 1000$.

Do Exercises 3–6.

METHOD 2: FINDING LCMS USING PRIME FACTORIZATIONS

6. 20, 40, 80

A second method for finding LCMs uses prime factorizations. Consider again 20 and 30. Their prime factorizations are $20 = 2 \cdot 2 \cdot 5$ and $30 = 2 \cdot 3 \cdot 5$. Let's look at these prime factorizations in order to find the LCM. Any multiple of 20 will have to have *two* 2's as factors and *one* 5 as a factor. Any multiple of 30 will have to have *one* 2, *one* 3, and *one* 5 as factors. The smallest number satisfying these conditions is

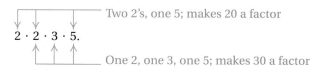
Two 2's, one 5; makes 20 a factor

$$2 \cdot 2 \cdot 3 \cdot 5.$$

One 2, one 3, one 5; makes 30 a factor

Answers on page A-6

The LCM must have all the factors of 20 and all the factors of 30, but the factors need not be repeated when they are common to both numbers.

The greatest number of times that a 2 occurs as a factor of either 20 or 30 is two, and the LCM has 2 as a factor twice. The greatest number of times that a 3 occurs as a factor of either 20 or 30 is one, and the LCM has 3 as a factor once. The greatest number of times that 5 occurs as a factor of either 20 or 30 is one, and the LCM has 5 as a factor once. The LCM is the product $2 \cdot 2 \cdot 3 \cdot 5$, or 60.

> *Method 2.* To find the LCM of a set of numbers using prime factorizations:
>
> **a)** Find the prime factorization of each number.
> **b)** Create a product of factors, using each factor the greatest number of times that it occurs in any one factorization.

EXAMPLE 6 Find the LCM of 6 and 8.

a) Find the prime factorization of each number.

$$6 = 2 \cdot 3, \qquad 8 = 2 \cdot 2 \cdot 2$$

b) Create a product by writing factors, using each the greatest number of times that it occurs in any one factorization.

Consider the factor 2. The greatest number of times that 2 occurs in any one factorization is three. We write 2 as a factor three times.

$$2 \cdot 2 \cdot 2 \cdot ?$$

Consider the factor 3. The greatest number of times that 3 occurs in any one factorization is one. We write 3 as a factor one time.

$$2 \cdot 2 \cdot 2 \cdot 3 \cdot ?$$

Since there are no other prime factors in either factorization, the

LCM is $2 \cdot 2 \cdot 2 \cdot 3$, or 24.

EXAMPLE 7 Find the LCM of 24 and 36.

a) Find the prime factorization of each number.

$$24 = 2 \cdot 2 \cdot 2 \cdot 3, \qquad 36 = 2 \cdot 2 \cdot 3 \cdot 3$$

b) Create a product by writing factors, using each the greatest number of times that it occurs in any one factorization.

Consider the factor 2. The greatest number of times that 2 occurs in any one factorization is three. We write 2 as a factor three times:

$$2 \cdot 2 \cdot 2 \cdot ?$$

Consider the factor 3. The greatest number of times that 3 occurs in any one factorization is two. We write 3 as a factor two times:

$$2 \cdot 2 \cdot 2 \cdot 3 \cdot 3 \cdot ?$$

Since there are no other prime factors in either factorization, the

LCM is $2 \cdot 2 \cdot 2 \cdot 3 \cdot 3$, or 72.

Do Exercises 7–9.

Use prime factorizations to find the LCM.

7. 8, 10

8. 18, 40

9. 32, 54

Answers on page A-6

10. Find the LCM of 24, 35, and 45.

11. Use exponents to find the LCM of 24, 35, and 45.

12. Redo Margin Exercises 7–9 using exponents.

EXAMPLE 8 Find the LCM of 27, 90, and 84.

a) Find the prime factorization of each number.

$$27 = 3 \cdot 3 \cdot 3, \qquad 90 = 2 \cdot 3 \cdot 3 \cdot 5, \qquad 84 = 2 \cdot 2 \cdot 3 \cdot 7$$

b) Create a product by writing factors, using each the greatest number of times that it occurs in any one factorization.

Consider the factor 2. The greatest number of times that 2 occurs in any one factorization is two. We write 2 as a factor two times:

$$2 \cdot 2 \cdot \text{?}$$

Consider the factor 3. The greatest number of times that 3 occurs in any one factorization is three. We write 3 as a factor three times:

$$2 \cdot 2 \cdot 3 \cdot 3 \cdot 3 \cdot \text{?}$$

Consider the factor 5. The greatest number of times that 5 occurs in any one factorization is one. We write 5 as a factor one time:

$$2 \cdot 2 \cdot 3 \cdot 3 \cdot 3 \cdot 5 \cdot \text{?}$$

Consider the factor 7. The greatest number of times that 7 occurs in any one factorization is one. We write 7 as a factor one time:

$$2 \cdot 2 \cdot 3 \cdot 3 \cdot 3 \cdot 5 \cdot 7 \cdot \text{?}$$

Since there are no other prime factors in any of the factorizations, the

LCM is $2 \cdot 2 \cdot 3 \cdot 3 \cdot 3 \cdot 5 \cdot 7$, or 3780.

Do Exercise 10.

The use of exponents might be helpful to you as an extension of the factorization method. Let's reconsider Example 8. We want to find the LCM of 27, 90, and 84. We factor and then convert to exponential notation:

$$27 = 3 \cdot 3 \cdot 3 = 3^3,$$
$$90 = 2 \cdot 3 \cdot 3 \cdot 5 = 2^1 \cdot 3^2 \cdot 5^1, \quad \text{and}$$
$$84 = 2 \cdot 2 \cdot 3 \cdot 7 = 2^2 \cdot 3^1 \cdot 7^1.$$

Thus the

LCM is $2^2 \cdot 3^3 \cdot 5^1 \cdot 7^1$, or 3780.

Note that in 84, the 2 in 2^2 is the largest exponent of 2 in any of the factorizations. It is also the exponent of 2 in the LCM. It indicates the greatest number of times that 2 occurs as a factor of any of the numbers. Similarly, in 27, the 3 in 3^3 is the largest exponent of 3 in any of the factorizations. It is also the exponent of 3 in the LCM. Likewise, the 1's in 5^1 and 7^1 tell us the exponents of 5 and 7 in the LCM. They indicate the greatest number of times that 5 and 7 occur as factors.

Thus the answers to Examples 6–8 can be expressed with exponents as $2^3 \cdot 3$, $2^3 \cdot 3^2$, and $2^2 \cdot 3^3 \cdot 5 \cdot 7$, respectively.

Do Exercises 11 and 12.

EXAMPLE 9 Find the LCM of 7 and 21.

We find the prime factorization of each number. Because 7 is prime, it has no prime factorization.

$$7 = 7, \qquad 21 = 3 \cdot 7$$

Note that 7 is a factor of 21. We stated earlier that if one number is a factor of another, the LCM is the larger of the numbers. Thus the LCM is $7 \cdot 3$, or 21.

Do Exercises 13 and 14.

EXAMPLE 10 Find the LCM of 8 and 9.

We find the prime factorization of each number.

$$8 = 2 \cdot 2 \cdot 2, \qquad 9 = 3 \cdot 3$$

Note that the two numbers, 8 and 9, have no common prime factor. When this is the case, the LCM is just the product of the two numbers. Thus the LCM is $2 \cdot 2 \cdot 2 \cdot 3 \cdot 3$, or 72.

Do Exercises 15 and 16.

Let's compare the two methods considered so far for finding LCMs: the multiples method and the factorization method.

Method 1, the **multiples method,** can be longer than the factorization method when the LCM is large or when there are more than two numbers. But this method can be faster and easier to use mentally for two numbers.

Method 2, the **factorization method,** works well for several numbers. It is just like a method used in algebra. If you are going to study algebra, you should definitely learn the factorization method.

METHOD 3: FINDING LCMS USING DIVISION BY PRIMES

Here is another method for finding LCMs that is especially useful for three or more numbers. For example, to find the LCM of 48, 72, and 80, we first look for any prime that divides any two of the numbers with no remainder. Then we divide as follows.

$$
\begin{array}{r|rrr}
2 & 48 & 72 & 80 \\
\hline
& 24 & 36 & 40
\end{array}
$$

We repeat the process, bringing down any numbers not divisible by the prime, until we can divide no more, that is, until there are no two numbers divisible by the same prime:

$$
\begin{array}{r|rrr}
2 & 48 & 72 & 80 \\
3 & 24 & 36 & 40 \\
2 & 8 & 12 & 40 \\
2 & 4 & 6 & 20 \\
2 & 2 & 3 & 10 \\
\hline
& 1 & 3 & 5
\end{array}
$$

40 is not divisible by 3.

3 is not divisible by 2.

The LCM is

$2 \cdot 3 \cdot 2 \cdot 2 \cdot 2 \cdot 1 \cdot 3 \cdot 5$, or 720.

Method 3: To find the LCM using division by primes:

a) First look for any prime that divides at least two of the numbers with no remainder. Then divide, bringing down any numbers not divisible by the prime.

b) Repeat the process until you can divide no more, that is, until there are no two numbers divisible by the same prime.

Find the LCM.

13. 3, 18

14. 12, 24

Find the LCM.

15. 4, 9

16. 5, 6, 7

Answers on page A-6

Find the LCM using division by primes.

17. 12, 75, 120

18. 27, 90, 84

19. 12, 24, 75, 120

Answers on page A-6

EXAMPLE 11 Find the LCM of 24, 35, and 45.

$$
\begin{array}{c|ccc}
5 & 24 & 35 & 45 \\
3 & 24 & 7 & 9 \\
& 8 & 7 & 3
\end{array}
\quad
\begin{array}{l}
\text{24 is not divisible by 5.} \\
\text{7 is not divisible by 3.}
\end{array}
$$

The LCM is $5 \cdot 3 \cdot 8 \cdot 7 \cdot 3$, or 2520.

EXAMPLE 12 Find the LCM of 12, 18, 20, and 21.

$$
\begin{array}{c|cccc}
3 & 12 & 18 & 20 & 21 \\
2 & 4 & 6 & 20 & 7 \\
2 & 2 & 3 & 10 & 7 \\
& 1 & 3 & 5 & 7
\end{array}
$$

The LCM is $3 \cdot 2 \cdot 2 \cdot 1 \cdot 3 \cdot 5 \cdot 7$, or 1260.

Do Exercises 17–19.

Study Tips

TIPS FROM A FORMER STUDENT

A former student of Professor Bittinger, Mike Rosenborg earned a master's degree in mathematics and now teaches mathematics. Here are some of his study tips.

- Because working problems is the best way to learn math, instructors generally assign lots of problems. Never let yourself get behind in your math homework.
- If you are struggling with a math concept, do not give up. Ask for help from your friends and your instructor. Since each concept is built on previous concepts, any gaps in your understanding will follow you through the entire course, so make sure you understand each concept as you go along.
- Read your textbook! It will often contain the help and tips you need to solve any problem with which you are struggling. It may also bring out points that you missed in class or that your instructor may not have covered.
- Learn to use scratch paper to jot down your thoughts and to draw pictures. Don't try to figure everything out "in your head." You will think more clearly and accurately this way.
- When preparing for a test, it is often helpful to work at least two problems per section as practice: one easy and one difficult. Write out all the new rules and procedures your test will cover, and then read through them twice. Doing so will enable you to both learn and retain them better.
- Most schools have classrooms set up where you can get free help from math tutors. Take advantage of this, but be sure you do the work first. Don't let your tutor do all the work for you—otherwise you'll never learn the material.
- In math, as in many other areas of life, patience and persistence are virtues—cultivate them. "Cramming" for an exam will not help you learn and retain the material.

3.1

EXERCISE SET

a   Find the LCM of the set of numbers.

1. 2, 4

2. 3, 15

3. 10, 25

4. 10, 15

5. 20, 40

6. 8, 12

7. 18, 27

8. 9, 11

9. 30, 50

10. 24, 36

11. 30, 40

12. 21, 27

13. 18, 24

14. 12, 18

15. 60, 70

16. 35, 45

17. 16, 36

18. 18, 20

19. 32, 36

20. 36, 48

21. 2, 3, 5

22. 5, 18, 3

23. 3, 5, 7

24. 6, 12, 18

25. 24, 36, 12

26. 8, 16, 22

27. 5, 12, 15

28. 12, 18, 40

29. 9, 12, 6

30. 8, 16, 12

31. 180, 100, 450, 60

32. 18, 30, 50, 48

33. 8, 48

34. 16, 32

35. 5, 50

36. 12, 72

37. 11, 13

38. 13, 14

39. 12, 35

40. 23, 25

41. 54, 63

42. 56, 72

43. 81, 90

44. 75, 100

Applications of LCMs: Planet Orbits. The earth, Jupiter, Saturn, and Uranus all revolve around the sun. The earth takes 1 yr, Jupiter 12 yr, Saturn 30 yr, and Uranus 84 yr to make a complete revolution. On a certain night, you look at those three distant planets and wonder how many years it will take before they have the same position again. (*Hint:* To find out, you find the LCM of 12, 30, and 84. It will be that number of years.)
Source: *The Handy Science Answer Book*

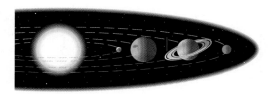

45. How often will Jupiter and Saturn appear in the same direction in the night sky as seen from the earth?

46. How often will Jupiter and Uranus appear in the same direction in the night sky as seen from the earth?

47. How often will Saturn and Uranus appear in the same direction in the night sky as seen from the earth?

48. How often will Jupiter, Saturn, and Uranus appear in the same direction in the night sky as seen from the earth?

49. ^{D}W Use both Methods 1 and 2 to find the LCM of each of the following sets of numbers.

a) 6, 8   **b)** 6, 7   **c)** 6, 21   **d)** 24, 36

Which method do you consider more efficient? Explain why.

50. ^{D}W Is the LCM of two numbers always larger than either number? Why or why not?

SKILL MAINTENANCE

Solve.

51. Joy uses $\frac{1}{2}$ yd of dental floss each day. How long will a 45-yd container of dental floss last for Joy?   [2.7d]

52. *Vehicle Expense.* The most expensive cities in which to own an automobile are Los Angeles, where the yearly cost is $9254, and Philadelphia, where the yearly cost is $8715. How much more does it cost in Los Angeles than in Philadelphia?   [1.8a]
Source: Runzheimer International

53. Add: $23,456 + 5677 + 4002$.   [1.2a]

54. Subtract: $10,007 - 3068$.   [1.3b]

55. Multiply and simplify: $\frac{4}{5} \cdot \frac{10}{12}$.   [2.6a]

56. Divide and simplify: $\frac{4}{5} \div \frac{7}{10}$.   [2.7b]

SYNTHESIS

57. Find the LCM of 27, 90, 84, 210, 108, and 50.

58. Find the LCM of 18, 21, 24, 36, 63, 56, and 20.

59. A pencil company uses two sizes of boxes, 5 in. by 6 in. and 5 in. by 8 in. These boxes are packed in bigger cartons for shipping. Find the width and the length of the smallest carton that will accommodate boxes of either size without any room left over. (Each carton can contain only one type of box and all boxes must point in the same direction.)

60. Consider 8 and 12. Determine whether each of the following is the LCM of 8 and 12. Tell why or why not.

a) $2 \cdot 2 \cdot 3 \cdot 3$
b) $2 \cdot 2 \cdot 3$
c) $2 \cdot 3 \cdot 3$
d) $2 \cdot 2 \cdot 2 \cdot 3$

3.2

ADDITION AND APPLICATIONS

a Addition Using Fraction Notation

Objectives

 a Add using fraction notation.

 b Solve applied problems involving addition with fraction notation.

LIKE DENOMINATORS

Addition using fraction notation corresponds to combining or putting like things together, just as addition with whole numbers does. For example,

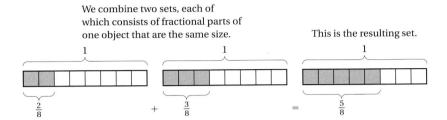

We combine two sets, each of which consists of fractional parts of one object that are the same size.

This is the resulting set.

$$2 \text{ eighths} + 3 \text{ eighths} = 5 \text{ eighths},$$

or $2 \cdot \dfrac{1}{8} + 3 \cdot \dfrac{1}{8} = 5 \cdot \dfrac{1}{8},$ or $\dfrac{2}{8} + \dfrac{3}{8} = \dfrac{5}{8}.$

We see that to add when denominators are the same, we add the numerators, keep the denominator, and simplify, if possible.

Do Exercise 1.

To add when denominators are the same,

a) add the numerators,

b) keep the denominator, and

c) simplify, if possible.

$$\frac{2}{6} + \frac{5}{6} = \frac{2+5}{6} = \frac{7}{6}$$

EXAMPLES Add and simplify.

1. $\dfrac{2}{4} + \dfrac{1}{4} = \dfrac{2+1}{4} = \dfrac{3}{4}$ No simplifying is possible.

2. $\dfrac{11}{6} + \dfrac{3}{6} = \dfrac{11+3}{6} = \dfrac{14}{6} = \dfrac{2 \cdot 7}{2 \cdot 3} = \dfrac{2}{2} \cdot \dfrac{7}{3} = 1 \cdot \dfrac{7}{3} = \dfrac{7}{3}$ Here we simplified.

3. $\dfrac{3}{12} + \dfrac{5}{12} = \dfrac{3+5}{12} = \dfrac{8}{12} = \dfrac{4 \cdot 2}{4 \cdot 3} = \dfrac{4}{4} \cdot \dfrac{2}{3} = 1 \cdot \dfrac{2}{3} = \dfrac{2}{3}$

Do Exercises 2–4.

DIFFERENT DENOMINATORS

What do we do when denominators are different? We can find a common denominator by multiplying by 1. Consider adding $\frac{1}{6}$ and $\frac{3}{4}$. There are many common denominators that can be obtained. Let's look at two possibilities.

1. Find $\dfrac{1}{5} + \dfrac{3}{5}$.

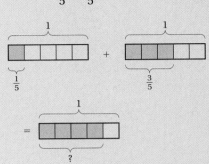

Add and simplify.

2. $\dfrac{1}{3} + \dfrac{2}{3}$

3. $\dfrac{5}{12} + \dfrac{1}{12}$

4. $\dfrac{9}{16} + \dfrac{3}{16}$

Answers on page A-6

173

5. Add. (Find the least common denominator.)

$$\frac{2}{3} + \frac{1}{6}$$

A. $\dfrac{1}{6} + \dfrac{3}{4} = \dfrac{1}{6} \cdot 1 + \dfrac{3}{4} \cdot 1$

$\qquad\quad = \dfrac{1}{6} \cdot \dfrac{4}{4} + \dfrac{3}{4} \cdot \dfrac{6}{6}$

$\qquad\quad = \dfrac{4}{24} + \dfrac{18}{24}$

$\qquad\quad = \dfrac{22}{24}$

$\qquad\quad = \dfrac{11}{12}$

B. $\dfrac{1}{6} + \dfrac{3}{4} = \dfrac{1}{6} \cdot 1 + \dfrac{3}{4} \cdot 1$

$\qquad\quad = \dfrac{1}{6} \cdot \dfrac{2}{2} + \dfrac{3}{4} \cdot \dfrac{3}{3}$

$\qquad\quad = \dfrac{2}{12} + \dfrac{9}{12}$

$\qquad\quad = \dfrac{11}{12}$

We had to simplify in (A). We didn't have to simplify in (B). In (B), we used the least common multiple of the denominators, 12. That number is called the **least common denominator,** or **LCD.**

> To add when denominators are different:
> **a)** Find the least common multiple of the denominators. That number is the least common denominator, LCD.
> **b)** Multiply by 1, using an appropriate notation, n/n, to express each number in terms of the LCD.
> **c)** Add the numerators, keeping the same denominator.
> **d)** Simplify, if possible.

6. Add: $\dfrac{3}{8} + \dfrac{5}{6}$.

EXAMPLE 4 Add: $\dfrac{3}{4} + \dfrac{1}{8}$.

The LCD is 8.   4 is a factor of 8 so the LCM of 4 and 8 is 8.

$\dfrac{3}{4} + \dfrac{1}{8} = \dfrac{3}{4} \cdot 1 + \dfrac{1}{8}$ ← This fraction already has the LCD as its denominator.

$\qquad\quad = \dfrac{3}{4} \cdot \dfrac{2}{2} + \dfrac{1}{8}$ *Think*: $4 \times \square = 8$. The answer is 2, so we multiply by 1, using $\frac{2}{2}$.

$\qquad\quad = \dfrac{6}{8} + \dfrac{1}{8} = \dfrac{7}{8}$

Do Exercise 5.

EXAMPLE 5 Add: $\dfrac{1}{9} + \dfrac{5}{6}$.

The LCD is 18. $9 = 3 \cdot 3$ and $6 = 2 \cdot 3$, so the LCM of 9 and 6 is $2 \cdot 3 \cdot 3$, or 18.

$\dfrac{1}{9} + \dfrac{5}{6} = \dfrac{1}{9} \cdot 1 + \dfrac{5}{6} \cdot 1 = \dfrac{1}{9} \cdot \dfrac{2}{2} + \dfrac{5}{6} \cdot \dfrac{3}{3}$

$\qquad\qquad$ *Think*: $6 \times \square = 18$. The answer is 3, so we multiply by 1 using $\frac{3}{3}$.

$\qquad\qquad$ *Think*: $9 \times \square = 18$. The answer is 2, so we multiply by 1 using $\frac{2}{2}$.

$\qquad\quad = \dfrac{2}{18} + \dfrac{15}{18} = \dfrac{17}{18}$

Do Exercise 6.

Answers on page A-6

EXAMPLE 6 Add: $\dfrac{5}{9} + \dfrac{11}{18}$.

The LCD is 18.

$$\frac{5}{9} + \frac{11}{18} = \frac{5}{9} \cdot \frac{2}{2} + \frac{11}{18} = \frac{10}{18} + \frac{11}{18}$$

$$= \frac{21}{18}$$

$$= \frac{7}{6}$$

We may still have to simplify, but it is usually easier if we have used the LCD.

Do Exercise 7.

EXAMPLE 7 Add: $\dfrac{1}{10} + \dfrac{3}{100} + \dfrac{7}{1000}$.

Since 10 and 100 are factors of 1000, the LCD is 1000. Then

$$\frac{1}{10} + \frac{3}{100} + \frac{7}{1000} = \frac{1}{10} \cdot \frac{100}{100} + \frac{3}{100} \cdot \frac{10}{10} + \frac{7}{1000}$$

$$= \frac{100}{1000} + \frac{30}{1000} + \frac{7}{1000} = \frac{137}{1000}.$$

Do Exercise 8.

When denominators are large, we most often use the prime factorization of each denominator. This is shown in Example 8. Using the prime factorization in this manner is similar to what is done in algebra.

EXAMPLE 8 Add: $\dfrac{13}{70} + \dfrac{11}{21} + \dfrac{6}{15}$.

We have

$$\frac{13}{70} + \frac{11}{21} + \frac{6}{15} = \frac{13}{2 \cdot 5 \cdot 7} + \frac{11}{3 \cdot 7} + \frac{6}{3 \cdot 5}. \quad \text{Factoring denominators}$$

The LCD is $2 \cdot 3 \cdot 5 \cdot 7$, or 210. Then

$$\frac{13}{70} + \frac{11}{21} + \frac{6}{15} = \frac{13}{2 \cdot 5 \cdot 7} \cdot \frac{3}{3} + \frac{11}{3 \cdot 7} \cdot \frac{2 \cdot 5}{2 \cdot 5} + \frac{6}{3 \cdot 5} \cdot \frac{7 \cdot 2}{7 \cdot 2}$$

> The LCD of 70, 21, and 15 is $2 \cdot 3 \cdot 5 \cdot 7$. In each case, think of which factors are needed to get the LCD. Then multiply by 1 to obtain the LCD in each denominator.

$$= \frac{13 \cdot 3}{2 \cdot 5 \cdot 7 \cdot 3} + \frac{11 \cdot 2 \cdot 5}{3 \cdot 7 \cdot 2 \cdot 5} + \frac{6 \cdot 7 \cdot 2}{3 \cdot 5 \cdot 7 \cdot 2}$$

$$= \frac{39}{3 \cdot 5 \cdot 7 \cdot 2} + \frac{110}{3 \cdot 5 \cdot 7 \cdot 2} + \frac{84}{3 \cdot 5 \cdot 7 \cdot 2}$$

$$= \frac{233}{3 \cdot 5 \cdot 7 \cdot 2}$$

$$= \frac{233}{210}. \quad \text{We left 210 factored until we knew we could not simplify.}$$

Do Exercises 9 and 10.

7. Add: $\dfrac{1}{6} + \dfrac{7}{18}$.

8. Add: $\dfrac{4}{10} + \dfrac{1}{100} + \dfrac{3}{1000}$.

Add.

9. $\dfrac{7}{10} + \dfrac{2}{21} + \dfrac{1}{7}$

10. $\dfrac{7}{18} + \dfrac{5}{24} + \dfrac{11}{36}$

Answers on page A-6

11. Sally jogs for $\frac{4}{5}$ mi, rests, and then jogs for another $\frac{1}{10}$ mi. How far does she jog in all?

$\frac{1}{10}$ mi

$\frac{4}{5}$ mi

D

b Applications and Problem Solving

EXAMPLE 9 *Carpentry.* Dick Bonewitz, a master carpenter, makes special pieces of furniture for his family and friends. To cut expenses, he sometimes glues two kinds of plywood together. He glues a $\frac{1}{4}$-in. $\left(\frac{1}{4}''\right)$ piece of walnut plywood to a $\frac{3}{8}$-in. $\left(\frac{3}{8}''\right)$ piece of less expensive plywood. What is the total thickness of these pieces?

1. **Familiarize.** We first make a drawing. We let $T =$ the total thickness of the plywood.

$\frac{1''}{4}$

$\frac{3''}{8}$

2. **Translate.** The problem can be translated to an equation as follows.

$$\underbrace{\text{Less expensive plywood}}_{} \quad \underset{+}{\text{plus}} \quad \underbrace{\text{Walnut plywood}}_{} \quad \underset{=}{\text{is}} \quad \underbrace{\text{Total thickness}}_{}$$

$$\frac{3}{8} \qquad + \qquad \frac{1}{4} \qquad = \qquad T$$

3. **Solve.** To solve the equation, we carry out the addition. The LCM of the denominators is 8 because 4 is a factor of 8. We multiply by 1 in order to obtain the LCD:

$$\frac{3}{8} + \frac{1}{4} = T$$

$$\frac{3}{8} + \frac{1}{4} \cdot \frac{2}{2} = T$$

$$\frac{3}{8} + \frac{2}{8} = T$$

$$\frac{5}{8} = T.$$

4. **Check.** We check by repeating the calculation. We also note that the sum should be larger than either of the individual measurements, which it is. This gives us a partial check on the reasonableness of the answer.

5. **State.** The total thickness of the plywood is $\frac{5}{8}$ in.

Do Exercise 11.

Answer on page A-6

3.2

EXERCISE SET

For Extra Help

Digital Video
Tutor CD 2
Videotape 5

InterAct
Math

Math Tutor
Center

MathXL

MyMathLab

a Add and simplify.

1. $\dfrac{7}{8} + \dfrac{1}{8}$

2. $\dfrac{2}{5} + \dfrac{3}{5}$

3. $\dfrac{1}{8} + \dfrac{5}{8}$

4. $\dfrac{3}{10} + \dfrac{3}{10}$

5. $\dfrac{2}{3} + \dfrac{5}{6}$

6. $\dfrac{5}{6} + \dfrac{1}{9}$

7. $\dfrac{1}{8} + \dfrac{1}{6}$

8. $\dfrac{1}{6} + \dfrac{3}{4}$

9. $\dfrac{4}{5} + \dfrac{7}{10}$

10. $\dfrac{3}{4} + \dfrac{1}{12}$

11. $\dfrac{5}{12} + \dfrac{3}{8}$

12. $\dfrac{7}{8} + \dfrac{1}{16}$

13. $\dfrac{3}{20} + \dfrac{3}{4}$

14. $\dfrac{2}{15} + \dfrac{2}{5}$

15. $\dfrac{5}{6} + \dfrac{7}{9}$

16. $\dfrac{5}{8} + \dfrac{5}{6}$

17. $\dfrac{3}{10} + \dfrac{1}{100}$

18. $\dfrac{9}{10} + \dfrac{3}{100}$

19. $\dfrac{5}{12} + \dfrac{4}{15}$

20. $\dfrac{3}{16} + \dfrac{1}{12}$

21. $\dfrac{9}{10} + \dfrac{99}{100}$

22. $\dfrac{3}{10} + \dfrac{27}{100}$

23. $\dfrac{7}{8} + \dfrac{0}{1}$

24. $\dfrac{0}{1} + \dfrac{5}{6}$

25. $\dfrac{3}{8} + \dfrac{1}{6}$

26. $\dfrac{5}{8} + \dfrac{1}{6}$

27. $\dfrac{5}{12} + \dfrac{7}{24}$

28. $\dfrac{1}{18} + \dfrac{7}{12}$

29. $\dfrac{3}{16} + \dfrac{5}{16} + \dfrac{4}{16}$

30. $\dfrac{3}{8} + \dfrac{1}{8} + \dfrac{2}{8}$

31. $\dfrac{8}{10} + \dfrac{7}{100} + \dfrac{4}{1000}$

32. $\dfrac{1}{10} + \dfrac{2}{100} + \dfrac{3}{1000}$

33. $\dfrac{3}{8} + \dfrac{5}{12} + \dfrac{8}{15}$

34. $\dfrac{1}{2} + \dfrac{3}{8} + \dfrac{1}{4}$

35. $\dfrac{15}{24} + \dfrac{7}{36} + \dfrac{91}{48}$

36. $\dfrac{5}{7} + \dfrac{25}{52} + \dfrac{7}{4}$

 Solve.

37. Rene bought $\frac{1}{3}$ lb of orange pekoe tea and $\frac{1}{2}$ lb of English cinnamon tea. How many pounds of tea did he buy?

38. Stan bought $\frac{1}{4}$ lb of gumdrops and $\frac{1}{2}$ lb of caramels. How many pounds of candy did he buy?

39. Russ walked $\frac{7}{6}$ mi to a friend's dormitory, and then $\frac{3}{4}$ mi to class. How far did he walk?

40. Elaine walked $\frac{7}{8}$ mi to the student union, and then $\frac{2}{5}$ mi to class. How far did she walk?

41. *Concrete Mix.* A cubic meter of concrete mix contains 420 kilograms (kg) of cement, 150 kg of stone, and 120 kg of sand. What is the total weight of the cubic meter of concrete mix? What part is cement? stone? sand? Add these amounts. What is the result?

42. *Punch Recipe.* A recipe for strawberry punch calls for $\frac{1}{5}$ quart (qt) of ginger ale and $\frac{3}{5}$ qt of strawberry soda. How much liquid is needed? If the recipe is doubled, how much liquid is needed? If the recipe is halved, how much liquid is needed?

43. A tile $\frac{5}{8}$ in. thick is glued to a board $\frac{7}{8}$ in. thick. The glue is $\frac{3}{32}$ in. thick. How thick is the result?

44. A baker used $\frac{1}{2}$ lb of flour for rolls, $\frac{1}{4}$ lb for donuts, and $\frac{1}{3}$ lb for cookies. How much flour was used?

45. ᴰW Explain the role of multiplication when adding using fraction notation with different denominators.

46. ᴰW To add numbers with different denominators, a student consistently uses the product of the denominators as a common denominator. Is this correct? Why or why not?

Multiply. [1.5a]

47. 408 · 516   **48.** 1125 · 3728   **49.** 423 · 8009   **50.** 2025 · 174

Holiday Expenditures.   The following chart shows average expenditures per person of consumers during the Christmas holidays of 1999 and 2000. Use these data for Exercises 51–56.   [1.8a]

HOLIDAY EXPENDITURES	1999	2000
Gifts	$1088	$1161
Entertainment	188	197
Travel	151	154
Decorations / cards	77	88
Other holiday expenses	54	84
Total	?	?

Source: *2000 American Express Retail Index*

51. How much more was spent on gifts in 2000 than in 1999?

52. How much more was spent on decorations and cards in 2000 than in 1999?

53. How much more was spent on travel in 2000 than in 1999?

54. How much more was spent on entertainment in 2000 than in 1999?

55. What was the total expenditure in 1999?

56. What was the total expenditure in 2000?

57. Elsa has $9 to spend on ride tickets at the fair. If the tickets cost 75¢, or $\frac{3}{4}$, each, how many tickets can she purchase? [2.7d]

58. The Bingham community garden is to be split into 16 equally sized plots. If the garden occupies $\frac{3}{4}$ acre of land, how large will each plot be? [2.7d]

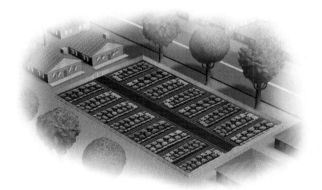

59. A guitarist's band is booked for Friday and Saturday nights at a local club. The guitarist is part of a trio on Friday and part of a quintet on Saturday. Thus the guitarist is paid one-third of one-half the weekend's pay for Friday and one-fifth of one-half the weekend's pay for Saturday. What fractional part of the band's pay did the guitarist receive for the weekend's work? If the band was paid $1200, how much did the guitarist receive?

Objectives

a Subtract using fraction notation.

b Use $<$ or $>$ with fraction notation to write a true sentence.

c Solve equations of the type $x + a = b$ and $a + x = b$, where a and b may be fractions.

d Solve applied problems involving subtraction with fraction notation.

Subtract and simplify.

1. $\dfrac{7}{8} - \dfrac{3}{8}$

2. $\dfrac{10}{16} - \dfrac{4}{16}$

3. $\dfrac{8}{10} - \dfrac{3}{10}$

Answers on page A-6

180

CHAPTER 3: Fraction Notation and Mixed Numerals

---

a Subtraction Using Fraction Notation

LIKE DENOMINATORS

We can consider the difference $\frac{4}{8} - \frac{3}{8}$ as we did before, as either "take away" or "missing addend." Let's consider "take away."

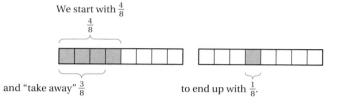

We start with $\frac{4}{8}$

and "take away" $\frac{3}{8}$ to end up with $\frac{1}{8}$.

We start with 4 eighths and take away 3 eighths:

$$4 \text{ eighths} - 3 \text{ eighths} = 1 \text{ eighth},$$

or $\quad 4 \cdot \dfrac{1}{8} - 3 \cdot \dfrac{1}{8} = \dfrac{1}{8}, \quad$ or $\quad \dfrac{4}{8} - \dfrac{3}{8} = \dfrac{1}{8}.$

> To subtract when denominators are the same,
> **a)** subtract the numerators,
> **b)** keep the denominator, and
> **c)** simplify, if possible.
>
> $$\dfrac{7}{10} - \dfrac{4}{10} = \dfrac{7 - 4}{10} = \dfrac{3}{10}$$

EXAMPLES Subtract and simplify.

1. $\dfrac{7}{10} - \dfrac{3}{10} = \dfrac{7 - 3}{10} = \dfrac{4}{10} = \dfrac{2 \cdot 2}{5 \cdot 2} = \dfrac{2}{5} \cdot \dfrac{2}{2} = \dfrac{2}{5} \cdot 1 = \dfrac{2}{5}$

2. $\dfrac{8}{9} - \dfrac{2}{9} = \dfrac{8 - 2}{9} = \dfrac{6}{9} = \dfrac{2 \cdot 3}{3 \cdot 3} = \dfrac{2}{3} \cdot \dfrac{3}{3} = \dfrac{2}{3} \cdot 1 = \dfrac{2}{3}$

3. $\dfrac{32}{12} - \dfrac{25}{12} = \dfrac{32 - 25}{12} = \dfrac{7}{12}$

Do Exercises 1–3.

DIFFERENT DENOMINATORS

> To subtract when denominators are different:
> **a)** Find the least common multiple of the denominators. That number is the least common denominator, LCD.
> **b)** Multiply by 1, using an appropriate notation, n/n, to express each number in terms of the LCD.
> **c)** Subtract the numerators, keeping the same denominator.
> **d)** Simplify, if possible.

EXAMPLE 4 Subtract: $\dfrac{2}{5} - \dfrac{3}{8}$.

The LCM of 5 and 8 is 40. The LCD is 40.

$$\frac{2}{5} - \frac{3}{8} = \frac{2}{5} \cdot \frac{8}{8} - \frac{3}{8} \cdot \frac{5}{5}$$

Think: $8 \times \square = 40$. The answer is 5, so we multiply by 1, using $\frac{5}{5}$.

Think: $5 \times \square = 40$. The answer is 8, so we multiply by 1, using $\frac{8}{8}$.

$$= \frac{16}{40} - \frac{15}{40} = \frac{16 - 15}{40} = \frac{1}{40}$$

Do Exercise 4.

EXAMPLE 5 Subtract: $\dfrac{5}{6} - \dfrac{7}{12}$.

Since 12 is a multiple of 6, the LCM of 6 and 12 is 12. The LCD is 12.

$$\frac{5}{6} - \frac{7}{12} = \frac{5}{6} \cdot \frac{2}{2} - \frac{7}{12}$$

$$= \frac{10}{12} - \frac{7}{12} = \frac{10 - 7}{12} = \frac{3}{12}$$

$$= \frac{3 \cdot 1}{3 \cdot 4} = \frac{3}{3} \cdot \frac{1}{4} = \frac{1}{4}$$

Do Exercises 5 and 6.

EXAMPLE 6 Subtract: $\dfrac{17}{24} - \dfrac{4}{15}$.

We have

$$\frac{17}{24} - \frac{4}{15} = \frac{17}{3 \cdot 2 \cdot 2 \cdot 2} - \frac{4}{5 \cdot 3}.$$

The LCD is $3 \cdot 2 \cdot 2 \cdot 2 \cdot 5$, or 120. Then

$$\frac{17}{24} - \frac{4}{15} = \frac{17}{3 \cdot 2 \cdot 2 \cdot 2} \cdot \frac{5}{5} - \frac{4}{5 \cdot 3} \cdot \frac{2 \cdot 2 \cdot 2}{2 \cdot 2 \cdot 2}$$

The LCD of 24 and 15 is $2 \cdot 2 \cdot 2 \cdot 3 \cdot 5$. In each case, we multiply by 1 to obtain the LCD.

$$= \frac{17 \cdot 5}{3 \cdot 2 \cdot 2 \cdot 2 \cdot 5} - \frac{4 \cdot 2 \cdot 2 \cdot 2}{5 \cdot 3 \cdot 2 \cdot 2 \cdot 2}$$

$$= \frac{85}{120} - \frac{32}{120} = \frac{53}{120}.$$

Do Exercise 7.

b Order

We see from this figure that $\frac{4}{5} > \frac{3}{5}$, and $\frac{3}{5} < \frac{4}{5}$. That is, $\frac{4}{5}$ is greater than $\frac{3}{5}$, and $\frac{3}{5}$ is less than $\frac{4}{5}$.

$\frac{4}{5}$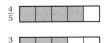

$\frac{3}{5}$

---

4. Subtract: $\dfrac{3}{4} - \dfrac{2}{3}$.

Subtract.

5. $\dfrac{5}{6} - \dfrac{1}{9}$

6. $\dfrac{4}{5} - \dfrac{3}{10}$

7. Subtract: $\dfrac{11}{28} - \dfrac{5}{16}$.

8. Use < or > for $\square$ to write a true sentence:

$$\frac{3}{8} \;\square\; \frac{5}{8}.$$

Answers on page A-6

9. Use $<$ or $>$ for $\square$ to write a true sentence:

$$\frac{7}{10} \square \frac{6}{10}.$$

Use $<$ or $>$ for $\square$ to write a true sentence.

10. $\frac{2}{3} \square \frac{5}{8}$

11. $\frac{3}{4} \square \frac{8}{12}$

12. $\frac{5}{6} \square \frac{7}{8}$

Solve.

13. $x + \frac{2}{3} = \frac{5}{6}$

14. $\frac{3}{5} + t = \frac{7}{8}$

To determine which of two numbers is greater when there is a common denominator, compare the numerators:

$$\frac{4}{5}, \quad \frac{3}{5}, \qquad 4 > 3 \qquad \frac{4}{5} > \frac{3}{5}.$$

Do Exercises 8 and 9. (Exercise 8 is on the preceding page.)

When denominators are different, we cannot compare numerators. We multiply by 1 to make the denominators the same.

EXAMPLE 7 Use $<$ or $>$ for $\square$ to write a true sentence:

$$\frac{2}{5} \square \frac{3}{4}.$$

We have

$$\frac{2}{5} \cdot \frac{4}{4} = \frac{8}{20}; \qquad \text{We multiply by 1 using } \tfrac{4}{4} \text{ to get the LCD.}$$

$$\frac{3}{4} \cdot \frac{5}{5} = \frac{15}{20}. \qquad \text{We multiply by 1 using } \tfrac{5}{5} \text{ to get the LCD.}$$

Now that the denominators are the same, 20, we can compare the numerators. Since $8 < 15$, it follows that $\frac{8}{20} < \frac{15}{20}$, so

$$\frac{2}{5} < \frac{3}{4}.$$

EXAMPLE 8 Use $<$ or $>$ for $\square$ to write a true sentence:

$$\frac{9}{10} \square \frac{89}{100}.$$

The LCD is 100.

$$\frac{9}{10} \cdot \frac{10}{10} = \frac{90}{100} \qquad \text{We multiply by } \tfrac{10}{10} \text{ to get the LCD.}$$

Since $90 > 89$, it follows that $\frac{90}{100} > \frac{89}{100}$, so

$$\frac{9}{10} > \frac{89}{100}.$$

Do Exercises 10–12.

C   **Solving Equations**

Now let's solve equations of the form $x + a = b$ or $a + x = b$, where a and b may be fractions. Proceeding as we have before, we subtract a on both sides of the equation.

EXAMPLE 9 Solve: $x + \dfrac{1}{4} = \dfrac{3}{5}$.

$$x + \frac{1}{4} - \frac{1}{4} = \frac{3}{5} - \frac{1}{4} \qquad \text{Subtracting } \tfrac{1}{4} \text{ on both sides}$$

$$x + 0 = \frac{3}{5} \cdot \frac{4}{4} - \frac{1}{4} \cdot \frac{5}{5} \qquad \begin{array}{l}\text{The LCD is 20. We multiply by 1}\\ \text{to get the LCD.}\end{array}$$

$$x = \frac{12}{20} - \frac{5}{20} = \frac{7}{20}$$

Do Exercises 13 and 14 on the preceding page.

d Applications and Problem Solving

EXAMPLE 10 *Pendant Necklace.* Coldwater Creek offers the pendant necklace illustrated at right. The sterling silver capping at the top measures $\frac{11}{32}$ in. and the total length of the pendant is $\frac{7}{8}$ in. Find the length, or diameter, w of the pearl ball on the pendant.

1. **Familiarize.** We let $w =$ the length of the pearl ball on the pendant.

2. **Translate.** We see that this is a "missing addend" situation. We can translate to an equation.

Length of silver capping	plus	Length of pearl ball	is	Total length of pendant
$\frac{11}{32}$	$+$	w	$=$	$\frac{7}{8}$

3. **Solve.** To solve the equation, we subtract $\frac{11}{32}$ on both sides:

$$\frac{11}{32} + w = \frac{7}{8}$$

$$\frac{11}{32} + w - \frac{11}{32} = \frac{7}{8} - \frac{11}{32} \qquad \text{Subtracting } \tfrac{11}{32} \text{ on both sides}$$

$$w + 0 = \frac{7}{8} \cdot \frac{4}{4} - \frac{11}{32} \qquad \begin{array}{l}\text{The LCD is 32. We multiply}\\ \text{by 1 to obtain the LCD.}\end{array}$$

$$w = \frac{28}{32} - \frac{11}{32}$$

$$= \frac{17}{32}.$$

4. **Check.** To check, we return to the original problem and add:

$$\frac{11}{32} + \frac{17}{32} = \frac{28}{32} = \frac{7}{8} \cdot \frac{4}{4} = \frac{7}{8}.$$

5. **State.** The length of the pearl ball on the pendant is $\frac{17}{32}$ in.

Do Exercise 15.

Source: ©Coldwater Creek Inc. www.coldwatercreek.com

15. Natasha has run for $\frac{2}{3}$ mi and will stop when she has run for $\frac{7}{8}$ mi. How much farther does she have to go?

Answer on page A-6

a Subtract and simplify.

1. $\dfrac{5}{6} - \dfrac{1}{6}$

2. $\dfrac{5}{8} - \dfrac{3}{8}$

3. $\dfrac{11}{12} - \dfrac{2}{12}$

4. $\dfrac{17}{18} - \dfrac{11}{18}$

5. $\dfrac{3}{4} - \dfrac{1}{8}$

6. $\dfrac{2}{3} - \dfrac{1}{9}$

7. $\dfrac{1}{8} - \dfrac{1}{12}$

8. $\dfrac{1}{6} - \dfrac{1}{8}$

9. $\dfrac{4}{3} - \dfrac{5}{6}$

10. $\dfrac{7}{8} - \dfrac{1}{16}$

11. $\dfrac{3}{4} - \dfrac{3}{28}$

12. $\dfrac{2}{5} - \dfrac{2}{15}$

13. $\dfrac{3}{4} - \dfrac{3}{20}$

14. $\dfrac{5}{6} - \dfrac{1}{2}$

15. $\dfrac{3}{4} - \dfrac{1}{20}$

16. $\dfrac{3}{4} - \dfrac{4}{16}$

17. $\dfrac{5}{12} - \dfrac{2}{15}$

18. $\dfrac{9}{10} - \dfrac{11}{16}$

19. $\dfrac{6}{10} - \dfrac{7}{100}$

20. $\dfrac{9}{10} - \dfrac{3}{100}$

21. $\dfrac{7}{15} - \dfrac{3}{25}$

22. $\dfrac{18}{25} - \dfrac{4}{35}$

23. $\dfrac{99}{100} - \dfrac{9}{10}$

24. $\dfrac{78}{100} - \dfrac{11}{20}$

25. $\dfrac{2}{3} - \dfrac{1}{8}$

26. $\dfrac{3}{4} - \dfrac{1}{2}$

27. $\dfrac{3}{5} - \dfrac{1}{2}$

28. $\dfrac{5}{6} - \dfrac{2}{3}$

29. $\dfrac{5}{12} - \dfrac{3}{8}$

30. $\dfrac{7}{12} - \dfrac{2}{9}$

31. $\dfrac{7}{8} - \dfrac{1}{16}$

32. $\dfrac{5}{12} - \dfrac{5}{16}$

33. $\dfrac{17}{25} - \dfrac{4}{15}$

34. $\dfrac{11}{18} - \dfrac{7}{24}$

35. $\dfrac{23}{25} - \dfrac{112}{150}$

36. $\dfrac{89}{90} - \dfrac{53}{120}$

b Use < or > for □ to write a true sentence.

37. $\dfrac{5}{8} \,\square\, \dfrac{6}{8}$

38. $\dfrac{7}{9} \,\square\, \dfrac{5}{9}$

39. $\dfrac{1}{3} \,\square\, \dfrac{1}{4}$

40. $\dfrac{1}{8} \,\square\, \dfrac{1}{6}$

41. $\dfrac{2}{3} \,\square\, \dfrac{5}{7}$

42. $\dfrac{3}{5} \,\square\, \dfrac{4}{7}$

43. $\dfrac{4}{5} \,\square\, \dfrac{5}{6}$

44. $\dfrac{3}{2} \,\square\, \dfrac{7}{5}$

45. $\dfrac{19}{20} \,\square\, \dfrac{4}{5}$

46. $\dfrac{5}{6} \,\square\, \dfrac{13}{16}$

47. $\dfrac{19}{20} \,\square\, \dfrac{9}{10}$

48. $\dfrac{3}{4} \,\square\, \dfrac{11}{15}$

49. $\dfrac{31}{21} \,\square\, \dfrac{41}{13}$

50. $\dfrac{12}{7} \,\square\, \dfrac{132}{49}$

CHAPTER 3: Fraction Notation and Mixed Numerals

c Solve.

51. $x + \dfrac{1}{30} = \dfrac{1}{10}$

52. $y + \dfrac{9}{12} = \dfrac{11}{12}$

53. $\dfrac{2}{3} + t = \dfrac{4}{5}$

54. $\dfrac{2}{3} + p = \dfrac{7}{8}$

55. $x + \dfrac{1}{3} = \dfrac{5}{6}$

56. $m + \dfrac{5}{6} = \dfrac{9}{10}$

d Solve.

57. Monica spent $\frac{3}{4}$ hr listening to tapes of *NSync and the Backstreet Boys. She spent $\frac{1}{3}$ hr listening to *NSync. How many hours were spent listening to the Backstreet Boys?

58. As part of a fitness program, Deb swims $\frac{1}{2}$ mi every day. She has already swum $\frac{1}{5}$ mi. How much farther should Deb swim?

59. *Tire Tread.*   A new long-life tire has a tread depth of $\frac{3}{8}$ in. instead of a more typical $\frac{11}{32}$ in. How much deeper is the new tread depth?
Source: *Popular Science*

60. From a $\frac{4}{5}$-lb wheel of cheese, a $\frac{1}{4}$-lb piece was served. How much cheese remained on the wheel?

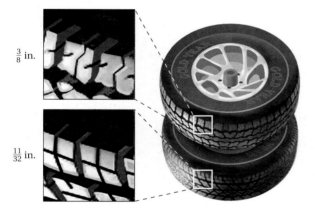

$\frac{3}{8}$ in.

$\frac{11}{32}$ in.

61. An Arby's franchise is owned by three people. One owns $\frac{7}{12}$ of the business and the second owns $\frac{1}{6}$. What part of the business does the third person own?

62. An estate was left to four children. One received $\frac{1}{4}$ of the estate, the second $\frac{1}{16}$, and the third $\frac{3}{8}$. How much did the fourth receive?

63. A server has a bottle containing $\frac{11}{12}$ cup of olive oil. He serves $\frac{1}{4}$ cup on a plate to a customer for bread dipping. How much remains in the bottle?

64. Jovan has an $\frac{11}{10}$-lb mixture of cashews and peanuts that includes $\frac{3}{5}$ lb of cashews. How many pounds of peanuts are in the mixture?

65. **D**w A fellow student made the following error:
$$\frac{8}{5} - \frac{8}{2} = \frac{8}{3}.$$
Find at least two ways to convince him of the mistake.

66. **D**w Explain how one could use pictures to convince someone that $\frac{7}{29}$ is larger than $\frac{13}{57}$.

Divide, if possible. If not possible, write "not defined." [1.6b], [2.3b]

67. $\dfrac{38}{38}$

68. $\dfrac{38}{0}$

69. $\dfrac{124}{0}$

70. $\dfrac{124}{31}$

Divide and simplify. [2.7b]

71. $\dfrac{3}{7} \div \dfrac{9}{4}$

72. $\dfrac{9}{10} \div \dfrac{3}{5}$

73. $7 \div \dfrac{1}{3}$

74. $\dfrac{1}{4} \div 8$

Solve. [2.6b]

75. A small box of Kellogg's cornflakes weighs $\frac{3}{4}$ lb. How much does 8 small boxes of cornflakes weigh?

76. A batch of fudge requires $\frac{3}{4}$ cup of sugar. How much sugar is needed to make 12 batches?

Solve.

77. ▦ $x + \dfrac{16}{323} = \dfrac{10}{187}$

78. ▦ $x + \dfrac{7}{253} = \dfrac{12}{299}$

79. A mountain climber, beginning at sea level, climbs $\frac{3}{5}$ km, descends $\frac{1}{4}$ km, climbs $\frac{1}{3}$ km, and then descends $\frac{1}{7}$ km. At what elevation does the climber finish?

Simplify. Use the rules for order of operations given in Section 1.9.

80. $\dfrac{2}{5} + \dfrac{1}{6} \div 3$

81. $\dfrac{7}{8} - \dfrac{1}{10} \times \dfrac{5}{6}$

82. $5 \times \dfrac{3}{7} - \dfrac{1}{7} \times \dfrac{4}{5}$

83. $\left(\dfrac{2}{3}\right)^2 + \left(\dfrac{3}{4}\right)^2$

84. A VCR can record up to 6 hr on one tape. It can also fill that same tape in either 4 hr or 2 hr when running at faster speeds. A tape is placed in the machine, which records for $\frac{1}{2}$ hr at the 4-hr speed and $\frac{3}{4}$ hr at the 2-hr speed. How much time is left on the tape to record at the 6-hr speed?

85. As part of a rehabilitation program, an athlete must swim and then walk a total of $\frac{9}{10}$ km each day. If one lap in the swimming pool is $\frac{3}{80}$ km, how far must the athlete walk after swimming 10 laps?

Use $<$, $>$, or $=$ for $\square$ to write a true sentence.

86. ▦ $\dfrac{12}{97} + \dfrac{67}{139} \; \square \; \dfrac{8167}{13{,}289}$

87. ▦ $\dfrac{37}{157} + \dfrac{19}{107} \; \square \; \dfrac{6941}{16{,}799}$

88. *Microsoft Interview.* The following is a question taken from an employment interview with Microsoft. Try to answer it.

"Given a gold bar that can be cut exactly twice and a contractor who must be paid one-seventh of a gold bar every day for seven days, how should the bar be cut?"
Source: *Fortune Magazine*, January 22, 2001

3.4 MIXED NUMERALS

Objectives

a. Convert between mixed numerals and fraction notation.

b. Divide whole numbers, writing the quotient as a mixed numeral.

a. Mixed Numerals

The following figure illustrates the use of a **mixed numeral** in daily life. The bolt shown is $2\frac{3}{8}$ in. long. The length is given as a whole-number part, 2, and a fractional part less than 1, $\frac{3}{8}$. We can represent the measurement of the bolt with fraction notation as $\frac{19}{8}$, but the meaning or interpretation of such a symbol is less understandable or visual.

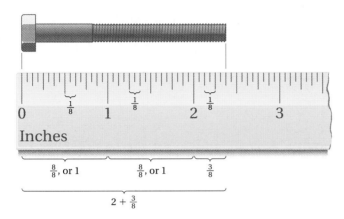

$$\frac{8}{8}, \text{ or } 1 \qquad \frac{8}{8}, \text{ or } 1 \qquad \frac{3}{8}$$

$$2 + \frac{3}{8}$$

A mixed numeral $2\frac{3}{8}$ represents a sum:

$$2\frac{3}{8} \quad \text{means} \quad 2 + \frac{3}{8}$$

This is a whole number.   This is a fraction less than 1.

EXAMPLES Convert to a mixed numeral.

1. $7 + \frac{2}{5} = 7\frac{2}{5}$

2. $4 + \frac{3}{10} = 4\frac{3}{10}$

Do Exercises 1–4.

The notation $2\frac{3}{4}$ has a plus sign left out. To aid in understanding, we sometimes write the missing plus sign.

EXAMPLES Convert to fraction notation.

3. $2\frac{3}{4} = 2 + \frac{3}{4}$ Inserting the missing plus sign

$= \frac{2}{1} + \frac{3}{4}$ $2 = \frac{2}{1}$

$= \frac{2}{1} \cdot \frac{4}{4} + \frac{3}{4}$ Finding a common denominator

$= \frac{8}{4} + \frac{3}{4} = \frac{11}{4}$

4. $4\frac{3}{10} = 4 + \frac{3}{10} = \frac{4}{1} + \frac{3}{10} = \frac{4}{1} \cdot \frac{10}{10} + \frac{3}{10} = \frac{40}{10} + \frac{3}{10} = \frac{43}{10}$

Convert to a mixed numeral.

1. $1 + \frac{2}{3} = \square\frac{\square}{\square}$

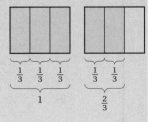

2. $2 + \frac{3}{4} = \square\frac{\square}{\square}$

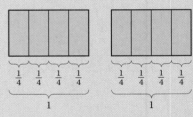

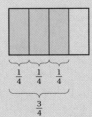

3. $8 + \frac{3}{4}$ **4.** $12 + \frac{2}{3}$

Answers on page A-6

187

Convert to fraction notation.

5. $4\dfrac{2}{5}$ **6.** $6\dfrac{1}{10}$

Convert to fraction notation. Use the faster method.

7. $4\dfrac{5}{6}$

8. $9\dfrac{1}{4}$

9. $20\dfrac{2}{3}$

Answers on page A-6

Do Exercises 5 and 6.

Let's now consider a faster method for converting a mixed numeral to fraction notation.

> To convert from a mixed numeral to fraction notation:
>
> (a) Multiply the whole number by the denominator: $4 \cdot 10 = 40$.
>
> (b) Add the result to the numerator: $40 + 3 = 43$.
>
> (c) Keep the denominator.
>
> $4\dfrac{3}{10} = \dfrac{43}{10}$

EXAMPLES   Convert to fraction notation.

5. $6\dfrac{2}{3} = \dfrac{20}{3}$ $6 \cdot 3 = 18,\ 18 + 2 = 20$

6. $8\dfrac{2}{9} = \dfrac{74}{9}$

7. $10\dfrac{7}{8} = \dfrac{87}{8}$

Do Exercises 7–9.

WRITING MIXED NUMERALS

We can find a mixed numeral for $\frac{5}{3}$ as follows:

$$\dfrac{5}{3} = \dfrac{3}{3} + \dfrac{2}{3} = 1 + \dfrac{2}{3} = 1\dfrac{2}{3}.$$

In terms of objects, we can think of $\frac{5}{3}$ as $\frac{3}{3}$, or 1, plus $\frac{2}{3}$, as shown below.

$$\dfrac{5}{3} = \qquad \dfrac{3}{3},\ \text{or } 1 \qquad + \qquad \dfrac{2}{3}$$

Fraction symbols like $\frac{5}{3}$ also indicate division; $\frac{5}{3}$ means $5 \div 3$. Let's divide the numerator by the denominator.

$$
\begin{array}{r}
1 \\
3\overline{)5} \\
3 \\
\hline
2
\end{array}
\longleftarrow 2 \div 3 = \tfrac{2}{3}
$$

Thus, $\frac{5}{3} = 1\frac{2}{3}$.

> To convert from fraction notation to a mixed numeral, divide.

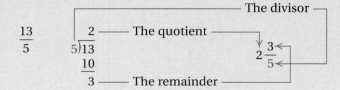

EXAMPLES   Convert to a mixed numeral.

8. $\dfrac{69}{10}$

$$10\overline{)69} \quad \dfrac{69}{10} = 6\dfrac{9}{10}$$
$$\underline{60}$$
$$9$$

9. $\dfrac{122}{8}$

$$8\overline{)122} \quad \dfrac{122}{8} = 15\dfrac{2}{8} = 15\dfrac{1}{4}$$
$$\underline{80}$$
$$42$$
$$\underline{40}$$
$$2$$

Do Exercises 10–12.

b   Writing Mixed Numerals for Quotients

It is quite common when dividing whole numbers to write the quotient using a mixed numeral. The remainder is the numerator of the fractional part of the mixed numeral.

EXAMPLE 10   Divide. Write a mixed numeral for the quotient.

$$7\overline{)6\ 3\ 4\ 1}$$

We first divide as usual.

$$
\begin{array}{r}
9\ 0\ 5 \\
7\overline{)6\ 3\ 4\ 1} \\
\underline{6\ 3\ 0\ 0} \\
4\ 1 \\
\underline{3\ 5} \\
6
\end{array}
$$

The answer is 905 R 6. We write a mixed numeral for the answer as follows:

$$905\dfrac{6}{7}.$$

The division $6341 \div 7$ can be expressed using fraction notation or a mixed numeral as follows:

$$\dfrac{6341}{7} = 905\dfrac{6}{7}.$$

Convert to a mixed numeral.

10. $\dfrac{7}{3}$

11. $\dfrac{11}{10}$

12. $\dfrac{110}{6}$

Answers on page A-6

Study Tips

FORMING A STUDY GROUP

Consider forming a study group with some of your fellow students. Exchange e-mail addresses, telephone numbers, and schedules so that you can coordinate study time for homework and tests.

Divide. Write a mixed numeral for the answer.

13. $6\overline{)4\ 8\ 4\ 6}$

14. $4\ 5\overline{)6\ 0\ 5\ 3}$

EXAMPLE 11   Divide. Write a mixed numeral for the answer.

$$4\ 2\overline{)8\ 9\ 1\ 5}$$

We first divide as usual.

$$
\begin{array}{r}
2\ 1\ 2 \\
4\ 2\overline{)8\ 9\ 1\ 5} \\
8\ 4\ 0\ 0 \\
\hline
5\ 1\ 5 \\
4\ 2\ 0 \\
\hline
9\ 5 \\
8\ 4 \\
\hline
1\ 1
\end{array}
\qquad
\frac{8915}{42} = 212\frac{11}{42}
$$

The answer is $212\frac{11}{42}$.

Do Exercises 13 and 14.

A fraction larger than 1, such as $\frac{27}{8}$, is sometimes referred to as an "improper" fraction. We will not use this terminology because notation such as $\frac{27}{8}$, $\frac{11}{9}$, and $\frac{89}{10}$ is quite "proper" and very common in algebra.

As we will see in subsequent sections, mixed numerals have many real-world applications, especially in the areas of garment manufacturing and carpentry, as well as many other fields.

CALCULATOR CORNER

Writing Quotients as Mixed Numerals   When using a calculator to divide whole numbers, we can express the result using a mixed numeral. To do so, we first find the quotient and the remainder as shown in the Calculator Corner on p. 56. The quotient is the whole-number part of the mixed numeral, the remainder is the numerator of the fractional part, and the divisor is the denominator of the fractional part. For example, on p. 56, we saw that $453 \div 15 = 30$ R 3. Then we can also write

$$453 \div 15 = 30\frac{3}{15} = 30\frac{1}{5}.$$

Exercises: Use a calculator to divide. Write the result as a mixed numeral.

1. $6\overline{)8\ 8\ 5\ 7}$ **6.** $3\ 2\overline{)2\ 3\ 4{,}5\ 6\ 7}$

2. $9\overline{)6\ 0\ 8\ 8}$ **7.** $4\ 5\overline{)6\ 0\ 3\ 3}$

3. $5\ 6\overline{)4\ 4{,}8\ 5\ 1}$ **8.** $2\ 1\ 3\overline{)5\ 6\ 7{,}9\ 8\ 8}$

4. $1\ 8\overline{)2\ 3\ 4{,}5\ 6\ 7}$ **9.** $1\ 1\ 2\overline{)4\ 0\ 0{,}0\ 0\ 3}$

5. $1\ 1\overline{)5\ 6\ 7{,}8\ 9\ 5}$ **10.** $9\ 0\ 8\overline{)1\ 1{,}2\ 3\ 4}$

a

1. *Garment Manufacturing.* A tailoring shop determines that for a certain size dress, it must use $3\frac{5}{8}$ yd of fabric that is 45 in. wide. To make the same dress with fabric that is 60 in. wide, it needs $2\frac{3}{4}$ yd. Convert $3\frac{5}{8}$ and $2\frac{3}{4}$ to fraction notation.

2. *Carpentry.* Dick Bonewitz, master carpenter, is making a display case according to the design below. Convert each mixed numeral to fraction notation.

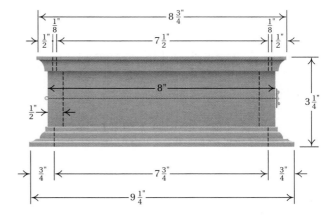

Convert to fraction notation.

3. $5\frac{2}{3}$

4. $3\frac{4}{5}$

5. $3\frac{1}{4}$

6. $6\frac{1}{2}$

7. $10\frac{1}{8}$

8. $20\frac{1}{5}$

9. $5\frac{1}{10}$

10. $9\frac{1}{10}$

11. $20\frac{3}{5}$

12. $30\frac{4}{5}$

13. $9\frac{5}{6}$

14. $8\frac{7}{8}$

15. $7\frac{3}{10}$

16. $6\frac{9}{10}$

17. $1\frac{5}{8}$

18. $1\frac{3}{5}$

19. $12\frac{3}{4}$

20. $15\frac{2}{3}$

21. $4\frac{3}{10}$

22. $5\frac{7}{10}$

23. $2\frac{3}{100}$

24. $5\frac{7}{100}$

25. $66\frac{2}{3}$

26. $33\frac{1}{3}$

27. $5\frac{29}{50}$

28. $84\frac{3}{8}$

Convert to a mixed numeral.

29. $\frac{18}{5}$

30. $\frac{17}{4}$

31. $\frac{14}{3}$

32. $\frac{39}{8}$

33. $\frac{27}{6}$

34. $\frac{30}{9}$

35. $\frac{57}{10}$

36. $\frac{89}{10}$

37. $\frac{53}{7}$

38. $\frac{59}{8}$

39. $\frac{45}{6}$

40. $\frac{50}{8}$

41. $\frac{46}{4}$

42. $\frac{39}{9}$

43. $\frac{12}{8}$

44. $\frac{28}{6}$

45. $\frac{757}{100}$

46. $\frac{467}{100}$

47. $\frac{345}{8}$

48. $\frac{223}{4}$

b   Divide. Write a mixed numeral for the answer.

49. $8\overline{)869}$

50. $3\overline{)2126}$

51. $5\overline{)3091}$

52. $9\overline{)9110}$

53. $21\overline{)852}$

54. $85\overline{)7672}$

55. $102\overline{)5612}$

56. $46\overline{)1081}$

57. ^{D}W Describe in your own words a method for rewriting a fraction as a mixed numeral.

58. ^{D}W Are the numbers $2\frac{1}{3}$ and $2 \cdot \frac{1}{3}$ equal? Why or why not?

SKILL MAINTENANCE

Multiply and simplify.   [2.4a], [2.6a]

59. $\frac{6}{5} \cdot 15$

60. $\frac{5}{12} \cdot 6$

61. $\frac{7}{10} \cdot \frac{5}{14}$

62. $\frac{1}{10} \cdot \frac{20}{5}$

Divide and simplify.   [2.7b]

63. $\frac{2}{3} \div \frac{1}{36}$

64. $28 \div \frac{4}{7}$

65. $200 \div \frac{15}{64}$

66. $\frac{3}{4} \div \frac{9}{16}$

SYNTHESIS

Write a mixed numeral.

67. ▦ $\frac{128,236}{541}$

68. ▦ $\frac{103,676}{349}$

69. $\frac{56}{7} + \frac{2}{3}$

70. $\frac{72}{12} + \frac{5}{6}$

71. There are $\frac{366}{7}$ weeks in a leap year.

72. There are $\frac{365}{7}$ weeks in a year.

ADDITION AND SUBTRACTION USING MIXED NUMERALS; APPLICATIONS

Objectives

a Add using mixed numerals.

b Subtract using mixed numerals.

c Solve applied problems involving addition and subtraction with mixed numerals.

a Addition Using Mixed Numerals

To find the sum $1\frac{5}{8} + 3\frac{1}{8}$, we first add the fractions. Then we add the whole numbers.

$$
\begin{array}{r}
1\dfrac{5}{8} = \\[2mm]
+\,3\dfrac{1}{8} = \\[2mm]
\hline
\dfrac{6}{8}
\end{array}
\qquad
\begin{array}{r}
1\dfrac{5}{8} \\[2mm]
+\,3\dfrac{1}{8} \\[2mm]
\hline
4\dfrac{6}{8} = 4\dfrac{3}{4}
\end{array}
$$

Simplifying

↑ Add the fractions.   ↑ Add the whole numbers.

Do Exercise 1.

1. Add.

$$
\begin{array}{r}
2\dfrac{3}{10} \\[2mm]
+\,5\dfrac{1}{10} \\
\hline
\end{array}
$$

EXAMPLE 1   Add: $5\frac{2}{3} + 3\frac{5}{6}$. Write a mixed numeral for the answer.

The LCD is 6.

$$
\begin{array}{r}
5\dfrac{2}{3}\cdot\dfrac{2}{2} = \quad 5\dfrac{4}{6} \\[2mm]
+\,3\dfrac{5}{6} = +\,3\dfrac{5}{6} \\[2mm]
\hline
8\dfrac{9}{6} = 8 + \dfrac{9}{6} \\[2mm]
= 8 + 1\dfrac{1}{2} \\[2mm]
= 9\dfrac{1}{2}
\end{array}
$$

To find a mixed numeral for $\frac{9}{6}$, we divide:

$$
\begin{array}{r}
1 \\
6\overline{)9} \\
6 \\
\hline
3
\end{array}
\qquad \frac{9}{6} = 1\frac{3}{6} = 1\frac{1}{2}
$$

$\frac{19}{2}$ is also a correct answer, but it is not a mixed numeral, which is what we are working with in Sections 3.4, 3.5, and 3.6.

Do Exercise 2.

2. Add.

$$
\begin{array}{r}
8\dfrac{2}{5} \\[2mm]
+\,3\dfrac{7}{10} \\
\hline
\end{array}
$$

EXAMPLE 2   Add: $10\frac{5}{6} + 7\frac{3}{8}$.

The LCD is 24.

$$
\begin{array}{r}
10\dfrac{5}{6}\cdot\dfrac{4}{4} = \quad 10\dfrac{20}{24} \\[2mm]
+\,7\dfrac{3}{8}\cdot\dfrac{3}{3} = +\,7\dfrac{9}{24} \\[2mm]
\hline
17\dfrac{29}{24} = 18\dfrac{5}{24}
\end{array}
$$

Do Exercise 3.

3. Add.

$$
\begin{array}{r}
9\dfrac{3}{4} \\[2mm]
+\,3\dfrac{5}{6} \\
\hline
\end{array}
$$

Answers on page A-7

b Subtraction Using Mixed Numerals

EXAMPLE 3 Subtract: $7\frac{3}{4} - 2\frac{1}{4}$.

$$
\begin{array}{r}
7\dfrac{3}{4} = \\[2mm]
-\ 2\dfrac{1}{4} = \\[2mm]
\hline
\dfrac{2}{4}
\end{array}
\qquad
\begin{array}{r}
7\dfrac{3}{4} \\[2mm]
-\ 2\dfrac{1}{4} \\[2mm]
\hline
5\dfrac{2}{4} = 5\dfrac{1}{2}
\end{array}
$$

↑   ↑   ↑ Simplifying

Subtract the   Subtract the
fractions.   whole numbers.

EXAMPLE 4 Subtract: $9\frac{4}{5} - 3\frac{1}{2}$.

The LCD is 10.

$$
\begin{array}{r}
9\dfrac{4}{5} \cdot \dfrac{2}{2} = 9\dfrac{8}{10} \\[3mm]
-\ 3\dfrac{1}{2} \cdot \dfrac{5}{5} = -\ 3\dfrac{5}{10} \\[3mm]
\hline
6\dfrac{3}{10}
\end{array}
$$

Do Exercises 4 and 5.

EXAMPLE 5 Subtract: $7\frac{1}{6} - 2\frac{1}{4}$.

The LCD is 12.

$$
\begin{array}{r}
7\dfrac{1}{6} \cdot \dfrac{2}{2} = 7\dfrac{2}{12} \\[3mm]
-\ 2\dfrac{1}{4} \cdot \dfrac{3}{3} = -\ 2\dfrac{3}{12}
\end{array}
$$

We cannot subtract $\frac{3}{12}$ from $\frac{2}{12}$.
We borrow 1, or $\frac{12}{12}$, from 7:
$7\frac{2}{12} = 6 + 1 + \frac{2}{12} = 6 + \frac{12}{12} + \frac{2}{12} = 6\frac{14}{12}$.

We can write this as

$$
\begin{array}{r}
7\dfrac{2}{12} = 6\dfrac{14}{12} \\[3mm]
-\ 2\dfrac{3}{12} = -\ 2\dfrac{3}{12} \\[3mm]
\hline
4\dfrac{11}{12}
\end{array}
$$

Do Exercise 6.

Subtract.

4. $10\dfrac{7}{8}$

    $-\ 9\dfrac{3}{8}$

5. $8\dfrac{2}{3}$

    $-\ 5\dfrac{1}{2}$

6. Subtract.

    $8\dfrac{1}{9}$

    $-\ 4\dfrac{5}{6}$

Answers on page A-7

7. Subtract.

$$5$$
$$-1\frac{1}{3}$$

EXAMPLE 6 Subtract: $12 - 9\frac{3}{8}$.

$$
\begin{array}{rcl}
12 & = & 11\dfrac{8}{8} \\[2mm]
-\ 9\dfrac{3}{8} & = & -\ 9\dfrac{3}{8} \\[2mm]
\hline
& & 2\dfrac{5}{8}
\end{array}
$$

$12 = 11 + 1 = 11 + \dfrac{8}{8} = 11\dfrac{8}{8}$

Do Exercise 7.

C Applications and Problem Solving

EXAMPLE 7 *Travel Distance.* Jim Lawler is a college textbook salesman for Addison-Wesley. On two business days, Jim drove $144\frac{9}{10}$ mi and $87\frac{1}{4}$ mi. What was the total distance driven?

1. Familiarize. We let $d =$ the total distance driven.

2. Translate. We translate as follows.

Distance driven first day	+	Distance driven second day	=	Total distance driven
$144\dfrac{9}{10}$	+	$87\dfrac{1}{4}$	=	d

3. Solve. The sentence tells us what to do. We add. The LCD is 20.

$$
\begin{array}{rcccl}
144\dfrac{9}{10} & = & 144\ \dfrac{9}{10}\cdot\dfrac{2}{2} & = & 144\dfrac{18}{20} \\[3mm]
+\ 87\dfrac{1}{4} & = & +\ \ 87\ \dfrac{1}{4}\cdot\dfrac{5}{5} & = & +\ \ 87\dfrac{5}{20} \\[3mm]
\hline
& & & & 231\dfrac{23}{20} = 232\dfrac{3}{20}
\end{array}
$$

Thus, $d = 232\frac{3}{20}$ mi.

4. Check. We check by repeating the calculation. We also note that the answer is larger than either of the distances driven, which means that the answer is reasonable.

5. State. The total distance driven was $232\frac{3}{20}$ mi.

Do Exercise 8.

8. Executive Car Care sold two pieces of synthetic leather, one $6\frac{1}{4}$ yd long and the other $10\frac{5}{6}$ yd long. What was the total length of the leather?

Answers on page A-7

EXAMPLE 8 *NCAA Football Goalposts.* In college football, the distance between goalposts was reduced from $23\frac{1}{3}$ ft to $18\frac{1}{2}$ ft. By how much was it reduced?

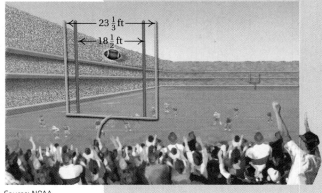

Source: NCAA

1. **Familiarize.** We let d = the amount of reduction and make a drawing to illustrate the situation.

2. **Translate.** We translate as follows.

$$
\underbrace{\text{Former} \atop \text{distance}} \quad - \quad \underbrace{\text{New} \atop \text{distance}} \quad = \quad \underbrace{\text{Amount of} \atop \text{reduction}}
$$

$$
23\frac{1}{3} \quad - \quad 18\frac{1}{2} \quad = \quad d
$$

3. **Solve.** To solve the equation, we carry out the subtraction. The LCD is 6.

$$
23\frac{1}{3} = \quad 23\,\frac{1}{3}\cdot\frac{2}{2} = \quad 23\frac{2}{6} = \quad 22\frac{8}{6}
$$

$$
- \,18\frac{1}{2} = -\,18\,\frac{1}{2}\cdot\frac{3}{3} = -\,18\frac{3}{6} = -\,18\frac{3}{6}
$$

$$
4\frac{5}{6}
$$

Thus, $d = 4\frac{5}{6}$ ft.

4. **Check.** To check, we add the reduction to the new distance:

$$
18\frac{1}{2} + 4\frac{5}{6} = 18\frac{3}{6} + 4\frac{5}{6}
$$

$$
= 22\frac{8}{6}
$$

$$
= 23\frac{2}{6}
$$

$$
= 23\frac{1}{3}.
$$

This checks.

5. **State.** The reduction in the goalpost distance was $4\frac{5}{6}$ ft.

Do Exercise 9.

9. **Damascus Blade.** The Damascus blade of a folding knife is $3\frac{3}{4}$ in. long. The same blade in an ATS-34 is $4\frac{1}{8}$ in. long. How many inches longer is the ATS-34 blade?
Source: *Blade Magazine* 23, no. 10, October 1996: 26–27

$\leftarrow\!\!-\; 3\frac{3}{4}\text{ in.}\;-\!\!\rightarrow$

$\leftarrow\!\!-\; 4\frac{1}{8}\text{ in.}\;-\!\!\rightarrow$

Answer on page A-7

10. There are $20\frac{1}{3}$ gal of water in a rainbarrel; $5\frac{3}{4}$ gal are poured out and $8\frac{2}{3}$ gal are returned after a heavy rainfall. How many gallons of water are then in the barrel?

MULTISTEP PROBLEMS

EXAMPLE 9 *Carpentry.* The following diagram shows the layout for the construction of a desk drawer. Find the missing length a.

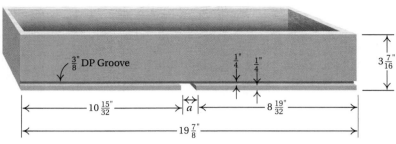

Middle Drawer / Back Layout

1. **Familiarize.** The length a is shown in the drawing.

2. **Translate.** From the drawing, we see that the length a is the full length of the drawer minus the sum of the smaller lengths, $10\frac{15}{32}''$ and $8\frac{19}{32}''$. Thus we have

$$a = 19\frac{7}{8} - \left(10\frac{15}{32} + 8\frac{19}{32}\right).$$

3. **Solve.** This is a two-step problem.

   a) We first add the two lengths, $10\frac{15}{32}$ and $8\frac{19}{32}$.

$$10\frac{15}{32}$$
$$+\ 8\frac{19}{32}$$
$$\overline{\qquad}$$
$$18\frac{34}{32} = 18\frac{17}{16} = 19\frac{1}{16}$$

   b) Next, we subtract $19\frac{1}{16}$ from $19\frac{7}{8}$.

$$19\frac{7}{8} = \quad 19\frac{14}{16}$$
$$-19\frac{1}{16} = -19\frac{1}{16}$$
$$\overline{\qquad}$$
$$\frac{13}{16} = a$$

4. **Check.** We check by repeating the calculation, or adding the three measures:

$$10\frac{15}{32} + 8\frac{19}{32} + \frac{13}{16} = 19\frac{7}{8}.$$

5. **State.** The length a in the diagram is $\frac{13}{16}''$.

Do Exercise 10.

Answer on page A-7

3.5
EXERCISE SET

For Extra Help

Digital Video
Tutor CD 2
Videotape 6

InterAct
Math

Math Tutor
Center

MathXL

MyMathLab

a Add. Write a mixed numeral for the answer.

1. $\begin{array}{r} 20 \\ +\ 8\frac{3}{4} \\ \hline \end{array}$

2. $\begin{array}{r} 37 \\ +\ 18\frac{2}{3} \\ \hline \end{array}$

3. $\begin{array}{r} 129\frac{7}{8} \\ +\ \ 56 \\ \hline \end{array}$

4. $\begin{array}{r} 2003\frac{4}{11} \\ +\ \ \ 59 \\ \hline \end{array}$

5. $\begin{array}{r} 2\frac{7}{8} \\ +\ 3\frac{5}{8} \\ \hline \end{array}$

6. $\begin{array}{r} 4\frac{5}{6} \\ +\ 3\frac{5}{6} \\ \hline \end{array}$

7. $1\frac{1}{4} + 1\frac{2}{3}$

8. $4\frac{1}{3} + 5\frac{2}{9}$

9. $\begin{array}{r} 8\frac{3}{4} \\ +\ 5\frac{5}{6} \\ \hline \end{array}$

10. $\begin{array}{r} 4\frac{3}{8} \\ +\ 6\frac{5}{12} \\ \hline \end{array}$

11. $\begin{array}{r} 3\frac{2}{5} \\ +\ 8\frac{7}{10} \\ \hline \end{array}$

12. $\begin{array}{r} 5\frac{1}{2} \\ +\ 3\frac{7}{10} \\ \hline \end{array}$

13. $\begin{array}{r} 5\frac{3}{8} \\ +\ 10\frac{5}{6} \\ \hline \end{array}$

14. $\begin{array}{r} \frac{5}{8} \\ +\ 1\frac{5}{6} \\ \hline \end{array}$

15. $\begin{array}{r} 12\frac{4}{5} \\ +\ \ 8\frac{7}{10} \\ \hline \end{array}$

16. $\begin{array}{r} 15\frac{5}{8} \\ +\ 11\frac{3}{4} \\ \hline \end{array}$

17. $\begin{array}{r} 14\frac{5}{8} \\ +\ 13\frac{1}{4} \\ \hline \end{array}$

18. $\begin{array}{r} 16\frac{1}{4} \\ +\ 15\frac{7}{8} \\ \hline \end{array}$

19. $\begin{array}{r} 7\frac{1}{8} \\ 9\frac{2}{3} \\ +\ 10\frac{3}{4} \\ \hline \end{array}$

20. $\begin{array}{r} 45\frac{2}{3} \\ 31\frac{3}{5} \\ +\ 12\frac{1}{4} \\ \hline \end{array}$

b Subtract. Write a mixed numeral for the answer.

21. $\begin{array}{r} 4\frac{1}{5} \\ -\ 2\frac{3}{5} \\ \hline \end{array}$

22. $\begin{array}{r} 5\frac{1}{8} \\ -\ 2\frac{3}{8} \\ \hline \end{array}$

23. $6\frac{3}{5} - 2\frac{1}{2}$

24. $7\frac{2}{3} - 6\frac{1}{2}$

25. $\begin{array}{r} 34\frac{1}{3} \\ -\ 12\frac{5}{8} \\ \hline \end{array}$

26. $\begin{array}{r} 23\frac{5}{16} \\ -\ 16\frac{3}{4} \\ \hline \end{array}$

27. $\begin{array}{r} 21 \\ -\ 8\frac{3}{4} \\ \hline \end{array}$

28. $\begin{array}{r} 42 \\ -\ 3\frac{7}{8} \\ \hline \end{array}$

29.
$$\begin{array}{r} 34 \\ -\ 18\frac{5}{8} \\ \hline \end{array}$$

30.
$$\begin{array}{r} 23 \\ -\ 19\frac{3}{4} \\ \hline \end{array}$$

31.
$$\begin{array}{r} 21\frac{1}{6} \\ -\ 13\frac{3}{4} \\ \hline \end{array}$$

32.
$$\begin{array}{r} 42\frac{1}{10} \\ -\ 23\frac{7}{12} \\ \hline \end{array}$$

33.
$$\begin{array}{r} 14\frac{1}{8} \\ -\ \ \ \frac{3}{4} \\ \hline \end{array}$$

34.
$$\begin{array}{r} 28\frac{1}{6} \\ -\ \ 5 \\ \hline \end{array}$$

35.
$$\begin{array}{r} 25\frac{1}{9} \\ -\ 13\frac{5}{6} \\ \hline \end{array}$$

36.
$$\begin{array}{r} 23\frac{5}{16} \\ -\ 14\frac{7}{12} \\ \hline \end{array}$$

C Solve.

37. *Sewing from a Pattern.* Suppose you want to make an outfit in size 8. Using 45-in. fabric, you need $1\frac{3}{8}$ yd for the dress, $\frac{5}{8}$ yd of contrasting fabric for the band at the bottom, and $3\frac{3}{8}$ yd for the jacket. How many yards in all of 45-in. fabric are needed to make the outfit?

38. *Sewing from a Pattern.* Suppose you want to make an outfit in size 12. Using 45-in. fabric, you need $2\frac{3}{4}$ yd for the dress and $3\frac{1}{2}$ yd for the jacket. How many yards in all of 45-in. fabric are needed to make the outfit?

39. For a family barbecue, Jason bought packages of hamburger weighing $1\frac{2}{3}$ lb and $5\frac{3}{4}$ lb. What was the total weight of the meat?

40. Marsha's Butcher Shop sold packages of sliced turkey breast weighing $1\frac{1}{3}$ lb and $4\frac{3}{5}$ lb. What was the total weight of the meat?

CHAPTER 3: Fraction Notation and
Mixed Numerals

41. Tara is 66 in. tall and her son is $59\frac{7}{12}$ in. tall. How much taller is Tara?

42. Nicholas is $73\frac{2}{3}$ in. tall and his daughter is $71\frac{5}{16}$ in. tall. How much taller is Nicholas?

43. A plumber uses pipes of lengths $10\frac{5}{16}$ in. and $8\frac{3}{4}$ in. in the installation of a sink. How much pipe was used?

$8\frac{3}{4}$ in.

$10\frac{5}{16}$ in.

44. *Writing Supplies.* The standard pencil is $6\frac{7}{8}$ in. wood and $\frac{1}{2}$ in. eraser. What is the total length of the standard pencil?
Source: Eberhard Faber American

$6\frac{7}{8}$ in.

$\frac{1}{2}$ in.

45. Kim Park is a computer technician. One day, she drove $180\frac{7}{10}$ mi away from Los Angeles for a service call. The next day, she drove $85\frac{1}{2}$ mi back toward Los Angeles for another service call. How far was she from Los Angeles?

46. Pilar is $4\frac{1}{2}$ in. taller than her daughter Teresa. Teresa is $66\frac{2}{3}$ in. tall. How tall is Pilar?

47. *Book Size.* One standard book size is $8\frac{1}{2}$ in. by $9\frac{3}{4}$ in. What is the total distance around (perimeter of) the front cover of such a book?

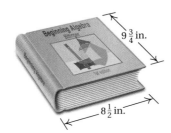

$9\frac{3}{4}$ in.

$8\frac{1}{2}$ in.

48. *Copier Paper.* A standard sheet of copier paper is $8\frac{1}{2}$ in. by 11 in. What is the total distance around (perimeter of) the paper?

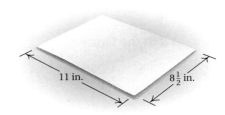

11 in.

$8\frac{1}{2}$ in.

49. *Carpentry.* When cutting wood with a saw, a carpenter must take into account the thickness of the saw blade. Suppose that from a piece of wood 36 in. long, a carpenter cuts a $15\frac{3}{4}$-in. length with a saw blade that is $\frac{1}{8}$ in. in thickness. How long is the piece that remains?

50. *Painting.* When redecorating, a painter used $1\frac{3}{4}$ gal of paint for the living room and $1\frac{1}{3}$ gal for the family room. How much paint was used in all?

51. Rene is $5\frac{1}{4}$ in. taller than his son, who is $72\frac{5}{6}$ in. tall. How tall is Rene?

52. A Boeing 767 flew 640 mi on a nonstop flight. On the return flight, it landed after having flown $320\frac{3}{10}$ mi. How far was the plane from its original point of departure?

53. *Interior Design.* Sue, an interior designer, worked $10\frac{1}{2}$ hr over a three-day period. If Sue worked $2\frac{1}{2}$ hr on the first day and $4\frac{1}{5}$ hr on the second, how many hours did Sue work on the third day?

54. *Painting.* Geri had $3\frac{1}{2}$ gal of paint. It took $2\frac{3}{4}$ gal to paint the family room. It was estimated that it would take $2\frac{1}{4}$ gal to paint the living room. How much more paint was needed?

Find the perimeter of (distance around) the figure.

55.

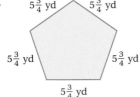

$5\frac{3}{4}$ yd $5\frac{3}{4}$ yd

$5\frac{3}{4}$ yd $5\frac{3}{4}$ yd

$5\frac{3}{4}$ yd

56.

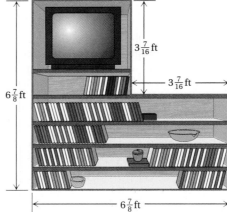

$3\frac{7}{16}$ ft

$3\frac{7}{16}$ ft

$6\frac{7}{8}$ ft

$6\frac{7}{8}$ ft

Find the length *d* in the figure.

57.

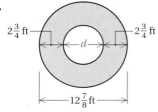

$2\frac{3}{4}$ ft   *d* $2\frac{3}{4}$ ft

$12\frac{7}{8}$ ft

58.

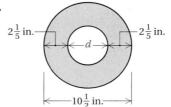

$2\frac{1}{5}$ in.   *d* $2\frac{1}{5}$ in.

$10\frac{1}{2}$ in.

59. Find the smallest length of a bolt that will pass through a piece of tubing with an outside diameter of $\frac{1}{2}$ in., a washer $\frac{1}{16}$ in. thick, a piece of tubing with a $\frac{3}{4}$-in. outside diameter, another washer, and a nut $\frac{3}{16}$ in. thick.

60. The front of the stage at the Lagrange Town Hall is $6\frac{1}{2}$ yd long. If renovation work succeeds in adding $2\frac{3}{4}$ yd in length, how long is the renovated stage?

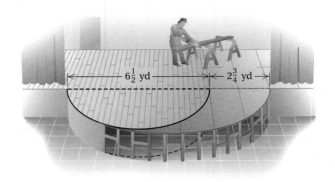

61. ^{D}W Write a problem for a classmate to solve. Design the problem so the solution is "The larger package holds $4\frac{1}{2}$ oz more than the smaller package."

62. ^{D}W Is the sum of two mixed numerals always a mixed numeral? Why or why not?

SKILL MAINTENANCE

Solve.

63. Rick's Market prepackages Swiss cheese in $\frac{3}{4}$-lb packages. How many packages can be made from a 12-lb slab of cheese? [2.7d]

64. Holstein's Dairy produced 4578 oz of milk one morning. How many 16-oz cartons were filled? How much milk was left over? [1.8a]

Determine whether the first number is divisible by the second. [2.2a]

65. 9993 by 3

66. 9993 by 9

67. 2345 by 9

68. 2345 by 5

69. 2335 by 10

70. 7764 by 6

71. 18,888 by 8

72. 18,888 by 4

73. Multiply and simplify: $\dfrac{15}{9} \cdot \dfrac{18}{39}$. [2.6a]

74. Divide and simplify: $\dfrac{12}{25} \div \dfrac{24}{5}$. [2.7b]

SYNTHESIS

Calculate each of the following. Write the result as a mixed numeral.

75. ▦ $3289\frac{1047}{1189} + 5278\frac{32}{41}$

76. ▦ $5798\frac{17}{53} - 3909\frac{1957}{2279}$

77. A post for a pier is 29 ft long. Half of the post extends above the water's surface and $8\frac{3}{4}$ ft of the post is buried in mud. How deep is the water at that point?

78. Solve: $47\frac{2}{3} + n = 56\frac{1}{4}$.

Objectives

a Multiply using mixed numerals.

b Divide using mixed numerals.

c Solve applied problems involving multiplication and division with mixed numerals.

1. Multiply: $6 \cdot 3\frac{1}{3}$.

2. Multiply: $2\frac{1}{2} \cdot \frac{3}{4}$.

3. Multiply: $2 \cdot 6\frac{2}{5}$.

4. Multiply: $3\frac{1}{3} \cdot 2\frac{1}{2}$.

Answers on page A-7

a Multiplication Using Mixed Numerals

Carrying out addition and subtraction with mixed numerals is usually easier if the numbers are left as mixed numerals. With multiplication and division, however, it is easier to convert the numbers first to fraction notation.

> ### MULTIPLICATION USING MIXED NUMERALS
>
> To multiply using mixed numerals, first convert to fraction notation. Then multiply with fraction notation and convert the answer back to a mixed numeral, if appropriate.

EXAMPLE 1 Multiply: $6 \cdot 2\frac{1}{2}$.

$$6 \cdot 2\frac{1}{2} = \frac{6}{1} \cdot \frac{5}{2} = \frac{6 \cdot 5}{1 \cdot 2} = \frac{2 \cdot 3 \cdot 5}{2 \cdot 1} = \frac{2}{2} \cdot \frac{3 \cdot 5}{1} = 15$$

Note that fraction notation is needed to carry out the multiplication.

Do Exercise 1.

EXAMPLE 2 Multiply: $3\frac{1}{2} \cdot \frac{3}{4}$.

$$3\frac{1}{2} \cdot \frac{3}{4} = \frac{7}{2} \cdot \frac{3}{4} = \frac{21}{8} = 2\frac{5}{8}$$

Here we write fraction notation.

Do Exercise 2.

EXAMPLE 3 Multiply: $8 \cdot 4\frac{2}{3}$.

$$8 \cdot 4\frac{2}{3} = \frac{8}{1} \cdot \frac{14}{3} = \frac{112}{3} = 37\frac{1}{3}$$

Do Exercise 3.

EXAMPLE 4 Multiply: $2\frac{1}{4} \cdot 3\frac{2}{5}$.

$$2\frac{1}{4} \cdot 3\frac{2}{5} = \frac{9}{4} \cdot \frac{17}{5} = \frac{153}{20} = 7\frac{13}{20}$$

> **CAUTION!**
>
> $2\frac{1}{4} \cdot 3\frac{2}{5} \neq 6\frac{2}{20}$. A common error is to multiply the whole numbers and then the fractions. This does not give the correct answer, $7\frac{13}{20}$, which is found by converting first to fraction notation.

Do Exercise 4.

b Division Using Mixed Numerals

The division $1\frac{1}{2} \div \frac{1}{6}$ is shown here. *Think*: "How many $\frac{1}{6}$'s are in $1\frac{1}{2}$?"

$$1\frac{1}{2} \div \frac{1}{6} = \frac{3}{2} \div \frac{1}{6} = \frac{3}{2} \cdot 6$$

$$= \frac{3 \cdot 6}{2} = \frac{3 \cdot 3 \cdot 2}{2 \cdot 1} = \frac{3 \cdot 3}{1} \cdot \frac{2}{2} = \frac{3 \cdot 3}{1} \cdot 1 = 9$$

DIVISION USING MIXED NUMERALS

To divide using mixed numerals, first write fraction notation. Then divide with fraction notation and convert the answer back to a mixed numeral, if appropriate.

EXAMPLE 5 Divide: $32 \div 3\frac{1}{5}$.

$$32 \div 3\frac{1}{5} = \frac{32}{1} \div \frac{16}{5}$$

$$= \frac{32}{1} \cdot \frac{5}{16} = \frac{32 \cdot 5}{1 \cdot 16} = \frac{2 \cdot 16 \cdot 5}{1 \cdot 16} = \frac{16}{16} \cdot \frac{2 \cdot 5}{1} = 10$$

⬆ Remember to multiply by the reciprocal.

Do Exercise 5.

EXAMPLE 6 Divide: $35 \div 4\frac{1}{3}$.

$$35 \div 4\frac{1}{3} = \frac{35}{1} \div \frac{13}{3} = \frac{35}{1} \cdot \frac{3}{13} = \frac{105}{13} = 8\frac{1}{13}$$

Do Exercise 6.

EXAMPLE 7 Divide: $2\frac{1}{3} \div 1\frac{3}{4}$.

$$2\frac{1}{3} \div 1\frac{3}{4} = \frac{7}{3} \div \frac{7}{4} = \frac{7}{3} \cdot \frac{4}{7} = \frac{7 \cdot 4}{7 \cdot 3} = \frac{7}{7} \cdot \frac{4}{3} = 1 \cdot \frac{4}{3} = \frac{4}{3} = 1\frac{1}{3}$$

(**CAUTION!**)

The reciprocal of $1\frac{3}{4}$ is *not* $1\frac{4}{3}$!

EXAMPLE 8 Divide: $1\frac{3}{5} \div 3\frac{1}{3}$.

$$1\frac{3}{5} \div 3\frac{1}{3} = \frac{8}{5} \div \frac{10}{3} = \frac{8}{5} \cdot \frac{3}{10} = \frac{2 \cdot 4 \cdot 3}{5 \cdot 2 \cdot 5} = \frac{2}{2} \cdot \frac{4 \cdot 3}{5 \cdot 5} = 1 \cdot \frac{4 \cdot 3}{5 \cdot 5} = \frac{12}{25}$$

Do Exercises 7 and 8.

5. Divide: $84 \div 5\frac{1}{4}$.

6. Divide: $26 \div 3\frac{1}{2}$.

Divide.

7. $2\frac{1}{4} \div 1\frac{1}{5}$

8. $1\frac{3}{4} \div 2\frac{1}{2}$

Answers on page A-7

9. Kyle's pickup truck travels on an interstate highway at 65 mph for $3\frac{1}{2}$ hr. How far does it travel?

Margaret Grace Bittinger, age 2

10. Holly's minivan travels 302 mi on $15\frac{1}{10}$ gal of gas. How many miles per gallon did it get?

Answers on page A-7

C   Applications and Problem Solving

EXAMPLE 9   *Cassette Tape Music.*   The tape in an audio cassette is played at a rate of $1\frac{7}{8}$ in. per second. A child has destroyed 30 in. of tape. How many seconds of music have been lost?

1. Familiarize.   We can make a drawing to help us visualize the situation.

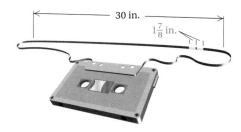

Since each $1\frac{7}{8}$ in. of tape represents 1 sec of lost music, the question can be regarded as asking how many times 30 can be divided by $1\frac{7}{8}$. We let $t =$ the number of seconds of music lost.

2. Translate.   The situation corresponds to a division sentence:

$$t = 30 \div 1\frac{7}{8}.$$

3. Solve.   To solve the equation, we perform the division:

$$t = 30 \div 1\frac{7}{8}$$
$$= \frac{30}{1} \div \frac{15}{8}$$
$$= \frac{30}{1} \cdot \frac{8}{15}$$
$$= \frac{15 \cdot 2 \cdot 8}{1 \cdot 15}$$
$$= \frac{15}{15} \cdot \frac{2 \cdot 8}{1}$$
$$= 16.$$

4. Check.   We check by multiplying. If 16 sec of music were lost, then

$$16 \cdot 1\frac{7}{8} = \frac{16}{1} \cdot \frac{15}{8}$$
$$= \frac{8 \cdot 2 \cdot 15}{1 \cdot 8}$$
$$= \frac{8}{8} \cdot \frac{2 \cdot 15}{1} = 30 \text{ in.}$$

of tape were destroyed. A quicker, but less precise, check can be made by noting that $1\frac{7}{8} \approx 2$. Then $16 \cdot 1\frac{7}{8} \approx 16 \cdot 2 = 32 \approx 30$. Our answer checks.

5. State.   The cassette has lost 16 sec of music.

Do Exercises 9 and 10.

EXAMPLE 10 *Mirror Area.* The mirror-backed candle shelf, shown below with a carpenter's diagram, was designed and built by Harry Cooper. Such shelves were popular in Colonial times because the mirror provided extra lighting from the candle. A rectangular walnut board is used to make the back of the shelf. Find the area of the original board and the amount left over after the mirror has been cut out.

Source: Popular Science Woodworking Projects

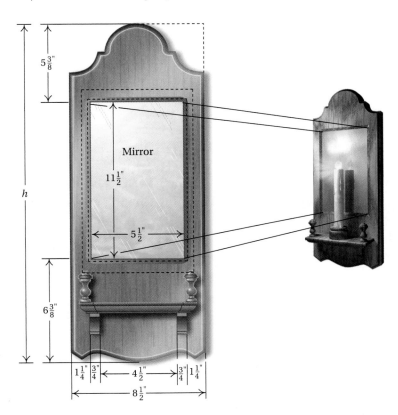

11. A room measures $22\frac{1}{2}$ ft by $15\frac{1}{2}$ ft. A 9-ft by 12-ft Oriental rug is placed in the center of the room. How much area is not covered by the rug?

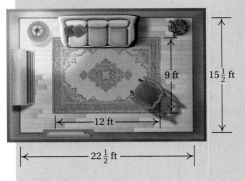

1. **Familiarize.** Refer to the figure above. We let $h =$ the height of the back of the shelf and $B =$ the area of the original board. We know the width of the original board, $8\frac{1}{2}$". (Remember, $8\frac{1}{2}$" means $8\frac{1}{2}$ in.) We let $A =$ the area left over after the mirror has been cut out.

2. **Translate.** This is a multistep problem. To find B, which equals $8\frac{1}{2} \cdot h$, we first need to calculate h. We read the dimensions $5\frac{3}{8}$", $11\frac{1}{2}$", and $6\frac{3}{8}$" from the diagram and add them to find h:

$$h = 5\frac{3}{8} + 11\frac{1}{2} + 6\frac{3}{8}.$$

The dimensions of the mirror are $11\frac{1}{2}$" and $5\frac{1}{2}$". Then A is the area of the original board minus the area of the mirror. That is,

$$A = B - 11\frac{1}{2} \cdot 5\frac{1}{2}.$$

Answer on page A-7

3. Solve. We carry out each calculation as follows:

$$h = 5\frac{3}{8} + 11\frac{1}{2} + 6\frac{3}{8} \qquad\qquad B = 8\frac{1}{2} \cdot h$$

$$= 5\frac{3}{8} + 11\frac{4}{8} + 6\frac{3}{8} \qquad\qquad = 8\frac{1}{2} \cdot 23\frac{1}{4} = \frac{17}{2} \cdot \frac{93}{4}$$

$$= 22\frac{10}{8} = 22\frac{5}{4} = 23\frac{1}{4}; \qquad\qquad = \frac{1581}{8} = 197\frac{5}{8};$$

$$A = B - 11\frac{1}{2} \cdot 5\frac{1}{2}$$

$$= 197\frac{5}{8} - 11\frac{1}{2} \cdot 5\frac{1}{2}$$

$$= 197\frac{5}{8} - \frac{23}{2} \cdot \frac{11}{2} = 197\frac{5}{8} - \frac{253}{4}$$

$$= 197\frac{5}{8} - 63\frac{1}{4} = 197\frac{5}{8} - 63\frac{2}{8} = 134\frac{3}{8}.$$

4. Check. We perform a check by repeating the calculations.

5. State. The area of the original board is $197\frac{5}{8}$ in². The area left over is $134\frac{3}{8}$ in².

Do Exercise 11 on the preceding page.

CALCULATOR CORNER

Operations on Fractions and Mixed Numerals Fraction calculators can add, subtract, multiply, and divide fractions and mixed numerals. The $\boxed{a_{b/c}}$ key is used to enter fractions and mixed numerals. To find $\frac{3}{4} + \frac{1}{2}$, for example, we press $\boxed{3}\ \boxed{a_{b/c}}\ \boxed{4}\ \boxed{+}\ \boxed{1}\ \boxed{a_{b/c}}\ \boxed{2}\ \boxed{=}$. Note that 3/4 and 1/2 appear on the display as $\boxed{\quad 3\ \lrcorner 4\quad}$ and $\boxed{\quad 1\ \lrcorner 2\quad}$, respectively. The result is given as the mixed numeral $1\frac{1}{4}$ and is displayed as $\boxed{\quad 1\ \lrcorner 1\ \lrcorner 4\quad}$. Fraction results that are greater than 1 are always displayed as mixed numerals. To express this result as a fraction, we press $\boxed{\text{SHIFT}}$ $\boxed{d/c}$. We get $\boxed{\quad 5\ \lrcorner 4\quad}$, or 5/4.

To find $3\frac{2}{3} \cdot 4\frac{1}{5}$, we press $\boxed{3}\ \boxed{a_{b/c}}\ \boxed{2}\ \boxed{a_{b/c}}\ \boxed{3}\ \boxed{\times}\ \boxed{4}\ \boxed{a_{b/c}}\ \boxed{1}\ \boxed{a_{b/c}}\ \boxed{5}\ \boxed{=}$. The calculator displays $\boxed{\quad 15\ \lrcorner 2\ \lrcorner 5\quad}$, so the product is $15\frac{2}{5}$.

Some calculators are capable of displaying mixed numerals in the way in which we write them, as shown below.

Exercises: Perform each calculation. Give the answer in fraction notation.

1. $\dfrac{1}{3} + \dfrac{1}{4}$ **2.** $\dfrac{7}{5} - \dfrac{3}{10}$

3. $\dfrac{15}{4} \cdot \dfrac{7}{12}$ **4.** $\dfrac{4}{5} \div \dfrac{8}{3}$

Perform each calculation. Give the answer as a mixed numeral.

5. $4\dfrac{1}{3} + 5\dfrac{4}{5}$ **6.** $9\dfrac{2}{7} - 8\dfrac{1}{4}$

7. $2\dfrac{1}{3} \cdot 4\dfrac{3}{5}$ **8.** $10\dfrac{7}{10} \div 3\dfrac{5}{6}$

For Extra Help

EXERCISE SET

a   Multiply. Write a mixed numeral for the answer.

1. $8 \cdot 2\frac{5}{6}$

2. $5 \cdot 3\frac{3}{4}$

3. $3\frac{5}{8} \cdot \frac{2}{3}$

4. $6\frac{2}{3} \cdot \frac{1}{4}$

5. $3\frac{1}{2} \cdot 2\frac{1}{3}$

6. $4\frac{1}{5} \cdot 5\frac{1}{4}$

7. $3\frac{2}{5} \cdot 2\frac{7}{8}$

8. $2\frac{3}{10} \cdot 4\frac{2}{5}$

9. $4\frac{7}{10} \cdot 5\frac{3}{10}$

10. $6\frac{3}{10} \cdot 5\frac{7}{10}$

11. $20\frac{1}{2} \cdot 10\frac{1}{5} \cdot 4\frac{2}{3}$

12. $21\frac{1}{3} \cdot 11\frac{1}{3} \cdot 3\frac{5}{8}$

b   Divide. Write a mixed numeral for the answer.

13. $20 \div 3\frac{1}{5}$

14. $18 \div 2\frac{1}{4}$

15. $8\frac{2}{5} \div 7$

16. $3\frac{3}{8} \div 3$

17. $4\frac{3}{4} \div 1\frac{1}{3}$

18. $5\frac{4}{5} \div 2\frac{1}{2}$

19. $1\frac{7}{8} \div 1\frac{2}{3}$

20. $4\frac{3}{8} \div 2\frac{5}{6}$

21. $5\dfrac{1}{10} \div 4\dfrac{3}{10}$　　　　**22.** $4\dfrac{1}{10} \div 2\dfrac{1}{10}$　　　　**23.** $20\dfrac{1}{4} \div 90$　　　　**24.** $12\dfrac{1}{2} \div 50$

C　Solve.

25. *Home Furnishings.* Each shelf in June's entertainment center is 27 in. long. A videocassette is $1\dfrac{1}{8}$ in. thick. How many cassettes can she place on each shelf?

26. *Exercise.* At one point during an aerobics class at Ray's health club, Kea's bicycle wheel was completing $76\dfrac{2}{3}$ revolutions per minute. How many revolutions did the wheel complete in 6 min?

27. *Sodium Consumption.* The average American woman consumes $1\dfrac{1}{3}$ tsp of sodium each day. How much sodium do 10 average American women consume in one day?
Source: *Nutrition Action Health Letter,* March 1994, p. 6. 1875 Connecticut Ave., N.W., Washington, DC 20009-5728

28. *Aeronautics.* Most space shuttles orbit the earth once every $1\dfrac{1}{2}$ hr. How many orbits are made every 24 hr?

29. *Weight of Water.* The weight of water is $62\dfrac{1}{2}$ lb per cubic foot. What is the weight of $5\dfrac{1}{2}$ cubic feet of water?

30. *Weight of Water.* The weight of water is $62\dfrac{1}{2}$ lb per cubic foot. What is the weight of $2\dfrac{1}{4}$ cubic feet of water?

31. *Fruit Cocktail Bars.* Listed below is the recipe for fruit cocktail bars. What are the ingredients for $\frac{1}{2}$ recipe? for 3 recipes?

Source: Reprinted with permission from Taste of Home, Greendale WI, www.tasteofhome.com

Fruit Cocktail Bars

1-1/2	cups sugar
2	eggs
1	can (17 ounces) fruit cocktail, undrained
1	teaspoon vanilla extract
2-1/4	cups all-purpose flour
1-1/2	teaspoons baking soda
1	teaspoon salt
1-1/3	cups flaked coconut
1	cup chopped walnuts

Glaze:

1/2	cup sugar
1/4	cup butter **or** margarine
2	tablespoons milk
1/4	teaspoon vanilla extract

32. *Cookie Sheet Apple Pie.* Listed below is the recipe for cookie sheet apple pie. What are the ingredients for $\frac{1}{2}$ recipe? for 5 recipes?

Source: Reprinted with permission from Taste of Home, Greendale WI, www.tasteofhome.com

Cookie Sheet Apple Pie

3-3/4	cups all-purpose flour
1-1/2	teaspoons salt
3/4	cup shortening
3	eggs, lightly beaten
1/3	cup milk
8	cups sliced peeled tart apples
1-1/2	cups sugar
1	teaspoon ground cinnamon
1/2	teaspoon ground nutmeg

1	cup crushed cornflakes
1	egg white, beaten

33. *Temperatures.* Fahrenheit temperature can be obtained from Celsius (centigrade) temperature by multiplying by $1\frac{4}{5}$ and adding $32°$. What Fahrenheit temperature corresponds to a Celsius temperature of $20°$?

34. *Temperature.* Fahrenheit temperature can be obtained from Celsius (centigrade) temperature by multiplying by $1\frac{4}{5}$ and adding $32°$. What Fahrenheit temperature corresponds to the Celsius temperature of boiling water, which is $100°$?

35. *Video Recording.* The tape in a VCR operating in the short-play mode travels at a rate of $1\frac{3}{8}$ in. per second. How many inches of tape are used to record for 60 sec in the short-play mode?

36. *Audio Recording.* The tape in an audio cassette is played at the rate of $1\frac{7}{8}$ in. per second. How many inches of tape are used when a cassette is played for $5\frac{1}{2}$ sec?

37. A car traveled 213 mi on $14\frac{2}{10}$ gal of gas. How many miles per gallon did it get?

38. A car traveled 385 mi on $15\frac{4}{10}$ gal of gas. How many miles per gallon did it get?

39. *Weight of Water.* The weight of water is $62\frac{1}{2}$ lb per cubic foot. How many cubic feet would be occupied by 250 lb of water?

40. *Weight of Water.* The weight of water is $62\frac{1}{2}$ lb per cubic foot. How many cubic feet would be occupied by 375 lb of water?

41. *Servings of Flounder.* A serving of filleted fish is generally considered to be about $\frac{1}{3}$ lb. How many servings can be prepared from $5\frac{1}{2}$ lb of flounder fillet?

42. *Servings of Tuna.* A serving of fish steak (cross section) is generally $\frac{1}{2}$ lb. How many servings can be prepared from a cleaned $18\frac{3}{4}$-lb tuna?

Find the area of the shaded region.

43.

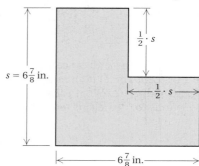

$s = 6\frac{7}{8}$ in.

$\frac{1}{2} \cdot s$

$\frac{1}{2} \cdot s$

$6\frac{7}{8}$ in.

44.

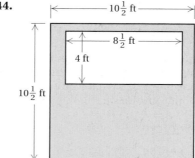

$10\frac{1}{2}$ ft

$8\frac{1}{2}$ ft

4 ft

$10\frac{1}{2}$ ft

45. *Construction.* A rectangular lot has dimensions of $302\frac{1}{2}$ ft by $205\frac{1}{4}$ ft. A building with dimensions of 100 ft by $25\frac{1}{2}$ ft is built on the lot. How much area is left over?

46. *Word Processing.* Kelly wants to create a table using Microsoft® Word software for word processing. She needs to have two columns, each $1\frac{1}{2}$ in. wide, and five columns, each $\frac{3}{4}$ in. wide. Will this table fit on a piece of standard paper that is $8\frac{1}{2}$ in. wide? If so, how wide will each margin be if her margins on each side are to be of equal width?

47. **D**_W Write a problem for a classmate to solve. Design the problem so that its solution is found by performing the multiplication $4\frac{1}{2} \cdot 33\frac{1}{3}$.

48. **D**_W Under what circumstances is a pair of mixed numerals more easily added than multiplied?

49. Multiply.   [1.5a]

$$\begin{array}{r} 6\ 7\ 0\ 9 \\ \times\quad 2\ 1\ 3 \\ \hline \end{array}$$

50. Round to the nearest hundred: 45,765.   [1.4a]

51. Solve: $\dfrac{5}{7} \cdot t = 420.$ [2.7c]

52. Divide and simplify: $\dfrac{4}{5} \div \dfrac{6}{5}.$ [2.7b]

53. Multiply and simplify: $\dfrac{3}{8} \cdot \dfrac{4}{9}.$ [2.6a]

54. Round to the nearest ten: 45,765.   [1.4a]

Simplify.   [2.5b]

55. $\dfrac{200}{375}$

56. $\dfrac{63}{75}$

57. $\dfrac{160}{270}$

58. $\dfrac{6996}{8028}$

Multiply. Write the answer as a mixed numeral whenever possible.

59. ▦ $15\dfrac{2}{11} \cdot 23\dfrac{31}{43}$

60. ▦ $17\dfrac{23}{31} \cdot 19\dfrac{13}{15}$

Simplify.

61. $8 \div \dfrac{1}{2} + \dfrac{3}{4} + \left(5 - \dfrac{5}{8}\right)^2$

62. $\left(\dfrac{5}{9} - \dfrac{1}{4}\right) \times 12 + \left(4 - \dfrac{3}{4}\right)^2$

63. $\dfrac{1}{3} \div \left(\dfrac{1}{2} - \dfrac{1}{5}\right) \times \dfrac{1}{4} + \dfrac{1}{6}$

64. $\dfrac{7}{8} - 1\dfrac{1}{8} \times \dfrac{2}{3} + \dfrac{9}{10} \div \dfrac{3}{5}$

65. $4\dfrac{1}{2} \div 2\dfrac{1}{2} + 8 - 4 \div \dfrac{1}{2}$

66. $6 - 2\dfrac{1}{3} \times \dfrac{3}{4} + \dfrac{5}{8} \div \dfrac{2}{3}$

Objectives

a Simplify expressions using the rules for order of operations.

b Estimate with fraction and mixed–numeral notation.

Simplify.

1. $\dfrac{2}{5} \cdot \dfrac{5}{8} + \dfrac{1}{4}$

2. $\dfrac{1}{3} \cdot \dfrac{3}{4} \div \dfrac{5}{8} - \dfrac{1}{10}$

3. Simplify: $\dfrac{3}{4} \cdot 16 + 8\dfrac{2}{3}$.

Answers on page A-7

a Order of Operations; Fraction Notation and Mixed Numerals

The rules for order of operations that we use with whole numbers (see Section 1.9) apply when we are simplifying expressions involving fraction notation and mixed numerals. For review, these rules are listed below.

RULES FOR ORDER OF OPERATIONS

1. Do all calculations within parentheses before operations outside.
2. Evaluate all exponential expressions.
3. Do all multiplications and divisions in order from left to right.
4. Do all additions and subtractions in order from left to right.

EXAMPLE 1 Simplify: $\dfrac{2}{3} \div \dfrac{1}{2} \cdot \dfrac{5}{8} + \dfrac{1}{6}$.

$$\dfrac{2}{3} \div \dfrac{1}{2} \cdot \dfrac{5}{8} + \dfrac{1}{6} = \dfrac{2}{3} \cdot \dfrac{2}{1} \cdot \dfrac{5}{8} + \dfrac{1}{6}$$ Doing the division first by multiplying by the reciprocal of $\frac{1}{2}$

$$= \dfrac{4 \cdot 5}{3 \cdot 8} + \dfrac{1}{6}$$ Doing the multiplication

$$= \dfrac{4 \cdot 5}{3 \cdot 4 \cdot 2} + \dfrac{1}{6}$$ Factoring in order to simplify

$$= \dfrac{5}{3 \cdot 2} + \dfrac{1}{6}$$ Removing a factor of 1: $\dfrac{4}{4} = 1$

$$= \dfrac{5}{6} + \dfrac{1}{6}$$

$$= \dfrac{6}{6}, \quad \text{or } 1$$ Doing the addition

Do Exercises 1 and 2.

EXAMPLE 2 Simplify: $\dfrac{2}{3} \cdot 24 - 11\dfrac{1}{2}$.

$$\dfrac{2}{3} \cdot 24 - 11\dfrac{1}{2} = \dfrac{2 \cdot 24}{3} - 11\dfrac{1}{2}$$ Doing the multiplication first

$$= \dfrac{2 \cdot 3 \cdot 8}{3} - 11\dfrac{1}{2}$$ Factoring the numerator

$$= 2 \cdot 8 - 11\dfrac{1}{2}$$ Removing a factor of 1: $\dfrac{3}{3} = 1$

$$= 16 - 11\dfrac{1}{2}$$ Completing the multiplication

$$= 4\dfrac{1}{2}, \quad \text{or } \dfrac{9}{2}$$ Doing the subtraction

Do Exercise 3 on the preceding page.

EXAMPLE 3 Melody has had three children. Their birth weights were $7\frac{1}{2}$ lb, $7\frac{3}{4}$ lb, and $6\frac{3}{4}$ lb. What was the average weight of her babies?

Recall that to compute an **average,** we add the numbers and then divide the sum by the number of addends (see Section 1.9). We have

$$\frac{7\frac{1}{2} + 7\frac{3}{4} + 6\frac{3}{4}}{3}.$$

We first add:

$$7\frac{1}{2} + 7\frac{3}{4} + 6\frac{3}{4} = 7\frac{2}{4} + 7\frac{3}{4} + 6\frac{3}{4}$$

$$= 20\frac{8}{4} = 22. \qquad 20\frac{8}{4} = 20 + \frac{8}{4} = 20 + 2$$

Then we divide:

$$\frac{7\frac{1}{2} + 7\frac{3}{4} + 6\frac{3}{4}}{3} = \frac{22}{3} = 7\frac{1}{3}. \qquad \text{Dividing by 3}$$

The average weight of the three babies is $7\frac{1}{3}$ lb.

Do Exercises 4–6.

EXAMPLE 4 Simplify: $\left(\dfrac{7}{8} - \dfrac{1}{3}\right) \times 48 + \left(13 + \dfrac{4}{5}\right)^2$.

$$\left(\frac{7}{8} - \frac{1}{3}\right) \times 48 + \left(13 + \frac{4}{5}\right)^2$$

$$= \left(\frac{7}{8} \cdot \frac{3}{3} - \frac{1}{3} \cdot \frac{8}{8}\right) \times 48 + \left(13 \cdot \frac{5}{5} + \frac{4}{5}\right)^2$$

Carrying out operations inside parentheses first. To do so, we first multiply by 1 to obtain the LCD.

$$= \left(\frac{21}{24} - \frac{8}{24}\right) \times 48 + \left(\frac{65}{5} + \frac{4}{5}\right)^2$$

$$= \frac{13}{24} \times 48 + \left(\frac{69}{5}\right)^2 \qquad \text{Completing the operations within parentheses}$$

$$= \frac{13}{24} \times 48 + \frac{4761}{25} \qquad \text{Evaluating exponential expressions next}$$

$$= 26 + \frac{4761}{25} \qquad \text{Doing the multiplication}$$

$$= 26 + 190\frac{11}{25} \qquad \text{Converting to a mixed numeral}$$

$$= 216\frac{11}{25}, \quad \text{or} \quad \frac{5411}{25} \qquad \text{Adding}$$

Answers can be given using either fraction notation or mixed numerals as desired. Consult with your instructor.

Do Exercise 7.

---

4. After two weeks, Kurt's tomato seedlings measure $9\frac{1}{2}$ in., $10\frac{3}{4}$ in., $10\frac{1}{4}$ in., and 9 in. tall. Find their average height.

5. Find the average of

$$\frac{1}{2}, \frac{1}{3}, \quad \text{and} \quad \frac{5}{6}.$$

6. Find the average of $\dfrac{3}{4}$ and $\dfrac{4}{5}$.

7. Simplify:

$$\left(\frac{2}{3} + \frac{3}{4}\right) \div 2\frac{1}{3} - \left(\frac{1}{2}\right)^3.$$

Answers on page A-7

Estimate each of the following as 0, $\frac{1}{2}$, or 1.

8. $\frac{3}{59}$

9. $\frac{61}{59}$

10. $\frac{29}{59}$

11. $\frac{57}{59}$

Find a number for the blank so that the fraction is close to but less than 1.

12. $\frac{11}{\square}$

13. $\frac{\square}{33}$

Answers on page A-7

Study Tips

SKILL MAINTENANCE EXERCISES

It is never too soon to begin reviewing for the final examination. The Skill Maintenance exercises found in each exercise set review and reinforce skills taught in earlier sections. Include all of these exercises in your weekly preparation. Answers to both odd-numbered and even-numbered exercises along with section references appear at the back of the book.

b   Estimation with Fraction Notation and Mixed Numerals

We now estimate with fraction notation and mixed numerals.

EXAMPLES   Estimate each of the following as 0, $\frac{1}{2}$, or 1.

5. $\frac{2}{17}$

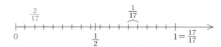

A fraction is very close to 0 when the numerator is very small in comparison to the denominator. Thus, 0 is an estimate for $\frac{2}{17}$ because 2 is very small in comparison to 17. Thus, $\frac{2}{17} \approx 0$.

6. $\frac{11}{23}$

A fraction is very close to $\frac{1}{2}$ when the denominator is about twice the numerator. Thus, $\frac{1}{2}$ is an estimate for $\frac{11}{23}$ because $2 \cdot 11 = 22$ and 22 is close to 23. Thus, $\frac{11}{23} \approx \frac{1}{2}$.

7. $\frac{37}{38}$

A fraction is very close to 1 when the numerator is nearly equal to the denominator. Thus, 1 is an estimate for $\frac{37}{38}$ because 37 is nearly equal to 38. Thus, $\frac{37}{38} \approx 1$.

8. $\frac{43}{41}$

As in the preceding example, the numerator 43 is very close to the denominator 41. Thus, $\frac{43}{41} \approx 1$.

Do Exercises 8–11.

EXAMPLE 9   Find a number for the blank so that $\frac{9}{\square}$ is close to but less than 1. Answers may vary.

If the number in the blank were 9, we would have 1, so we increase 9 to 10. The answer is 10; $\frac{9}{10}$ is close to 1. The number 11 would also be a correct answer; $\frac{9}{11}$ is close to 1.

Do Exercises 12 and 13.

EXAMPLE 10 Find a number for the blank so that $\dfrac{9}{\Box}$ is close to but less than $\frac{1}{2}$. Answers may vary.

If we double 9 to get 18 and use it for the blank, we have $\frac{1}{2}$. If we increase that denominator by 1, to get 19, and use it for the blank, we get a number less than $\frac{1}{2}$ but close to $\frac{1}{2}$. Thus, $\dfrac{9}{19} \approx \frac{1}{2}$.

Do Exercises 14 and 15.

EXAMPLE 11 Find a number for the blank so that $\dfrac{\Box}{50}$ is close to but greater than 0.

Since 50 is rather large, any small number such as 1, 2, or 3 will make the fraction close to 0. For example, $\dfrac{1}{50} \approx 0$.

Do Exercises 16 and 17.

EXAMPLE 12 Estimate $16\frac{8}{9} + 11\frac{2}{13} - 4\frac{22}{43}$ as a whole number or as a mixed numeral where the fractional part is $\frac{1}{2}$.

We estimate each fraction as 0, $\frac{1}{2}$, or 1. Then we calculate:

$$16\frac{8}{9} + 11\frac{2}{13} - 4\frac{22}{43} \approx 17 + 11 - 4\frac{1}{2}$$

$$= 28 - 4\frac{1}{2}$$

$$= 23\frac{1}{2}.$$

Do Exercises 18–20.

Find a number for the blank so that the fraction is close to but less than $\frac{1}{2}$.

14. $\dfrac{13}{\Box}$ **15.** $\dfrac{\Box}{31}$

Find a number for the blank so that the fraction is close to but greater than 0.

16. $\dfrac{\Box}{37}$ **17.** $\dfrac{13}{\Box}$

Estimate each part of the following as a whole number or as a mixed numeral where the fractional part is $\frac{1}{2}$.

18. $5\dfrac{9}{10} + 26\dfrac{1}{2} - 10\dfrac{3}{29}$

19. $10\dfrac{7}{8} \cdot \left(25\dfrac{11}{13} - 14\dfrac{1}{9}\right)$

20. $\left(10\dfrac{4}{5} + 7\dfrac{5}{9}\right) \div \dfrac{17}{30}$

Answers on page A-7

a   Simplify.

1. $\dfrac{1}{2} \cdot \dfrac{1}{3} \cdot \dfrac{1}{4}$

2. $\dfrac{1}{3} \cdot \dfrac{1}{4} \cdot \dfrac{1}{5}$

3. $6 \div 3 \div 5$

4. $12 \div 4 \div 8$

5. $\dfrac{2}{3} \div \dfrac{4}{3} \div \dfrac{7}{8}$

6. $\dfrac{5}{6} \div \dfrac{3}{4} \div \dfrac{2}{5}$

7. $\dfrac{5}{8} \div \dfrac{1}{4} - \dfrac{2}{3} \cdot \dfrac{4}{5}$

8. $\dfrac{4}{7} \cdot \dfrac{7}{15} + \dfrac{2}{3} \div 8$

9. $\dfrac{3}{4} - \dfrac{2}{3} \cdot \left(\dfrac{1}{2} + \dfrac{2}{5} \right)$

10. $\dfrac{3}{4} \div \dfrac{1}{2} \cdot \left(\dfrac{8}{9} - \dfrac{2}{3} \right)$

11. $28\dfrac{1}{8} - 5\dfrac{1}{4} + 3\dfrac{1}{2}$

12. $10\dfrac{3}{5} - 4\dfrac{1}{10} - 1\dfrac{1}{2}$

13. $\dfrac{7}{8} \div \dfrac{1}{2} \cdot \dfrac{1}{4}$

14. $\dfrac{7}{10} \cdot \dfrac{4}{5} \div \dfrac{2}{3}$

15. $\left(\dfrac{2}{3} \right)^2 - \dfrac{1}{3} \cdot 1\dfrac{1}{4}$

16. $\left(\dfrac{3}{4} \right)^2 + 3\dfrac{1}{2} \div 1\dfrac{1}{4}$

17. $\dfrac{1}{2} - \left(\dfrac{1}{2} \right)^2 + \left(\dfrac{1}{2} \right)^3$

18. $1 + \dfrac{1}{4} + \left(\dfrac{1}{4} \right)^2 - \left(\dfrac{1}{4} \right)^3$

19. Find the average of $\dfrac{2}{3}$ and $\dfrac{7}{8}$.

20. Find the average of $\dfrac{1}{4}$ and $\dfrac{1}{5}$.

21. Find the average of $\dfrac{1}{6}$, $\dfrac{1}{8}$, and $\dfrac{3}{4}$.

22. Find the average of $\dfrac{4}{5}$, $\dfrac{1}{2}$, and $\dfrac{1}{10}$.

23. Find the average of $3\dfrac{1}{2}$ and $9\dfrac{3}{8}$.

24. Find the average of $10\dfrac{2}{3}$ and $24\dfrac{5}{6}$.

25. *Birth Weights.* The Piper quadruplets of Great Britain weighed $2\frac{9}{16}$ lb, $2\frac{9}{32}$ lb, $2\frac{1}{8}$ lb, and $2\frac{5}{16}$ lb at birth. Find their average birth weight.
Source: *The Guinness Book of Records,* 1998

26. *Vertical Leaps.* Eight-year-old Zachary registered vertical leaps of $12\frac{3}{4}$ in., $13\frac{3}{4}$ in., $13\frac{1}{2}$ in., and 14 in. Find his average vertical leap.

27. *Manufacturing.* A test of five light bulbs showed that they burned for the lengths of time given on the graph below. For how many days, on average, did the bulbs burn?

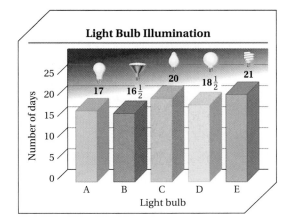

28. *Packaging.* A sample of four bags of beef jerky showed the weights given on the graph below. What was the average weight?

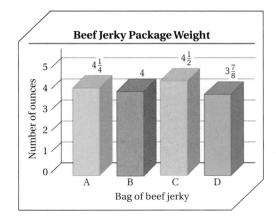

Simplify.

29. $\left(\dfrac{2}{3} + \dfrac{3}{4}\right) \div \left(\dfrac{5}{6} - \dfrac{1}{3}\right)$

30. $\left(\dfrac{3}{5} - \dfrac{1}{2}\right) \div \left(\dfrac{3}{4} - \dfrac{3}{10}\right)$

31. $\left(\dfrac{1}{2} + \dfrac{1}{3}\right)^2 \cdot 144 - \dfrac{5}{8} \div 10\dfrac{1}{2}$

32. $\left(3\dfrac{1}{2} - 2\dfrac{1}{3}\right)^2 + 6 \cdot 2\dfrac{1}{2} \div 32$

b Estimate each of the following as 0, $\frac{1}{2}$, or 1.

33. $\dfrac{2}{47}$ **34.** $\dfrac{4}{5}$ **35.** $\dfrac{1}{13}$ **36.** $\dfrac{7}{8}$ **37.** $\dfrac{6}{11}$ **38.** $\dfrac{10}{13}$

39. $\dfrac{7}{15}$ **40.** $\dfrac{1}{16}$ **41.** $\dfrac{7}{100}$ **42.** $\dfrac{5}{9}$ **43.** $\dfrac{19}{20}$ **44.** $\dfrac{5}{12}$

Find a number for the blank so that the fraction is close to but greater than $\frac{1}{2}$. Answers may vary.

45. $\dfrac{\Box}{11}$ **46.** $\dfrac{\Box}{8}$ **47.** $\dfrac{\Box}{23}$ **48.** $\dfrac{\Box}{35}$

49. $\dfrac{10}{\Box}$ **50.** $\dfrac{51}{\Box}$

Find a number for the blank so that the fraction is close to but greater than 1. Answers may vary.

51. $\dfrac{7}{\square}$ **52.** $\dfrac{11}{\square}$ **53.** $\dfrac{13}{\square}$ **54.** $\dfrac{27}{\square}$ **55.** $\dfrac{\square}{15}$ **56.** $\dfrac{\square}{100}$

Estimate each part of the following as a whole number, $\frac{1}{2}$, or as a mixed numeral where the fractional part is $\frac{1}{2}$.

57. $2\dfrac{7}{8}$

58. $1\dfrac{1}{3}$

59. $12\dfrac{5}{6}$

60. $26\dfrac{6}{13}$

61. $\dfrac{4}{5} + \dfrac{7}{8}$

62. $\dfrac{1}{12} \cdot \dfrac{7}{15}$

63. $\dfrac{2}{3} + \dfrac{7}{13} + \dfrac{5}{9}$

64. $\dfrac{8}{9} + \dfrac{4}{5} + \dfrac{11}{12}$

65. $\dfrac{43}{100} + \dfrac{1}{10} - \dfrac{11}{1000}$

66. $\dfrac{23}{24} + \dfrac{37}{39} + \dfrac{51}{50}$

67. $7\dfrac{29}{60} + 10\dfrac{12}{13} \cdot 24\dfrac{2}{17}$

68. $5\dfrac{13}{14} - 1\dfrac{5}{8} + 1\dfrac{23}{28} \cdot 6\dfrac{35}{74}$

69. $24 \div 7\dfrac{8}{9}$

70. $43\dfrac{16}{17} \div 11\dfrac{2}{13}$

71. $76\dfrac{3}{14} + 23\dfrac{19}{20}$

72. $76\dfrac{13}{14} \cdot 23\dfrac{17}{20}$

73. $16\dfrac{1}{5} \div 2\dfrac{1}{11} + 25\dfrac{9}{10} - 4\dfrac{11}{23}$

74. $96\dfrac{2}{13} \div 5\dfrac{19}{20} + 3\dfrac{1}{7} \cdot 5\dfrac{18}{21}$

75. ^{D}W A student insists that $3\frac{2}{5} \cdot 1\frac{3}{7} = 3\frac{6}{35}$. What mistake is he making and how should he have proceeded?

76. ^{D}W A student insists that $5 \cdot 3\frac{2}{7} = (5 \cdot 3) \cdot \left(5 \cdot \frac{2}{7}\right)$. What mistake is she making and how should she have proceeded?

⎯⎯(**SKILL MAINTENANCE**)⎯⎯⎯⎯⎯⎯⎯⎯⎯⎯⎯⎯⎯⎯⎯⎯⎯⎯⎯⎯⎯⎯⎯⎯⎯⎯⎯⎯

77. Multiply: $27 \cdot 126$. [1.5a]

78. Multiply: $132 \cdot 7865$. [1.5a]

79. Divide: $7865 \div 132$. [1.6b]

Multiply and simplify. [2.4a], [2.6a]

80. $\dfrac{2}{3} \cdot 522$

81. $\dfrac{3}{2} \cdot 522$

Divide and simplify. [2.7b]

82. $\dfrac{4}{5} \div \dfrac{3}{10}$

83. $\dfrac{3}{10} \div \dfrac{4}{5}$

84. Classify the given numbers as prime, composite, or neither. [2.1c]

      1, 5, 7, 9, 14, 23, 43

Solve.

85. *Luncheon Servings.* Ian purchased 6 lb of cold cuts for a luncheon. If Ian is to allow $\frac{3}{8}$ lb per person, how many people can he invite to the luncheon? [2.7d]

86. *Cholesterol.* A 3-oz serving of crabmeat contains 85 milligrams (mg) of cholesterol. A 3-oz serving of shrimp contains 128 mg of cholesterol. How much more cholesterol is in the shrimp? [1.8a]

SYNTHESIS

87. a) Find an expression for the sum of the areas of the two rectangles shown here.
    b) Simplify the expression.
    c) How is the computation in part (b) related to the rules for order of operations?

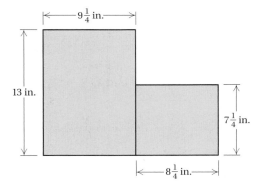

88. Find r if

$$\frac{1}{r} = \frac{1}{100} + \frac{1}{150} + \frac{1}{200}.$$

89. ▦ In the sum below, a and b are digits. Find a and b.

$$\frac{a}{17} + \frac{1b}{23} = \frac{35a}{391}$$

90. ▦ Consider only the numbers 3, 4, 5, and 6. Assume each can be placed in a blank in the following.

$$\Box + \frac{\Box}{\Box} \cdot \Box = ?$$

What placement of the numbers in the blanks yields the largest number?

91. ▦ Consider only the numbers 2, 3, 4, and 5. Assume each is placed in a blank in the following.

$$\frac{\Box}{\Box} + \frac{\Box}{\Box} = ?$$

What placement of the numbers in the blanks yields the largest sum?

92. ▦ Use a standard calculator. Arrange the following in order from smallest to largest.

$$\frac{3}{4}, \frac{17}{21}, \frac{13}{15}, \frac{7}{9}, \frac{15}{17}, \frac{13}{12}, \frac{19}{22}$$

The review that follows is meant to prepare you for a chapter exam. It consists of two parts. The first part is a checklist of some of the Study Tips referred to in this text. The second part is the Review Exercises. These provide practice exercises for the exam, together with references to section objectives so you can go back and review. Before beginning, stop and look back over the skills you have obtained. What skills in mathematics do you have now that you did not have before studying this chapter?

STUDY TIPS CHECKLIST

The foundation of all your study skills is TIME!

☐ Have you formed a study group with some of your fellow students?

☐ Are you doing some skill maintenance exercises as part of your daily assignment whether they have been recommended or not?

☐ Have you found adequate time to study?

☐ Are you stopping to work the margin exercises when directed to do so?

☐ Are you doing your homework as soon as possible after class?

REVIEW EXERCISES

Find the LCM.   [3.1a]

1. 12 and 18

2. 18 and 45

3. 3, 6, and 30

4. 26, 36, and 54

Add and simplify.   [3.2a]

5. $\dfrac{6}{5} + \dfrac{3}{8}$

6. $\dfrac{5}{16} + \dfrac{1}{12}$

7. $\dfrac{6}{5} + \dfrac{11}{15}$

8. $\dfrac{5}{16} + \dfrac{1}{8}$

Subtract and simplify.   [3.3a]

9. $\dfrac{5}{9} - \dfrac{2}{9}$

10. $\dfrac{7}{8} - \dfrac{3}{4}$

11. $\dfrac{11}{27} - \dfrac{2}{9}$

12. $\dfrac{5}{6} - \dfrac{2}{9}$

Use $<$ or $>$ for ☐ to write a true sentence.   [3.3b]

13. $\dfrac{4}{7} \square \dfrac{5}{9}$

14. $\dfrac{8}{9} \square \dfrac{11}{13}$

Solve.   [3.3c]

15. $x + \dfrac{2}{5} = \dfrac{7}{8}$

16. $\dfrac{1}{2} + y = \dfrac{9}{10}$

Convert to fraction notation.   [3.4a]

17. $7\dfrac{1}{2}$

18. $8\dfrac{3}{8}$

19. $4\dfrac{1}{3}$

20. $10\dfrac{5}{7}$

Convert to a mixed numeral.   [3.4a]

21. $\dfrac{7}{3}$

22. $\dfrac{27}{4}$

23. $\dfrac{63}{5}$

24. $\dfrac{7}{2}$

Divide. Write a mixed numeral for the answer.
[3.4b]

25. $9 \overline{)7896}$ **26.** $23 \overline{)10,493}$

Add. Write a mixed numeral for the answer. [3.5a]

27. $5\frac{3}{5}$
$+ 4\frac{4}{5}$

28. $8\frac{1}{3}$
$+ 3\frac{2}{5}$

29. $5\frac{5}{6}$
$+ 4\frac{5}{6}$

30. $2\frac{3}{4}$
$+ 5\frac{1}{2}$

Subtract. Write a mixed numeral for the answer where appropriate. [3.5b]

31. 12
$- 4\frac{2}{9}$

32. $9\frac{3}{5}$
$- 4\frac{13}{15}$

33. $10\frac{1}{4}$
$- 6\frac{1}{10}$

34. 24
$- 10\frac{5}{8}$

Multiply. Write a mixed numeral for the answer where appropriate. [3.6a]

35. $6 \cdot 2\frac{2}{3}$

36. $5\frac{1}{4} \cdot \frac{2}{3}$

37. $2\frac{1}{5} \cdot 1\frac{1}{10}$

38. $2\frac{2}{5} \cdot 2\frac{1}{2}$

Divide. Write a mixed numeral for the answer where appropriate. [3.6b]

39. $27 \div 2\frac{1}{4}$

40. $2\frac{2}{5} \div 1\frac{7}{10}$

41. $3\frac{1}{4} \div 26$

42. $4\frac{1}{5} \div 4\frac{2}{3}$

Solve. [3.2b], [3.5c], [3.6c]

43. *Sewing from a Pattern.* Suppose you want to make an outfit in size 12. On the back of the pattern envelope, it states that for size 12, using 60-in. fabric, you need $1\frac{5}{8}$ yd of fabric for the dress and $2\frac{5}{8}$ yd for the jacket. How many yards in all are needed to make the outfit?

44. *Turkey Servings.* Turkey contains $1\frac{1}{3}$ servings per pound. How many pounds are needed for 32 servings?

45. *Weightlifting.* In 1998, Sun Tianni of China snatched 111 kg. This amount was about $1\frac{3}{5}$ times her body weight. How much did Tianni weigh?
Source: *The Guinness Book of Records, 2000*

46. *Carpentry.* A board $\frac{9}{10}$ in. thick is glued to a board $\frac{8}{10}$ in. thick. The glue is $\frac{3}{100}$ in. thick. How thick is the result?

47. What is the sum of the areas in the figure below?

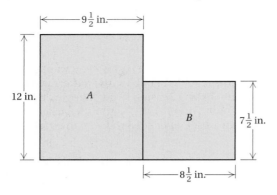

48. In the figure above, how much larger is the area of rectangle *A* than the area of rectangle *B*?

49. *Cake Recipe.* A wedding-cake recipe requires 12 cups of shortening. Being calorie-conscious, the wedding couple decides to reduce the shortening by $3\frac{5}{8}$ cups and replace it with prune purée. How many cups of shortening are used in their new recipe?

50. *Firefighters' Pie Sale.* Green River's Volunteer Fire Department recently hosted its annual ice cream social. Each of the donated 83 pies was cut into 6 pieces. At the end of the evening, the cashier said they had sold 382 pieces of pie. How many pies did they sell? How many were left over? Express your answers in mixed numerals.

51. Simplify this expression using the rules for order of operations: [3.7a]

$$\frac{1}{8} \div \frac{1}{4} + \frac{1}{2}.$$

52. Find the average of $\frac{1}{2}, \frac{1}{4}, \frac{1}{3}$, and $\frac{1}{5}$. [3.7a]

Estimate each of the following as 0, $\frac{1}{2}$, or 1. [3.7b]

53. $\frac{29}{59}$ **54.** $\frac{2}{59}$ **55.** $\frac{61}{59}$

Estimate each of the following as a whole number or as a mixed numeral where the fractional part is $\frac{1}{2}$. [3.7b]

56. $6\frac{7}{8}$ **57.** $10\frac{2}{17}$

58. $\frac{3}{10} + \frac{5}{6} + \frac{31}{29}$

59. $32\frac{14}{15} + 27\frac{3}{4} - 4\frac{25}{28} \cdot 6\frac{37}{76}$

60. ^{D}w Discuss the role of least common multiples in adding and subtracting with fraction notation. [3.2a], [3.3a]

61. ^{D}w Find a real-world situation that fits this equation: [3.5c], [3.6c]

$$2 \cdot 15\frac{3}{4} + 2 \cdot 28\frac{5}{8} = 88\frac{3}{4}.$$

SKILL MAINTENANCE

Certain objectives from four particular sections will be retested on the chapter test. The objectives are listed with the practice problems that follow.

62. Multiply and simplify: $\frac{9}{10} \cdot \frac{4}{3}$. [2.6a]

63. Divide and simplify: $\frac{5}{4} \div \frac{5}{6}$. [2.7b]

64. Multiply: $176 \cdot 4023$. [1.5a]

65. *Digital Tire Gauges.* A factory produces 3885 digital tire gauges per day. How long will it take to fill an order for 66,045 tire gauges? [1.8a]

Tire gauge

SYNTHESIS

66. *Orangutan Circus Act.* Yuri and Olga are orangutans who perform in a circus by riding bicycles around a circular track. It takes Yuri 6 min and Olga 4 min to make one trip around the track. Suppose they start at the same point and then complete their act when they again reach the same point. How long is their act? [3.1a]

67. Place the numbers 3, 4, 5, and 6 in the boxes in order to make a true equation: [3.5a]

$$\frac{\square}{\square} + \frac{\square}{\square} = 3\frac{1}{4}.$$

Chapter Test

1. Find the LCM of 12 and 16.

Add and simplify.

2. $\dfrac{1}{2} + \dfrac{5}{2}$

3. $\dfrac{7}{8} + \dfrac{2}{3}$

4. $\dfrac{7}{10} + \dfrac{9}{100}$

Subtract and simplify.

5. $\dfrac{5}{6} - \dfrac{3}{6}$

6. $\dfrac{5}{6} - \dfrac{3}{4}$

7. $\dfrac{17}{24} - \dfrac{5}{8}$

8. Use < or > for □ to write a true sentence:

$\dfrac{6}{7} \; \square \; \dfrac{21}{25}$.

9. Solve: $x + \dfrac{2}{3} = \dfrac{11}{12}$.

Convert to fraction notation.

10. $3\dfrac{1}{2}$

11. $9\dfrac{7}{8}$

Convert to a mixed numeral.

12. $\dfrac{9}{2}$

13. $\dfrac{74}{9}$

Divide. Write a mixed numeral for the answer.

14. $1\,1\,)\,\overline{1\,7\,8\,9}$

Add. Write a mixed numeral for the answer.

15. $\begin{array}{r} 6\frac{2}{5} \\ + 7\frac{4}{5} \\ \hline \end{array}$

16. $\begin{array}{r} 9\frac{1}{4} \\ + 5\frac{1}{6} \\ \hline \end{array}$

Subtract. Write a mixed numeral for the answer.

17. $\begin{array}{r} 10\frac{1}{6} \\ - \; 5\frac{7}{8} \\ \hline \end{array}$

18. $\begin{array}{r} 14 \\ - \; 7\frac{5}{6} \\ \hline \end{array}$

Multiply. Write a mixed numeral for the answer.

19. $9 \cdot 4\dfrac{1}{3}$

20. $6\dfrac{3}{4} \cdot \dfrac{2}{3}$

Divide. Write a mixed numeral for the answer.

21. $2\dfrac{1}{3} \div 1\dfrac{1}{6}$

22. $2\dfrac{1}{12} \div 75$

23. *Weightlifting.* In 1999, Hossein Rezazadeh of Iran did a clean and jerk of $262\frac{1}{2}$ kg. This amount was $2\frac{1}{2}$ times his body weight. How much did Rezazadeh weigh?
Source: *The Guinness Book of Records, 2000*

24. *Book Order.* An order of books for a math course weighs 220 lb. Each book weighs $2\frac{3}{4}$ lb. How many books are in the order?

225

25. *Carpentry.* The following diagram shows a middle drawer support guide for a cabinet drawer. Find each of the following.

a) The short length a across the top
b) The length b across the bottom

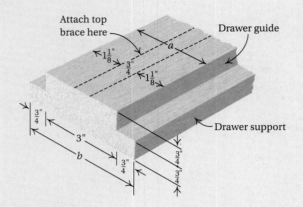

26. *Carpentry.* In carpentry, some pieces of plywood that are called "$\frac{3}{4}$-inch" plywood are actually $\frac{11}{16}$-in. thick. How much thinner is such a piece than its name indicates?

27. *Women's Dunks.* Only three women in the history of college basketball have been able to dunk a basketball. Their names, heights, and universities are:

Michelle Snow, $6\frac{5}{12}$ ft, Tennessee;

Charlotte Smith, $5\frac{11}{12}$ ft, North Carolina;

Georgeann Wells, $6\frac{7}{12}$ ft, West Virginia.

Find the average height of these women.
Source: *USA Today,* 11/30/00. p. 3C

28. Simplify: $\dfrac{2}{3} + 1\dfrac{1}{3} \cdot 2\dfrac{1}{8}$.

Estimate each of the following as 0, $\frac{1}{2}$, or 1.

29. $\dfrac{3}{82}$

30. $\dfrac{93}{91}$

Estimate each of the following as a whole number or as a mixed numeral where the fractional part is $\frac{1}{2}$.

31. $3\dfrac{8}{9}$

32. $18\dfrac{9}{17}$

33. $256 \div 15\dfrac{19}{21}$

34. $43\dfrac{15}{31} \cdot 27\dfrac{3}{4} - 9\dfrac{15}{28} + 6\dfrac{5}{76}$

SKILL MAINTENANCE

35. Multiply:
$$\begin{array}{r} 4\ 5\ 6\ 1 \\ \times\quad 7\ 6 \\ \hline \end{array}$$

36. Divide and simplify: $\dfrac{4}{3} \div \dfrac{5}{6}$.

37. Multiply and simplify: $\dfrac{4}{3} \cdot \dfrac{5}{6}$.

38. A container has 8570 oz of beverage with which to fill 16-oz bottles. How many of these bottles can be filled? How much beverage will be left over?

SYNTHESIS

39. The students in a math class can be organized into study groups of 8 each so that no students are left out. The same class of students can also be organized into groups of 6 so that no students are left out.

a) Find some class sizes for which this will work.
b) Find the smallest such class size.

40. Dolores runs 17 laps at her health club. Terence runs 17 laps at his health club. If the track at Dolores's health club is $\frac{1}{7}$ mi long, and the track at Terence's is $\frac{1}{8}$ mi long, who runs farther? How much farther?

Cumulative Review

Solve.

1. *Excelsior Made from Aspen.* Excelsior consists of slender, curved wood shavings used especially for packing. Shown below are examples of excelsior and the saw blades used to cut it, as made by Western Excelsior Corporation of Mancos, CO. The width of strips for craft decoration is either $\frac{1}{16}$ in. or $\frac{1}{8}$ in. The width for erosion control mats used for stabilizing soil and nourishing young crops is $\frac{1}{24}$ in.

   a) How much wider is the $\frac{1}{16}$-in. craft decoration excelsior than the erosion control excelsior?
   b) How much wider is the $\frac{1}{8}$-in. craft decoration excelsior than the erosion control excelsior?

   Source: Western Excelsior Corporation

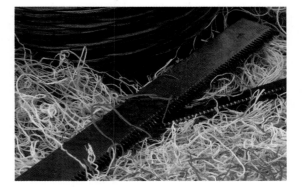

2. *DVD Storage.* Gregory is making a home entertainment center. He is planning a 27-in. shelf that holds DVDs that are each $\frac{7}{16}$ in. thick. How many DVDs will the shelf hold?

3. *Electronics Purchases.* Juanita goes to H. H. Gregg's Appliances to make the following purchases for her sorority:

   KOSS, 3-Piece CD/Radio Cassette Recorder, $69

   JVC, 4-Head Hi-Fi Stereo VCR with Super VHS, $199

   SONY, DVD Player, $249

   a) Find the total cost of the purchases.
   b) Find the average cost of these purchases. Express your answer as a mixed numeral.

   Sources: Koss, JVC, SONY

4. *Room Carpeting.* The Chandlers have an L-shaped family room consisting of a rectangle that is $8\frac{1}{2}$ ft by 11 ft and one that is $6\frac{1}{2}$ ft by $7\frac{1}{2}$ ft. They carpet the room.

   a) Find the area of the carpet.
   b) Find the perimeter of the carpet.

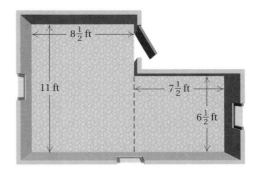

5. How many people can get equal $16 shares from a total of $496?

6. An emergency food pantry fund contains $423. From this fund, $148 and $167 are withdrawn for expenses. How much is left in the fund?

7. A recipe calls for $\frac{4}{5}$ tsp of salt. How much salt should be used for $\frac{1}{2}$ recipe? for 5 recipes?

8. A book weighs $2\frac{3}{5}$ lb. How much do 15 books weigh?

9. How many pieces, each $2\frac{3}{8}$ ft long, can be cut from a piece of wire 38 ft long?

10. In a walkathon, one person walked $\frac{9}{10}$ mi and another walked $\frac{75}{100}$ mi. What was the total distance walked?

11. In the number 2753, what digit names tens?

12. Write expanded notation for 6075.

13. Write a word name for the number in the following sentence: The diameter of Uranus is 29,500 miles.

14. What part is shaded?

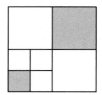

Calculate and simplify.

15.
$$\begin{array}{r} 6\ 2\ 8 \\ +\ 2\ 7\ 1 \\ \hline \end{array}$$

16.
$$\begin{array}{r} 3\ 7\ 0\ 4 \\ +\ 5\ 2\ 7\ 8 \\ \hline \end{array}$$

17. $\dfrac{3}{8} + \dfrac{1}{24}$

18. $2\dfrac{3}{4}$
$+ 5\dfrac{1}{2}$

19.
$$\begin{array}{r} 7\ 4\ 6\ 9 \\ -\ 2\ 3\ 4\ 5 \\ \hline \end{array}$$

20.
$$\begin{array}{r} 7\ 6\ 0\ 5 \\ -\ 3\ 0\ 8\ 7 \\ \hline \end{array}$$

21. $\dfrac{3}{4} - \dfrac{1}{3}$

22. $2\dfrac{1}{3}$
$- 1\dfrac{1}{6}$

23.
$$\begin{array}{r} 2\ 7\ 8 \\ \times\quad 1\ 8 \\ \hline \end{array}$$

24.
$$\begin{array}{r} 8\ 9\ 4 \\ \times\ 3\ 2\ 8 \\ \hline \end{array}$$

25. $\dfrac{9}{10} \cdot \dfrac{5}{3}$

26. $18 \cdot \dfrac{5}{6}$

27. $2\dfrac{1}{3} \cdot 3\dfrac{1}{7}$

Divide. Write the answer with the remainder in the form 34 R 7.

28. $6 \overline{)\ 4\ 2\ 9\ 0}$

29. $4\ 5 \overline{)\ 2\ 5\ 3\ 1}$

30. In Question 29, write a mixed numeral for the answer.

Divide and simplify, where appropriate.

31. $\dfrac{2}{5} \div \dfrac{7}{10}$

32. $2\dfrac{1}{5} \div \dfrac{3}{10}$

33. Round 38,478 to the nearest hundred.

34. Find the LCM of 18 and 24.

35. Determine whether 3718 is divisible by 8.

36. Find all factors of 16.

Use $<$, $>$, or $=$ for $\square$ to write a true sentence.

37. $\dfrac{4}{5} \square \dfrac{4}{6}$

38. $\dfrac{5}{12} \square \dfrac{3}{7}$

Simplify.

39. $\dfrac{36}{45}$

40. $\dfrac{320}{10}$

41. Convert to fraction notation: $4\dfrac{5}{8}$.

42. Convert to a mixed numeral: $\dfrac{17}{3}$.

Solve.

43. $x + 24 = 117$ 　　　**44.** $x + \dfrac{7}{9} = \dfrac{4}{3}$ 　　　**45.** $\dfrac{7}{9} \cdot t = \dfrac{4}{3}$ 　　　**46.** $y = 32{,}580 \div 36$

Estimate each of the following as 0, $\dfrac{1}{2}$, or 1.

47. $\dfrac{29}{30}$ 　　　　　**48.** $\dfrac{15}{29}$ 　　　　　**49.** $\dfrac{2}{43}$

Estimate each of the following as a whole number or as a mixed numeral where the fractional part is $\dfrac{1}{2}$.

50. $30\dfrac{4}{53}$ 　　　**51.** $\dfrac{9}{10} - \dfrac{7}{8} + \dfrac{41}{39}$ 　　　**52.** $78\dfrac{14}{15} - 28\dfrac{7}{8} - 7\dfrac{25}{28} \div \dfrac{65}{66}$

53. *Matching.* Match each item in the first column with the appropriate item in the second column by drawing connecting lines. Note that some numbers in the right column will be used more than once, and some numbers will not be used at all.

Multiply: $\dfrac{3}{5} \cdot 4\dfrac{1}{6}$.

Simplify: $\dfrac{72}{48}$.

Add: $1\dfrac{1}{8} + 1\dfrac{3}{16}$.

Divide: $1\dfrac{1}{2} \div \dfrac{2}{3}$.

The reciprocal of $\dfrac{4}{9}$.

Simplify: $\dfrac{108}{54}$.

Simplify: $\dfrac{24}{54}$.

Subtract: $14\dfrac{1}{8} - 11\dfrac{5}{8}$.

Divide: $3^2 \div 2^2$.

Simplify: $\dfrac{216}{96}$.

2

$\dfrac{37}{16}$

$\dfrac{4}{9}$

$1\dfrac{1}{3}$

1

$\dfrac{9}{4}$

$10\dfrac{1}{4}$

$\dfrac{3}{2}$

$2\dfrac{1}{2}$

For each of Exercises 54 and 55, choose the correct answer from the selections given.

54. A gallon of ice cream provides $28\dfrac{1}{2}$ servings. How many gallons of ice cream are needed to serve 228 guests at a wedding reception?

a) 8 　　　　**b)** $8\dfrac{1}{2}$ 　　　　**c)** 10
d) $10\dfrac{1}{2}$ 　　**e)** None

55. For a certain type of load of dishes in a Kitchen-Aid® dishwasher, $1\dfrac{3}{4}$ oz of Cascade® detergent are required. How many ounces of detergent are needed for 8 loads?

a) 14 　　　　**b)** $9\dfrac{3}{4}$ 　　　　**c)** $4\dfrac{4}{7}$
d) $7\dfrac{3}{32}$ 　　**e)** None

⸻ SYNTHESIS ⸻

56. a) Simplify each of the following, using fraction notation for your answers.

$$\dfrac{1}{1 \cdot 2}$$

$$\dfrac{1}{1 \cdot 2} + \dfrac{1}{2 \cdot 3}$$

$$\dfrac{1}{1 \cdot 2} + \dfrac{1}{2 \cdot 3} + \dfrac{1}{3 \cdot 4}$$

$$\dfrac{1}{1 \cdot 2} + \dfrac{1}{2 \cdot 3} + \dfrac{1}{3 \cdot 4} + \dfrac{1}{4 \cdot 5}$$

b) Look for a pattern in your answers to part (a). Then find the following without carrying out the computations.

$$\dfrac{1}{1 \cdot 2} + \dfrac{1}{2 \cdot 3} + \dfrac{1}{3 \cdot 4} + \dfrac{1}{4 \cdot 5} + \dfrac{1}{5 \cdot 6}$$
$$+ \dfrac{1}{6 \cdot 7} + \dfrac{1}{7 \cdot 8} + \dfrac{1}{8 \cdot 9} + \dfrac{1}{9 \cdot 10}$$

57. Find the smallest prime number that is larger than 2000.

Decimal Notation

100

90

80

70

60

50

JAN

40

Gateway to Chapter 4

In this chapter, we consider the operations of addition, subtraction, multiplication, and division with decimal notation. These skills will allow us to solve applied problems like the one about changing values of The Quaker Oats Company stock shown here. We will also study estimating sums, differences, products, and quotients. Conversion between fraction and decimal notation in which the decimal notation may be repeating will be discussed as well.

Real-World Application

Over the entire 52 weeks of 2000, the price per share of The Quaker Oats Company stock ranged in value from a low of $45.81 to a high of $98.94. By how much did the high value differ from the low value?

Sources: The New York Stock Exchange; The Quaker Oats Company

This problem appears as Example 1 in Section 4.7.

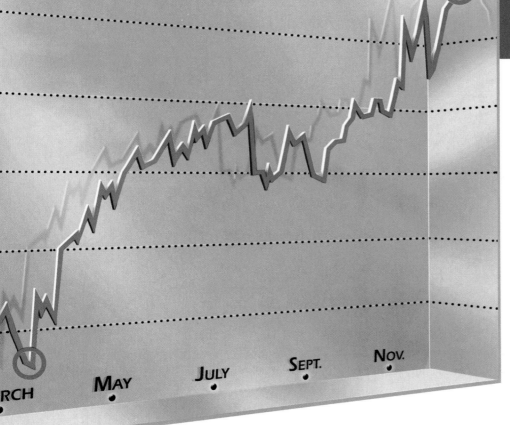

THE QUAKER OATS COMPANY, 2000

MARCH   MAY   JULY   SEPT.   NOV.

1. Write a word name: 2.347.   [4.1a]

2. Write a word name, as on a check, for $3264.78.
   [4.1a]

Write fraction notation.   [4.1b]

3. 0.21

4. 5.408

Write decimal notation.   [4.1b]

5. $\dfrac{379}{1000}$

6. $28\dfrac{439}{1000}$

Which number is larger?   [4.1c]

7. 3.2, 0.321

8. 0.099, 0.091

Round 21.0448 to the nearest: [4.1d]

9. Tenth.

10. Thousandth.

11. Add: 6 0 1.3   [4.2a]
               5.8 1
        +   0.1 0 9

12. Subtract: 4 0.0   [4.2b]
        −   0.9 0 9 9

Multiply.   [4.3a]

13.   0.8 3 5
     ×   0.7 4

14. 0.001×324.56

Divide.   [4.4a]

15. $6.6 \overline{)\, 2\,0\,0.6\,4}$

16. $\dfrac{576.98}{1000}$

Solve.

17. $9.6 \cdot y = 808.896$ [4.4b]

18. $54.96 + q = 6400.117$ [4.2c]

Solve.   [4.7a]

19. **Travel Distance.**   On a three-day trip, a traveler drove these distances: 432.6 mi, 179.2 mi, and 469.8 mi. What is the total number of miles driven?

20. A checking account contained $434.19. After a purchase of $148.24 was made using a debit card, how much was left in the account?

21. **DVD Purchase.**   Tanya bought 8 DVDs of the movie *The Matrix*. She paid $19.98 for each copy. What was the total cost?

22. A developer paid $47,567.89 for 14 acres of land. How much was paid for 1 acre? Round to the nearest cent.

23. Estimate the product 6.92×32.458 by rounding to the nearest one.   [4.6a]

Find decimal notation. Use multiplying by 1.
[4.5a]

24. $\dfrac{7}{5}$

25. $\dfrac{37}{40}$

Find decimal notation. Use division.   [4.5a]

26. $\dfrac{11}{4}$

27. $\dfrac{29}{7}$

Round $4.\overline{61}$ to the nearest: [4.5b]

28. Tenth.

29. Hundredth.

30. Thousandth.

31. Convert from cents to dollars: 949 cents.   [4.3b]

32. Convert to standard notation: 490 trillion.   [4.3b]

Calculate.

33. $(1 - 0.06)^2 + 8[5(12.1 - 7.8) + 20(17.3 - 8.7)]$
   [4.4c]

34. $\dfrac{2}{3} \times 89.95 - \dfrac{5}{9} \times 3.234$ [4.5c]

4.1

DECIMAL NOTATION, ORDER, AND ROUNDING

The set of **arithmetic numbers,** or **nonnegative rational numbers,** consists of the whole numbers 0, 1, 2, 3, 4, 5, 6, 7, 8, 9, 10, and so on, and fractions like $\frac{1}{2}, \frac{2}{3}, \frac{7}{8}, \frac{17}{10}$, and so on. We studied the use of fraction notation for arithmetic numbers in Chapters 2 and 3. In Chapter 4, we will study the use of *decimal notation.* The word *decimal* comes from the Latin word *decima,* meaning a tenth part. Although we are using different notation, we are still considering the same set of numbers. For example, instead of using fraction notation for $\frac{7}{8}$, we use decimal notation, 0.875, and instead of $48\frac{97}{100}$, we use 48.97.

Objectives

a Given decimal notation, write a word name, and write a word name for an amount of money.

b Convert between fraction notation and decimal notation.

c Given a pair of numbers in decimal notation, tell which is larger.

d Round decimal notation to the nearest thousandth, hundredth, tenth, one, ten, hundred, or thousand.

a Decimal Notation and Word Names

The Razor Kick Scooter® costs $148.97. The dot in $148.97 is called a **decimal point.** Since 0.97, or 97¢, is $\frac{97}{100}$ of a dollar, it follows that

$$\$148.97 = 148 + \frac{97}{100} \text{ dollars.}$$

Also, since $0.97, or 97¢, has the same value as

9 dimes + 7 cents

and 1 dime is $\frac{1}{10}$ of a dollar and 1 cent is $\frac{1}{100}$ of a dollar, we can write

$$148.97 = 1 \cdot 100 + 4 \cdot 10 + 8 \cdot 1 + 9 \cdot \frac{1}{10} + 7 \cdot \frac{1}{100}.$$

This is an extension of the expanded notation for whole numbers that we used in Chapter 1. The place values are 100, 10, 1, $\frac{1}{10}$, $\frac{1}{100}$, and so on. We can see this on a **place-value chart.** The value of each place is $\frac{1}{10}$ as large as the one to its left.

Let's see how to understand decimal notation using a place-value chart, using the following:

Source: Razor USA

PLACE-VALUE CHART							
Hundreds	Tens	Ones	Tenths	Hundredths	Thousandths	Ten-Thousandths	Hundred-Thousandths
100	10	1	$\frac{1}{10}$	$\frac{1}{100}$	$\frac{1}{1000}$	$\frac{1}{10,000}$	$\frac{1}{100,000}$
	2	9 .	5	2	9	7	

The women's record, held by Junxia Wang of China, in the 10,000-meter run is 29.5297 min.

Study Tips

QUIZ–TEST FOLLOW-UP

You may have just completed a chapter quiz or test. Immediately after each chapter quiz or test, write out a step-by-step solution of the questions you missed. Visit your instructor or tutor for help with problems that are still giving you trouble. When the week of the final examination arrives, you will be glad to have the excellent study guide these corrected tests provide.

Write a word name for the number.

1. Each person in the United States consumed an average of 15.3 lb of seafood in a recent year.
Source: National Oceanographic and Atmospheric Administration

2. In 1999, the race horse *Charismatic* won the Kentucky Derby in a time of 2.05333 min.
Source: *The New York Times Almanac, 2000*

3. 245.89

4. 34.0064

5. 31,079.764

Answers on page A-8

The decimal notation 29.5297 means

$$2 \text{ tens} + 9 \text{ ones} + 5 \text{ tenths} + 2 \text{ hundredths} + 9 \text{ thousandths} + 7 \text{ ten-thousandths}$$

or

$$2 \cdot 10 + 9 \cdot 1 + 5 \cdot \frac{1}{10} + 2 \cdot \frac{1}{100} + 9 \cdot \frac{1}{1000} + 7 \cdot \frac{1}{10,000}$$

or

$$20 + 9 + \frac{5}{10} + \frac{2}{100} + \frac{9}{1000} + \frac{7}{10,000}.$$

We read both 29.5297 and $29\frac{5297}{10,000}$ as

"Twenty-nine and five thousand two hundred ninety-seven ten-thousandths."

When we come to the decimal point, we read it as "and." We can also read 29.5297 as

"Two nine *point* five two nine seven."

To write a word name from decimal notation,

a) write a word name for the whole number (the number named to the left of the decimal point),

397.685 → Three hundred ninety-seven

b) write the word "and" for the decimal point, and

397.685 Three hundred ninety-seven and

c) write a word name for the number named to the right of the decimal point, followed by the place value of the last digit.

397.685 Three hundred ninety-seven and six hundred eighty-five *thousandths*

EXAMPLE 1 Write a word name for the number in this sentence: Arnold Schwarzenegger's body mass index is 27.7.

Twenty-seven and seven tenths

EXAMPLE 2 Write a word name for 410.87.

Four hundred ten and eighty-seven hundredths

EXAMPLE 3 Write a word name for the number in this sentence: The world record in the men's 800-meter run is 1.6852 min, held by Wilson Kipketer of Denmark.

One and six thousand eight hundred fifty-two ten-thousandths

EXAMPLE 4 Write a word name for 1788.405.

One thousand, seven hundred eighty-eight and four hundred five thousandths

Do Exercises 1–5 on the preceding page.

Decimal notation is also used with money. It is common on a check to write "and ninety-five cents" as "and $\frac{95}{100}$ dollars."

EXAMPLE 5 Write a word name for the amount on the check, $5876.95.

Five thousand, eight hundred seventy-six and $\frac{95}{100}$ dollars

Do Exercises 6 and 7.

Write a word name as on a check.

6. $4217.56

b   Converting Between Decimal Notation and Fraction Notation

We can find fraction notation as follows:

$$9.875 = 9 + \frac{8}{10} + \frac{7}{100} + \frac{5}{1000}$$

$$= 9 \cdot \frac{1000}{1000} + \frac{8}{10} \cdot \frac{100}{100} + \frac{7}{100} \cdot \frac{10}{10} + \frac{5}{1000}$$

$$= \frac{9000}{1000} + \frac{800}{1000} + \frac{70}{1000} + \frac{5}{1000} = \frac{9875}{1000}.$$

7. $13.98

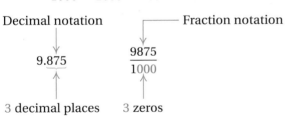

Decimal notation — Fraction notation

9.875 $\frac{9875}{1000}$

3 decimal places   3 zeros

To convert from decimal to fraction notation,

a) count the number of decimal places,   4.98

                                                   2 places

b) move the decimal point that many   4.98,   Move
   places to the right, and   2 places.

c) write the answer over a denominator $\frac{498}{100}$
   with a 1 followed by that number of zeros.   2 zeros

Answers on page A-8

EXAMPLE 6 Write fraction notation for 0.876. Do not simplify.

$$0.876 \qquad 0.876. \qquad 0.876 = \frac{876}{1000}$$

3 places   3 zeros

For a number like 0.876, we generally write a 0 before the decimal point to avoid forgetting or omitting it.

EXAMPLE 7 Write fraction notation for 56.23. Do not simplify.

$$56.23 \qquad 56.23. \qquad 56.23 = \frac{5623}{100}$$

2 places   2 zeros

EXAMPLE 8 Write fraction notation for 1.5018. Do not simplify.

$$1.5018 \qquad 1.5018. \qquad 1.5018 = \frac{15{,}018}{10{,}000}$$

4 places   4 zeros

Do Exercises 8–11.

If fraction notation has a denominator that is a power of ten, such as 10, 100, 1000, and so on, we reverse the procedure we used before.

> To convert from fraction notation to decimal notation when the denominator is 10, 100, 1000, and so on,
>
> a) count the number of zeros, and
>
> $$\frac{8679}{1000}$$
>
> 3 zeros
>
> b) move the decimal point that number of places to the left. Leave off the denominator.
>
> 8.679. Move 3 places.
>
> $$\frac{8679}{1000} = 8.679$$

EXAMPLE 9 Write decimal notation for $\frac{47}{10}$.

$$\frac{47}{10} \qquad\qquad 4.7. \qquad \frac{47}{10} = 4.7$$

1 zero   1 place

EXAMPLE 10 Write decimal notation for $\frac{123,067}{10,000}$.

$$\frac{123,067}{10,000}$$
$$\underset{\text{4 zeros}}{\uparrow}$$

$$12.3067.$$
$$\text{4 places}$$

$$\frac{123,067}{10,000} = 12.3067$$

EXAMPLE 11 Write decimal notation for $\frac{13}{1000}$.

$$\frac{13}{1000}$$
$$\underset{\text{3 zeros}}{\uparrow}$$

$$0.013.$$
$$\text{3 places}$$

$$\frac{13}{1000} = 0.013$$

EXAMPLE 12 Write decimal notation for $\frac{570}{100,000}$.

$$\frac{570}{100,000}$$
$$\underset{\text{5 zeros}}{\uparrow}$$

$$0.00570.$$
$$\text{5 places}$$

$$\frac{570}{100,000} = 0.0057$$

Do Exercises 12–17.

When denominators are numbers other than 10, 100, and so on, we will use another method for conversion. It will be considered in Section 4.5.

If a mixed numeral has a fractional part with a denominator that is a power of ten, such as 10, 100, or 1000, and so on, we first write the mixed numeral as a sum of a whole number and a fraction. Then we convert to decimal notation.

EXAMPLE 13 Write decimal notation for $23\frac{59}{100}$.

$$23\frac{59}{100} = 23 + \frac{59}{100} = 23 \text{ and } \frac{59}{100} = 23.59$$

EXAMPLE 14 Write decimal notation for $772\frac{129}{10,000}$.

$$772\frac{129}{10,000} = 772 + \frac{129}{10,000} = 772 \text{ and } \frac{129}{10,000} = 772.0129$$

Do Exercises 18–20.

Write decimal notation.

12. $\frac{743}{100}$

13. $\frac{406}{1000}$

14. $\frac{67,089}{10,000}$

15. $\frac{9}{10}$

16. $\frac{57}{1000}$

17. $\frac{830}{10,000}$

Write decimal notation.

18. $4\frac{3}{10}$

19. $283\frac{71}{100}$

20. $456\frac{13}{1000}$

Answers on page A-8

Which number is larger?

21. 2.04,   2.039

22. 0.06,   0.008

23. 0.5,   0.58

24. 1,   0.9999

25. 0.8989,   0.09898

26. 21.006,   21.05

c Order

To understand how to compare numbers in decimal notation, consider 0.85 and 0.9. First note that $0.9 = 0.90$ because $\frac{9}{10} = \frac{90}{100}$. Then $0.85 = \frac{85}{100}$ and $0.90 = \frac{90}{100}$. Since $\frac{85}{100} < \frac{90}{100}$, it follows that $0.85 < 0.90$. This leads us to a quick way to compare two numbers in decimal notation.

> **COMPARING NUMBERS IN DECIMAL NOTATION**
>
> To compare two numbers in decimal notation, start at the left and compare corresponding digits moving from left to right. If two digits differ, the number with the larger digit is the larger of the two numbers. To ease the comparison, extra zeros can be written to the right of the last decimal place.

EXAMPLE 15   Which of 2.109 and 2.1 is larger?

$$\begin{array}{l} \quad\quad\quad\quad \textit{Think.} \\ 2.109 \xrightarrow{\quad\quad} 2.109 \\ 2.1 \quad\quad\quad\quad 2.100 \\ \quad\quad\quad\quad\quad \text{Same} \quad \text{Different; } 9 > 0 \end{array}$$

Thus, 2.109 is larger than 2.1. That is, $2.109 > 2.1$.

EXAMPLE 16   Which of 0.09 and 0.108 is larger?

$$\begin{array}{l} \quad\quad\quad\quad \textit{Think.} \\ 0.09 \xrightarrow{\quad\quad} 0.090 \\ 0.108 \quad\quad\quad\quad 0.108 \\ \quad\quad\quad \text{Same} \quad \text{Different; } 1 > 0 \end{array}$$

Thus, 0.108 is larger than 0.09. That is, $0.108 > 0.09$.

Do Exercises 21–26.

d Rounding

Rounding is done as for whole numbers. To understand, we first consider an example using a number line. It might help to review Section 1.4.

EXAMPLE 17   Round 0.37 to the nearest tenth.

Here is part of a number line.

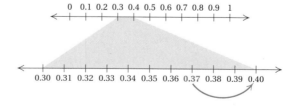

We see that 0.37 is closer to 0.40 than to 0.30. Thus, 0.37 rounded to the nearest tenth is 0.4.

ROUNDING DECIMAL NOTATION

To round to a certain place:

a) Locate the digit in that place.

b) Consider the next digit to the right.

c) If the digit to the right is 5 or higher, round up; if the digit to the right is 4 or lower, round down.

EXAMPLE 18 Round 3872.2459 to the nearest tenth.

a) Locate the digit in the tenths place, 2.

3 8 7 2 . 2̲ 4 5 9

b) Consider the next digit to the right, 4.

3 8 7 2 . 2 4̲ 5 9

c) Since that digit, 4, is less than 5, round down.

3 8 7 2 . 2 ← This is the answer.

> **CAUTION!**
>
> 3872.3 is not a correct answer to Example 18. It is *incorrect* to round from the ten-thousandths digit over to the tenths digit, as follows:
>
> 3872.246 → 3872.25 → 3872.3.

EXAMPLE 19 Round 3872.2459 to the nearest thousandth, hundredth, tenth, one, ten, hundred, and thousand.

Thousandth:	3872.246	Ten:	3870
Hundredth:	3872.25	Hundred:	3900
Tenth:	3872.2	Thousand:	4000
One:	3872		

EXAMPLE 20 Round 14.8973 to the nearest hundredth.

a) Locate the digit in the hundredths place, 9.

1 4 . 8 9̲ 7 3

b) Consider the next digit to the right, 7.

1 4 . 8 9 7̲ 3

c) Since that digit, 7, is 5 or higher, round up. When we make the hundredths digit a 10, we carry 1 to the tenths place.

The answer is 14.90. Note that the 0 in 14.90 indicates that the answer is correct to the nearest hundredth.

EXAMPLE 21 Round 0.008 to the nearest tenth.

a) Locate the digit in the tenths place, 0.

0 . 0̲ 0 8

b) Consider the next digit to the right, 0.

0 . 0 0̲ 8

c) Since that digit, 0, is less than 5, round down.

The answer is 0.0.

Do Exercises 27–45.

Round to the nearest tenth.

27. 2.76

28. 13.85

29. 234.448

30. 7.009

Round to the nearest hundredth.

31. 0.636

32. 7.834

33. 34.675

34. 0.025

Round to the nearest thousandth.

35. 0.9434

36. 8.0038

37. 43.1119

38. 37.4005

Round 7459.3548 to the nearest:

39. Thousandth.

40. Hundredth.

41. Tenth.

42. One.

43. Ten. (*Caution:* "Tens" are not "tenths.")

44. Hundred.

45. Thousand.

Answers on page A-8

a   Write a word name for the number in the sentence.

1. *MP3.* The MP3 audio format is changing the way we obtain and listen to music, allowing us to download digital music from the Internet. The cost of a Creative Labs NOMAD II Digital Audio Player is $249.94.
Source: Creative Labs

2. *Microsoft.* Recently, the stock of Microsoft sold for $43.9375 per share.

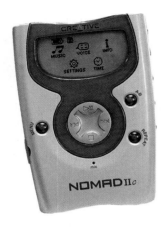

NASDAQ
NASDAQ COMPOSITE INDEX
MOST ACTIVE: SHARE VOLUME

	Vol.(000s)	Last	Change
SunMicro s	101,978	9.87	+.58
Cisco	83,072	14.94	+.52
Intel	47,702	21.96	+.41
DellCptr	42,577	22.56	+.24
Microsft	39,588	43.9375	+1.25
Oracle s	38,682	14.20	+.41
Qualcom	38,606	38.46	-3.54
JnprNtw	32,009	15.08	+1.81
WorldCom	30,837	13.33	-.60
JDS Uniph	27,145	6.92	-.10

3. *Quaker Oats.* Recently, the stock of Quaker Oats sold for $96.4375 per share.

4. *Water Weight.* One gallon of water weighs 8.35 lb.

Write a word name.

5. 34.891

6. 27.1245

Write a word name as on a check.

7. $326.48

8. $125.99

9. $36.72

10. $0.67

b   Write fraction notation. Do not simplify.

11. 8.3

12. 0.17

13. 3.56

14. 203.6

15. 46.03

16. 1.509

17. 0.00013

18. 0.0109

19. 1.0008

20. 2.0114

21. 20.003

22. 4567.2

Write decimal notation.

23. $\dfrac{8}{10}$

24. $\dfrac{51}{10}$

25. $\dfrac{889}{100}$

26. $\dfrac{92}{100}$

27. $\dfrac{3798}{1000}$

28. $\dfrac{780}{1000}$

29. $\dfrac{78}{10,000}$

30. $\dfrac{56,788}{100,000}$

31. $\dfrac{19}{100,000}$

32. $\dfrac{2173}{100}$

33. $\dfrac{376,193}{1,000,000}$

34. $\dfrac{8,953,074}{1,000,000}$

35. $99\dfrac{44}{100}$

36. $4\dfrac{909}{1000}$

37. $3\dfrac{798}{1000}$

38. $67\dfrac{83}{100}$

39. $2\dfrac{1739}{10,000}$

40. $9243\dfrac{1}{10}$

41. $8\dfrac{953,073}{1,000,000}$

42. $2256\dfrac{3059}{10,000}$

c Which number is larger?

43. 0.06,   0.58

44. 0.008,   0.8

45. 0.905,   0.91

46. 42.06,   42.1

47. 0.0009,   0.001

48. 7.067,   7.054

49. 234.07,   235.07

50. 0.99999,   1

51. 0.004, $\dfrac{4}{100}$

52. $\dfrac{73}{10}$,   0.73

53. 0.432,   0.4325

54. 0.8437,   0.84384

d Round to the nearest tenth.

55. 0.11

56. 0.85

57. 0.49

58. 0.5794

59. 2.7449

60. 4.78

61. 123.65

62. 36.049

Round to the nearest hundredth.

63. 0.893

64. 0.675

65. 0.6666

66. 6.529

67. 0.995

68. 207.9976

69. 0.094

70. 11.4246

Round to the nearest thousandth.

71. 0.3246

72. 0.6666

73. 17.0015

74. 123.4562

75. 10.1011

76. 0.1161

77. 9.9989

78. 67.100602

Round 809.4732 to the nearest:

79. Hundred.

80. Tenth.

81. Thousandth.

82. Hundredth.

83. One.

84. Ten.

Round 34.54389 to the nearest:

85. Ten-thousandth.

86. Thousandth.

87. Hundredth.

88. Tenth.

89. One.

90. Ten.

91. D_W Describe in your own words a procedure for converting from decimal notation to fraction notation.

92. D_W A fellow student rounds 236.448 to the nearest one and gets 237. Explain the possible error.

SKILL MAINTENANCE

Round 6172 to the nearest: [1.4a]

93. Ten.

94. Hundred.

95. Thousand.

Add or subtract.

96. $\begin{array}{r} 6\ 8\ 1 \\ +\ 1\ 4\ 9 \end{array}$ [1.2a]

97. $\dfrac{681}{1000} + \dfrac{149}{1000}$ [3.2a]

98. $\begin{array}{r} 2\ 6\ 7 \\ -\ \ \ 8\ 5 \end{array}$ [1.3b]

99. $\dfrac{267}{100} - \dfrac{85}{100}$ [3.3a]

Find the prime factorization. [2.1d]

100. 2000

101. 1530

102. 2002

103. 4312

SYNTHESIS

104. Arrange the following numbers in order from smallest to largest.

0.99, 0.099, 1, 0.9999, 0.89999, 1.00009, 0.909, 0.9889

105. Arrange the following numbers in order from smallest to largest.

2.1, 2.109, 2.108, 2.018, 2.0119, 2.0302, 2.000001

Truncating. There are other methods of rounding decimal notation. A computer often uses a method called **truncating.** To round using truncating, we drop off all decimal places past the rounding place, which is the same as changing all digits to the right to zeros. For example, rounding 6.78093456285102 to the ninth decimal place, using truncating, gives us 6.780934562. Use truncating to round each of the following to the fifth decimal place, that is, the hundred thousandth.

106. 6.78346123

107. 6.783461902

108. 99.999999999

109. 0.030303030303

4.2 ADDITION AND SUBTRACTION

Objectives

a Add using decimal notation.

b Subtract using decimal notation.

c Solve equations of the type $x + a = b$ and $a + x = b$, where a and b may be in decimal notation.

d Balance a checkbook.

a Addition

Adding with decimal notation is similar to adding whole numbers. First we line up the decimal points so that we can add corresponding place-value digits. Then we add digits from the right. For example, we add the thousandths, then the hundredths, and so on, carrying if necessary. If desired, we can write extra zeros to the right of the decimal point so that the number of places is the same.

EXAMPLE 1 Add: $56.314 + 17.78$.

$$\begin{array}{r} 5\;6\;.\;3\;1\;4 \\ +\;1\;7\;.\;7\;8\;0 \\ \hline \end{array}$$

Lining up the decimal points in order to add
Writing an extra zero to the right of the decimal point

$$\begin{array}{r} 5\;6\;.\;3\;1\;\mathbf{4} \\ +\;1\;7\;.\;7\;8\;\mathbf{0} \\ \hline \mathbf{4} \end{array}$$

Adding thousandths

$$\begin{array}{r} 5\;6\;.\;3\;\mathbf{1}\;4 \\ +\;1\;7\;.\;7\;\mathbf{8}\;0 \\ \hline \mathbf{9}\;4 \end{array}$$

Adding hundredths

$$\begin{array}{r} 1 \\ 5\;6\;.\;\mathbf{3}\;1\;4 \\ +\;1\;7\;.\;\mathbf{7}\;8\;0 \\ \hline .\;\mathbf{0}\;9\;4 \end{array}$$

Adding tenths
Write a decimal point in the answer.

We get 10 tenths = 1 one + 0 tenths, so we carry the 1 to the ones column.

$$\begin{array}{r} 1\;\;1 \\ 5\;\mathbf{6}\;.\;3\;1\;4 \\ +\;1\;\mathbf{7}\;.\;7\;8\;0 \\ \hline \mathbf{4}\;.\;0\;9\;4 \end{array}$$

Adding ones

We get 14 ones = 1 ten + 4 ones, so we carry the 1 to the tens column.

$$\begin{array}{r} 1\;\;1 \\ \mathbf{5}\;6\;.\;3\;1\;4 \\ +\;\mathbf{1}\;7\;.\;7\;8\;0 \\ \hline \mathbf{7}\;4\;.\;0\;9\;4 \end{array}$$

Adding tens

Do Exercises 1 and 2.

EXAMPLE 2 Add: $3.42 + 0.237 + 14.1$.

$$\begin{array}{r} 3.4\;2\;0 \\ 0.2\;3\;7 \\ +\;1\;4.1\;0\;0 \\ \hline 1\;7.7\;5\;7 \end{array}$$

Lining up the decimal points and writing extra zeros

Adding

Do Exercises 3–5.

Add.

1.
$$\begin{array}{r} 0.8\;4\;7 \\ +\;1\;0.0\;7 \\ \hline \end{array}$$

2.
$$\begin{array}{r} 2.1 \\ 0.7\;3\;9 \\ +\;3\;1.3\;6\;8\;9 \\ \hline \end{array}$$

Add.

3. $0.02 + 4.3 + 0.649$

4. $0.12 + 3.006 + 0.4357$

5. $0.4591 + 0.2374 + 8.70894$

Answers on page A-8

Add.

6. 789 + 123.67

7. 45.78 + 2467 + 1.993

Subtract.

8. 37.428 − 26.674

9.   0.3 4 7
       − 0.0 0 8

Answers on page A-8

Consider the addition 3456 + 19.347. Keep in mind that any whole number has an "unwritten" decimal point at the right, with 0 fractional parts. For example, 3456 can also be written 3456.000. When adding, we can always write in that decimal point and extra zeros if desired.

EXAMPLE 3   Add: 3456 + 19.347.

$$
\begin{array}{r}
\overset{1}{}\ 3\ 4\ 5\ 6.0\ 0\ 0 \\
+\ 1\ 9.3\ 4\ 7 \\
\hline
3\ 4\ 7\ 5.3\ 4\ 7
\end{array}
$$

Writing in the decimal point and extra zeros
Lining up the decimal points
Adding

Do Exercises 6 and 7.

b   Subtraction

Subtracting with decimal notation is similar to subtracting whole numbers. First we line up the decimal points so that we can subtract corresponding place-value digits. Then we subtract digits from the right. For example, we subtract the thousandths, then the hundredths, the tenths, and so on, borrowing if necessary.

EXAMPLE 4   Subtract: 56.314 − 17.78.

$$
\begin{array}{r}
5\ 6.3\ 1\ 4 \\
-\ 1\ 7.7\ 8\ 0 \\
\end{array}
$$

Lining up the decimal points in order to subtract
Writing an extra 0

$$
\begin{array}{r}
5\ 6.3\ 1\ 4 \\
-\ 1\ 7.7\ 8\ 0 \\
\hline
4
\end{array}
$$

Subtracting thousandths

$$
\begin{array}{r}
\overset{2\ \ 11}{5\ 6.3\ \cancel{1}\ 4} \\
-\ 1\ 7.7\ 8\ 0 \\
\hline
3\ 4
\end{array}
$$

Borrowing tenths to subtract hundredths

$$
\begin{array}{r}
\overset{\ \ \ \ 12}{\overset{5\ \ 2\ 11}{5\ \cancel{6}.\cancel{3}\ \cancel{1}\ 4}} \\
-\ 1\ 7.7\ 8\ 0 \\
\hline
.5\ 3\ 4
\end{array}
$$

Borrowing ones to subtract tenths
Writing a decimal point

$$
\begin{array}{r}
\overset{15\ 12}{\overset{4\ \ 5\ \ 2\ 11}{\cancel{5}\ \cancel{6}.\cancel{3}\ \cancel{1}\ 4}} \\
-\ 1\ 7.7\ 8\ 0 \\
\hline
8.5\ 3\ 4
\end{array}
$$

Borrowing tens to subtract ones

$$
\begin{array}{r}
\overset{15\ 12}{\overset{4\ \ 5\ \ 2\ 11}{\cancel{5}\ \cancel{6}.\cancel{3}\ \cancel{1}\ 4}} \\
-\ 1\ 7.7\ 8\ 0 \\
\hline
3\ 8.5\ 3\ 4
\end{array}
$$

Subtracting tens

CHECK:
$$
\begin{array}{r}
\overset{1\ \ \ 1\ \ \ 1}{3\ 8.5\ 3\ 4} \\
+\ 1\ 7.7\ 8\ 0 \\
\hline
5\ 6.3\ 1\ 4
\end{array}
$$

Do Exercises 8 and 9.

EXAMPLE 5   Subtract: 13.07 − 9.205.

$$
\begin{array}{r}
\overset{12}{} \\
\overset{\cancel{2}\ \ \overset{10}{\cancel{6}}\ \overset{10}{}}{1\ 3.0\ 7\ 0} \\
-\quad 9.2\ 0\ 5 \\
\hline
3.8\ 6\ 5
\end{array}
$$

Writing an extra zero

Subtracting

EXAMPLE 6   Subtract: 23.08 − 5.0053.

$$
\begin{array}{r}
\overset{1\ \ 13\ \ \ \ \ 7\ \ 9\ \ 10}{2\ 3.0\ 8\ 0\ 0} \\
-\quad 5.0\ 0\ 5\ 3 \\
\hline
1\ 8.0\ 7\ 4\ 7
\end{array}
$$

Writing two extra zeros

Subtracting

Do Exercises 10–12.

When subtraction involves a whole number, again keep in mind that there is an "unwritten" decimal point that can be written in if desired. Extra zeros can also be written in to the right of the decimal point.

EXAMPLE 7   Subtract: 456 − 2.467.

$$
\begin{array}{r}
\overset{5\ \ 9\ \ 9\ \ 10}{4\ 5\ 6.0\ 0\ 0} \\
-\qquad 2.4\ 6\ 7 \\
\hline
4\ 5\ 3.5\ 3\ 3
\end{array}
$$

Writing in the decimal point and extra zeros

Subtracting

Do Exercises 13 and 14.

CALCULATOR CORNER

Addition and Subtraction with Decimal Notation   To use a calculator to add and subtract with decimal notation, we use the $\boxed{\cdot}$, $\boxed{+}$, $\boxed{-}$, and $\boxed{=}$ keys. To find 47.046 − 28.193, for example, we press $\boxed{4}\boxed{7}\boxed{\cdot}\boxed{0}\boxed{4}\boxed{6}\boxed{-}\boxed{2}\boxed{8}\boxed{\cdot}\boxed{1}\boxed{9}\boxed{3}\boxed{=}$. The display reads $\boxed{18.853}$, so 47.046 − 28.193 = 18.853.

Exercises:

Use a calculator to add.

1. $\begin{array}{r} 2\ 7\ 4.1\ 5\ 9 \\ +\quad 4\ 3.4\ 8\ 6 \\ \hline \end{array}$

2. $\begin{array}{r} 1\ 9.8\ 0\ 5 \\ +\ 4\ 8\ 6.7\ 4\ 8 \\ \hline \end{array}$

3. 1.7 + 14.56 + 0.89

4. 3.4 + 45 + 0.68

Use a calculator to subtract.

5. $\begin{array}{r} 9.2 \\ -\ 4.8 \\ \hline \end{array}$

6. $\begin{array}{r} 5\ 2.3\ 4 \\ -\ 1\ 8.5\ 1 \\ \hline \end{array}$

7. 489 − 34.26

8. 6.09 − 5.1

Subtract.

10. 1.2345 − 0.7

11. 0.9564 − 0.4392

12. 7.37 − 0.00008

Subtract.

13. 1277 − 82.78

14. 5 − 0.0089

Answers on page A-8

Solve.

15. $x + 17.78 = 56.314$

[C] Solving Equations

Now let's solve equations $x + a = b$ and $a + x = b$, where a and b may be in decimal notation. Proceeding as we have before, we subtract a on both sides.

EXAMPLE 8 Solve: $x + 28.89 = 74.567$.

We have

$$x + 28.89 - 28.89 = 74.567 - 28.89 \qquad \text{Subtracting 28.89 on both sides}$$
$$x = 45.677.$$

$$\begin{array}{r} \overset{6\ \ 13\ 14\ 16}{7\ 4.5\ 6\ 7} \\ -\ 2\ 8.8\ 9\ 0 \\ \hline 4\ 5.6\ 7\ 7 \end{array}$$

The solution is 45.677.

EXAMPLE 9 Solve: $0.8879 + y = 9.0026$.

We have

$$0.8879 + y - 0.8879 = 9.0026 - 0.8879 \qquad \text{Subtracting 0.8879 on both sides}$$
$$y = 8.1147.$$

$$\begin{array}{r} \overset{8\ \ 9\ \ 9\ 11\ 16}{9.0\ 0\ 2\ 6} \\ -\ 0.8\ 8\ 7\ 9 \\ \hline 8.1\ 1\ 4\ 7 \end{array}$$

The solution is 8.1147.

16. $8.906 + t = 23.07$

Do Exercises 15 and 16.

EXAMPLE 10 Solve: $120 + x = 4380.6$.

We have

$$120 + x - 120 = 4380.6 - 120 \qquad \text{Subtracting 120 on both sides}$$
$$x = 4260.6$$

$$\begin{array}{r} 4\ 3\ 8\ 0.6 \\ -\ \ 1\ 2\ 0.0 \\ \hline 4\ 2\ 6\ 0.6 \end{array}$$

17. Solve: $241 + y = 2374.5$.

The solution is 4260.6.

Do Exercise 17.

[d] Balancing a Checkbook

Let's use addition and subtraction with decimals to balance a checkbook.

EXAMPLE 11 Find the errors, if any, in the balances in this checkbook.

20___		RECORD ALL CHARGES OR CREDITS THAT AFFECT YOUR ACCOUNT					
DATE	CHECK NUMBER	TRANSACTION DESCRIPTION	√ T	(−) PAYMENT/ DEBIT	(+ OR −) OTHER	(+) DEPOSIT/ CREDIT	BALANCE FORWARD
							8767 73
8/16	432	Burch Laundry		23 56			8744 16
8/19	433	Rogers TV		20 49			8764 65
8/20		Deposit				85 00	8848 65
8/21	434	Galaxy Records		48 60			8801 05
8/22	435	Electric Works		267 95			8533 09

There are two ways to determine whether there are errors. We assume that the amount $8767.73 in the "Balance forward" column is correct. If we can determine that the ending balance is correct, we have some assurance that the checkbook is correct. But two errors could offset each other to give us that balance.

METHOD 1

a) We add the debits:

$$23.56 + 20.49 + 48.60 + 267.95 = 360.60.$$

b) We add the deposits/credits. In this case, there is only one deposit, 85.00.

c) We add the total of the deposits to the balance brought forward:

$$8767.73 + 85.00 = 8852.73.$$

d) We subtract the total of the debits:

$$8852.73 - 360.60 = 8492.13.$$

The result should be the ending balance, 8533.09. We see that $8492.13 \neq 8533.09$. Since the numbers are not equal, we proceed to method 2.

METHOD 2   We successively add or subtract deposit/credits and debits, and check the result in the "Balance forward" column.

$$8767.73 - 23.56 = 8744.17.$$

We have found our first error. The subtraction was incorrect. We correct it and continue, using 8744.17 as the corrected balance forward:

$$8744.17 - 20.49 = 8723.68.$$

It looks as though 20.49 was added instead of subtracted. Actually, we would have to correct this line even if it had been subtracted, because the error of 1¢ in the first step has been carried through successive calculations. We correct that balance line and continue, using 8723.68 as the balance and adding the deposit 85.00:

$$8723.68 + 85.00 = 8808.68.$$

We make the correction and continue subtracting the last two debits:

$$8808.68 - 48.60 = 8760.08.$$

Then

$$8760.08 - 267.95 = 8492.13.$$

The corrected checkbook is below.

20___		RECORD ALL CHARGES OR CREDITS THAT AFFECT YOUR ACCOUNT						
DATE	CHECK NUMBER	TRANSACTION DESCRIPTION	√T	(−) PAYMENT/ DEBIT	(+ OR −) OTHER	(+) DEPOSIT/ CREDIT	BALANCE FORWARD 8767 73	
8/16	432	Burch Laundry		23 56			8744 16	→ 8744.17
8/19	433	Rogers TV		20 49			8764 65	→ 8723.68
8/20		Deposit				85 00	8848 65	→ 8808.68
8/21	434	Galaxy Records		48 60			8801 05	→ 8760.08
8/22	435	Electric Works		267 95			8533 09	→ 8492.13

Do Exercise 18.

There are other ways in which errors can be made in checkbooks, such as forgetting to record a transaction or writing the amounts incorrectly, but we will not consider those here.

18. Find the errors, if any, in this checkbook.

20	DATE	CHECK NUMBER	TRANSACTION DESCRIPTION	√T	(−) PAYMENT/ DEBIT	(+ OR −) OTHER	(+) DEPOSIT/ CREDIT	BALANCE FORWARD
								3078 92
	12/1	888	HH Gregg Appliances		340 69			2738 23
	12/3	889	Marie Callendar's Pies		78 54			2659 66
	12/5		Deposit <Paycheck>				230 80	2890 46
	12/6	890	Chili's Restaurant		13 14			2877 32
	12/8	891	Stonecreek Golf Course		48 00			2829 32
	12/8		Deposit <Molly>				39 58	2868 90
	12/10	892	Galyan's Trading Post		102 87			2766 83
	12/14	893	Goody's Music		48 59			2697 45
	12/15	894	Salvation Army		100 00			2497 45

Answer on page A-9

a   Add.

1.
```
  3 1 6.2 5
+   1 8.1 2
```

2.
```
  6 4 1.8 0 3
+   1 4.9 3 5
```

3.
```
  6 5 9.4 0 3
+ 9 1 6.8 1 2
```

4.
```
  4 2 0 3.2 8
+       3.3 9
```

5.
```
      9.1 0 4
+ 1 2 3.4 5 6
```

6.
```
  6.1 5 2 8
+ 5.2 7 7 7
```

7.
```
  8 1.0 0 8
+   3.4 0 9
```

8. $0.8096 + 0.7856$

9. $20.0124 + 30.0124$

10. $0.687 + 0.9$

11. $39 + 1.007$

12. $0.845 + 10.02$

13. $0.34 + 3.5 + 0.127 + 768$

14. $2.3 + 0.729 + 23$

15. $17 + 3.24 + 0.256 + 0.3689$

16.
```
      4 7.8
  2 1 9.8 5 2
    4 3.5 9
+ 6 6 6.7 1 3
```

17.
```
      2.7 0 3
     7 8.3 3
    2 8.0 0 0 9
+ 1 1 8.4 3 4 1
```

18.
```
        1 3.7 2
         9.1 1 2
  6 5 4 2.7 9 0 8
+      2 3.9 0 1
```

19. $99.6001 + 7285.18 + 500.042 + 870$

20. $65.987 + 9.4703 + 6744.02 + 1.0003 + 200.895$

b   Subtract.

21.
```
  5.2
- 3.9
```

22.
```
  4 4.3 4 5
-    3.1 0 5
```

23.
```
  5 1.3 1
-    2.2 9
```

24.
```
  8 7.4 6
-    6.3 2
```

25.
```
  4 8.7 6
-    3.1 5
```

26.
```
  9 7.0 1
-    3.1 5
```

27.
```
  9 2.3 4 1
-      6.4 2
```

28.
```
  0.8 4 6 8
- 0.0 3 4
```

29.
```
   2.5
-0.0025
```

30.
```
  3 9.0
-    0.2 8
```

31.
```
   3.4
-0.0 0 3
```

32.
```
   2.8
- 2.0 8
```

33. $28.2 - 19.35$

34. $100.16 - 0.118$

35. $34.07 - 30.7$

36. $36.2 - 16.28$

37. $8.45 - 7.405$

38. $3.801 - 2.81$

39. $6.003 - 2.3$

40. $9.087 - 8.807$

41. $1 - 0.0098$

42. $2 - 1.0908$

43. $100 - 0.34$

44. $624 - 18.79$

45. $7.48 - 2.6$

46. $18.4 - 5.92$

47. $3 - 2.006$

48. $263.7 - 102.08$

49. $19 - 1.198$

50. $2548.98 - 2.007$

51. $65 - 13.87$

52. $45 - 0.999$

53. $3.907 - 1.416$

54. $70.0009 - 23.0567$

55.
$$\begin{array}{r} 3\,2.7\,9\,7\,8 \\ -\quad 0.0\,5\,9\,2 \\ \hline \end{array}$$

56.
$$\begin{array}{r} 0.4\,9\,6\,3\,4 \\ -\ 0.1\,2\,6\,7\,8 \\ \hline \end{array}$$

57.
$$\begin{array}{r} 3.0\,0\,7\,4 \\ -\ 1.3\,4\,0\,8 \\ \hline \end{array}$$

58.
$$\begin{array}{r} 6.0\,7 \\ -\ 2.0\,0\,7\,8 \\ \hline \end{array}$$

59.
$$\begin{array}{r} 2\,3\,4\,5.9\,0\,7\,8\,6 \\ -\qquad\quad 0.9\,9\,9 \\ \hline \end{array}$$

60.
$$\begin{array}{r} 1.0 \\ -\ 0.9\,9\,9\,9 \\ \hline \end{array}$$

---

c Solve.

61. $x + 17.5 = 29.15$

62. $t + 50.7 = 54.07$

63. $3.205 + m = 22.456$

64. $4.26 + q = 58.32$

65. $17.95 + p = 402.63$

66. $w + 1.3004 = 47.8$

67. $13{,}083.3 = x + 12{,}500.33$

68. $100.23 = 67.8 + z$

69. $x + 2349 = 17{,}684.3$

70. $1830.4 + t = 23{,}067$

---

d Find the errors, if any, in each checkbook.

71.

20____		RECORD ALL CHARGES OR CREDITS THAT AFFECT YOUR ACCOUNT						
DATE	CHECK NUMBER	TRANSACTION DESCRIPTION	√ T	(−) PAYMENT/ DEBIT	(+ OR −) OTHER	(+) DEPOSIT/ CREDIT	\multicolumn{2}{c}{BALANCE FORWARD}
							9704	56
8/8	342	Bill Rydman		27 44			9,677	12
8/9		Deposit <Beauty Contest>				1000 00	10,677	12
8/12	343	Jason Jordan		123 95			10,553	17
8/14	344	Jennifer Crum		124 02			10,677	19
8/22	345	Neon Johnny's Pizza		12 43			10,664	76
8/24		Deposit <Bowling Tournament>				2500 00	13,164	76
8/29	346	Border's Bookstore		137 78			13,302	54
9/2		Deposit <Bodybuilder Contest>				18 88	13,283	66
9/3	347	Fireman's Fund		2800 00			10,483	66

72.

				20___ RECORD ALL CHARGES OR CREDITS THAT AFFECT YOUR ACCOUNT			
DATE	CHECK NUMBER	TRANSACTION DESCRIPTION	√T	(−) PAYMENT/ DEBIT	(+ OR −) OTHER	(+) DEPOSIT/ CREDIT	BALANCE FORWARD 1876 43
4/1	500	Ed Moura		500 12			1376 31
4/3	501	Jim Lawler		28 56			1347 75
4/3		Deposit <State Lottery>				10,000 00	11,347 75
4/3	502	Victoria Montoya		464 00			10,883 75
4/3		Deposit <Jewelry Sale>				2500 00	8383 75
4/4	503	Baskin & Robbins		1600 00			6783 75
4/8	504	Golf Galaxy		1349 98			5433 77
4/12	505	Don Mitchell Pro Shops		658 97			4774 80
4/13		Deposit <Publisher's Clearing House>				100000 00	104,774 80
4/15	506	American Airlines		6885 58			98,889 22

73. D_W Explain the error in the following:

Add.

$$\begin{array}{r} 1\ 3.0\ 7 \\ +\ \ \ 9.2\ 0\ 5 \\ \hline 1\ 0.5\ 1\ 2 \end{array}$$

74. D_W Explain the error in the following:

Subtract.

$$\begin{array}{r} 7\ 3.0\ 8\ 9 \\ -\ \ \ 5.0\ 0\ 6\ 1 \\ \hline 2.3\ 0\ 2\ 8 \end{array}$$

75. Round 34,567 to the nearest thousand. [1.4a]

76. Round 34,496 to the nearest thousand. [1.4a]

Subtract.

77. $\dfrac{13}{24} - \dfrac{3}{8}$ [3.3a]

78. $\dfrac{8}{9} - \dfrac{2}{15}$ [3.3a]

79. $8805 - 2639$ [1.3b]

80. $8005 - 2639$ [1.3b]

Solve.

81. A serving of filleted fish is generally considered to be about $\frac{1}{3}$ lb. How many servings can be prepared from $5\frac{1}{2}$ lb of flounder fillet? [3.6c]

82. A photocopier technician drove $125\frac{7}{10}$ mi away from Scottsdale for a repair call. The next day he drove $65\frac{1}{2}$ mi back toward Scottsdale for another service call. How far was the technician from Scottsdale? [3.5c]

83. A student presses the wrong button when using a calculator and adds 235.7 instead of subtracting it. The incorrect answer is 817.2. What is the correct answer?

Objectives

a   Multiply using decimal notation.

b   Convert from notation like 45.7 million to standard notation, and from dollars to cents and cents to dollars.

a   Multiplication

Let's find the product

$$2.3 \times 1.12.$$

To understand how we find such a product, we first convert each factor to fraction notation. Next, we multiply the whole numbers 23 and 112, and then divide by 1000.

$$2.3 \times 1.12 = \frac{23}{10} \times \frac{112}{100} = \frac{23 \times 112}{10 \times 100} = \frac{2576}{1000} = 2.576$$

Note the number of decimal places.

$$
\begin{array}{r}
1.1\ 2 \quad \text{(2 decimal places)} \\
\times \quad 2.3 \quad \text{(1 decimal place)} \\
\hline
2.5\ 7\ 6 \quad \text{(3 decimal places)}
\end{array}
$$

Now consider

$$0.011 \times 15.0002 = \frac{11}{1000} \times \frac{150{,}002}{10{,}000} = \frac{1{,}650{,}022}{10{,}000{,}000} = 0.1650022.$$

Note the number of decimal places.

$$
\begin{array}{r}
1\ 5.0\ 0\ 0\ 2 \quad \text{(4 decimal places)} \\
\times \quad 0.0\ 1\ 1 \quad \text{(3 decimal places)} \\
\hline
0.1\ 6\ 5\ 0\ 0\ 2\ 2 \quad \text{(7 decimal places)}
\end{array}
$$

To multiply using decimals: 0.8×0.43

a) Ignore the decimal points and multiply as though both factors were whole numbers.

$$
\begin{array}{r}
{\scriptstyle 2} \\
0.4\ 3 \\
\times \quad 0.8 \\
\hline
3\ 4\ 4
\end{array}
$$
Ignore the decimal points for now.

b) Then place the decimal point in the result. The number of decimal places in the product is the sum of the numbers of places in the factors (count places from the right).

$$
\begin{array}{r}
0.4\ 3 \quad \text{(2 decimal places)} \\
\times \quad 0.8 \quad \text{(1 decimal place)} \\
\hline
0.3\ 4\ 4 \quad \text{(3 decimal places)}
\end{array}
$$

EXAMPLE 1   Multiply: 8.3×74.6.

a) Ignore the decimal points and multiply as though factors were whole numbers:

$$
\begin{array}{r}
{\scriptstyle 3}\ \ {\scriptstyle 4} \\
{\scriptstyle 1}\ \ {\scriptstyle 1} \\
7\ 4.6 \\
\times \quad 8.3 \\
\hline
2\ 2\ 3\ 8 \\
5\ 9\ 6\ 8\ 0 \\
\hline
6\ 1\ 9\ 1\ 8
\end{array}
$$

b) Place the decimal point in the result. The number of decimal places in the product is the sum, $1 + 1$, of the number of places in the factors.

$$
\begin{array}{r}
7\ 4.6 \quad \text{(1 decimal place)} \\
\times \qquad 8.3 \quad \text{(1 decimal place)} \\
\hline
2\ 2\ 3\ 8 \\
5\ 9\ 6\ 8\ 0 \\
\hline
6\ 1\ 9.1\ 8 \quad \text{(2 decimal places)}
\end{array}
$$

Do Exercise 1.

EXAMPLE 2 Multiply: 0.0032×2148.

As we catch on to the skill, we can combine the two steps.

$$
\begin{array}{r}
2\ 1\ 4\ 8 \quad \text{(0 decimal places)} \\
\times\ 0.0\ 0\ 3\ 2 \quad \text{(4 decimal places)} \\
\hline
4\ 2\ 9\ 6 \\
6\ 4\ 4\ 4\ 0 \\
\hline
6.8\ 7\ 3\ 6 \quad \text{(4 decimal places)}
\end{array}
$$

EXAMPLE 3 Multiply: 0.14×0.867.

$$
\begin{array}{r}
0.8\ 6\ 7 \quad \text{(3 decimal places)} \\
\times \qquad 0.1\ 4 \quad \text{(2 decimal places)} \\
\hline
3\ 4\ 6\ 8 \\
8\ 6\ 7\ 0 \\
\hline
0.1\ 2\ 1\ 3\ 8 \quad \text{(5 decimal places)}
\end{array}
$$

Do Exercises 2 and 3.

MULTIPLYING BY 0.1, 0.01, 0.001, AND SO ON

Now let's consider some special kinds of products. The first involves multiplying by a tenth, hundredth, thousandth, or ten-thousandth. Let's look at those products.

$$0.1 \times 38 = \frac{1}{10} \times 38 = \frac{38}{10} = 3.8$$

$$0.01 \times 38 = \frac{1}{100} \times 38 = \frac{38}{100} = 0.38$$

$$0.001 \times 38 = \frac{1}{1000} \times 38 = \frac{38}{1000} = 0.038$$

$$0.0001 \times 38 = \frac{1}{10,000} \times 38 = \frac{38}{10,000} = 0.0038$$

Note in each case that the product is *smaller* than 38.

---

1. Multiply.

$$
\begin{array}{r}
8\ 5.4 \\
\times \qquad 6.2 \\
\hline
\end{array}
$$

Multiply.

2.
$$
\begin{array}{r}
1\ 2\ 3\ 4 \\
\times\ 0.0\ 0\ 4\ 1 \\
\hline
\end{array}
$$

3.
$$
\begin{array}{r}
4\ 2.6\ 5 \\
\times\ 0.8\ 0\ 4 \\
\hline
\end{array}
$$

Answers on page A-9

Multiply.

4. 0.1×3.48

5. 0.01×3.48

6. 0.001×3.48

7. 0.0001×3.48

Multiply.

8. 10×3.48

9. 100×3.48

10. 1000×3.48

11. $10,000 \times 3.48$

Answers on page A-9

To multiply any number by 0.1, 0.01, 0.001, and so on,

a) count the number of decimal places in the tenth, hundredth, or thousandth, and so on, and

b) move the decimal point that many places to the left.

$$\underbrace{0.001}_{} \times 34.45678$$
$$\longrightarrow 3 \text{ places}$$

$$0.001 \times 34.45678 = 0.034.45678$$

Move 3 places to the left.

$$0.001 \times 34.45678 = 0.03445678$$

EXAMPLES Multiply.

4. $0.1 \times 14.605 = 1.4605 \qquad 1.4.605$

5. $0.01 \times 14.605 = 0.14605$

6. $0.001 \times 14.605 = 0.014605$

— We write an extra zero.

7. $0.0001 \times 14.605 = 0.0014605$

— We write two extra zeros.

Do Exercises 4–7.

MULTIPLYING BY 10, 100, 1000, AND SO ON

Next, let's consider multiplying by 10, 100, 1000, and so on. Let's look at those products.

$$10 \times 97.34 = 973.4$$
$$100 \times 97.34 = 9734$$
$$1000 \times 97.34 = 97,340$$
$$10,000 \times 97.34 = 973,400$$

Note in each case that the product is *larger* than 97.34.

To multiply any number by 10, 100, 1000, and so on,

a) count the number of zeros, and

b) move the decimal point that many places to the right.

$$\underbrace{1000}_{} \times 34.45678$$
$$\longrightarrow 3 \text{ zeros}$$

$$1000 \times 34.45678 = 34.456.78$$

Move 3 places to the right.

$$1000 \times 34.45678 = 34,456.78$$

EXAMPLES Multiply.

8. $10 \times 14.605 = 146.05 \qquad 14.6.05$

9. $100 \times 14.605 = 1460.5$

10. $1000 \times 14.605 = 14,605$

11. $10,000 \times 14.605 = 146,050 \qquad 14.6050.$

Do Exercises 8–11.

b Applications Using Multiplication with Decimal Notation

NAMING LARGE NUMBERS

We often see notation like the following in newspapers and magazines and on television.

> The largest building in the world is the Pentagon, which has 3.7 million square feet of floor space.
>
> By 2004, it is expected that $7.3 trillion dollars worth of business will be transacted over the Internet.
>
> In 1999, the U. S. Mint produced 11.6 billion pennies.

11.6 billion — Pennies
4.4 billion — Quarters
3.6 billion — Dimes
2.3 billion — Nickels

Source: U.S. Mint

To understand such notation, consider the information in the following table.

NAMING LARGE NUMBERS

1 hundred = 100 = 10^2 └→ 2 zeros

1 thousand = 1000 = 10^3 └→ 3 zeros

1 million = 1,000,000 = 10^6 └→ 6 zeros

1 billion = 1,000,000,000 = 10^9 └→ 9 zeros

1 trillion = 1,000,000,000,000 = 10^{12} └→ 12 zeros

CALCULATOR CORNER

Multiplication with Decimal Notation To use a calculator to multiply with decimal notation, we use the $\boxed{\cdot}$, $\boxed{\times}$, and $\boxed{=}$ keys. To find 4.78 × 0.34, for example, we press $\boxed{4}\ \boxed{\cdot}\ \boxed{7}\ \boxed{8}\ \boxed{\times}\ \boxed{\cdot}$ $\boxed{3}\ \boxed{4}\ \boxed{=}$. The display reads $\boxed{1.6252}$, so 4.78 × 0.34 = 1.6252.

Exercises: Use a calculator to multiply.

1. $\begin{array}{r} 5.4 \\ \times\ \ \ 9 \\ \hline \end{array}$

2. $\begin{array}{r} 4\ 1\ 5 \\ \times\ 1\ 6.7 \\ \hline \end{array}$

3. $\begin{array}{r} 1\ 7.6\ 3 \\ \times\ \ \ \ \ \ 8.1 \\ \hline \end{array}$

4. 0.04 × 12.69

5. 586.4 × 13.5

6. 4.003 × 5.1

Convert the number in the sentence to standard notation.

12. In 1999, the U.S. Mint produced 4.4 billion quarters.
Source: U.S. Mint

13. The largest building in the world is the Pentagon, which has 3.7 million square feet of floor space.

Convert from dollars to cents.

14. $15.69

15. $0.17

Convert from cents to dollars.

16. 35¢

17. 577¢

Answers on page A-9

To convert a large number to standard notation, we proceed as follows.

EXAMPLE 12 Convert the number in this sentence to standard notation: In 1999, the U.S. Mint produced 11.6 billion pennies.
Source: U.S. Mint

$$11.6 \text{ billion} = 11.6 \times 1 \text{ billion}$$
$$= 11.6 \times 1,\underbrace{000,000,000}_{9 \text{ zeros}}$$
$$= 11,600,000,000$$

Do Exercises 12 and 13.

MONEY CONVERSION

Converting from dollars to cents is like multiplying by 100. To see why, consider $19.43.

$19.43 = 19.43 \times \$1$	We think of $19.43 as 19.43 × 1 dollar, or 19.43 × $1.
$= 19.43 \times 100¢$	Substituting 100¢ for $1: $1 = 100¢
$= 1943¢$	Multiplying

> **DOLLARS TO CENTS**
>
> To convert from dollars to cents, move the decimal point two places to the right and change from the $ sign in front to the ¢ sign at the end.

EXAMPLES Convert from dollars to cents.

13. $189.64 = 18,964¢

14. $0.75 = 75¢

Do Exercises 14 and 15.

Converting from cents to dollars is like multiplying by 0.01. To see why, consider 65¢.

$65¢ = 65 \times 1¢$	We think of 65¢ as 65 × 1 cent, or 65 × 1¢.
$= 65 \times \$0.01$	Substituting $0.01 for 1¢: 1¢ = $0.01
$= \$0.65$	Multiplying

> **CENTS TO DOLLARS**
>
> To convert from cents to dollars, move the decimal point two places to the left and change from the ¢ sign at the end to the $ sign in front.

EXAMPLES Convert from cents to dollars.

15. 395¢ = $3.95

16. 8503¢ = $85.03

Do Exercises 16 and 17.

a Multiply.

1. $\begin{array}{r} 8.6 \\ \times\ \ \ 7 \\ \hline \end{array}$

2. $\begin{array}{r} 5.7 \\ \times\ 0.8 \\ \hline \end{array}$

3. $\begin{array}{r} 0.8\,4 \\ \times\ \ \ \ \ 8 \\ \hline \end{array}$

4. $\begin{array}{r} 9.4 \\ \times\ 0.6 \\ \hline \end{array}$

5. $\begin{array}{r} 6.3 \\ \times\ 0.0\,4 \\ \hline \end{array}$

6. $\begin{array}{r} 9.8 \\ \times\ 0.0\,8 \\ \hline \end{array}$

7. $\begin{array}{r} 8\,7 \\ \times\ 0.0\,0\,6 \\ \hline \end{array}$

8. $\begin{array}{r} 1\,8.4 \\ \times\ 0.0\,7 \\ \hline \end{array}$

9. 10×23.76

10. 100×3.8798

11. 1000×583.686852

12. 0.34×1000

13. 7.8×100

14. 0.00238×10

15. 0.1×89.23

16. 0.01×789.235

17. 0.001×97.68

18. 8976.23×0.001

19. 78.2×0.01

20. 0.0235×0.1

21. $\begin{array}{r} 3\,2.6 \\ \times\ \ \ 1\,6 \\ \hline \end{array}$

22. $\begin{array}{r} 9.2\,8 \\ \times\ \ \ 8.6 \\ \hline \end{array}$

23. $\begin{array}{r} 0.9\,8\,4 \\ \times\ \ \ \ \ 3.3 \\ \hline \end{array}$

24. $\begin{array}{r} 8.4\,8\,9 \\ \times\ \ \ \ \ 7.4 \\ \hline \end{array}$

25. $\begin{array}{r} 3\,7\,4 \\ \times\ \ \ 2.4 \\ \hline \end{array}$

26. $\begin{array}{r} 8\,6\,5 \\ \times\ \ \ 1.0\,8 \\ \hline \end{array}$

27. $\begin{array}{r} 7\,4\,9 \\ \times\ 0.4\,3 \\ \hline \end{array}$

28. $\begin{array}{r} 9\,7\,8 \\ \times\ 2\,0.5 \\ \hline \end{array}$

29. $\begin{array}{r} 0.8\,7 \\ \times\ \ \ 6\,4 \\ \hline \end{array}$

30. $\begin{array}{r} 7.2\,5 \\ \times\ \ \ 6\,0 \\ \hline \end{array}$

31. $\begin{array}{r} 4\,6.5\,0 \\ \times\ \ \ \ \ 7\,5 \\ \hline \end{array}$

32. $\begin{array}{r} 8.2\,4 \\ \times\ 7\,0\,3 \\ \hline \end{array}$

33.
$$\begin{array}{r} 8\,1.7 \\ \times\ 0.6\,1\,2 \\ \hline \end{array}$$

34.
$$\begin{array}{r} 3\,1.8\,2 \\ \times\ \ \ 7.1\,5 \\ \hline \end{array}$$

35.
$$\begin{array}{r} 1\,0.1\,0\,5 \\ \times\ 1\,1.3\,2\,4 \\ \hline \end{array}$$

36.
$$\begin{array}{r} 1\,5\,1.2 \\ \times\ 4.5\,5\,5 \\ \hline \end{array}$$

37.
$$\begin{array}{r} 1\,2.3 \\ \times\ 1.0\,8 \\ \hline \end{array}$$

38.
$$\begin{array}{r} 7.8\,2 \\ \times\ 0.0\,2\,4 \\ \hline \end{array}$$

39.
$$\begin{array}{r} 3\,2.4 \\ \times\ \ \ 2.8 \\ \hline \end{array}$$

40.
$$\begin{array}{r} 8.0\,9 \\ \times\ 0.0\,0\,7\,5 \\ \hline \end{array}$$

41.
$$\begin{array}{r} 0.0\,0\,3\,4\,2 \\ \times\ \ \ \ \ 0.8\,4 \\ \hline \end{array}$$

42.
$$\begin{array}{r} 2.0\,0\,5\,6 \\ \times\ \ \ \ \ 3.8 \\ \hline \end{array}$$

43.
$$\begin{array}{r} 0.3\,4\,7 \\ \times\ \ \ 2.0\,9 \\ \hline \end{array}$$

44.
$$\begin{array}{r} 2.5\,3\,2 \\ \times\ 1.0\,6\,7 \\ \hline \end{array}$$

45.
$$\begin{array}{r} 3.0\,0\,5 \\ \times\ 0.6\,2\,3 \\ \hline \end{array}$$

46.
$$\begin{array}{r} 1\,6.3\,4 \\ \times\ 0.0\,0\,0\,5\,1\,2 \\ \hline \end{array}$$

47. 1000×45.678

48. 0.001×45.678

---

b Convert from dollars to cents.

49. $28.88

50. $67.43

51. $0.66

52. $1.78

Convert from cents to dollars.

53. 34¢

54. 95¢

55. 3445¢

56. 933¢

Convert the number in the sentence to standard notation.

57. The average distance from the earth to the sun is 93 million miles. (This was a $1 million question on the TV quiz show "Who Wants to Be a Millionaire?")

58. In 2001, 3.5 million sport utility vehicles were sold.
Source: Autodata

59. In 1999, total box office sales at the movies was $7.2 billion.
Source: Motion Picture Association of America

60. By 2003, it is expected that $3.5 trillion dollars worth of business will be transacted over the Internet.

61. D_W If two rectangles have the same perimeter, will they also have the same area? Experiment with different dimensions. Be sure to use decimals. Explain your answer.

62. D_W A student insists that $346.708 \times 0.1 = 3467.08$. How could you convince him that a mistake had been made without checking on a calculator?

──────────────── SKILL MAINTENANCE ────────────────

Calculate.

63. $2\frac{1}{3} \cdot 4\frac{4}{5}$ [3.6a]

64. $2\frac{1}{3} \div 4\frac{4}{5}$ [3.6b]

65. $4\frac{4}{5} - 2\frac{1}{3}$ [3.5b]

66. $4\frac{4}{5} + 2\frac{1}{3}$ [3.5a]

Divide.   [1.6b]

67. $2\,4\,\overline{)\,8\,2\,0\,8}$

68. $4\,\overline{)\,3\,4\,8}$

69. $7\,\overline{)\,3\,1,9\,6\,2}$

70. $1\,8\,\overline{)\,2\,2,6\,2\,6}$

71. $4\,0\,\overline{)\,3\,4\,8\,0}$

72. $1\,7\,\overline{)\,2\,0,0\,0\,6}$

──────────────── SYNTHESIS ────────────────

Consider the following names for large numbers in addition to those already discussed in this section:

$$1 \text{ quadrillion} = 1,000,000,000,000,000 = 10^{15};$$
$$1 \text{ quintillion} = 1,000,000,000,000,000,000 = 10^{18};$$
$$1 \text{ sextillion} = 1,000,000,000,000,000,000,000 = 10^{21};$$
$$1 \text{ septillion} = 1,000,000,000,000,000,000,000,000 = 10^{24}.$$

Find each of the following. Express the answer with a name that is a power of 10.

73. (1 trillion) · (1 billion)

74. (1 million) · (1 billion)

75. (1 trillion) · (1 trillion)

76. Is a billion millions the same as a million billions? Explain.

Objectives

a Divide using decimal notation.

b Solve equations of the type $a \cdot x = b$, where a and b may be in decimal notation.

c Simplify expressions using the rules for order of operations.

Divide.

1. $9 \overline{)\, 5.4}$

2. $1\,5 \overline{)\, 2\,2.5}$

3. $8\,2 \overline{)\, 3\,8.5\,4}$

4.4 DIVISION

a Division

WHOLE-NUMBER DIVISORS

Compare these divisions by a whole number.

$$\frac{588}{7} = 84$$

$$\frac{58.8}{7} = 8.4$$

$$\frac{5.88}{7} = 0.84$$

$$\frac{0.588}{7} = 0.084$$

When we are dividing by a whole number, the number of decimal places in the *quotient* is the same as the number of decimal places in the *dividend*.

These examples lead us to this method for dividing by a whole number.

To divide by a whole number,

a) place the decimal point directly above the decimal point in the dividend, and

b) divide as though dividing whole numbers.

$$
\begin{array}{r}
0.8\;4 \leftarrow \text{Quotient}\\
\text{Divisor} \rightarrow 7 \overline{)\, 5.8\;8} \leftarrow \text{Dividend}\\
\underline{5\;6\;0}\\
2\;8\\
\underline{2\;8}\\
0 \leftarrow \text{Remainder}
\end{array}
$$

EXAMPLE 1 Divide: $379.2 \div 8$.

Place the decimal point.

$$
\begin{array}{r}
4\;7.4\\
8 \overline{)\, 3\;7\;9.2}\\
\underline{3\;2\;0\;0}\\
5\;9\;2\\
\underline{5\;6\;0}\\
3\;2\\
\underline{3\;2}\\
0
\end{array}
$$

Divide as though dividing whole numbers.

EXAMPLE 2 Divide: $82.08 \div 24$.

Place the decimal point.

$$
\begin{array}{r}
3.4\;2\\
2\,4 \overline{)\, 8\;2.0\;8}\\
\underline{7\;2\;0\;0}\\
1\;0\;0\;8\\
\underline{9\;6\;0}\\
4\;8\\
\underline{4\;8}\\
0
\end{array}
$$

Divide as though dividing whole numbers.

Do Exercises 1–3 on the preceding page.

Sometimes it helps to write some extra zeros to the right of the decimal point. They don't change the number.

EXAMPLE 3 Divide: $30 \div 8$.

$$
\begin{array}{r}
3. \\
8\,\overline{)\,3\,0.} \\
2\,4 \\
\hline
6
\end{array}
$$
Place the decimal point and divide to find how many ones

$$
\begin{array}{r}
3. \\
8\,\overline{)\,3\,0.0} \\
2\,4\downarrow \\
\hline
6\,0
\end{array}
$$
Write an extra zero.

$$
\begin{array}{r}
3.7 \\
8\,\overline{)\,3\,0.0} \\
2\,4 \\
\hline
6\,0 \\
5\,6 \\
\hline
4
\end{array}
$$
Divide to find how many tenths.

$$
\begin{array}{r}
3.7 \\
8\,\overline{)\,3\,0.0\,0} \\
2\,4 \\
\hline
6\,0 \\
5\,6\downarrow \\
\hline
4\,0
\end{array}
$$
Write an extra zero.

$$
\begin{array}{r}
3.7\,5 \\
8\,\overline{)\,3\,0.0\,0} \\
2\,4 \\
\hline
6\,0 \\
5\,6 \\
\hline
4\,0 \\
4\,0 \\
\hline
0
\end{array}
$$
Divide to find how many hundredths.

EXAMPLE 4 Divide: $4 \div 25$.

$$
\begin{array}{r}
0.1\,6 \\
2\,5\,\overline{)\,4.0\,0} \\
2\,5 \\
\hline
1\,5\,0 \\
1\,5\,0 \\
\hline
0
\end{array}
$$

Do Exercises 4–6.

Divide.

4. $2\,5\,\overline{)\,8}$

5. $4\,\overline{)\,1\,5}$

6. $8\,6\,\overline{)\,2\,1.5}$

Answers on page A-9

7. a) Complete.

$$\frac{3.75}{0.25} = \frac{3.75}{0.25} \times \frac{100}{100}$$

$$= \frac{(\quad)}{25}$$

b) Divide.

$$0.2\,5\,\overline{)\,3.7\,5}$$

Divide.

8. $0.8\,3\,\overline{)\,4.0\,6\,7}$

9. $3.5\,\overline{)\,4\,4.8}$

DIVISORS THAT ARE NOT WHOLE NUMBERS

Consider the division

$$0.2\,4\,\overline{)\,8.2\,0\,8}$$

We write the division as $\dfrac{8.208}{0.24}$. Then we multiply by 1 to change to a whole-number divisor:

The division $0.24\overline{)8.208}$ is the same as $24\overline{)820.8}$.

$$\frac{8.208}{0.24} = \frac{8.208}{0.24} \times \frac{100}{100} = \frac{820.8}{24}.$$

The divisor is now a whole number.

To divide when the divisor is not a whole number,

a) move the decimal point (multiply by 10, 100, and so on) to make the divisor a whole number;

$$0.2\,4\,\overline{)\,8.2\,0\,8}$$
Move 2 places to the right.

b) move the decimal point (multiply the same way) in the dividend the same number of places; and

$$0.2\,4\,\overline{)\,8.2\,0\,8}$$
Move 2 places to the right.

c) place the decimal point directly above the new decimal point in the dividend and divide as though dividing whole numbers.

$$
\begin{array}{r}
3\,4.2 \\
0.2\,4\,\overline{)\,8.2\,0{\scriptstyle\wedge}8} \\
7\,2\,0\,0 \\
\hline
1\,0\,0\,8 \\
9\,6\,0 \\
\hline
4\,8 \\
4\,8 \\
\hline
0
\end{array}
$$

(The new decimal point in the dividend is indicated by a caret.)

EXAMPLE 5 Divide: $5.848 \div 8.6$.

$$8.6\,\overline{)\,5.8\,4\,8}$$

Multiply the divisor by 10 (move the decimal point 1 place). Multiply the same way in the dividend (move 1 place).

$$
\begin{array}{r}
0.6\,8 \\
8.6\,\overline{)\,5.8{\scriptstyle\wedge}4\,8} \\
5\,1\,6\,0 \\
\hline
6\,8\,8 \\
6\,8\,8 \\
\hline
0
\end{array}
$$

Place a decimal point above the new decimal point and then divide.
Note: $\dfrac{5.848}{8.6} = \dfrac{5.848}{8.6} \cdot \dfrac{10}{10} = \dfrac{58.48}{86}.$

Do Exercises 7–9.

Suppose the dividend is a whole number. We can think of it as having a decimal point at the end with as many 0's as we wish after the decimal point. For example,

$$12 = 12. = 12.0 = 12.00 = 12.000, \text{ and so on.}$$

EXAMPLE 6 Divide: $12 \div 0.64$.

$$0.6\,4\,\overline{)\,1\,2.}$$

Place a decimal point at the end of the whole number.

$$0.6\,4\,\overline{)\,1\,2.0\,0}$$

Multiply the divisor by 100 (move the decimal point 2 places). Multiply the same way in the dividend (move 2 places).

$$
\begin{array}{r}
1\,8.7\,5 \\
0.6\,4\,\overline{)\,1\,2.0\,0\,0\,0} \\
6\,4\,0 \\
\hline
5\,6\,0 \\
5\,1\,2 \\
\hline
4\,8\,0 \\
4\,4\,8 \\
\hline
3\,2\,0 \\
3\,2\,0 \\
\hline
0
\end{array}
$$

Place a decimal point above and then divide.

Do Exercise 10.

DIVIDING BY 10, 100, 1000, AND SO ON

It is often helpful to be able to divide quickly by a ten, hundred, or thousand, or by a tenth, hundredth, or thousandth. Each procedure we use is based on multiplying by 1. Consider the following example:

$$\frac{23.789}{1000} = \frac{23.789}{1000} \cdot \frac{1000}{1000} = \frac{23{,}789}{1{,}000{,}000} = 0.023789.$$

We are dividing by a number greater than 1: The result is *smaller* than 23.789.

To divide by 10, 100, 1000, and so on,

a) count the number of zeros in the divisor, and

$$\frac{713.49}{100}$$

⤷ 2 zeros

b) move the decimal point that number of places to the left.

$$\frac{713.49}{100}, \qquad 7\underset{\curvearrowleft}{.}13.49 \qquad \frac{713.49}{100} = 7.1349$$

2 places to the left

EXAMPLE 7 Divide: $\dfrac{0.0104}{10}$.

$$\frac{0.0104}{10}, \qquad 0\underset{\curvearrowleft}{.}0.0104, \qquad \frac{0.0104}{10} = 0.00104$$

1 zero   1 place to the left

---

10. Divide.

$$1.6\,\overline{)\,2\,5}$$

CALCULATOR CORNER

Division with Decimal Notation To use a calculator to divide with decimal notation, we use the $\boxed{\cdot}$, $\boxed{\div}$, and $\boxed{=}$ keys. To find $237.12 \div 5.2$, for example, we press

$\boxed{2}\boxed{3}\boxed{7}\boxed{\cdot}\boxed{1}\boxed{2}\boxed{\div}$
$\boxed{5}\boxed{\cdot}\boxed{2}\boxed{=}$. The display reads $\boxed{45.6}$, so $237.12 \div 5.2 = 45.6$.

Exercises: Use a calculator to divide.

1. $1\,2.4\,\overline{)\,1\,7\,7.3\,2}$

2. $4\,9\,\overline{)\,1\,2\,5.4\,4}$

3. $3.2\,\overline{)\,6\,4\,0}$

4. $1\,6\,\overline{)\,1\,2}$

5. $14 \div 0.7$

6. $1.6 \div 25$

7. $474.14 \div 30.2$

8. $518.472 \div 6.84$

Answer on page A-9

Study Tips

HOMEWORK TIPS

Prepare for your homework assignment by reading the explanations of concepts and following the step-by-step solutions of examples in the text. The time you spend preparing will save valuable time when you do your assignment.

263

4.4 Division

Divide.

11. $\dfrac{0.1278}{0.01}$

12. $\dfrac{0.1278}{100}$

13. $\dfrac{98.47}{1000}$

14. $\dfrac{6.7832}{0.1}$

DIVIDING BY 0.1, 0.01, 0.001, AND SO ON

Now consider the following example:

$$\frac{23.789}{0.01} = \frac{23.789}{0.01} \cdot \frac{100}{100} = \frac{2378.9}{1} = 2378.9.$$

We are dividing by a number less than 1: The result is *larger* than 23.789. We use the following procedure.

> To divide by 0.1, 0.01, 0.001, and so on,
>
> **a)** count the number of decimal places in the divisor, and
>
> **b)** move the decimal point that number of places to the right.
>
> $$\frac{713.49}{0.001}, \qquad 713.490. \qquad \frac{713.49}{0.001} = 713{,}490$$
>
> 3 places to the right
>
> $\dfrac{713.49}{0.001}$
>
> 3 places

EXAMPLE 8   Divide: $\dfrac{23.738}{0.001}$.

$$\frac{23.738}{0.001}, \qquad 23.738. \qquad \frac{23.738}{0.001} = 23{,}738$$

3 places   3 places to the right to change 0.001 to 1

Do Exercises 11–14.

b   Solving Equations

Now let's solve equations of the type $a \cdot x = b$, where a and b may be in decimal notation. Proceeding as before, we divide by a on both sides.

EXAMPLE 9   Solve: $8 \cdot x = 27.2$.

We have

$$\frac{8 \cdot x}{8} = \frac{27.2}{8} \qquad \text{Dividing by 8 on both sides}$$

$$x = 3.4.$$

$$\begin{array}{r} 3.4 \\ 8\,)\,\overline{2\ 7.2} \\ 2\ 4\ 0 \\ \hline 3\ 2 \\ 3\ 2 \\ \hline 0 \end{array}$$

The solution is 3.4.

EXAMPLE 10 Solve: $2.9 \cdot t = 0.14616$.

We have

$$\frac{2.9 \cdot t}{2.9} = \frac{0.14616}{2.9} \qquad \text{Dividing by 2.9 on both sides}$$

$$t = 0.0504.$$

$$
\begin{array}{r}
0.0\,5\,0\,4 \\
2.9\,)\,\overline{0.1_\wedge4\,6\,1\,6} \\
1\,4\,5\,0\,0 \\
\hline
1\,1\,6 \\
1\,1\,6 \\
\hline
0
\end{array}
$$

The solution is 0.0504.

Do Exercises 15 and 16.

Solve.
15. $100 \cdot x = 78.314$

C   Order of Operations: Decimal Notation

The same rules for order of operations used with whole numbers and fraction notation apply when simplifying expressions with decimal notation.

> **RULES FOR ORDER OF OPERATIONS**
>
> **1.** Do all calculations within grouping symbols before operations outside.
> **2.** Evaluate all exponential expressions.
> **3.** Do all multiplications and divisions in order from left to right.
> **4.** Do all additions and subtractions in order from left to right.

16. $0.25 \cdot y = 276.4$

EXAMPLE 11 Simplify: $2.56 \times 25.6 \div 25{,}600 \times 256$.

There are no exponents or parentheses, so we multiply and divide from left to right:

$2.56 \times 25.6 \div 25{,}600 \times 256 = 65.536 \div 25{,}600 \times 256$ Doing all multiplications and divisions in order from left to right

$$= 0.00256 \times 256$$
$$= 0.65536.$$

EXAMPLE 12 Simplify: $(5 - 0.06) \div 2 + 3.42 \times 0.1$.

$(5 - 0.06) \div 2 + 3.42 \times 0.1 = 4.94 \div 2 + 3.42 \times 0.1$ Carrying out operations inside parentheses

$$= 2.47 + 0.342 \qquad \text{Doing all multiplications and divisions in order from left to right}$$
$$= 2.812$$

Answers on page A-9

Simplify.

17. $625 \div 62.5 \times 25 \div 6250$

18. $0.25 \cdot (1 + 0.08) - 0.0274$

19. $20^2 - 3.4^2 + \{2.5[20(9.2 - 5.6)] + 5(10 - 5)\}$

20. Mountains in Peru. Refer to the figure in Example 14. Find the average height of the mountains, in meters.

EXAMPLE 13   Simplify: $10^2 \times \{[(3 - 0.24) \div 2.4] - (0.21 - 0.092)\}$.

$10^2 \times \{[(3 - 0.24) \div 2.4] - (0.21 - 0.092)\}$

$= 10^2 \times \{[2.76 \div 2.4] - 0.118\}$ Doing the calculations in the innermost parentheses first

$= 10^2 \times \{1.15 - 0.118\}$ Again, doing the calculations in the innermost parentheses

$= 10^2 \times 1.032$ Subtracting inside the parentheses

$= 100 \times 1.032$ Evaluating the exponential expression

$= 103.2$

Do Exercises 17–19.

EXAMPLE 14   *Mountains in Peru.*   The following figure shows a range of very high mountains in Peru, together with their altitudes, given both in feet and in meters. Find the average height of these mountains, in feet.
Source: *National Geographic,* July 1968, p. 130

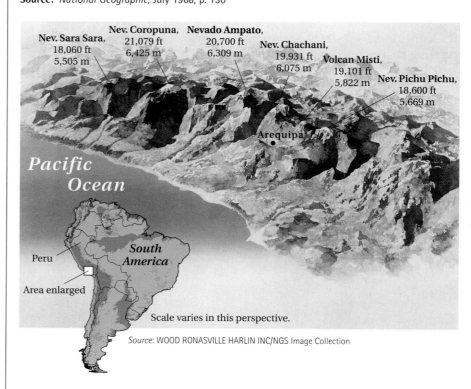

Nev. Sara Sara, 18,060 ft 5,505 m
Nev. Coropuna, 21,079 ft 6,425 m
Nevado Ampato, 20,700 ft 6,309 m
Nev. Chachani, 19,931 ft 6,075 m
Volcan Misti, 19,101 ft 5,822 m
Nev. Pichu Pichu, 18,600 ft 5,669 m
Arequipa
Pacific Ocean
Peru
South America
Area enlarged
Scale varies in this perspective.
Source: WOOD RONASVILLE HARLIN INC/NGS Image Collection

The **average** of a set of numbers is the sum of the numbers divided by the number of addends. (See Section 1.9.) We find the sum of the heights divided by the number of addends, 6:

$$\frac{18,060 + 21,079 + 20,700 + 19,931 + 19,101 + 18,600}{6} = \frac{117,471}{6} = 19,578.5.$$

Thus the average height of these mountains is 19,578.5 ft.

Do Exercise 20.

Answers on page A-9

a Divide.

1. $2 \overline{)5.9\,8}$

2. $5 \overline{)1\,8}$

3. $4 \overline{)9\,5.1\,2}$

4. $8 \overline{)2\,5.9\,2}$

5. $1\,2 \overline{)8\,9.7\,6}$

6. $2\,3 \overline{)2\,5.0\,7}$

7. $3\,3 \overline{)2\,3\,7.6}$

8. $12.4 \div 4$

9. $9.144 \div 8$

10. $4.5 \div 9$

11. $12.123 \div 3$

12. $7 \overline{)5.6}$

13. $5 \overline{)0.3\,5}$

14. $0.0\,4 \overline{)1.6\,8}$

15. $0.1\,2 \overline{)8.4}$

16. $0.3\,6 \overline{)2.8\,8}$

17. $3.4 \overline{)6\,8}$

18. $0.2\,5 \overline{)5}$

19. $1\,5 \overline{)6}$

20. $1\,2 \overline{)1.8}$

21. $3\,6 \overline{)1\,4.7\,6}$

22. $5\,2 \overline{)1\,1\,9.6}$

23. $3.2 \overline{)2\,7.2}$

24. $8.5 \overline{)2\,7.2}$

25. $4.2 \overline{)3\,9.0\,6}$

26. $4.8 \overline{)0.1\,1\,0\,4}$

27. $8 \overline{)5}$

28. $8 \overline{)3}$

29. $0.4\,7 \overline{)0.1\,2\,2\,2}$

30. $1.0\,8 \overline{)0.5\,4}$

31. $4.8 \overline{)7\,5}$

32. $0.2\,8\,\overline{)\,6\ 3}$

33. $0.0\,3\,2\,\overline{)\,0.0\ 7\ 4\ 8\ 8}$

34. $0.0\,1\,7\,\overline{)\,1.5\ 8\ 1}$

35. $8\,2\,\overline{)\,3\ 8.5\ 4}$

36. $3\,4\,\overline{)\,0.1\ 4\ 6\ 2}$

37. $\dfrac{213.4567}{1000}$

38. $\dfrac{213.4567}{100}$

39. $\dfrac{213.4567}{10}$

40. $\dfrac{100.7604}{0.1}$

41. $\dfrac{1.0237}{0.001}$

42. $\dfrac{1.0237}{0.01}$

b   Solve.

43. $4.2 \cdot x = 39.06$

44. $36 \cdot y = 14.76$

45. $1000 \cdot y = 9.0678$

46. $789.23 = 0.25 \cdot q$

47. $1048.8 = 23 \cdot t$

48. $28.2 \cdot x = 423$

c   Simplify.

49. $14 \times (82.6 + 67.9)$

50. $(26.2 - 14.8) \times 12$

51. $0.003 + 3.03 \div 0.01$

52. $9.94 + 4.26 \div (6.02 - 4.6) - 0.9$

53. $42 \times (10.6 + 0.024)$

54. $(18.6 - 4.9) \times 13$

55. $4.2 \times 5.7 + 0.7 \div 3.5$

56. $123.3 - 4.24 \times 1.01$

57. $9.0072 + 0.04 \div 0.1^2$

58. $12 \div 0.03 - 12 \times 0.03^2$

59. $(8 - 0.04)^2 \div 4 + 8.7 \times 0.4$

60. $(5 - 2.5)^2 \div 100 + 0.1 \times 6.5$

61. $86.7 + 4.22 \times (9.6 - 0.03)^2$

62. $2.48 \div (1 - 0.504) + 24.3 - 11 \times 2$

63. $4 \div 0.4 + 0.1 \times 5 - 0.1^2$

64. $6 \times 0.9 + 0.1 \div 4 - 0.2^3$

65. $5.5^2 \times [(6 - 4.2) \div 0.06 + 0.12]$

66. $12^2 \div (12 + 2.4) - [(2 - 1.6) \div 0.8]$

67. $200 \times \{[(4 - 0.25) \div 2.5] - (4.5 - 4.025)\}$

68. $0.03 \times \{1 \times 50.2 - [(8 - 7.5) \div 0.05]\}$

69. Find the average of $1276.59, $1350.49, $1123.78, and $1402.58.

70. Find the average weight of two wrestlers who weigh 308 lb and 296.4 lb.

71. *Porsche Sales.* Because of the exchange rate of the dollar to the Euro (European Monetary Unit), sales of Porsches have soared in the United States, as shown in the bar graph below. Find the average number of sales per year over the 5-yr period.

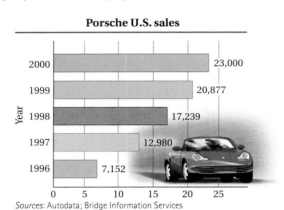

Porsche U.S. sales

Year	Sales
2000	23,000
1999	20,877
1998	17,239
1997	12,980
1996	7,152

Sources: Autodata; Bridge Information Services

72. *Apples.* Americans are growing and eating more apples each year. The following graph shows the number of apples, in millions of bushels, in storage on May 1 of four recent years. Find the average number of apples in storage per year.

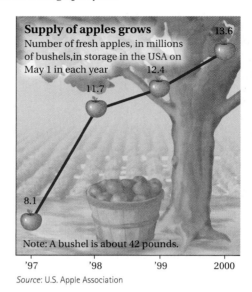

Supply of apples grows
Number of fresh apples, in millions of bushels, in storage in the USA on May 1 in each year

13.6
12.4
11.7
8.1

Note: A bushel is about 42 pounds.

'97   '98   '99   2000

Source: U.S. Apple Association

73. D_W How is division with decimal notation similar to division of whole numbers? How is it different?

74. D_W A student made these two computational mistakes:
$$0.247 \div 0.1 = 0.0247; \quad 0.247 \div 10 = 2.47.$$
In each case, how could you convince her that a mistake has been made?

SKILL MAINTENANCE

Simplify. [2.5b]

75. $\dfrac{36}{42}$ **76.** $\dfrac{56}{64}$ **77.** $\dfrac{38}{146}$ **78.** $\dfrac{114}{438}$

Find the prime factorization. [2.1d]

79. 684   **80.** 162   **81.** 2007   **82.** 2005

83. Add: $10\frac{1}{2} + 4\frac{5}{8}$. [3.5a]

84. Subtract: $10\frac{1}{2} - 4\frac{5}{8}$. [3.5b]

SYNTHESIS

Simplify.

85. 🖩 $9.0534 - 2.041^2 \times 0.731 \div 1.043^2$

86. 🖩 $23.042(7 - 4.037 \times 1.46 - 0.932^2)$

In Exercises 87–90, find the missing value.

87. $439.57 \times 0.01 \div 1000 \times \square = 4.3957$

88. $5.2738 \div 0.01 \times 1000 \div \square = 52.738$

89. $0.0329 \div 0.001 \times 10^4 \div \square = 3290$

90. $0.0047 \times 0.01 \div 10^4 \times \square = 4.7$

4.5 CONVERTING FROM FRACTION NOTATION TO DECIMAL NOTATION

Objectives

a Convert from fraction notation to decimal notation.

b Round numbers named by repeating decimals in problem solving.

c Calculate using fraction and decimal notation together.

a Fraction Notation to Decimal Notation

When a denominator has no prime factors other than 2's and 5's, we can find decimal notation by multiplying by 1. We multiply to get a denominator that is a power of ten, like 10, 100, or 1000.

EXAMPLE 1 Find decimal notation for $\frac{3}{5}$.

$$\frac{3}{5} = \frac{3}{5} \cdot \frac{2}{2} = \frac{6}{10} = 0.6 \qquad \text{We use } \tfrac{2}{2} \text{ for 1 to get a denominator of 10.}$$

EXAMPLE 2 Find decimal notation for $\frac{7}{20}$.

$$\frac{7}{20} = \frac{7}{20} \cdot \frac{5}{5} = \frac{35}{100} = 0.35 \qquad \text{We use } \tfrac{5}{5} \text{ for 1 to get a denominator of 100.}$$

EXAMPLE 3 Find decimal notation for $\frac{87}{25}$.

$$\frac{87}{25} = \frac{87}{25} \cdot \frac{4}{4} = \frac{348}{100} = 3.48 \qquad \text{We use } \tfrac{4}{4} \text{ for 1 to get a denominator of 100.}$$

EXAMPLE 4 Find decimal notation for $\frac{9}{40}$.

$$\frac{9}{40} = \frac{9}{40} \cdot \frac{25}{25} = \frac{225}{1000} = 0.225 \qquad \text{We use } \tfrac{25}{25} \text{ for 1 to get a denominator of 1000.}$$

Do Exercises 1–4.

We can also divide to find decimal notation.

EXAMPLE 5 Find decimal notation for $\frac{3}{5}$.

$$\frac{3}{5} = 3 \div 5 \qquad \begin{array}{r} 0.6 \\ 5\overline{)3.0} \\ \underline{3\ 0} \\ 0 \end{array} \qquad \frac{3}{5} = 0.6$$

EXAMPLE 6 Find decimal notation for $\frac{7}{8}$.

$$\frac{7}{8} = 7 \div 8 \qquad \begin{array}{r} 0.8\ 7\ 5 \\ 8\overline{)7.0\ 0\ 0} \\ \underline{6\ 4} \\ 6\ 0 \\ \underline{5\ 6} \\ 4\ 0 \\ \underline{4\ 0} \\ 0 \end{array} \qquad \frac{7}{8} = 0.875$$

Do Exercises 5 and 6.

Find decimal notation. Use multiplying by 1.

1. $\dfrac{4}{5}$

2. $\dfrac{9}{20}$

3. $\dfrac{11}{40}$

4. $\dfrac{33}{25}$

Find decimal notation.

5. $\dfrac{2}{5}$

6. $\dfrac{3}{8}$

Answers on page A-9

Find decimal notation.

7. $\dfrac{1}{6}$

8. $\dfrac{2}{3}$

Find decimal notation.

9. $\dfrac{5}{11}$

10. $\dfrac{12}{11}$

Answers on page A-9

In Examples 5 and 6, the division *terminated,* meaning that eventually we got a remainder of 0. A **terminating decimal** occurs when the denominator has only 2's or 5's, or both, as factors, as in $\frac{17}{25}$, $\frac{5}{8}$, or $\frac{83}{100}$. This assumes that the fraction notation has been simplified.

Consider a different situation:

$$\frac{5}{6}, \quad \text{or} \quad \frac{5}{2 \cdot 3}.$$

Since 6 has a 3 as a factor, the division will not terminate. Although we can still use division to get decimal notation, the answer will be a **repeating decimal,** as follows.

EXAMPLE 7 Find decimal notation for $\frac{5}{6}$.

$$\frac{5}{6} = 5 \div 6 \qquad \begin{array}{r} 0.8\ 3\ 3 \\ 6\ )\ \overline{5.0\ 0\ 0} \\ \underline{4\ 8} \\ 2\ 0 \\ \underline{1\ 8} \\ 2\ 0 \\ \underline{1\ 8} \\ 2 \end{array}$$

Since 2 keeps reappearing as a remainder, the digits repeat and will continue to do so; therefore,

$$\frac{5}{6} = 0.83333\ldots.$$

The red dots indicate an endless sequence of digits in the quotient. When there is a repeating pattern, the dots are often replaced by a bar to indicate the repeating part—in this case, only the 3:

$$\frac{5}{6} = 0.8\overline{3}.$$

Do Exercises 7 and 8.

EXAMPLE 8 Find decimal notation for $\frac{4}{11}$.

$$\frac{4}{11} = 4 \div 11 \qquad \begin{array}{r} 0.3\ 6\ 3\ 6 \\ 1\ 1\ )\ \overline{4.0\ 0\ 0\ 0} \\ \underline{3\ 3} \\ 7\ 0 \\ \underline{6\ 6} \\ 4\ 0 \\ \underline{3\ 3} \\ 7\ 0 \\ \underline{6\ 6} \\ 4 \end{array}$$

Since 7 and 4 keep repeating as remainders, the sequence of digits "36" repeats in the quotient, and

$$\frac{4}{11} = 0.363636\ldots, \quad \text{or} \quad 0.\overline{36}.$$

Do Exercises 9 and 10.

EXAMPLE 9 Find decimal notation for $\frac{5}{7}$.

$$
\begin{array}{r}
0.7\ 1\ 4\ 2\ 8\ 5 \\
7\ )\ \overline{5.0\ 0\ 0\ 0\ 0\ 0} \\
\underline{4\ 9} \\
1\ 0 \\
\underline{7} \\
3\ 0 \\
\underline{2\ 8} \\
2\ 0 \\
\underline{1\ 4} \\
6\ 0 \\
\underline{5\ 6} \\
4\ 0 \\
\underline{3\ 5} \\
5
\end{array}
$$

Since 5 appears as a remainder, the sequence of digits "714285" repeats in the quotient, and

$$\frac{5}{7} = 0.714285714285\ldots, \quad \text{or} \quad 0.\overline{714285}.$$

The length of a repeating part can be very long—too long to find on a calculator. An example is $\frac{5}{97}$, which has a repeating part of 96 digits.

Do Exercise 11.

b Rounding in Problem Solving

In applied problems, repeating decimals are rounded to get approximate answers. To round a repeating decimal, we can extend the decimal notation at least one place past the rounding digit, and then round as before.

EXAMPLES Round each of the following to the nearest tenth, hundredth, and thousandth.

	Nearest tenth	*Nearest hundredth*	*Nearest thousandth*
10. $0.8\overline{3} = 0.83333\ldots$	0.8	0.83	0.833
11. $0.\overline{09} = 0.090909\ldots$	0.1	0.09	0.091
12. $0.\overline{714285} = 0.714285714285\ldots$	0.7	0.71	0.714

Do Exercises 12–14.

CONVERTING RATIOS TO DECIMAL NOTATION

When solving applied problems, we often convert ratios to decimal notation.

EXAMPLE 13 *Forest Fires.* The National Forest Service reports that in a recent year, 6.4 million acres were burned by 73,000 fires. Find the ratio of number of acres burned to number of fires and convert it to decimal notation. Round to the nearest thousandth.
Source: National Forest Service

11. Find decimal notation for $\frac{3}{7}$.

Round each to the nearest tenth, hundredth, and thousandth.

12. $0.\overline{6}$

13. $0.\overline{80}$

14. $6.2\overline{45}$

Answers on page A-9

15. Coin Tossing. A coin is tossed 51 times. It lands heads 26 times. Find the ratio of heads to tosses and convert it to decimal notation. Round to the nearest thousandth. (This is also the experimental probability of getting heads.)

Heads   Tails

16. Gas Mileage. A car goes 380 mi on 15.7 gal of gasoline. Find the gasoline mileage and convert the ratio to decimal notation rounded to the nearest tenth.

17. SUV Models. Refer to the data in bold on the bar graph in Example 15. Find the average number of SUV models available per year for the 5-yr period. Round to the nearest tenth.

Answers on page A-9

We have

$$\frac{\text{Acres burned}}{\text{Number of fires}} = \frac{6{,}400{,}000 \text{ acres}}{73{,}000 \text{ fires}} \approx 87.671.$$

There were about 87.671 acres burned per fire.

EXAMPLE 14 *Gas Mileage.* A car goes 457 mi on 16.4 gal of gasoline. The ratio of number of miles driven to amount of gasoline used is *gas mileage*. Find the gas mileage and convert the ratio to decimal notation rounded to the nearest tenth.

$$\frac{\text{Miles driven}}{\text{Gasoline used}} = \frac{457}{16.4} \approx 27.9$$

The gas mileage is 27.9 miles to the gallon.

Do Exercises 15 and 16.

AVERAGES

When finding an average, we may at times need to round an answer.

EXAMPLE 15 *Sport Utility Vehicles.* Sport utility vehicles have experienced a great explosion in sales. The following bar graph shows total sales, in millions, and the number of models available (in bold). Find the average number of vehicles sold per year for the period from 1995 to 2001. Round the answer to the nearest hundredth.

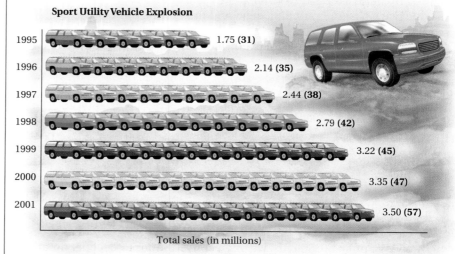

Sport Utility Vehicle Explosion

1995	1.75 (31)
1996	2.14 (35)
1997	2.44 (38)
1998	2.79 (42)
1999	3.22 (45)
2000	3.35 (47)
2001	3.50 (57)

Total sales (in millions)

Source: Autodata

We add the sales totals shown on the bar graph and divide by the number of addends, 7. Since all the units are in millions, we need not convert them to standard notation. The average is

$$\frac{1.75 + 2.14 + 2.44 + 2.79 + 3.22 + 3.35 + 3.50}{7} = \frac{19.19}{7} = 2.7414\ldots \approx 2.74.$$

The average number of SUVs sold per year for the 7-yr period is about 2.74 million.

Do Exercise 17.

C. Calculations with Fraction and Decimal Notation Together

In certain kinds of calculations, fraction and decimal notation might occur together. In such cases, there are at least three ways in which we might proceed.

EXAMPLE 16 Calculate: $\frac{2}{3} \times 0.576$.

METHOD 1 One way to do this calculation is to convert the fraction notation to decimal notation so that both numbers are in decimal notation. Since $\frac{2}{3}$ converts to repeating decimal notation, it is first rounded to some chosen decimal place. We choose three decimal places. Then, using decimal notation, we multiply.

$$\frac{2}{3} \times 0.576 = 0.\overline{6} \times 0.576 \approx 0.667 \times 0.576 = 0.384192$$

METHOD 2 A second way to do this calculation is to convert the decimal notation to fraction notation so that both numbers are in fraction notation. The answer can be left in fraction notation and simplified, or we can convert back to decimal notation and round, if appropriate.

$$\frac{2}{3} \times 0.576 = \frac{2}{3} \cdot \frac{576}{1000} = \frac{2 \cdot 576}{3 \cdot 1000}$$

$$= \frac{2 \cdot 2 \cdot 2 \cdot 2 \cdot 2 \cdot 2 \cdot 2 \cdot 3 \cdot 3}{2 \cdot 2 \cdot 2 \cdot 3 \cdot 5 \cdot 5 \cdot 5}$$

$$= \frac{2 \cdot 2 \cdot 2 \cdot 3}{2 \cdot 2 \cdot 2 \cdot 3} \cdot \frac{2 \cdot 2 \cdot 2 \cdot 3}{5 \cdot 5 \cdot 5}$$

$$= 1 \cdot \frac{2 \cdot 2 \cdot 2 \cdot 3}{5 \cdot 5 \cdot 5}$$

$$= \frac{2 \cdot 2 \cdot 2 \cdot 3}{5 \cdot 5 \cdot 5} = \frac{48}{125}, \text{ or } 0.384$$

METHOD 3 A third way to do this calculation is to treat 0.576 as $\frac{0.576}{1}$. Then we multiply 0.576 by 2, and divide the result by 3.

$$\frac{2}{3} \times 0.576 = \frac{2}{3} \times \frac{0.576}{1} = \frac{2 \times 0.576}{3} = \frac{1.152}{3} = 0.384$$

Do Exercise 18.

EXAMPLE 17 Calculate: $\frac{2}{3} \times 0.576 + 3.287 \div \frac{4}{5}$.

We use the rules for order of operations, doing first the multiplication and then the division. Then we add.

$$\frac{2}{3} \times 0.576 + 3.287 \div \frac{4}{5} = 0.384 + 3.287 \cdot \frac{5}{4}$$

Method 3:
$\frac{2}{3} \times \frac{0.576}{1} = 0.384$;
$\frac{3.287}{1} \times \frac{5}{4} = 4.10875$

$$= 0.384 + 4.10875$$

$$= 4.49275$$

Do Exercises 19 and 20.

---

18. Calculate: $\frac{5}{6} \times 0.864$.

Calculate.

19. $\frac{1}{3} \times 0.384 + \frac{5}{8} \times 0.6784$

20. $\frac{5}{6} \times 0.864 + 14.3 \div \frac{8}{5}$

Answers on page A-9

a   Find decimal notation.

1. $\dfrac{23}{100}$ **2.** $\dfrac{9}{100}$ **3.** $\dfrac{3}{5}$ **4.** $\dfrac{19}{20}$ **5.** $\dfrac{13}{40}$ **6.** $\dfrac{3}{16}$

7. $\dfrac{1}{5}$ **8.** $\dfrac{4}{5}$ **9.** $\dfrac{17}{20}$ **10.** $\dfrac{11}{20}$ **11.** $\dfrac{3}{8}$ **12.** $\dfrac{7}{8}$

13. $\dfrac{39}{40}$ **14.** $\dfrac{31}{40}$ **15.** $\dfrac{13}{25}$ **16.** $\dfrac{61}{125}$ **17.** $\dfrac{2502}{125}$ **18.** $\dfrac{181}{200}$

19. $\dfrac{1}{4}$ **20.** $\dfrac{1}{2}$ **21.** $\dfrac{29}{25}$ **22.** $\dfrac{37}{25}$ **23.** $\dfrac{19}{16}$ **24.** $\dfrac{5}{8}$

25. $\dfrac{4}{15}$ **26.** $\dfrac{7}{9}$ **27.** $\dfrac{1}{3}$ **28.** $\dfrac{1}{9}$ **29.** $\dfrac{4}{3}$ **30.** $\dfrac{8}{9}$

31. $\dfrac{7}{6}$ **32.** $\dfrac{7}{11}$ **33.** $\dfrac{4}{7}$ **34.** $\dfrac{14}{11}$ **35.** $\dfrac{11}{12}$ **36.** $\dfrac{5}{12}$

b

37.–47. Odds. Round each answer of the odd-numbered Exercises 25–35 to the nearest tenth, hundredth, and thousandth.

38.–48. Evens. Round each answer of the even-numbered Exercises 26–36 to the nearest tenth, hundredth, and thousandth.

Round each to the nearest tenth, hundredth, and thousandth.

49. $0.\overline{18}$ **50.** $0.\overline{83}$ **51.** $0.2\overline{7}$ **52.** $3.5\overline{4}$

53. For this set of people, what is the ratio, in decimal notation rounded to the nearest thousandth, where appropriate, of:

a) women to the total number of people?
b) women to men?
c) men to the total number of people?
d) men to women?

54. For this set of nuts and bolts, what is the ratio, in decimal notation rounded to the nearest thousandth, where appropriate, of:

a) nuts to bolts?
b) bolts to nuts?
c) nuts to the total?
d) total number to nuts?

Gas Mileage. In each of Exercises 55–58, find the gas mileage rounded to the nearest tenth.

55. 285 mi; 18 gal

56. 396 mi; 17 gal

57. 324.8 mi; 18.2 gal

58. 264.8 mi; 12.7 gal

59. *Windy Cities.* Although nicknamed the Windy City, Chicago is not the windiest city in the United States. Listed in the table below are the six windiest cities and their average wind speeds. Find the average of these wind speeds and round your answer to the nearest tenth.
Source: *The Handy Geography Answer Book*

CITY	AVERAGE WIND SPEED (in miles per hour)
Mt. Washington, NH	35.3
Boston, MA	12.5
Honolulu, HI	11.3
Dallas, TX	10.7
Kansas City, MO	10.7
Chicago, IL	10.4

60. *Areas of the New England States.* The table below lists the areas of the New England states. Find the average area and round your answer to the nearest tenth.
Source: *The New York Times Almanac*

STATE	TOTAL AREA (in square miles)
Maine	33,265
New Hampshire	9,279
Vermont	9,614
Massachusetts	8,284
Connecticut	5,018
Rhode Island	1,211

Stock Prices. At one time stock prices were given using mixed numerals involving halfs, fourths, eighths, and, more recently, sixteenths. The Securities and Exchange Commission has mandated the use of decimal notation. Thus a price of 23\frac{13}{16}$ is now converted to decimal notation rounded to the nearest hundredth, that is, $23.81. Complete the following table.

Sources: *The Indianapolis Star,* 1/30/01; www.yahoo.com

	STOCK	PRICE PER SHARE	DECIMAL NOTATION	ROUNDED TO NEAREST HUNDREDTH
61.	General Mills	41\frac{11}{16}$		
62.	Quaker Oats	98\frac{15}{16}$		
63.	Kellogg	25\frac{7}{8}$		
64.	Dillard's	20\frac{5}{8}$		
65.	Hudson's Bay	19\frac{3}{64}$		
66.	Abercrombie & Fitch	31\frac{47}{64}$		

C Calculate.

67. $\dfrac{7}{8} \times 12.64$

68. $\dfrac{4}{5} \times 384.8$

69. $2\dfrac{3}{4} + 5.65$

70. $4\dfrac{4}{5} + 3.25$

71. $\dfrac{47}{9} \times 79.95$

72. $\dfrac{7}{11} \times 2.7873$

73. $\dfrac{1}{2} - 0.5$

74. $3\dfrac{1}{8} - 2.75$

75. $4.875 - 2\dfrac{1}{16}$

76. $55\dfrac{3}{5} - 12.22$

77. $\dfrac{5}{6} \times 0.0765 + \dfrac{5}{4} \times 0.1124$

78. $\dfrac{3}{5} \times 6384.1 - \dfrac{3}{8} \times 156.56$

79. $\dfrac{4}{5} \times 384.8 + 24.8 \div \dfrac{8}{3}$

80. $102.4 \div \dfrac{2}{5} - 12 \times \dfrac{5}{6}$

81. $\dfrac{7}{8} \times 0.86 - 0.76 \times \dfrac{3}{4}$

82. $17.95 \div \dfrac{5}{8} + \dfrac{3}{4} \times 16.2$

83. $3.375 \times 5\dfrac{1}{3}$

84. $2.5 \times 3\dfrac{5}{8}$

85. $6.84 \div 2\dfrac{1}{2}$

86. $8\dfrac{1}{2} \div 2.125$

87. $\mathbf{D_W}$ When is long division *not* the fastest way to convert from fraction notation to decimal notation?

88. $\mathbf{D_W}$ Examine Example 16 of this section. How could the problem be changed so that method 1 would give a result that is completely accurate?

SKILL MAINTENANCE

Multiply.   [3.6a]

89. $9 \cdot 2\dfrac{1}{3}$

90. $10\dfrac{1}{2} \cdot 22\dfrac{3}{4}$

Divide.   [3.6b]

91. $84 \div 8\dfrac{2}{5}$

92. $8\dfrac{3}{5} \div 10\dfrac{2}{5}$

Add.   [3.5a]

93. $17\dfrac{5}{6} + 32\dfrac{3}{8}$

94. $14\dfrac{3}{5} + 16\dfrac{1}{10}$

Subtract.   [3.5b]

95. $16\dfrac{1}{10} - 14\dfrac{3}{5}$

96. $32\dfrac{3}{8} - 17\dfrac{5}{6}$

Solve.   [3.5c]

97. A recipe for bread calls for $\dfrac{2}{3}$ cup of water, $\dfrac{1}{4}$ cup of milk, and $\dfrac{1}{8}$ cup of oil. How many cups of liquid ingredients does the recipe call for?

98. A board $\dfrac{7}{10}$ in. thick is glued to a board $\dfrac{3}{5}$ in. thick. The glue is $\dfrac{3}{100}$ in. thick. How thick is the result?

SYNTHESIS

▦ Find decimal notation.

99. $\dfrac{1}{7}$

100. $\dfrac{2}{7}$

101. $\dfrac{3}{7}$

102. $\dfrac{4}{7}$

103. $\dfrac{5}{7}$

104. ▦ From the pattern of Exercises 99–103, guess the decimal notation for $\dfrac{6}{7}$. Check on your calculator.

▦ Find decimal notation.

105. $\dfrac{1}{9}$

106. $\dfrac{1}{99}$

107. $\dfrac{1}{999}$

108. ▦ From the pattern of Exercises 105–107, guess the decimal notation for $\dfrac{1}{9999}$. Check on your calculator.

ESTIMATING

1. Estimate by rounding to the nearest ten the total cost of one TV and one vacuum cleaner. Which of the following is an appropriate estimate?
   a) $5700   b) $570
   c) $500   d) $57

2. About how much more does the TV cost than the vacuum cleaner? Estimate by rounding to the nearest ten. Which of the following is an appropriate estimate?
   a) $130   b) $1300
   c) $580   d) $13

Answers on page A-10

Study Tips

a **Estimating Sums, Differences, Products, and Quotients**

Estimating has many uses. It can be done before a problem is even attempted in order to get an idea of the answer. It can be done afterward as a check, even when we are using a calculator. In many situations, an estimate is all we need. We usually estimate by rounding the numbers so that there are one or two nonzero digits, depending on how accurate we want our estimate. Consider the following advertisements for Examples 1–4.

EXAMPLE 1 Estimate by rounding to the nearest ten the total cost of one fax machine and one TV.

We are estimating the sum

$149.95 + $346.95 = Total cost.

The estimate found by rounding the addends to the nearest ten is

$150 + $350 = $500.   (Estimated total cost)

Do Exercise 1.

EXAMPLE 2 About how much more does the TV cost than the fax machine? Estimate by rounding to the nearest ten.

We are estimating the difference

$346.95 − $149.95 = Price difference.

The estimate to the nearest ten is

$350 − $150 = $200.   (Estimated price difference)

Do Exercise 2.

EXAMPLE 3 Estimate the total cost of 4 vacuum cleaners.

We are estimating the product

$$4 \times \$219.95 = \text{Total cost.}$$

The estimate is found by rounding $219.95 to the nearest ten:

$$4 \times \$220 = \$880.$$

Do Exercise 3.

EXAMPLE 4 About how many fax machines can be purchased for $1480?

We estimate the quotient

$$\$1480 \div \$149.95.$$

Since we want a whole-number estimate, we choose our rounding appropriately. Rounding $149.95 to the nearest one, we get $150. Since $1480 is close to $1500, which is a multiple of 150, we estimate

$$\$1500 \div \$150,$$

so the answer is 10.

Do Exercise 4.

EXAMPLE 5 Estimate: 4.8×52. Do not find the actual product. Which of the following is an appropriate estimate?

a) 25 b) 250 c) 2500 d) 360

We have

$$5 \times 50 = 250. \text{(Estimated product)}$$

We rounded 4.8 to the nearest one and 52 to the nearest ten. Thus an appropriate estimate is (b).

Compare these estimates for the product 4.94×38:

$$5 \times 40 = 200, 5 \times 38 = 190, 4.9 \times 40 = 196.$$

The first estimate was the easiest. You could probably do it mentally. The others had more nonzero digits.

Do Exercises 5–10.

---

3. Estimate the total cost of 6 TVs. Which of the following is an appropriate estimate?
   a) $4400   b) $350
   c) $21,000   d) $2100

4. About how many vacuum cleaners can be purchased for $1100? Which of the following is an appropriate estimate?
   a) 8   b) 5
   c) 11   d) 124

Estimate the product. Do not find the actual product. Which of the following is an appropriate estimate?

5. 2.4×8
   a) 16   b) 34
   c) 125   d) 5

6. 24×0.6
   a) 200   b) 5
   c) 110   d) 20

7. 0.86×0.432
   a) 0.04   b) 0.4
   c) 1.1   d) 4

8. 0.82×0.1
   a) 800   b) 8
   c) 0.08   d) 80

9. 0.12×18.248
   a) 180   b) 1.8
   c) 0.018   d) 18

10. 24.234×5.2
    a) 200   b) 125
    c) 12.5   d) 234

Answers on page A-10

Estimate the quotient. Which of the following is an appropriate estimate?

11. 59.78 ÷ 29.1
a) 200   b) 20
c) 2   d) 0.2

12. 82.08 ÷ 2.4
a) 40   b) 4.0
c) 400   d) 0.4

13. 0.1768 ÷ 0.08
a) 8   b) 10
c) 2   d) 20

14. Estimate: 0.0069 ÷ 0.15. Which of the following is an appropriate estimate?
a) 0.5   b) 50
c) 0.05   d) 0.004

EXAMPLE 6 Estimate: 82.08 ÷ 24. Which of the following is an appropriate estimate?

a) 400   b) 16   c) 40   d) 4

This is about 80 ÷ 20, so the answer is about 4. Thus an appropriate estimate is (d).

EXAMPLE 7 Estimate: 94.18 ÷ 3.2. Which of the following is an appropriate estimate?

a) 30   b) 300   c) 3   d) 60

This is about 90 ÷ 3, so the answer is about 30. Thus an appropriate estimate is (a).

EXAMPLE 8 Estimate: 0.0156 ÷ 1.3. Which of the following is an appropriate estimate?

a) 0.2   b) 0.002   c) 0.02   d) 20

This is about 0.02 ÷ 1, so the answer is about 0.02. Thus an appropriate estimate is (c).

Do Exercises 11–13.

In some cases, it is easier to estimate a quotient directly rather than by rounding the divisor and the dividend.

EXAMPLE 9 Estimate: 0.0074 ÷ 0.23. Which of the following is an appropriate estimate?

a) 0.3   b) 0.03   c) 300   d) 3

We estimate 3 for a quotient. We check by multiplying.

$$0.23 \times 3 = 0.69$$

We make the estimate smaller. We estimate 0.3 and check by multiplying.

$$0.23 \times 0.3 = 0.069$$

We make the estimate smaller. We estimate 0.03 and check by multiplying.

$$0.23 \times 0.03 = 0.0069$$

This is about 0.0074, so the quotient is about 0.03. Thus an appropriate estimate is (b).

Do Exercise 14.

Answers on page A-10

4.6

EXERCISE SET

For Extra Help

Digital Video
Tutor CD 3
Videotape 8

InterAct
Math

Math Tutor
Center

MathXL

MyMathLab

a Consider the following advertisements for Exercises 1–8. Estimate the sums, differences, products, or quotients involved in these problems. Indicate which of the choices is an appropriate estimate.

Entertainment Center
2 adjustable shelves for electronic equipment. Storage behind doors.

19" COLOR TV
• Remote Control
• Matrix Picture Tube
• Stereo Sound
• On-Screen Displays

SALE $299
Reg. $349

All Mini Systems ON SALE!

$249⁹⁵

CD Mini System
• Digital tuner
• Dual cassette deck
• 2-way speakers
• 32 watts/channel
Reg. $299.95

Reg. $134.95 **$109.95**
Ready to assemble.

1. Estimate the total cost of one entertainment center and one sound system.

  a) $36   **b)** $72   **c)** $3.60   **d)** $360

2. Estimate the total cost of one entertainment center and one TV.

  a) $410   **b)** $820   **c)** $41   **d)** $4.10

3. About how much more does the TV cost than the sound system?

  a) $500   **b)** $80   **c)** $50   **d)** $5

4. About how much more does the TV cost than the entertainment center?

  a) $100   **b)** $190   **c)** $250   **d)** $150

5. Estimate the total cost of 9 TVs.

  a) $2700   **b)** $27   **c)** $270   **d)** $540

6. Estimate the total cost of 16 sound systems.

  a) $5010   **b)** $4000   **c)** $40   **d)** $410

7. About how many TVs can be purchased for $1700?

  a) 600   **b)** 72   **c)** 6   **d)** 60

8. About how many sound systems can be purchased for $1300?

  a) 10   **b)** 5   **c)** 50   **d)** 500

Estimate by rounding as directed.

9. 0.02 + 1.31 + 0.34;
   nearest tenth

10. 0.88 + 2.07 + 1.54;
    nearest one

11. 6.03 + 0.007 + 0.214;
    nearest one

12. 1.11 + 8.888 + 99.94;
    nearest one

13. 52.367 + 1.307 + 7.324;
    nearest one

14. 12.9882 + 1.0115;
    nearest tenth

15. 2.678 − 0.445; nearest tenth

16. 12.9882 − 1.0115; nearest one

17. 198.67432 − 24.5007; nearest ten

Estimate. Choose a rounding digit that gives one or two nonzero digits. Indicate which of the choices is an appropriate estimate.

18. 234.12321 − 200.3223

    a) 600   **b)** 60
    c) 300   **d)** 30

19. 49 × 7.89

    a) 400   **b)** 40
    c) 4   **d)** 0.4

20. 7.4 × 8.9

    a) 95   **b)** 63
    c) 124   **d)** 6

21. 98.4 × 0.083

    a) 80   **b)** 12
    c) 8   **d)** 0.8

22. 78 × 5.3

    a) 400   **b)** 800
    c) 40   **d)** 8

23. 3.6 ÷ 4

    a) 10   **b)** 1
    c) 0.1   **d)** 0.01

24. 0.0713 ÷ 1.94

    a) 4   **b)** 0.4
    c) 0.04   **d)** 40

25. 74.68 ÷ 24.7

    a) 9   **b)** 3
    c) 12   **d)** 120

26. 914 ÷ 0.921

    a) 10   **b)** 100
    c) 1000   **d)** 1

27. *Palm VIIxe and the Sears Tower.* The Palm VIIxe PDA (Personal Digital Assistant) is 4.7 in. (about 0.39167 ft) high. Estimate how many PDAs it would take, if placed end to end, to reach from the ground to the top of the Sears Tower, which is 1454 ft tall. Round to the nearest one.
Source: www. yahoo.com

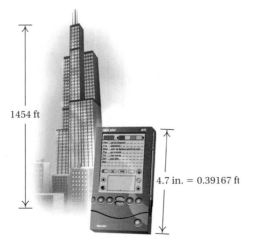

1454 ft

4.7 in. = 0.39167 ft

28. *Ticketmaster.* Recently, Ticketmaster stock sold for $8.63 per share. Estimate how many shares can be purchased for $27,000.

29. D_W Describe a situation in which an estimation is made by rounding to the nearest 10,000 and then multiplying.

30. D_W A roll of fiberglass insulation costs $21.95. Describe two situations involving estimating and the cost of fiberglass insulation. Devise one situation so that $21.95 is rounded to $22. Devise the other situation so that $21.95 is rounded to $20.

SKILL MAINTENANCE

Find the prime factorization. [2.1d]

31. 108

32. 400

33. 325

34. 666

35. 1728

Simplify. [2.5b]

36. $\dfrac{125}{400}$

37. $\dfrac{3225}{6275}$

38. $\dfrac{72}{81}$

39. $\dfrac{325}{625}$

40. $\dfrac{625}{475}$

SYNTHESIS

The following were done on a calculator. Estimate to determine whether the decimal point was placed correctly.

41. $178.9462 \times 61.78 = 11{,}055.29624$

42. $14{,}973.35 \div 298.75 = 501.2$

43. $19.7236 - 1.4738 \times 4.1097 = 1.366672414$

44. $28.46901 \div 4.9187 - 2.5081 = 3.279813473$

45. ▦ Use one of $+$, $-$, $\times$, and $\div$ in each blank to make a true sentence.

    a) $(0.37 \,\square\, 18.78) \,\square\, 2^{13} = 156{,}876.8$

    b) $2.56 \,\square\, 6.4 \,\square\, 51.2 \,\square\, 17.4 = 312.84$

46. ▦ In the subtraction below, a and b are digits. Find a and b.

$$\begin{array}{r} b876.a4321 \\ -\,1234.a678b \\ \hline 8641.b7a32 \end{array}$$

Exercise Set 4.6

Objective

a Solve applied problems involving decimals.

a Solving Applied Problems

Solving applied problems with decimals is like solving applied problems with whole numbers. We translate first to an equation that corresponds to the situation. Then we solve the equation.

EXAMPLE 1 *Quaker Oats Stock Prices.* The Quaker Oats Company is a manufacturer of hot cereals, pancake syrups, grain-based snacks, pancake mixes, and pasta products. Over the entire 52 weeks in 2000, the price per share of its stock ranged in value from a low of $45.81 to a high of $98.94. By how much did the high value differ from the low value?

Sources: The New York Stock Exchange; The Quaker Oats Company

1. **Familiarize.** The stock prices are charted in the graph above. We let $c =$ the amount that the price per share rose over the 52-week period.

2. **Translate.** This is a "missing-addend" situation. We translate as follows, using the given information.

Price at the start	plus	Amount of increase	is	Price at the end
$45.81	+	c	=	$98.94

3. **Solve.** We solve the equation by subtracting $45.81 on both sides:

$$45.81 + c = 98.94$$
$$45.81 + c - 45.81 = 98.94 - 45.81$$
$$c = 53.13.$$

$$\begin{array}{r} 9\ 8.9\ 4 \\ -\ 4\ 5.8\ 1 \\ \hline 5\ 3.1\ 3 \end{array}$$

4. **Check.** We can check by adding 53.13 to 45.81 to get 98.94.

5. **State.** The stock rose by $53.13 per share over the 52-week period.

Do Exercise 1.

---

1. **Body Temperature.** Normal body temperature is 98.6°F. When fevered, most people will die if their bodies reach 107°F. This is a rise of how many degrees?

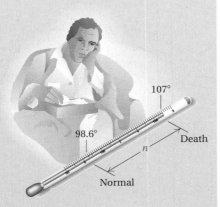

Answer on page A-10

EXAMPLE 2 *Injections of Medication.* A patient was given injections of 2.8 mL, 1.35 mL, 2.0 mL, and 1.88 mL over a 24-hr period. What was the total amount of the injections?

1. **Familiarize.** We make a drawing or at least visualize the situation. We let t = the amount of the injections.

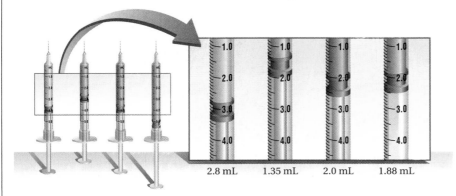

2.8 mL   1.35 mL   2.0 mL   1.88 mL

2. **Translate.** Amounts are being combined. We translate to an equation:

First plus second plus third plus fourth is total.

$$2.8 + 1.35 + 2.0 + 1.88 = t$$

3. **Solve.** To solve, we carry out the addition.

```
  2 1
  2.8 0
  1.3 5
  2.0 0
+ 1.8 8
  8.0 3
```

Thus, $t = 8.03$.

4. **Check.** We can check by repeating our addition. We can also see whether our answer is reasonable by first noting that it is indeed larger than any of the numbers being added. We can also partially check by rounding:

$$2.8 + 1.35 + 2.0 + 1.88 \approx 3 + 1 + 2 + 2$$
$$= 8 \approx 8.03.$$

If we had gotten an answer like 80.3 or 0.803, then our estimate, 8, would have told us that we did something wrong, like not lining up the decimal points.

5. **State.** The total amount of the injections was 8.03 mL.

Do Exercise 2.

---

2. **Liquid Consumption.** Each year, the average American drinks about 49.0 gal of soft drinks, 41.2 gal of water, 25.3 gal of milk, 24.8 gal of coffee, and 7.8 gal of fruit juice. What is the total amount that the average American drinks?

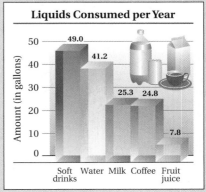

Liquids Consumed per Year

Source: U.S. Department of Agriculture

Answer on page A-10

3. Printing Costs. At a printing company, the cost of copying is 11 cents per page. How much, in dollars, would it cost to make 466 copies?

EXAMPLE 3 *IRS Driving Allowance.* In 2001, the Internal Revenue Service allowed a tax deduction of 34.5¢ per mile for mileage driven for business purposes. What deduction, in dollars, would be allowed for driving 127 mi?
Source: Internal Revenue Service

1. **Familiarize.** We first make a drawing or at least visualize the situation. Repeated addition fits this situation. We let $d = $ the deduction, in dollars, allowed for driving 127 mi.

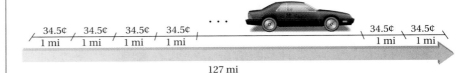

127 mi

2. **Translate.** We translate as follows.

Deduction for each mile	times	Number of miles driven	is	Total deduction
↓	↓	↓	↓	↓
$0.345	×	127	=	d

Converting 34.5 cents to dollars gives us $0.345.

3. **Solve.** To solve the equation, we carry out the multiplication.

$$
\begin{array}{r}
1\ 2\ 7 \\
\times\ \ 0.3\ 4\ 5 \\
\hline
6\ 3\ 5 \\
5\ 0\ 8\ 0 \\
3\ 8\ 1\ 0\ 0 \\
\hline
4\ 3.8\ 1\ 5
\end{array}
$$

Thus, $d = 43.815 \approx \$43.82$.

4. **Check.** We can obtain a partial check by rounding and estimating:
$$127 \times 0.345 \approx 130 \times 0.3$$
$$= 39 \approx 43.82.$$

5. **State.** The total allowable deduction would be $43.82.

Do Exercise 3.

EXAMPLE 4 *Loan Payments.* A car loan of $7382.52 is to be paid off in 36 monthly payments. How much is each payment?

1. **Familiarize.** We first make a drawing. We let $n = $ the amount of each payment.

There may be some fractional part of $1.

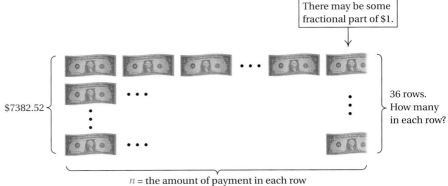

$7382.52

36 rows. How many in each row?

$n = $ the amount of payment in each row

Answer on page A-10

2. Translate. The problem can be translated to the following equation, thinking that

(Total loan) ÷ (Number of payments) = Amount of each payment

$$\$7382.52 \div 36 = n.$$

3. Solve. To solve the equation, we carry out the division.

```
        2 0 5.0 7
3 6 ) 7 3 8 2.5 2
      7 2 0 0 0 0
      1 8 2 5 2
      1 8 0 0 0
          2 5 2
          2 5 2
              0
```

Thus, $n = 205.07$.

4. Check. A partial check can be obtained by estimating the quotient: $\$7382.56 \div 36 \approx 8000 \div 40 = 200 \approx 205.07$. The estimate checks.

5. State. Each payment is $205.07.

Do Exercise 4.

EXAMPLE 5 *Jackie Robinson Poster.* A special limited-edition poster was painted by sports artist Leroy Neiman. Commissioned by Barton L. Kaufman, it commemorates the entrance of the first African-American, Jackie Robinson, into major league baseball in 1947. The dimensions of the poster are 19.3 in. by 27.4 in. Find the area.
Source: Barton L. Kaufman, private collection

1. Familiarize. We first make a drawing. We let $A =$ the area.

|← 19.3 in. →|

27.4 in.

4. Loan Payments. A loan of $4425 is to be paid off in 12 monthly payments. How much is each payment?

Answer on page A-10

Study Tips

FIVE STEPS FOR PROBLEM SOLVING

Are you remembering to use the following five steps for problem solving that were developed in Section 1.8?

1. **Familiarize** yourself with the situation.
   a) Carefully read and reread until you understand *what* you are being asked to find.
   b) Draw a diagram or see if there is a formula that applies.
   c) Assign a letter, or *variable*, to the unknown.
2. **Translate** the problem to an equation using the letter or variable.
3. **Solve** the equation.
4. **Check** the answer in the original wording of the problem.
5. **State** the answer to the problem clearly with appropriate units.

5. Index Cards. A standard-size index card measures 12.7 cm by 7.6 cm. Find its area.

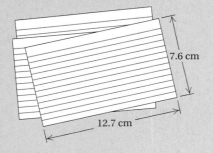

7.6 cm

12.7 cm

6. One pound of lean boneless ham contains 4.5 servings. It costs $5.99 per pound. What is the cost per serving? Round to the nearest cent.

2. Translate. We use the formula $A = l \cdot w$ and substitute.

$$A = l \cdot w$$
$$A = 27.4 \times 19.3.$$

3. Solve. We solve by carrying out the multiplication.

```
        2 7.4
    ×   1 9.3
    ─────────
        8 2 2
    2 4 6 6 0
    2 7 4 0 0
    ─────────
    5 2 8.8 2
```

4. Check. We obtain a partial check by estimating the product:

$$A = 27.4 \times 19.3 \approx 30 \times 20 = 600.$$

Since this estimate is not too close to 528.82, we might repeat our calculation or change our estimate by rounding to the nearest one. This is left to the student. We see that 528.82 checks.

5. State. The area of the Jackie Robinson poster is 528.82 in².

Do Exercise 5.

EXAMPLE 6 *Digital Camera Purchase.* Kelly Real Estate spends $11,998.80 on a set of 24 Olympus D-490 Digital Zoom cameras, so that its realtors can make instant photos and place them on the firm's website. How much did each camera cost?
Source: d-store™: Olympus Camera and Accessories Store

1. Familiarize. We let c = the cost of each camera.

2. Translate. We translate as follows.

Cost of each camera	is	Total cost of purchase	divided by	Number of cameras purchased
↓	↓	↓	↓	↓
c	=	$11,998.80	÷	24

3. Solve. To solve, we carry out the division.

```
              4 9 9.9 5
      2 4 ) 1 1,9 9 8.8 0
              9 6
            ─────
              2 3 9
              2 1 6
              ─────
                2 3 8
                2 1 6
                ─────
                  2 2 8
                  2 1 6
                  ─────
                    1 2 0
                    1 2 0
                    ─────
                        0
```

4. Check. We check by estimating $11,998.80 \div 24 \approx 12,000 \div 25 = 480$. Since 480 is close to 499.95, the answer is probably correct.

5. State. The cost of each camera was $499.95.

Do Exercise 6.

Answers on page A-10

Multistep Problems

EXAMPLE 7 *Gas Mileage.* A driver filled the gasoline tank and noted that the odometer read 67,507.8. After the next filling, the odometer read 68,006.1. It took 16.5 gal to fill the tank. How many miles per gallon did the driver get?

1. Familiarize. We first make a drawing.

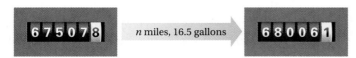

This is a two-step problem. First, we find the number of miles that have been driven between fillups. We let n = the number of miles driven.

2., 3. Translate and **Solve.** This is a "missing-addend" situation. We translate and solve as follows.

First odometer reading   plus   Number of miles driven   is   Second odometer reading

$$67{,}507.8 \quad + \quad n \quad = \quad 68{,}006.1$$

To solve the equation, we subtract 67,507.8 on both sides:

$$n = 68{,}006.1 - 67{,}507.8$$
$$= 498.3.$$

$$\begin{array}{r} 6\ 8{,}0\ 0\ 6.1 \\ -\ 6\ 7{,}5\ 0\ 7.8 \\ \hline 4\ 9\ 8.3 \end{array}$$

Second, we divide the total number of miles driven by the number of gallons. This gives us m = the number of miles per gallon—that is, the mileage. The division that corresponds to the situation is

$$498.3 \div 16.5 = m.$$

To find the number m, we divide.

$$\begin{array}{r} 3\ 0.2 \\ 1\ 6.5\ )\overline{\ 4\ 9\ 8.3_{\wedge}0} \\ 4\ 9\ 5\ 0 \\ \hline 3\ 3\ 0 \\ 3\ 3\ 0 \\ \hline 0 \end{array}$$

Thus, $m = 30.2$.

4. Check. To check, we first multiply the number of miles per gallon times the number of gallons:

$$16.5 \times 30.2 = 498.3.$$

Then we add 498.3 to 67,507.8:

$$67{,}507.8 + 498.3 = 68{,}006.1.$$

The mileage 30.2 checks.

5. State. The driver got 30.2 miles per gallon.

Do Exercise 7.

7. Gas Mileage. A driver filled the gasoline tank and noted that the odometer read 38,320.8. After the next filling, the odometer read 38,735.5. It took 14.5 gal to fill the tank. How many miles per gallon did the driver get?

Answer on page A-10

EXAMPLE 8 *Home-Cost Comparison.* Suppose you own a home like the one shown here and it is valued at $250,000 in Indianapolis, Indiana. What would it cost to buy a similar (replacement) home in Palo Alto, California? To find out, we can use an index table prepared by Coldwell Banker Real Estate Corporation. (For a complete index table, contact your local representative.) We use the following formula:

$$\begin{pmatrix} \text{Cost of your} \\ \text{home in new city} \end{pmatrix} = \begin{pmatrix} \text{Value of} \\ \text{your home} \end{pmatrix} \div \begin{pmatrix} \text{Index of} \\ \text{your city} \end{pmatrix} \times \begin{pmatrix} \text{Index of} \\ \text{new city} \end{pmatrix}.$$

Find the cost of your Indianapolis home in Palo Alto. Round to the nearest one.

Source: Coldwell Banker Real Estate Corporation

STATE	CITY	INDEX
California	San Francisco	310
	Palo Alto	398
	Hollywood Hills	271
Indiana	Indianapolis	63
	Fort Wayne	52
Arizona	Phoenix	79
	Tucson	77
Illinois	Barrington	184
	Naperville	99
Texas	Austin	88
	Dallas	73
	Houston	80
Florida	Miami	112
	Orlando	77
	Tampa	74
Minnesota	Minneapolis	112
	St. Paul	94
Georgia	Atlanta	97
New York	Queens/North Shore	159
	Albany	80

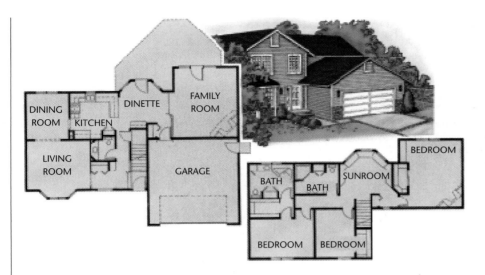

Refer to the table in Example 8 to answer Margin Exercises 8 and 9.

8. Home-Cost Comparison. Find the cost of a $250,000 home in Indianapolis if you were to try to replace it when moving to Dallas. Round to the nearest one.

1. **Familiarize.** We let C = the cost of the home in Palo Alto. We use the table and look up the indexes of the city in which you now live and the city to which you are moving.

2. **Translate.** Using the formula, we translate to the following equation:

   $C = \$250{,}000 \div 63 \times 398.$

3. **Solve.** To solve, we carry out the computations using the rules for order of operations (see Section 4.4):

   $C = \$250{,}000 \div 63 \times 398$

   $\approx \$3968.254 \times 398$ Carrying out the division first

   $\approx \$1{,}579{,}365.$ Carrying out the multiplication and rounding to the nearest one

   On a calculator, the computation could be done in one step.

4. **Check.** We can repeat our computations.

5. **State.** A home that sells for $250,000 in Indianapolis would cost about $1,579,365 in Palo Alto.

Do Exercises 8 and 9.

9. Find the cost of a $250,000 home in Phoenix if you were to try to replace it when moving to Barrington. Round to the nearest one.

Answers on page A-10

a Solve.

1. *Sherwin-Williams® Stock.* Sherwin-Williams Company specializes in many kinds of home-improvement items. Over the 52 weeks in 2000, the price per share of its stock ranged in value from a low of $17.13 to a high of $27.63. By how much did the high value differ from the low value?
Sources: The New York Stock Exchange; Sherwin-Williams Company

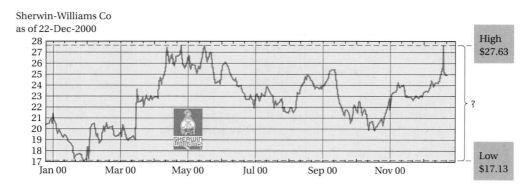

2. *Intel Corporation® Stock.* Intel is a corporation that specializes in microprocessors and other semiconductor products, such as computer chips. Over the 52 weeks in 2000, the price per share of its stock ranged in value from a low of $31.25 to a high of $75.81. By how much did the high value differ from the low value?
Sources: NasdaqNM; Intel Corporation

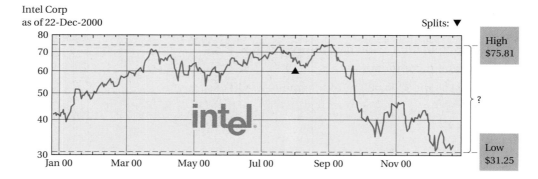

3. Roberto bought the CD "No Strings Attached" by *NSYNC for $14.99 plus $1.14 sales tax. He paid for it with a $20 bill. How much change did he receive?

4. Hannah bought a DVD of the movie *Gladiator* for $25.87 plus $1.55 sales tax. She paid for it with a $50 bill. How much change did she receive?

Russell Crowe holds his Oscar for Best Actor for his role in *Gladiator* at the 73rd Annual Academy Awards.

5. *Body Temperature.*   Normal body temperature is 98.6°F. During an illness, a patient's temperature rose 4.2°. What was the new temperature?

6. *Gasoline Cost.*   What is the cost, in dollars, of 20.4 gal of gasoline at 159.9 cents per gallon? Round the answer to the nearest cent.

7. *Lottery Winnings.*   In Texas, one of the state lotteries is called "Cash 5." In a recent weekly game, the lottery prize of $127,315 was shared equally by 6 winners. How much was each winner's share? Round to the nearest cent.
Source: Texas Lottery

8. *Lunch Costs.*   A group of 4 students pays $47.84 for lunch. What is each person's share?

9. *Stamp.*   Find the area and the perimeter of the stamp shown here.

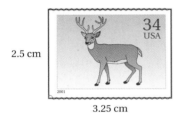

2.5 cm

3.25 cm

10. *Pole Vault Pit.*   Find the area and the perimeter of the landing area and the pole vault pit shown here.

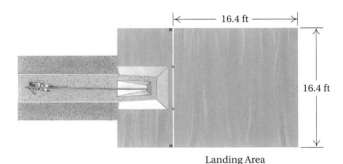

16.4 ft

16.4 ft

Landing Area

11. *Odometer Reading.*   A family checked the odometer before starting a trip. It read 22,456.8 and they know that they will be driving 234.7 mi. What will the odometer read at the end of the trip?

12. *Miles Driven.*   Petra bought gasoline when the odometer read 14,296.3. At the next gasoline purchase, the odometer read 14,515.8. How many miles had been driven?

13. *Gas Mileage.*   Peggy filled her van's gas tank and noted that the odometer read 26,342.8. After the next filling, the odometer read 26,736.7. It took 19.5 gal to fill the tank. How many miles per gallon did the van get?

14. *Gas Mileage.*   Peter filled his Honda's gas tank and noted that the odometer read 18,943.2. After the next filling, the odometer read 19,306.2. It took 13.2 gal to fill the tank. How many miles per gallon did the car get?

15. *Cost of Video Game.*   A certain video game costs 25 cents and runs for 1.5 min. Assuming a player does not win any free games and plays continuously, how much money, in dollars, does it cost to play the video game for 1 hr?

16. *Property Taxes.*   The Colavitos own a house with an assessed value of $184,500. For every $1000 of assessed value, they pay $7.68 in taxes. How much do they pay in taxes?

17. *Chemistry.* The water in a filled tank weighs 748.45 lb. One cubic foot of water weighs 62.5 lb. How many cubic feet of water does the tank hold?

18. *Highway Routes.* You can drive from home to work using either of two routes:

> *Route A:* Via interstate highway, 7.6 mi, with a speed limit of 65 mph.
> *Route B:* Via a country road, 5.6 mi, with a speed limit of 50 mph.

Assuming you drive at the posted speed limit, which route takes less time? (Use the formula *Distance = Speed × Time.*)

Find the distance around (perimeter of) the figure.

19.

8.9 cm
23.8 cm
4.7 cm
18.6 cm
22.1 cm

20.

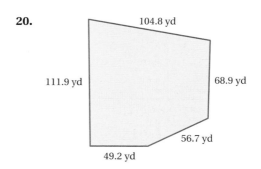

104.8 yd
111.9 yd
68.9 yd
56.7 yd
49.2 yd

21.

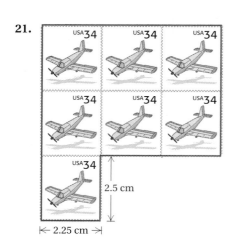

2.5 cm
2.25 cm

22.

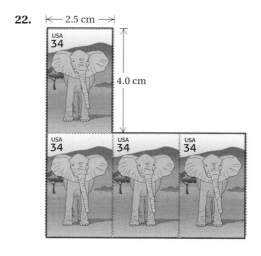

2.5 cm
4.0 cm

Find the length *d* in the figure.

23.

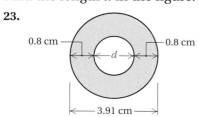

0.8 cm
0.8 cm
d
3.91 cm

24.

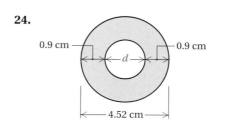

0.9 cm
0.9 cm
d
4.52 cm

25. *Calories Burned Mowing.* A person weighing 150 lb burns 7.3 calories per minute while mowing a lawn with a power lawnmower. How many calories would be burned in 2 hr of mowing?
Source: *The Handy Science Answer Book*

26. Lot A measures 250.1 ft by 302.7 ft. Lot B measures 389.4 ft by 566.2 ft. What is the total area of the two lots?

27. Holly had $1123.56 in her checking account. She used her debit card to pay bills of $23.82, $507.88, and $98.32. She then deposited a bonus check of $678.20. How much is in her account after these changes?

28. Natalie had $185.00 to spend for fall clothes: $44.95 was spent on shoes, $71.95 for a jacket, and $55.35 for pants. How much was left?

29. A rectangular yard is 20 ft by 15 ft. The yard is covered with grass except for an 8.5-ft square flower garden. How much grass is in the yard?

30. Rita earns a gross paycheck (before deductions) of $495.72. Her deductions are $59.60 for federal income tax, $29.00 for FICA, and $29.00 for medical insurance. What is her take-home paycheck?

31. *Batting Averages.* For the 2000 season, Todd Helton of the Colorado Rockies won the National League batting title with 216 hits in 580 times at bat. What part of his at-bats were hits? Give decimal notation to the nearest thousandth. (This is a player's *batting average.*)
Source: Major League Baseball

32. *Batting Averages.* For the 2000 season, Nomar Garciaparra of the Boston Red Sox won the American League batting title with 197 hits in 529 times at bat. What part of his at-bats were hits? Give decimal notation to the nearest thousandth.
Source: Major League Baseball

33. *CellularOne® Rates.* One recent plan for a cellular phone in Indiana was called "Indiana 400." The charge was $39.99 per month, and it included up to 400 min of statewide calling time. Minutes over 400 were charged at a rate of $0.25 per minute. One month Maggie used her cell phone for 517 min. What was the charge?
Source: CellularOne® from Bell South

34. *CellularOne® Rates.* One recent plan for a cellular phone in Indiana was called "Indiana 700." The charge was $64.99 per month, and it included up to 700 min of calling time. Minutes over 700 were charged at a rate of $0.25 per minute. One month Dave used his cell phone for 946 min. What was the charge?
Source: CellularOne® from Bell South

35. *Construction Pay.* A construction worker is paid $18.50 per hour for the first 40 hr of work, and time and a half, or $27.75 per hour, for any overtime exceeding 40 hr per week. One week she works 46 hr. How much is her pay?

36. *Summer Work.* Zachary worked 53 hr during a week one summer. He earned $6.50 per hour for the first 40 hr and $9.75 per hour for overtime (hours exceeding 40). How much did Zachary earn during the week?

37. *Egg Costs.*   A restaurant owner bought 20 dozen eggs for $13.80. Find the cost of each egg to the nearest tenth of a cent (thousandth of a dollar).

38. *Weight Loss.*   A person weighing 170 lb burns 8.6 calories per minute while mowing a lawn. One must burn about 3500 calories in order to lose 1 lb. How many pounds would be lost by mowing for 2 hr? Round to the nearest tenth.

39. *Field Dimensions.*   The dimensions of a World Cup soccer field are 114.9 yd by 74.4 yd. The dimensions of a standard football field are 120 yd by 53.3 yd. How much greater is the area of a World Cup soccer field?

40. *Loan Payment.*   In order to make money on loans, financial institutions are paid back more money than they loan. You borrow $120,000 to buy a house and agree to make monthly payments of $880.52 for 30 yr. How much do you pay back altogether? How much more do you pay back than the amount of the loan?

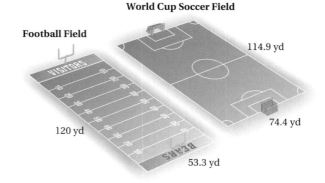

41. *World Population.*   Using the information in the following bar graph, determine the average population of the world for the years 1950 through 2000.

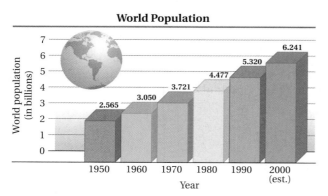

Source: Francis Urban and Philip Rose. *World Population by Country and Region, 1950–86,* and *Projections to 2050,* U.S. Dept. of Agriculture.

42. *Sleep Aid Prescriptions.*   The following bar graph shows the number of sleep aid prescriptions written for recent years. Find the average number of prescriptions written each year for that period.
Source: IMS Health

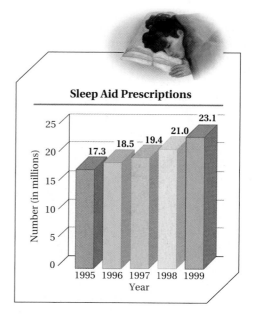

43. *Body Temperature.* Normal body temperature is 98.6°F. A baby's bath water should be 100°F. How many degrees above normal body temperature is this?

44. *Body Temperature.* Normal body temperature is 98.6°F. The lowest temperature at which a patient has survived is 69°F. How many degrees below normal is this?

Home-Cost Comparison. Use the table and formula from Example 8. In each of the following cases, find the value of the house in the new location.
Source: Coldwell Banker Real Estate Corporation

	VALUE	PRESENT LOCATION	NEW LOCATION	NEW VALUE
45.	$125,000	Hollywood Hills	San Francisco	
46.	$180,000	Barrington	Palo Alto	
47.	$96,000	Indianapolis	Tampa	
48.	$300,000	Miami	Queens/North Shore	
49.	$240,000	San Francisco	Atlanta	
50.	$160,000	St. Paul	Phoenix	

Comparison Shopping. The Internet now provides many sites to shop for a product rather than shopping in a local store. The following lists various web sites for purchasing the recent best-selling novel *The Rescue* by Nicholas Sparks.

51. ^{D}w Complete the total costs in the table below for each merchant. Then decide which site you would use to make a purchase. Discuss possible ways in which answers might vary.

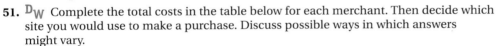

MERCHANT	RETAIL PRICE	OUR PRICE	SHIPPING AND HANDLING	SALES TAX	TOTAL COST
Amazon.com www.amazon.com	$22.95	$13.77	$4.48	$0	?
Barnes & Noble www.bn.com	$22.95	$13.77	$3.49 per order, plus $0.99 per item	$0	?
Powell's Books www.powells.com	$22.95	$15.00	$4.00 per order, plus $1.00 per item	$0	?
Costco Wholesale www.costco.com	$22.95	$12.99	$2.45	$0	?
1bookstreet.com www.1bookstreet.com	$22.95	$16.07	Free if the order is over $15	$0	?
Borders Bookstore, local store	$22.95	$22.95	$0	$1.38	?

Sources: www.yahoo.com; Borders Bookstore, Indianapolis

52. D_W *Internet Project.* Consider using the Internet to buy a copy of the book *The Bear and the Dragon* by Tom Clancy. Then compare your costs with buying it at a local bookstore. Decide which way you would make a purchase. Discuss possible ways in which answers might vary.

SKILL MAINTENANCE

Add.

53. $4569 + 1766$ [1.2a]

54. $\dfrac{2}{3} + \dfrac{5}{8}$ [3.2a]

55. $4\dfrac{1}{3} + 2\dfrac{1}{2}$ [3.5a]

56. $\dfrac{5}{6} + \dfrac{7}{10}$ [3.2a]

57. $8099 + 5667$ [1.2a]

Subtract.

58. $4569 - 1766$ [1.3b]

59. $\dfrac{2}{3} - \dfrac{5}{8}$ [3.3a]

60. $4\dfrac{1}{3} - 2\dfrac{1}{2}$ [3.5b]

61. $\dfrac{5}{6} - \dfrac{7}{10}$ [3.3a]

62. $8099 - 5667$ [1.3b]

Solve. [3.6c]

63. If a water wheel made 469 revolutions at a rate of $16\dfrac{3}{4}$ revolutions per minute, how long did it rotate?

64. If a bicycle wheel made 480 revolutions at a rate of $66\dfrac{2}{3}$ revolutions per minute, how long did it rotate?

SYNTHESIS

65. You buy a half-dozen packs of basketball cards with a dozen cards in each pack. The cost is twelve dozen cents for each half-dozen cards. How much do you pay for the cards?

Summary and Review

The review that follows is meant to prepare you for a chapter exam. It consists of two parts. The first part is a checklist of some of the Study Tips referred to in this text. The second part is the Review Exercises. These provide practice exercises for the exam, together with references to section objectives so you can go back and review. Before beginning, stop and look back over the skills you have obtained. What powers in mathematics do you have now that you did not have before studying this chapter?

STUDY TIPS CHECKLIST

The foundation of all your study skills is TIME!

☐ Have you been trying to take the primary responsibility for your learning?

☐ Have you been asking questions in class?

☐ Have you been using the five steps for problem solving?

☐ Have you established a learning relationship with your instructor?

☐ Are you doing your homework as soon as possible after class?

REVIEW EXERCISES

Convert the number in the sentence to standard notation.

1. Russia has the largest total area of any country in the world, at 6.59 million square miles. [4.3b]

2. The total weight of the turkeys consumed by Americans during the Thanksgiving holidays is about 6.9 million pounds. [4.3b]

Write a word name. [4.1a]

3. 3.47

4. 0.031

Write a word name as on a check. [4.1a]

5. $597.25

6. $0.96

Write fraction notation. [4.1b]

7. 0.09

8. 4.561

9. 0.089

10. 3.0227

Write decimal notation. [4.1b]

11. $\dfrac{34}{1000}$

12. $\dfrac{42,603}{10,000}$

13. $27\dfrac{91}{100}$

14. $867\dfrac{6}{1000}$

Which number is larger? [4.1c]

15. 0.034, 0.0185

16. 0.91, 0.19

17. 0.741, 0.6943

18. 1.038, 1.041

Round 17.4287 to the nearest: [4.1d]

19. Tenth.

20. Hundredth.

21. Thousandth.

22. One.

Add. [4.2a]

23.
```
      2.0 4 8
    6 5.3 7 1
  + 5 0 7.1
```

24.
```
    0.6
    0.0 0 4
    0.0 7
  +0.0 0 9 8
```

25. 219.3 + 2.8 + 7

26. 0.41 + 4.1 + 41 + 0.041

Subtract. [4.2b]

27.
```
    3 0.0
  −    0.7 9 0 8
```

28.
```
    8 4 5.0 8
  −     5 4.7 9
```

29. 37.645 − 8.497

30. 70.8 − 0.0109

Multiply. [4.3a]

31.
```
      4 8
  × 0.2 7
```

32.
```
    0.1 7 4
  ×   0.8 3
```

33. 100 × 0.043

34. 0.001 × 24.68

Divide. [4.4a]

35. 8) 6 0

36. 5 2) 2 3.4

37. 2.6) 1 1 7.5 2

38. 2.1 4) 2.1 8 7 0 8

39. $\dfrac{276.3}{1000}$

40. $\dfrac{13.892}{0.01}$

Solve. [4.2c], [4.4b]

41. $x + 51.748 = 548.0275$

42. $3 \cdot x = 20.85$

43. $10 \cdot y = 425.4$

44. $0.0089 + y = 5$

Solve. [4.7a]

45. *Tea Consumption.* The average person drinks about 3.48 cups of tea per day. How many cups of tea does the average person drink in a week? in a 30-day month?
Source: Tom Parker, *In One Day.* Boston: Houghton Mifflin, 1984.

46. Stacia, a coronary intensive care nurse, earned $620.74 during a recent 40-hr week. What was her hourly wage? Round to the nearest cent.

47. Derek had $6274.35 in his checking account. He used $485.79 to buy a Palm Digital Assistant with his debit card. How much was left in his account?

48. *CellularOne® Rates.* One recent plan for a cellular phone in Indiana was called "Indiana 1600." The charge was $124.99 per month, and it included up to 1600 min of calling time. Minutes over 1600 were charged at a rate of $0.25 per minute. One month Maria used her cell phone for 2000 min. What was the charge?
Source: CellularOne® from Bell South

49. *Gas Mileage.* A driver wants to estimate gas mileage per gallon. At 36,057.1 mi, the tank is filled with 10.7 gal. At 36,217.6 mi, the tank is filled with 11.1 gal. Find the mileage per gallon. Round to the nearest tenth.

50. *Seafood Consumption.* The following graph shows the annual consumption, in pounds, of seafood per person in the United States in recent years. [4.4c]

a) Find the total per capita consumption for the four years.

b) Find the average per capita consumption.

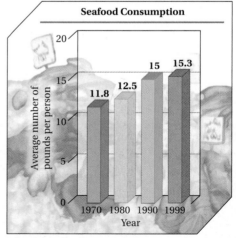

Source: National Oceanographic and Atmospheric Administration

Estimate each of the following. [4.6a]

51. The product 7.82×34.487 by rounding to the nearest one

52. The difference $219.875 - 4.478$ by rounding to the nearest one

53. The quotient $82.304 \div 17.287$ by rounding to the nearest ten

54. The sum $\$45.78 + \$78.99$ by rounding to the nearest one

Find decimal notation. Use multiplying by 1. [4.5a]

55. $\dfrac{13}{5}$ **56.** $\dfrac{32}{25}$ **57.** $\dfrac{11}{4}$

Find decimal notation. Use division. [4.5a]

58. $\dfrac{13}{4}$ **59.** $\dfrac{7}{6}$ **60.** $\dfrac{17}{11}$

Round the answer to Exercise 60 to the nearest:
[4.5b]

61. Tenth. **62.** Hundredth. **63.** Thousandth.

Convert from cents to dollars. [4.3b]

64. 8273 cents **65.** 487 cents

Convert from dollars to cents. [4.3b]

66. $24.93 **67.** $9.86

Calculate. [4.4c], [4.5c]

68. $(8 - 1.23) \div 4 + 5.6 \times 0.02$

69. $(1 + 0.07)^2 + 10^3 \div 10^2$
$+ [4(10.1 - 5.6) + 8(11.3 - 7.8)]$

70. $\dfrac{3}{4} \times 20.85$

71. $\dfrac{1}{3} \times 123.7 + \dfrac{4}{9} \times 0.684$

72. $\mathbf{D_W}$ Consider finding decimal notation for $\frac{44}{125}$. Discuss as many ways as you can for finding such notation and give the answer. [4.5a]

73. $\mathbf{D_W}$ Explain how we can use fraction notation to understand why we count decimal places when multiplying with decimal notation. [4.3a]

SKILL MAINTENANCE

Certain objectives from four particular sections will be retested on the chapter test. The objectives are listed with the practice problems that follow.

74. Multiply: $8\dfrac{1}{3} \cdot 5\dfrac{1}{4}$. [3.6a]

75. Divide: $20 \div 5\dfrac{1}{3}$. [3.6b]

76. Add: $12\dfrac{1}{2} + 7\dfrac{3}{10}$. [3.5a]

77. Subtract: $24 - 17\dfrac{2}{5}$. [3.5b]

78. Simplify: $\dfrac{28}{56}$. [2.5b]

79. Find the prime factorization of 192. [2.1d]

SYNTHESIS

80. ▦ In each of the following, use one of $+$, $-$, $\times$, and $\div$ in each blank to make a true sentence. [4.4c]
a) $2.56 \,\square\, 6.4 \,\square\, 51.2 \,\square\, 17.4 \,\square\, 89.7 = 72.62$
b) $(11.12 \,\square\, 0.29) \,\square\, 3^4 = 877.23$

81. Find repeating decimal notation for 1 and explain. Use the following hints. [4.5a]

$$\dfrac{1}{3} = 0.33333333\ldots,$$

$$\dfrac{2}{3} = 0.66666666\ldots$$

82. Find repeating decimal notation for 2. [4.5a]

Chapter Test

Convert the number in the sentence to standard notation.

1. The annual sales of antibiotics in the United States is $8.9 billion.
Source: IMS Health

2. There are 3.756 million people enrolled in bowling organizations in the United States.
Source: *Bowler's Journal International,* December 2000

3. Write a word name: 2.34.

4. Write a word name, as on a check, for $1234.78.

Write fraction notation.

5. 0.91

6. 2.769

Write decimal notation.

7. $\dfrac{74}{1000}$

8. $\dfrac{37,047}{10,000}$

9. $756\dfrac{9}{100}$

10. $91\dfrac{703}{1000}$

Which number is larger?

11. 0.07, 0.162

12. 0.078, 0.06

13. 0.09, 0.9

Round 5.6783 to the nearest:

14. One.

15. Hundredth.

16. Thousandth.

17. Tenth.

Calculate.

18.
$$\begin{array}{r} 0.7 \\ 0.0\ 8 \\ 0.0\ 0\ 9 \\ +\ 0.0\ 0\ 1\ 2 \\ \hline \end{array}$$

19. 102.4 + 6.1 + 78

20. 0.93 + 9.3 + 93 + 930

21.
$$\begin{array}{r} 5\ 2.6\ 7\ 8 \\ -\ \ \ \ 4.3\ 2\ 1 \\ \hline \end{array}$$

22.
$$\begin{array}{r} 2\ 0.0 \\ -\ \ \ \ 0.9\ 0\ 9\ 9 \\ \hline \end{array}$$

23. 234.6788 − 81.7854

24.
$$\begin{array}{r} 0.1\ 2\ 5 \\ \times\ \ \ \ 0.2\ 4 \\ \hline \end{array}$$

25. 0.001 × 213.45

26. 1000 × 73.962

27. $4\overline{)1\ 9}$

28. $3.3\overline{)100.32}$

29. $82\overline{)15.58}$

30. $\dfrac{346.89}{1000}$

31. $\dfrac{346.89}{0.01}$

Solve.

32. $4.8 \cdot y = 404.448$

33. $x + 0.018 = 9$

34. *CellularOne® Rates.* One recent plan for a cellular phone in Indiana was called "Indiana 1000." The charge was $84.99 per month, and it included up to 1000 min of calling time. Minutes over 1000 were charged at a rate of $0.25 per minute. One month Ramon used his cell phone for 1142 min. What was the charge?
Source: CellularOne® from Bell South

35. *Gas Mileage.* Tina wants to estimate the gas mileage per gallon in her economy car. At 76,843 mi, the tank is filled with 14.3 gal of gasoline. At 77,310 mi, the tank is filled with 16.5 gal of gasoline. Find the mileage per gallon. Round to the nearest tenth.

36. *Checking Account Balance.* Nicholas has a balance of $10,200 in his checking account before making purchases of $123.89, $56.68, and $3446.98 with his debit card. What was the balance after making the purchases?

37. *MP3 Players.* Matt buys 6 Compaq iPAQ 64 Mb Personal Audio Players at $199.99 each. What is the total cost?
Source: Compaq

38. *Airport Passengers.* The following graph shows the number of passengers in a recent year who traveled through the country's busiest airports. Find the average number of passengers through these airports.

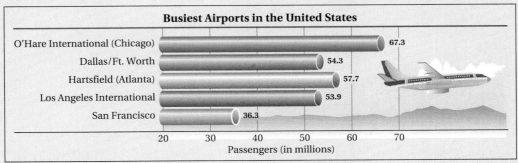

Busiest Airports in the United States

O'Hare International (Chicago) — 67.3
Dallas/Ft. Worth — 54.3
Hartsfield (Atlanta) — 57.7
Los Angeles International — 53.9
San Francisco — 36.3

Passengers (in millions)

Source: Air Transport Association of America

Estimate each of the following.

39. The product 8.91×22.457 by rounding to the nearest one

40. The quotient $78.2209 \div 16.09$ by rounding to the nearest ten

Find decimal notation. Use multiplying by 1.

41. $\dfrac{8}{5}$

42. $\dfrac{22}{25}$

43. $\dfrac{21}{4}$

Find decimal notation. Use division.

44. $\dfrac{3}{4}$

45. $\dfrac{11}{9}$

46. $\dfrac{15}{7}$

Round the answer to Question 46 to the nearest:

47. Tenth.

48. Hundredth.

49. Thousandth.

50. Convert from cents to dollars: 949 cents.

Calculate.

51. $256 \div 3.2 \div 2 - 1.56 + 78.325 \times 0.02$

52. $(1 - 0.08)^2 + 6[5(12.1 - 8.7) + 10(14.3 - 9.6)]$

53. $\dfrac{7}{8} \times 345.6$

54. $\dfrac{2}{3} \times 79.95 - \dfrac{7}{9} \times 1.235$

SKILL MAINTENANCE

55. Multiply: $2\dfrac{1}{10} \cdot 6\dfrac{2}{3}$.

56. Add: $2\dfrac{3}{16} + \dfrac{1}{2}$.

57. Subtract: $28\dfrac{2}{3} - 2\dfrac{1}{6}$.

58. Divide. $3\dfrac{3}{8} \div 3$.

59. Simplify: $\dfrac{33}{54}$.

60. Find the prime factorization of 360.

SYNTHESIS

61. The Silver's Health Club generally charges a $79 membership fee and $42.50 a month. Allise has a coupon that will allow her to join the club for $299 for six months. How much will Allise save if she uses the coupon?

62. ▦ Arrange from smallest to largest.

$$\dfrac{2}{3}, \ \dfrac{15}{19}, \ \dfrac{11}{13}, \ \dfrac{5}{7}, \ \dfrac{13}{15}, \ \dfrac{17}{20}$$

Convert to fraction notation.

1. $2\frac{2}{9}$

2. 3.052

Find decimal notation.

3. $\frac{7}{5}$

4. $\frac{6}{11}$

5. Determine whether 43 is prime, composite, or neither.

6. Determine whether 2,053,752 is divisible by 4.

Calculate.

7. $48 + 12 \div 4 - 10 \times 2 + 6892 \div 4$

8. $4.7 - \{0.1[1.2(3.95 - 1.65) + 1.5 \div 2.5]\}$

Round to the nearest hundredth.

9. 584.903

10. $218.\overline{5}$

11. Estimate the product 16.392×9.715 by rounding to the nearest one.

12. Estimate by rounding to the nearest tenth:
$2.714 + 4.562 - 3.31 - 0.0023$.

13. Estimate the product 6418×1984 by rounding to the nearest hundred.

14. Estimate the quotient $717.832 \div 124.998$ by rounding to the nearest ten.

Global Warming. The following table lists the global average temperature for the years 1990 through 1999. Use the table for Exercises 15 and 16.

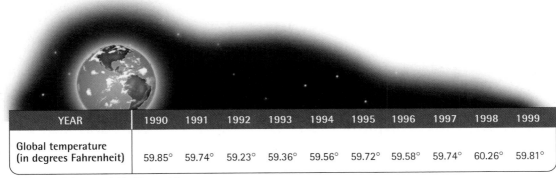

YEAR	1990	1991	1992	1993	1994	1995	1996	1997	1998	1999
Global temperature (in degrees Fahrenheit)	59.85°	59.74°	59.23°	59.36°	59.56°	59.72°	59.58°	59.74°	60.26°	59.81°

Sources: Lester R. Brown et al., *Vital Signs*, and the Council of Environmental Quality

15. Find the average global temperature for the years 1990 through 1994. Round the answer to the nearest hundredth.

16. Find the average global temperature for the years 1995 through 1999. Round the answer to the nearest hundredth.

17. *Medical Transplants.* In a recent year, there were 1952 heart transplants, 9004 kidney transplants, 3229 liver transplants, and 89 pancreas transplants. How many transplants of these four organs were performed that year?
Source: U.S. Department of Health and Human Services

18. *Abercrombie & Fitch.* Operating more than 350 stores, Abercrombie & Fitch is a company principally engaged in the purchase, distribution, and sale of men's, women's, and kids' casual apparel. Over the 52 weeks in 2000, the price per share of its stock ranged in value from a low of $8 to a high of $31.31. By how much did the high value differ from the low value?
Sources: The New York Stock Exchange; Abercrombie & Fitch

19. *ABC Carpet Sales.* ABC Warehouse Outlet has a $12\frac{1}{6}$-ft by 16-ft Red Bokhara rug on sale marked down from $5199 to $3499.

a) Find the area and the perimeter of the rug.
b) By how much has the price been reduced?
Sources: The New York Times, 7/16/00; www.abccarpet.com

20. After making a $150 down payment on a sofa, $\frac{3}{10}$ of the total cost was paid. How much did the sofa cost?

21. A clerk in a delicatessen sold $1\frac{1}{2}$ lb of ham, $2\frac{3}{4}$ lb of turkey, and $2\frac{1}{4}$ lb of roast beef. How many pounds of meat were sold?

22. A baker used $\frac{1}{2}$ lb of sugar for cookies, $\frac{2}{3}$ lb of sugar for pie, and $\frac{5}{6}$ lb of sugar for cake. How much sugar was used in all?

Divide and simplify.

23. $16.5 \overline{)35.013}$

24. $26 \overline{)47,918}$

25. $13.8621 \div 0.001$

26. $\dfrac{4}{9} \div \dfrac{8}{15}$

Solve.

27. $8.32 + x = 9.1$

28. $75 \cdot x = 2100$

29. $y \cdot 9.47 = 81.6314$

30. $1062 + y = 368,313$

31. $t + \dfrac{5}{6} = \dfrac{8}{9}$

32. $\dfrac{7}{8} \cdot t = \dfrac{7}{16}$

Add and simplify.

33.
$$2\frac{1}{4}$$
$$+3\frac{4}{5}$$

34.
$$\begin{array}{r} 34,921 \\ 93,092 \\ +11,103 \\ \hline \end{array}$$

35. $\dfrac{1}{6} + \dfrac{2}{3} + \dfrac{8}{9}$

36. $143.9 + 2.053$

Subtract and simplify.

37. $723,041 - 12,904$

38. $19 - 5.903$

39. $5\dfrac{1}{7} - 4\dfrac{3}{7}$

40. $\dfrac{10}{11} - \dfrac{9}{10}$

Multiply and simplify.

41. $\dfrac{3}{8} \cdot \dfrac{4}{9}$

42.
$$\begin{array}{r} 2532 \\ \times\ 2100 \\ \hline \end{array}$$

43.
$$\begin{array}{r} 23.9 \\ \times\ 0.2 \\ \hline \end{array}$$

44.
$$\begin{array}{r} 27.9431 \\ \times\ 0.001 \\ \hline \end{array}$$

45. Simplify: $\left(\dfrac{3}{4}\right)^2 - \dfrac{1}{8} \cdot \left(3 - 1\dfrac{1}{2}\right)^2$.

Look for a pattern in the following sequences, and find the missing numbers.

46. $22.22, $33.34, $44.46, $55.58, ____ , ____ , ____ , ____ , ____ , ____ , ____ .

47. $2344.78, $2266, $2187.22, $2108.44, ____ , ____ , ____ , ____ .

Each of the following is called a *magic square*. The sum along each row, column, or diagonal is the same. Find the missing numbers.

48.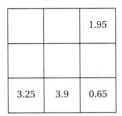

		1.95
3.25	3.9	0.65

Magic sum = ?

49.

2.16		1.08
1.62		0.54

Magic sum = 4.05

50.

6.16		34.72	16.24
38.64	12.32		
	34.16	44.24	
40.32			37.52

Magic sum = 100.24

For each of Exercises 51–54, choose the correct answer from the selections given.

51. In the equation $11 - 4 = 7$, what is the subtrahend?

a) 11 **b)** 7 **c)** 4
d) "=" sign **e)** None

52. Find decimal notation: $\dfrac{35}{27}$.

a) 1.259259 **b)** 0.7714285 **c)** 1.296
d) 1.333333 **e)** None

53. An athletic coach paid $2974.36 for 30 pairs of running shoes. Find the cost of each pair. Round to the nearest cent.

a) $75.24 **b)** $77.32 **c)** $76.26
d) $76.27 **e)** None

54. Find decimal notation: $\dfrac{716{,}605}{10{,}000}$.

a) 716,605.0000 **b)** 71.6605 **c)** 0.716605
d) 0.0716605 **e)** None

SYNTHESIS

55. A customer in a grocery store used a manufacturer's coupon to buy juice. With the coupon, if 5 cartons of juice were purchased, the sixth carton was free. The price of each carton was $1.09. What was the cost per carton with the coupon? Round to the nearest cent.

Ratio and Proportion

Gateway to Chapter 5

In Chapter 2, we introduced fraction notation as a ratio.
Here we use ratios and proportions to solve problems.
We also consider such topics as rates, unit prices, and
geometric applications.

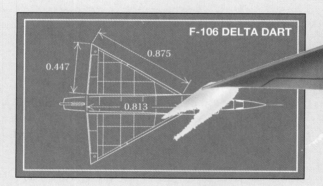

F-106 DELTA DART

0.447   0.875

0.813

Real-World Application

A blueprint is a scale drawing. Each wing of an F-106 Delta Dart military
airplane has a triangular shape. The blueprint for this airplane shows similar
triangles. Find the length a of a side of the wing.

This problem appears as Example 3 in Section 5.5.

19.2 ft

a

Write fraction notation for the ratio. [5.1a]

1. 35 to 43

2. 0.079 to 1.043

Solve. [5.3b]

3. $\dfrac{5}{6} = \dfrac{x}{27}$

4. $\dfrac{y}{0.25} = \dfrac{0.3}{0.1}$

5. $\dfrac{3\frac{1}{2}}{4\frac{1}{3}} = \dfrac{6\frac{3}{4}}{x}$

6. What is the rate in miles per gallon? [5.2a]

408 miles, 16 gallons

7. A student picked 10 qt (quarts) of strawberries in 45 min. What is the rate in quarts per minute? [5.2a]

8. Folger's Coffee. For each package listed below, find the unit price rounded to the nearest hundredth. Then determine which package has the lower unit price. [5.2b]

PACKAGE	PRICE	UNIT PRICE
11.5 oz	$2.32	
34.5 oz	$5.67	

Solve.

9. Todd traveled 216 km in 6 hr. At this rate, how far will he travel in 54 hr? [5.4a]

10. If 4 packs of gum cost $5.16, how many packs of gum can you buy for $28.38? [5.4a]

11. Juan's digital car clock loses 5 min in 10 hr. At this rate, how much will it lose in 24 hr? [5.4a]

12. On a map, 4 in. represents 225 mi. If two cities are 7 in. apart on the map, how far apart are they in reality? [5.4a]

13. These triangles are similar. Find the missing lengths. [5.5a]

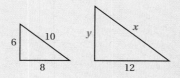

14. New York Commuters. Of every 5 people who commute to work in New York City, 2 spend more than 90 min a day commuting. Find the ratio of people whose daily commute to New York exceeds 90 min a day to those whose commute is 90 min or less. [5.4a]

Source: *The Amicus Journal*

5.1

INTRODUCTION TO RATIOS

a   Ratios

Objectives

a Find fraction notation for ratios.

b Simplify ratios.

> **RATIO**
>
> A **ratio** is the quotient of two quantities.

In January 2001, the Atlanta Hawks basketball team averaged 86.5 points per game and allowed their opponents an average of 90.8 points per game. The *ratio* of points earned to points allowed is given by the fraction notation

Points earned $\longrightarrow$ $\dfrac{86.5}{90.8}$ or by the colon notation   86.5 : 90.8.
Points allowed $\longrightarrow$

We read both forms of notation as "the ratio of 86.5 to 90.8," listing the numerator first and the denominator second.

> **RATIO NOTATION**
>
> The **ratio** of a to b is given by the fraction notation $\dfrac{a}{b}$, where a is the numerator and b is the denominator, or by the colon notation $a:b$.

EXAMPLE 1   Find the ratio of 7 to 8.

The ratio is $\dfrac{7}{8}$,   or $7:8$.

EXAMPLE 2   Find the ratio of 31.4 to 100.

The ratio is $\dfrac{31.4}{100}$,   or $31.4:100$.

EXAMPLE 3   Find the ratio of $4\dfrac{2}{3}$ to $5\dfrac{7}{8}$. You need not simplify.

The ratio is $\dfrac{4\frac{2}{3}}{5\frac{7}{8}}$,   or $4\frac{2}{3} : 5\frac{7}{8}$.

Do Exercises 1–3.

In most of our work, we will use fraction notation for ratios.

EXAMPLE 4   *Wind Speeds.*   The average wind speed in Chicago is 10.4 mph. The average wind speed in Boston is 12.5 mph. Find the ratio of the wind speed in Chicago to the wind speed in Boston.
Source: *The Handy Geography Answer Book*

The ratio is $\dfrac{10.4}{12.5}$.

1. Find the ratio of 5 to 11.

2. Find the ratio of 57.3 to 86.1.

3. Find the ratio of $6\dfrac{3}{4}$ to $7\dfrac{2}{5}$.

4. **Rainfall.**   The greatest amount of rainfall ever recorded for a 12-month period was 739 in. in Kukui, Maui, Hawaii, from December 1981 to December 1982. Find the ratio of rainfall to time in months.
Source: *The Handy Science Answer Book*

Answers on page A-11

313

5. Fat Grams. In one serving ($\frac{1}{2}$-cup) of fried scallops, there is 12 g of fat. In one serving ($\frac{1}{2}$-cup) of fried oysters, there is 14 g of fat. What is the ratio of grams of fat in one serving of scallops to grams of fat in one serving of oysters?
Source: *Better Homes and Gardens: A New Cook Book*

6. Earned Runs. In the 2000 season, Randy Johnson of the Arizona Diamondbacks gave up 73 earned runs in $248\frac{2}{3}$ innings pitched. What was the ratio of earned runs to innings pitched? of innings pitched to earned runs?
Source: Major League Baseball

7. In the triangle below, what is the ratio of the length of the shortest side to the length of the longest side?

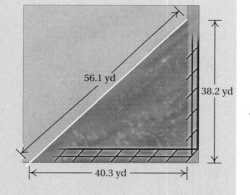

EXAMPLE 5 *Batting.* In the 2000 season, Gary Sheffield of the Los Angeles Dodgers got 163 hits in 501 at-bats. What was the ratio of hits to at-bats? of at-bats to hits?
Source: Major League Baseball

The ratio of hits to at-bats is

$$\frac{163}{501}.$$

The ratio of at-bats to hits is

$$\frac{501}{163}.$$

Do Exercises 4–6. (Exercise 4 is on the preceding page.)

EXAMPLE 6 Refer to the triangle below.

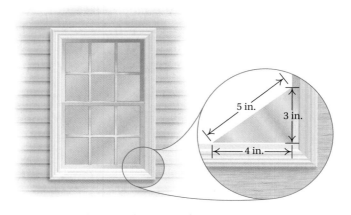

a) What is the ratio of the length of the longest side to the length of the shortest side?

$$\frac{5}{3}$$

b) What is the ratio of the length of the shortest side to the length of the longest side?

$$\frac{3}{5}$$

Do Exercise 7.

b Simplifying Notation for Ratios

Sometimes a ratio can be simplified. This provides a means of finding other numbers with the same ratio.

EXAMPLE 7 Find the ratio of 6 to 8. Then simplify and find two other numbers in the same ratio.

Answers on page A-11

We write the ratio in fraction notation and then simplify:

$$\frac{6}{8} = \frac{2 \cdot 3}{2 \cdot 4} = \frac{2}{2} \cdot \frac{3}{4} = 1 \cdot \frac{3}{4} = \frac{3}{4}.$$

Thus, 3 and 4 have the same ratio as 6 and 8. We can express this by saying "6 is to 8" as "3 is to 4."

Do Exercise 8.

EXAMPLE 8 Find the ratio of 2.4 to 10. Then simplify and find two other numbers in the same ratio.

We first write the ratio in fraction notation. Next, we multiply by 1 to clear the decimal from the numerator. Then we simplify.

$$\frac{2.4}{10} = \frac{2.4}{10} \cdot \frac{10}{10} = \frac{24}{100} = \frac{4 \cdot 6}{4 \cdot 25} = \frac{4}{4} \cdot \frac{6}{25} = \frac{6}{25}$$

Thus, 2.4 is to 10 as 6 is to 25.

Do Exercises 9 and 10.

EXAMPLE 9 A standard television screen with a width of 32 in. has a height of 24 in. Find the ratio of width to height and simplify.

The ratio is $\dfrac{32}{24} = \dfrac{8 \cdot 4}{8 \cdot 3} = \dfrac{8}{8} \cdot \dfrac{4}{3} = \dfrac{4}{3}$.

Thus we can say that the ratio of width to height is 4 to 3.

Do Exercise 11.

8. Find the ratio of 18 to 27. Then simplify and find two other numbers in the same ratio.

9. Find the ratio of 3.6 to 12. Then simplify and find two other numbers in the same ratio.

10. Find the ratio of 1.2 to 1.5. Then simplify and find two other numbers in the same ratio.

11. In Example 9, find the ratio of the width of the shortest side of the television screen to the width of the longest side and simplify.

Answers on page A-11

Answers on page A-11

Study Tips

TIME MANAGEMENT (PART 2)

Here are some additional tips to help you with time management. (See also the Study Tips on time management in Sections 1.5 and 5.4.)

■ **Avoid "time killers."** We live in a media age, and the Internet, e-mail, television, and movies all are time killers. Allow yourself a break to enjoy some college and outside activities. But keep track of the time you spend on such activities and compare it to the time you spend studying.

■ **Prioritize your tasks.** Be careful about taking on too many college activities that fall outside of academics. Examples of such activities are decorating a homecoming float, joining a fraternity or sorority, and participating on a student council committee. Any of these is important but keep them to a minimum to be sure that you have enough time for your studies.

■ **Be aggressive about your study tasks.** Instead of worrying over your math homework or test preparation, do something to get yourself started. Work a problem here and a problem there, and before long you will accomplish the task at hand. If the task is large, break it down into smaller parts, and do one at a time. You will be surprised at how quickly the large task can then be completed.

5.1

EXERCISE SET

For Extra Help

Digital Video
Tutor CD 3
Videotape 9

InterAct
Math

Math Tutor
Center

MathXL

MyMathLab

a Find fraction notation for the ratio. You need not simplify.

1. 4 to 5

2. 3 to 2

3. 178 to 572

4. 329 to 967

5. 0.4 to 12

6. 2.3 to 22

7. 3.8 to 7.4

8. 0.6 to 0.7

9. 56.78 to 98.35

10. 456.2 to 333.1

11. $8\frac{3}{4}$ to $9\frac{5}{6}$

12. $10\frac{1}{2}$ to $43\frac{1}{4}$

13. *Corvette Accidents.* Of every 5 fatal accidents involving a Corvette, 4 do not involve another vehicle. Find the ratio of fatal accidents involving just a Corvette to those involving a Corvette and at least one other vehicle.
Source: *Harper's Magazine*

14. *The Rescue.* The recent novel *The Rescue* had a list price of $22.95 but was sold by Amazon.com for $13.77. What was the ratio of the sale price to the list price? of the list price to the sale price?
Source: www.amazon.com

15. *Batting.* In the 2000 season, Derek Jeter of the New York Yankees got 201 hits in 593 at-bats. What was the ratio of hits to at-bats? of at-bats to hits?
Source: Major League Baseball

16. *Batting.* In the 2000 season, Alex Rodriguez then of the Seattle Mariners got 175 hits in 554 at-bats. What was the ratio of hits to at-bats? of at-bats to hits?
Source: Major League Baseball

17. *Heart Disease.* In the state of Minnesota, of every 1000 people, 93.2 will die of heart disease. Find the ratio of those who die of heart disease to every 1000 people.
Source: "Reforming the Health Care System; State Profiles 1999," AARP

18. *Cancer Deaths.* In the state of Texas, of every 1000 people, 122.8 will die of cancer. Find the ratio of those who die of cancer to every 1000 people.
Source: "Reforming the Health Care System; State Profiles 1999," AARP

19. *Silicon in the Earth's Crust.* Of every 100 tons of the earth's crust, there will be about 28 tons of silicon in its content. What is the ratio of silicon to the weight of crust? of the weight of crust to the weight of silicon?
Source: *The Handy Science Answer Book*

20. *Smokers.* In 1998, of every 100 people 18 years or older, 24.1 smoked cigarettes. Find the ratio of smokers to every 100 people.
Source: U.S. Centers for Disease Control

21. *Field Hockey.* A diagram of the playing area for field hockey is shown below. What is the ratio of width to length? of length to width?
Source: *Sports: The Complete Visual Reference*

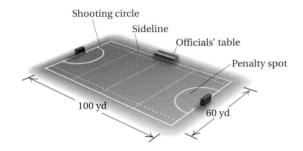

Shooting circle
Sideline
Officials' table
Penalty spot
100 yd
60 yd

22. *The Leaning Tower of Pisa.* At the time of this writing, the Leaning Tower of Pisa is still standing. It is 184.5 ft tall but leans about 17 ft out from its base. What is the ratio of the distance it leans to its height? its height to the distance it leans?
Source: *The Handy Science Answer Book*

184.5 ft

17 ft

b Find the ratio of the first number to the second and simplify.

23. 4 to 6

24. 6 to 10

25. 18 to 24

26. 28 to 36

27. 4.8 to 10

28. 5.6 to 10

29. 2.8 to 3.6

30. 4.8 to 6.4

31. 20 to 30   **32.** 40 to 60   **33.** 56 to 100   **34.** 42 to 100

35. 128 to 256   **36.** 232 to 116   **37.** 0.48 to 0.64   **38.** 0.32 to 0.96

39. In this rectangle, find the ratios of length to width and of width to length.

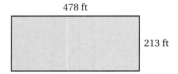

478 ft

213 ft

40. In this right triangle, find the ratios of shortest length to longest length and of longest length to shortest length.

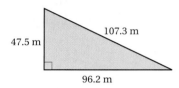

47.5 m

107.3 m

96.2 m

41. ^{D}W Can every ratio be written as the ratio of some number to 1? Why or why not?

42. ^{D}W What can be concluded about a rectangle's width if the ratio of length to perimeter is 1 to 3? Make some sketches and explain your reasoning.

SKILL MAINTENANCE

Use = or ≠ for ☐ to write a true sentence.   [2.5c]

43. $\frac{12}{8}$ ☐ $\frac{6}{4}$ **44.** $\frac{4}{7}$ ☐ $\frac{5}{9}$ **45.** $\frac{7}{2}$ ☐ $\frac{31}{9}$ **46.** $\frac{17}{25}$ ☐ $\frac{68}{100}$

Divide. Write decimal notation for the answer.   [4.4a]

47. 200 ÷ 4   **48.** 95 ÷ 10   **49.** 232 ÷ 16   **50.** 342 ÷ 2.25

Solve.   [3.5c]

51. Rocky is $187\frac{1}{10}$ cm tall and his daughter is $180\frac{3}{4}$ cm tall. How much taller is Rocky?

52. Aunt Louise is $168\frac{1}{4}$ cm tall and her son is $150\frac{7}{10}$ cm tall. How much taller is Aunt Louise?

SYNTHESIS

53. Find the ratio of $3\frac{3}{4}$ to $5\frac{7}{8}$ and simplify.

Fertilizer.   Exercises 54 and 55 refer to a common lawn fertilizer known as "5, 10, 15." This mixture contains 5 parts of potassium for every 10 parts of phosphorus and 15 parts of nitrogen (this is often denoted 5 : 10 : 15).

54. Find the ratio of potassium to nitrogen and of nitrogen to phosphorus.

55. Simplify the ratio 5 : 10 : 15.

5.2

RATES AND UNIT PRICES

Objectives

 a Give the ratio of two different measures as a rate.

 b Find unit prices and use them to compare purchases.

a Rates

A 2001 Honda Civic can go 464 miles on 16 gallons of gasoline. Let's consider the ratio of miles to gallons:

Source: *Consumer Reports*

$$\frac{464 \text{ mi}}{16 \text{ gal}} = \frac{464}{16} \frac{\text{miles}}{\text{gallon}} = \frac{29}{1} \frac{\text{miles}}{\text{gallon}}$$

$$= 29 \text{ miles per gallon} = 29 \text{ mpg.}$$

"per" means "division," or "for each."

The ratio

$$\frac{464 \text{ mi}}{16 \text{ gal}}, \quad \text{or} \quad \frac{464}{16} \frac{\text{mi}}{\text{gal}}, \quad \text{or } 29 \text{ mpg}$$

is called a **rate.**

RATE

When a ratio is used to compare two different kinds of measure, we call it a **rate.**

Suppose Alyssa says her car goes 462.4 mi on 15.8 gal of gasoline. Is the mpg (mileage) of her car better than that of the Civic above? To determine this, it helps to convert to decimal notation and perhaps round. Then we have

$$\frac{462.4 \text{ miles}}{15.8 \text{ gallons}} = \frac{462.4}{15.8} \text{ mpg} \approx 29.266 \text{ mpg.}$$

Since $29.266 > 29$, Alyssa's car gets better mileage than the Civic does.

EXAMPLE 1 It takes 60 oz of grass seed to seed 3000 sq ft of lawn. What is the rate in ounces per square foot?

$$\frac{60 \text{ oz}}{3000 \text{ sq ft}} = \frac{1}{50} \frac{\text{oz}}{\text{sq ft}}, \quad \text{or} \quad 0.02 \frac{\text{oz}}{\text{sq ft}}$$

EXAMPLE 2 A cook buys 10 lb of potatoes for $3.69. What is the rate in cents per pound?

$$\frac{\$3.69}{10 \text{ lb}} = \frac{369 \text{ cents}}{10 \text{ lb}}, \quad \text{or} \quad 36.9 \frac{\text{cents}}{\text{lb}}$$

Study Tips

TAPING YOUR LECTURES

Consider recording your notes and playing them back when convenient, say, while commuting to campus. It can even be advantageous to record math lectures. (Be sure to get permission from your instructor before doing so, however.) Important points can be emphasized verbally. We consider this idea so worthwhile that we provide a series of audiotapes that accompany the book. (See the Preface for more information.)

A ratio of distance traveled to time is called *speed*. What is the rate, or speed, in miles per hour?

1. 45 mi, 9 hr

2. 120 mi, 10 hr

3. 89 km, 13 hr (Round to the nearest hundredth.)

What is the rate, or speed, in feet per second?

4. 2200 ft, 2 sec

5. 52 ft, 13 sec

6. 242 ft, 16 sec

7. Ratio of Home Runs to Strikeouts. Referring to Example 4, determine Kent's home-run to strikeout rate.
Source: Major League Baseball

Answers on page A-11

EXAMPLE 3 A pharmacy student working as a pharmacist's assistant earned $3690 for working 3 months one summer. What was the rate of pay per month?

The rate of pay is the ratio of money earned per length of time worked, or

$$\frac{\$3690}{3 \text{ mo}} = 1230 \frac{\text{dollars}}{\text{month}}, \quad \text{or}$$
$1230 per month.

EXAMPLE 4 *Ratio of Strikeouts to Home Runs.* In the 2000 season, Jeff Kent of the San Francisco Giants had 107 strikeouts and 33 home runs. What was his strikeout to home-run rate?
Source: Major League Baseball

$$\frac{107 \text{ strikeouts}}{33 \text{ home runs}} = \frac{107}{33} \frac{\text{strikeouts}}{\text{home runs}} = \frac{107}{33} \text{ strikeouts per home run}$$
$$\approx 3.24 \text{ strikeouts per home run}$$

Do Exercises 1–8. (Exercise 8 is on the following page.)

b   Unit Pricing

UNIT PRICE

A **unit price,** or **unit rate,** is the ratio of price to the number of units.

EXAMPLE 5 *Unit Price of Pears.* A consumer bought a $15\frac{1}{4}$-oz can of pears for $1.07. What is the unit price in cents per ounce?

Often it is helpful to change the cost to cents so we can compare unit prices more easily:

$$\text{Unit price} = \frac{\text{Price}}{\text{Number of units}}$$
$$= \frac{\$1.07}{15\frac{1}{4} \text{ oz}} = \frac{107 \text{ cents}}{15.25 \text{ oz}} = \frac{107}{15.25} \frac{\text{cents}}{\text{oz}}$$
$$\approx 7.016 \text{ cents per ounce.}$$

Do Exercise 9 on the following page.

To do comparison shopping, it helps to compare unit prices.

EXAMPLE 6 *Unit price of Heinz Ketchup.* Many factors can contribute to determining unit pricing in food, such as variations in store pricing and special discounts. Heinz produces ketchup in containers of various sizes. The table below lists several examples of pricing for these packages from a Meijer store. Starting with the price given for each package, compute the unit prices and decide which is the best purchase on the basis of unit price per ounce alone.
Source: Meijer Stores

PACKAGE	PRICE	UNIT PRICE
14 oz	$1.19	8.5 ¢/oz
24 oz	1.39	5.792 ¢/oz
36 oz	1.99	5.528 ¢/oz
46 oz	2.67	5.804 ¢/oz
101-oz twin pack (two 50$\frac{1}{2}$-oz packages)	4.99	4.941 ¢/oz ← Lowest unit price

We compute the unit price for the 24-oz package and leave the remaining prices to the student to check. The unit price for the 24 oz, $1.39 package is given by

$$\frac{\$1.39}{24\ oz} = \frac{139\ cents}{24\ oz} = \frac{139}{24}\frac{cents}{oz} \approx 5.792 \text{ cents per ounce} = 5.792 \text{ ¢/oz.}$$

On the basis of unit price alone, we see that the 101-oz twin pack is the best buy.

Sometimes, as you will see in Margin Exercise 10, a larger size may not have the lower unit price. It is also worth noting that "bigger" is not always "cheaper." (For example, you may not have room for larger packages or the food may go to waste before it is used.)

Do Exercise 10.

8. Babe Ruth. In his entire career, Babe Ruth had 1330 strikeouts and 714 home runs. What was his home-run to strikeout rate? How does it compare to Kent's?
Source: Major League Baseball

9. Unit Price of Olives. A consumer bought a 5$\frac{3}{4}$-oz jar of olives for $1.39. What is the unit price in cents per ounce?

10. Meijer Brand Olives. Complete the following table of unit prices for Meijer Brand olives. Which package has the better unit price?
Source: Meijer Stores

PACKAGE	PRICE	UNIT PRICE
7 oz	$1.69	
10 oz	$2.59	
5$\frac{3}{4}$ oz	$1.39	

Answers on page A-11

a   In Exercises 1–8, find the rate, or speed, as a ratio of distance to time. Round to the nearest hundredth where appropriate.

1. 120 km, 3 hr

2. 18 mi, 9 hr

3. 217 mi, 29 sec

4. 443 m, 48 sec

5. *Mercedes Cabriolet—City Driving.* A 2001 Mercedes-Benz Cabriolet will go 297 miles on 16.5 gallons of gasoline in city driving. What is the rate in miles per gallon?
Source: Mercedes-Benz

6. *Mercedes Cabriolet—Highway Driving.* A 2001 Mercedes-Benz Cabriolet will go 396 miles on 16.5 gallons of gasoline in highway driving. What is the rate in miles per gallon?
Source: Mercedes-Benz

7. *BMW 330Ci Convertible.* A 2001 BMW 330Ci Convertible will go 434 mi on 15.5 gal of gasoline in highway driving. What is the rate in miles per gallon?
Source: BMW

8. *BMW 330Ci Convertible.* A 2001 BMW 330Ci Convertible will go 310 mi on 15.5 gal of gasoline in city driving. What is the rate in miles per gallon?
Source: BMW

9. *Heavenly Ham.* A bone-in, 14-lb ham contains 45 luncheon servings. What is the rate in servings per pound of ham?
Source: Heavenly Ham

45 servings

14 lb

10. *Population Density of Monaco.* Monaco is a tiny country on the Mediterranean coast of France. It has an area of 1.21 square miles and a population of 32,149 people. What is the rate of number of people per square mile? The rate per square mile is called the *population density*. Monaco has the highest population density in the world.
Sources: *The New York Times Almanac; The Handy Geography Answer Book*

11. A car is driven 500 mi in 20 hr. What is the rate in miles per hour? in hours per mile?

12. A student eats 3 hamburgers in 15 min. What is the rate in hamburgers per minute? in minutes per hamburger?

13. *Points Per Game.*   At one point in the 2001 season, Allen Iverson of the Philadelphia 76ers had scored 884 points in 33 games. What was the rate in points per game?
Source: National Basketball Association

14. *Points Per Game.*   At one point in the 2001 season, Shaquille O'Neal of the Los Angeles Lakers had scored 826 points in 32 games. What was the rate in points per game?
Source: National Basketball Association

15. *Lawn Watering.*   To water a lawn adequately requires 623 gal of water for every 1000 ft^2. What is the rate in gallons per square foot?

16. A car is driven 200 km on 40 L of gasoline. What is the rate in kilometers per liter?

17. *Speed of Light.*   Light travels 186,000 mi in 1 sec. What is its rate, or speed, in miles per second?
Source: *The Handy Science Answer Book*

18. *Speed of Sound.*   Sound travels 1100 ft in 1 sec. What is its rate, or speed, in feet per second?
Source: *The Handy Science Answer Book*

19. Impulses in nerve fibers travel 310 km in 2.5 hr. What is the rate, or speed, in kilometers per hour?

20. A black racer snake can travel 4.6 km in 2 hr. What is its rate, or speed, in kilometers per hour?

21. *Elephant Heartbeat.*   The heart of an elephant, at rest, will beat an average of 1500 beats in 60 min. What is the rate in beats per minute?
Source: *The Handy Science Answer Book*

22. *Human Heartbeat.*   The heart of a human, at rest, will beat an average of 4200 beats in 60 min. What is the rate in beats per minute?
Source: *The Handy Science Answer Book*

b Find each unit price in each of Exercises 23–32. Then determine which size has the lower unit price.

23. *Scope Mouthwash.*

PACKAGE	PRICE	UNIT PRICE
33 fl oz	$3.97	
50 fl oz	$5.78	

24. *Roll-on Deodorant.*

PACKAGE	PRICE	UNIT PRICE
2.25 oz	$2.19	
2.5 oz	$2.89	

25. *Crest Toothpaste.*

PACKAGE	PRICE	UNIT PRICE
6.2 oz	$2.97	
8.0 oz	$3.47	

26. *Colgate Toothpaste.*

PACKAGE	PRICE	UNIT PRICE
6.0 oz	$2.97	
7.8 oz	$3.47	

27. *Meijer Coffee.*

PACKAGE	PRICE	UNIT PRICE
11.5 oz	$2.09	
34.5 oz	$5.27	

28. *Maxwell House Coffee.*

PACKAGE	PRICE	UNIT PRICE
13 oz	$2.28	
26 oz	$4.88	

29. *Paper Towels.*

PACKAGE	PRICE	UNIT PRICE
53.7 sq ft	$0.85	
59.5 sq ft	$1.97	
61.8 sq ft	$0.95	
80.6 sq ft	$1.95	
90 sq ft	$1.39	
94 sq ft	$1.79	

30. *Downy Fabric Softener.*

PACKAGE	PRICE	UNIT PRICE
20 oz	$2.69	
40 oz	$3.87	
64 oz	$3.57	
90 oz	$8.69	
120 oz	$10.99	

31. *Tide Liquid Laundry Detergent.*

PACKAGE	PRICE	UNIT PRICE
50 fl oz	$3.97	
100 fl oz	$4.99	
200 fl oz	$12.24	
300 fl oz	$17.97	

32. *"All" Liquid Laundry Detergent.*

PACKAGE	PRICE	UNIT PRICE
100 fl oz	$3.97	
200 fl oz	$9.49	
300 fl oz	$13.99	

Use the unit prices listed in Exercises 23–32 when doing Exercises 33 and 34.

33. ^{D}W Look over the unit prices for each size package of Downy fabric softener. What seems to violate common sense about these unit prices? Why do you think the products are sold this way?

34. ^{D}W Compare the prices and unit prices for the 8.0-oz tube of Crest toothpaste and the 7.8-oz tube of Colgate toothpaste. What seems unusual about these prices? Explain why you think this has happened.

SKILL MAINTENANCE

Solve.

35. There are 20.6 million people in this country who play the piano and 18.9 million who play the guitar. How many more play the piano than the guitar?   [4.7a]

36. A serving of fish steak (cross section) is generally $\frac{1}{2}$ lb. How many servings can be prepared from a cleaned $18\frac{3}{4}$-lb tuna?   [3.6c]

37. *Surf Expo.*   In a swimwear showing at Surf Expo, a trade show for retailers of beach supplies, each swimsuit test takes 8 minutes (min). If the show runs for 240 min, how many tests can be scheduled?   [1.8a]

38. *Eating Habits.*   Each year, Americans eat 24.8 billion hamburgers and 15.9 billion hot dogs. How many more hamburgers than hot dogs do Americans eat?   [4.7a]

Multiply.   [4.3a]

39. $\begin{array}{r} 4\ 5.6\ 7 \\ \times \quad\quad 2.4 \\ \hline \end{array}$

40. $\begin{array}{r} 6\ 7\ 8.1\ 9 \\ \times \quad\quad 1\ 0\ 0 \\ \hline \end{array}$

41. 84.3×69.2

42. 1002.56×465

SYNTHESIS

43. Recently, certain manufacturers have been changing the size of their containers in such a way that the consumer thinks the price of a product has been lowered when, in reality, a higher unit price is being charged.

Some aluminum juice cans are now concave (curved in) on the bottom. Suppose the volume of the can in the figure has been reduced from a fluid capacity of 6 oz to 5.5 oz, and the price of each can has been reduced from 65¢ to 60¢. Find the unit price of each container in cents per ounce.

PROPORTIONS

During the 2000 season, Peyton Manning of the Indianapolis Colts completed 357 passes out of 571 attempts. His pass completion rate was

$$\text{Completion rate} = \frac{357 \text{ completions}}{571 \text{ attempts}} = \frac{357}{571} \frac{\text{completions}}{\text{attempt}}$$

$$\approx 0.625 \frac{\text{completions}}{\text{attempt}}.$$

The rate was 0.625 completions per attempt.

Daunte Culpepper of the Minnesota Vikings completed 297 passes out of 474 attempts. His pass-completion rate was

$$\text{Completion rate} = \frac{297 \text{ completions}}{474 \text{ attempts}} = \frac{297}{474} \frac{\text{completions}}{\text{attempt}}$$

$$\approx 0.627 \frac{\text{completions}}{\text{attempt}}.$$

The rate was 0.627 completions per attempt. We can see that the rates are not equal.

Source: National Football League

Instead of comparing the rates in decimal notation, we can compare the ratios

$$\frac{357}{571} \quad \text{and} \quad \frac{297}{474}$$

using the test for equality considered in Section 2.5. We compare cross products.

$$357 \cdot 474 \overset{?}{=} \frac{357}{571} \overset{?}{=} \frac{297}{474} = 571 \cdot 297$$
$$169{,}218 \qquad\qquad\qquad 169{,}587$$

Since

169,218 $\neq$ 169,587, we know that

$$\frac{357}{571} \neq \frac{297}{474}.$$

Thus the ratios are not equal. If the ratios had been equal, we would say they are proportional.

a   Proportions

When two pairs of numbers (such as 3, 2 and 6, 4) have the same ratio, we say that they are **proportional.** The equation

$$\frac{3}{2} = \frac{6}{4}$$

states that the pairs 3, 2 and 6, 4 are proportional. Such an equation is called a **proportion.** We sometimes read $\frac{3}{2} = \frac{6}{4}$ as "3 is to 2 as 6 is to 4."

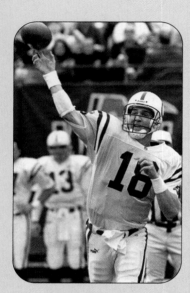

Peyton Manning

Dauente Culpepper

EXAMPLE 1   Determine whether 1, 2, and 3, 6 are proportional.

We can use cross products:

$$1 \cdot 6 = 6 \qquad \overset{?}{\underset{}{}} \quad \frac{1}{2} \overset{?}{=} \frac{3}{6} \qquad 2 \cdot 3 = 6.$$

Since the cross products are the same, $6 = 6$, we know that $\frac{1}{2} = \frac{3}{6}$, so the numbers are proportional.

EXAMPLE 2   Determine whether 2, 5 and 4, 7 are proportional.

We can use cross products:

$$2 \cdot 7 = 14 \qquad \frac{2}{5} \overset{?}{=} \frac{4}{7} \qquad 5 \cdot 4 = 20.$$

Since the cross products are not the same, $14 \neq 20$, we know that $\frac{2}{5} \neq \frac{4}{7}$, so the numbers are not proportional.

Do Exercises 1–3.

EXAMPLE 3   Determine whether 3.2, 4.8 and 0.16, 0.24 are proportional.

We can use cross products:

$$3.2 \times 0.24 = 0.768 \qquad \frac{3.2}{4.8} \overset{?}{=} \frac{0.16}{0.24} \qquad 4.8 \times 0.16 = 0.768.$$

Since the cross products are the same, $0.768 = 0.768$, we know that $\frac{3.2}{4.8} = \frac{0.16}{0.24}$, so the numbers are proportional.

Do Exercises 4 and 5.

EXAMPLE 4   Determine whether $4\frac{2}{3}$, $5\frac{1}{2}$ and $8\frac{7}{8}$, $16\frac{1}{3}$ are proportional.

We can use cross products:

$$4\frac{2}{3} \cdot 16\frac{1}{3} = \frac{14}{3} \cdot \frac{49}{3} \qquad \frac{4\frac{2}{3}}{5\frac{1}{2}} \overset{?}{=} \frac{8\frac{7}{8}}{16\frac{1}{3}} \qquad 5\frac{1}{2} \cdot 8\frac{7}{8} = \frac{11}{2} \cdot \frac{71}{8}$$

$$= \frac{686}{9} \qquad\qquad\qquad\qquad = \frac{781}{16}$$

$$= 76\frac{2}{9} \qquad\qquad\qquad\qquad = 48\frac{13}{16}.$$

Since the cross products are not the same, $76\frac{2}{9} \neq 48\frac{13}{16}$, we know that the numbers are not proportional.

Do Exercise 6.

b   Solving Proportions

Let's now look at solving proportions. Consider the proportion

$$\frac{x}{3} = \frac{4}{6}.$$

One way to solve a proportion is to use cross products. Then we can divide on

---

Determine whether the two pairs of numbers are proportional.

1. 3, 4 and 6, 8

2. 1, 4 and 10, 39

3. 1, 2 and 20, 39

Determine whether the two pairs of numbers are proportional.

4. 6.4, 12.8 and 5.3, 10.6

5. 6.8, 7.4 and 3.4, 4.2

6. Determine whether $4\frac{2}{3}$, $5\frac{1}{2}$ and 14, $16\frac{1}{2}$ are proportional.

Answers on page A-11

7. Solve: $\dfrac{x}{63} = \dfrac{2}{9}$.

8. Solve: $\dfrac{x}{9} = \dfrac{5}{4}$.

Answers on page A-11

Study Tips

WRITING ALL THE STEPS

Take the time to include all the steps when working your homework problems. Doing so will help you organize your thinking and avoid computational errors. If you find a wrong answer, having all the steps allows easier checking of your work. It will also give you complete, step-by-step solutions of the exercises that can be used to study for an exam.

Writing down all the steps and keeping your work organized may also give you a better chance of getting partial credit.

"Success comes before work only in the dictionary."

Anonymous

both sides to get the variable alone:

$$x \cdot 6 = 3 \cdot 4 \qquad \text{Equating cross products (finding cross products and setting them equal)}$$

$$\frac{x \cdot 6}{6} = \frac{3 \cdot 4}{6} \qquad \text{Dividing by 6 on both sides}$$

$$x = \frac{3 \cdot 4}{6} = \frac{12}{6} = 2.$$

We can check that 2 is the solution by replacing x with 2 and using cross products:

$$2 \cdot 6 = 12 \qquad \frac{2}{3} \overset{?}{=} \frac{4}{6} \qquad 3 \cdot 4 = 12$$

Since the cross products are the same, it follows that $\frac{2}{3} = \frac{4}{6}$; so the numbers 2, 3 and 4, 6 are proportional, and 2 is the solution of the equation.

> **SOLVING PROPORTIONS**
>
> To solve $\dfrac{x}{a} = \dfrac{c}{d}$, equate *cross products* and divide on both sides to get x alone.

Do Exercise 7.

EXAMPLE 5 Solve: $\dfrac{x}{7} = \dfrac{5}{3}$. Write a mixed numeral for the answer.

We have

$$\frac{x}{7} = \frac{5}{3}$$

$$x \cdot 3 = 7 \cdot 5 \qquad \text{Equating cross products}$$

$$\frac{x \cdot 3}{3} = \frac{7 \cdot 5}{3} \qquad \text{Dividing by 3}$$

$$x = \frac{7 \cdot 5}{3}$$

$$= \frac{35}{3}, \text{ or } 11\frac{2}{3}.$$

The solution is $11\frac{2}{3}$.

Do Exercise 8.

EXAMPLE 6 Solve: $\dfrac{7.7}{15.4} = \dfrac{y}{2.2}$.

We have

$$\frac{7.7}{15.4} = \frac{y}{2.2}$$

$$7.7 \times 2.2 = 15.4 \times y \qquad \text{Equating cross products}$$

$$\frac{7.7 \times 2.2}{15.4} = \frac{15.4 \times y}{15.4}. \qquad \text{Dividing by 15.4}$$

Then

$$\frac{7.7 \times 2.2}{15.4} = y$$

$$\frac{16.94}{15.4} = y \qquad \text{Multiplying}$$

$$1.1 = y. \qquad \text{Dividing:} \quad 15.4\overline{)16.9\wedge 4}$$

$$\begin{array}{r} 1.1 \\ 15.4\overline{)16.9\wedge 4} \\ \underline{1540} \\ 154 \\ \underline{154} \\ 0 \end{array}$$

The solution is 1.1.

Do Exercise 9.

EXAMPLE 7 Solve: $\dfrac{8}{x} = \dfrac{5}{3}$. Write decimal notation for the answer.

We have

$$\frac{8}{x} = \frac{5}{3}$$

$$8 \cdot 3 = x \cdot 5 \qquad \text{Equating cross products}$$

$$\frac{8 \cdot 3}{5} = \frac{x \cdot 5}{5} \qquad \text{Dividing by 5}$$

$$\frac{8 \cdot 3}{5} = x$$

$$\frac{24}{5} = x \qquad \text{Multiplying}$$

$$4.8 = x. \qquad \text{Simplifying}$$

The solution is 4.8.

Do Exercise 10.

EXAMPLE 8 Solve: $\dfrac{3.4}{4.93} = \dfrac{10}{n}$.

We have

$$\frac{3.4}{4.93} = \frac{10}{n}$$

$$3.4 \times n = 4.93 \times 10 \qquad \text{Equating cross products}$$

$$\frac{3.4 \times n}{3.4} = \frac{4.93 \times 10}{3.4} \qquad \text{Dividing by 3.4}$$

$$n = \frac{4.93 \times 10}{3.4}$$

$$= \frac{49.3}{3.4} \qquad \text{Multiplying}$$

$$= 14.5. \qquad \text{Dividing}$$

The solution is 14.5.

Do Exercise 11.

9. Solve: $\dfrac{21}{5} = \dfrac{n}{2.5}$.

10. Solve: $\dfrac{6}{x} = \dfrac{25}{11}$.

11. Solve: $\dfrac{0.4}{0.9} = \dfrac{4.8}{t}$.

Answers on page A-11

12. Solve:

$$\frac{8\frac{1}{3}}{x} = \frac{10\frac{1}{2}}{3\frac{3}{4}}.$$

Answer on page A-11

EXAMPLE 9 Solve: $\frac{4\frac{2}{3}}{5\frac{1}{2}} = \frac{14}{x}$. Write a mixed numeral for the answer.

We have

$$\frac{4\frac{2}{3}}{5\frac{1}{2}} = \frac{14}{x}$$

$$4\frac{2}{3} \cdot x = 14 \cdot 5\frac{1}{2} \qquad \text{Equating cross products}$$

$$\frac{14}{3} \cdot x = 14 \cdot \frac{11}{2} \qquad \text{Converting to fraction notation}$$

$$\frac{\frac{14}{3} \cdot x}{\frac{14}{3}} = \frac{14 \cdot \frac{11}{2}}{\frac{14}{3}} \qquad \text{Dividing by } \frac{14}{3}$$

$$x = 14 \cdot \frac{11}{2} \div \frac{14}{3}$$

$$= 14 \cdot \frac{11}{2} \cdot \frac{3}{14} \qquad \text{Multiplying by the reciprocal of the divisor}$$

$$= \frac{11 \cdot 3}{2} \qquad \text{Simplifying by removing a factor of 1: } \frac{14}{14} = 1$$

$$= \frac{33}{2}, \text{ or } 16\frac{1}{2}.$$

The solution is $16\frac{1}{2}$.

Do Exercise 12.

CALCULATOR CORNER

Solving Proportions   Note in Examples 5–9 that when we solve a proportion, we equate cross products and then we divide on both sides to isolate the variable on one side of the equation. We can use a calculator to do the calculations in this situation. In Example 8, for instance, after equating cross products and dividing by 3.4 on both sides, we have

$$n = \frac{4.93 \times 10}{3.4}.$$

To find n on a calculator, we can press [4] [.] [9] [3] [×] [1] [0] [÷] [3] [.] [4] [=]. The result is 14.5, so $n = 14.5$.

Exercises

1. Use a calculator to solve each of the proportions in Examples 5–7.

2. Use a calculator to solve each of the proportions in Margin Exercises 7–11.

Solve each proportion.

3. $\frac{15.75}{20} = \frac{a}{35}$

4. $\frac{32}{x} = \frac{25}{20}$

5. $\frac{t}{57} = \frac{17}{64}$

6. $\frac{71.2}{a} = \frac{42.5}{23.9}$

7. $\frac{29.6}{3.15} = \frac{x}{4.23}$

8. $\frac{a}{3.01} = \frac{1.7}{0.043}$

a   Determine whether the two pairs of numbers are proportional.

1. 5, 6 and 7, 9

2. 7, 5 and 6, 4

3. 1, 2 and 10, 20

4. 7, 3 and 21, 9

5. 2.4, 3.6 and 1.8, 2.7

6. 4.5, 3.8 and 6.7, 5.2

7. $5\frac{1}{3}, 8\frac{1}{4}$ and $2\frac{1}{5}, 9\frac{1}{2}$

8. $2\frac{1}{3}, 3\frac{1}{2}$ and 14, 21

Pass Completion Rates.   The table below lists the records of four NFL quarterbacks from the 2000 season.

PLAYER	TEAM	NUMBER OF PASSES COMPLETED	NUMBER OF PASSES ATTEMPTED	NUMBER OF COMPLETIONS PER ATTEMPT (COMPLETION RATE)
Kerry Collins	New York Giants	311	529	
Trent Dilfer	Baltimore Ravens	134	226	
Brian Griese	Denver Broncos	216	336	
Rich Gannon	Oakland Raiders	284	473	

Source: National Football League

9. Find each pass completion rate rounded to the nearest hundredth. Are any the same?

10. Use cross products to determine whether any quarterback completion rates are the same.

b   Solve.

11. $\dfrac{18}{4} = \dfrac{x}{10}$

12. $\dfrac{x}{45} = \dfrac{20}{25}$

13. $\dfrac{x}{8} = \dfrac{9}{6}$

14. $\dfrac{8}{10} = \dfrac{n}{5}$

15. $\dfrac{t}{12} = \dfrac{5}{6}$

16. $\dfrac{12}{4} = \dfrac{x}{3}$

17. $\dfrac{2}{5} = \dfrac{8}{n}$

18. $\dfrac{10}{6} = \dfrac{5}{x}$

19. $\dfrac{n}{15} = \dfrac{10}{30}$

20. $\dfrac{2}{24} = \dfrac{x}{36}$

21. $\dfrac{16}{12} = \dfrac{24}{x}$

22. $\dfrac{7}{11} = \dfrac{2}{x}$

23. $\dfrac{6}{11} = \dfrac{12}{x}$

24. $\dfrac{8}{9} = \dfrac{32}{n}$

25. $\dfrac{20}{7} = \dfrac{80}{x}$

26. $\dfrac{5}{x} = \dfrac{4}{10}$

27. $\dfrac{12}{9} = \dfrac{x}{7}$

28. $\dfrac{x}{20} = \dfrac{16}{15}$

29. $\dfrac{x}{13} = \dfrac{2}{9}$

30. $\dfrac{1.2}{4} = \dfrac{x}{9}$

31. $\dfrac{t}{0.16} = \dfrac{0.15}{0.40}$

32. $\dfrac{x}{11} = \dfrac{7.1}{2}$

33. $\dfrac{100}{25} = \dfrac{20}{n}$

34. $\dfrac{35}{125} = \dfrac{7}{m}$

35. $\dfrac{7}{\frac{1}{4}} = \dfrac{28}{x}$

36. $\dfrac{x}{6} = \dfrac{1}{6}$

37. $\dfrac{\frac{1}{4}}{\frac{1}{2}} = \dfrac{\frac{1}{2}}{x}$

38. $\dfrac{1}{7} = \dfrac{x}{4\frac{1}{2}}$

39. $\dfrac{1}{2} = \dfrac{7}{x}$

40. $\dfrac{x}{3} = \dfrac{0}{9}$

41. $\dfrac{\frac{2}{7}}{\frac{3}{4}} = \dfrac{\frac{5}{6}}{y}$

42. $\dfrac{\frac{5}{4}}{\frac{5}{8}} = \dfrac{\frac{3}{2}}{Q}$

43. $\dfrac{2\frac{1}{2}}{3\frac{1}{3}} = \dfrac{x}{4\frac{1}{4}}$

44. $\dfrac{5\frac{1}{5}}{6\frac{1}{6}} = \dfrac{y}{3\frac{1}{2}}$

45. $\dfrac{1.28}{3.76} = \dfrac{4.28}{y}$

46. $\dfrac{10.4}{12.4} = \dfrac{6.76}{t}$

47. $\dfrac{10\frac{3}{8}}{12\frac{2}{3}} = \dfrac{5\frac{3}{4}}{y}$

48. $\dfrac{12\frac{7}{8}}{20\frac{3}{4}} = \dfrac{5\frac{2}{3}}{y}$

CHAPTER 5: Ratio and Proportion

49. D_W Instead of equating cross products, a student solves $\frac{x}{7} = \frac{5}{3}$ (see Example 5) by multiplying on both sides by the least common denominator, 21. Is his approach a good one? Why or why not?

50. D_W An instructor predicts that a student's test grade will be proportional to the amount of time the student spends studying. What is meant by this? Write an example of a proportion that involves the grades of two students and their study times.

SKILL MAINTENANCE

Use = or ≠ for ☐ to write a true sentence.   [2.5c]

51. $\frac{3}{4} \square \frac{5}{6}$

52. $\frac{18}{24} \square \frac{36}{48}$

53. $\frac{7}{8} \square \frac{7}{9}$

54. $\frac{19}{37} \square \frac{15}{19}$

Divide. Write decimal notation for the answer.   [4.4a]

55. $260 \div 4$

56. $395 \div 10$

57. $4648 \div 16$

58. $3427 \div 2.25$

Divide. Write decimal notation rounded to the nearest thousandth for the answer.   [4.1d], [4.4a]

59. $311 \div 529$

60. $134 \div 226$

61. $216 \div 336$

62. $284 \div 473$

SYNTHESIS

▦ Solve.

63. $\dfrac{1728}{5643} = \dfrac{836.4}{x}$

64. $\dfrac{328.56}{627.48} = \dfrac{y}{127.66}$

65. *Strikeouts per Home Run.*   Baseball Hall-of-Famer Babe Ruth had 1330 strikeouts and 714 home runs in his career. Hall-of-Famer Mike Schmidt had 1883 strikeouts and 548 home runs in his career.

a) Find the unit rate of each player in terms of strikeouts per home run. (These rates were considered among the highest in the history of the game and yet each made the Hall of Fame.)

b) Which player had the higher rate?

Objective

a Solve applied problems involving proportions.

1. Calories Burned. Your author generally exercises for 2 hr each day. The readout on an exercise machine tells him that if he exercises for 24 min, he will burn 356 calories. How many calories will he burn if he exercises for 30 min?
Source: Star Trac Treadmill

a Applications and Problem Solving

Proportions have applications in such diverse fields as business, chemistry, health sciences, and home economics, as well as to many areas of daily life. Proportions are useful in making predictions.

EXAMPLE 1 *Predicting Total Distance.* Donna drives her delivery van 800 mi in 3 days. At this rate, how far will she drive in 15 days?

1. **Familiarize.** We let d = the distance traveled in 15 days.

2. **Translate.** We translate to a proportion. We make each side the ratio of distance to time, with distance in the numerator and time in the denominator.

$$\text{Distance in 15 days} \rightarrow \frac{d}{15} = \frac{800}{3} \leftarrow \text{Distance in 3 days}$$
$$\text{Time} \rightarrow \phantom{\frac{d}{15}} \phantom{\frac{800}{3}} \leftarrow \text{Time}$$

It may help to verbalize the proportion above as "the unknown distance d is to 15 days as the known distance 800 miles is to 3 days."

3. **Solve.** Next, we solve the proportion:

$$3 \cdot d = 15 \cdot 800 \qquad \text{Equating cross products}$$

$$\frac{3 \cdot d}{3} = \frac{15 \cdot 800}{3} \qquad \text{Dividing by 3 on both sides}$$

$$d = \frac{15 \cdot 800}{3}$$

$$d = 4000. \qquad \text{Multiplying and dividing}$$

4. **Check.** We substitute into the proportion and check cross products:

$$\frac{4000}{15} = \frac{800}{3};$$

$$4000 \cdot 3 = 12{,}000; \qquad 15 \cdot 800 = 12{,}000.$$

The cross products are the same.

5. **State.** Donna will drive 4000 mi in 15 days.

Do Exercise 1.

Problems involving proportion can be translated in more than one way. In Example 1, any one of the following is an appropriate translation:

$$\frac{800}{3} = \frac{d}{15}, \qquad \frac{15}{d} = \frac{3}{800}, \qquad \frac{15}{3} = \frac{d}{800}, \qquad \frac{800}{d} = \frac{3}{15}.$$

Equating the cross products in each equation gives us the equation $3 \cdot d = 15 \cdot 800$.

Answer on page A-12

EXAMPLE 2 *Recommended Dosage.* To control a fever, a doctor suggests that a child who weighs 28 kg be given 420 mg of Tylenol. If the dosage is proportional to the child's weight, how much Tylenol is recommended for a child who weighs 35 kg?

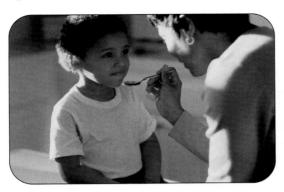

1. **Familiarize.** We let t = the number of milligrams of Tylenol.
2. **Translate.** We translate to a proportion, keeping the amount of Tylenol in the numerators.

$$\text{Tylenol suggested} \rightarrow \frac{420}{28} = \frac{t}{35} \leftarrow \text{Tylenol suggested}$$
$$\text{Child's weight} \rightarrow \qquad\qquad \leftarrow \text{Child's weight}$$

3. **Solve.** Next, we solve the proportion:

$$420 \cdot 35 = 28 \cdot t \qquad \text{Equating cross products}$$

$$\frac{420 \cdot 35}{28} = \frac{28 \cdot t}{28} \qquad \text{Dividing by 28 on both sides}$$

$$\frac{420 \cdot 35}{28} = t$$

$$525 = t. \qquad \text{Multiplying and dividing}$$

4. **Check.** We substitute into the proportion and check cross products:

$$\frac{420}{28} = \frac{525}{35};$$

$$420 \cdot 35 = 14{,}700; \qquad 28 \cdot 525 = 14{,}700.$$

The cross products are the same.

5. **State.** The dosage for a child who weighs 35 kg is 525 mg.

Do Exercise 2.

EXAMPLE 3 *Purchasing Tickets.* Carey bought 8 tickets to an international food festival for $52. How many tickets could she purchase with $90?

1. **Familiarize.** We let n = the number of tickets that can be purchased with $90.
2. **Translate.** We translate to a proportion, keeping the number of tickets in the numerators.

$$\text{Tickets} \rightarrow \frac{8}{52} = \frac{n}{90} \leftarrow \text{Tickets}$$
$$\text{Cost} \rightarrow \qquad\qquad \leftarrow \text{Cost}$$

2. Determining Paint Needs.
Lowell and Chris run a summer painting company to support their college expenses. They can paint 1600 ft² of clapboard with 4 gal of paint. How much paint would be needed for a building with 6000 ft² of clapboard?

Answer on page A-12

3. Purchasing Shirts. If 2 shirts can be bought for $47, how many shirts can be bought with $200?

3. Solve. Next, we solve the proportion:

$$52 \cdot n = 8 \cdot 90 \qquad \text{Equating cross products}$$

$$\frac{52 \cdot n}{52} = \frac{8 \cdot 90}{52} \qquad \text{Dividing by 52 on both sides}$$

$$n = \frac{8 \cdot 90}{52}$$

$$n = 13.8. \qquad \text{Multiplying and dividing}$$

Because it is impossible to buy a fractional part of a ticket, we must round our answer *down* to 13.

4. Check. As a check, we use a different approach: We find the cost per ticket and then divide $90 by that price. Since $52 \div 8 = 6.50$ and $90 \div 6.50 \approx 13.8$, we have a check.

5. State. Carey could purchase 13 tickets with $90.

Do Exercise 3.

EXAMPLE 4 *Women's Hip Measurements.* For improved health, it is recommended that a woman's waist-to-hip ratio be 0.85 (or lower). Marta's hip measurement is 40 in. To meet the recommendation, what should Marta's waist measurement be?
Source: David Schmidt, "Lifting Weight Myths," *Nutrition Action Newsletter* 20, no. 4, October 1993

4. Men's Hip Measurements. It is recommended that a man's waist-to-hip ratio be 0.95 (or lower). Malcolm's hip measurement is 40 in. To meet the recommendation, what should Malcolm's waist measurement be?
Source: David Schmidt, "Lifting Weight Myths," *Nutrition Action Newsletter* 20, no. 4, October 1993

 Hip measurement is the largest measurement around the widest part of the buttocks.

 Waist measurement is the smallest measurement below the ribs but above the navel.

1. Familiarize. Note that $0.85 = \frac{85}{100}$. We let $w =$ Marta's waist measurement.

2. Translate. We translate to a proportion as follows:

$$\begin{array}{l} \text{Waist measurement} \rightarrow \\ \text{Hip measurement} \rightarrow \end{array} \frac{w}{40} = \frac{85}{100}. \begin{array}{l} \leftarrow \text{Recommended} \\ \leftarrow \text{waist-to-hip ratio} \end{array}$$

3. Solve. Next, we solve the proportion:

$$100 \cdot w = 40 \cdot 85 \qquad \text{Equating cross products}$$

$$\frac{100 \cdot w}{100} = \frac{40 \cdot 85}{100} \qquad \text{Dividing by 100 on both sides}$$

$$w = \frac{40 \cdot 85}{100}$$

$$w = 34. \qquad \text{Multiplying and dividing}$$

4. Check. As a check, we divide 34 by 40: $34 \div 40 = 0.85$. This is the desired ratio.

5. State. Marta's recommended waist measurement is 34 in. (or less).

Do Exercise 4.

Answers on page A-12

EXAMPLE 5 *Construction Plans.* Architects make blueprints of projects being constructed. These are scale drawings in which lengths are in proportion to actual sizes. The Hennesseys are constructing a rectangular deck just outside their house. The architectural blueprints are rendered such that $\frac{3}{4}$ in. on the drawing is actually 2.25 ft on the deck. The width of the deck on the drawing is 4.3 in. How wide is the deck in reality?

5. **Construction Plans.** In Example 5, the length of the actual deck is 28.5 ft. What is the length of the deck on the blueprints?

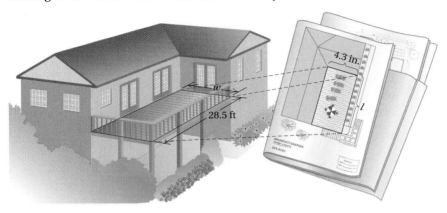

1. **Familiarize.** We let w = the width of the deck.

2. **Translate.** Then we translate to a proportion, using 0.75 for $\frac{3}{4}$ in.

$$\begin{array}{c} \text{Measure on drawing} \rightarrow \\ \text{Measure on deck} \rightarrow \end{array} \frac{0.75}{2.25} = \frac{4.3}{w} \begin{array}{c} \leftarrow \text{Width of drawing} \\ \leftarrow \text{Width of deck} \end{array}$$

3. **Solve.** Next, we solve the proportion:

$$0.75 \times w = 2.25 \times 4.3 \qquad \text{Equating cross products}$$

$$\frac{0.75 \times w}{0.75} = \frac{2.25 \times 4.3}{0.75} \qquad \text{Dividing by 0.75 on both sides}$$

$$w = \frac{2.25 \times 4.3}{0.75}$$

$$w = 12.9.$$

4. **Check.** We substitute into the proportion and check cross products:

$$\frac{0.75}{2.25} = \frac{4.3}{12.9};$$

$$0.75 \times 12.9 = 9.675; \qquad 2.25 \times 4.3 = 9.675.$$

The cross products are the same.

5. **State.** The width of the deck is 12.9 ft.

Do Exercise 5.

Answer on page A-12

6. Estimating a Deer Population.
To determine the number of deer in a forest, a conservationist catches 612 deer, tags them, and releases them. Later, 244 deer are caught, and it is found that 72 of them are tagged. Estimate how many deer are in the forest.

EXAMPLE 6 *Estimating a Wildlife Population.* To determine the number of fish in a lake, a conservationist catches 225 fish, tags them, and throws them back into the lake. Later, 108 fish are caught, and it is found that 15 of them are tagged. Estimate how many fish are in the lake.

1. **Familiarize.** We let $F =$ the number of fish in the lake.

2. **Translate.** We translate to a proportion as follows:

$$\text{Fish tagged originally} \rightarrow \frac{225}{F} = \frac{15}{108} \leftarrow \text{Tagged fish caught later}$$
$$\text{Fish in lake} \rightarrow \phantom{\frac{225}{F}} \phantom{\frac{15}{108}} \leftarrow \text{Fish caught later}$$

3. **Solve.** Next, we solve the proportion:

$$225 \cdot 108 = F \cdot 15 \qquad \text{Equating cross products}$$

$$\frac{225 \cdot 108}{15} = \frac{F \cdot 15}{15} \qquad \text{Dividing by 15 on both sides}$$

$$\frac{225 \cdot 108}{15} = F$$

$$1620 = F. \qquad \text{Multiplying and dividing}$$

4. **Check.** We substitute into the proportion and check cross products:

$$\frac{225}{1620} = \frac{15}{108};$$

$$225 \cdot 108 = 24{,}300; \qquad 1620 \cdot 15 = 24{,}300.$$

The cross products are the same.

5. **State.** We estimate that there are 1620 fish in the lake.

Do Exercise 6.

Answer on page A-12

5.4

EXERCISE SET

For Extra Help

Digital Video
Tutor CD 3
Videotape 9

InterAct
Math

Math Tutor
Center

MathXL

MyMathLab

a   Solve.

1. *Study Time and Test Grades.*   An English instructor asserted that students' test grades are directly proportional to the amount of time spent studying. Lisa studies 9 hr for a particular test and gets a score of 75. At this rate, how many hours would she have had to study to get a score of 92?

2. *Study Time and Test Grades.*   A mathematics instructor asserted that students' test grades are directly proportional to the amount of time spent studying. Brent studies 15 hr for a particular test and gets a score of 85. At this rate, what score would he have received if he had studied 16 hr?

3. *Complete™ Cereal.*   The nutritional chart on the side of a box of Kellogg's Complete™ Cereal states that there are 90 calories in a $\frac{3}{4}$-cup serving. How many calories are there in 5 cups of the cereal?

4. *Coco Wheats® Cereal.*   The nutritional chart on the side of a box of Little Crow Foods' Coco Wheats® Cereal states that there are 200 calories in $\frac{1}{3}$ cup of precooked mix. How many calories are there in 4 cups of the mix?

Source: Kellogg's

Source: Little Crow Foods

5. *Overweight Americans.*   A study recently confirmed that of every 100 Americans, 60 are considered overweight. There were 281 million Americans in 2001. How many would be considered overweight?
Source: U.S. Centers for Disease Control

6. *Cancer Death Rate in Illinois.*   It is predicted that for every 1000 people in the state of Illinois, 130.9 will die of cancer. The population of Chicago is about 2,721,547. How many of these people will die of cancer?
Source: 2001 New York Times Almanac

7. *Gasoline Mileage.*   Nancy's van traveled 84 mi on 6.5 gal of gasoline. At this rate, how many gallons would be needed to travel 126 mi?

8. *Bicycling.*   Roy bicycled 234 mi in 14 days. At this rate, how far would Roy travel in 42 days?

9. *Quality Control.* A quality-control inspector examined 100 lightbulbs and found 7 of them to be defective. At this rate, how many defective bulbs will there be in a lot of 2500?

10. *Grading.* A professor must grade 32 essays in a literature class. She can grade 5 essays in 40 min. At this rate, how long will it take her to grade all 32 essays?

11. *Painting.* Fred uses 3 gal of paint to cover 1275 ft^2 of siding. How much siding can Fred paint with 7 gal of paint?

12. *Waterproofing.* Bonnie can waterproof 450 ft^2 of decking with 2 gal of sealant. How many gallons should Bonnie buy for a 1200-ft^2 deck?

13. *Publishing.* Every 6 pages of an author's manuscript corresponds to 5 published pages. How many published pages will a 540-page manuscript become?

14. *Turkey Servings.* An 8-lb turkey breast contains 36 servings of meat. How many pounds of turkey breast would be needed for 54 servings?

15. *Exchanging Money.* On 22 December 2000, 1 U.S. dollar was worth about 1.80 Australian dollars.

   a) How much would 250 U.S. dollars be worth in Australian dollars?

   b) Derek was traveling in Australia and bought a sweatshirt that cost 50 Australian dollars. How much would it cost in U.S. dollars?

16. *Exchanging Money.* On 22 December 2000, 1 U.S. dollar was worth about 0.676453 British pound.

   a) How much would 250 U.S. dollars be worth in British pounds?

   b) Brittany was traveling in England and bought a watch that cost 320 British pounds. How much would it cost in U.S. dollars?

17. *Gas Mileage.* A 2001 BMW 330Ci Convertible will go 434 mi on 15.5 gal of gasoline in highway driving.

   a) How many gallons of gasoline will it take to drive 2690 mi from Boston to Phoenix?

   b) How far can the car be driven on 140 gal of gasoline?
Source: BMW

18. *Gas Mileage.* A 2001 Mercedes-Benz Cabriolet will go 396 mi on 16.5 gal of gasoline in highway driving.

   a) How many gallons of gasoline will it take to drive 1650 mi from Pittsburgh to Albuquerque?

   b) How far can the car be driven on 130 gal of gasoline?
Source: Mercedes-Benz

19. *Lefties.*   In a class of 40 students, on average, 6 will be left-handed. If a class includes 9 "lefties," how many students would you estimate are in the class?

20. *Sugaring.*   When 38 gal of maple sap are boiled down, the result is 2 gal of maple syrup. How much sap is needed to produce 9 gal of syrup?

21. *Mileage.*   Jean bought a new car. In the first 8 months, it was driven 9000 mi. At this rate, how many miles will the car be driven in 1 yr?

22. *Coffee Production.*   Coffee beans from 14 trees are required to produce the 17 lb of coffee that the average person in the United States drinks each year. How many trees are required to produce 375 lb of coffee?

23. *Metallurgy.*   In a metal alloy, the ratio of zinc to copper is 3 to 13. If there are 520 lb of copper, how many pounds of zinc are there?

24. *Class Size.*   A college advertises that its student-to-faculty ratio is 14 to 1. If 56 students register for Introductory Spanish, how many sections of the course would you expect to see offered?

25. *Painting.*   Helen can paint 950 ft^2 with 2 gal of paint. How many 1-gal cans does she need in order to paint a 30,000-ft^2 wall?

26. *Snow to Water.*   Under typical conditions, $1\frac{1}{2}$ ft of snow will melt to 2 in. of water. To how many inches of water will $5\frac{1}{2}$ ft of snow melt?

27. *Grass-Seed Coverage.* It takes 60 oz of grass seed to seed 3000 ft² of lawn. At this rate, how much would be needed for 5000 ft² of lawn?

28. *Grass-Seed Coverage.* In Exercise 27, how much seed would be needed for 7000 ft² of lawn?

29. *Estimating a Deer Population.* To determine the number of deer in a game preserve, a forest ranger catches 318 deer, tags them, and releases them. Later, 168 deer are caught, and it is found that 56 of them are tagged. Estimate how many deer are in the game preserve.

30. *Estimating a Trout Population.* To determine the number of trout in a lake, a conservationist catches 112 trout, tags them, and throws them back into the lake. Later, 82 trout are caught, and it is found that 32 of them are tagged. Estimate how many trout there are in the lake.

31. *Map Scaling.* On a road atlas map, 1 in. represents 16.6 mi. If two cities are 3.5 in. apart on the map, how far apart are they in reality?

32. *Map Scaling.* On a map, $\frac{1}{4}$ in. represents 50 mi. If two cities are $3\frac{1}{4}$ in. apart on the map, how far apart are they in reality?

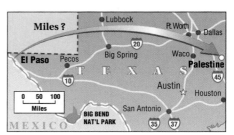

33. *Points per Game.* At one point in the 2000–2001 NBA season, Allen Iverson of the Philadelphia 76ers had scored 884 points in 33 games.

   a) At this rate, how many games would it take him to score 1500 points?

   b) There are 82 games in an entire NBA season. At this rate, how many points would Iverson score in the entire season?

Source: National Basketball Association

34. *Points per Game.* At one point in the 2000–2001 NBA season, Shaquille O'Neal of the Los Angeles Lakers had scored 826 points in 32 games.

   a) At this rate, how many games would it take him to score 2000 points?

   b) There are 82 games in an entire NBA season. At this rate, how many points would O'Neal score in the entire season?

35. D_W Can unit prices be used to solve proportions that involve money? Explain why or why not.

36. D_W *Earned Run Average.* In baseball, the average number of runs given up by a pitcher in nine innings is his *earned run average,* or *ERA.* Set up a formula for determining a player's ERA. Then verify it using the fact that in the 2000 season, Daryl Kile of the St. Louis Cardinals gave up 101 earned runs in $232\frac{1}{3}$ innings to compile an ERA of 3.91. Then use your formula to determine the ERA of Randy Johnson of the Arizona Diamondbacks, who gave up 73 earned runs in $248\frac{2}{3}$ innings. Is a low ERA considered good or bad?

Source: Major League Baseball

SKILL MAINTENANCE

Determine whether each number is prime, composite, or neither.   [2.1c]

37. 1   **38.** 28   **39.** 83   **40.** 93   **41.** 47

Find the prime factorization of each number.   [2.1d]

42. 808   **43.** 28   **44.** 866   **45.** 93   **46.** 2020

SYNTHESIS

47. ▦ Carney College is expanding from 850 to 1050 students. To avoid any rise in the student-to-faculty ratio, the faculty of 69 professors must also increase. How many new faculty positions should be created?

48. ▦ In recognition of her outstanding work, Sheri's salary has been increased from $26,000 to $29,380. Tim is earning $23,000 and is requesting a proportional raise. How much more should he ask for?

49. *Baseball Statistics.* Cy Young, one of the greatest baseball pitchers of all time, gave up an average of 2.63 earned runs every 9 innings. Young pitched 7356 innings, more than anyone in the history of baseball. How many earned runs did he give up?

50. ▦ *Real-Estate Values.* According to Coldwell Banker Real Estate Corporation, a home selling for $189,000 in Austin, Texas, would sell for $665,795 in San Francisco. How much would a $450,000 home in San Francisco sell for in Austin? Round to the nearest $1000.

Source: Coldwell Banker Real Estate Corporation

51. ▦ The ratio $1:3:2$ is used to estimate the relative costs of a CD player, receiver, and speakers when shopping for a stereo. That is, the receiver should cost three times the amount spent on the CD player and the speakers should cost twice as much as the amount spent on the CD player. If you had $900 to spend, how would you allocate the money, using this ratio?

Objectives

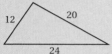

a Find lengths of sides of similar triangles using proportions.

b Use proportions to find lengths in pairs of figures that differ only in size.

---

1. This pair of triangles is similar. Find the missing length x.

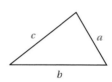

---

a Proportions and Similar Triangles

Look at the pair of triangles below. Note that they appear to have the same shape, but their sizes are different. These are examples of **similar triangles.** By using a magnifying glass, you could imagine enlarging the smaller triangle to get the larger. This process works because the corresponding sides of each triangle have the same ratio. That is, the following proportion is true.

$$\frac{a}{d} = \frac{b}{e} = \frac{c}{f}$$

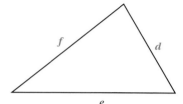

SIMILAR TRIANGLES

Similar triangles have the same shape. The lengths of their corresponding sides have the same ratio—that is, they are proportional.

EXAMPLE 1 The triangles below are similar triangles. Find the missing length x.

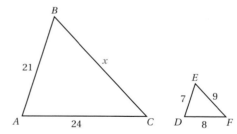

The ratio of x to 9 is the same as the ratio of 24 to 8 or 21 to 7. We get the proportions

$$\frac{x}{9} = \frac{24}{8} \quad \text{and} \quad \frac{x}{9} = \frac{21}{7}.$$

We can solve either one of these proportions. We use the first:

$$\frac{x}{9} = \frac{24}{8}$$

$x \cdot 8 = 24 \cdot 9$ Equating cross products

$\dfrac{x \cdot 8}{8} = \dfrac{24 \cdot 9}{8}$ Dividing by 8 on both sides

$x = 27.$ Simplifying

The missing length x is 27. Other proportions could also be used.

Answer on page A-12

Do Exercise 1 on the preceding page.

Similar triangles and proportions can often be used to find lengths that would ordinarily be difficult to measure. For example, we could find the height of a flagpole without climbing it or the distance across a river without crossing it.

EXAMPLE 2 How high is a flagpole that casts a 56-ft shadow at the same time that a 6-ft man casts a 5-ft shadow?

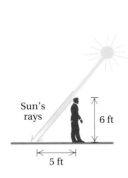

If we use the sun's rays to represent the third side of the triangle in our drawing of the situation, we see that we have similar triangles. Let $p =$ the height of the flagpole. The ratio of 6 to p is the same as the ratio of 5 to 56. Thus we have the proportion

$$\text{Height of man} \longrightarrow \frac{6}{p} = \frac{5}{56}. \longleftarrow \text{Length of shadow of man}$$
$$\text{Height of pole} \longrightarrow \phantom{\frac{6}{p} = \frac{5}{56}.} \longleftarrow \text{Length of shadow of pole}$$

Solve: $6 \cdot 56 = 5 \cdot p$ Equating cross products

$$\frac{6 \cdot 56}{5} = \frac{5 \cdot p}{5}$$ Dividing by 5 on both sides

$$\frac{6 \cdot 56}{5} = p$$ Simplifying

$$67.2 = p$$

The height of the flagpole is 67.2 ft.

Do Exercise 2.

EXAMPLE 3 *F-106 Blueprint.*   A blueprint is a scale drawing. Each wing of an F-106 Delta Dart military airplane has a triangular shape. The blueprint for this airplane shows similar triangles. Find the length a of a side of the wing.

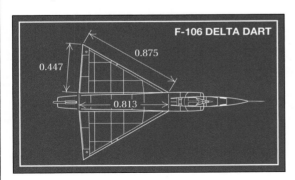

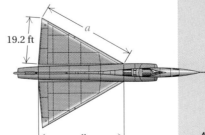

We let $a =$ the length of the wing.

2. How high is a flagpole that casts a 45-ft shadow at the same time that a 5.5-ft woman casts a 10-ft shadow?

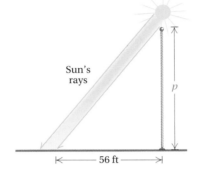

Answer on page A-12

3. F-106 Blueprint. Referring to Example 3, find the length x of the wing to the nearest tenth.

Thus we have the proportion

Length on the blueprint $\longrightarrow$ $\dfrac{0.447}{19.2} = \dfrac{0.875}{a}$ $\longleftarrow$ Length on the blueprint
Length of the wing $\longrightarrow$ $\phantom{\dfrac{0.447}{19.2} = \dfrac{0.875}{a}}$ $\longleftarrow$ Length of the wing

Solve: $\quad 0.447 \times a = 19.2 \times 0.875 \qquad$ Equating cross products

$$\frac{0.447 \times a}{0.447} = \frac{19.2 \times 0.875}{0.447} \qquad \text{Dividing by 0.447 on both sides}$$

$$a = \frac{19.2 \times 0.875}{0.447} \qquad \text{Simplifying}$$

$$a \approx 37.6 \text{ ft}$$

The length of side a of the wing is about 37.6 ft.

Do Exercise 3.

b Proportions and Other Geometric Shapes

When one geometric figure is a magnification of another, the figures are similar. Thus the corresponding lengths are proportional.

4. The sides in the photographs below are proportional. Find the width of the larger photograph.

EXAMPLE 4 The sides in the negative and photograph below are proportional. Find the width of the photograph.

We let $x = $ the width of the photograph. Then we translate to a proportion.

Photo width $\longrightarrow$ $\dfrac{x}{2.5} = \dfrac{10.5}{3.5}$ $\longleftarrow$ Photo length
Negative width $\longrightarrow$ $\phantom{\dfrac{x}{2.5} = \dfrac{10.5}{3.5}}$ $\longleftarrow$ Negative length

Solve: $\quad 3.5 \times x = 2.5 \times 10.5 \qquad$ Equating cross products

$$\frac{3.5 \times x}{3.5} = \frac{2.5 \times 10.5}{3.5} \qquad \text{Dividing by 3.5 on both sides}$$

$$x = \frac{2.5 \times 10.5}{3.5} \qquad \text{Simplifying}$$

$$x = 7.5$$

Thus the width of the photograph is 7.5 cm.

Do Exercise 4.

Answers on page A-12

EXAMPLE 5 A scale model of an addition to an athletic facility is 12 cm wide at the base and rises to a height of 15 cm. If the actual base is to be 116 ft, what will be the actual height of the addition?

5. Refer to the figures in Example 5. If a model skylight is 3 cm wide, how wide will the actual skylight be?

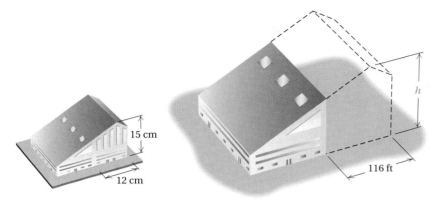

15 cm

12 cm

116 ft

We let h = the height of the addition. Then we translate to a proportion.

Width in model $\longrightarrow \dfrac{12}{116} = \dfrac{15}{h} \longleftarrow$ Height in model
Actual width $\longrightarrow$ $\longleftarrow$ Actual height

Solve: $\quad 12 \cdot h = 116 \cdot 15 \qquad$ Equating cross products

$$\frac{12 \cdot h}{12} = \frac{116 \cdot 15}{12} \qquad \text{Dividing by 12 on both sides}$$

$$h = \frac{116 \cdot 15}{12} = 145.$$

Thus the height of the addition will be 145 ft.

Do Exercise 5.

Answer on page A-12

Study Tips

TIME MANAGEMENT (PART 3)

Here are some additional tips to help you with time management. (See also the Study Tips on time management in Sections 1.5 and 5.1.)

■ **Are you a morning or an evening person?** If you are an evening person, it might be best to avoid scheduling early-morning classes. If you are a morning person, do the opposite, but go to bed earlier to compensate. Nothing can drain your study time and effectiveness like fatigue.

■ **Keep on schedule.** Your course syllabus provides a plan for the semester's schedule. Use a write-on calendar, daily planner, laptop computer, or personal digital assistant to outline your time for the semester. Be sure to note deadlines involving term papers and exams so you can begin a task early, breaking it down into smaller segments that can be accomplished more easily.

■ **Balance your class schedule.** You may be someone who prefers large blocks of time for study on the off days. In that case, it might be advantageous for you to take courses that meet only three days a week. Keep in mind, however, that this might be a problem when tests in more than one course are scheduled for the same day.

5.5

EXERCISE SET

For Extra Help

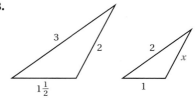

Digital Video   InterAct   Math Tutor   MathXL   MyMathLab
Tutor CD 3   Math   Center
Videotape 9

a The triangles in each exercise are similar. Find the missing lengths.

1.

2.

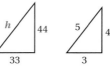

3.

4.

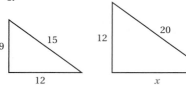

5.

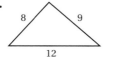

6.

7.

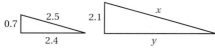

8.

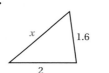

9. When a tree 8 m high casts a shadow 5 m long, how long a shadow is cast by a person 2 m tall?

10. How high is a flagpole that casts a 42-ft shadow at the same time that a $5\frac{1}{2}$-ft woman casts a 7-ft shadow?

11. How high is a tree that casts a 27-ft shadow at the same time that a 4-ft fence post casts a 3-ft shadow?

12. How high is a tree that casts a 32-ft shadow at the same time that an 8-ft light pole casts a 9-ft shadow?

13. Find the height h of the wall.

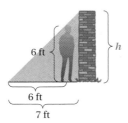

6 ft
h
6 ft
7 ft

14. Find the length L of the lake. Assume that the ratio of L to 120 yd is the same as the ratio of 720 yd to 30 yd.

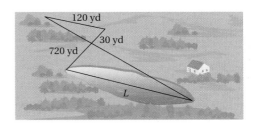

120 yd
30 yd
720 yd
L

15. Find the distance across the river. Assume that the ratio of d to 25 ft is the same as the ratio of 40 ft to 10 ft.

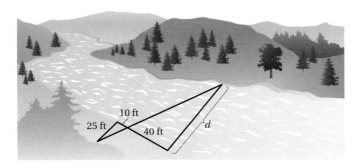

10 ft
25 ft
40 ft
d

16. To measure the height of a hill, a string is drawn tight from level ground to the top of the hill. A 3-ft stick is placed under the string, touching it at point P, a distance of 5 ft from point G, where the string touches the ground. The string is then detached and found to be 120 ft long. How high is the hill?

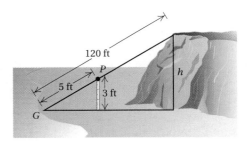

120 ft
P
5 ft
3 ft
h
G

b In each of Exercises 17–26, the sides in each pair of figures are proportional. Find the missing lengths.

17.

6
9
x
6

18.

5
x
7
14

19.

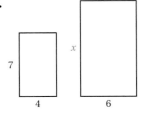

7
x
4
6

20.

x
11
4
3

21.

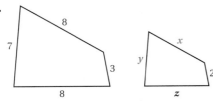

22.

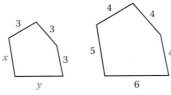

23.

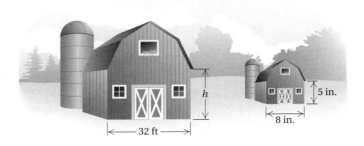

24.

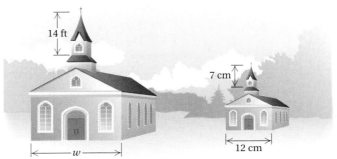

25.

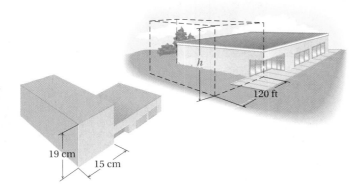

26.

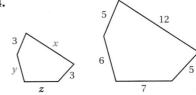

27. A scale model of an addition to an athletic facility is 15 cm wide at the base and rises to a height of 19 cm. If the actual base is to be 120 ft, what will be the height of the addition?

28. Refer to the figures in Exercise 27. If a model skylight is 3 cm wide, how wide will the actual skylight be?

29. **D_W** Is it possible for two triangles to have two pairs of sides that are proportional without the triangles being similar? Why or why not?

30. **D_W** Design for a classmate a problem involving similar triangles for which

$$\frac{18}{128.95} = \frac{x}{789.89}.$$

31. *Expense Needs.* A student has $34.97 to spend for a book at $49.95, a CD at $14.88, and a sweatshirt at $29.95. How much more money does the student need to make these purchases? [4.7a]

32. Divide: 80.892 ÷ 8.4. [4.4a]

Multiply. [4.3a]

33. 8.4 × 80.892

34. 0.01 × 274.568

35. 100 × 274.568

36. 0.002 × 274.568

Find decimal notation and round to the nearest thousandth, if appropriate. [4.5a, b]

37. $\dfrac{17}{20}$

38. $\dfrac{73}{40}$

39. $\dfrac{10}{11}$

40. $\dfrac{43}{51}$

Hockey Goals. An official hockey goal is 6 ft wide. To make scoring more difficult, goalies often locate themselves far in front of the goal to "cut down the angle." In Exercises 41 and 42, suppose that a slapshot from point *A* is attempted and that the goalie is 2.7 ft wide. Determine how far from the goal the goalie should be located if point *A* is the given distance from the goal. (*Hint:* First find how far the goalie should be from point *A*.)

41. 🔲 25 ft

42. 🔲 35 ft

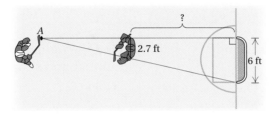

43. 🔲 A miniature basketball hoop is built for the model referred to in Exercise 27. An actual hoop is 10 ft high. How high should the model hoop be? Round to the nearest thousandth of a centimeter.

🔲 Solve. Round the answer to the nearest thousandth.

44. $\dfrac{8664.3}{10{,}344.8} = \dfrac{x}{9776.2}$

45. $\dfrac{12.0078}{56.0115} = \dfrac{789.23}{y}$

🔲 The triangles in each exercise are similar triangles. Find the lengths not given.

46.

47.

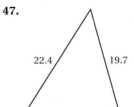

351

Summary and Review

The review that follows is meant to prepare you for a chapter exam. It consists of two parts. The first part is a checklist of some of the Study Tips referred to in this text. The second part is the Review Exercises. These provide practice exercises for the exam, together with references to the section objectives so you can go back and review. Before beginning, stop and look back over the skills you have obtained. What skills in mathematics do you have now that you did not have before studying this chapter?

STUDY TIPS CHECKLIST

The foundation of all your study skills is TIME!	☐ Are you avoiding time killers?
	☐ Are you being aggressive about your study tasks?
	☐ Are you keeping on schedule?
	☐ Have you been writing out *all* the steps when working your homework problems?
	☐ Have you been recording your lectures with your instructor's permission?

REVIEW EXERCISES

Write fraction notation for the ratio. Do not simplify. [5.1a]

1. 47 to 84

2. 46 to 1.27

3. 83 to 100

4. 0.72 to 197

5. *Kona Jack's Restaurants.* Kona Jack's is a seafood restaurant chain in Indianapolis. Each year they sell 12,480 lb of tuna and 16,640 lb of salmon. [5.1a]
  a) Write fraction notation for the ratio of tuna sold to salmon sold.
  b) Write fraction notation for the ratio of salmon sold to the total number of pounds of both kinds of fish.

Source: Kona Jack's Restaurants

Find the ratio of the first number to the second number and simplify. [5.1b]

6. 9 to 12

7. 3.6 to 6.4

8. *Gas Mileage.* The Chrysler PT Cruiser will go 377 mi on 14.5 gal of gasoline in highway driving. What is the rate in miles per gallon? [5.2a]
Source: DaimlerChrysler Corporation

9. *CD-ROM Spin Rate.* A 12x CD-ROM on a computer will spin 472,500 revolutions if left running for 75 min. What is the rate of its spin in revolutions per minute (rpm)? [5.2a]
Source: *Electronic Engineering Times*, June 1997

10. A lawn requires 319 gal of water for every 500 ft². What is the rate in gallons per square foot? [5.2a]

11. *Turkey Servings.* A 25-lb turkey serves 18 people. Find the rate in servings per pound. [5.2a]

12. *Calcium Supplement.* The price for a particular calcium supplement is $12.99 for 300 tablets. Find the unit price in cents per tablet. [5.2b]

13. *Pillsbury Orange Breakfast Rolls.* The price for these breakfast rolls is $1.97 for 13.9 oz. Find the unit price in cents per ounce. [5.2b]

In each of Exercises 14 and 15, find the unit prices. Then determine in each case which has the lower unit price. [5.2b]

14. *Cheer Liquid Laundry Detergent.*

PACKAGE	PRICE	UNIT PRICE
100 oz	$5.40	
150 oz	$9.69	

15. *Crisco Oil.*

PACKAGE	PRICE	UNIT PRICE
16 oz	$1.29	
32 oz	$1.73	
48 oz	$1.99	
64 oz	$2.64	
128 oz	$5.65	

Determine whether the two pairs of numbers are proportional. [5.3a]

16. 9, 15 and 36, 59

17. 24, 37 and 40, 46.25

Solve. [5.3b]

18. $\dfrac{8}{9} = \dfrac{x}{36}$

19. $\dfrac{6}{x} = \dfrac{48}{56}$

20. $\dfrac{120}{\frac{3}{7}} = \dfrac{7}{x}$

21. $\dfrac{4.5}{120} = \dfrac{0.9}{x}$

Solve. [5.4a]

22. If 3 dozen eggs cost $2.67, how much will 5 dozen eggs cost?

23. *Quality Control.* A factory manufacturing computer circuits found 39 defective circuits in a lot of 65 circuits. At this rate, how many defective circuits can be expected in a lot of 585 circuits?

24. *Exchanging Money.* On 22 December 2000, 1 U.S. dollar was worth about 1.08 European Monetary Units (Euros).
a) How much would 250 U.S. dollars be worth in Euros?
b) Jamal was traveling in France and saw a sweatshirt that cost 50 Euros. How much would it cost in U.S. dollars?

25. A train travels 448 mi in 7 hr. At this rate, how far will it travel in 13 hr?

26. Fifteen acres are required to produce 54 bushels of tomatoes. At this rate, how many acres are required to produce 97.2 bushels of tomatoes?

27. *Garbage Production.* It is known that 5 people produce 13 kg of garbage in one day. San Diego, California, has 1,220,666 people. How many kilograms of garbage are produced in San Diego in one day?

28. *Snow to Water.* Under typical conditions, $1\frac{1}{2}$ ft of snow will melt to 2 in. of water. To how many inches of water will $4\frac{1}{2}$ ft of snow melt?

29. *Lawyers in Michigan.* In Michigan, there are 2.3 lawyers for every 1000 people. The population of Detroit is 4,307,000. How many lawyers would you expect there to be in Detroit?
Source: U.S. Bureau of the Census

Each pair of triangles in Exercises 30 and 31 is similar. Find the missing length(s). [5.5a]

30.
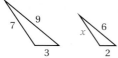

31.
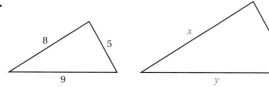

32. How high is a billboard that casts a 25-ft shadow at the same time that an 8-ft sapling casts a 5-ft shadow? [5.5a]

33. The lengths in the figures below are proportional. Find the missing lengths. [5.5b]

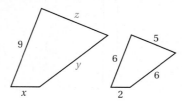

34. ^{D}W If you were a college president, which would you prefer: a low or high faculty-to-student ratio? Why? [5.1a]

35. ^{D}W Write a proportion problem for a classmate to solve. Design the problem so that the solution is "Leslie would need 16 gal of gasoline in order to travel 368 mi." [5.4a]

Certain objectives from four particular sections will be retested on the chapter test. The objectives are listed with the practice problems that follow.

Solve. [4.7a]

36. A family has $2347.89 in its checking account. Gwyneth uses her debit card to make purchases of $678.95 and $38.54. How much is left in the checking account?

37. What is the total cost of 8 CD players at $349.95 each?

Use $=$ or $\neq$ for $\square$ to write a true sentence. [2.5c]

38. $\dfrac{5}{2} \square \dfrac{10}{4}$ **39.** $\dfrac{4}{6} \square \dfrac{8}{10}$

40. Multiply. [4.3a]

$$\begin{array}{r} 4\ 5\ 6.1 \\ \times\quad 2\ 3.4 \\ \hline \end{array}$$

41. Divide. Write decimal notation for the answer. [4.4a]

$$5.6\,\overline{)\,2\ 5\ 4.8}$$

42. It takes Yancy Martinez 10 min to type two-thirds of a page of his term paper. At this rate, how long will it take him to type a 7-page term paper? [5.4a]

43. ▦ The following triangles are similar. Find the missing lengths. [5.5a]

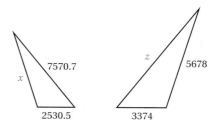

44. Shine-and-Glo Painters uses 2 gal of finishing paint for every 3 gal of primer. Each gallon of finishing paint covers 450 ft^2. If a surface of 4950 ft^2 needs both primer and finishing paint, how many gallons of each should be purchased? [5.4a]

Chapter Test

CHAPTER

5

Write fraction notation for the ratio. Do not simplify.

1. 85 to 97

2. 0.34 to 124

Find the ratio of the first number to the second number and simplify.

3. 18 to 20

4. 0.75 to 0.96

5. What is the rate in feet per second?

10 feet,   16 seconds

6. *Ham Servings.*   A 12-lb shankless ham contains 16 servings. What is the rate in servings per pound?

7. *Gas Mileage.*   The 2000 Volkswagen New Beetle GL will go 341 mi on 14.5 gal of gasoline in city driving. What is the rate in miles per gallon?
Source: Volkswagen of America, Inc.

8. *Laundry Detergent.*   A box of Cheer laundry detergent powder sells at $6.29 for 81 oz. Find the unit price in cents per ounce.

9. The following table lists prices for various packages of Tide laundry detergent powder. Find the unit price of each package. Then determine which has the lower unit price.

PACKAGE	PRICE	UNIT PRICE
33 oz	$3.69	
87 oz	$6.22	
131 oz	$10.99	
263 oz	$17.99	

Determine whether the two pairs of numbers are proportional.

10. 7, 8 and 63, 72

11. 1.3, 3.4 and 5.6, 15.2

Solve.

12. $\frac{9}{4} = \frac{27}{x}$

13. $\frac{150}{2.5} = \frac{x}{6}$

14. $\frac{x}{100} = \frac{27}{64}$

15. $\frac{68}{y} = \frac{17}{25}$

Solve.

16. *Distance Traveled.*   An ocean liner traveled 432 km in 12 hr. At this rate, how far would the boat travel in 42 hr?

17. *Time Loss.*   A watch loses 2 min in 10 hr. At this rate, how much will it lose in 24 hr?

18. *Map Scaling.* On a map, 3 in. represents 225 mi. If two cities are 7 in. apart on the map, how far are they apart in reality?

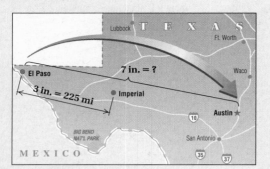

7 in. = ?
3 in. = 225 mi

19. *Tower Height.* A birdhouse built on a pole that is 3 m high casts a shadow 5 m long. At the same time, the shadow of a tower is 110 m long. How high is the tower?

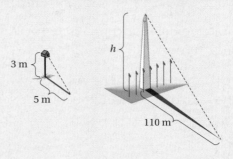

3 m
5 m
h
110 m

20. *Exchanging Money.* On 22 December 2000, it was known that 1 U.S. dollar was worth about 1.52 Canadian dollars.

a) How much would 450 U.S. dollars be worth in Canadian dollars?

b) Mitchell was traveling in Toronto and saw a DVD player that cost 560 Canadian dollars. How much would it cost in U.S. dollars?

21. *Automobile Violations.* In a recent year, the Indianapolis Police Department employed 1088 officers and made 37,493 arrests. At this rate, how many arrests could be made if the number of officers were increased to 2500?

Source: *Indianapolis Star*, 12-31-00

The lengths in each pair of figures are proportional. Find the missing lengths.

22.

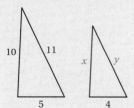

10   11
5
x   y
4

23.

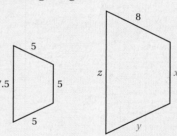

8
5
7.5   5
5
z   x
y

SKILL MAINTENANCE

24. In a recent year, Kellogg's sold 146.2 million lb of Corn Flakes and 120.4 million lb of Frosted Flakes. How many more pounds of Corn Flakes did it sell than Frosted Flakes?

25. Use = or ≠ for ☐ to write a true sentence:

$$\frac{6}{5} \,\square\, \frac{11}{9}.$$

26. Multiply:
$$\begin{array}{r} 2\ 3\ 4.1\ 1 \\ \times \quad\quad 7\ 4 \\ \hline \end{array}$$

27. Divide: $\dfrac{99.44}{100}$.

SYNTHESIS

28. Nancy Morano-Smith wants to win a season football ticket from the local bookstore. Her goal is to guess the number of marbles in an 8-gal jar. She knows that there are 128 oz in a gallon. She goes home and fills an 8-oz jar with 46 marbles. How many marbles should she guess are in the jar?

Cumulative Review

1. *Baseball Salaries.* In 2000, Alex Rodriguez signed a 10-yr contract for $252 million with the Texas Rangers.
   a) Find standard notation for the dollar amount of this contract.
   b) How many billion dollars was this contract?
   c) How much money did he make each year?
   d) In 2000, Rodriguez had 554 at-bats. How much money did he make for each at-bat?
   Source: Major League Baseball

2. *Gas Mileage.* The 2000 Volkswagen New Beetle GL will go 300 mi on 12.5 gal of gasoline in city driving. What is the rate in miles per gallon?
   Source: Volkswagen of America, Inc.

Add and simplify.

3.   2 7.6 8
          3.0 1 9
     + 4 8 3.2 9 7

4. $2\frac{1}{3}$
    $+4\frac{5}{12}$

5. $\frac{6}{35} + \frac{5}{28}$

Subtract and simplify.

6.   4 0.2
    $-$ 9.7 0 9

7. $73.82 - 0.908$

8. $\frac{4}{15} - \frac{3}{20}$.

Multiply and simplify.

9.   3 7.6 4
   $\times$ 5.9

10. 5.678×100

11. $2\frac{1}{3} \cdot 1\frac{2}{7}$

Divide and simplify.

12. $2.3 \overline{)\ 9\ 8.9}$

13. $5\ 4\ \overline{)\ 4\ 8,5\ 4\ 6}$

14. $\frac{7}{11} \div \frac{14}{33}$

15. Write expanded notation: 30,074.

16. Write a word name for 120.07.

Which number is larger?

17. 0.7,   0.698

18. 0.799,   0.8

19. Find the prime factorization of 144.

20. Find the LCM of 28 and 35.

21. What part is shaded?

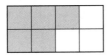

22. Simplify: $\frac{90}{144}$.

Calculate.

23. $\dfrac{3}{5} \times 9.53$

24. $\dfrac{1}{3} \times 0.645 - \dfrac{3}{4} \times 0.048$

25. Write fraction notation for the ratio 0.3 to 15.

26. Determine whether the pairs 3, 9 and 25, 75 are proportional.

27. What is the rate in meters per second?
     660 meters,   12 seconds

28. The following table lists prices for various brands of liquid dish soap. Find the unit price of each brand.

BRAND	PACKAGE	PRICE	UNIT PRICE
Palmolive	13 oz	$1.53	
Dawn	42.7 oz	$3.99	
Dawn	14.7 oz	$1.43	
Joy	28 oz	$1.78	
Joy	42.7 oz	$2.99	

Solve.

29. $\dfrac{14}{25} = \dfrac{x}{54}$

30. $423 = 16 \cdot t$

31. $\dfrac{2}{3} \cdot y = \dfrac{16}{27}$

32. $\dfrac{7}{16} = \dfrac{56}{x}$

33. $34.56 + n = 67.9$

34. $t + \dfrac{7}{25} = \dfrac{5}{7}$

Solve.

35. A particular kind of fettuccini alfredo has 520 calories in 1 cup. How many calories are there in $\frac{3}{4}$ cup?

36. *Exchanging Money.*   On 22 December 2000, 1 U.S. dollar was worth about 112.75 Japanese Yen.

   a) How much would 350 U.S. dollars be worth in Yen?
   b) Monica was traveling in Tokyo and saw a camera that cost 40,000 Yen. How much would it cost in U.S. dollars?

37. *Gas Mileage.*   A Greyhound tour bus traveled 347.6 mi, 249.8 mi, and 379.5 mi on three separate trips. What was the total mileage of the bus?

38. A machine can stamp out 925 washers in 5 min. The company owning the machine needs 1295 washers by the end of the morning. How long will it take to stamp them out?

39. A 46-oz juice can contains $5\frac{3}{4}$ cups of juice. A recipe calls for $3\frac{1}{2}$ cups of juice. How many cups are left over?

40. It takes a carpenter $\frac{2}{3}$ hr to hand a door. How many doors can the carpenter hang in 8 hr?

41. *The Leaning Tower of Pisa.*   At the time of this writing, the Leaning Tower of Pisa was still standing. It is 184.5 ft tall but leans about 17 ft out from its base. Each year, it leans about an additional $\frac{1}{20}$ in., or $\frac{1}{240}$ ft.
   a) After how many years will it lean the same length as it is tall?
   b) At most how many years do you think the Tower will stand?

42. *Airplane Tire Costs.*   A Boeing 747-400 jumbo jet has 2 nose tires and 16 rear tires. Each tire costs about $20,000.
   a) What is the total cost of a new set of tires for such a plane?
   b) Suppose an airline has a fleet of 400 such planes. What is the total cost of a new set of tires for all the planes?
   c) Suppose the airline has to change tires every month. What would be the total cost for tires for the airline for an entire year?
   Source: *World-Traveler,* October 2000

43. *Car Travel.*   A car travels 337.62 mi in 8 hr. How far does it travel in 1 hr?

44. *Shuttle Orbits.*   A recent space shuttle made 16 orbits a day during an 8.25-day mission. How many orbits were made during the entire mission?

For each of Exercises 45–47, choose the correct answer from the selections given.

45. How many even prime numbers are there?
   a) 5   **b)** 3   **c)** 2
   d) 1   **e)** None

46. The gas mileage of a car is 28.16 miles per gallon. How many gallons per mile is this?
   a) $\dfrac{704}{25}$ **b)** $\dfrac{25}{704}$ **c)** $\dfrac{2816}{100}$
   d) $\dfrac{250}{704}$ **e)** None

47. By what number do you multiply the side s of a square to find its perimeter?
   a) s itself   **b)** 4   **c)** 2
   d) 8   **e)** None

(**SYNTHESIS**)

48. A soccer goalie wishing to block an opponent's shot moves toward the shooter to reduce the shooter's view of the goal. If the goalie can only defend a region 10 ft wide, how far in front of the goal should the goalie be? (See the figure at right.)

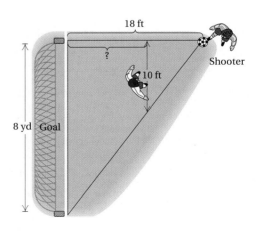

18 ft

?

10 ft

Shooter

8 yd   Goal

Percent Notation

Gateway to Chapter 6

This chapter introduces percent notation. We will see that $\frac{3}{8}$ (fraction notation), 0.375 (decimal notation), and 37.5% (percent notation) are all names for the same number. Percent notation has extensive applications in everyday life, in such diverse areas as business, sports, science, and medicine. We consider as well applications involving sales tax, commission, discount, interest, and interest rates on credit cards and loans.

Real-World Application

George W. Bush was inaugurated as the 43rd president of the United States in 2001. Since Grover Cleveland was both the 22nd and the 24th presidents, there have been only 42 different presidents. Of these 43 presidents, 8 have died in office: William Henry Harrison, Zachary Taylor, Abraham Lincoln, James A. Garfield, William McKinley, Warren G. Harding, Franklin D. Roosevelt, and John F. Kennedy. What percent have died in office?

This problem appears as Example 1 in Section 6.5.

CHAPTER

6

1. Insurance costs account for 13.3% of the annual cost of owning and operating an automobile. Find decimal notation for 13.3%. [6.1b]
   Source: Runzheimer International

2. Depreciation and interest charges on a car loan account for 0.504 of the annual cost of owning and operating an automobile. Find percent notation for 0.504. [6.1b]
   Source: Runzheimer International

Insurance 13.3%

Depreciation and interest 0.504

Tires $\frac{1}{25}$

Fuel 19%

3. Tire costs account for $\frac{1}{25}$ of the annual cost of owning and operating an automobile. Find percent notation for $\frac{1}{25}$. [6.2a]
   Source: Runzheimer International

4. Fuel costs account for 19% of the annual cost of owning and operating an automobile. Find fraction notation for 19%. [6.2b]
   Source: Runzheimer International

5. Translate to a percent equation. Then solve.
   What is 60% of 75? [6.3a, b]

6. Translate to a proportion. Then solve.
   What percent of 50 is 35? [6.4a, b]

Solve.

7. **Weight of Muscles.** The weight of muscles in a human body is 40% of total body weight. A person weighs 225 lb. What do the muscles weigh? [6.5a]

8. **Ticket Price Increase.** In 2001, the Indianapolis Colts raised the price of a ticket from $125 to $149 for a seat between the 30-yd lines on the lower level. What was the percent of increase? [6.5b]
   Source: The Indianapolis Colts

9. **Massachusetts Sales Tax.** The sales tax rate in Massachusetts is 5%. How much tax is charged on a purchase of $286? What is the total price? [6.6a]

10. A salesperson's commission rate is 28%. What is the commission from the sale of $18,400 worth of merchandise? [6.6b]

11. The marked price of a home theater system is $4450. The system is on sale at Lowland Appliances for 25% off. What are the discount and the sale price? [6.6c]

12. What is the simple interest on $1200 principal at the interest rate of 8.3% for 1 year? [6.7a]

13. What is the simple interest on $500 at 8% for $\frac{1}{2}$ year? [6.7a]

14. Interest is compounded annually. Find the amount in an account if $6000 is invested at 9% for 2 years. [6.7b]

15. The Beechers invest $7500 in an investment account paying 6%, compounded monthly. How much is in the account after 5 years? [6.7b]

6.1

PERCENT NOTATION

a Understanding Percent Notation

Of all the surface area of the earth, 70% of it is covered by water. What does this mean? It means that of every 100 square miles of the earth's surface area, 70 square miles are covered by water. Thus, 70% is a ratio of 70 to 100, or $\frac{70}{100}$.

Source: *The Handy Geography Answer Book*

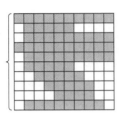

70 of 100 squares are shaded.

70% or $\frac{70}{100}$ or 0.70 of the large square is shaded.

Percent notation is used extensively in our everyday lives. Here are some examples:

63% of all aluminum used in the United States is recycled.

46% of the people at a major-league baseball game are women.

33% of all Americans say the day they dread the most is the day they go to the dentist.

20% of the time that people declare as sick leave is actually used for personal needs.

60% of the vehicles involved in a rollover fatality are sport utility vehicles.

0.08% blood alcohol level is a standard used by some states as the legal limit for drunk driving.

Percent notation is often represented in pie charts to show how the parts of a quantity are related. For example, the chart below relates the amounts of different kinds of juices that are sold.

Juices Sold

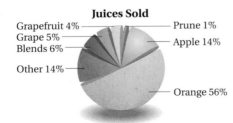

Grapefruit 4%
Grape 5%
Blends 6%
Other 14%
Prune 1%
Apple 14%
Orange 56%

Source: Beverage Marketing Corporation

PERCENT NOTATION

The notation **n%** means "*n* per hundred."

Objectives

a Write three kinds of notation for a percent.

b Convert between percent notation and decimal notation.

Write three kinds of notation as in Examples 1 and 2.

1. 70%

2. 23.4%

3. 100%

It is thought that the Roman emperor Augustus began percent notation by taxing goods sold at a rate of $\frac{1}{100}$. In time, the symbol "%" evolved by interchanging the parts of the symbol "100" to "0/0" and then to "%."

Answers on page A-13

From 1998 to 2008, the number of jobs for professional chefs will increase by 13.4%.
Source: *Handbook of U.S. Labor Statistics*

This definition leads us to the following equivalent ways of defining percent notation.

NOTATION FOR *n*%

Percent notation, *n*%, can be expressed using:

ratio $\rightarrow$ *n*% = the ratio of *n* to 100 = $\dfrac{n}{100}$,

fraction notation $\rightarrow$ *n*% = $n \times \dfrac{1}{100}$, or

decimal notation $\rightarrow$ *n*% = $n \times 0.01$.

EXAMPLE 1   Write three kinds of notation for 35%.

Using ratio: $35\% = \dfrac{35}{100}$ A ratio of 35 to 100

Using fraction notation: $35\% = 35 \times \dfrac{1}{100}$ Replacing % with $\times \dfrac{1}{100}$

Using decimal notation: $35\% = 35 \times 0.01$ Replacing % with $\times 0.01$

EXAMPLE 2   Write three kinds of notation for 67.8%.

Using ratio: $67.8\% = \dfrac{67.8}{100}$ A ratio of 67.8 to 100

Using fraction notation: $67.8\% = 67.8 \times \dfrac{1}{100}$ Replacing % with $\times \dfrac{1}{100}$

Using decimal notation: $67.8\% = 67.8 \times 0.01$ Replacing % with $\times 0.01$

Do Exercises 1–3 on the preceding page.

b   Converting Between Percent Notation and Decimal Notation

Consider 78%. To convert to decimal notation, we can think of percent notation as a ratio and write

$78\% = \dfrac{78}{100}$ Using the definition of percent as a ratio

$ = 0.78.$ Dividing

Similarly,

$4.9\% = \dfrac{4.9}{100}$ Using the definition of percent as a ratio

$ = 0.049.$ Dividing

We could also convert 78% to decimal notation by replacing "%" with "$\times 0.01$" and write

$78\% = 78 \times 0.01$ Replacing % with $\times 0.01$

$ = 0.78.$ Multiplying

Similarly,

$4.9\% = 4.9 \times 0.01$ Replacing % with $\times 0.01$

$ = 0.049.$ Multiplying

Dividing by 100 amounts to moving the decimal point two places to the left, which is the same as multiplying by 0.01. This leads us to a quick way to convert from percent notation to decimal notation: We drop the percent symbol and move the decimal point two places to the left.

To convert from percent notation to decimal notation,	36.5%
a) replace the percent symbol % with × 0.01, and	36.5 × 0.01
b) multiply by 0.01, which means move the decimal point two places to the left.	0.36.5 Move 2 places to the left.
	36.5% = 0.365

EXAMPLE 3 Find decimal notation for 99.44%.

a) Replace the percent symbol with × 0.01.   99.44 × 0.01

b) Move the decimal point two places to the left.   0.99.44

Thus, 99.44% = 0.9944.

EXAMPLE 4 The interest rate on a $2\frac{1}{2}$-year certificate of deposit is $6\frac{3}{8}\%$. Find decimal notation for $6\frac{3}{8}\%$.

a) Convert $6\frac{3}{8}$ to decimal notation and replace the percent symbol with × 0.01.

$$6\frac{3}{8}\%$$
$$6.375 \times 0.01$$

b) Move the decimal point two places to the left.

$$0.06.375$$

Thus, $6\frac{3}{8}\% = 0.06375$.

Do Exercises 4–8.

To convert 0.38 to percent notation, we can first write fraction notation, as follows:

$$0.38 = \frac{38}{100} \quad \text{Converting to fraction notation}$$

$$= 38\%. \quad \text{Using the definition of percent as a ratio}$$

Note that 100% = 100 × 0.01 = 1. Thus to convert 0.38 to percent notation, we can multiply by 1, using 100% as a symbol for 1. Then

$$0.38 = 0.38 \times 1$$
$$= 0.38 \times 100\%$$
$$= 0.38 \times 100 \times 0.01 \quad \text{Replacing 100\% with 100 × 0.01}$$
$$= (0.38 \times 100) \times 0.01 \quad \text{Using the associative law of multiplication}$$
$$= 38 \times 0.01$$
$$= 38\%. \quad \text{Replacing "× 0.01" with the \% symbol}$$

Even more quickly, since 0.38 = 0.38 × 100%, we can simply multiply 0.38 by 100 and write the % symbol.

Find decimal notation.

4. 34%

5. 78.9%

6. $6\frac{5}{8}\%$

Find decimal notation for the percent notation in the sentence.

7. Of all aluminum used in the United States, 63% is recycled.

8. A blood alcohol level of 0.08% is a standard used by some states as the legal limit for drunk driving.

Answers on page A-13

Find percent notation.

9. 0.24

10. 3.47

11. 1

Find percent notation for the decimal notation in the sentence.

12. Of all vehicles involved in a rollover fatality, 0.6 are sport utility vehicles.
Source: National Highway Traffic Safety Administration

13. Of those who play golf, 0.253 play 25–49 rounds per year.
Source: U.S. Golf Association

To convert from decimal notation to percent notation, we multiply by 100%—that is, we move the decimal point two places to the right and write a percent symbol.

To convert from decimal notation to percent notation, multiply by 100%. That is,	$0.675 = 0.675 \times 100\%$
a) move the decimal point two places to the right, and	0.67.5 Move 2 places to the right.
b) write a % symbol.	67.5%
	$0.675 = 67.5\%$

EXAMPLE 5   Find percent notation for 1.27.

a) Move the decimal point two places to the right.   1.27.

b) Write a % symbol.   127%

Thus, 1.27 = 127%.

EXAMPLE 6   Of the time that people declare as sick leave, 0.21 is actually used for family issues. Find percent notation for 0.21.
Source: CCH Inc.

a) Move the decimal point two places to the right.   0.21.

b) Write a % symbol.   21%

Thus, 0.21 = 21%.

EXAMPLE 7   Find percent notation for 5.6.

a) Move the decimal point two places to the right, adding an extra zero.   5.60.

b) Write a % symbol.   560%

Thus, 5.6 = 560%.

EXAMPLE 8   Of those who play golf, 0.149 play 8–24 rounds per year. Find percent notation for 0.149.
Source: U.S. Golf Association

a) Move the decimal point two places to the right.   0.14.9

b) Write a % symbol.   14.9%

Thus, 0.149 = 14.9%.

Do Exercises 9–13.

Answers on page A-13

6.1 EXERCISE SET

a Write three kinds of notation as in Examples 1 and 2 on p. 364.

1. 90% **2.** 58.7% **3.** 12.5% **4.** 130%

b Find decimal notation.

5. 67% **6.** 17% **7.** 45.6% **8.** 76.3%

9. 59.01% **10.** 30.02% **11.** 10% **12.** 80%

13. 1% **14.** 100% **15.** 200% **16.** 300%

17. 0.1% **18.** 0.4% **19.** 0.09% **20.** 0.12%

21. 0.18% **22.** 5.5% **23.** 23.19% **24.** 87.99%

25. $14\frac{7}{8}\%$ **26.** $93\frac{1}{8}\%$ **27.** $56\frac{1}{2}\%$ **28.** $61\frac{3}{4}\%$

Find decimal notation for the percent notation in the sentence.

29. Of the people who declare time off as sick leave, 40% actually have a personal illness.
Source: CCH, Inc.

30. Of those who play golf, 39% play 50–99 rounds per year.
Source: U.S. Golf Association

31. Of those who play golf, 18.6% play 100 or more rounds per year.
Source: U.S. Golf Association

32. Recently, the average interest rate on a 30-yr mortgage loan was 6.89%.
Source: Freddie Mac

33. According to a recent survey, 29% of those asked to name their favorite ice cream chose vanilla.
Source: International Ice Cream Association

34. According to a recent survey, 95.1% of those asked to name what sports they participate in chose swimming.
Source: Sporting Goods Manufacturers

Find percent notation.

35. 0.47

36. 0.87

37. 0.03

38. 0.01

39. 8.7

40. 4

41. 0.334

42. 0.889

43. 0.75

44. 0.99

45. 0.4

46. 0.5

47. 0.006

48. 0.008

49. 0.017

50. 0.024

51. 0.2718

52. 0.8911

53. 0.0239

54. 0.00073

Find percent notation for the decimal notation in the sentence.

55. According to a recent survey, 0.526 of those asked to name what sports they participate in chose bowling.
Source: Sporting Goods Manufacturers

56. On average, churchgoers donate 0.03 of their income to their churches.
Source: Lutheran Brotherhood

57. In 2000, the cost of college to a middle-income family was 0.17 of their income.
Source: College Board

58. About 0.69 of all newspapers are recycled.
Sources: American Forest and Paper Association; Newspaper Association of America

59. At one point in the 2000–2001 NBA season, Allen Iverson of the Philadelphia 76ers had made 0.411 of his field goals. His shooting percentage was 0.411.
Source: National Basketball Association

60. Of those people living in North Carolina, 0.1134 will die of heart disease.
Source: American Association of Retired Persons

61. **D**w *Winning Percentage.* During the 2000 regular baseball season, the New York Yankees won 87 of 162 games and went on to win the World Series. Find the ratio of number of wins to total number of games played in the regular season and convert it to decimal notation. Such a rate is often called a "winning percentage." Explain why.

62. **D**w Athletes sometimes speak of "giving 110%" effort. Does this make sense? Explain.

SKILL MAINTENANCE

Convert to a mixed numeral.   [3.4a]

63. $\dfrac{100}{3}$

64. $\dfrac{75}{2}$

65. $\dfrac{75}{8}$

66. $\dfrac{297}{16}$

67. $\dfrac{567}{98}$

68. $\dfrac{2345}{21}$

Convert to decimal notation.   [4.5a]

69. $\dfrac{2}{3}$

70. $\dfrac{1}{3}$

71. $\dfrac{5}{6}$

72. $\dfrac{17}{12}$

73. $\dfrac{8}{3}$

74. $\dfrac{15}{16}$

a Converting from Fraction Notation to Percent Notation

Consider the fraction notation $\frac{7}{8}$. To convert to percent notation, we use two skills we already have. We first find decimal notation by dividing:

$$\frac{7}{8} = 0.875$$

$$
\begin{array}{r}
0.8\ 7\ 5 \\
8\ \overline{)\ 7.0\ 0\ 0} \\
6\ 4 \\
\hline
6\ 0 \\
5\ 6 \\
\hline
4\ 0 \\
4\ 0 \\
\hline
0
\end{array}
$$

Then we convert the decimal notation to percent notation. We move the decimal point two places to the right

$$0.8\ 7.5$$

and write a % symbol:

$$\frac{7}{8} = 87.5\%, \text{ or } 87\frac{1}{2}\%.$$

To convert from fraction notation to percent notation,

a) find decimal notation by division, and

b) convert the decimal notation to percent notation.

$\frac{3}{5}$ Fraction notation

$$
\begin{array}{r}
0.6 \\
5\ \overline{)\ 3.0} \\
3\ 0 \\
\hline
0
\end{array}
$$

$0.6 = 0.60 = 60\%$ Percent

$\frac{3}{5} = 60\%$ notation

EXAMPLE 1 Find percent notation for $\frac{9}{16}$.

a) We first find decimal notation by division.

$$
\begin{array}{r}
0.5\ 6\ 2\ 5 \\
1\ 6\ \overline{)\ 9.0\ 0\ 0\ 0} \\
8\ 0 \\
\hline
1\ 0\ 0 \\
9\ 6 \\
\hline
4\ 0 \\
3\ 2 \\
\hline
8\ 0 \\
8\ 0 \\
\hline
0
\end{array}
$$

$$\frac{9}{16} = 0.5625$$

b) Next, we convert the decimal notation to percent notation. We move the decimal point two places to the right and write a % symbol.

$$0.56.25$$

$$\frac{9}{16} = 56.25\%, \text{ or } 56\tfrac{1}{4}\%$$

Don't forget the % symbol.

Do Exercises 1 and 2.

Find percent notation.

1. $\dfrac{1}{4}$ 2. $\dfrac{5}{8}$

CALCULATOR CORNER

Converting from Fraction Notation to Percent Notation   A calculator can be used to convert from fraction notation to percent notation. We simply perform the division on the calculator and then use the percent key. To convert $\dfrac{17}{40}$ to percent notation, for example, we press $\boxed{1}\,\boxed{7}\,\boxed{\div}\,\boxed{4}\,\boxed{0}\,\boxed{\text{2nd}}\,\boxed{\%}$, or $\boxed{1}\,\boxed{7}\,\boxed{\div}\,\boxed{4}\,\boxed{0}\,\boxed{\text{SHIFT}}\,\boxed{\%}$. The display reads $\boxed{42.5}$, so $\dfrac{17}{40} = 42.5\%$. Read the user's manual to determine whether your calculator can do this conversion.

Exercises: Use a calculator to find percent notation. Round to the nearest hundredth of a percent.

1. $\dfrac{13}{25}$ 4. $\dfrac{12}{7}$

2. $\dfrac{5}{13}$ 5. $\dfrac{217}{364}$

3. $\dfrac{43}{39}$ 6. $\dfrac{2378}{8401}$

EXAMPLE 2   *Death from Heart Attack.*   Of all those who suffer a heart attack, $\frac{1}{3}$ will die. Find percent notation for $\frac{1}{3}$.
Source: American Heart Association

a) Find decimal notation by division.

$$
\begin{array}{r}
0.3\ 3\ 3 \\
3\,\overline{)\,1.0\ 0\ 0} \\
\underline{9} \\
1\ 0 \\
\underline{9} \\
1\ 0 \\
\underline{9} \\
1
\end{array}
$$

We get a repeating decimal: $0.33\overline{3}$.

b) Convert the answer to percent notation.

$$0.33.\overline{3}$$

$$\frac{1}{3} = 33.\overline{3}\%, \text{ or } 33\frac{1}{3}\%$$

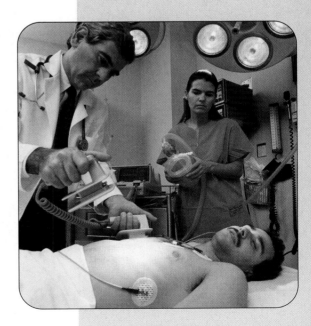

Answers on page A-13

3. Water is the single most abundant chemical in the body. The human body is about $\frac{2}{3}$ water. Find percent notation for $\frac{2}{3}$.

Do Exercises 3 and 4.

In some cases, division is not the fastest way to convert. The following are some optional ways in which conversion might be done.

EXAMPLE 3 Find percent notation for $\frac{69}{100}$.

We use the definition of percent as a ratio.

$$\frac{69}{100} = 69\%$$

EXAMPLE 4 Find percent notation for $\frac{17}{20}$.

We multiply by 1 to get 100 in the denominator. We think of what we have to multiply 20 by in order to get 100. That number is 5, so we multiply by 1 using $\frac{5}{5}$.

$$\frac{17}{20} \cdot \frac{5}{5} = \frac{85}{100} = 85\%$$

Note that this shortcut works only when the denominator is a factor of 100.

4. Find percent notation: $\frac{5}{6}$.

Do Exercises 5 and 6.

b   Converting from Percent Notation to Fraction Notation

To convert from percent notation to fraction notation,	30% Percent notation
a) use the definition of percent as a ratio, and	$\dfrac{30}{100}$
b) simplify, if possible.	$\dfrac{3}{10}$ Fraction notation

EXAMPLE 5 Find fraction notation for 75%.

$$75\% = \frac{75}{100} \qquad \text{Using the definition of percent}$$

$$= \frac{3 \cdot 25}{4 \cdot 25} = \frac{3}{4} \cdot \frac{25}{25} \left. \vphantom{\frac{3}{4}} \right\}$$

$$= \frac{3}{4} \qquad\qquad\qquad \text{Simplifying}$$

Find percent notation.

5. $\dfrac{57}{100}$

6. $\dfrac{19}{25}$

Answers on page A-13

CHAPTER 6: Percent Notation

EXAMPLE 6 Find fraction notation for 62.5%.

$$62.5\% = \frac{62.5}{100} \qquad \text{Using the definition of percent}$$

$$= \frac{62.5}{100} \times \frac{10}{10} \qquad \text{Multiplying by 1 to eliminate the decimal point in the numerator}$$

$$= \frac{625}{1000}$$

$$\left.\begin{array}{l} = \dfrac{5 \cdot 125}{8 \cdot 125} = \dfrac{5}{8} \cdot \dfrac{125}{125} \\[2mm] = \dfrac{5}{8} \end{array}\right\} \quad \text{Simplifying}$$

EXAMPLE 7 Find fraction notation for $16\frac{2}{3}\%$.

$$16\frac{2}{3}\% = \frac{50}{3}\% \qquad \text{Converting from the mixed numeral to fraction notation}$$

$$= \frac{50}{3} \times \frac{1}{100} \qquad \text{Using the definition of percent}$$

$$\left.\begin{array}{l} = \dfrac{50 \cdot 1}{3 \cdot 50 \cdot 2} = \dfrac{1}{6} \cdot \dfrac{50}{50} \\[2mm] = \dfrac{1}{6} \end{array}\right\} \quad \text{Simplifying}$$

The table on the inside front cover lists decimal, fraction, and percent equivalents used so often that it would speed up your work if you memorized them. For example, $\frac{1}{3} = 0.\overline{3}$, so we say that the **decimal equivalent** of $\frac{1}{3}$ is $0.\overline{3}$, or that $0.\overline{3}$ has the **fraction equivalent** $\frac{1}{3}$.

EXAMPLE 8 Find fraction notation for $16.\overline{6}\%$.

We can use the table on the inside front cover or recall that $16.\overline{6}\% = 16\frac{2}{3}\% = \frac{1}{6}$. We can also recall from our work with repeating decimals in Chapter 4 that $0.\overline{6} = \frac{2}{3}$. Then we have $16.\overline{6}\% = 16\frac{2}{3}\%$ and can proceed as in Example 7.

Do Exercises 7–10.

Find fraction notation.

7. 60%

8. 3.25%

9. $66\frac{2}{3}\%$

10. Complete this table.

Fraction Notation	$\dfrac{1}{5}$		
Decimal Notation		$0.83\overline{3}$	
Percent Notation			$37\frac{1}{2}\%$

Answers on page A-13

Applications of Ratio and Percent: The Price–Earnings Ratio and Stock Yields

The Price–Earnings Ratio   If the total earnings of a company one year were $5,000,000 and 100,000 shares of stock were issued, the earnings per share was $50. At one time, the price per share of Coca-Cola was $58.125 and the earnings per share was $0.76. The **price–earnings ratio,** P/E, is the price of the stock divided by the earnings per share. For the Coca-Cola stock, the price–earnings ratio, P/E, is given by

$$\frac{P}{E} = \frac{58.125}{0.76} \approx 76.48. \qquad \text{Dividing, using a calculator, and rounding to the nearest hundredth}$$

Stock Yields   At one time, the price per share of Coca-Cola stock was $58.125 and the company was paying a yearly dividend of $0.68 per share. It is helpful to those interested in stocks to know what percent the dividend is of the price of the stock. The percent is called the **yield.** For the Coca-Cola stock, the yield is given by

$$\text{Yield} = \frac{\text{Dividend}}{\text{Price per share}} = \frac{0.68}{58.125} \approx 0.0117 \qquad \text{Dividing and rounding to the nearest ten-thousandth}$$

$$= 1.17\% \qquad \text{Converting to percent notation}$$

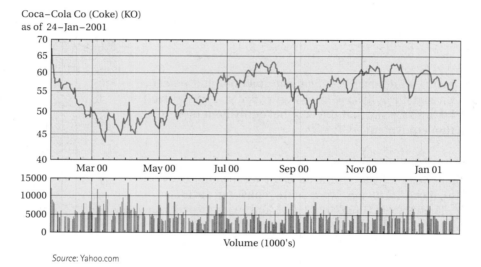

Coca–Cola Co (Coke) (KO) as of 24–Jan–2001

Volume (1000's)

Source: Yahoo.com

Exercises: Compute the price–earnings ratio and the yield for each stock listed below.

	STOCK	PRICE PER SHARE	EARNINGS	DIVIDEND	P/E	YIELD
1.	Pepsi (PEP)	$42.75	$1.40	$0.56		
2.	Pearson (PSO)	$25.00	$0.78	$0.30		
3.	Quaker Oats (OAT)	$92.375	$2.68	$1.10		
4.	Texas Insts (TEX)	$42.875	$1.62	$0.43		
5.	Ford Motor Co (F)	$27.5625	$2.30	$1.19		
6.	Wendy's Intl (WEN)	$25.75	$1.47	$0.23		

6.2

EXERCISE SET

a Find percent notation.

1. $\frac{41}{100}$ 2. $\frac{36}{100}$ 3. $\frac{5}{100}$ 4. $\frac{1}{100}$ 5. $\frac{2}{10}$ 6. $\frac{7}{10}$

7. $\frac{3}{10}$ 8. $\frac{9}{10}$ 9. $\frac{1}{2}$ 10. $\frac{3}{4}$ 11. $\frac{7}{8}$ 12. $\frac{1}{8}$

13. $\frac{4}{5}$ 14. $\frac{2}{5}$ 15. $\frac{2}{3}$ 16. $\frac{1}{3}$ 17. $\frac{1}{6}$ 18. $\frac{5}{6}$

19. $\frac{3}{16}$ 20. $\frac{11}{16}$ 21. $\frac{13}{16}$ 22. $\frac{7}{16}$ 23. $\frac{4}{25}$ 24. $\frac{17}{25}$

25. $\frac{1}{20}$ 26. $\frac{31}{50}$ 27. $\frac{17}{50}$ 28. $\frac{3}{20}$

Find percent notation for the fraction notation in the sentence.

29. Of all people, $\frac{2}{25}$ dread their birthday.
Source: Yankelovich Partners for Lutheran Brotherhood

30. Of all the water taken into the body, $\frac{3}{5}$ of it comes from beverages.

In Exercises 31–34, write percent notation for the fractions in this pie chart.

Engagement Times of Married Couples
Never engaged $\frac{1}{5}$
Less than 1 year $\frac{6}{25}$
1–2 years $\frac{21}{100}$
More than 2 years $\frac{7}{20}$

31. $\frac{21}{100}$ 32. $\frac{1}{5}$

33. $\frac{6}{25}$ 34. $\frac{7}{20}$

Find fraction notation. Simplify.

35. 85%

36. 55%

37. 62.5%

38. 12.5%

39. $33\frac{1}{3}\%$

40. $83\frac{1}{3}\%$

41. $16.\overline{6}\%$

42. $66.\overline{6}\%$

43. 7.25%

44. 4.85%

45. 0.8%

46. 0.2%

47. $25\frac{3}{8}\%$

48. $48\frac{7}{8}\%$

49. $78\frac{2}{9}\%$

50. $16\frac{5}{9}\%$

51. $64\frac{7}{11}\%$

52. $73\frac{3}{11}\%$

53. 150%

54. 110%

55. 0.0325%

56. 0.419%

57. $33.\overline{3}\%$

58. $83.\overline{3}\%$

Find fraction notation for the percent notation in the following bar graph.

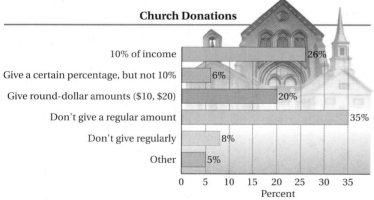

Church Donations

10% of income — 26%
Give a certain percentage, but not 10% — 6%
Give round-dollar amounts ($10, $20) — 20%
Don't give a regular amount — 35%
Don't give regularly — 8%
Other — 5%

0 5 10 15 20 25 30 35
Percent

Source: Lutheran Brotherhood

59. 26%

60. 20%

61. 5%

62. 8%

63. 6%

64. 35%

Find fraction notation for the percent notation in the sentence.

65. A 1.1-oz serving of Complete® cereal with $\frac{1}{2}$ cup of skim milk satisfies 45% of the minimum daily requirements for iron.
Source: Kellogg's USA, Inc.

66. A 1-cup serving of Wheaties® cereal with $\frac{1}{2}$ cup of skim milk satisfies 15% of the minimum daily requirements for calcium.
Source: General Mills Sales, Inc.

67. Of all those who are 85 or older, 47% have Alzheimer's disease.
Source: Alzheimer's Association

68. In 1998, 24.1% of Americans 18 and older smoked cigarettes.
Source: U.S. Centers for Disease Control

Complete the table.

69.

FRACTION NOTATION	DECIMAL NOTATION	PERCENT NOTATION
$\frac{1}{8}$		12.5%, or $12\frac{1}{2}\%$
$\frac{1}{6}$		
		20%
	0.25	
		$33.\overline{3}\%$, or $33\frac{1}{3}\%$
		37.5%, or $37\frac{1}{2}\%$
		40%
$\frac{1}{2}$		

70.

FRACTION NOTATION	DECIMAL NOTATION	PERCENT NOTATION
$\frac{3}{5}$		
	0.625	
$\frac{2}{3}$		
	0.75	75%
$\frac{4}{5}$		
$\frac{5}{6}$		$83.\overline{3}\%$, or $83\frac{1}{3}\%$
$\frac{7}{8}$		87.5%, or $87\frac{1}{2}\%$
		100%

71.

FRACTION NOTATION	DECIMAL NOTATION	PERCENT NOTATION
	0.5	
$\frac{1}{3}$		
		25%
		$16.\overline{6}\%$, or $16\frac{2}{3}\%$
	0.125	
$\frac{3}{4}$		
	$0.8\overline{3}$	
$\frac{3}{8}$		

72.

FRACTION NOTATION	DECIMAL NOTATION	PERCENT NOTATION
		40%
		62.5%, or $62\frac{1}{2}\%$
	0.875	
$\frac{1}{1}$		
	0.6	
	$0.\overline{6}$	
$\frac{1}{5}$		

73. D_W What do the following have in common? Explain.

$$\frac{23}{16}, \quad 1\frac{875}{2000}, \quad 1.4375, \quad \frac{207}{144}, \quad 1\frac{7}{16}, \quad 143.75\%, \quad 1\frac{4375}{10,000}$$

74. D_W Is it always best to convert from fraction notation to percent notation by first finding decimal notation? Why or why not?

Solve.

75. $13 \cdot x = 910$ [1.7b]

76. $15 \cdot y = 75$ [1.7b]

77. $0.05 \times b = 20$ [4.4b]

78. $3 = 0.16 \times b$ [4.4b]

79. $\dfrac{24}{37} = \dfrac{15}{x}$ [5.3b]

80. $\dfrac{17}{18} = \dfrac{x}{27}$ [5.3b]

Convert to a mixed numeral. [3.4a]

81. $\dfrac{100}{3}$

82. $\dfrac{75}{2}$

83. $\dfrac{250}{3}$

84. $\dfrac{123}{6}$

85. $\dfrac{345}{8}$

86. $\dfrac{373}{6}$

87. $\dfrac{75}{4}$

88. $\dfrac{67}{9}$

Write percent notation.

89. ▦ $\dfrac{41}{369}$

90. ▦ $\dfrac{54}{999}$

91. $2.5\overline{74631}$

92. $3.2\overline{93847}$

Write decimal notation.

93. $\dfrac{14}{9}\%$

94. $\dfrac{19}{12}\%$

95. $\dfrac{729}{7}\%$

96. $\dfrac{637}{6}\%$

6.3

SOLVING PERCENT PROBLEMS USING PERCENT EQUATIONS

Objectives

a Translate percent problems to percent equations.

b Solve basic percent problems.

a Translating to Equations

To solve a problem involving percents, it is helpful to translate first to an equation. To distinguish the method in Section 6.3 from that of Section 6.4, we will call these *percent equations*.

EXAMPLE 1 Translate:

$$\begin{array}{ccccc} 23\% & \text{of} & 5 & \text{is} & \text{what?} \\ \downarrow & \downarrow & \downarrow & \downarrow & \downarrow \\ 23\% & \cdot & 5 & = & a \end{array}$$

This is a *percent equation*.

KEY WORDS IN PERCENT TRANSLATIONS

"**Of**" translates to "·", or "×".   "**Is**" translates to "=".

"**What**" translates to any letter.   "**%**" translates to "$\times \frac{1}{100}$" or "$\times 0.01$".

EXAMPLE 2 Translate:

$$\begin{array}{ccccc} \text{What} & \text{is} & 11\% & \text{of} & 49? \\ \downarrow & \downarrow & \downarrow & \downarrow & \downarrow \\ a & = & 11\% & \cdot & 49 \end{array}$$

Any letter can be used.

Do Exercises 1 and 2.

EXAMPLE 3 Translate:

$$\begin{array}{ccccc} 3 & \text{is} & 10\% & \text{of} & \text{what?} \\ \downarrow & \downarrow & \downarrow & \downarrow & \downarrow \\ 3 & = & 10\% & \cdot & b \end{array}$$

EXAMPLE 4 Translate:

$$\begin{array}{ccccc} 45\% & \text{of} & \text{what} & \text{is} & 23? \\ \downarrow & \downarrow & \downarrow & \downarrow & \downarrow \\ 45\% & \times & b & = & 23 \end{array}$$

Do Exercises 3 and 4.

EXAMPLE 5 Translate:

$$\begin{array}{ccccc} 10 & \text{is} & \text{what percent} & \text{of} & 20? \\ \downarrow & \downarrow & \downarrow & \downarrow & \downarrow \\ 10 & = & p & \times & 20 \end{array}$$

Translate to an equation. Do not solve.

1. 12% of 50 is what?

2. What is 40% of 60?

Translate to an equation. Do not solve.

3. 45 is 20% of what?

4. 120% of what is 60?

Answers on page A-14

Translate to an equation. Do not solve.

5. 16 is what percent of 40?

6. What percent of 84 is 10.5?

7. Solve:

What is 12% of 50?

John's Used Cars currently has 60 used cars for sale; 15% of those cars are at least 8 yr old. How many cars are at least 8 yr old?

Answers on page A-14

EXAMPLE 6 Translate:

What percent of 50 is 7?
$$p \cdot 50 = 7$$

Do Exercises 5 and 6.

b Solving Percent Problems

In solving percent problems, we use the *Translate* and *Solve* steps in the problem-solving strategy used throughout this text.

Percent problems are actually of three different types. Although the method we present does *not* require that you be able to identify which type you are solving, it is helpful to know them.

We know that

15 is 25% of 60, or

$15 = 25\% \times 60.$

We can think of this as:

> Amount = Percent number × Base.

Each of the three types of percent problems depend on which of the three pieces of information is missing.

1. **Finding the *amount* (the result of taking the percent)**

   Example: What is 25% of 60?

   Translation: $a = 25\% \cdot 60$

2. **Finding the *base* (the number you are taking the percent of)**

   Example: 15 is 25% of what number?

   Translation: $15 = 25\% \cdot b$

3. **Finding the *percent number* (the percent itself)**

   Example: 15 is what percent of 60?

   Translation: $15 = p \cdot 60$

FINDING THE AMOUNT

EXAMPLE 7 What is 15% of 60?

Translate: $a = 15\% \times 60.$

Solve: The letter is by itself. To solve the equation, we just convert 15% to decimal notation and multiply:

$$a = 15\% \times 60 = 0.15 \times 60 = 9.$$

Thus, 9 is 15% of 60. The answer is 9.

Do Exercise 7.

EXAMPLE 8   120% of $42 is what?

Translate: $120\% \times 42 = a$.

Solve: The letter is by itself. To solve the equation, we carry out the calculation:

$$a = 120\% \times 42$$
$$= 1.2 \times 42$$
$$= 50.4.$$

Thus, 120% of $42 is $50.40. The answer is $50.40.

Do Exercise 8.

FINDING THE BASE

EXAMPLE 9   5% of what is 20?

Translate: $5\% \times b = 20$.

Solve: This time the letter is *not* by itself. To solve the equation, we divide by 5% on both sides:

$$\frac{5\% \times b}{5\%} = \frac{20}{5\%} \qquad \text{Dividing by 5\% on both sides}$$

$$b = \frac{20}{0.05} \qquad 5\% = 0.05$$

$$b = 400.$$

Thus, 5% of 400 is 20. The answer is 400.

EXAMPLE 10 $3 is 16% of what?

Translate:
$$\begin{array}{ccccc} \$3 & \text{is} & 16\% & \text{of} & \text{what?} \\ \downarrow & \downarrow & \downarrow & \downarrow & \downarrow \\ 3 & = & 16\% & \times & b. \end{array}$$

Solve: Again, the letter is *not* by itself. To solve the equation, we divide by 16% on both sides:

$$\frac{3}{16\%} = \frac{16\% \times b}{16\%} \qquad \text{Dividing by 16\% on both sides}$$

$$\frac{3}{0.16} = b \qquad 16\% = 0.16$$

$$18.75 = b.$$

Thus, $3 is 16% of $18.75. The answer is $18.75.

Do Exercises 9 and 10.

8. Solve:

$$64\% \text{ of } \$55 \text{ is what?}$$

In a survey of a group of people, it was found that 5%, or 20 people, chose strawberry as their favorite ice cream. How many people were surveyed?
Source: International Ice Cream Association

Solve.

9. 20% of what is 45?

10. $60 is 120% of what?

Answers on page A-14

Of every 20 people who travel, 10 will feel stress while waiting in line. What percent will feel stress?
Source: TNS Intersearch

11. Solve:

16 is what percent of 40?

12. Solve:

What percent of $84 is $10.50?

FINDING THE PERCENT NUMBER

In solving these problems, you *must* remember to convert to percent notation after you have solved the equation.

EXAMPLE 11 10 is what percent of 20?

Translate: 10 is what percent of 20?

$$10 = p \times 20.$$

Solve: To solve the equation, we divide by 20 on both sides and convert the result to percent notation:

$$p \cdot 20 = 10$$

$$\frac{p \cdot 20}{20} = \frac{10}{20} \qquad \text{Dividing by 20 on both sides}$$

$$p = 0.50 = 50\%. \qquad \text{Converting to percent notation}$$

Thus, 10 is 50% of 20. The answer is 50%.

Do Exercise 11.

EXAMPLE 12 What percent of $50 is $16?

Translate: What percent of $50 is $16?

$$p \times 50 = 16.$$

Solve: To solve the equation, we divide by 50 on both sides and convert the answer to percent notation:

$$\frac{p \times 50}{50} = \frac{16}{50} \qquad \text{Dividing by 50 on both sides}$$

$$p = \frac{16}{50}$$

$$p = 0.32$$

$$p = 32\%. \qquad \text{Converting to percent notation}$$

Thus, 32% of $50 is $16. The answer is 32%.

Do Exercise 12.

CAUTION!

When a question asks "what percent?", be sure to give the answer in percent notation.

Answers on page A-14

Using Percents in Computations   Many calculators have a $\boxed{\%}$ key that can be used in computations. (See the Calculator Corner on page 364.) For example, to find 11% of 49, we press $\boxed{1}\boxed{1}\boxed{\text{2nd}}\boxed{\%}\boxed{\times}\boxed{4}\boxed{9}\boxed{=}$ or $\boxed{4}\boxed{9}\boxed{\times}\boxed{1}\boxed{1}\boxed{\text{SHIFT}}\boxed{\%}$. The display reads $\boxed{\qquad 5.39}$, so 11% of 49 is 5.39.

In Example 9, we perform the computation 20/5%. To use the $\boxed{\%}$ key in this computation, we press $\boxed{2}\boxed{0}\boxed{\div}\boxed{5}\boxed{\text{2nd}}\boxed{\%}\boxed{=}$, or $\boxed{2}\boxed{0}\boxed{\div}\boxed{5}\boxed{\text{SHIFT}}\boxed{\%}$. The result is 400.

We can also use the $\boxed{\%}$ key to find the percent number in a problem. In Example 11, for instance, we answer the question "10 is what percent of 20?" On a calculator, we press $\boxed{1}\boxed{0}\boxed{\div}\boxed{2}\boxed{0}\boxed{\text{2nd}}\boxed{\%}\boxed{=}$, or $\boxed{1}\boxed{0}\boxed{\div}$ $\boxed{2}\boxed{0}\boxed{\text{SHIFT}}\boxed{\%}$. The result is 50, so 10 is 50% of 20.

Exercises: Use a calculator to find each of the following.

1. What is 5% of 24?

2. What is 12.6% of $40?

3. What is 19% of 256?

4. 140% of $16 is what?

5. 0.04% of 28 is what?

6. 33% of $90 is what?

7. Use the percent key on a calculator to perform the computations in Example 10 and Margin Exercises 9 and 10.

8. Use the percent key on a calculator to perform the computations in Example 12 and Margin Exercises 11 and 12.

a Translate to an equation. Do not solve.

1. What is 32% of 78?

2. 98% of 57 is what?

3. 89 is what percent of 99?

4. What percent of 25 is 8?

5. 13 is 25% of what?

6. 21.4% of what is 20?

b Solve.

7. What is 85% of 276?

8. What is 74% of 53?

9. 150% of 30 is what?

10. 100% of 13 is what?

11. What is 6% of $300?

12. What is 4% of $45?

13. 3.8% of 50 is what?

14. $33\frac{1}{3}\%$ of 480 is what?
$\left(\textit{Hint: } 33\frac{1}{3}\% = \frac{1}{3}.\right)$

15. $39 is what percent of $50?

16. $16 is what percent of $90?

17. 20 is what percent of 10?

18. 60 is what percent of 20?

19. What percent of $300 is $150?

20. What percent of $50 is $40?

21. What percent of 80 is 100?

22. What percent of 60 is 15?

23. 20 is 50% of what?

24. 57 is 20% of what?

25. 40% of what is $16?

26. 100% of what is $74?

27. 56.32 is 64% of what?

28. 71.04 is 96% of what?

29. 70% of what is 14?

30. 70% of what is 35?

31. What is $62\frac{1}{2}$% of 10?

32. What is $35\frac{1}{4}$% of 1200?

33. What is 8.3% of $10,200?

34. What is 9.2% of $5600?

35. ^{D}W Write a question that could be translated to the equation

$$25 = 4\% \times b.$$

36. ^{D}W Suppose we know that 40% of 92 is 36.8. What is a quick way to find 4% of 92? 400% of 92? Explain.

SKILL MAINTENANCE

Write fraction notation.   [4.1b]

37. 0.09

38. 1.79

39. 0.875

40. 0.125

41. 0.9375

42. 0.6875

Write decimal notation.   [4.1b]

43. $\dfrac{89}{100}$

44. $\dfrac{7}{100}$

45. $\dfrac{3}{10}$

46. $\dfrac{17}{1000}$

SYNTHESIS

Solve.

47. 🔲 What is 7.75% of $10,880?
Estimate _____
Calculate _____

48. 🔲 50,951.775 is what percent of 78,995?
Estimate _____
Calculate _____

49. 🔲 $2496 is 24% of what amount?
Estimate _____
Calculate _____

50. 🔲 What is 38.2% of $52,345.79?
Estimate _____
Calculate _____

51. 40% of $18\frac{3}{4}$% of $25,000 is what?

SOLVING PERCENT PROBLEMS USING PROPORTIONS*

Objectives

a Translate percent problems to proportions.

b Solve basic percent problems.

A survey has found that 75% of all people watch TV in bed before they go to sleep. The city of San Francisco has 745,780 people. This means that 559,335 of them watch TV in bed before they go to sleep.
Sources: Bruskin–Goldring Research for Serta; *The New York Times Almanac*

Note: This section presents an alternative method for solving basic percent problems. You can use either equations or proportions to solve percent problems, but you might prefer one method over the other, or your instructor may direct you to use one method over the other.

a Translating to Proportions

A percent is a ratio of some number to 100. For example, 75% is the ratio $\frac{75}{100}$. The numbers 559,335 and 745,780 have the same ratio as 75 and 100. The numbers 3 and 4 also have the same ratio.

$$\frac{75}{100} = \frac{559,335}{745,780} = \frac{3}{4}$$

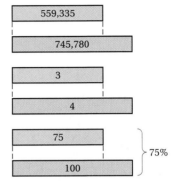

To solve a percent problem using a proportion, we translate as follows:

$$\text{Number} \rightarrow \frac{P}{100} = \frac{a}{b} \begin{array}{l} \leftarrow \text{Amount} \\ \leftarrow \text{Base} \end{array}$$

100 ⟶

You might find it helpful to read this as "part is to whole as part is to whole."

For example, 60% of 25 is 15 translates to

$$\frac{60}{100} = \frac{15}{25}. \begin{array}{l} \leftarrow \text{Amount} \\ \leftarrow \text{Base} \end{array}$$

A clue in translating is that the base, b, corresponds to 100 and usually follows the wording "percent of." Also, $P\%$ always translates to $P/100$. Another aid in translating is to make a comparison drawing. To do this, we start with the percent side and list 0% at the top and 100% near the bottom. Then we estimate where the specified percent—in this case, 60%—is located. The corresponding quantities are then filled in. The base—in this case, 25—always corresponds to 100% and the amount—in this case, 15—corresponds to the specified percent.

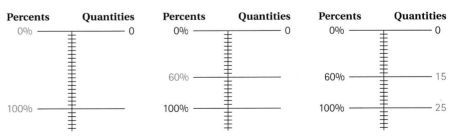

The proportion can then be read easily from the drawing: $\frac{60}{100} = \frac{15}{25}$.

EXAMPLE 1 Translate to a proportion.

23% of 5 is what?

$$\frac{23}{100} = \frac{a}{5}$$

Percents		Quantities
0%		0
23%		a
100%		5

EXAMPLE 2 Translate to a proportion.

What is 124% of 49?

$$\frac{124}{100} = \frac{a}{49}$$

Percents		Quantities
0%		0
100%		49
124%		a

Do Exercises 1–3.

EXAMPLE 3 Translate to a proportion.

3 is 10% of what?

$$\frac{10}{100} = \frac{3}{b}$$

Percents		Quantities
0%		0
10%		3
100%		b

EXAMPLE 4 Translate to a proportion.

45% of what is 23?

$$\frac{45}{100} = \frac{23}{b}$$

Percents		Quantities
0%		0
45%		23
100%		b

Do Exercises 4 and 5.

EXAMPLE 5 Translate to a proportion.

10 is what percent of 20?

$$\frac{P}{100} = \frac{10}{20}$$

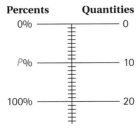

Percents		Quantities
0%		0
P%		10
100%		20

Translate to a proportion. Do not solve.

1. 12% of 50 is what?

2. What is 40% of 60?

3. 130% of 72 is what?

Translate to a proportion. Do not solve.

4. 45 is 20% of what?

5. 120% of what is 60?

Answers on page A-14

6.4 Solving Percent Problems
Using Proportions

Translate to a proportion. Do not solve.

6. 16 is what percent of 40?

7. What percent of 84 is 10.5?

8. Solve:

20% of what is $45?

Solve.

9. 64% of 55 is what?

10. What is 12% of 50?

Answers on page A-14

EXAMPLE 6   Translate to a proportion.

What percent of 50 is 7?

$$\frac{P}{100} = \frac{7}{50}$$

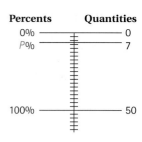

Percents	Quantities
0%	0
P%	7
100%	50

Do Exercises 6 and 7.

b   Solving Percent Problems

After a percent problem has been translated to a proportion, we solve as in Section 5.3.

EXAMPLE 7   5% of what is $20?

Translate: $\dfrac{5}{100} = \dfrac{20}{b}$

Solve: $5 \cdot b = 100 \cdot 20$ Equating cross products

$\dfrac{5 \cdot b}{5} = \dfrac{100 \cdot 20}{5}$ Dividing by 5

$b = \dfrac{2000}{5}$

$b = 400$ Simplifying

Thus, 5% of $400 is $20. The answer is $400.

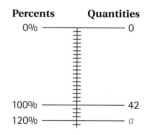

Percents	Quantities
0%	0
5%	20
100%	b

Do Exercise 8.

EXAMPLE 8   120% of 42 is what?

Translate: $\dfrac{120}{100} = \dfrac{a}{42}$

Solve: $120 \cdot 42 = 100 \cdot a$ Equating cross products

$\dfrac{120 \cdot 42}{100} = \dfrac{100 \cdot a}{100}$ Dividing by 100

$\dfrac{5040}{100} = a$

$50.4 = a$ Simplifying

Thus, 120% of 42 is 50.4. The answer is 50.4.

Percents	Quantities
0%	0
100%	42
120%	a

Do Exercises 9 and 10.

EXAMPLE 9 3 is 16% of what?

Translate: $\dfrac{3}{b} = \dfrac{16}{100}$

Percents	Quantities
0%	0
16%	3
100%	b

Solve: $3 \cdot 100 = b \cdot 16$ Equating cross products

$\dfrac{3 \cdot 100}{16} = \dfrac{b \cdot 16}{16}$ Dividing by 16

$\dfrac{300}{16} = b$ Multiplying and simplifying

$18.75 = b$ Dividing

Thus, 3 is 16% of 18.75. The answer is 18.75.

Do Exercise 11.

EXAMPLE 10 $10 is what percent of $20?

Translate: $\dfrac{10}{20} = \dfrac{P}{100}$

Percents	Quantities
0%	0
P%	$10
100%	$20

Solve: $10 \cdot 100 = 20 \cdot P$ Equating cross products

$\dfrac{10 \cdot 100}{20} = \dfrac{20 \cdot P}{20}$ Dividing by 20

$\dfrac{1000}{20} = P$ Multiplying and simplifying

$50 = P$ Dividing

Thus, $10 is 50% of $20. The answer is 50%.

Do Exercise 12.

EXAMPLE 11 What percent of 50 is 16?

Translate: $\dfrac{P}{100} = \dfrac{16}{50}$

Percents	Quantities
0%	0
P%	16
100%	50

Solve: $50 \cdot P = 100 \cdot 16$ Equating cross products

$\dfrac{50 \cdot P}{50} = \dfrac{100 \cdot 16}{50}$ Dividing by 50

$P = \dfrac{1600}{50}$ Multiplying and simplifying

$P = 32$ Dividing

Thus, 32% of 50 is 16. The answer is 32%.

Do Exercise 13.

11. Solve:

60 is 120% of what?

12. Solve:

$12 is what percent of $40?

13. Solve:

What percent of 84 is 10.5?

Answers on page A-14

6.4

EXERCISE SET

For Extra Help

Digital Video
Tutor CD 4
Videotape 10

InterAct
Math

Math Tutor
Center

MathXL

MyMathLab

a   Translate to a proportion. Do not solve.

1. What is 37% of 74?

2. 66% of 74 is what?

3. 4.3 is what percent of 5.9?

4. What percent of 6.8 is 5.3?

5. 14 is 25% of what?

6. 133% of what is 40?

b   Solve.

7. What is 76% of 90?

8. What is 32% of 70?

9. 70% of 660 is what?

10. 80% of 920 is what?

11. What is 4% of 1000?

12. What is 6% of 2000?

13. 4.8% of 60 is what?

14. 63.1% of 80 is what?

15. $24 is what percent of $96?

16. $14 is what percent of $70?

17. 102 is what percent of 100?

18. 103 is what percent of 100?

19. What percent of $480 is $120?

20. What percent of $80 is $60?

21. What percent of 160 is 150?

22. What percent of 33 is 11?

23. $18 is 25% of what?

24. $75 is 20% of what?

25. 60% of what is 54?

26. 80% of what is 96?

27. 65.12 is 74% of what?

28. 63.7 is 65% of what?

29. 80% of what is 16?

30. 80% of what is 10?

31. What is $62\frac{1}{2}$% of 40?

32. What is $43\frac{1}{4}$% of 2600?

33. What is 9.4% of $8300?

34. What is 8.7% of $76,000?

35. ^{D}w In your own words, list steps that a classmate could use to solve any percent problem in this section.

36. ^{D}w In solving Example 10, a student simplifies $\frac{10}{20}$ before solving. Is this a good idea? Why or why not?

SKILL MAINTENANCE

Solve. [5.3b]

37. $\dfrac{x}{188} = \dfrac{2}{47}$

38. $\dfrac{15}{x} = \dfrac{3}{800}$

39. $\dfrac{4}{7} = \dfrac{x}{14}$

40. $\dfrac{612}{t} = \dfrac{72}{244}$

41. $\dfrac{5000}{t} = \dfrac{3000}{60}$

42. $\dfrac{75}{100} = \dfrac{n}{20}$

43. $\dfrac{x}{1.2} = \dfrac{36.2}{5.4}$

44. $\dfrac{y}{1\frac{1}{2}} = \dfrac{2\frac{3}{4}}{22}$

Solve.

45. A recipe for muffins calls for $\frac{1}{2}$ qt of buttermilk, $\frac{1}{3}$ qt of skim milk, and $\frac{1}{16}$ qt of oil. How many quarts of liquid ingredients does the recipe call for? [3.2b]

46. The Ferristown School District purchased $\frac{3}{4}$ ton (T) of clay. If the clay is to be shared equally among the district's 6 art departments, how much will each art department receive? [2.7d]

SYNTHESIS

Solve.

47. ▦ What is 8.85% of $12,640?

Estimate _____

Calculate _____

48. ▦ 78.8% of what is 9809.024?

Estimate _____

Calculate _____

a   Solve applied problems involving percent.

b   Solve applied problems involving percent of increase or decrease.

6.5 APPLICATIONS OF PERCENT

a   **Applied Problems Involving Percent**

Applied problems involving percent are not always stated in a manner easily translated to an equation. In such cases, it is helpful to rephrase the problem before translating. Sometimes it also helps to make a drawing.

EXAMPLE 1   *Presidential Deaths in Office.*   George W. Bush was inaugurated as the 43rd President of the United States in 2001. Since Grover Cleveland was both the 22nd and the 24th presidents, there have been only 42 different presidents. Of the 42 presidents, 8 have died in office: William Henry Harrison, Zachary Taylor, Abraham Lincoln, James A. Garfield, William McKinley, Warren G. Harding, Franklin D. Roosevelt, and John F. Kennedy. What percent have died in office?

Harrison   Taylor   Garfield   McKinley

Harding   Roosevelt   Kennedy

1. **Familiarize.**   The question asks for a percent of the presidents who have died in office. We note that 42 is approximately 40 and 8 is $\frac{1}{5}$, or 20%, of 40, so our answer is close to 20%. We let p = the percent who have died in office.

2. **Translate.**   There are two ways in which we can translate this problem.

   Percent equation (see Section 6.3):

   8   is   what percent   of   42?

   8   = p ·   42

   Proportion (see Section 6.4):

   $$\frac{P}{100} = \frac{8}{42}$$

   For proportions, $P\% = p$.

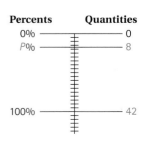

Percents	Quantities
0%	0
P%	8
100%	42

3. Solve. We now have two ways in which to solve this problem.

Percent equation (see Section 6.3):

$$8 = p \cdot 42$$

$$\frac{8}{42} = \frac{p \cdot 42}{42} \qquad \text{Dividing by 42 on both sides}$$

$$\frac{8}{42} = p$$

$$0.190 \approx p \qquad \text{Finding decimal notation and rounding to the nearest thousandth}$$

$$19.0\% \approx p \qquad \text{Remember to find percent notation.}$$

Note here that the solution, p, includes the % symbol.

Proportion (see Section 6.4):

$$\frac{P}{100} = \frac{8}{42}$$

$$P \cdot 42 = 100 \cdot 8 \qquad \text{Equating cross products}$$

$$\frac{P \cdot 42}{42} = \frac{800}{42} \qquad \text{Dividing by 42 on both sides}$$

$$P = \frac{800}{42}$$

$$P \approx 19.0 \qquad \text{Dividing and rounding to the nearest tenth}$$

We use the solution of the proportion to express the answer to the problem as 19.0%. Note that in the proportion method, $P\% = p$.

4. Check. To check, we note that the answer 19.0% is close to 20%, as estimated in the *Familiarize* step.

5. State. About 19.0% of the U.S. presidents have died in office.

Do Exercise 1.

EXAMPLE 2 *Water Ingested as Beverages.* The average adult ingests 2500 milliliters (mL) of water each day, that is, about 2.5 qt. About 60% of this comes from beverages. How many milliliters of water does the average adult ingest as beverages in one day?
Source: Elaine N. Meriab, *Essentials of Anatomy & Physiology*, 6th ed. San Francisco: Benjamin/Cummings Science Publishing, 2000

---

1. Presidential Assassinations in Office. Of the 43 U.S. presidents, 4 have been assassinated in office. These were Garfield, McKinley, Lincoln, and Kennedy. What percent have been assassinated in office?

Answer on page A-14

2. Water Ingested as Food. The average adult ingests 2500 mL of water each day. About 30% of this comes from foods. How many milliliters of water does the average adult ingest as food in one day?

Source: Elaine N. Meriab, *Essentials of Anatomy & Physiology,* 6th ed. San Francisco: Benjamin/Cummings Science Publishing, 2000

1. Familiarize. We can make a drawing of a pie chart to help familiarize ourselves with the problem. We let b = the total number of milliliters of water ingested through beverages.

Water Ingested

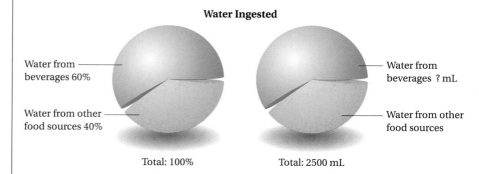

Water from beverages 60%

Water from other food sources 40%

Total: 100%

Water from beverages ? mL

Water from other food sources

Total: 2500 mL

2. Translate. There are two ways in which we can translate this problem.

Percent equation:

What number is 60% of 2500?

$$b = 60\% \cdot 2500$$

Proportion:

$$\frac{60}{100} = \frac{b}{2500}$$

3. Solve. We now have two ways in which to solve this problem.

Percent equation:

$$b = 60\% \cdot 2500$$

We convert 60% to decimal notation and multiply:

$$b = 60\% \cdot 2500 = 0.60 \times 2500 = 1500.$$

Proportion:

$$\frac{60}{100} = \frac{b}{2500}$$

$$60 \cdot 2500 = 100 \cdot b \qquad \text{Equating cross products}$$

$$\frac{60 \cdot 2500}{100} = \frac{100 \cdot b}{100} \qquad \text{Dividing by 100}$$

$$\frac{150,000}{100} = b$$

$$1500 = b \qquad \text{Simplifying}$$

Percents **Quantities**

0% — 0

60% — b

100% — 2500

4. Check. To check, we can repeat the calculations. We can also think about our answer. Since we are taking 60% of 2500, we would expect 1500 to be smaller than 2500 and exactly three-fifths of 2500, which it is.

5. State. The amount of water ingested through beverages by the average adult in one day is 1500 mL.

Do Exercise 2.

Answer on page A-14

b | Percent of Increase or Decrease

Percent is often used to state increase or decrease. Let's consider an example of each, using the price of a car as the original number.

PERCENT OF INCREASE

One year a car sold for $20,455. The manufacturer decides to raise the price of the following year's model by 6%. The increase is 0.06 × $20,455, or $1227.30. The new price is $20,455 + $1227.30, or $21,682.30. The *percent of increase* is 6%.

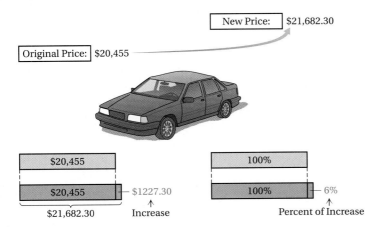

PERCENT OF DECREASE

Lisa buys the car listed above for $20,455. After one year, the car depreciates in value by 25%. This is 0.25 × $20,455, or $5113.75. This lowers the value of the car to

$20,455 − $5113.75,   or $15,341.25.

Note that the new price is thus 75% of the original price. If Lisa decides to sell the car after a year, $15,341.25 might be the most she could expect to get for it. The *percent of decrease* is 25%, and the decrease is $5113.75.

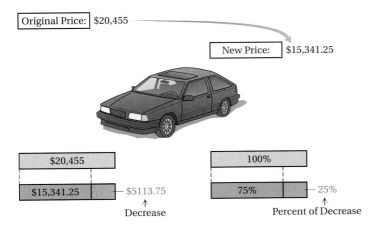

Do Exercises 3 and 4.

3. **Percent of Increase.** The value of a car is $36,875. The price is increased by 4%.

   a) How much is the increase?
   b) What is the new price?

4. **Percent of Decrease.** The value of a car is $36,875. The car depreciates in value by 25% after one year.

   a) How much is the decrease?
   b) What is the depreciated value of the car?

Answers on page A-14

When a quantity is decreased by a certain percent, we say we have **percent of decrease.**

EXAMPLE 3 *DVD Price.* Barnes & Noble recently sold the DVD *When Harry Met Sally* on its Web site. The retail price of $24.98 was decreased to a sale price of $19.98. What was the percent of decrease?
Source: Barnes & Noble

1. **Familiarize.** We find the amount of decrease and then make a drawing.

$$
\begin{array}{rl}
2\ 4.9\ 8 & \text{Retail price} \\
-\ 1\ 9.9\ 8 & \text{Sale price} \\
\hline
5.0\ 0 & \text{Decrease}
\end{array}
$$

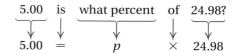

2. **Translate.** There are two ways in which we can translate this problem.

Percent equation:

$$
\underbrace{5.00}_{\displaystyle 5.00} \quad \underbrace{\text{is}}_{\displaystyle =} \quad \underbrace{\text{what percent}}_{\displaystyle p} \quad \underbrace{\text{of}}_{\displaystyle \times} \quad \underbrace{24.98?}_{\displaystyle 24.98}
$$

Proportion:

$$
\frac{P}{100} = \frac{5.00}{24.98}
$$

For proportions, $P\% = p$.

3. **Solve.** We have two ways in which to solve this problem.

Percent equation:

$$
5.00 = p \times 24.98
$$
$$
\frac{5.00}{24.98} = \frac{p \times 24.98}{24.98} \qquad \text{Dividing by 24.98 on both sides}
$$
$$
\frac{5.00}{24.98} = p
$$
$$
0.20 \approx p
$$
$$
20\% \approx p \qquad \text{Converting to percent notation}
$$

Proportion:

$$
\frac{P}{100} = \frac{5.00}{24.98}
$$
$$
24.98 \times P = 100 \times 5 \qquad \text{Equating cross products}
$$
$$
\frac{24.98 \times P}{24.98} = \frac{100 \times 5}{24.98} \qquad \text{Dividing by 24.98 on both sides}
$$
$$
P = \frac{500}{24.98}
$$
$$
P \approx 20
$$

We use the solution of the proportion to express the answer to the problem as 20%.

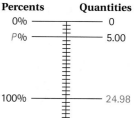

4. Check. To check, we note that, with a 20% decrease, the reduced (or sale) price should be 80% of the retail (or original) price. Since

$$80\% \times 24.98 = 0.80 \times 24.98 = 19.984 \approx 19.98,$$

our answer checks.

5. State. The percent of decrease in the price of the DVD was 20%.

Do Exercise 5 on the preceding page.

When a quantity is increased by a certain percent, we say we have **percent of increase.**

EXAMPLE 4 *Doctor Visits.* The average length of a doctor's visit paid for by the patient or insurance increased from 16.4 min in 1989 to 18.5 min in 1998. What was the percent of increase in the time of an average visit?
Source: *The New England Journal of Medicine*

1. Familiarize. We note that the increase in time was $18.5 - 16.4$, or 2.1 min. A drawing can help us visualize the situation. We let $p =$ the percent of increase.

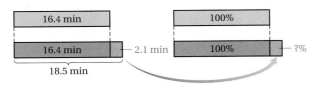

2. Translate. There are two ways in which we can translate this problem.

Percent equation:

$$
\underbrace{2.1 \text{ min}}_{2.1} \quad \underbrace{\text{is}}_{=} \quad \underbrace{\text{what percent}}_{p} \quad \underbrace{\text{of}}_{\cdot} \quad \underbrace{16.4?}_{16.4}
$$

Proportion:

$$\frac{P}{100} = \frac{2.1}{16.4}$$

For proportions, $P\% = p$.

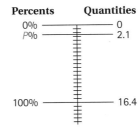

Percents	Quantities
0%	0
P%	2.1
100%	16.4

6. Doctor Visits. The average length of a visit paid for by an HMO increased from 15.4 min in 1989 to 17.9 min in 1998. What was the percent of increase in the time of an average visit?

Source: *The New England Journal of Medicine*

3. Solve. We have two ways in which to solve this problem.

Percent equation:

$$2.1 = p \cdot 16.4$$

$$\frac{2.1}{16.4} = \frac{p \times 16.4}{16.4} \qquad \text{Dividing by 16.4 on both sides}$$

$$\frac{2.1}{16.4} = p$$

$$0.128 \approx p$$

$$12.8\% \approx p \qquad \text{Converting to percent notation}$$

Proportion:

$$\frac{P}{100} = \frac{2.1}{16.4}$$

$$16.4 \times P = 100 \times 2.1 \qquad \text{Equating cross products}$$

$$\frac{16.4 \times P}{16.4} = \frac{100 \times 2.1}{16.4} \qquad \text{Dividing by 16.4 on both sides}$$

$$P = \frac{210}{16.4}$$

$$P \approx 12.8$$

We use the solution of the proportion to express the answer to the problem as 12.8%.

4. Check. To check, we take 12.8% of 16.4:

$$12.8\% \times 16.4 = 0.128 \times 16.4 = 2.0992.$$

Since we rounded the percent, this approximation is close enough to 2.1 to be a good check.

5. State. The percent of increase in the average length of a visit to a doctor was 12.8%

Do Exercise 6.

Answer on page A-14

a Solve.

1. *Panda Survival.* Breeding the much-loved panda bear in captivity has been quite difficult for zookeepers.

a) From 1964 to 1997, of 133 panda cubs born in captivity, only 90 lived to be one month old. What percent lived to be one month old?

b) In 1999, Mark Edwards of the San Diego Zoo developed a nutritional formula on which 18 of 20 newborns lived to be one month old. What percent lived to be one month old?

2. *Batting Averages.* Nomar Garciaparra of the Boston Red Sox won the 2000 American League baseball batting title with 196 hits in 586 at-bats. What percent of his at-bats were hits?
Source: Major League Baseball

3. *Pass Completions.* Trent Dilfer, quarterback of the Baltimore Ravens, completed 59.3% of his passes in the 2000 NFL season. He attempted 226 passes. How many did he complete?
Source: National Football League

4. *Pass Completions.* Kerry Collins, quarterback of the New York Giants, completed 58.8% of his passes in the 2000 NFL season. He attempted 529 passes. How many did he complete?
Source: National Football League

5. *Overweight and Obese.* Of the 281 million people in the United States, 60% are considered overweight and 25% are considered obese. How many are overweight? How many are obese?
Source: U.S. Centers for Disease Control

6. *Smoking and Diabetes.* Of the 281 million people in the United States, 25% are smokers and 6.5% have diabetes. How many are smokers? How many have diabetes?
Source: U.S. Centers for Disease Control

7. A lab technician has 680 mL of a solution of water and acid; 3% is acid. How many milliliters are acid? water?

8. A lab technician has 540 mL of a solution of alcohol and water; 8% is alcohol. How many milliliters are alcohol? water?

9. *Field Goals.* At one point in the 2000–2001 NBA season, Vince Carter of the Toronto Raptors had successfully completed 45.2% of his field goals. He made 288 field goals. How many did he attempt?
Source: National Basketball Association

10. *Field Goals.* At one point in the 2000–2001 NBA season, Glenn Robinson of the Milwaukee Bucks had completed 45.4% of his field goals. He made 269 field goals. How many did he attempt?
Source: National Basketball Association

11. *Test Results.* On a test of 80 items, Antonio got 76 correct. What percent were correct? incorrect?

12. *Test Results.* On a test of 40 items, Cole got 33 correct. What percent were correct? incorrect?

13. *Test Results.* On a test of 40 items, Christina got 91% correct. (There was partial credit on some items.) How many items did she get correct? incorrect?

14. *Test Results.* On a test of 80 items, Pedro got 93% correct. (There was partial credit on some items.) How many items did he get correct? incorrect?

15. *Test Results.* On a test, Maj Ling got 86%, or 81.7, of the items correct. (There was partial credit on some items.) How many items were on the test?

16. *Test Results.* On a test, Juan got 85%, or 119, of the items correct. How many items were on the test?

17. *TV Usage.* Of the 8760 hr in a year, most television sets are on for 2190 hr. What percent is this?

18. *Colds from Kissing.* In a medical study, it was determined that if 800 people kiss someone who has a cold, only 56 will actually catch a cold. What percent is this?
Source: U.S. Centers for Disease Control

19. *Maximum Heart Rate.*   Treadmill tests are often administered to diagnose heart ailments. A guideline in such a test is to try to get you to reach your *maximum heart rate,* in beats per minute. The maximum heart rate is found by subtracting your age from 220 and then multiplying by 85%. What is the maximum heart rate of someone whose age is 25? 36? 48? 55? 76? Round to the nearest one.

20. It costs an oil company $40,000 a day to operate two refineries. Refinery A accounts for 37.5% of the cost, and refinery B for the rest of the cost.

   a) What percent of the cost does it take to run refinery B?

   b) What is the cost of operating refinery A? refinery B?

---

b   Solve.

21. *Savings Increase.*   The amount in a savings account increased from $200 to $216. What was the percent of increase?

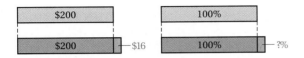

22. *Population Increase.*   The population of a small mountain town increased from 840 to 882. What was the percent of increase?

23. During a sale, a dress decreased in price from $90 to $72. What was the percent of decrease?

24. A person on a diet goes from a weight of 125 lb to a weight of 110 lb. What is the percent of decrease?

25. *Population Increase.*   The population of the state of Colorado increased from 3,294,394 in 1990 to 4,301,261 in 2000. What is the percent of increase?
Source: U.S. Bureau of the Census

26. *Population Increase.*   The population of the state of Utah increased from 1,722,850 in 1990 to 2,233,169 in 2000. What is the percent of increase?
Source: U.S. Bureau of the Census

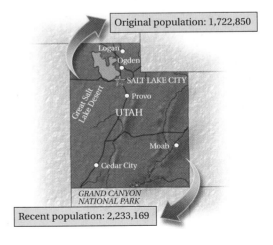

Original population: 1,722,850

Logan
Ogden
SALT LAKE CITY
Provo
Great Salt Lake Desert
UTAH
Moab
Cedar City
GRAND CANYON NATIONAL PARK

Recent population: 2,233,169

27. A person earns $28,600 one year and receives a 5% raise in salary. What is the new salary?

28. A person earns $20,400 one year and receives an 8% raise in salary. What is the new salary?

29. *Car Depreciation.* Irwin buys a car for $21,566. It depreciates 25% each year that he owns it. What is the depreciated value of the car after 1 yr? after 2 yr?

30. *Car Depreciation.* Janice buys a car for $22,688. It depreciates 25% each year that she owns it. What is the depreciated value of the car after 1 yr? after 2 yr?

31. *DVD Price.* The set of DVDs *Ken Burns: Jazz* has a retail price of $199.98. Barnes & Noble offers it on its Web site at a sale price of $149.99. What is the percent of decrease?
Source: Barnes & Noble

32. *Portable DVD Player.* A Sharp Portable DVD video player has a retail price of $1,499.95. Amazon.com offers it on its Web site at a sale price of $899.88. What is the percent of decrease?
Sources: Sharp Electronics Corporation; Amazon.com

Sharp DV–L70U Portable DVD–Video Player
Other products by **Sharp**
List Price: ~~$1,499.95~~
Our Price: $899.88
You Save: $600.07
Availability:
Usually ships within 24 hours

33. *Two-by-Four.* A cross-section of a standard or nominal "two-by-four" board actually measures $1\frac{1}{2}$ in. by $3\frac{1}{2}$ in. The rough board is 2 in. by 4 in. but is planed and dried to the finished size. What percent of the wood is removed in planing and drying?

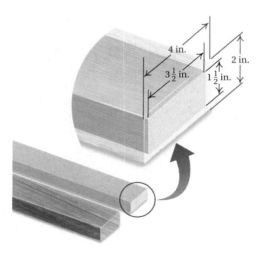

4 in.

$3\frac{1}{2}$ in.

2 in.

$1\frac{1}{2}$ in.

34. *Tipping.* Diners frequently add a 15% tip when charging a meal to a credit card. What is the total amount charged if the cost of the meal, without tip, is $18? $34? $49?

35. *Population Decrease.* Between 1990 and 2000, the population of Washington, D. C., decreased from 606,900 to 572,059.

    a) What is the percent of decrease?
    b) If this percent of decrease repeated itself in the following decade, what would the population be in 2010?
    Source: U.S. Bureau of the Census

36. *World Population.* World population is increasing by 1.6% each year. In 2000, it was 6.26 billion. How much will it be in 2003? 2005? 2008?

Life Insurance Rates for Smokers and Nonsmokers. The following table provides data showing how yearly rates (premiums) for a $500,000 term life insurance policy are increased for smokers. Complete the missing numbers in the table.

TYPICAL INSURANCE PREMIUMS (DOLLARS)

	AGE	RATE FOR NONSMOKER	RATE FOR SMOKER	PERCENT INCREASE FOR SMOKER
	35	$ 345	$ 630	83%
37.	40	$ 430	$ 735	
38.	45	$ 565		84%
39.	50	$ 780		100%
40.	55	$ 985		117%
41.	60	$1645	$2955	
42.	65	$2943	$5445	

Source: Pacific Life PL Protector Term Life Portfolio, OYT Rates

Population Increase. The following table provides data showing how the populations of various states increased from 1990 to 2000. Complete the missing numbers in the table.

	STATE	POPULATION IN 1990	POPULATION IN 2000	CHANGE	PERCENT CHANGE
43.	Vermont	562,758	608,827		
44.	Virginia	6,187,358	7,078,515		
45.	Washington	4,866,692	5,894,121		
46.	West Virginia	1,793,477		14,867	
47.	Wisconsin		5,363,675	471,906	
48.	Wyoming	453,588		40,194	

Source: U.S. Bureau of the Census

49. *Car Depreciation.* A car generally depreciates 25% of its original value in the first year. A car is worth $27,300 after the first year. What was its original cost?

50. *Car Depreciation.* Given normal use, an American-made car will depreciate 25% of its original cost the first year and 14% of its remaining value in the second year. What is the value of a car at the end of the second year if its original cost was $36,400? $28,400? $26,800?

51. *Strike Zone.* In baseball, the *strike zone* is normally a 17-in. by 30-in. rectangle. Some batters give the pitcher an advantage by swinging at pitches thrown out of the strike zone. By what percent is the area of the strike zone increased if a 2-in. border is added to the outside?
Source: Major League Baseball

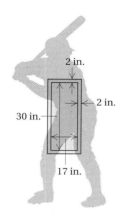

52. Tony is planting grass on a 24-ft by 36-ft area in his back yard. He installs a 6-ft by 8-ft garden. By what percent has he reduced the area he has to mow?

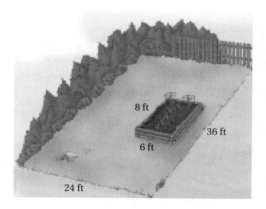

53. ᴰ_W Which is better for a wage earner, and why: a 10% raise followed by a 5% raise a year later, or a 5% raise followed by a 10% raise a year later?

54. ᴰ_W A worker receives raises of 3%, 6%, and then 9%. By what percent has the original salary increased? Explain.

SKILL MAINTENANCE

Convert to decimal notation.   [4.1b], [4.5a]

55. $\dfrac{25}{11}$

56. $\dfrac{11}{25}$

57. $\dfrac{27}{8}$

58. $\dfrac{43}{9}$

59. $\dfrac{23}{25}$

60. $\dfrac{20}{24}$

61. $\dfrac{14}{32}$

62. $\dfrac{2317}{1000}$

63. $\dfrac{34,809}{10,000}$

64. $\dfrac{27}{40}$

SYNTHESIS

65. *Adult Height.* It has been determined that at the age of 10, a girl has reached 84.4% of her final adult growth. Cynthia is 4 ft, 8 in. at the age of 10. What will be her final adult height?
Source: *Dunlop Illustrated Encyclopedia of Facts.* New York: Sterling Publishing, 1970.

66. *Adult Height.* It has been determined that at the age of 15, a boy has reached 96.1% of his final adult height. Claude is 6 ft, 4 in. at the age of 15. What will be his final adult height?
Source: *Dunlop Illustrated Encyclopedia of Facts.* New York: Sterling Publishing, 1970.

67. If *p* is 120% of *q*, then *q* is what percent of *p*?

68. A coupon allows a couple to have dinner and then have $10 subtracted from the bill. Before subtracting $10, however, the restaurant adds a tip of 15%. If the couple is presented with a bill for $44.05, how much would the dinner (without tip) have cost without the coupon?

6.6

SALES TAX, COMMISSION, AND DISCOUNT

Objectives

a Solve applied problems involving sales tax and percent.

b Solve applied problems involving commission and percent.

c Solve applied problems involving discount and percent.

a Sales Tax

Sales tax computations represent a special type of percent of increase problem. The sales tax rate in Maryland is 5%. This means that the tax is 5% of the purchase price. Suppose the purchase price on a coat is $124.95. The sales tax is then 5% of $124.95, or 0.05×124.95, or 6.2475, or about $6.25.

$124.95
+ 5% sales tax

Baltimore

Annapolis

BILL:

Purchase price	=	$124.95
Sales tax (5% of $124.95)	=	+ 6.25
Total price		$131.20

The total that you pay is the price plus the sales tax:

$124.95 + $6.25, or $131.20.

> **SALES TAX**
>
> **Sales tax** = Sales tax rate × Purchase price
> **Total price** = Purchase price + Sales tax

EXAMPLE 1 *Connecticut Sales Tax.* The sales tax rate in Connecticut is 6%. How much tax is charged on the purchase of 3 copies of the DVD *The Matrix* at $19.98 each? What is the total price?

a) We first find the cost of the DVDs. It is

$3 \times \$19.98 = \$59.94.$

b) The sales tax on items costing $59.94 is

$$\underbrace{\text{Sales tax rate}}_{6\%} \times \underbrace{\text{Purchase price}}_{\$59.94}$$

or 0.06×59.94, or 3.5964. Thus the tax is $3.60 (rounded to the nearest cent).

c) The total price is given by the purchase price plus the sales tax:

$59.94 + $3.60, or $63.54.

To check, note that the total price is the purchase price plus 6% of the purchase price. Thus the total price is 106% of the purchase price. Since $1.06 \times 59.94 \approx 63.54$, we have a check. The sales tax is $3.60 and the total price is $63.54.

Do Exercises 1 and 2.

1. California Sales Tax. The sales tax rate in California is 8%. How much tax is charged on the purchase of a refrigerator that sells for $668.95? What is the total price?

2. Illinois Sales Tax. Maggie buys 5 hardcover copies of Dean Koontz's novel *From the Corner of His Eye* for $26.95 each. The sales tax rate in Illinois is 7%. How much sales tax will be charged? What is the total price?

Answers on page A-14

3. The sales tax is $50.94 on the purchase of a night table that costs $849. What is the sales tax?

4. The sales tax on a television is $25.20 and the sales tax rate is 6%. Find the purchase price (the price before taxes are added).

High Definition TV **$?**
$25.20 tax at 6%

EXAMPLE 2   The sales tax is $83.96 on the purchase of this lingerie chest, which costs $2099. What is the sales tax rate?

Lingerie Chest
26" w × 18½" d × 54½" h
$2099
+ $83.96 sales tax
at ?% sales tax rate

Rephrase: Sales tax   is   what percent   of   purchase price?

Translate: $83.96   = r × $2099

To solve the equation, we divide by 2099 on both sides:

$$\frac{83.96}{2099} = \frac{r \times 2099}{2099}$$

$$\frac{83.96}{2099} = r$$

$$0.04 = r$$

$$4\% = r.$$

The sales tax rate is 4%.

Do Exercise 3.

EXAMPLE 3   The sales tax on an inkjet printer is $17.19 and the sales tax rate is 5%. Find the purchase price (the price before taxes are added).

Rephrase: Sales tax   is   5%   of   what?

Translate: 17.19   =   5%   × b,   or $17.19 = 0.05 \times b$.

To solve, we divide by 0.05 on both sides:

$$\frac{17.19}{0.05} = \frac{0.05 \times b}{0.05}$$

$$\frac{17.19}{0.05} = b$$

$$343.80 = b.$$

The purchase price is $343.80.

Do Exercise 4.

New Low Price

$?
$17.19
TAX
@5%

Answers on page A-14

5. Raul's commission rate is 30%. What is the commission from the sale of $18,760 worth of air conditioners?

b Commission

When you work for a **salary,** you receive the same amount of money each week or month. When you work for a **commission,** you are paid a percentage of the total sales for which you are responsible.

> **COMMISSION**
>
> **Commission** = Commission rate × Sales

EXAMPLE 4 *Stereo Equipment Sales.* A salesperson's commission rate is 20%. What is the commission from the sale of $25,560 worth of stereophonic equipment?

$$
\begin{array}{ccccc}
Commission & = & Commission\ rate & \times & Sales \\
C & = & 20\% & \times & 25{,}560 \\
C & = & 0.20 & \times & 25{,}560 \\
C & = & 5112 & &
\end{array}
$$

The commission is $5112.

Do Exercise 5.

EXAMPLE 5 *Farm Machinery Sales.* Dawn earns a commission of $30,000 selling $600,000 worth of farm machinery. What is the commission rate?

$$
\begin{array}{ccccc}
Commission & = & Commission\ rate & \times & Sales \\
30{,}000 & = & r & \times & 600{,}000
\end{array}
$$

Answer on page A-14

6. Liz earns a commission of $3000 selling $24,000 worth of *NSYNC concert tickets. What is the commission rate?

To solve this equation, we divide by 600,000 on both sides:

$$\frac{30,000}{600,000} = \frac{r \times 600,000}{600,000}$$

$$\frac{1}{20} = r$$

$$0.05 = r$$

$$5\% = r.$$

The commission rate is 5%.

Do Exercise 6.

EXAMPLE 6 *Motorcycle Sales.*   Joyce's commission rate is 25%. She receives a commission of $425 on the sale of a motorcycle. How much did the motorcycle cost?

$$
\begin{array}{ccccc}
\textit{Commission} & = & \textit{Commission rate} & \times & \textit{Sales} \\
425 & = & 25\% & \times & S, \quad \text{or } 425 = 0.25 \times S
\end{array}
$$

To solve this equation, we divide by 0.25 on both sides:

$$\frac{425}{0.25} = \frac{0.25 \times S}{0.25}$$

$$\frac{4.25}{0.25} = S$$

$$1700 = S.$$

The motorcycle cost $1700.

Do Exercise 7.

7. Ben's commission rate is 16%. He receives a commission of $268 from sales of clothing. How many dollars worth of clothing were sold?

C Discount

Suppose that the regular price of a rug is $60, and the rug is on sale at 25% off. Since 25% of $60 is $15, the sale price is $60 − $15, or $45. We call $60 the **original,** or **marked price,** 25% the **rate of discount,** $15 the **discount,** and $45 the **sale price.** Note that discount problems are a type of percent of decrease problem.

DISCOUNT AND SALE PRICE
Discount = Rate of discount × Original price
Sale price = Original price − Discount

EXAMPLE 7 A rug marked $240 is on sale at $33\frac{1}{3}\%$ off. What is the discount? the sale price?

a) *Discount* $=$ *Rate of discount* $\times$ *Original price*

$$D = 33\frac{1}{3}\% \times 240$$

$$D = \frac{1}{3} \times 240$$

$$D = \frac{240}{3} = 80$$

b) *Sale price* $=$ *Original price* $-$ *Discount*

$$S = 240 - 80$$

$$S = 160$$

The discount is $80 and the sale price is $160.

Do Exercise 8.

EXAMPLE 8 *Antique Pricing.* An antique table is marked down from $620 to $527. What is the rate of discount?

We first find the discount by subtracting the sale price from the original price:

$$620 - 527 = 93.$$

The discount is $93.

Next, we use the equation for discount:

Discount $=$ *Rate of discount* $\times$ *Original price*

$$93 = r \times 620.$$

To solve, we divide by 620 on both sides:

$$\frac{93}{620} = \frac{r \times 620}{620}$$

$$\frac{93}{620} = r$$

$$0.15 = r$$

$$15\% = r.$$

> To check, note that a 15% discount rate means that 85% of the original price is paid:
> $0.85 \times 620 = 527.$

The discount rate is 15%.

Do Exercise 9.

9. A pair of hiking boots is reduced from $75 to $60. Find the rate of discount.

Answers on page A-14

6.6

EXERCISE SET

For Extra Help

Digital Video
Tutor CD 4
Videotape 11

InterAct
Math

Math Tutor
Center

MathXL

MyMathLab

a Solve.

1. *Indiana Sales Tax.* The sales tax rate in Indiana is 5%. How much sales tax would be charged on an I-JAM MP-3 player that costs $219?

2. *Pennsylvania Sales Tax.* The sales tax rate in Pennsylvania is 6%. How much sales tax would be charged on an I-JAM MP-3 player that costs $219?

3. *Fort Worth, Texas, Sales Tax.* The sales tax rate in Fort Worth, Texas, is 8.25%. How much sales tax would be charged on a copy of J. K. Rowling's novel, *Harry Potter and the Goblet of Fire*, which sells for $25.95?

4. *Fort Worth, Texas, Sales Tax.* The sales tax rate in Fort Worth, Texas, is 8.25%. How much sales tax would be charged on a copy of John Grisham's novel *A Painted House*, which sells for $27.95?
   Source: Borders Bookstore; Andrea Sutcliffe, *Numbers*

5. *Illinois Sales Tax.* The sales tax rate in Illinois is 7%. How much tax is charged on a purchase of 5 telephones at $53 apiece? What is the total price?

6. *Kentucky Sales Tax.* The sales tax rate in Kentucky is 6%. How much tax is charged on a purchase of 5 teapots at $37.99 apiece? What is the total price?

7. The sales tax is $48 on the purchase of a dining room set that sells for $960. What is the sales tax rate?

8. The sales tax is $15 on the purchase of a diamond ring that sells for $500. What is the sales tax rate?

9. The sales tax is $35.80 on the purchase of a refrigerator–freezer that sells for $895. What is the sales tax rate?

10. The sales tax is $9.12 on the purchase of a patio set that sells for $456. What is the sales tax rate?

11. The sales tax on a used car is $100 and the sales tax rate is 5%. Find the purchase price (the price before taxes are added).

12. The sales tax on the purchase of a new boat is $112 and the sales tax rate is 2%. Find the purchase price.

13. The sales tax on a dining room set is $28 and the sales tax rate is 3.5%. Find the purchase price.

14. The sales tax on a portable CD player is $66 and the sales tax rate is 5.5%. Find the purchase price.

15. The sales tax rate in Austin is 2% for the city and county and 6.25% for the state. Find the total amount paid for 2 shower units at $332.50 apiece.

16. The sales tax rate in Omaha is 1.5% for the city and 5% for the state. Find the total amount paid for 3 air conditioners at $260 apiece.

17. The sales tax is $1030.40 on an automobile purchase of $18,400. What is the sales tax rate?

18. The sales tax is $979.60 on an automobile purchase of $15,800. What is the sales tax rate?

b Solve.

19. Katrina's commission rate is 6%. What is the commission from the sale of $45,000 worth of furnaces?

20. Jose's commission rate is 32%. What is the commission from the sale of $12,500 worth of sailboards?

21. Vince earns $120 selling $2400 worth of television sets in a consignment shop. What is the commission rate?

22. Donna earns $408 selling $3400 worth of shoes. What is the commission rate?

23. An art gallery's commission rate is 40%. They receive a commission of $392. How many dollars worth of artwork were sold?

24. A real estate agent's commission rate is 7%. She receives a commission of $5600 on the sale of a home. How much did the home sell for?

25. A real estate commission is 6%. What is the commission on the sale of a $98,000 home?

26. A real estate commission is 8%. What is the commission on the sale of a piece of land for $68,000?

27. Bonnie earns $280.80 selling $2340 worth of tee shirts. What is the commission rate?

28. Chuck earns $1147.50 selling $7650 worth of ski passes. What is the commission rate?

29. Miguel's commission is increased according to how much he sells. He receives a commission of 5% for the first $2000 and 8% on the amount over $2000. What is the total commission on sales of $6000?

30. Lucinda earns a salary of $500 a month, plus a 2% commission on sales. One month, she sold $990 worth of encyclopedias. What were her wages that month?

	MARKED PRICE	RATE OF DISCOUNT	DISCOUNT	SALE PRICE
31.	$300	10%		
32.	$2000	40%		
33.	$17	15%		
34.	$20	25%		
35.		10%	$12.50	
36.		15%	$65.70	
37.	$600		$240	
38.	$12,800		$1920	

39. Find the discount and the rate of discount for the ring in this ad.

40. Find the discount and the rate of discount for the calculator in this ad.

41. Find the marked price and the rate of discount for the camcorder in this ad.

42. Find the marked price and the rate of discount for the cedar chest in this ad.

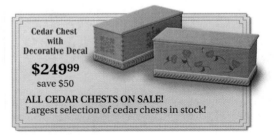

43. ^{D}W Is the following ad mathematically correct? Why or why not?

FAMOUS MAKER WATCHES

$6.95 Regularly $9.95

Choose from men's and ladies' casual or dress designs

Limited time offer

30% OFF

44. ^{D}W An item that is no longer on sale at "25% off" receives a price tag that is $33\frac{1}{3}\%$ more than the sale price. Has the item price been restored to its original price? Why or why not?

45. ^{D}W Which is better, a discount of 40% or a discount of 20% followed by another of 20%? Explain.

46. ^{D}W You take 40% of 50% of a number. What percent of the number could you take to obtain the same result making only one multiplication? Explain your answer.

Solve. [5.3b]

47. $\dfrac{x}{12} = \dfrac{24}{16}$

48. $\dfrac{7}{2} = \dfrac{11}{x}$

Solve. [4.4b]

49. $0.64 \cdot x = 170$

50. $28.5 = 25.6 \times y$

Find decimal notation. [4.5a]

51. $\dfrac{5}{9}$

52. $\dfrac{23}{11}$

53. $\dfrac{11}{12}$

54. $\dfrac{13}{7}$

55. $\dfrac{15}{7}$

56. $\dfrac{19}{12}$

Convert to standard notation. [4.3b]

57. $4.03 trillion

58. 5.8 million

59. 42.7 million

60. 6.09 trillion

61. 🖩 *Magazine Subscriptions.* In a recent subscription drive, *People* offered a subscription of 52 weekly issues for a price of $1.89 per issue. They advertised that this was a savings of 29.7% off the newsstand price. What was the newsstand price?
Source: *People Magazine*

62. 🖩 Gordon receives a 10% commission on the first $5000 in sales and 15% on all sales beyond $5000. If Gordon receives a commission of $2405, how much did he sell? Use a calculator and trial and error if you wish.

63. Herb collects baseball memorabilia. He bought two autographed plaques, but became short of funds and had to sell them quickly for $200 each. On one, he made a 20% profit and on the other, he lost 20%. Did he make or lose money on the sale?

64. Tee shirts are being sold at the mall for $5 each, or 3 for $10. If you buy three tee shirts, what is the rate of discount?

413

SIMPLE AND COMPOUND INTEREST

Objectives

a Solve applied problems involving simple interest.

b Solve applied problems involving compound interest.

1. What is the simple interest on $4300 invested at an interest rate of 14% for 1 year?

2. What is the simple interest on a principal of $4300 invested at an interest rate of 14% for $\frac{3}{4}$ year?

a Simple Interest

Suppose you put $1000 into an investment for 1 year. The $1000 is called the **principal.** If the **interest rate** is 8%, in addition to the principal, you get back 8% of the principal, which is

8% of $1000, or 0.08 × 1000, or $80.00.

The $80.00 is called the **simple interest.** It is, in effect, the price that a financial institution pays for the use of the money over time.

> **SIMPLE INTEREST FORMULA**
>
> The **simple interest** I on principal P, invested for t years at interest rate r, is given by
> $$I = P \cdot r \cdot t.$$

EXAMPLE 1 What is the simple interest on $2500 invested at an interest rate of 6% for 1 year?

We use the formula $I = P \cdot r \cdot t$:

$$I = P \cdot r \cdot t = \$2500 \times 6\% \times 1$$
$$= \$2500 \times 0.06$$
$$= \$150.$$

The simple interest for 1 year is $150.

Do Exercise 1.

EXAMPLE 2 What is the simple interest on a principal of $2500 invested at an interest rate of 6% for $\frac{1}{4}$ year?

We use the formula $I = P \cdot r \cdot t$:

$$I = P \cdot r \cdot t = \$2500 \times 6\% \times \frac{1}{4}$$
$$= \frac{\$2500 \times 0.06}{4}$$
$$= \$37.50.$$

We could instead have found $\frac{1}{4}$ of 6% and then multiplied by 2500.

The simple interest for $\frac{1}{4}$ year is $37.50.

Do Exercise 2.

Answers on page A-15

When time is given in days, we generally divide it by 365 to express the time as a fractional part of a year.

EXAMPLE 3 To pay for a shipment of tee shirts, New Wave Designs borrows $8000 at $9\frac{3}{4}$% for 60 days. Find (a) the amount of simple interest that is due and (b) the total amount that must be paid after 60 days.

a) We express 60 days as a fractional part of a year:

$$I = P \cdot r \cdot t = \$8000 \times 9\frac{3}{4}\% \times \frac{60}{365}$$

$$= \$8000 \times 0.0975 \times \frac{60}{365}$$

$$\approx \$128.22.$$

The interest due for 60 days is $128.22.

b) The total amount to be paid after 60 days is the principal plus the interest:

$$\$8000 + \$128.22 = \$8128.22$$

The total amount due is $8128.22.

Do Exercise 3.

b   Compound Interest

When interest is paid *on interest*, we call it **compound interest.** This is the type of interest usually paid on investments. Suppose you have $5000 in a savings account at 6%. In 1 year, the account will contain the original $5000 plus 6% of $5000. Thus the total in the account after 1 year will be

106% of $5000,   or   1.06 × $5000,   or $5300.

Now suppose that the total of $5300 remains in the account for another year. At the end of this second year, the account will contain the $5300 plus 6% of $5300. The total in the account would thus be

106% of $5300,   or   1.06 × $5300,   or $5618.

Note that in the second year, interest is earned on the first year's interest. When this happens, we say that interest is **compounded annually.**

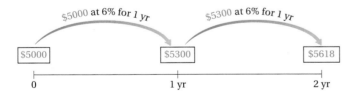

3. The Glass Nook borrows $4800 at $9\frac{1}{2}$% for 30 days. Find (a) the amount of simple interest due and (b) the total amount that must be paid after 30 days.

Answer on page A-15

4. Find the amount in an account if $2000 is invested at 11%, compounded annually, for 2 years.

EXAMPLE 4 Find the amount in an account if $2000 is invested at 8%, compounded annually, for 2 years.

a) After 1 year, the account will contain 108% of $2000:

$$1.08 \times \$2000 = \$2160.$$

b) At the end of the second year, the account will contain 108% of $2160:

$$1.08 \times \$2160 = \$2332.80.$$

The amount in the account after 2 years is $2332.80.

Do Exercise 4.

Suppose that the interest in Example 4 were **compounded semi-annually**—that is, every half year. Interest would then be calculated twice a year at a rate of 8% ÷ 2, or 4% each time. The approach used in Example 4 can then be adapted, as follows.

After the first $\frac{1}{2}$ year, the account will contain 104% of $2000:

$$1.04 \times \$2000 = \$2080.$$

After a second $\frac{1}{2}$ year (1 full year), the account will contain 104% of $2080:

$$1.04 \times \$2080 = \$2163.20.$$

After a third $\frac{1}{2}$ year $\left(1\frac{1}{2} \text{ full years}\right)$, the account will contain 104% of $2163.20:

$$1.04 \times \$2163.20 = \$2249.728$$
$$\approx \$2249.73. \qquad \text{Rounding to the nearest cent}$$

Finally, after a fourth $\frac{1}{2}$ year (2 full years), the account will contain 104% of $2249.73:

$$1.04 \times \$2249.73 = \$2339.7192$$
$$\approx \$2339.72. \qquad \text{Rounding to the nearest cent}$$

Let's summarize our results and look at them another way:

End of 1st $\frac{1}{2}$ year $\rightarrow 1.04 \times 2000 = 2000 \times (1.04)^1$;
End of 2nd $\frac{1}{2}$ year $\rightarrow 1.04 \times (1.04 \times 2000) = 2000 \times (1.04)^2$;
End of 3rd $\frac{1}{2}$ year $\rightarrow 1.04 \times (1.04 \times 1.04 \times 2000) = 2000 \times (1.04)^3$;
End of 4th $\frac{1}{2}$ year $\rightarrow 1.04 \times (1.04 \times 1.04 \times 1.04 \times 2000) = 2000 \times (1.04)^4$.

Note that each multiplication was by 1.04 and that

$$\$2000 \times 1.04^4 \approx \$2339.72. \qquad \text{Using a calculator and rounding to the nearest cent}$$

We have illustrated the following result.

COMPOUND INTEREST FORMULA

If a principal P has been invested at interest rate r, compounded n times a year, in t years it will grow to an amount A given by

$$A = P \cdot \left(1 + \frac{r}{n}\right)^{n \cdot t}.$$

Answer on page A-15

Let's apply this formula to confirm our preceding discussion, where the amount invested is $P = \$2000$, the number of years is $t = 2$, and the number of compounding periods each year is $n = 2$. Substituting into the compound interest formula, we have

$$A = P \cdot \left(1 + \frac{r}{n}\right)^{n \cdot t} = 2000 \cdot \left(1 + \frac{8\%}{2}\right)^{2 \cdot 2}$$

$$= 2000 \cdot \left(1 + \frac{0.08}{2}\right)^4 = 2000(1.04)^4$$

$$= 2000 \times 1.16985856 \approx \$2339.72.$$

If you are using a calculator, you could perform this computation in one step.

EXAMPLE 5 The Ibsens invest \$4000 in an account paying $8\frac{5}{8}\%$, compounded quarterly. Find the amount in the account after $2\frac{1}{2}$ years.

The compounding is quarterly, so n is 4. We substitute \$4000 for P, $8\frac{5}{8}\%$, or 0.08625, for r, 4 for n, and $2\frac{1}{2}$, or $\frac{5}{2}$, for t and compute A:

$$A = P \cdot \left(1 + \frac{r}{n}\right)^{n \cdot t} = \$4000 \cdot \left(1 + \frac{8\frac{5}{8}\%}{4}\right)^{4 \cdot 5/2}$$

$$= \$4000 \cdot \left(1 + \frac{0.08625}{4}\right)^{10}$$

$$= \$4000(1.0215625)^{10}$$

$$\approx \$4951.19.$$

The amount in the account after $2\frac{1}{2}$ years is \$4951.19.

Do Exercise 5.

5. A couple invests \$7000 in an account paying $10\frac{3}{8}\%$, compounded semiannually. Find the amount in the account after $1\frac{1}{2}$ years.

Answer on page A-15

CALCULATOR CORNER

Compound Interest A calculator is useful in computing compound interest. Not only does it do computations quickly but it also eliminates the need to round until the computation is completed. This minimizes "round-off errors" that occur when rounding is done at each stage of the computation. We must keep order of operations in mind when computing compound interest.

To find the amount due on a \$20,000 loan made for 25 days at 11% interest, compounded daily, we would compute $20,000\left(1 + \frac{0.11}{365}\right)^{25}$. To do this on a calculator, we press $\boxed{2}\boxed{0}\boxed{0}\boxed{0}\boxed{0} \boxed{\times} \boxed{(} \boxed{1} \boxed{+} \boxed{.}\boxed{1}\boxed{1} \boxed{\div} \boxed{3}\boxed{6}\boxed{5} \boxed{)} \boxed{y^x}$

(or $\boxed{x^y}$) $\boxed{2}\boxed{5} \boxed{=}$. Without parentheses, we would first find $1 + \frac{0.11}{365}$, raise this result to the 25th power, and then

multiply by 20,000. To do this, we press $\boxed{1} \boxed{+} \boxed{.}\boxed{1}\boxed{1} \boxed{\div} \boxed{3}\boxed{6}\boxed{5} \boxed{=} \boxed{y^x}$ (or $\boxed{x^y}$) $\boxed{2}\boxed{5} \boxed{=} \boxed{\times} \boxed{2}\boxed{0}\boxed{0}\boxed{0}\boxed{0}$.
In either case, the result is 20,151.23, rounded to the nearest cent.

Some calculators have business keys that allow such computations to be done more quickly.

Exercises:

1. Find the amount due on a \$16,000 loan made for 62 days at 13% interest, compounded daily.

2. An investment of \$12,500 is made for 90 days at 8.5% interest, compounded daily. How much is the investment worth after 90 days?

a Find the simple interest.

	PRINCIPAL	RATE OF INTEREST	TIME	SIMPLE INTEREST
1.	$200	8%	1 year	
2.	$450	6%	1 year	
3.	$2,000	8.4%	$\frac{1}{2}$ year	
4.	$200	7.7%	$\frac{1}{2}$ year	
5.	$4,300	10.56%	$\frac{1}{4}$ year	
6.	$8,000	9.42%	$\frac{1}{6}$ year	
7.	$20,000	$7\frac{5}{8}\%$	1 year	
8.	$100,000	$8\frac{7}{8}\%$	1 year	
9.	$50,000	$5\frac{3}{8}\%$	$\frac{1}{4}$ year	
10.	$80,000	$6\frac{3}{4}\%$	$\frac{1}{12}$ year	

Solve. Assume that simple interest is being calculated in each case.

11. CopiPix, Inc. borrows $10,000 at 9% for 60 days. Find (a) the amount of interest due and (b) the total amount that must be paid after 60 days.

12. Sal's Laundry borrows $8000 at 10% for 90 days. Find (a) the amount of interest due and (b) the total amount that must be paid after 90 days.

13. Animal Instinct, a pet supply shop, borrows $6500 at 8% for 90 days. Find (a) the amount of interest due and (b) the total amount that must be paid after 90 days.

14. Andante's Cafe borrows $4500 at 9% for 60 days. Find (a) the amount of interest due and (b) the total amount that must be paid after 60 days.

15. Jean's Garage borrows $5600 at 10% for 30 days. Find (a) the amount of interest due and (b) the total amount that must be paid after 30 days.

16. Shear Delights, a hair salon, borrows $3600 at 8% for 30 days. Find (a) the amount of interest due and (b) the total amount that must be paid after 30 days.

b Interest is compounded annually. Find the amount in the account after the given length of time. Round to the nearest cent.

	PRINCIPAL	RATE OF INTEREST	TIME	AMOUNT IN THE ACCOUNT
17.	$400	10%	2 years	
18.	$450	8%	2 years	
19.	$2,000	8.8%	4 years	
20.	$4,000	7.7%	4 years	
21.	$4,300	10.56%	6 years	
22.	$8,000	9.42%	6 years	
23.	$20,000	$7\frac{5}{8}\%$	25 years	
24.	$100,000	$8\frac{7}{8}\%$	30 years	

Interest is compounded semiannually. Find the amount in the account after the given length of time. Round to the nearest cent.

	PRINCIPAL	RATE OF INTEREST	TIME	AMOUNT IN THE ACCOUNT
25.	$4,000	7%	1 year	
26.	$1,000	5%	1 year	
27.	$20,000	8.8%	4 years	
28.	$40,000	7.7%	4 years	
29.	$5,000	10.56%	6 years	
30.	$8,000	9.42%	8 years	
31.	$20,000	$7\frac{5}{8}\%$	25 years	
32.	$100,000	$8\frac{7}{8}\%$	30 years	

Solve.

33. A family invests $4000 in an account paying 6%, compounded monthly. How much is in the account after 5 months?

34. A couple invests $2500 in an account paying 9%, compounded monthly. How much is in the account after 6 months?

35. A couple invests $1200 in an account paying 10%, compounded quarterly. How much is in the account after 1 year?

36. The O'Hares invest $6000 in an account paying 8%, compounded quarterly. How much is in the account after 18 months?

37. D_W Which is a better investment and why: $1000 invested at $14\frac{3}{4}$% simple interest for 1 year, or $1000 invested at 14% compounded monthly for 1 year?

38. D_W A firm must choose between borrowing $5000 at 10% for 30 days and borrowing $10,000 at 8% for 60 days. Give arguments in favor of and against each option.

Solve. [5.3b]

39. $\dfrac{9}{10} = \dfrac{x}{5}$

40. $\dfrac{7}{x} = \dfrac{4}{5}$

41. $\dfrac{3}{4} = \dfrac{6}{x}$

42. $\dfrac{7}{8} = \dfrac{x}{100}$

Convert to a mixed numeral. [3.4a]

43. $\dfrac{100}{3}$

44. $\dfrac{64}{17}$

45. $\dfrac{38}{3}$

46. $\dfrac{38}{11}$

Convert from a mixed numeral to fraction notation. [3.4a]

47. $1\dfrac{1}{17}$

48. $20\dfrac{9}{10}$

49. $101\dfrac{1}{2}$

50. $32\dfrac{3}{8}$

SYNTHESIS

Effective Yield. The *effective yield* is the yearly rate of simple interest that corresponds to a rate for which interest is compounded two or more times a year. For example, if P is invested at 12%, compounded quarterly, we would multiply P by $(1 + 0.12/4)^4$, or 1.03^4. Since $1.03^4 \approx 1.126$, the 12% compounded quarterly corresponds to an effective yield of approximately 12.6%. In Exercises 51 and 52, find the effective yield for the indicated account.

51. ▦ The account pays 9% compounded monthly.

52. ▦ The account pays 10% compounded daily.

CHAPTER 6: Percent Notation

Copyright © 2003 Pearson Education, Inc.

6.8 INTEREST RATES ON CREDIT CARDS AND LOANS

Objective

a Solve applied problems involving interest rates on credit cards and loans.

a Credit Cards and Loans

Look at the following graphs. They offer good reason for a study of the real-world applications of percent, interest, loans, and credit cards.

The True Cost of "Free" Money

Many 18–35-year-olds are borrowing and mortgaging their futures, according to debt and consumer advocates.

Average credit card debt among undergraduate students

| 1998 | $1879 |
| 2000 | $2748 |

Americans younger than 35 who sought bankruptcy protection from creditors

| 1991 | 380,000 |
| 1999 | 461,000 |

Percentage of undergraduate college students with a credit card

1998 — 67%   2000 — 78%

Home ownership rate for heads of households younger than 35

1982 — 41.2%   1999 — 39.7%

Sources: Nellie Mae, U.S. Census Bureau, and Harvard Law School

Comparing interest rates is essential if one is to become financially responsible. A small change in an interest rate can make a *large* difference in the cost of a loan. When you make a payment on a loan, do you know how much of that payment is interest and how much is applied to reducing the principal?

We begin with an example involving credit cards. A balance carried on a credit card is a type of loan. Last year in the United States, 100,000 young adults declared bankruptcy because of excessive credit card debt. The money you obtain through the use of a credit card is not "free" money. There is a price (interest) to be paid for the privilege.

EXAMPLE 1 *Credit Cards.* After the holidays, Sarah has a balance of $3216.28 on a credit card with an annual percentage rate (APR) of 19.7%. She decides to not make any additional purchases with this card until she has paid off the balance.

a) Many credit cards require a minimum monthly payment of 2% of the balance. What is Sarah's minimum payment on a balance of $3216.28? Round the answer to the nearest dollar.

b) Find the amount of interest and the amount applied to reduce the principal in the minimum payment found in part (a).

c) If Sarah had transferred her balance to a card with an APR of 12.5%, how much of her first payment would be interest and how much would be applied to reduce the principal?

d) Compare the amounts for 12.5% from part (c) with the amounts for 19.7% from part (b).

1. Credit Cards. After the holidays, Jamal has a balance of $4867.59 on a credit card with an annual percentage rate (APR) of 21.3%. He decides to not make any additional purchases with this card until he has paid off the balance.

a) Many credit cards require a minimum monthly payment of 2% of the balance. What is Jamal's minimum payment on a balance of $4867.59? Round the answer to the nearest dollar.

b) Find the amount of interest and the amount applied to reduce the principal in the minimum payment found in part (a).

c) If Jamal had transferred his balance to a card with an APR of 13.6%, how much of his first payment would be interest and how much would be applied to reduce the principal?

We solve as follows.

a) We multiply the balance of $3216.28 by 2%:

$$0.02 \times \$3216.28 = \$64.3256.$$ Sarah's minimum payment, rounded to the nearest dollar, is $64.

b) The amount of interest on $3216.28 at 19.7% for one month* is given by

$$I = P \cdot r \cdot t = \$3216.28 \times 0.197 \times \frac{1}{12} \approx \$52.80.$$

We subtract to find the amount applied to reduce the principal in the first payment:

$$\text{Amount applied to reduce the principal} = \text{Minimum payment} - \text{Interest for the month}$$
$$= \$64 - \$52.80$$
$$= \$11.20.$$

Thus the principal of $3216.28 is decreased by only $11.20 with the first payment. (Sarah still owes $3205.08.)

c) The amount of interest on $3216.28 at 12.5% for one month is

$$I = P \cdot r \cdot t = \$3216.28 \times 0.125 \times \frac{1}{12} \approx \$33.50.$$

We subtract to find the amount applied to reduce the principal in the first payment:

$$\text{Amount applied to reduce the principal} = \text{Minimum payment} - \text{Interest for the month}$$
$$= \$64 - \$33.50$$
$$= \$30.50.$$

Thus the principal of $3216.28 is decreased by $30.50 with the first payment. (Sarah still owes $3185.78.)

d) Let's organize the information for both rates in the following table.

BALANCE BEFORE FIRST PAYMENT	FIRST MONTH'S PAYMENT	% APR	AMOUNT OF INTEREST	AMOUNT APPLIED TO PRINCIPAL	BALANCE AFTER FIRST PAYMENT
$3216.28	$64	19.7%	$52.80	$11.20	$3205.08
3216.28	64	12.5	33.50	30.50	3185.78

Difference in balance after first payment → $19.30

d) Compare the amounts for 13.6% from part (c) with the amounts for 21.3% from part (b).

At 19.7%, the interest is $52.80 and the principal is decreased by $11.20. At 12.5%, the interest is $33.50 and the principal is decreased by $30.50. Thus the principal is decreased by $30.50 − $11.20, or $19.30 more with the 12.5% rate than with the 19.7% rate. Thus the interest at 19.7% is $19.30 greater than the interest at 12.5%.

---

*Actually, the interest on a credit card is computed daily with a rate called a daily percentage rate (DPR). The DPR for Example 1 would be 19.7%/365 ≈ 0.054%. When no payments or additional purchases are made during the month, the difference in total interest for the month is minimal and we will not deal with it here.

Answers on page A-15

Do Exercise 1 on the preceding page.

Even though the mathematics of the information in the chart below is beyond the scope of this text, it is interesting to compare how long it takes to pay off the balance of Example 1 if Sarah continues to pay $64 for each payment with how long it takes if she pays double that amount, $128, for each payment. Financial consultants frequently tell clients that if they want to take control of their debt, they should pay double the minimum payment.

RATE	PAYMENT PER MONTH	NUMBER OF PAYMENTS TO PAY OFF DEBT	TOTAL PAID BACK	ADDITIONAL COST OF PURCHASES
19.7%	$64	107, or 8 yr 11 mo	$6848	$3631.72
19.7	128	33, or 2 yr 9 mo	4224	1007.72
12.5	64	72, or 6 yr	4608	1391.72
12.5	128	29, or 2 yr 5 mo	3712	495.72

As with most loans, if you pay an extra amount toward the principal with each payment, the length of the loan can be greatly reduced. Note that at the rate of 19.7%, it will take Sarah almost 9 yr to pay off her debt if she pays only $64 per month and does not make additional purchases. If she transfers her balance to a card with a 12.5% rate and pays $128 per month, she could eliminate her debt in approximately $2\frac{1}{2}$ yr. You can see how debt can get out of control if you continue to make purchases and pay only the minimum payment. The debt will never be eliminated.

The Federal Stafford Loan program provides educational loans to students at interest rates (usually from 6.5% to 8.5%) that are much lower than those on credit cards. Payments on a loan do not begin until 6 months after graduation. At that time, the student has 10 years, or 120 monthly payments, to pay off the loan.

EXAMPLE 2 *Federal Stafford Loans.* After graduation, the balance on Taylor's Stafford loan is $28,650. If the rate on his loan is 7%, he will make 120 payments of approximately $333 each to pay off the loan.

a) Find the amount of interest and the amount of principal in the first payment.

b) If the interest rate were 8.25%, he would make 120 monthly payments of approximately $351 each. How much more of the first payment is interest if the loan is 8.25% rather than 7%?

c) Compare the total amount of interest on the loan at 7% with the amount on the loan at 8.25%. How much more would Taylor pay in interest on the 8.25% loan than on the 7% loan?

We solve as follows.

a) We use the formula $I = P \cdot r \cdot t$, substituting $28,650 for P, 0.07 for r, and $1/12$ for t:

$$I = \$28{,}650 \times 0.07 \times \frac{1}{12}$$

$$\approx \$167.13.$$

The amount of interest in the first payment is $167.13. The payment is $333. We subtract to determine the amount applied to the principal:

$$\$333 - \$167.13 = \$165.87.$$

With the first payment, the principal will be reduced by $165.87.

2. Federal Stafford Loans. After graduation, the balance on Maggie's Stafford loan is $32,680. To pay off the loan at 7.25%, she will make 120 payments of approximately $384 each.

a) Find the amount of interest and the amount of principal in the first payment.

b) If the interest rate were 8.5%, she would make 120 payments of approximately $405 each. How much more of the first payment is interest if the loan is 8.5% rather than 7.25%?

c) Compare the total amount of interest on the loan at 7.25% with the amount of interest on the loan at 8.5%. How much more would Maggie pay in interest on the 8.5% loan than on the 7.25% loan?

Answers on page A-15

b) The interest at 8.25% would be

$$I = \$28{,}650 \times 0.0825 \times \frac{1}{12}$$

$$\approx \$196.97.$$

At the rate of 8.25%, the additional interest in the first payment is

$$\$196.97 - \$167.13 = \$29.84.$$

The higher interest rate results in an additional $29.84 in interest in the first payment.

c) For the 7% loan, there will be 120 payments of $333 each:

$$120 \times \$333 = \$39{,}960.$$

The total amount of interest at this rate is

$$\$39{,}960 - \$28{,}650 = \$11{,}310.$$

For the 8.25% loan, there will be 120 payments of $351 each:

$$120 \times \$351 = \$42{,}120.$$

The total amount of interest at this rate is

$$\$42{,}120 - \$28{,}650 = \$13{,}470.$$

At the rate of 8.25%, Taylor would pay

$$\$13{,}470 - \$11{,}310 = \$2160$$

more in interest than at the rate of 7%.

Do Exercise 2.

EXAMPLE 3 *Home Loans.* The Sawyers recently purchased their first home. They borrowed $123,000 at $8\frac{7}{8}$% for 30 years (360 payments). Their monthly payment (excluding insurance and taxes) is $978.64.

a) How much of the first payment is interest and how much is applied to reduce the principal?

b) If the Sawyers pay the entire 360 payments, how much interest will be paid on the loan?

We solve as follows.

a) To find the amount of interest paid in the first payment, we use the formula $I = P \cdot r \cdot t$:

$$I = P \cdot r \cdot t = \$123{,}000 \times 0.08875 \times \frac{1}{12} \approx \$909.69.$$

The amount applied to the principal is

$$\$978.64 - \$909.69, \text{ or } \$68.95.$$

b) Over the 30-year period, the total paid will be

$$360 \times \$978.64, \text{ or } \$352{,}310.40.$$

The total amount of interest paid over the lifetime of the loan is

$$\$352{,}310.40 - \$123{,}000, \text{ or } \$229{,}310.40.$$

Do Exercises 3 and 4 on the following page.

AMORTIZATION TABLES

If we make 360 calculations as in Example 3(a) and continue with a decreased principal as in Margin Exercise 4, we can create an *amortization table*, part of which is shown below. Such tables are also found in reference books. The beginning, middle, and last part of the loan described are shown. Look over the table and note how small a portion of the payment reduces the principal at the beginning of the loan and how that portion increases throughout the lifetime of the loan. Do you see again why a loan is not "free"?

MORTGAGE AMORTIZATION PROGRAM

MORTGAGE AMOUNT: $123,000
INTEREST RATE: 8.875%
NUMBER OF YEARS: 30
MONTHLY PAYMENTS ARE: $978.64

PAYMENT	PRINCIPAL	INTEREST	BALANCE	
1	$ 68.95	$909.69	$122,931.05	← Example 3
2	69.46	909.18	122,861.59	← Margin
3	69.98	908.66	122,791.61	Exercise 4
4	70.49	908.15	122,721.12	
5	71.02	907.62	122,650.10	
6	71.54	907.10	122,578.56	
7	72.07	906.57	122,506.49	
8	72.60	906.04	122,433.89	
9	73.14	905.50	122,360.75	
10	73.68	904.96	122,287.07	
11	74.23	904.41	122,212.84	
12	74.77	903.87	122,138.07	
⋮	⋮	⋮	Interest for 12 periods = $10,881.75	
175	248.53	730.11	98,470.83	
176	250.37	728.27	98,220.46	
177	252.22	726.42	97,968.24	
178	254.08	724.56	97,714.16	
179	255.96	722.68	97,458.20	
180	257.86	720.78	97,200.34	
181	259.76	718.88	96,940.58	
182	261.68	716.96	96,678.90	
183	263.62	715.02	96,415.28	
184	265.57	713.07	96,149.71	
185	267.53	711.11	95,882.18	
186	269.51	709.13	95,612.67	
⋮	⋮	⋮	Interest for 12 periods = $8636.99	
349	895.78	82.86	10,307.76	
350	902.41	76.23	9,405.35	
351	909.08	69.56	8,496.27	
352	915.80	62.84	7,580.47	
353	922.58	56.06	6,657.89	
354	929.40	49.24	5,728.49	
355	936.27	42.37	4,792.22	
356	943.20	35.44	3,849.02	
357	950.17	28.47	2,898.85	
358	957.20	21.44	1,941.65	
359	964.28	14.36	977.37	
360	971.41	7.23	5.96	

Interest for 12 periods = $546.10

Total Interest for 360 Periods = $229,316.36

Refer to Example 3 for Margin Exercises 3 and 4.

3. **Home Loans.** Since the principal has been reduced by the first payment, at the time of the second payment of the Sawyers' 30-year loan, the new principal is the decreased principal

$123,000 − $68.95,

or

$122,931.05.

Use $122,931.05 as the principal, and determine how much of the second payment is interest and how much is applied to reduce the principal. (In effect, repeat Example 3(a) using the new principal.)

4. **Home Loans.** The Sawyers decide to change the period of their home loan from 30 years to 15 years. Their monthly payment increases to $1238.42.

a) How much of the first payment is interest and how much is applied to reduce the principal?

b) If the Sawyers pay the entire 180 payments, how much interest will be paid on this loan?

c) Compare the amount of interest to pay off the 15-yr loan with the amount of interest to pay off the 30-yr loan.

Answers on page A-15

5. Refinancing a Home Loan.
Consider Example 4 for a 15-yr loan. The new monthly payment is $1071.46.

5. Refinancing a Home Loan.
Consider Example 4 for a 15-yr loan. The new monthly payment is $1071.46.

a) How much of the first payment is interest and how much is applied to reduce the principal?

b) If the Sawyers pay the entire 180 payments, how much interest will be paid on this loan?

c) Compare the amount of interest to pay off the 15-yr loan at $6\frac{1}{2}\%$ with the amount of interest to pay off the 15-yr loan at $8\frac{7}{8}\%$ in Margin Exercise 4.

Answer on page A-15

EXAMPLE 4 *Refinancing a Home Loan.* Refer to Example 3. Ten months after the Sawyers buy their home financed at a rate of $8\frac{7}{8}\%$, the rates drop to $6\frac{1}{2}\%$. After much consideration, they decide to refinance even though the new loan will cost them $1200 in refinance charges. They have reduced the principal a small amount in the 10 payments they have made, but they decide to again borrow $123,000 for 30 years at the new rate. Their new monthly payment is $777.44.

a) How much of the first payment is interest and how much is applied to the principal?

b) Compare the amounts at $6\frac{1}{2}\%$ found in part (a) with the amounts at $8\frac{7}{8}\%$ found in Example 3(a).

c) With the lower house payment, how long will it take the Sawyers to recoup the refinance charge of $1200?

d) If the Sawyers pay the entire 360 payments, how much interest will be paid on this loan? How much less is the total interest at $6\frac{1}{2}\%$ than at $8\frac{7}{8}\%$?

We solve as follows.

a) To find the interest paid in the first payment, we use the formula $I = P \cdot r \cdot t$:

$$I = P \cdot r \cdot t = \$123,000 \times 0.065 \times \frac{1}{12} = \$662.25.$$

The amount applied to the principal is

$$\$777.44 - \$666.25, \text{ or } \$111.19.$$

b) We compare the amount found in part (a) with the amount found in Example 3(a):

Rate	Monthly payment	Interest in first payment	Amount applied to principal
$8\frac{7}{8}\%$	$978.64	$909.69	$68.95
$6\frac{1}{2}\%$	$777.44	$666.25	$111.19

At $6\frac{1}{2}\%$, the amount of interest in the payment is $909.69 − $666.25, or $243.44, less than at $8\frac{7}{8}\%$. The amount applied to the principal is $111.19 − $68.95, or $42.24, more.

c) The monthly payment at $6\frac{1}{2}\%$ is $978.64 − $777.44, or $201.20 less than the payment at $8\frac{7}{8}\%$. The total savings each month is approximately $200. We can divide the cost of the refinancing by this monthly savings to determine the number of months it will take to recoup the $1200 refinancing charge: $1200 ÷ $200 = 6. It will take the Sawyers approximately 6 months to break even.

d) Over the 30-year period, the total paid will be

$$360 \times \$777.44, \text{ or } \$279,878.40.$$

The total amount of interest paid over the lifetime of the loan is

$$\$279,878.40 - \$123,000, \text{ or } \$156,878.40.$$

The total interest paid at $6\frac{1}{2}\%$ is

$$\$229,310.40 \text{ (see Example 3)} - \$156,878.40, \text{ or } \$72,432$$

less than the total interest paid at $8\frac{7}{8}\%$. Thus the $6\frac{1}{2}\%$ loan saves the Sawyers approximately $70,000 in interest charges over the 30 years.

Do Exercise 5.

6.8

EXERCISE SET

For Extra Help

Digital Video
Tutor CD 4
Videotape 11

InterAct
Math

Math Tutor
Center

MathXL

MyMathLab

a   Solve.

1. *Credit Cards.*   At the end of his freshman year of college, Antonio has a balance of $4876.54 on a credit card with an annual percentage rate (APR) of 21.3%. He decides to not make any additional purchases with his card until he has paid off the balance.

a) Many credit cards require a minimum monthly payment of 2% of the balance. What is Antonio's minimum payment on a balance of $4876.54? Round the answer to the nearest dollar.

b) Find the amount of interest and the amount applied to reduce the principal in the minimum payment found in part (a).

c) If Antonio had transferred his balance to a card with an APR of 12.6%, how much of his first payment would be interest and how much would be applied to reduce the principal?

d) Compare the amounts for 12.6% from part (c) with the amounts for 21.3% from part (b).

2. *Credit Cards.*   At the end of her junior year of college, Becky had a balance of $5328.88 on a credit card with an annual percentage rate (APR) of 18.7%. She decides to not make any additional purchases with this card until she has paid off the balance.

a) Many credit cards require a minimum monthly payment of 2% of the balance. What is Becky's minimum payment on a balance of $5328.88? Round the answer to the nearest dollar.

b) Find the amount of interest and the amount applied to reduce the principal in the minimum payment found in part (a).

c) If Becky had transferred her balance to a card with an APR of 13.2%, how much of her first payment would be interest and how much would be applied to reduce the principal?

d) Compare the amounts for 13.2% from part (c) with the amounts for 18.7% from part (b).

3. *Federal Stafford Loans.*   After graduation, the balance on Grace's Stafford loan is $44,560. To pay off the loan at 6.5%, she will make 120 payments of approximately $505.97 each.

a) Find the amount of interest and the amount applied to reduce the principal in the first payment.

b) If the interest rate were 8.5%, she would make 120 monthly payments of approximately $552.48 each. How much more of the first payment is interest if the loan is 8.5% rather than 6.5%?

c) Compare the total amount of interest on a loan at 6.5% with the amount on the loan at 8.5%. How much more would Grace pay on the 8.5% loan than on the 6.5% loan?

4. *Federal Stafford Loans.*   After graduation, the balance on Ricky's Stafford loan is $38,970. To pay off the loan at 8.2%, he will make 120 payments of approximately $476.94 each.

a) Find the amount of interest and the amount applied to reduce the principal in the first payment.

b) If the interest rate were 7.4%, he would make 120 monthly payments of approximately $460.55 each. How much less of the first payment is interest if the loan is 7.4% rather than 8.2%?

c) Compare the total amount of interest on the loan at 8.2% with the amount on the loan at 7.4%. How much more would Ricky pay on the 8.2% loan than on the 7.4% loan?

5. *Home Loan.*   The Martinez family recently purchased a home. They borrowed $150,000 at 6.98% for 30 years (360 payments). Their monthly payment (excluding insurance and taxes) is $995.94.

a) How much of the first payment is interest and how much is applied to reduce the principal?

b) If this family pays the entire 360 payments, how much interest will be paid on the loan?

c) Determine the new principal after the first payment. Use that new principal to determine how much of the second payment is interest and how much is applied to reduce the principal.

6. *Home Loan.*   The Kaufmans recently purchased a home. They borrowed $180,000 at 8.36% for 30 years (360 payments). Their monthly payment (excluding insurance and taxes) is $1366.22.

a) How much of the first payment is interest and how much is applied to reduce the principal?

b) If the Kaufmans pay the entire 360 payments, how much interest will be paid on the loan?

c) Determine the new principal after the first payment. Use that new principal to determine how much of the second payment is interest and how much is applied to reduce the principal.

7. *Refinancing a Home Loan.* Refer to Exercise 5. The Martinez decide to change the period of their home loan to 15 years. Their monthly payment increased to $1346.57.

a) How much of the first payment is interest and how much is applied to reduce the principal?

b) If the Martinez pay the entire 180 payments, how much interest will be paid on the loan?

c) Compare the amount of interest to pay off the 15-yr loan with the amount of interest to pay off the 30-yr loan.

8. *Refinancing a Home Loan.* Refer to Exercise 6. The Kaufmans decide to change the period of their home loan to 15 years. Their monthly payment increased to $1757.79.

a) How much of the first payment is interest and how much is applied to reduce the principal?

b) If the Kaufmans pay the entire 180 payments, how much interest will be paid on the loan?

c) Compare the amount of interest to pay off the 15-yr loan with the amount of interest to pay off the 30-yr loan.

Complete the following table, assuming monthly payments as given.

	INTEREST RATE	HOME MORTGAGE	TIME OF LOAN	MONTHLY PAYMENT	PRINCIPAL AFTER FIRST PAYMENT	PRINCIPAL AFTER SECOND PAYMENT
9.	6.98%	$100,000	360 mos	$663.96		
10.	6.98%	$100,000	180 mos	$897.71		
11.	8.04%	$100,000	180 mos	$957.96		
12.	8.04%	$100,000	360 mos	$736.55		
13.	7.24%	$150,000	360 mos	$1022.25		
14.	7.24%	$75,000	180 mos	$684.22		
15.	7.24%	$200,000	180 mos	$1824.60		
16.	7.24%	$180,000	360 mos	$1226.70		

17. *New-Car Loan.* After working at her first job for 2 years, Janice buys a new Saturn for $16,385. She makes a down payment of $1385 and finances $15,000 for 4 years at a new-car loan rate of 8.99%. Her monthly payment is $373.20.

a) How much of her first payment is interest and how much is applied to reduce the principal?

b) Find the principal balance at the beginning of the second month and determine how much less interest she will pay in the second payment than in the first.

c) What is the total interest cost of the loan if she pays all of the 48 payments?

18. *Manufacturer's Car Loan Offer.* For a trip to Colorado, Michael and Rebecca buy a new Jeep Cherokee Classic whose selling price is $22,085. For financing, they accept the promotion from the manufacturer that offers a 36-month loan at 1.9% with 20% down. Their monthly payment is $505.29.

a) What is the down payment? the amount borrowed?

b) How much of the first payment is interest and how much is applied to reduce the principal?

c) What is the total interest cost of the loan if they pay all of the 36 payments?

19. *Used-Car Loan.* Twin brothers, Jerry and Terry, each take a job at the college cafeteria in order to have the money to make payments on the purchase of a used 1997 Ford Taurus for $7900. They make a down payment of 10% and finance the remainder at 12.49% for 3 years. (Used-car loan rates are generally higher than new-car loan rates.) Their monthly payment is $237.82.

a) What is the down payment? the amount borrowed?
b) How much of the first payment is interest and how much is applied to reduce the principal?
c) If they pay all 36 payments, how much interest will they pay for the loan?

20. *Used-Car Loan.* For his construction job, Clint buys a 1994 Chevrolet S-10 truck for $5350. He makes a down payment of $550 and finances the remainder for 2 years at 11.3%. The monthly payment is $224.39.

a) How much is financed?
b) How much of the first payment is interest and how much is applied to reduce the principal?
c) If he pays all 24 payments, how much interest will he pay for the loan?

21. ^{D}w Based on the skills of mathematics you have obtained in this section, discuss the significant new ideas you now have about interest rates and credit cards that you didn't have before.

22. ^{D}w Examine the information in the graphs at the beginning of the section. Discuss how a knowledge of this section might have been of help to some of these students.

23. ^{D}w Compare the following two purchases and describe a situation in which each purchase is the best choice.

Purchase A: A new car for $15,145. The loan is for $14,500 at 6.9% for 4 years. The monthly payment is $346.55.

Purchase B: A used car for $10,600. The loan is for $9300 at 12.5% for 3 years. The monthly payment is $311.12.

24. ^{D}w Look over the examples and exercises in this section. What seems to happen to the monthly payment on a loan if the time of payment changes from 30 years to 15 years, assuming the interest rate stays the same? Discuss the pros and cons of both time periods.

⸺(**SKILL MAINTENANCE**)⸺

Solve. Round the answer to the nearest hundredth where appropriate. [5.3b]

25. $\dfrac{5}{8} = \dfrac{x}{28}$

26. $\dfrac{5}{8} = \dfrac{17.5}{y}$

27. $\dfrac{13}{16} = \dfrac{81.25}{N}$

28. $\dfrac{9}{16} = \dfrac{p}{100}$

29. $\dfrac{1284}{t} = \dfrac{3456}{5000}$

30. $\dfrac{12.8}{32.5} = \dfrac{x}{2000}$

31. $\dfrac{56.3}{78.4} = \dfrac{t}{100}$

32. $\dfrac{28}{x} = \dfrac{8}{5}$

33. $\dfrac{16}{9} = \dfrac{100}{p}$

34. $\dfrac{t}{1284} = \dfrac{5000}{3456}$

The review that follows is meant to prepare you for a chapter exam. It consists of two parts. The first part is a checklist of some of the Study Tips referred to in this and preceding chapters, as well as a list of important properties and formulas. The second part is the Review Exercises. These provide practice exercises for the exam, together with references to section objectives so you can go back and review. Before beginning, stop and look back over the skills you have obtained. What skills in mathematics do you have now that you did not have before studying this chapter?

STUDY TIPS CHECKLIST

The foundation of all your study skills is TIME!	☐ Have you tried *memorizing* the formulas and the three types of percent problems?
	☐ Have you tried being a tutor to a fellow student?
	☐ Have you found any applications of percent in a newspaper or magazine?
	☐ Are you stopping to work the margin exercises when directed to do so?
	☐ Are you doing your homework as soon as possible after class?

IMPORTANT PROPERTIES AND FORMULAS

Commission = Commission rate × Sales
Discount = Rate of discount × Original price
Sale price = Original price − Discount

Simple Interest: $I = P \cdot r \cdot t$

Compound Interest: $A = P \cdot \left(1 + \dfrac{r}{n}\right)^{n \cdot t}$

REVIEW EXERCISES

Find percent notation for the decimal notation in the sentence in Exercises 1 and 2.   [6.1b]

1. Of all the vehicles in Mexico City, 0.017 of them are taxis.
Source: *The Handy Geography Answer Book*

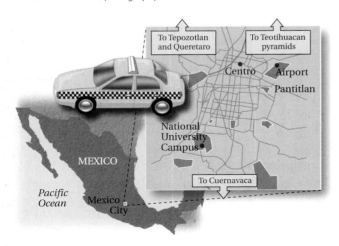

2. Of all the snacks eaten on Super Bowl Sunday, 0.56 of them are chips and salsa.
Source: Korbel Research and Pace Foods

Find percent notation.   [6.2a]

3. $\dfrac{3}{8}$

4. $\dfrac{1}{3}$

Find decimal notation.   [6.1b]

5. 73.5%

6. $6\dfrac{1}{2}\%$

Find fraction notation.   [6.2b]

7. 24%

8. 6.3%

Translate to a percent equation. Then solve.
[6.3a, b]

9. 30.6 is what percent of 90?

10. 63 is 84 percent of what?

11. What is $38\frac{1}{2}\%$ of 168?

Translate to a proportion. Then solve. [6.4a, b]

12. 24 percent of what is 16.8?

13. 42 is what percent of 30?

14. What is 10.5% of 84?

Solve. [6.5a, b]

15. *Favorite Ice Creams.* According to a recent survey, 8.9% of those interviewed chose chocolate as their favorite ice cream flavor and 4.2% chose butter pecan. Of the 2500 students in a freshman class, how many would choose chocolate as their favorite ice cream? butter pecan?
Source: International Ice Cream Association

16. *Prescriptions.* Of the 281 million people in the United States, 123.64 million take at least one kind of prescription drug per day. What percent take at least one kind of prescription drug per day?
Source: American Society of Health-System Pharmacies

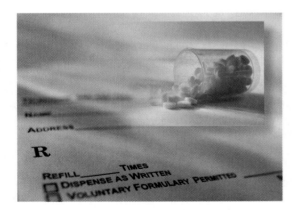

17. *Water Output.* The average person expels 200 mL of water per day by sweating. This is 8% of the total output of water from the body. How much is the total output of water?
Source: Elaine N. Marieb, *Essentials of Human Anatomy and Physiology*, 6th ed. Boston: Addison Wesley Longman, Inc., 2000

18. *Test Scores.* Jason got a 75 on a math test. He was allowed to go to the math lab and take a retest. He increased his score to 84. What was the percent of increase?

19. *Test Scores.* Jenny got an 81 on a math test. By taking a retest in the math lab, she increased her score by 15%. What was her new score?

Solve. [6.6a, b, c]

20. A state charges a meals tax of $4\frac{1}{2}\%$. What is the meals tax charged on a dinner party costing $320?

21. In a certain state, a sales tax of $378 is collected on the purchase of a used car for $7560. What is the sales tax rate?

22. Kim earns $753.50 selling $6850 worth of televisions. What is the commission rate?

23. An air conditioner has a marked price of $350. It is placed on sale at 12% off. What are the discount and the sale price?

24. A fax machine priced at $305 is discounted at the rate of 14%. What are the discount and the sale price?

25. An insurance salesperson receives a 7% commission. If $42,000 worth of life insurance is sold, what is the commission?

26. Find the rate of discount.

Solve. [6.7a, b]

27. What is the simple interest on $1800 at 6% for $\frac{1}{3}$ year?

28. The Dress Shack borrows $24,000 at 10% simple interest for 60 days. Find (a) the amount of interest due and (b) the total amount that must be paid after 60 days.

29. What is the simple interest on $2200 principal at the interest rate of 5.5% for 1 year?

30. The Kleins invest $7500 in an investment account paying an annual interest rate of 12%, compounded monthly. How much is in the account after 3 months?

31. Find the amount in an investment account if $8000 is invested at 9%, compounded annually, for 2 years.

Solve. [6.8a]

32. *Credit Cards.* At the end of her junior year of college, Judy has a balance of $6428.74 on a credit card with an annual percentage rate (APR) of 18.7%. She decides to not make any additional purchases with this card until she has paid off the balance.

a) Many credit cards require a minimum payment of 2% of the balance. What is Judy's minimum payment on a balance of $6428.74? Round the answer to the nearest dollar.
b) Find the amount of interest and the amount applied to reduce the principal in the minimum payment found in part (a).
c) If Judy had transferred her balance to a card with an APR of 13.2%, how much of her first payment would be interest and how much would be applied to reduce the principal?
d) Compare the amounts for 13.2% from part (c) with the amounts for 18.7% from part (b).

Certain objectives from four particular sections will be retested on the chapter test. The objectives are listed with the practice problems that follow.

Solve. [5.3b]

33. $\frac{3}{8} = \frac{7}{x}$

34. $\frac{1}{6} = \frac{7}{x}$

Solve. [4.4b]

35. $10.4 \times y = 665.6$

36. $100 \cdot x = 761.23$

Convert to decimal notation. [4.5a]

37. $\frac{11}{3}$

38. $\frac{11}{7}$

Convert to a mixed numeral. [3.4a]

39. $\frac{11}{3}$

40. $\frac{121}{7}$

41. $^{D}\mathbf{W}$ Ollie buys a microwave oven during a 10%-off sale. The sale price that Ollie paid was $162. To find the original price, Ollie calculates 10% of $162 and adds that to $162. Is this correct? Why or why not? [6.6c]

42. $^{D}\mathbf{W}$ Which is the better deal for a consumer and why: a discount of 40% or a discount of 20% followed by another of 22%? [6.6c]

43. ▦ *Land Area of the United States.* After Hawaii and Alaska became states, the total land area of the United States increased from 2,963,681 mi^2 to 3,540,939 mi^2. What was the percent of increase? [5.5b]

44. Rhonda's Dress Shop reduces the price of a dress by 40% during a sale. By what percent must the store increase the sale price, after the sale, to get back to the original price? [6.6c]

45. A $200 coat is marked up 20%. After 30 days, it is marked down 30% and sold. What was the final selling price of the coat? [6.6c]

1. During a recent month, 0.905 of all the flights of Aloha Airlines arrived on time. This was the highest percentage in the airline industry. Find percent notation for 0.905.
   Source: U.S. Department of Transportation

2. Stephen King's novel *Dream Catcher* was sold on the Barnes & Noble Web site at a 20% discount. Find decimal notation for 20%.
   Source: Barnes & Noble

3. Find percent notation for $\frac{11}{8}$.

4. Find fraction notation for 65%.

5. Translate to a percent equation. Then solve.

   What is 40% of 55?

6. Translate to a proportion. Then solve.

   What percent of 80 is 65?

Solve.

7. *Cruise Ship Passengers.* Of the passengers on a typical cruise ship, 16% are in the 25–34 age group and 23% are in the 35–44 age group. A cruise ship has 2500 passengers. How many are in the 25–34 age group? the 35–44 age group?
   Source: Polk

8. *Batting Averages.* Luis Castillo, second baseman for the Florida Marlins, got 180 hits during the 2000 baseball season. This was about 33.4% of his at-bats. How many at-bats did he have?
   Source: Major League Baseball

9. *Airline Profits.* Profits of the entire U. S. Airline industry decreased from $5.5 billion in 1999 to $2.7 billion in 2000. Find the percent of decrease.
   Source: Air Transport Association

10. There are 6.6 billion people living in the world today. It is estimated that the total number who have ever lived is about 120 billion. What percent of people who have ever lived are alive today?
    Source: *The Handy Geography Answer Book*

11. *Maine Sales Tax.* The sales tax rate in Maine is 5%. How much tax is charged on a purchase of $324? What is the total price?

12. Gwen's commission rate is 15%. What is the commission from the sale of $4200 worth of merchandise?

13. The marked price of a CD player is $200 and the item is on sale at 20% off. What are the discount and the sale price?

14. What is the simple interest on a principal of $120 at the interest rate of 7.1% for 1 year?

15. The Burnham Parents–Teachers Association invests $5200 at 6% simple interest. How much is in the account after $\frac{1}{2}$ year?

16. Find the amount in an account if $1000 is invested at $5\frac{3}{8}$%, compounded annually, for 2 years.

17. The Suarez family invests $10,000 at an annual interest rate of 9%, compounded monthly. How much is in the account after 3 years?

18. *Job Opportunities.* The table below lists job opportunities, in thousands, in 1998 and projected increases to 2008. Find the missing numbers.

OCCUPATION	NUMBER OF JOBS IN 1998 (in thousands)	NUMBER OF JOBS IN 2008 (in thousands)	CHANGE	PERCENT OF INCREASE
Restaurant waitstaff	2019	2322	303	15.0%
Dental assistant	229		97	
Nurse psychiatric aide	1461	1794		
Child-care worker		1141	236	
Hairdresser/hairstylist/cosmetologist		670		10.2%

Source: Handbook of U.S. Labor Statistics

19. Find the discount and the discount rate of the bed in this ad.

WHITE IRON DAYBED
WITH BRASS ACCENTS
100 TO SELL
FANTASTIC VALUE!
MARKET VALUE
$249.95
CHOICE OF FINISH!
$118 Springs Included!

20. *Home Loan.* Complete the following table, assuming the monthly payment as given.

Interest Rate	7.4%
Mortgage	$120,000
Time of Loan	360 mos
Monthly Payment	$830.86
Principal after First Payment	
Principal after Second Payment	

SKILL MAINTENANCE

Solve.

21. $8.4 \times y = 1864.8$

22. $\dfrac{5}{8} = \dfrac{10}{x}$

23. Convert to decimal notation: $\dfrac{17}{12}$.

24. Convert to a mixed numeral: $\dfrac{153}{44}$.

SYNTHESIS

25. By selling a home without using a realtor, Juan and Marie can avoid paying a 7.5% commission. They receive an offer of $180,000 from a potential buyer. In order to give a comparable offer, for what price would a realtor need to sell the house? Round to the nearest hundred.

26. Karen's commission rate is 16%. She invests her commission from the sale of $15,000 worth of merchandise at the interest rate of 12%, compounded quarterly. How much is Karen's investment worth after 6 months?

1. It is expected that by 2010, 53% of all food expenses will occur away from home. Find decimal notation for 53%.
Source: National Restaurant Association

2. During the 2000 baseball season, Shawn Green of the Los Angeles Dodgers had a 0.269 batting average. Find percent notation for 0.269.
Source: Major League Baseball

3. Find percent notation: $\frac{9}{8}$.

4. Find decimal notation: $\frac{13}{6}$.

5. Write fraction notation for the ratio 5 to 0.5.

6. Find the rate in kilometers per hour: 350 km, 15 hr.

Use $<$, $>$, or $=$ for $\square$ to write a true sentence.

7. $\frac{5}{7} \square \frac{6}{8}$

8. $\frac{6}{14} \square \frac{15}{25}$

Estimate the sum or difference by rounding to the nearest hundred.

9. $263,961 + 32,090 + 127.89$

10. $73,510 - 23,450$

Calculate.

11. $46 - [4(6 + 4 \div 2) + 2 \times 3 - 5]$

12. $[0.8(1.5 - 9.8 \div 49) + (1 + 0.1)^2] \div 1.5$

Compute and simplify.

13. $\frac{6}{5} + 1\frac{5}{6}$

14. $46.9 + 2.84$

15.
$$
\begin{array}{r}
4\ 8\ 7{,}0\ 9\ 4 \\
6{,}9\ 3\ 6 \\
+\ \ \ \ 2\ 1{,}1\ 2\ 0 \\
\hline
\end{array}
$$

16. $35 - 34.98$

17. $3\frac{1}{3} - 2\frac{2}{3}$

18. $\frac{8}{9} - \frac{6}{7}$

19. $\frac{7}{9} \cdot \frac{3}{14}$

20.
$$
\begin{array}{r}
2\ 3\ 6{,}9\ 8\ 4 \\
\times\ \ \ \ \ 3{,}6\ 0\ 0 \\
\hline
\end{array}
$$

21.
$$
\begin{array}{r}
4\ 6.0\ 1\ 2 \\
\times\ \ \ \ 0.0\ 3 \\
\hline
\end{array}
$$

22. $6\frac{3}{5} \div 4\frac{2}{5}$

23. $431.2 \div 35.2$

24. $15\overline{)1\ 8\ 5\ 0}$

Solve.

25. $36 \cdot x = 3420$

26. $y + 142.87 = 151$

27. $\frac{2}{15} \cdot t = \frac{6}{5}$

28. $\frac{3}{4} + x = \frac{5}{6}$

29. $\frac{y}{25} = \frac{24}{15}$

30. $\frac{16}{n} = \frac{21}{11}$

Solve.

31. *Box-Office Revenue.* The table below shows the weekend gross (box-office revenue), in millions, of the movie *The Family Man.*

WEEKEND	GROSS (in millions)
1	$15.1
2	16.4
3	9.1
4	6.6
5	3.3

a) Convert each amount to standard notation.
b) Find the total amount earned for the five weekends.
c) Find the average amount earned.
d) Find the percent of increase from the first weekend to the second.
e) Find the percent of decrease from the second weekend to the third.
Source: Exhibitor Relations

32. *Box-Office Revenue.* North Americans spent $7.45 billion at the movies in 2000. This was an increase of $0.14 billion over the amount spent in 1999.

a) How much was spent in 1999?
b) What was the percent of increase?
Source: A. C. Nielsen EDI

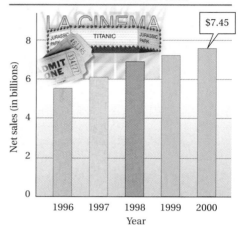

Movie Box Office Revenue

$7.45

33. *e-mails.* The typical Internet user receives 17,000 e-mail messages per year. How many messages are received by the typical user each day? Use 365 days in one year.
Source: Dave Barry, *The Miami Herald*

34. At one point in the 2000–2001 NBA season, the Utah Jazz had won 26 out of 40 games. At this rate, how many games would they win in the entire season of 82 games?
Source: National Basketball Association

35. *Neckties.* A total of $212.50 was paid for 5 neckties at an upscale men's store. How much did each necktie cost?

$212.50

36. *Unit Price.* A 200-oz bottle of Gain liquid laundry detergent costs $9.99. What is the unit price?

37. Patty walked $\frac{7}{10}$ mi to school and then $\frac{8}{10}$ mi to the library. How far did she walk?

38. On a map, 1 in. represents 80 mi. How much does $\frac{3}{4}$ in. represent?

39. *Compound Interest.* The Bakers invest $8500 in an investment account paying 8%, compounded monthly. How much is in the account after 5 years?

40. *Ribbons.* How many pieces of ribbon $1\frac{4}{5}$ yd long can be cut from a length of ribbon 9 yd long?

41. *Job Opportunities.* The table below shows job opportunities, in thousands, in 1998 and projected to 2008. Find the missing numbers.

OCCUPATION	NUMBER OF JOBS IN 1998 (in thousands)	NUMBER OF JOBS IN 2008 (in thousands)	CHANGE	% INCREASE
Court clerk	100	112	12	12%
Office manager	1611		313	
Office clerk	3021	3484		
Teacher assistant		1567	375	
Host/hostess		351		18.2%

Source: Handbook of U.S. Labor Statistics

For each of Exercises 42–45, choose the correct answer from the selections given.

42. The population of the state of Kentucky increased from 3,685,296 in 1990 to 4,041,769 in 2000. What was the percent increase?
Source: U.S. Bureau of the Census

a) 9.04%   **b)** 9.7%   **c)** 7.9%
d) 8.8%   **e)** None

43. Find decimal notation: $\frac{8}{13}$.

a) 0.615   **b)** 0.615384   **c)** 0.615385
d) 0.6153846154   **e)** None

44. Antonio bought a sweater vest for $59.95 and a pair of blue jeans for $39.50 and paid for them with a $100 bill. How much change did he receive?

a) $95   **b)** $55   **c)** $99.45
d) $0.55   **e)** None

45. Subtract and simplify: $\frac{14}{25} - \frac{3}{20}$.

a) $\frac{11}{500}$ **b)** $\frac{11}{5}$ **c)** $\frac{41}{100}$
d) $\frac{205}{500}$ **e)** None

(SYNTHESIS)

46. *Nutrition Facts.* Food companies are required by law to provide nutrition facts on packaging. But when choosing a product, one must be careful that a proper comparison is made. Consider, for example, the following nutrition information on a box of Wheaties compared to a box of Kellogg's Complete. Note that Wheaties defines 1 serving as 1 cup and Complete defines 1 serving as $\frac{3}{4}$ cup.
Sources: General Mills; Kellogg's

a) Use proportions to rewrite the nutrition facts for a box of Complete so that it is based on a serving size being 1 cup. Then compare the nutrition facts between the cereals.
b) Which cereal has the most calories per serving?
c) Which cereal has the most fat per serving?
d) Which cereal has the most sodium per serving?
e) Which cereal has the most potassium per serving?

Wheaties

Nutrition Facts		
Serving Size	1 cup (30g)	
Servings Per Container	About 17	

Amount Per Serving	Wheaties	with 1/2 cup skim milk
Calories	110	150
Calories from Fat	10	10
	% Daily Value **	
Total Fat 1g*	1%	2%
Saturated Fat 0g	0%	0%
Polyunsaturated Fat 0g		
Monounsaturated Fat 0g		
Cholesterol 0mg	0%	1%
Sodium 220mg	9%	12%
Potassium 110mg	3%	9%
Total Carbohydrate 24g	8%	10%
Dietary Fiber 3g	12%	12%
Sugars 4g		
Other Carbohydrate 17g		
Protein 3g		

Not the same

Complete

Nutrition Facts		
Serving Size	3/4 cup (29g/1.1 oz.)	
Servings Per Container	About 17	

Amount Per Serving	Cereal	Cereal with 1/2 Cup Vitamins A & D Fat Free Milk
Calories	90	130
Calories from Fat	5	5
	% Daily Value **	
Total Fat 0.5g*	1%	1%
Saturated Fat 0g	0%	0%
Polyunsaturated Fat 0g		
Monounsaturated Fat 0g		
Cholesterol 0mg	0%	0%
Sodium 210mg	9%	11%
Potassium 170mg	5%	11%
Total Carbohydrate 23g	8%	10%
Dietary Fiber 5g	20%	20%
Soluble Fiber 1g		
Insoluble Fiber 4g		
Sugars 5g		
Other Carbohydrate 13g		
Protein 3g		

437

Data, Graphs, and Statistics

Gateway to Chapter 7

There are many ways in which we can describe and analyze data. The data in the media or the world around us might be presented in a table. The data in a table might then be used to draw a graph. In this chapter, we consider many kinds of graphs: pictographs, bar graphs, line graphs, and circle graphs.

One way to analyze data is to look at statistics. Statistics that we consider in this chapter are averages (or means), medians, and modes.

Real-World Application

Tricia adds one slice of chocolate cake with fudge frosting (560 calories) to her diet each day for one year (365 days) and makes no other changes in her eating or exercise habits. The consumption of 3500 calories will add about 1 pound to her body weight. How many pounds will she have gained at the end of the year?

This problem appears as Exercise 12 in Section 7.3.

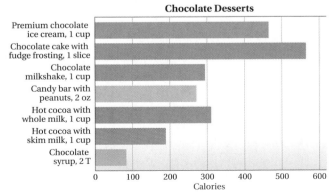

Chocolate Desserts

Premium chocolate ice cream, 1 cup
Chocolate cake with fudge frosting, 1 slice
Chocolate milkshake, 1 cup
Candy bar with peanuts, 2 oz
Hot cocoa with whole milk, 1 cup
Hot cocoa with skim milk, 1 cup
Chocolate syrup, 2 T

0   100   200   300   400   500   600
Calories

Source: *Better Homes and Gardens*, December 1996

In Questions 1–3, find (a) the average, (b) the median, and (c) the mode.   [7.1a, b, c]

1. 46, 50, 53, 55

2. 3, 1, 2, 8, 8

3. 50, 55, 46, 53, 50, 46, 50

4. A car was driven 660 mi in 12 hr. What was the average number of miles per hour?   [7.1a]

5. To get a C in chemistry, Delia must average 70 on four tests. Scores on the first three tests were 68, 71, and 65. What is the lowest score that she can make on the last test and still get a C?   [7.1a]

6. Teenage Spending.   The following data show how typical teenagers spend their money. Make a circle graph of the data.   [7.4b]

CATEGORY	PERCENT
Clothing	34%
Entertainment	22
Food	22
Other	22

Source: Rand Youth Poll, eMarketer

7. Cost of Life Insurance.   The following table shows the comparison of the cost of a $100,000 life insurance policy for female smokers and nonsmokers at certain ages.   [7.2a]

a) How much does it cost a female nonsmoker, age 32, for insurance?

b) How much more does it cost a female smoker, age 35, than a nonsmoker at the same age?

LIFE INSURANCE: FEMALE		
AGE	COST (Smoker)	COST (Nonsmoker)
31	$294	$170
32	298	172
33	302	176
34	310	178
35	316	182

Source: State Farm Insurance

8. Using the data in Question 7, draw a vertical bar graph showing the cost of insurance for a female smoker at various ages. Use age on the horizontal scale and cost on the vertical scale.   [7.3b]

9. Using the data in Question 7, draw a line graph showing the cost of insurance for a female smoker at various ages. Use age on the horizontal scale and cost on the vertical scale.   [7.3d]

Risk of Heart Disease.   The line graph below shows the relationship between blood cholesterol level and incidence of coronary heart disease.   [7.3c]

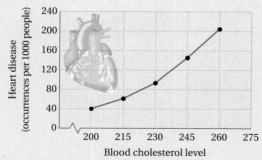

Source: American Heart Association

10. At what cholesterol level is the risk highest?

11. About how much higher is the risk at 260 than at 200?

12. Study Time vs. Grades.   An English instructor asked his students to keep track of how much time each spent studying for a chapter test. He collected the information together with the test scores. The data are given in the table below. Draw a line graph of the data.   [7.3d]

STUDY TIME (in hours)	TEST GRADE (in percent)
9	75
11	83
13	80
15	85
17	80
18	86
21	87
23	92

7.1

AVERAGES, MEDIANS, AND MODES

Data are often available regarding some kind of application involving mathematics. We can use tables and graphs of various kinds to show information about the data and to extract information from the data that can lead us to make analyses and predictions. Graphs allow us to communicate a message from the data.

For example, the following two pages show data regarding the number of alcohol-related traffic deaths in recent years. Examine each method of presentation. Which method, if any, do you like the best and why? Which do you like the least and why?

Paragraph The National Highway Traffic Safety Administration has recently released data regarding the number of alcohol-related traffic deaths for various years. In 1990, there were 22,084 deaths; in 1991, there were 19,887 deaths; in 1992, there were 17,859 deaths; in 1993, there were 17,473 deaths; in 1994, there were 16,589 deaths; in 1995, there were 17,274 deaths; in 1996, there were 17,126 deaths; in 1997, there were 16,189 deaths; and finally, in 1998, there were 15,936 deaths.

Objectives

a Find the average of a set of numbers and solve applied problems involving averages.

b Find the median of a set of numbers and solve applied problems involving medians.

c Find the mode of a set of numbers and solve applied problems involving modes.

d Compare two sets of data using their means.

Pictograph

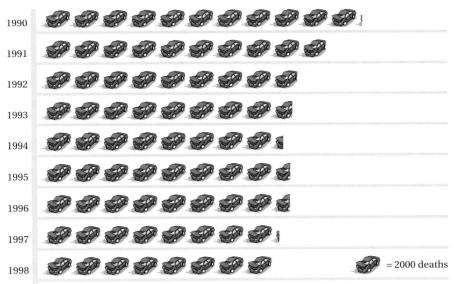

Bar Graph

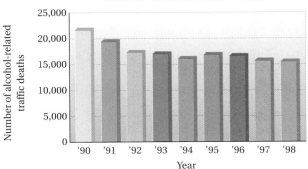

Table

YEAR	ALCOHOL–RELATED TRAFFIC DEATHS
1990	22,084
1991	19,887
1992	17,859
1993	17,473
1994	16,589
1995	17,274
1996	17,126
1997	16,189
1998	15,936

Source: National Highway Traffic Safety Administration

Line Graph

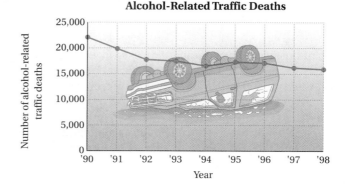

Alcohol-Related Traffic Deaths

Circle, or Pie, Graph

Alcohol-Related Traffic Deaths

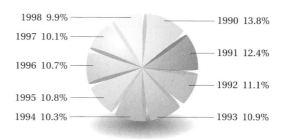

1998 9.9%
1997 10.1%
1996 10.7%
1995 10.8%
1994 10.3%
1990 13.8%
1991 12.4%
1992 11.1%
1993 10.9%

Most people would not find the paragraph method for displaying the data most useful. It takes time to read, and it is hard to look for a trend. The circle, or pie, graph might be used to compare what part of the total number of alcohol-related traffic deaths, 160,417, over the nine years each individual year represents. But since most sectors are almost the same size, the comparison is difficult. The bar and line graphs might be more worthwhile if we want to see the overall trend of decreased deaths and to make a prediction about years to come.

In this chapter, we will learn not only how to extract information from various kinds of tables and graphs, but also how to create various kinds of graphs.

a   Averages

A **statistic** is a number describing a set of data. One statistic is a *center point,* or *measure of central tendency,* that characterizes the data. The most common kind of center point is the *mean,* or *average,* of a set of numbers. We first considered averages in Section 1.9 and extended the coverage in Sections 3.7 and 4.4.

Let's consider the data of the number of alcohol-related traffic deaths:

> 22,084, 19,887, 17,859, 17,473, 16,589, 17,274, 17,126, 16,189, 15,936.

What is the average of this set of numbers? First, we add the numbers:

> 22,084 + 19,887 + 17,859 + 17,473 + 16,589 + 17,274 + 17,126 + 16,189 + 15,936 = 160,417.

Study Tips

THE AW MATH TUTOR CENTER

The AW Math Tutor Center is staffed by highly qualified mathematics instructors who provide students with tutoring on text examples and odd-numbered exercises. Tutoring is provided free to students who have bought a new textbook with a special access card bound with the book. Tutoring is available by telephone (toll-free), fax, and e-mail. White-board technology allows tutors and students to actually see problems worked while they talk live during the tutoring sessions. If you purchased a book without this card, you can purchase an access code through your bookstore using ISBN# 0-201-72170-8. (This is also discussed in the Preface.)

Next, we divide by the number of data items, 9:

$$\frac{160{,}417}{9} \approx 17{,}824. \qquad \text{Rounding to the nearest one}$$

Note that if the traffic deaths had been the average (same) for 9 years, we would have:

17,824 + 17,824 + 17,824 + 17,824 + 17,824 + 17,824 + 17,824 + 17,824 + 17,824

= 160,416 ≈ 160,417.

The number 17,824 is called the *average* of the set of numbers. It is also called the **arithmetic** (pronounced ăr´ ĭth-mĕt´-ĭk) **mean** or simply the **mean.**

> ## AVERAGE
>
> To find the **average** of a set of numbers, add the numbers and then divide by the number of items of data.

EXAMPLE 1 On a 4-day trip, a car was driven the following number of miles each day: 240, 302, 280, 320. What was the average number of miles per day?

$$\frac{240 + 302 + 280 + 320}{4} = \frac{1142}{4}, \quad \text{or} \quad 285.5$$

The car was driven an average of 285.5 mi per day. Had the car been driven exactly 285.5 mi each day, the same total distance (1142 mi) would have been traveled.

Do Exercises 1–4.

EXAMPLE 2 *Scoring Average.*
Kareem Abdul-Jabbar is the all-time leading scorer in the history of the National Basketball Association. He scored 38,387 points in 1560 games. What was the average number of points scored per game? Round to the nearest tenth.
Source: National Basketball Association

We already know the total number of points, 38,387, and the number of games, 1560. We divide and round to the nearest tenth:

$$\frac{38{,}387}{1560} = 24.60705\ldots \approx 24.6.$$

Abdul-Jabbar's average was 24.6 points per game.

Do Exercise 5.

Find the average.

1. 14, 175, 36

2. 75, 36.8, 95.7, 12.1

3. A student scored the following on five tests; 68, 85, 82, 74, 96. What was the average score?

4. In the first five games, a basketball player scored points as follows: 26, 21, 13, 14, 23. Find the average number of points scored per game.

5. Home-Run Batting Average.
At the end of the 2000 baseball season, Mark McGwire had the most career home runs of any active player in the major leagues, 554 in 15 seasons. What was his average number of home runs per season? Round to the nearest tenth.
Source: Major League Baseball

Answers on page A-16

6. Gas Mileage. The Honda Insight gets 816 miles of highway driving on 12 gallons of gasoline. What is the average number of miles per gallon—that is, what is its gas mileage for highway driving?

Source: *ACEE Green Book: The Environmental Guide to Cars & Trucks, Model Year 2001*

Answer on page A-16

EXAMPLE 3 *Gas Mileage.* The Honda Insight, shown here, is a hybrid car powered by gas and electricity. It is estimated to go 732 miles of city driving on 12 gallons of gasoline. What is the expected average number of miles per gallon—that is, what is its gas mileage for city driving?

Source: *ACEE Green Book: The Environmental Guide to Cars & Trucks, Model Year 2001*

We divide the total number of miles, 732, by the total number of gallons, 12:

$$\frac{732}{12} = 61 \text{ mpg.}$$

The Honda Insight's expected average is 61 miles per gallon for city driving.

Do Exercise 6.

EXAMPLE 4 *Grade Point Average.* In most colleges, students are assigned grade point values for grades obtained. The **grade point average,** or **GPA,** is the average of the grade point values for each credit hour taken. At most colleges, grade point values are assigned as follows:

A:	4.0	D:	1.0
B:	3.0	F:	0.0
C:	2.0		

Meg earned the following grades for one semester. What was her grade point average?

COURSE	GRADE	NUMBER OF CREDIT HOURS IN COURSE
Colonial History	B	3
Basic Mathematics	A	4
English Literature	A	3
French	C	4
Physical Education	D	1

To find the GPA, we first multiply the grade point value (in color below) by the number of credit hours in the course to determine the number of *quality points,* and then add, as follows:

Course	Calculation	
Colonial History	$3.0 \cdot 3 =$	9
Basic Mathematics	$4.0 \cdot 4 =$	16
English Literature	$4.0 \cdot 3 =$	12
French	$2.0 \cdot 4 =$	8
Physical Education	$1.0 \cdot 1 =$	1

46 (Total)

The total number of credit hours taken is $3 + 4 + 3 + 4 + 1$, or 15. We divide the number of quality points, 46, by the number of credits, 15, and round to the nearest tenth:

$$\text{GPA} = \frac{46}{15} \approx 3.1.$$

Meg's grade point average was 3.1.

Do Exercise 7.

EXAMPLE 5 *Grading.* To get a B in math, Geraldo must score an average of 80 on the tests. On the first four tests, his scores were 79, 88, 64, and 78. What is the lowest score that Geraldo can get on the last test and still get a B?

We can find the total of the five scores needed as follows:

$$80 + 80 + 80 + 80 + 80 = 5 \cdot 80, \quad \text{or} \quad 400.$$

The total of the scores on the first four tests is

$$79 + 88 + 64 + 78 = 309.$$

Thus Geraldo needs to get at least

$$400 - 309, \quad \text{or} \quad 91$$

in order to get a B. We can check this as follows:

$$\frac{79 + 88 + 64 + 78 + 91}{5} = \frac{400}{5}, \quad \text{or} \quad 80.$$

Do Exercise 8.

CALCULATOR CORNER

Computing Averages Averages can be computed easily on a calculator. We must keep the rules for order of operations in mind when doing this. For example, to calculate

$$\frac{84 + 92 + 79}{3}$$

on a calculator with parenthesis keys, we press ⎡ (⎤ ⎡ 8 ⎤ ⎡ 4 ⎤ ⎡ + ⎤ ⎡ 9 ⎤ ⎡ 2 ⎤ ⎡ + ⎤ ⎡ 7 ⎤ ⎡ 9 ⎤ ⎡) ⎤ ⎡ ÷ ⎤ ⎡ 3 ⎤ ⎡ = ⎤. On a calculator without parenthesis keys, we first add the numbers in the numerator and then divide that result by 3. To do this, we press ⎡ 8 ⎤ ⎡ 4 ⎤ ⎡ + ⎤ ⎡ 9 ⎤ ⎡ 2 ⎤ ⎡ + ⎤ ⎡ 7 ⎤ ⎡ 9 ⎤ ⎡ = ⎤ ⎡ ÷ ⎤ ⎡ 3 ⎤ ⎡ = ⎤. In either case, the result is 85.

Exercises:

1. Use a calculator to perform the computation in Example 1.

2. Use a calculator to perform the computations in Margin Exercises 1–4.

3. What would the result have been if we had not used parentheses in the first set of keystrokes above? (Keep the rules for order of operations in mind.)

---

7. Grade Point Average. Alex earned the following grades one semester.

GRADE	NUMBER OF CREDIT HOURS IN COURSE
B	3
C	4
C	4
A	2

What was Alex's grade point average? Assume that the grade point values are 4.0 for an A, 3.0 for a B, and so on. Round to the nearest tenth.

8. Grading. To get an A in math, Rosa must score an average of 90 on the tests. On the first three tests, her scores were 80, 100, and 86. What is the lowest score that Rosa can get on the last test and still get an A?

Answers on page A-16

Find the median.

9. 17, 13, 18, 14, 19

10. 20, 14, 13, 19, 16, 18, 17

11. 78, 81, 83, 91, 103, 102, 122, 119, 88

b Medians

Another type of center-point statistic is the *median*. Medians are useful when we wish to de-emphasize unusually extreme scores. For example, suppose a small class scored as follows on an exam.

Phil:	78	Pat:	56
Jill:	81	Olga:	84
Matt:	82		

Let's first list the scores in order from smallest to largest:

56, 78, 81, 82, 84.
        ↑
    Middle score

The middle score—in this case, 81—is called the **median.** Note that because of the extremely low score of 56, the average of the scores is 76.2. In this example, the median may be a more appropriate center-point statistic.

EXAMPLE 6 What is the median of this set of numbers?

99, 870, 91, 98, 106, 90, 98

We first rearrange the numbers in order from smallest to largest. Then we locate the middle number, 98.

90, 91, 98, 98, 99, 106, 870
                ↑
          Middle number

The median is 98.

Do Exercises 9–11.

> **MEDIAN**
>
> Once a set of data is listed in order, from smallest to largest, the **median** is the middle number if there is an odd number of data items. If there is an even number of items, the median is the number that is the average of the two middle numbers.

EXAMPLE 7 What is the median of this set of numbers?

69, 80, 61, 63, 62, 65

We first rearrange the numbers in order from smallest to largest. There is an even number of numbers. We look for the middle two, which are 63 and 65. The median is halfway between 63 and 65, the number 64.

61, 62, 63, 65, 69, 80   The average of the middle numbers is
            ↑ $\frac{63 + 65}{2}$, or 64.
            └──── The median is 64.

EXAMPLE 8 *Salaries.* The following are the salaries of the top four employees of Verducci's Dress Company. What is the median of the salaries?

$85,000, $100,000, $78,000, $84,000

We rearrange the numbers in order from smallest to largest. The two middle numbers are $84,000 and $85,000. Thus the median is halfway between $84,000 and $85,000 (the average of $84,000 and $85,000):

$78,000, $84,000, $85,000, $100,000

$$\text{Median} = \frac{\$84,000 + \$85,000}{2} = \frac{\$169,000}{2} = \$84,500.$$

Do Exercises 12 and 13.

C   Modes

The final type of center-point statistic is the *mode*.

> **MODE**
>
> The **mode** of a set of data is the number or numbers that occur most often. If each number occurs the same number of times, there is *no* mode.

EXAMPLE 9   Find the mode of these data.

13,   14,   17,   17,   18,   19

The number that occurs most often is 17. Thus the mode is 17.

A set of data has just one average (mean) and just one median, but it can have more than one mode. It is also possible for a set of data to have no mode—when all numbers are equally represented. For example, the set of data 5, 7, 11, 13, 19 has no mode.

EXAMPLE 10   Find the modes of these data.

33,   34,   34,   34,   35,   36,   37,   37,   37,   38,   39,   40

There are two numbers that occur most often, 34 and 37. Thus the modes are 34 and 37.

Do Exercises 14–17.

Which statistic is best for a particular situation? If someone is bowling, the *average* from several games is a good indicator of that person's ability. If someone is applying for a job, the *median* salary at that business is often most indicative of what people are earning there because although executives tend to make a lot more money, there are fewer of them. Finally, if someone is re-ordering for a clothing store, the *mode* of the sizes sold is probably the most important statistic.

Find the median.

12. Salaries of Part-Time Typists.
$3300, $4000, $3900, $3600, $3800, $3400

13. 68, 34, 67, 69, 34, 70

Find the modes of these data.

14. 23, 45, 45, 45, 78

15. 34, 34, 67, 67, 68, 70

16. 13, 24, 27, 28, 67, 89

17. In a lab, Gina determined the mass, in grams, of each of five eggs:

15 g, 19 g, 19 g, 14 g, 18 g.

a) What is the mean?
b) What is the median?
c) What is the mode?

Answers on page A-16

18. Growth of Wheat. Rudy experiments to see which of two kinds of wheat is better. (In this situation, the shorter wheat is considered "better.") He grows both kinds under similar conditions and measures stalk heights, in inches, as follows. Which kind is better?

WHEAT A STALK HEIGHTS (in inches)			
16.2	42.3	19.5	25.7
25.6	18.0	15.6	41.7
22.6	26.4	18.4	12.6
41.5	13.7	42.0	21.6

WHEAT B STALK HEIGHTS (in inches)			
19.7	18.4	19.7	17.2
19.7	14.6	32.0	25.7
14.0	21.6	42.5	32.6
22.6	10.9	26.7	22.8

d Comparing Two Sets of Data

We have seen how to calculate averages, medians, and modes from data. A way to analyze two sets of data is to make a determination about which of two groups is "better." One way to do so is by comparing the averages.

EXAMPLE 11 *Battery Testing.* An experiment is performed to compare battery quality. Two kinds of battery were tested to see how long, in hours, they kept a portable CD player running. On the basis of this test, which battery is better?

BATTERY A: ETERNREADY TIMES (in hours)			BATTERY B: STURDYCELL TIMES (in hours)		
27.9	28.3	27.4	28.3	27.6	27.8
27.6	27.9	28.0	27.4	27.6	27.9
26.8	27.7	28.1	26.9	27.8	28.1
28.2	26.9	27.4	27.9	28.7	27.6

Note that it is difficult to analyze the data at a glance because the numbers are close together. We need a way to compare the two groups. Let's compute the average of each set of data.

Battery A: Average

$$= \frac{27.9 + 28.3 + 27.4 + 27.6 + 27.9 + 28.0 + 26.8 + 27.7 + 28.1 + 28.2 + 26.9 + 27.4}{12}$$

$$= \frac{332.2}{12} \approx 27.68$$

Battery B: Average

$$= \frac{28.3 + 27.6 + 27.8 + 27.4 + 27.6 + 27.9 + 26.9 + 27.8 + 28.1 + 27.9 + 28.7 + 27.6}{12}$$

$$= \frac{333.6}{12} = 27.8$$

We see that the average time of battery B is higher than that of battery A and thus conclude that battery B is "better." (It should be noted that statisticians might question whether these differences are what they call "significant." The answer to that question belongs to a later math course.)

Do Exercise 18.

Answer on page A-17

Digital Video Tutor CD 4 Videotape 12   InterAct Math   Math Tutor Center   MathXL   MyMathLab

EXERCISE SET

a, **b**, **c** For each set of numbers, find the average, the median, and any modes that exist.

1. 17, 19, 29, 18, 14, 29

2. 72, 83, 85, 88, 92

3. 5, 37, 20, 20, 35, 5, 25

4. 13, 32, 25, 27, 13

5. 4.3, 7.4, 1.2, 5.7, 7.4

6. 13.4, 13.4, 12.6, 42.9

7. 234, 228, 234, 229, 234, 278

8. $29.95, $28.79, $30.95, $29.95

9. *Atlantic Storms and Hurricanes.* The following bar graph shows the number of Atlantic storms or hurricanes that formed in various months from 1980 to 2000. What is the average number for the 9 months given? the median? the mode?

10. *PBA Scores.* Chris Barnes rolled scores of 224, 224, 254, and 187 in a recent tournament of the Professional Bowlers Association. What was his average? his median? his mode?
Source: Professional Bowlers Association

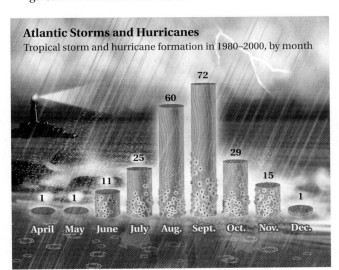

Atlantic Storms and Hurricanes
Tropical storm and hurricane formation in 1980–2000, by month

72 — Sept.
60 — Aug.
29 — Oct.
25 — July
15 — Nov.
11 — June
1 — April
1 — May
1 — Dec.

Source: Colorado State University

11. *Gas Mileage.* The Saturn SW gets 342 miles of highway driving on 9 gallons of gasoline. What is the average number of miles expected per gallon—that is, what is its gas mileage?
Source: *ACEE Green Book: The Environmental Guide to Cars & Trucks, Model Year 2001*

12. *Gas Mileage.* The Toyota Camry gets 322 miles of city driving on 14 gallons of gasoline. What is the average number of miles expected per gallon—that is, what is its gas mileage?
Source: *ACEE Green Book: The Environmental Guide to Cars & Trucks, Model Year 2001*

Grade Point Average. The tables in Exercises 13 and 14 show the grades of a student for one semester. In each case, find the grade point average. Assume that the grade point values are 4.0 for an A, 3.0 for a B, and so on. Round to the nearest tenth.

13.

GRADE	NUMBER OF CREDIT HOURS IN COURSE
B	4
A	5
D	3
C	4

14.

GRADE	NUMBER OF CREDIT HOURS IN COURSE
A	5
C	4
F	3
B	5

15. *Salmon Prices.* The following prices per pound of Atlantic salmon were found at five fish markets:

$6.99, $8.49, $8.99, $6.99, $9.49.

What was the average price per pound? the median price? the mode?

16. *Cheddar Cheese Prices.* The following prices per pound of sharp cheddar cheese were found at five supermarkets:

$5.99, $6.79, $5.99, $6.99, $6.79.

What was the average price per pound? the median price? the mode?

17. *Grading.* To get a B in math, Rich must score an average of 80 on five tests. Scores on the first four tests were 80, 74, 81, and 75. What is the lowest score that Rich can get on the last test and still receive a B?

18. *Grading.* To get an A in math, Cybil must score an average of 90 on five tests. Scores on the first four tests were 90, 91, 81, and 92. What is the lowest score that Cybil can get on the last test and still receive an A?

19. *Length of Pregnancy.* Marta was pregnant 270 days, 259 days, and 272 days for her first three pregnancies. In order for Marta's average pregnancy to equal the worldwide average of 266 days, how long must her fourth pregnancy last?
Source: David Crystal (ed.), *The Cambridge Factfinder.* Cambridge CB2 1RP: Cambridge University Press, 1993, p. 84.

20. *Male Height.* Jason's brothers are 174 cm, 180 cm, 179 cm, and 172 cm tall. The average male is 176.5 cm tall. How tall is Jason if he and his brothers have an average height of 176.5 cm?

d Solve.

21. *Light-Bulb Testing.* An experiment is performed to compare the lives of two types of light bulb. Several bulbs of each type were tested and the results are listed in the following table. On the basis of this test, which bulb is better?

BULB A: HOTLIGHT TIMES (in hours)			BULB B: BRIGHTBULB TIMES (in hours)		
983	964	1214	979	1083	1344
1417	1211	1521	984	1445	975
1084	1075	892	1492	1325	1283
1423	949	1322	1325	1352	1432

22. *Cola Testing.* An experiment is conducted to determine which of two colas tastes better. Students drank each cola and gave it a rating from 1 to 10. The results are given in the following table. On the basis of this test, which cola tastes better?

COLA A: VERVCOLA				COLA B: COLA-COLA			
6	8	10	7	10	9	9	6
7	9	9	8	8	8	10	7
5	10	9	10	8	7	4	3
9	4	7	6	7	8	10	9

23. D_W You are applying for an entry-level job at a large firm. You can be informed of the mean, median, or mode salary. Which of the three figures would you request? Why?

24. D_W Is it possible for a driver to average 20 mph on a 30-mi trip and still receive a ticket for driving 75 mph? Why or why not?

Multiply.

25. $14 \cdot 14$ [1.5a]

26. $\dfrac{2}{3} \cdot \dfrac{2}{3}$ [2.6a]

27. 1.4×1.4 [4.3a]

28. 1.414×1.414 [4.3a]

29. 12.86×17.5 [4.3a]

30. 222×0.5678 [4.3a]

31. $\dfrac{4}{5} \cdot \dfrac{3}{28}$ [2.6a]

32. $\dfrac{28}{45} \cdot \dfrac{3}{2}$ [2.6a]

Bowling Averages.   Bowling averages are always computed by rounding down to the nearest integer. For example, suppose a bowler gets a total of 599 for 3 games. To find the average, we divide 599 by 3 and drop the amount to the right of the decimal point:

$$\frac{599}{3} \approx 199.67 \qquad \text{The bowler's average is 199.}$$

In each case, find the bowling average.

33. ▦ 547 in 3 games

34. ▦ 4621 in 27 games

35. *Hank Aaron.*   Hank Aaron averaged $34\frac{7}{22}$ home runs per year over a 22-yr career. After 21 yr, Aaron had averaged $35\frac{10}{21}$ home runs per year. How many home runs did Aaron hit in his final year?

36. The ordered set of data 18, 21, 24, a, 36, 37, b has a median of 30 and an average of 32. Find a and b.

37. *Grades.*   Because of a poor grade on the fifth of five tests, Chris's average test grade fell from 90.5 to 84.0. What did Chris score on the fifth test? Assume that all tests are equally important.

38. *Price Negotiations.*   Amy offers $3200 for a used Ford Taurus advertised at $4000. The first offer from Jim, the car's owner, is to "split the difference" and sell the car for $(3200 + 4000) \div 2$, or $3600. Amy's second offer is to split the difference between Jim's offer and her first offer. Jim's second offer is to split the difference between Amy's second offer and his first offer. If this pattern continues and Amy accepts Jim's third (and final) offer, how much will she pay for the car?

451

TABLES AND PICTOGRAPHS

Objectives

a Extract and interpret data from tables.

b Extract and interpret data from pictographs.

a Reading and Interpreting Tables

A **table** is often used to present data in rows and columns.

EXAMPLE 1 *Nutrition Information.* The following table lists nutrition information for a 1-cup serving of five name-brand cereals (it does not consider the use of milk, sugar, or sweetener).

CEREAL	CALORIES	FAT (in grams)	TOTAL CARBOHYDRATES (in grams)	SODIUM (in milligrams)
Cinnamon Life	160	1.3	34.7	200
Life (Regular)	160	2.0	33.3	213.3
Lucky Charms	120	1.0	25.0	210
Kellogg's Complete	120	0.7	30.7	280
Wheaties	110	1.0	24.0	220

Sources: Quaker Oats; General Mills; Kellogg's

Use the table in Example 1 to answer Margin Exercises 1–7.

1. Which cereal has the most total carbohydrates?

2. Which cereal has the least total carbohydrates?

3. Which cereal has the least number of calories?

4. Find the average number of grams of fat in the cereals listed.

a) Which cereal has the least amount of sodium per serving?

b) Which cereal has the greatest amount of fat?

c) Which cereal has the least amount of fat?

d) Find the average total carbohydrates in the cereals listed.

Careful examination of the table will give the answers.

a) To determine which cereal has the least amount of sodium, look down the column headed "Sodium" and find the smallest number. That number is 200 mg. Then look across that row to find the brand of cereal, Cinnamon Life.

b) To determine which cereal has the greatest amount of fat, look down the column headed "Fat" and find the largest number. That number is 2.0 g. Then look across that row to find the cereal, Life (Regular).

c) To determine which cereal has the least amount of fat, look down the column headed "Fat" and find the smallest number. That number is 0.7 g. Then look across that row to find the cereal, Kellogg's Complete.

d) Find the average of all the numbers in the column headed "Total Carbohydrates":

$$\frac{34.7 + 33.3 + 25.0 + 30.7 + 24.0}{5} = \frac{147.7}{5} = 29.54 \text{ g.}$$

The average total carbohydrates is 29.54 g.

Do Exercises 1–7. (Exercises 5–7 are on the following page.)

Answers on page A-17

EXAMPLE 2 *Wheaties Nutrition Facts.* Most foods are required by law to provide factual information regarding nutrition, as shown in the following table of Nutrition Facts from a box of Wheaties cereal. Although this can be very helpful to the consumer, one must be careful in interpreting the data. The % Daily Value figures shown here are based on a 2000-calorie diet. Your daily values may be higher or lower, depending on your calorie needs or intake.

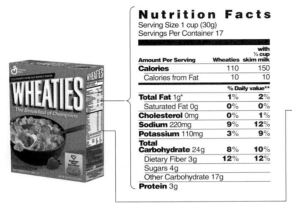

Nutrition Facts

Serving Size 1 cup (30g)
Servings Per Container 17

Amount Per Serving	Wheaties	with ½ cup skim milk
Calories	110	150
Calories from Fat	10	10
	% Daily value**	
Total Fat 1g*	1%	2%
Saturated Fat 0g	0%	0%
Cholesterol 0mg	0%	1%
Sodium 220mg	9%	12%
Potassium 110mg	3%	9%
Total Carbohydrate 24g	8%	10%
Dietary Fiber 3g	12%	12%
Sugars 4g		
Other Carbohydrate 17g		
Protein 3g		

Vitamin A	10%	15%
Vitamin C	10%	10%
Calcium	0%	15%
Iron	45%	45%
Vitamin D	10%	25%
Thiamin	50%	50%
Riboflavin	50%	60%
Niacin	50%	50%
Vitamin B$_6$	50%	50%
Folic Acid	50%	50%
Phosphorus	10%	20%
Magnesium	8%	10%
Zinc	50%	50%
Copper	4%	4%

*Amount in Cereal. A serving of cereal plus skim milk provides 1g fat, less than 5mg cholesterol, 280mg sodium, 310mg potassium, 30g carbohydrate (10g sugars), and 7g protein.
**Percent Daily Values are based on a 2,000 calorie diet. Your daily values may be higher or lower depending on your calorie needs.

Source: General Mills

Suppose your morning bowl of cereal consists of 2 cups of Wheaties with 1 cup of skim milk, with artificial sweetener containing 0 calories.

a) How many calories have you consumed?

b) What percent of the daily value of total fat have you consumed?

c) A nutritionist recommends that you look for foods that provide 10% or more of the daily value for vitamin C. Do you get that with your bowl of Wheaties?

d) Suppose you are trying to limit your daily caloric intake to 2500 calories. How many bowls of cereal would it take to exceed the 2500 calories, even though you probably would not eat just cereal?

Careful examination of the table of nutrition facts will give the answers.

a) Look at the column marked "with 1/2 cup skim milk" and note that 1 cup of cereal with 1/2 cup skim milk contains 150 calories. Since you are having twice that amount, you are consuming

$$2 \times 150, \quad \text{or} \quad 300 \text{ calories.}$$

b) Read across from "Total Fat" and note that in 1 cup of cereal with 1/2 cup skim milk, you get 2% of the daily value of fat. Since you are doubling that, you get 4% of the daily value of fat.

c) Find the row labeled "Vitamin C" on the left and look under the column labeled "with 1/2 cup skim milk." Note that you get 10% of the daily value for "1 cup with 1/2 cup of skim milk," and since you are doubling that, you are more than satisfying the 10% requirement.

d) From part (a), we know that you are consuming 300 calories per bowl. Dividing 2500 by 300 gives $\frac{2500}{300} \approx 8.33$. Thus if you eat 9 bowls of cereal in this manner, you will exceed the 2500 calories.

Do Exercises 8–12.

---

5. Find the average amount of sodium in the cereals.

6. Find the median of the amount of sodium in the cereals.

7. Find the average, the median, and the mode of the number of calories in the cereals.

Use the Nutrition Facts data from the Wheaties box and the bowl of cereal described in Example 2 to answer Margin Exercises 8–12.

8. How many calories from fat are in your bowl of cereal?

9. A nutritionist recommends that you look for foods that provide 10% or more of the daily value for iron. Do you get that with your bowl of Wheaties?

10. How much sodium have you consumed?

11. What daily value of sodium have you consumed?

12. How much protein have you consumed?

Answers on page A-17

Use the pictograph in Example 3 to answer Margin Exercises 13–15.

13. How many elephants are there in Tanzania?

14. How does the elephant population of Zimbabwe compare to that of Cameroon?

15. What is the average number of elephants in these six countries?

b   Reading and Interpreting Pictographs

Pictographs (or *picture graphs*) are another way to show information. Instead of actually listing the amounts to be considered, a **pictograph** uses symbols to represent the amounts. In addition, a *key* is given telling what each symbol represents.

EXAMPLE 3   *Elephant Population.*   The following pictograph shows the elephant population of various countries in Africa. Located on the graph is a key that tells you that each symbol represents 10,000 elephants.

Elephant Population

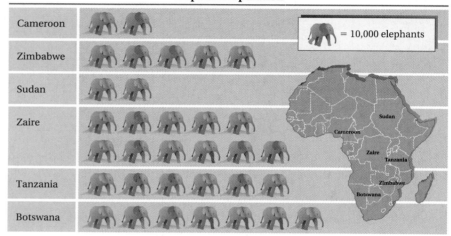

Source: National Geographic

a) Which country has the greatest number of elephants?

b) Which country has the least number of elephants?

c) How many more elephants are there in Zaire than in Botswana?

We can compute the answers by first reading the pictograph.

a) The country with the most symbols has the greatest number of elephants: Zaire, with $11 \times 10{,}000$, or 110,000 elephants.

b) The countries with the fewest symbols have the least number of elephants: Cameroon and Sudan, each with $2 \times 10{,}000$, or 20,000 elephants.

c) From part (a), we know that there are 110,000 elephants in Zaire. In Botswana there are $7 \times 10{,}000$, or 70,000 elephants. Thus there are $110{,}000 - 70{,}000$, or 40,000 more elephants in Zaire than in Botswana.

Do Exercises 13–15.

You have probably noticed that, although they seem to be very easy to read, pictographs are difficult to draw accurately because whole symbols reflect loose approximations due to significant rounding. In pictographs, you also need to use some mathematics to find the actual amounts.

Answers on page A-17

EXAMPLE 4 *Coffee Consumption.* For selected countries, the following pictograph shows approximately how many cups of coffee each person (per capita) drinks annually.

Coffee Consumption

Germany	
United States	
Switzerland	
France	= 100 cups
Italy	

Source: Beverage Marketing Corporation

a) Determine the approximate annual coffee consumption per capital of Germany.

b) Which two countries have the greatest difference in coffee consumption? Estimate that difference.

We use the data from the pictograph as follows.

a) Germany's consumption is represented by 11 whole symbols (1100 cups) and, though it is visually debatable, about $\frac{1}{8}$ of another symbol (about 13 cups), for a total of 1113 cups.

b) Visually, we see that Switzerland has the most consumption and that the United States has the least consumption. Switzerland's annual coffee consumption per capita is represented by 12 whole symbols (1200 cups) and about $\frac{1}{5}$ of another symbol (20 cups), for a total of 1220 cups. U.S. consumption is represented by 6 whole symbols (600 cups) and about $\frac{1}{10}$ of another symbol (10 cups), for a total of 610 cups. The difference between these amounts is $1220 - 610$, or 610 cups.

One advantage of pictographs is that the appropriate choice of a symbol will tell you, at a glance, the kind of measurement being made. Another advantage is that the comparison of amounts represented in the graph can be expressed more easily by just counting symbols. For instance, in Example 3, the ratio of elephants in Zaire to those in Cameroon is $11:2$.

There are at least three disadvantages of pictographs:

1. To make a pictograph easy to read, the amounts must be rounded significantly to the unit that a symbol represents. This makes it difficult to accurately represent an amount.

2. It is difficult to determine very accurately how much a partial symbol represents.

3. Some mathematics is required to finally compute the amount represented, since there is usually no explicit statement of the amount.

Do Exercises 16–18.

Use the pictograph in Example 4 to answer Margin Exercises 16–18.

16. Determine the approximate coffee consumption per capita of France.

17. Determine the approximate coffee consumption per capita of Italy.

18. The approximate coffee consumption of Finland is about the same as the combined coffee consumptions of Switzerland and the United States. What is the approximate coffee consumption of Finland?

Answers on page A-17

a   *Planets.*   Use the following table, which lists information about the planets, for Exercises 1–10.

PLANET	AVERAGE DISTANCE FROM SUN (in miles)	DIAMETER (in miles)	LENGTH OF PLANET'S DAY IN EARTH TIME (in days)	TIME OF REVOLUTION IN EARTH TIME (in years)
Mercury	35,983,000	3,031	58.82	0.24
Venus	67,237,700	7,520	224.59	0.62
Earth	92,955,900	7,926	1.00	1.00
Mars	141,634,800	4,221	1.03	1.88
Jupiter	483,612,200	88,846	0.41	11.86
Saturn	888,184,000	74,898	0.43	29.46
Uranus	1,782,000,000	31,763	0.45	84.01
Neptune	2,794,000,000	31,329	0.66	164.78
Pluto	3,666,000,000	1,423	6.41	248.53

Source: *The Handy Science Answer Book*, Gale Research, Inc.

1. Find the average distance from the sun to Jupiter.

2. How long is a day on Venus?

3. Which planet has a time of revolution of 164.78 yr?

4. Which planet has a diameter of 4221 mi?

5. Which planets have an average distance from the sun that is greater than 1,000,000 mi?

6. Which planets have a diameter that is less than 100,000 mi?

7. About how many earth diameters would it take to equal one Jupiter diameter?

8. How much longer is the longest time of revolution than the shortest?

9. What are the average, the median, and the mode of the diameters of the planets?

10. What are the average, the median, and the mode of the average distances from the sun of the planets?

Heat Index.   In warm weather, a person can feel hotter due to reduced heat loss from the skin caused by higher humidity. The **temperature–humidity index,** or **apparent temperature,** is what the temperature would have to be with no humidity in order to give the same heat effect. The following table lists the apparent temperatures for various actual temperatures and relative humidities. Use this table for Exercises 11–22.

ACTUAL TEMPERATURE (°F)	RELATIVE HUMIDITY									
	10%	20%	30%	40%	50%	60%	70%	80%	90%	100%
	APPARENT TEMPERATURE (°F)									
75°	75	77	79	80	82	84	86	88	90	92
80°	80	82	85	87	90	92	94	97	99	102
85°	85	88	91	94	97	100	103	106	108	111
90°	90	93	97	100	104	107	111	114	118	121
95°	95	99	103	107	111	115	119	123	127	131
100°	100	105	109	114	118	123	127	132	137	141
105°	105	110	115	120	125	131	136	141	146	151

In Exercises 11–14, find the apparent temperature for the given actual temperature and humidity combinations.

11. 80°, 60%

12. 90°, 70%

13. 85°, 90%

14. 95°, 80%

15. How many listed temperature–humidity combinations give an apparent temperature of 100°?

16. How many listed temperature–humidity combinations given an apparent temperature of 111°?

17. At a relative humidity of 50%, what actual temperatures give an apparent temperature above 100°?

18. At a relative humidity of 90%, what actual temperatures give an apparent temperature above 100°?

19. At an actual temperature of 95°, what relative humidities give an apparent temperature above 100°?

20. At an actual temperature of 85°, what relative humidities give an apparent temperature above 100°?

21. At an actual temperature of 85°, by how much would the humidity have to increase in order to raise the apparent temperature from 94° to 108°?

22. At an actual temperature of 80°, by how much would the humidity have to increase in order to raise the apparent temperature from 87° to 102°?

Global Warming. Ecologists are increasingly concerned about global warming, that is, the trend of average global temperatures to rise over recent years. One possible effect is the melting of the polar icecaps. Use the following table for Exercises 23–26.

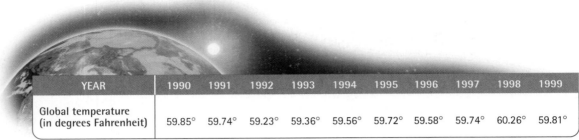

YEAR	1990	1991	1992	1993	1994	1995	1996	1997	1998	1999
Global temperature (in degrees Fahrenheit)	59.85°	59.74°	59.23°	59.36°	59.56°	59.72°	59.58°	59.74°	60.26°	59.81°

Sources: Lester R. Brown et al., *Vital Signs*; the Council of Environmental Quality

23. Find the average global temperatures in 1997 and 1998. What was the percent of increase in the temperature from 1997 to 1998?

24. Find the average global temperatures in 1998 and 1999. What was the percent of decrease in the temperature from 1998 to 1999?

25. Find the average of the average global temperatures for the years 1990 to 1993. Find the average of the average global temperatures for the years 1997 to 1999. By how many degrees does the latter average exceed the former?

26. Find the average of the average global temperatures for the years 1994 to 1996. Find the ten-year average of the average global temperatures for the years 1990 to 1999. By how many degrees does the ten-year average exceed the average for the years 1994 to 1996?

b *World Population Growth.* The following pictograph shows world population in various years. Use the pictograph for Exercises 27–34.

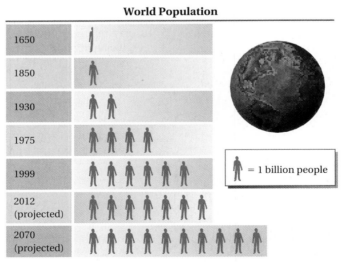

World Population

1650	
1850	
1930	
1975	
1999	
2012 (projected)	
2070 (projected)	

= 1 billion people

Source: U.S. Census Bureau, International Data Base

27. What was the world population in 1850?

28. What was the world population in 1975?

29. In which year will the population be the greatest?

30. In which year was the population the least?

31. Between which two years was the amount of growth the least?

32. Between which two years was the amount of growth the greatest?

33. How much greater will the world population in 2012 be than in 1975? What is the percent of increase?

34. How much greater was the world population in 1999 than in 1930? What is the percent of increase?

Water Consumption. The following pictograph shows water consumption, per person, in different regions of the world in a recent year. Use the pictograph for Exercises 35–40.

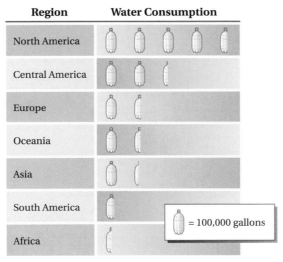

Region	Water Consumption
North America	
Central America	
Europe	
Oceania	
Asia	
South America	
Africa	

= 100,000 gallons

Sources: World Resources Institute; U.S. Energy Information Administration

35. What region consumes the least water?

36. Which region consumes the most water?

37. About how many gallons are consumed per person in North America?

38. About how many gallons are consumed per person in Europe?

39. Approximately how many more gallons are consumed per person in North America than in Asia?

40. Approximately how many more gallons are consumed per person in Central America than in Africa?

41. D_W Loreena is drawing a pictograph in which dollar bills are used as symbols to represent the tuition at various private colleges. Should each dollar bill represent $8000, $4000, or $400? Why?

42. D_W What advantage(s) does a table have over a pictograph?

Solve.   [6.5a]

43. *Kitchen Costs.*   The average cost of a kitchen is $26,888. Some of the cost percentages are as follows.

Cabinets: 50%
Appliances: 8%
Countertops: 15%
Fixtures: 3%

Find the costs for each part of a kitchen.
Source: National Kitchen and Bath Association

44. *Bathroom Costs.*   The average cost of a bathroom is $11,605. Some of the cost percentages are as follows.

Cabinets: 31%
Countertops: 11%
Labor: 25%
Flooring: 6%

Find the costs for each part of a bathroom.
Source: National Kitchen and Bath Association

Convert to fraction notation and simplify.   [6.2b]

45. 24%

46. 45%

47. 4.8%

48. 6.4%

49. 53.1%

50. 87.3%

51. 100%

52. 2%

53. Redraw the pictograph appearing in Example 4 as one in which each symbol represents 150 cups of coffee.

BAR GRAPHS AND LINE GRAPHS

Objectives

a	Extract and interpret data from bar graphs.
b	Draw bar graphs.
c	Extract and interpret data from line graphs.
d	Draw line graphs.

A **bar graph** is convenient for showing comparisons because you can tell at a glance which amount represents the largest or smallest quantity. Of course, since a bar graph is a more abstract form of pictograph, this is true of pictographs as well. However, with bar graphs, a *second scale* is usually included so that a more accurate determination of the amount can be made.

a Reading and Interpreting Bar Graphs

EXAMPLE 1   *Fat Content in Fast Foods.*   Wendy's Hamburgers is a national food franchise. The following bar graph shows the fat content of various sandwiches sold by Wendy's.

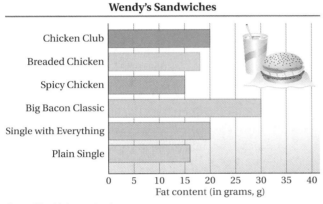

Wendy's Sandwiches

Fat content (in grams, g)

Source: Wendy's International

a) About how much fat is in a Plain Single sandwich?

b) Which sandwich contains the least amount of fat?

c) Which sandwiches contain about 20 g of fat?

We look at the graph to answer the questions.

a) We move to the right along the bar representing Plain Single sandwiches. We can read, fairly accurately, that there is approximately 16 g of fat in the Plain Single sandwich.

b) The shortest bar is for the Spicy Chicken sandwich. Thus that sandwich contains the least amount of fat.

c) We locate the line representing 20 g and then go up until we reach a bar that ends at approximately 20 g. We then go across to the left and read the name of the sandwich. This happens twice, for the Single with Everything and the Chicken Club.

Do Exercises 1–3.

Use the bar graph in Example 1 to answer Margin Exercises 1–3.

1. About how much fat is in the Breaded Chicken sandwich?

2. Which sandwich contains the greatest amount of fat?

3. Which sandwiches contain 15 g or more of fat?

Answers on page A-17

Use the bar graph in Example 2 to answer Margin Exercises 4–7.

4. Approximately how many women, per 100,000, develop breast cancer between the ages of 35 and 39?

5. In what age group is the mortality rate the highest?

6. In what age group do about 350 out of every 100,000 women develop breast cancer?

7. Does the breast-cancer mortality rate seem to increase from the youngest to the oldest age group?

Bar graphs are often drawn vertically and sometimes a double bar graph is used to make comparisons.

EXAMPLE 2 *Breast Cancer.* The following graph indicates the incidence and mortality rates of breast cancer for women of various age groups.

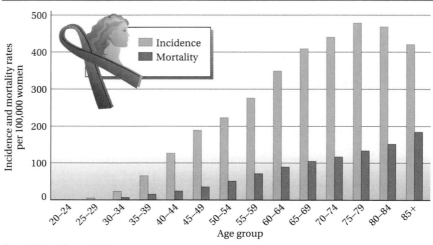

When Breast Cancer Strikes

Source: National Cancer Institute

a) Approximately how many women, per 100,000, develop breast cancer between the ages of 40 and 44?

b) In what age range is the mortality rate for breast cancer approximately 100 for every 100,000 women?

c) In what age range is the incidence of breast cancer the highest?

d) Does the incidence of breast cancer seem to increase from the youngest to the oldest age group?

We look at the graph to answer the questions.

a) We go to the right, across the bottom, to the green bar above the age group 40–44. Next, we go up to the top of that bar and, from there, back to the left to read approximately 130 on the vertical scale. About 130 out of every 100,000 women develop breast cancer between the ages of 40 and 44.

b) We read up the vertical scale to the number 100. From there we move to the right until we come to the top of a red bar. Moving down that bar, we find that in the 65–69 age group, about 100 out of every 100,000 women die of breast cancer.

c) We look for the tallest green bar and read the age range below it. The incidence of breast cancer is highest for women in the 75–79 age group.

d) Looking at the heights of the bars, we see that the incidence of breast cancer increases to a high point in the 75–79 age group and then decreases.

Do Exercises 4–7.

Answers on page A-17

b   Drawing Bar Graphs

EXAMPLE 3   *Police Officers.*   Listed below are the numbers of police officers per 10,000 people in various cities. Make a vertical graph of the data.

CITY	POLICE OFFICERS, PER 10,000 PEOPLE
Cincinnati	31
Cleveland	37
Minneapolis	26
Pittsburgh	30
Columbus	26
St. Louis	44

Sources: U.S. Census Bureau; FBI Uniform Crime Reports

First, we indicate the different names of the cities in six equally spaced intervals on the horizontal scale and give the horizontal scale the title "Cities." (See the figure on the left below.)

Next, we scale the vertical axis. To do so, we look over the data and note that it ranges from 26 to 44. We start the vertical scaling at 0, labeling the marks by 5's from 0 to 50. We give the vertical scale the title "Number of Officers (per 10,000 people)."

Finally, we draw vertical bars to show the various numbers, as shown in the figure at the right. We give the graph an overall title, "Police Officers."

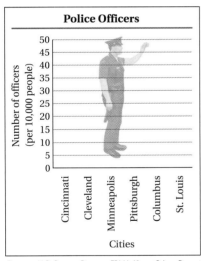

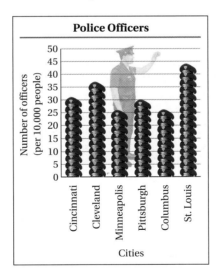

Sources: U.S. Census Bureau; FBI Uniform Crime Reports

Do Exercise 8.

c   Reading and Interpreting Line Graphs

Line graphs are often used to show a change over time as well as to indicate patterns or trends.

EXAMPLE 4   *New Home Sales.*   The following line graph shows the number of new home sales, in thousands, over a twelve-month period. The jagged line

---

8. Planetary Moons.   Make a horizontal bar graph to show the number of moons orbiting the various planets.

PLANET	MOONS
Earth	1
Mars	2
Jupiter	17
Saturn	28
Uranus	21
Neptune	8
Pluto	1

Source: National Aeronautics and Space Administration

Answer on page A-17

Use the line graph in Example 4 to answer Margin Exercises 9–11.

9. For which month were new home sales lowest?

at the base of the vertical scale indicates an unnecessary portion of the scale. Note that the vertical scale differs from the horizontal scale so that the data can be shown reasonably.

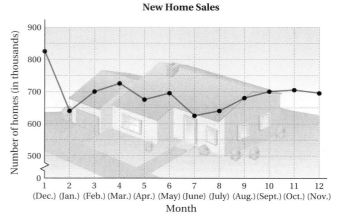

New Home Sales

Source: U.S. Department of Commerce

a) For which month were new home sales the greatest?

b) Between which months did new home sales increase?

c) For which months were new home sales about 700 thousand?

We look at the graph to answer the questions.

a) The greatest number of new home sales was about 825 thousand in month 1, December.

b) Reading the graph from left to right, we see that new home sales increased from month 2 to month 3, from month 3 to month 4, from month 5 to month 6, from month 7 to month 8, from month 8 to month 9, from month 9 to month 10, and from month 10 to month 11.

c) We look from left to right along the line at 700.

10. Between which months did new home sales decrease?

11. For which months were new home sales about 650 thousand?

New Home Sales

We see that points are close to 700 thousand at months 3, 6, 10, 11, and 12.

Do Exercises 9–11.

EXAMPLE 5 *Monthly Loan Payment.* Suppose that you borrow $110,000 at an interest rate of 9% to buy a home. The following graph shows the monthly payment required to pay off the loan, depending on the length of the loan.

Answers on page A-17

(*Caution:* A low monthly payment means that you will pay more interest over the duration of the loan.)

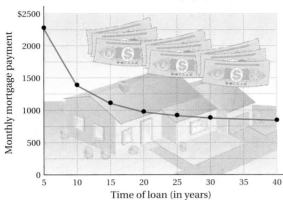

$110,000 Loan Repayment

a) Estimate the monthly payment for a loan of 15 yr.

b) What time period corresponds to a monthly payment of about $1400?

c) By how much does the monthly payment decrease when the loan period is increased from 10 yr to 20 yr?

We look at the graph to answer the questions.

a) We find the time period labeled "15" on the bottom scale and move up from that point to the line. We then go straight across to the left and find that the monthly payment is about $1100.

b) We locate $1400 on the vertical axis. Then we move to the right until we hit the line. The point $1400 is on the line at the 10-yr time period.

c) The graph shows that the monthly payment for 10 yr is about $1400; for 20 yr, it is about $990. Thus the monthly payment is decreased by $1400 − $990, or $410. (It should be noted that you will pay back $990 · 20 · 12 − $1400 · 10 · 12, or $69,600, more in interest for a 20-yr loan.)

Do Exercises 12–14.

d | Drawing Line Graphs

EXAMPLE 6 *Cell Phones with Internet Access.* Listed below are projections on the use of cell phones with access to the Internet. Make a line graph of the data.

YEAR	CELL PHONES WITH WEB ACCESS (in millions)
2001	29.4
2002	69.6
2003	120.1
2004	152.4
2005	171.1

Source: Forrester Research

Use the line graph in Example 5 to answer Margin Exercises 12–14.

12. Estimate the monthly payment for a loan of 25 yr.

13. What time period corresponds to a monthly payment of about $850?

14. By how much does the monthly payment decrease when the loan period is increased from 5 yr to 20 yr?

Answers on page A-17

15. Cell Phones. Listed below are projections on the use of cell phones with or without access to the Internet. Make a line graph of the data.

YEAR	NUMBER OF CELL PHONES (in millions)
2001	119.8
2002	135.3
2003	150.2
2004	163.8
2005	176.9

Source: Forrester Research

First, we indicate the different years on the horizontal scale and give the horizontal scale the title "Year." (See the figure on the left below.) Next, we scale the vertical axis by 25's to show the number of phones, in millions, and give the vertical scale the title "Number of cell phones (in millions)". We also give the graph the overall title "Cell Phones with Web Access."

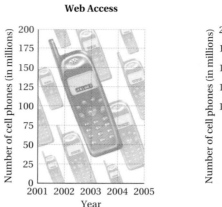
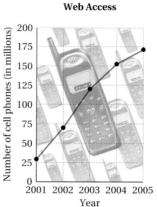

Next, we mark the number of phones at the appropriate level above each year. Then we draw line segments connecting the points. The dramatic change over time can now be observed easily from the graph.

Do Exercise 15.

Answer on page A-17

Answer on page A-17

Study Tips

ATTITUDE AND THE POWER OF YOUR CHOICES

Making the right choices can give you the power to succeed in learning mathematics.

You can choose to improve your attitude and raise the academic goals that you have set for yourself. Projecting a positive attitude toward your study of mathematics and expecting a positive outcome can make it easier for you to learn and to perform well in this course.

Here are some positive choices you can make:

- Choose to allocate the proper amount of time to learn.
- Choose to place the primary responsibility for learning on yourself.
- Choose to establish a learning relationship with your instructor.
- Choose to make a strong commitment to learning.

Well-known American psychologist William James once said, "The one thing that will guarantee the successful conclusion of a doubtful undertaking is faith in the beginning that you can do it." Having a positive attitude and making the right choices will add to your confidence in this course and multiply your successes.

7.3

EXERCISE SET

For Extra Help

Digital Video
Tutor CD 4
Videotape 12

InterAct
Math

Math Tutor
Center

MathXL

MyMathLab

a *Chocolate Desserts.*   The following horizontal bar graph shows the average caloric content of various kinds of chocolate desserts. Use the bar graph for Exercises 1–12.

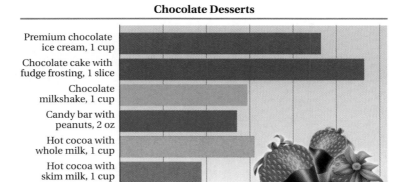

Chocolate Desserts

Source: *Better Homes and Gardens*, December 1996

1. Estimate how many calories there are in 1 cup of hot cocoa with skim milk.

2. Estimate how many calories there are in 1 cup of premium chocolate ice cream.

3. Which dessert has the highest caloric content?

4. Which dessert has the lowest caloric content?

5. Which dessert contains about 460 calories?

6. Which desserts contain about 300 calories?

7. How many more calories are there in 1 cup of hot cocoa made with whole milk than in 1 cup of hot cocoa made with skim milk?

8. Fred generally drinks a 4-cup chocolate milkshake. How many calories does he consume?

9. Kristin likes to eat 2 cups of premium chocolate ice cream at bedtime. How many calories does she consume?

10. Barney likes to eat a 6-oz chocolate bar with peanuts for lunch. How many calories does he consume?

11. Paul adds a 2-oz chocolate bar with peanuts to his diet each day for 1 yr (365 days) and makes no other changes in his eating or exercise habits. Consumption of 3500 extra calories will add about 1 lb to his body weight. How many pounds will he gain?

12. Tricia adds one slice of chocolate cake with fudge frosting to her diet each day for one year (365 days) and makes no other changes in her eating or exercise habits. The consumption of 3500 extra calories will add about 1 lb to her body weight. How many pounds will she have gained at the end of the year?

Education and Earnings. Side-by-side bar graphs allow for comparisons. The one shown at right provides data on the effect of education on earning power for men and women from 1970 to 1997. Use the bar graph in Exercises 13–20.

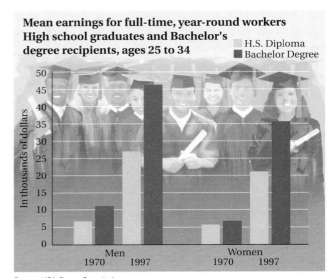

Source: USA Group Foundation

13. How much was the mean earnings for men with bachelor's degrees in 1970? in 1997? How much had it increased? What was the percent of increase?

14. How much was the mean earnings for women with bachelor's degrees in 1970? in 1997? How much had it increased? What was the percent of increase?

15. How much was the mean earnings for women who had ended their education at high school graduation in 1970? in 1997? How much had it increased? What was the percent of increase?

16. How much was the mean earnings for men who had ended their education at high school graduation in 1970? in 1997? How much had it increased? What was the percent of increase?

17. In 1970, how much more did men with bachelor's degrees earn than men who ended their education at high school graduation?

18. In 1997, how much more did men with bachelor's degrees earn than men who ended their education at high school graduation?

19. In 1997, how much more did women with bachelor's degrees earn than men who ended their education at high school graduation?

20. In 1970, how much more did men with bachelor's degrees earn than women who ended their education at high school graduation?

21. *Commuting Time.* The following table lists the average commuting time in six metropolitan areas with more than 1 million people. Make a vertical bar graph to illustrate the data.

CITY	COMMUTING TIME (in minutes)
New York	30.6
Los Angeles	26.4
Phoenix	23.0
Dallas	24.1
Indianapolis	21.9
Orlando	22.9

Source: U.S. Census Bureau

Use the data and the bar graph in Exercise 21 to do Exercises 22–25.

22. Which city has the greatest commuting time?

23. Which city has the least commuting time?

24. What was the median commuting time for all six cities?

25. What was the average commuting time for the six cities?

26. *Airline Net Profits.* The net profits (the amount remaining after all deductions like expenses have been made) of U.S. airlines in various years are listed in the table below. Make a horizontal bar graph illustrating the data.

YEAR	NET PROFIT (in billions)
1995	$2.3
1996	2.8
1997	5.2
1998	4.9
1999	5.5
2000	2.7

Source: Air Transportation Association

Use the data and the bar graph in Exercise 26 to do Exercises 27–32.

27. Between which pairs of years was there an increase in profit?

28. Between which pairs of years was there a decrease in profit?

29. What was the percent of decrease between 1999 and 2000?

30. What was the percent of increase between 1996 and 1997?

31. What was the average net profit for all 6 yr?

32. What was the median net profit for all 6 yr?

C *Golf Distances.* In recent years, new equipment and technology have had a tremendous impact on the distance a golfer can hit a golf ball. The line graph below shows the average driving distances for years from 1980 to 1999. Use the graph for Exercises 33–36.

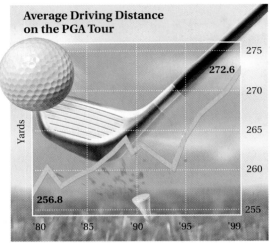

Average Driving Distance on the PGA Tour

Yards

275
272.6
270
265
260
256.8
255

'80 '85 '90 '95 '99

Source: U.S. Golf Association

33. How much farther was the driving distance in 1999 than in 1980?

34. What was the percent of increase in the driving distance from 1980 to 1999?

35. In what years was the average driving distance about 264 yd?

36. In what year was the average driving distance about 270 yd?

d Make a line graph of the data in the following tables (Exercises 37 and 42), using the horizontal axis to scale "Year."

37. *Longevity Beyond Age 65.* These data tell us the number of years a 65-year-old male in the given year can expect to live. Draw a line graph.

YEAR	AVERAGE NUMBER OF YEARS MEN ARE ESTIMATED TO LIVE BEYOND AGE 65
1980	14
1990	15
2000	15.9
2010	16.4
2020	16.9
2030	17.5

Source: 2000 Social Security Report

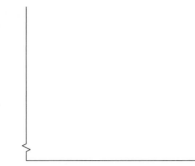

38. What was the percent of increase in longevity (years beyond 65) between 1980 and 2000?

39. What is the expected percent of increase in longevity between 1980 and 2030?

40. What will be the percent of increase in longevity between 2020 and 2030?

41. What will be the percent of increase in longevity between 2000 and 2030?

42. *Homicide Rate in Baltimore.* These data indicate the number of homicides in Baltimore, Maryland, for several years. Draw a line graph.

YEAR	NUMBER OF MURDERS COMMITTED
1995	325
1996	331
1997	311
1998	313
1999	305
2000	262

Source: Baltimore Police Department

43. Between which two years was the increase in murders the greatest?

44. Between which two years was the increase in murders the least?

45. What is the average number of murders committed per year from 1995 to 2000?

46. What is the median number of murders committed?

47. What was the percent of decrease in the murders from 1999 to 2000?

48. What was the percent of increase in the number of murders from 1995 to 1996?

49. D_W Can bar graphs always, sometimes, or never be converted to line graphs? Why?

50. D_W Compare bar graphs and line graphs. Discuss why you might use one rather than the other to graph a particular set of data.

SKILL MAINTENANCE

Solve.

51. A clock loses 3 min every 12 hr. At this rate, how much time will the clock lose in 72 hr?   [5.4a]

52. Managers of pizza restaurants know that if 50 pizzas are ordered in an evening, people will request extra cheese on 9 of them. What percent of the pizzas sold are ordered with extra cheese?   [6.5a]

53. 110% of 75 is what?   [6.3b], [6.4b]

54. 34 is what percent of 51?   [6.3b[, [6.4b]

Convert to percent notation.   [6.2a]

55. $\dfrac{17}{32}$

56. $\dfrac{11}{16}$

57. $\dfrac{673}{1000}$

58. $\dfrac{9781}{10,000}$

59. $\dfrac{19}{16}$

60. $\dfrac{33}{32}$

61. $\dfrac{64}{125}$

62. $\dfrac{249}{250}$

SYNTHESIS

63. *Movie Theater Screens.*   Draw a line graph of the data presented in the following bar graph. Use the line graph and the data to make the best estimate you can of the number of screens in 2000, 2001, and 2003. Explain.

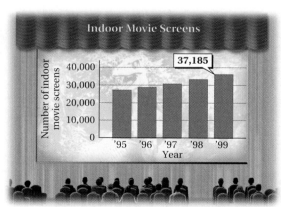

Source: Motion Picture Association of America

7.4 CIRCLE GRAPHS

Objectives

 a Extract and interpret data from circle graphs.

b Draw circle graphs.

We often use **circle graphs,** also called **pie charts,** to show the percent of a quantity used in different categories. Circle graphs can also be used very effectively to show visually the *ratio* of one category to another. In either case, it is quite often necessary to use mathematics to find the actual amounts represented for each specific category.

a Reading and Interpreting Circle Graphs

EXAMPLE 1 *Costs of Owning a Dog.* The following circle graph shows the relative costs of raising a dog from birth to death.

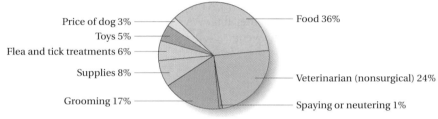

Costs of Owning a Dog

Price of dog 3%
Toys 5%
Flea and tick treatments 6%
Supplies 8%
Grooming 17%
Food 36%
Veterinarian (nonsurgical) 24%
Spaying or neutering 1%

Source: The American Pet Products Manufacturers Association

a) Which item costs the most?

b) What percent of the total cost is spent on grooming?

c) Which item involves 24% of the cost?

d) The American Pet Products Manufacturers Association estimates that the total cost of owning a dog for its lifetime is $6600. How much of that amount is spent for food?

e) What percent of the expense is for grooming and flea and tick treatments?

We look at the sections of the graph to find the answers.

a) The largest section (or sector) of the graph, 36%, is for food.

b) We see that grooming is 17% of the cost.

c) Nonsurgical veterinarian bills account for 24% of the cost.

d) The section of the graph representing food costs is 36%; 36% of $6600 is $2376.

e) We add the percents corresponding to grooming and flea and tick treatments. We have

17% (grooming) + 6% (flea and tick treatments) = 23%.

Do Exercises 1–4.

Use the circle graph in Example 1 to answer Margin Exercises 1–4.

1. Which item costs the least?

2. What percent is not spent on either toys or supplies?

3. How much of the $6600 lifetime cost of owning a dog is for grooming?

4. What part of the expense is for supplies and for buying the dog?

Answers on page A-18

5. Lengths of Engagement of Married Couples. The data below relate the percent of married couples who were engaged for a certain time period before marriage. Use this information to draw a circle graph.

ENGAGEMENT PERIOD	PERCENT
Less than 1 year	24
1–2 years	21
More than 2 years	35
Never engaged	20

Source: Bruskin Goldring Research

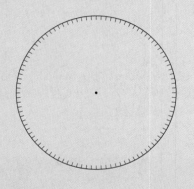

Answer on page A-18

Study Tips

TUNE OUT DISTRACTIONS

Are the places you generally study noisy? If there is constant noise in your home, dorm, or sorority/fraternity house, consider finding a quiet place in the library—maybe a place where the flow of people is minimized and you are not distracted with people-watching!

b Drawing Circle Graphs

To draw a circle graph, or pie chart, like the one in Example 1, think of a pie cut into 100 equally sized pieces. We would then shade in a wedge equal in size to 36 of these pieces to represent 36% for food. We shade a wedge equal in size to 5 of these pieces to represent 5% for toys, and so on.

EXAMPLE 2 *Fruit Juice Sales.* The percents of various kinds of fruit juice sold are given in the list at right. Use this information to draw a circle graph.
Source: Beverage Marketing Corporation

Apple:	14%
Orange:	56%
Blends:	6%
Grape:	5%
Grapefruit:	4%
Prune:	1%
Other:	14%

Using a circle with 100 equally spaced tick marks, we start with the 14% given for apple juice. We draw a line from the center to any tick mark. Then we count off 14 ticks and draw another line. We shade the wedge with a color—in this case, red—and label the wedge as shown in the figure on the left below.

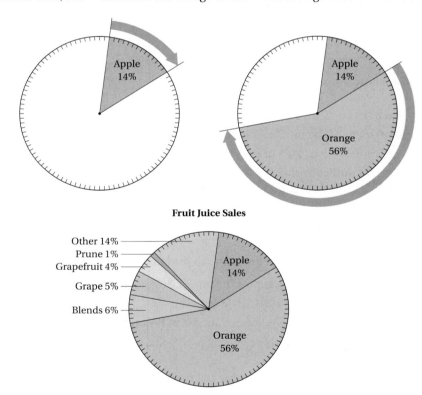

Fruit Juice Sales

To shade a wedge for orange juice, at 56%, we count off 56 ticks and draw another line. We shade the wedge with a different color—in this case, orange—and label the wedge as shown in the figure on the right above. Continuing in this manner and choosing different colors, we obtain the graph shown above. Finally, we give the graph the overall title "Fruit Juice Sales."

Do Exercise 5.

a *Musical Recordings.*   This circle graph, in the shape of a CD, shows music preferences of customers on the basis of music store sales. Use the graph for Exercises 1–6.

Musical Recordings

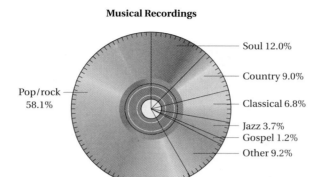

Soul 12.0%

Country 9.0%

Classical 6.8%

Jazz 3.7%
Gospel 1.2%
Other 9.2%

Pop/rock
58.1%

Source: National Association of Recording Merchandisers

1. What percent of all recordings sold are jazz?

2. Together, what percent of all recordings sold are either soul or pop/rock?

3. Camelot Music Store sells 3000 recordings a month. How many are country?

4. Sam's Music Store sells 2500 recordings a month. How many are gospel?

5. What percent of all recordings are neither soul nor jazz?

6. What percent of all recordings are not pop/rock?

Family Expenses.   This circle graph shows expenses as a percent of income for a family of four. Use the graph for Exercises 7–10.

Family Expenses

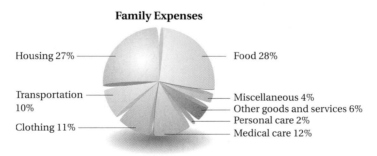

Housing 27%

Food 28%

Transportation
10%

Miscellaneous 4%
Other goods and services 6%
Personal care 2%
Medical care 12%

Clothing 11%

Source: U.S. Bureau of Labor Statistics

7. Which item accounts for the greatest expense?

8. In a family with a $4000 monthly income, how much is spent for transportation?

9. Some surveys combine medical care with personal care. What percent would be spent on those two items combined?

10. In a family with a $2000 monthly income, what is the ratio of the amount spent on medical care to the amount spent on personal care?

b Use the given information to complete a circle graph. Note that each circle is divided into 100 sections.

11. *Holiday Baking.* The table below lists the percentages of when people do their holiday baking.

PREFERENCE	PERCENT
Pre-dawn, before 6 A.M.	28.7%
Late night, 9 P.M. to midnight	11.7%
Overnight, midnight to 5 A.M.	11.5%
Marathon style, all day/night	10.5%
Take day off work	3.4%
Don't bake for the holidays	23.4%
Don't know	10.8%

Source: Land O'Lakes Holiday Bakeline

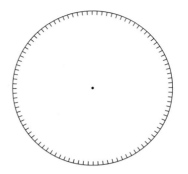

12. *Wealthy Givers.* The table below lists the amounts that people with a net worth of over $1 million make in charitable contributions.

DONATIONS	PERCENT
Under $1000	18%
$1000–$2499	21%
$2500–$4999	20%
$5000–$9999	17%
$10,000 or more	18%
Don't contribute/no answer	6%

Source: Yankelovich Partners

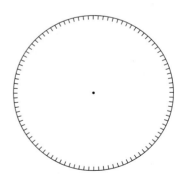

13. *Pregnancy Weight Gain.* The table below lists the amounts of weight gain during pregnancy.

WEIGHT GAIN (in pounds)	PERCENT
Less than 20	22%
21–30	32%
31–40	27%
41 or more	19%

Source: National Vital Statistics Report

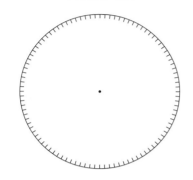

14. *e-mail.* The table below lists the numbers of e-mails people get per day at work.

NUMBER OF E-MAILS PER DAY	PERCENT
Less than 1	28%
1–5	20%
6–10	12%
11–20	9%
21 or more	31%

Source: John J. Heldrich Center for Workforce Development

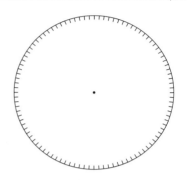

15. *Causes of Spinal Cord Injuries.* The table below lists the causes of spinal cord injury.

CAUSES	PERCENT
Motor vehicle accidents	44%
Acts of violence	24%
Falls	22%
Sports	8%
Other	2%

Source: National Spinal Cord Injury Association

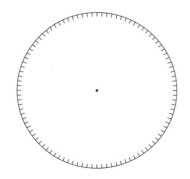

16. *Kids in Foster Care.* There are approximately one-half million children in foster care in the United States. Most of these children are under the age of 10. The table below lists the percentages by ages of children in foster care.

AGE GROUP	PERCENT
Under 1	3%
1–5	25%
6–10	27%
11–15	27%
16+	18%

Source: The Administration for Children and Families

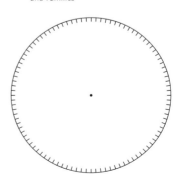

17. ^{D}W Discuss the advantages of being able to read a circle graph.

18. ^{D}W Compare circle graphs to bar graphs.

Convert each percent to fraction notation. Then simplify.   [6.2b], [2.5b]

	CAUSES	PERCENT	FRACTION NOTATION
19.	Motor vehicle accidents	44	
20.	Acts of violence	24	
21.	Falls	22	
22.	Sports	8	
23.	Other	2	

Sources: National Spinal Cord Injury Association, Purdue University

	NUMBER OF E-MAILS	PERCENT	FRACTION NOTATION
24.	Less than 1	28	
25.	1–5	20	
26.	6–10	12	
27.	11–20	9	
28.	21 or more	31	

Source: John J. Heldrich Center for Workforce Development

The review that follows is meant to prepare you for a chapter exam. It consists of two parts. The first part is a checklist of some of the Study Tips referred to in this and preceding chapters. The second part is the Review Exercises. These provide practice exercises for the exam, together with references to section objectives so you can go back and review. Before beginning, stop and look back over the skills you have obtained. What skills in mathematics do you have now that you did not have before studying this chapter?

STUDY TIPS CHECKLIST

The foundation of all your study skills is TIME!	☐ Have you tried using the AWL Math Tutor Center?
	☐ Are you keeping one section ahead in your syllabus?
	☐ Are you working even-numbered exercises to better prepare you to take exams?
	☐ Have you done all your homework on time and been on time to all your classes?

REVIEW EXERCISES

FedEx Mailing Costs. Federal Express has three types of delivery service for packages of various weights within a certain distance, as shown in the following table. Use this table for Exercises 1–6. [7.2a]

Delivery by 10:00 a.m. next business day	Delivery by 3:00 p.m. next business day	Delivery by 4:30 p.m. second business day	
	FEDEX PRIORITY OVERNIGHT®	FEDEX STANDARD OVERNIGHT®	FEDEX 2DAY®
FEDEX LETTER			

All other packaging / weight in lbs.	FEDEX PRIORITY OVERNIGHT®	FEDEX STANDARD OVERNIGHT®	FEDEX 2DAY®
up to 8 oz.	$ 16.50	$ 14.25	$ n/a
1 lb.	$ 23.75	$ 20.00	$ 9.25
2 lbs.	26.75	22.50	10.75
3	29.50	25.25	12.00
4	32.25	27.50	13.75
5	35.00	30.00	15.25
6	38.00	32.00	17.25
7	40.50	34.50	19.25
8	43.25	36.50	21.25
9	46.25	38.75	23.00
10	49.25	41.00	25.00
11	51.75	43.00	26.50

Source: Federal Express Corporation

1. Find the cost of a 3-lb FedEx Priority Overnight delivery.

2. Find the cost of a 10-lb FedEx Standard Overnight delivery.

3. How much would you save by sending the package listed in Exercise 1 by FedEx 2Day delivery?

4. How much would you save by sending the package in Exercise 2 by FedEx 2Day delivery?

5. Is there any difference in price between sending a 5-oz package FedEx Priority Overnight and sending an 8-oz package in the same way?

6. An author has a 4-lb manuscript to send by FedEx Standard Overnight delivery to her publisher. She calls and the package is picked up. Later that day she completes work on another part of her manuscript that weighs 5 lb. She calls and sends it by FedEx Standard Overnight delivery to the same address. How much could she have saved if she had waited and sent both packages as one?

U.S. Police Forces. This pictograph shows the number of officers in the largest U.S. police forces. Use the graph for Exercises 7–10.

America's Largest Police Forces

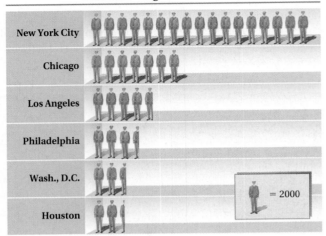

Source: International Association of Chiefs of Police

7. About how many officers are in the Chicago police force? [7.2b]

8. Which city has about 9000 officers on its force? [7.2b]

9. Of the cities listed, which has the smallest police force? [7.2b]

10. Estimate the average size of these six police forces. [7.1a], [7.2b]

Find the mode. [7.1c]

11. 26, 34, 43, 26, 51

12. 17, 7, 11, 11, 14, 17, 18

13. 0.2, 0.2, 1.7, 1.9, 2.4, 0.2

14. 700, 700, 800, 2700, 800

15. $14, $17, $21, $29, $17, $2

16. 20, 20, 20, 20, 20, 500

17. One summer, a student earned the following amounts over a four-week period: $102, $112, $130, and $98. What was the average amount earned per week? the median? [7.1a, b]

18. *Gas Mileage.* A 2001 Ford Focus gets 528 miles of highway driving on 16 gallons of gasoline. What is the gas mileage? [7.1a]
Source: Ford Motor Company

19. To get an A in math, a student must score an average of 90 on four tests. Scores on the first three tests were 94, 78, and 92. What is the lowest score that the student can make on the last test and still get an A? [7.1a]

Calorie Content in Fast Foods. Wendy's Hamburgers is a national food franchise. The following bar graph shows the caloric content of various sandwiches sold by Wendy's. Use the graph for Exercises 20–27. [7.3a]

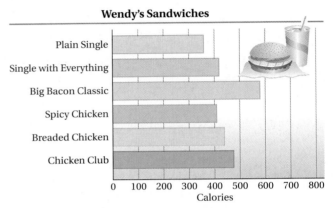

Source: Wendy's International

20. How many calories are in a Single with Everything?

21. How many calories are in a Breaded Chicken sandwich?

22. Which sandwich has the highest caloric content?

23. Which sandwich has the lowest caloric content?

24. Which sandwich contains about 360 calories?

25. Which sandwich contains about 470 calories?

26. How many more calories are in a Chicken Club than in a Single with Everything?

27. How many more calories are in a Big Bacon Classic than in a Plain Single?

Accidents by Driver Age. The following line graph shows the number of accidents per 100 drivers, by age. Use the graph for Exercises 28–33. [7.3c]

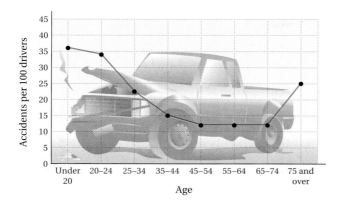

28. Which age group has the most accidents per 100 drivers?

29. What is the fewest number of accidents per 100 in any age group?

30. How many more accidents do people over 75 yr of age have than those in the age range of 65–74?

31. Between what ages does the number of accidents stay basically the same?

32. How many fewer accidents do people 25–34 yr of age have than those 20–24 yr of age?

33. Which age group has accidents more than three times as often as people 55–64 yr of age?

Hotel Preferences. This circle graph shows hotel preferences for travelers. Use the graph for Exercises 34–37. [7.4a]

Types of Hotels

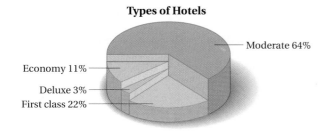

Moderate 64%
Economy 11%
Deluxe 3%
First class 22%

34. What percent of travelers prefer a first-class hotel?

35. What percent of travelers prefer an economy hotel?

36. Suppose 2500 travelers arrive in a city one day. How many of them might seek a moderate room?

37. What percent of travelers prefer either a first-class or a deluxe hotel?

First-Class Postage. The following table shows the cost of first-class postage in various years. Use the table for Exercises 38 and 39.

YEAR	FIRST-CLASS POSTAGE
1983	20¢
1989	25¢
1991	29¢
1995	32¢
1999	33¢
2001	34¢

Source: U.S. Postal Service

38. Make a vertical bar graph of the data. [7.3b]

39. Make a line graph of the data. [7.3d]

40. *Battery Testing.* An experiment is performed to compare battery quality. Two kinds of battery were tested to see how long, in hours, they kept a hand radio running. On the basis of this test, which battery is better? [7.1d]

BATTERY A: TIMES (in hours)		
38.9	39.3	40.4
53.1	41.7	38.0
36.8	47.7	48.1
38.2	46.9	47.4

BATTERY B: TIMES (in hours)		
39.3	38.6	38.8
37.4	47.6	37.9
46.9	37.8	38.1
47.9	50.1	38.2

Find the average. [7.1a]

41. 26, 34, 43, 51

42. 11, 14, 17, 18, 7

43. 0.2, 1.7, 1.9, 2.4

44. 700, 2700, 3000, 900, 1900

45. $2, $14, $17, $17, $21, $29

46. 20, 190, 280, 470, 470, 500

Find the median. [7.1b]

47. 26, 34, 43, 51

48. 7, 11, 14, 17, 18

49. 0.2, 1.7, 1.9, 2.4

50. 700, 900, 1900, 2700, 3000

51. $2, $17, $21, $29, $14, $17

52. 470, 20, 190, 280, 470, 500

53. *Grade Point Average.* Find the grade point average for one semester given the following grades. Assume the grade point values are 4.0 for A, 3.0 for B, and so on. Round to the nearest tenth. [7.1a]

COURSE	GRADE	NUMBER OF CREDIT HOURS IN COURSE
Basic math	A	5
English	B	3
Computer applications	C	4
Russian	B	3
College skills	B	1

54. D_W Compare and contrast averages, medians, and modes. Discuss why you might use one over the others to analyze a set of data. [7.1a, b, c]

55. D_W Find a real-world situation that fits this equation: [7.1a]

$$T = \frac{(20{,}500 + 22{,}800 + 23{,}400 + 26{,}000)}{4}.$$

Certain objectives from four particular sections will be retested on the chapter test. The objectives are listed with the practice problems that follow.

Solve.

56. A company car was driven 4200 mi in the first 4 months of a year. At this rate, how far will it be driven in 12 months? [5.4a]

57. 92% of the world population does not have a telephone. The population is about 6.26 billion. How many do not have a telephone? [6.5a]

58. 789 is what percent of 355.05? [6.3b], [6.4b]

59. What percent of 98 is 49? [6.3b], [6.4b]

Divide and simplify. [2.7b]

60. $\dfrac{3}{4} \div \dfrac{5}{6}$

61. $\dfrac{5}{8} \div \dfrac{3}{2}$

62. The ordered set of data 298, 301, 305, *a*, 323, *b*, 390 has a median of 316 and an average of 326. Find *a* and *b*. [7.1a, b]

Chapter Test

Desirable Body Weights. The following tables list the desirable body weights for men and women over age 25. Use the tables for Exercises 1–4.

DESIRABLE WEIGHT OF MEN

Height	Small Frame (in pounds)	Medium Frame (in pounds)	Large Frame (in pounds)
5 ft, 7 in.	138	152	166
5 ft, 9 in.	146	160	174
5 ft, 11 in.	154	169	184
6 ft, 1 in.	163	179	194
6 ft, 3 in.	172	188	204

DESIRABLE WEIGHT OF WOMEN

Height	Small Frame (in pounds)	Medium Frame (in pounds)	Large Frame (in pounds)
5 ft, 1 in.	105	113	122
5 ft, 3 in.	111	120	130
5 ft, 5 in.	118	128	139
5 ft, 7 in.	126	137	147
5 ft, 9 in.	134	144	155

Source: U.S. Department of Agriculture

1. What is the desirable weight for a 6 ft, 1 in. man with a medium frame?

2. What is the desirable weight for a 5 ft, 3 in. woman with a small frame?

3. What size woman has a desirable weight of 120 lb?

4. What size man has a desirable weight of 169 lb?

TV News Magazine Programs. The number of network news magazine programs has increased dramatically since the early 1980s when there were just two—ABC's "20/20" and CBS's "60 Minutes." In the pictograph at right, each symbol represents a 1-hr prime-time news magazine in the network's weekly fall schedule. Use the pictograph for Exercises 5–8.

5. In which year was there exactly 7 hr of prime-time news magazine programming per week?

6. In which year was there exactly 6 hr of prime-time news magazine programming per week?

7. How many hours per week of prime-time news magazine programming were there in 1994?

8. How many hours per week of prime-time news magazine programming were there in 1998?

1992 1993 1994 1995 1996 1997 1998

Sources: "Total Television," by Alex McNeil. Penguin Books; *The Hollywood Reporter;* Fox Broadcasting.

The New York Times, June 8, 1998

Find the average.

9. 45, 49, 52, 52

10. 1, 1, 3, 5, 3

11. 3, 17, 17, 18, 18, 20

Find the median and the mode.

12. 45, 49, 52, 52

13. 1, 1, 3, 5, 3

14. 3, 17, 17, 18, 18, 20

15. *Gas Mileage.* A 2001 Saturn S gets 608 miles of highway driving on 16 gallons of gasoline. What is the gas mileage?
Source: Saturn

16. *Grades.* To get a C in chemistry, a student must score an average of 70 on four tests. Scores on the first three tests were 68, 71, and 65. What is the lowest score that the student can make on the last test and still get a C?

Food Dollars Spent Away from Home. The line graph below shows the percentage of food dollars spent away from home for various years, and projected to 2010. Use the graph for Exercises 17–20.

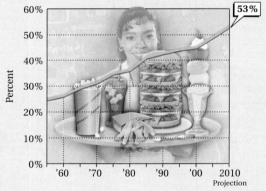

Food dollars spent away from home

Sources: U.S. Bureau of Labor Statistics; National Restaurant Association

17. What percent of meals will be eaten away from home in 2010?

18. What percent of meals were eaten away from home in 1985?

19. In what year was the percent of meals eaten away from home about 30%?

20. In what year will the percent of meals eaten away from home be about 50%?

21. *Animal Speeds.*   The following table lists maximum speeds of movement for various animals, in miles per hour, compared to the speed of the fastest human. Make a vertical bar graph of the data.

ANIMAL	SPEED (in miles per hour)
Antelope	61
Peregrine falcon	225
Cheetah	70
Fastest human	28
Greyhound	42
Golden eagle	150
Grant's gazelle	47

Source: Barbara Ann Kipfer, *The Order of Things.*
New York: Random House, 1998.

Refer to the table and the graph in Exercise 21 for Exercises 22–25.

22. By how much does the fastest speed exceed the slowest speed?

23. Does a human have a chance of outrunning a greyhound? Explain.

24. Find the average of all the speeds.

25. Find the median of all the speeds.

26. *Shoplifting and Employee Theft.*   The following table lists ways in which American retailers lost money recently. Construct a circle graph representing these data.

TYPE OF LOSS	PERCENT
Employee theft	44
Shoplifting	32.7
Administrative error	17.5
Vendor fraud	5.1
Other	0.7

Source: University of Florida, Department of Sociology
for Sensormatic Electronics Corporation

27. In reference to Exercise 26, it is known that retailers lost $23 billion dollars. Using the percents from the table and the circle graph, find the amount of money lost from each type of loss.

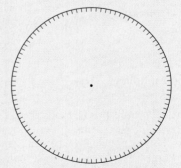

Porsche Sales. The table below lists the number of Porsche sales in the United States for various years. Use the table for Exercises 28 and 29.

YEAR	PORSCHE U.S. SALES
1996	7,152
1997	12,980
1998	17,239
1999	20,877
2000	23,000

Sources: Autodata, Bridge Information Systems

28. Make a bar graph of the data.

29. Make a line graph of the data.

30. *Chocolate Bars.* An experiment is performed to compare the quality of new Swiss chocolate bars being introduced in the United States. People were asked to taste the candies and rate them on a scale of 1 to 10. On the basis of this test, which chocolate bar is better?

BAR A: SWISS PECAN			BAR B: SWISS HAZELNUT		
9	10	8	10	6	8
10	9	7	9	10	10
6	9	10	8	7	6
7	8	8	9	10	8

31. *Grade Point Average.* Find the grade point average for one semester given the following grades. Assume the grade point values are 4.0 for A, 3.0 for B, and so on. Round to the nearest tenth.

COURSE	GRADE	NUMBER OF CREDIT HOURS IN COURSE
Introductory algebra	B	3
English	A	3
Business	C	4
Spanish	B	3
Typing	B	2

SKILL MAINTENANCE

32. Divide and simplify: $\dfrac{3}{5} \div \dfrac{12}{125}$.

33. 17 is 25% of what number?

34. On a particular Sunday afternoon, 78% of the television sets that were on were tuned to one of the major networks. Suppose 20,000 TV sets in a town are being watched. How many are tuned to a major network?

35. A baseball player gets 7 hits in the first 20 times at bat. At this rate, how many times at bat will it take to get 119 hits?

SYNTHESIS

36. The ordered set of data 69, 71, 73, a, 78, 98, b has a median of 74 and a mean of 82. Find a and b.

CHAPTER 7: Data, Graphs, and Statistics

1. *Net Worth.* In 1998, Bill Gates, of Microsoft fame, was worth $51 billion. Write standard notation for $51 billion.
Source: Forbes

2. *Gas Mileage.* A 2001 Saturn S gets 324 miles of city driving on 12 gallons of gasoline. What is the gas mileage?
Source: Saturn

3. In 402,513, what does the digit 5 mean?

4. Evaluate: $3 + 5^3$.

5. Find all the factors of 60.

6. Round 52.045 to the nearest tenth.

7. Convert to fraction notation: $3\frac{3}{10}$.

8. Convert from cents to dollars: 210¢.

9. Convert to standard notation: $3.25 trillion.

10. Determine whether 11, 30 and 4, 12 are proportional.

Compute and simplify.

11. $2\frac{2}{5} + 4\frac{3}{10}$

12. $41.063 + 3.5721$

13. $\frac{14}{15} - \frac{3}{5}$

14. $350 - 24.57$

15. $3\frac{3}{7} \cdot 4\frac{3}{8}$

16. $12{,}456 \times 220$

17. $\frac{13}{15} \div \frac{26}{27}$

18. $104{,}676 \div 24$

Solve.

19. $\frac{5}{8} = \frac{6}{x}$

20. $\frac{2}{5} \cdot y = \frac{3}{10}$

21. $21.5 \cdot y = 146.2$

22. $x = 398{,}112 \div 26$

Solve.

23. Tortilla chips cost $2.99 for 14.5 oz. Find the unit price rounded to the nearest tenth of a cent, in cents per ounce.

24. A college has a student body of 6000 students. Of these, 55.4% own a car. How many students own a car?

25. A piece of fabric $1\frac{3}{4}$ yd long is cut into 7 equal strips. What is the length of each strip?

26. A recipe calls for $\frac{3}{4}$ cup of sugar. How much sugar should be used for $\frac{1}{2}$ of the recipe?

27. *Peanut Products.* In any given year, the average American eats 2.7 lb of peanut butter, 1.5 lb of salted peanuts, 1.2 lb of peanut candy, 0.7 lb of in-shell peanuts, and 0.1 lb of peanuts in other forms. How many pounds of peanuts and products containing peanuts does the average American eat in one year?

28. *Energy Consumption.* In a recent year, American utility companies generated 1464 billion kilowatt-hours of electricity using coal, 455 billion using nuclear power, 273 billion using natural gas, 250 billion using hydroelectric plants, 118 billion using petroleum, and 12 billion using geothermal technology and other methods. How many kilowatt-hours of electricity were produced that year?

29. *Heart Disease.* Of the 281 million people in the United States, 7 million have coronary heart disease and 500,000 die of heart attacks each year. What percent have coronary heart disease? What percent die of heart attacks? Round your answers to the nearest tenth of a percent.
Source: U.S. Centers for Disease Control

30. *Billionaires.* In 1996, the median net worth of U.S. billionaires was $2.5 billion. By 1998, this figure had increased to $6.8 billion. What was the percent of increase?
Source: Forbes

31. *Football Fields.* The Arena Football League (AFL) is a professional league playing indoors, mostly on converted basketball and/or hockey rinks. The figure shows the AFL field compared with the larger field of the National Football League (NFL).

a) Find the area of an AFL field. Include the end zones in your calculation.
b) Find the area of an NFL field. Include the end zones in your calculation.
c) How much larger is an NFL field than an AFL field?

FedEx. The following table lists the cost of delivering a package by FedEx Priority Overnight shipping. Use the table for Questions 32–34.

WEIGHT (in pounds)	COST
1	$23.75
2	26.75
3	29.50
4	32.25
5	35.00
6	38.00
7	40.50
8	43.25
9	46.25
10	49.25
11	51.75

Source: Federal Express Corporation

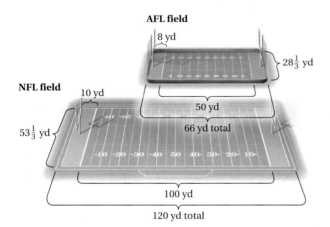

AFL field
8 yd
$28\frac{1}{3}$ yd
NFL field
10 yd
50 yd
66 yd total
$53\frac{1}{3}$ yd
100 yd
120 yd total

32. Find the average and the median of these costs.

33. Make a vertical bar graph of the data.

34. Make a line graph of the data.

35. A business is owned by four people. One owns $\frac{1}{3}$, the second owns $\frac{1}{4}$, and the third owns $\frac{1}{6}$. How much does the fourth person own?

36. In manufacturing valves for engines, a factory was discovered to have made 4 defective valves in a lot of 18 valves. At this rate, how many defective valves can be expected in a lot of 5049 valves?

37. A landscaper bought 22 evergreen trees for $210. What was the cost of each tree? Round to the nearest cent.

38. A salesperson earns $182 selling $2600 worth of electronic equipment. What is the commission rate?

Teen Spending. Teenagers are big spenders. More and more retailers are catering to the 13–19 year-old crowd. Those in this group who shop regularly spend an average of $381 per month. Use the table to answer Exercises 41–44.

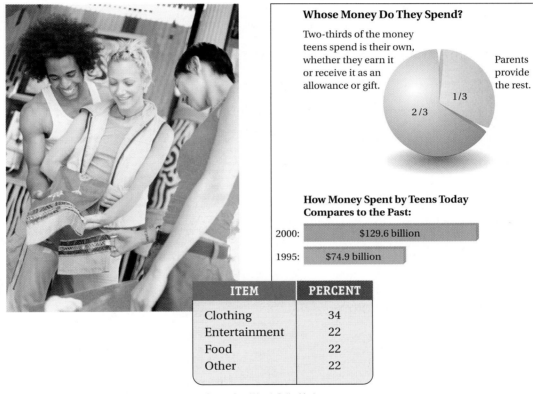

Whose Money Do They Spend?

Two-thirds of the money teens spend is their own, whether they earn it or receive it as an allowance or gift.

Parents provide the rest.

1/3

2/3

How Money Spent by Teens Today Compares to the Past:

2000: $129.6 billion

1995: $74.9 billion

ITEM	PERCENT
Clothing	34
Entertainment	22
Food	22
Other	22

Source: Rand Youth Poll, eMarketer

39. How is their monthly spending allocated?

40. How much of their monthly spending, whether they earn it or receive it as an allowance or gift, is their own? The rest of their spending money comes from their parents. How much of the monthly spending money comes from their parents?

41. In 2000, teenagers spent $129.6 billion. How was it allocated?

42. By what percent has teenage spending increased from 1995 to 2000?

SYNTHESIS

43. A photography club meets four times a month. In September, the attendance figures were 28, 23, 26, and 23. In October, the attendance figures were 26, 20, 14, and 28. What was the percent of increase or decrease in average attendance from September to October?

Measurement

Gateway to Chapter 8

In this chapter, we introduce American and metric systems used to measure length, weight, mass, capacity, and temperature. We present conversion from one unit to another within, as well as between, each system. We then apply these systems to everyday situations and to the field of medicine. We also consider units of time and their conversion.

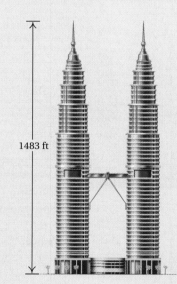

1483 ft

Real-World Application

The height of the Petronas Towers in Kuala Lumpur, Malaysia, is 1483 ft. Find the height in meters.

This problem appears as Example 5 in Section 8.3.

Complete.

1. 8 ft = _____ in.   [8.1a]

2. 5 in. = _____ ft   [8.1a]

3. 8.46 km = _____ m   [8.2a]

4. 9.2 mm = _____ cm   [8.2a]

Complete the following table.   [8.2a]

	OBJECT	MILLIMETERS (mm)	CENTIMETERS (cm)	METERS (m)
5.	Length of a videotape box		19	
6.	Height of Shanghai World Financial Center			457

Complete.   [8.3a]

7. 6.7 cm = _____ in.
(The width of a dollar bill)

8. 93,000,000 mi = _____ km
(The distance from the earth to the sun)

Complete the following table.   [8.1a], [8.2a], [8.3a]

	OBJECT	YARDS (yd)	CENTIMETER (cm)	INCHES (in.)	METERS (m)	MILLIMETERS (mm)
9.	Width of a videotape box		10.6			
10.	Length of a football field	100				

Complete.

11. 2304 mL = _____ L   [8.5a]

12. 2.4 L = _____ mL   [8.5a]

13. 5 lb = _____ oz   [8.4a]

14. 4.4 T = _____ lb   [8.4a]

15. 4.8 kg = _____ g   [8.4b]

16. 6.2 mg = _____ cg   [8.4b]

17. 3400 mg = _____ g   [8.4b]

18. 7 hr = _____ min   [8.6a]

19. 16 days = _____ hr   [8.6a]

20. 128 pt = _____ qt   [8.5a]

21. 20 gal = _____ oz   [8.5a]

22. 3 cups = _____ oz   [8.5a]

23. Convert 77°F to Celsius.   [8.6b]

24. Convert 37°C to Fahrenheit.   [8.6b]

Solve.

25. Medical Dosage. Nitroglycerin sublingual tablets come in 0.4-mg tablets. How many micrograms are in each tablet?   [8.4c]

26. Medical Dosage. For conversion, a pharmacist knows that 1 oz ≈ 29.57 mL. A prescription calls for 3 oz of theophylline. For how many milliliters is the prescription?   [8.5b]

27. A bundle of concert programs weighs 1 kg. How many grams does 1 bundle weigh?   [8.4b]

Complete.

28. $1 \text{ ft}^2 =$ _____ in^2 [8.7a]

29. $2 \text{ km}^2 =$ _____ m^2 [8.7b]

8.1

LINEAR MEASURES: AMERICAN UNITS

Length, or distance, is one kind of measure. To find lengths, we start with some **unit segment** and assign to it a measure of 1. Suppose $\overline{AB}$ below is a unit segment.

Let's measure segment $\overline{CD}$ below, using $\overline{AB}$ as our unit segment.

Since we can place 4 unit segments end to end along $\overline{CD}$, the measure of $\overline{CD}$ is 4.

Sometimes we have to use parts of units, called **subunits.** For example, the measure of the segment $\overline{MN}$ below is $1\frac{1}{2}$. We place one unit segment and one half-unit segment end to end.

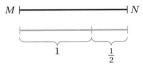

Do Exercises 1–4.

a American Measures

American units of length are related as follows.

(Actual size, in inches)

AMERICAN UNITS OF LENGTH	
12 inches (in.) = 1 foot (ft)	3 feet = 1 yard (yd)
36 inches = 1 yard	5280 feet = 1 mile (mi)

Use the unit below to measure the length of each segment or object.

1.

2.

3.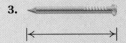

4.

Answers on page A-20

Complete.

5. 8 yd = _____ in.

6. $2\frac{5}{6}$ yd = _____ ft

7. 3.8 mi = _____ in.

We can visualize comparisons of the units as follows:

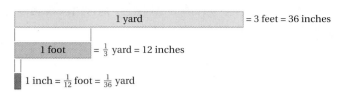

The symbolism 13 in. = 13″ and 27 ft = 27′ is also used for inches and feet. American units have also been called "English," or "British–American," because at one time they were used by both countries. Today, both Canada and England have officially converted to the metric system. However, if you travel in England, you will still see units such as "miles" on road signs.

To change from certain American units to others, we make substitutions. Such a substitution is usually helpful when we are converting from a *larger* unit to a *smaller* one.

EXAMPLE 1 Complete: 1 yd = _____ in.

$$
\begin{aligned}
1 \text{ yd} &= 3 \text{ ft} \\
&= 3 \times 1 \text{ ft} \qquad \text{We think of 3 ft as } 3 \times \text{ft, or } 3 \times 1 \text{ ft.} \\
&= 3 \times 12 \text{ in.} \qquad \text{Substituting 12 in. for 1 ft} \\
&= 36 \text{ in.} \qquad \text{Multiplying}
\end{aligned}
$$

EXAMPLE 2 Complete: $7\frac{1}{3}$ yd = _____ in.

$$
\begin{aligned}
7\frac{1}{3} \text{ yd} &= 7\frac{1}{3} \times 1 \text{ yd} \\
&= 7\frac{1}{3} \times 3 \text{ ft} \qquad \text{Substituting 3 ft for 1 yd} \\
&= 7\frac{1}{3} \times 3 \times 1 \text{ ft} \\
&= \frac{22}{3} \times 3 \times 12 \text{ in.} \qquad \text{Substituting 12 in. for 1 ft} \\
&= 264 \text{ in.}
\end{aligned}
$$

Do Exercises 5–7.

Sometimes it helps to use multiplying by 1 in making conversions. For example, 12 in. = 1 ft, so

$$\frac{12 \text{ in.}}{1 \text{ ft}} = 1 \quad \text{and} \quad \frac{1 \text{ ft}}{12 \text{ in.}} = 1.$$

If we divide 12 in. by 1 ft or 1 ft by 12 in., we get 1 because the lengths are the same. Let's first convert from *smaller* to *larger* units.

Answers on page A-20

EXAMPLE 3 Complete: 48 in. = _____ ft.

We want to convert from "in." to "ft." We multiply by 1 using a symbol for 1 with "in." on the bottom and "ft" on the top to eliminate inches and to convert to feet:

$$48 \text{ in.} = \frac{48 \text{ in.}}{1} \times \frac{1 \text{ ft}}{12 \text{ in.}} \qquad \text{Multiplying by 1 using } \frac{1 \text{ ft}}{12 \text{ in.}} \text{ to eliminate in.}$$

$$= \frac{48 \text{ in.}}{12 \text{ in.}} \times 1 \text{ ft}$$

$$= \frac{48}{12} \times \frac{\text{in.}}{\text{in.}} \times 1 \text{ ft}$$

$$= 4 \times 1 \text{ ft} \qquad \text{The } \frac{\text{in.}}{\text{in.}} \text{ acts like 1, so we can omit it.}$$

$$= 4 \text{ ft.}$$

We can also look at this conversion as "canceling" units:

$$48 \text{ in.} = \frac{48 \cancel{\text{ in.}}}{1} \times \frac{1 \text{ ft}}{12 \cancel{\text{ in.}}} = \frac{48}{12} \times 1 \text{ ft} = 4 \text{ ft.}$$

This method is not used only in mathematics, as here, but also in fields such as medicine, chemistry, and physics.

Do Exercises 8 and 9.

EXAMPLE 4 Complete: 25 ft = _____ yd.

Since we are converting from "ft" to "yd," we choose a symbol for 1 with "yd" on the top and "ft" on the bottom:

$$25 \text{ ft} = 25 \text{ ft} \times \frac{1 \text{ yd}}{3 \text{ ft}} \qquad 3 \text{ ft} = 1 \text{ yd, so } \frac{3 \text{ ft}}{1 \text{ yd}} = 1 \text{, and } \frac{1 \text{ yd}}{3 \text{ ft}} = 1. \text{ We}$$

$$\text{use } \frac{1 \text{ yd}}{3 \text{ ft}} \text{ to eliminate ft.}$$

$$= \frac{25}{3} \times \frac{\text{ft}}{\text{ft}} \times 1 \text{ yd}$$

$$= 8\frac{1}{3} \times 1 \text{ yd} \qquad \text{The } \frac{\text{ft}}{\text{ft}} \text{ acts like 1, so we can omit it.}$$

$$= 8\frac{1}{3} \text{ yd, or } 8.\overline{3} \text{ yd.}$$

Again, in this example, we can consider conversion from the point of view of canceling:

$$25 \text{ ft} = 25 \cancel{\text{ ft}} \times \frac{1 \text{ yd}}{3 \cancel{\text{ ft}}}$$

$$= \frac{25}{3} \times 1 \text{ yd} = 8\frac{1}{3} \text{ yd, or } 8.\overline{3} \text{ yd.}$$

Do Exercises 10 and 11.

Complete.

8. 72 in. = _____ ft

9. 17 in. = _____ ft

Complete.

10. 24 ft = _____ yd

11. 35 ft = _____ yd

Answers on page A-20

Complete.

12. 26,400 ft = _____ mi

13. 2640 ft = _____ mi

14. Complete. Use multiplying by 1.

8 yd = _____ in.

Answers on page A-20

EXAMPLE 5 Complete: 23,760 ft = _____ mi.

We choose a symbol for 1 with "mi" on the top and "ft" on the bottom:

$$23{,}760 \text{ ft} = 23{,}760 \text{ ft} \times \frac{1 \text{ mi}}{5280 \text{ ft}} \qquad 5280 \text{ ft} = 1 \text{ mi, so } \frac{1 \text{ mi}}{5280 \text{ ft}} = 1.$$

$$= \frac{23{,}760}{5280} \times \frac{\text{ft}}{\text{ft}} \times 1 \text{ mi}$$

$$= 4.5 \times 1 \text{ mi} \qquad \text{Dividing}$$

$$= 4.5 \text{ mi.}$$

Let's also consider this example using canceling:

$$23{,}760 \text{ ft} = 23{,}760 \text{ ft} \times \frac{1 \text{ mi}}{5280 \text{ ft}}$$

$$= \frac{23{,}760}{5280} \times 1 \text{ mi}$$

$$= 4.5 \times 1 \text{ mi} = 4.5 \text{ mi.}$$

Do Exercises 12 and 13.

We can also use multiplying by 1 to convert from larger to smaller units. Let's redo Example 2.

EXAMPLE 6 Complete: $7\frac{1}{3}$ yd = _____ in.

$$7\frac{1}{3} \text{ yd} = \frac{22 \text{ yd}}{3} \times \frac{36 \text{ in.}}{1 \text{ yd}} = \frac{22 \times 36}{3} \times 1 \text{ in.} = 264 \text{ in.}$$

Do Exercise 14.

Study Tips

TAKING NOTES

Effective note-taking can greatly enhance your learning of mathematics. Here are some suggestions.

■ This textbook has been written and designed so that it represents a quality set of notes at the same time that it teaches. Thus you might not need to take notes in class. Just watch, listen, and ask yourself questions as the class moves along, rather than racing to keep up your note-taking.

However, if you still feel more comfortable taking your own notes, consider using the following two-column method. Divide your page in half vertically so that you have two columns side by side. Write down what is on the board in the left column; then, in the right column, write clarifying comments or questions.

■ If you have any difficulty keeping up with the instructor, use abbreviations to speed up your note-taking. Consider standard abbreviations like "Ex" for "Example," "≈" for "is approximately equal to," or "∴" for "therefore." Create your own abbreviations as well.

■ Another shortcut for note-taking is to write only the beginning of a word, leaving space for the rest. Be sure you write enough of the word to know what it means later on!

a Complete.

1. 1 ft = _____ in.

2. 1 yd = _____ ft

3. 1 in. = _____ ft

4. 1 mi = _____ yd

5. 1 mi = _____ ft

6. 1 ft = _____ yd

7. 3 yd = _____ in.

8. 10 yd = _____ ft

9. 84 in. = _____ ft

10. 48 ft = _____ yd

11. 18 in. = _____ ft

12. 29 ft = _____ yd

13. 5 mi = _____ ft

14. 5 mi = _____ yd

15. 63 in. = _____ ft

16. 11,616 ft = _____ mi

17. 10 ft = _____ yd

18. 9.6 yd = _____ ft

19. 7.1 mi = _____ ft

20. 31,680 ft = _____ mi

21. $4\frac{1}{2}$ ft = _____ yd

22. 48 in. = _____ ft

23. 45 in. = _____ yd

24. $6\frac{1}{3}$ yd = _____ in.

25. 330 ft = _____ yd

26. 5280 yd = _____ mi

27. 3520 yd = _____ mi

28. 25 mi = _____ ft

29. 100 yd = _____ ft

30. 480 in. = _____ ft

31. 360 in. = _____ ft

32. 720 in. = _____ yd

33. 1 in. = _____ yd

34. 25 in. = _____ ft

35. 2 mi = _____ in.

36. 63,360 in. = _____ mi

37. D_W A student makes the following error:

$$23 \text{ in.} = 23 \cdot (12 \text{ ft}) = 276 \text{ ft}.$$

Explain the error in at least two ways.

38. D_W Describe two methods of making unit conversions discussed in this section.

SKILL MAINTENANCE

Convert to fraction notation.   [6.2b]

39. 9.25%

40. $87\frac{1}{2}\%$

Find fraction notation for the percent notation in the sentence.   [6.2b]

41. Of all 18-year olds, 27.5% are registered to vote.

42. Of all those who buy CDs, 57% are in the 20–39 age group.

Convert to percent notation.   [6.2a]

43. $\dfrac{11}{8}$

44. $\dfrac{2}{3}$

45. $\dfrac{1}{4}$

46. $\dfrac{7}{16}$

Find fraction notation for the ratio.   [5.1a]

47. *Heart Disease.*   In one year, in California, there are 373 deaths due to heart disease for every 100,000 women. What is the ratio of number of deaths due to heart disease to number of women? What is the ratio of number of women to number of deaths due to heart disease?
Source: U.S. Centers for Disease Control

48. *Wildfires.*   In 2000, 73,000 forest fires burned 6.4 million acres in the United States. What was the ratio of number of fires to number of acres? What was the ratio of number of acres to number of fires?
Source: National Forest Service

SYNTHESIS

49. *Noah's Ark.*   In biblical measures, it is thought that 1 cubit ≈ 18 in. The dimensions of Noah's ark are given as follows: "The length of the ark shall be three hundred cubits, the breadth of it fifty cubits, and the height of it thirty cubits." What were the dimensions of Noah's ark in inches? in feet?
Source: *Holy Bible, King James Version*, Gen. 6:15

50. *Goliath's Height.*   In biblical measures, a span was considered to be half of a cubit (1 cubit = 18 in.; see Exercise 49). The giant Goliath's height "was six cubits and a span." What was the height of Goliath in inches? in feet?
Source: *Holy Bible, King James Version*, 1 Sam. 17:4

8.2 LINEAR MEASURES: THE METRIC SYSTEM

Objective

a Convert from one metric unit of length to another.

The **metric system** is used in most countries of the world, but very little in the United States. The metric system does not use inches, feet, pounds, and so on, although units for time and electricity are the same as those used now in the United States.

An advantage of the metric system is that it is easier to convert from one unit to another. That is because the metric system is based on the number 10.

The basic unit of length is the **meter.** It is just over a yard. In fact, 1 meter ≈ 1.1 yd.

(Comparative sizes are shown.)

1 Meter

1 Yard

The other units of length are multiples of the length of a meter:

10 times a meter,   100 times a meter,   1000 times a meter,   and so on,

or fractions of a meter:

$\frac{1}{10}$ of a meter, $\frac{1}{100}$ of a meter, $\frac{1}{1000}$ of a meter,   and so on.

METRIC UNITS OF LENGTH

1 *kilo*meter (km) = 1000 meters (m)

1 *hecto*meter (hm) = 100 meters (m)

1*deka*meter (dam) = 10 meters (m)

1 meter (m)

1 *deci*meter (dm) = $\frac{1}{10}$ meter (m)

1 *centi*meter (cm) = $\frac{1}{100}$ meter (m)

1 *milli*meter (mm) = $\frac{1}{1000}$ meter (m)

dam and *dm* are not used often.

You should memorize these names and abbreviations. Think of *kilo*- for 1000, *hecto*- for 100, *deka*- for 10, *deci*- for $\frac{1}{10}$, *centi*- for $\frac{1}{100}$, and *milli*- for $\frac{1}{1000}$. We will also use these prefixes when considering units of area, capacity, and mass.

Study Tips

WORKING WITH A CLASSMATE

If you are finding it difficult to master a particular topic or concept, try talking about it with a classmate. Verbalizing your questions about the material might help clarify it. If your classmate is also finding the material difficult, it is possible that the majority of the people in your class are confused and you can ask your instructor to explain the concept again.

Use a centimeter ruler. Measure each object.

1.

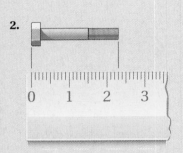

2.

3.

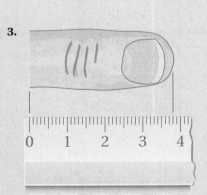

THINKING METRIC

To familiarize yourself with metric units, consider the following.

1 kilometer (1000 meters)	is slightly more than $\frac{1}{2}$ mile (0.6 mi).
1 meter	is just over a yard (1.1 yd).
1 centimeter (0.01 meter)	is a little more than the width of a paperclip (about 0.3937 inch).

1 cm

1 cm

1 inch is about 2.54 centimeters.

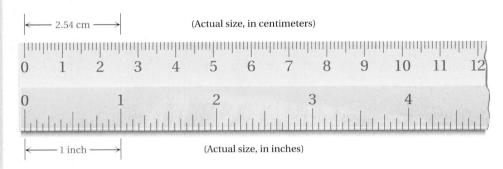

(Actual size, in centimeters)

(Actual size, in inches)

1 millimeter is about the diameter of a paperclip wire.

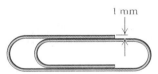

1 mm

The millimeter (mm) is used to measure small distances, especially in industry.

3 mm

Answers on page A-20

In many countries, the centimeter (cm) is used for body dimensions and clothing sizes.

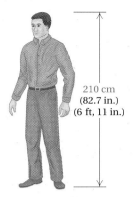

210 cm
(82.7 in.)
(6 ft, 11 in.)

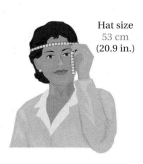

Hat size
53 cm
(20.9 in.)

Do Exercises 1–3 on the preceding page.

The meter (m) is used for expressing dimensions of larger objects—say, the length of a building—and for shorter distances, such as the length of a rug.

25 m (82.0 ft)

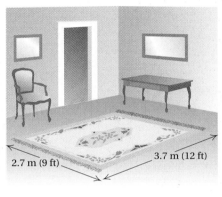

2.7 m (9 ft)   3.7 m (12 ft)

The kilometer (km) is used for longer distances, mostly in cases where miles are now being used.

1 mile is about 1.6 km.

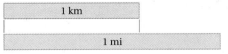

1 km

1 mi

Albuquerque
80 MI

Albuquerque
128 KM

Do Exercises 4–9.

Complete with mm, cm, m, or km.

4. A stick of gum is 7 _____ long.

5. Minneapolis is 3213 _____ from San Francisco.

6. A penny is 1 _____ thick.

7. The halfback ran 7 _____ .

8. The book is 3 _____ thick.

9. The desk is 2 _____ long.

Answers on page A-20

Complete.

10. 23 km = _____ m

11. 4 hm = _____ m

Complete.

12. 1.78 m = _____ cm

13. 9.04 m = _____ mm

a **Changing Metric Units**

As with American units, when changing from a *larger* unit to a *smaller* unit, we usually make substitutions.

EXAMPLE 1   Complete: 4 km = _____ m.

We want to convert from "km" to "m." Since we are converting from a *larger* to a *smaller* unit, we use substitution.

$$4 \text{ km} = 4 \times 1 \text{ km}$$
$$= 4 \times 1000 \text{ m} \qquad \text{Substituting 1000 m for 1 km}$$
$$= 4000 \text{ m}$$

Do Exercises 10 and 11.

Since

$$\frac{1}{10} \text{ m} = 1 \text{ dm}, \qquad \frac{1}{100} \text{ m} = 1 \text{ cm}, \quad \text{and} \quad \frac{1}{1000} \text{ m} = 1 \text{ mm},$$

it follows that

$$1 \text{ m} = 10 \text{ dm}, \qquad 1 \text{ m} = 100 \text{ cm}, \quad \text{and} \quad 1 \text{ m} = 1000 \text{ mm}.$$

EXAMPLE 2   Complete: 93.4 m = _____ cm.

We want to convert from "m" to "cm." Since we are converting from a *larger* to a *smaller* unit, we use substitution.

$$93.4 \text{ m} = 93.4 \times 1 \text{ m}$$
$$= 93.4 \times 100 \text{ cm} \qquad \text{Substituting 100 cm for 1 m}$$
$$= 9340 \text{ cm}$$

EXAMPLE 3   Complete: 0.248 m = _____ mm.

We want to convert from "m" to "mm." Since we are again converting from a *larger* to a *smaller* unit, we use substitution.

$$0.248 \text{ m} = 0.248 \times 1 \text{ m}$$
$$= 0.248 \times 1000 \text{ mm} \qquad \text{Substituting 1000 mm for 1 m}$$
$$= 248 \text{ mm}$$

Do Exercises 12 and 13.

We now convert from "m" to "km." Since we are converting from a *smaller* unit to a *larger* unit, we use multiplying by 1. We choose a symbol for 1 with "km" in the numerator and "m" in the denominator.

EXAMPLE 4   Complete: 2347 m = _____ km.

$$2347 \text{ m} = 2347 \text{ m} \times \frac{1 \text{ km}}{1000 \text{ m}} \qquad \text{Multiplying by 1 using } \frac{1 \text{ km}}{1000 \text{ m}}$$

$$= \frac{2347}{1000} \times \frac{\text{m}}{\text{m}} \times 1 \text{ km} \qquad \text{The } \frac{\text{m}}{\text{m}} \text{ acts like 1, so we omit it.}$$

$$= 2.347 \text{ km} \qquad \text{Dividing by 1000 moves the decimal point three places to the left.}$$

Using canceling, we can work this example as follows:

$$2347 \text{ m} = 2347 \cancel{\text{m}} \times \frac{1 \text{ km}}{1000 \cancel{\text{m}}}$$

$$= \frac{2347}{1000} \times 1 \text{ km} = 2.347 \text{ km}.$$

Do Exercises 14 and 15.

Sometimes we multiply by 1 more than once.

EXAMPLE 5   Complete: 8.42 mm = _____ cm.

$$8.42 \text{ mm} = 8.42 \text{ mm} \times \frac{1 \text{ m}}{1000 \text{ mm}} \times \frac{100 \text{ cm}}{1 \text{ m}} \qquad \begin{array}{l}\text{Multiplying by 1 using}\\ \frac{1 \text{ m}}{1000 \text{ mm}} \text{ and } \frac{100 \text{ cm}}{1 \text{ m}}\end{array}$$

$$= \frac{8.42 \times 100}{1000} \times \frac{\text{mm}}{\text{mm}} \times \frac{\text{m}}{\text{m}} \times 1 \text{ cm}$$

$$= \frac{842}{1000} \text{ cm} = 0.842 \text{ cm}$$

Do Exercises 16 and 17.

MENTAL CONVERSION

Look back over the examples and exercises done so far and you will see that changing from one unit to another in the metric system amounts to only the movement of a decimal point. That is because the metric system is based on 10. Let's find a faster way to convert. Look at the following table.

1000 m	100 m	10 m	1 m	0.1 m	0.01 m	0.001 m
1 km	1 hm	1 dam	1 m	1 dm	1 cm	1 mm

Each place in the table has a value $\frac{1}{10}$ that to the left or 10 times that to the right. Thus moving one place in the table corresponds to moving one decimal place.

Complete.

14. 7814 m = _____ km

15. 7814 m = _____ dam

Complete.

16. 9.67 mm = _____ cm

17. 89 km = _____ cm

Answers on page A-20

Complete. Try to do this mentally using the table.

18. 6780 m = _____ km

19. 9.74 cm = _____ mm

20. 1 mm = _____ cm

21. 845.1 mm = _____ dm

Let's convert mentally.

■ **EXAMPLE 6** Complete: 8.42 mm = _____ cm.

Think: To go from mm to cm in the table is a move of one place to the left. Thus we move the decimal point one place to the left.

1000 m	100 m	10 m	1 m	0.1 m	0.01 m	0.001 m
1 km	1 hm	1 dam	1 m	1 dm	1 cm	1 mm

1 place to the left

8.42   0.8.42   8.42 mm = 0.842 cm

■ **EXAMPLE 7** Complete: 1.886 km = _____ cm.

Think: To go from km to cm is a move of five places to the right. Thus we move the decimal point five places to the right.

1000 m	100 m	10 m	1 m	0.1 m	0.01 m	0.001 m
1 km	1 hm	1 dam	1 m	1 dm	1 cm	1 mm

5 places to the right

1.886   1.88600.   1.886 km = 188,600 cm

■ **EXAMPLE 8** Complete: 3 m = _____ cm.

Think: To go from m to cm in the table is a move of two places to the right. Thus we move the decimal point two places to the right.

1000 m	100 m	10 m	1 m	0.1 m	0.01 m	0.001 m
1 km	1 hm	1 dam	1 m	1 dm	1 cm	1 mm

2 places to the right

3   3.00.   3 m = 300 cm

You should try to make metric conversions mentally as much as possible. The fact that conversions can be done so easily is an important advantage of the metric system. The most commonly used metric units of length are km, m, cm, and mm. We have purposely used these more often than the others in the exercises.

Do Exercises 18–21.

a   Complete. Do as much as possible mentally.

1. a) 1 km = _____ m
   b) 1 m = _____ km

2. a) 1 hm = _____ m
   b) 1 m = _____ hm

3. a) 1 dam = _____ m
   b) 1 m = _____ dam

4. a) 1 dm = _____ m
   b) 1 m = _____ dm

5. a) 1 cm = _____ m
   b) 1 m = _____ cm

6. a) 1 mm = _____ m
   b) 1 m = _____ mm

7. a) 6.7 km = _____ m
   b) This conversion is from a larger unit to a smaller. Did you substitute or multiply by 1?

8. 27 km = _____ m

9. a) 98 cm = _____ m
   b) This conversion is from a smaller unit to a larger. Did you substitute or multiply by 1?

10. 0.789 cm = _____ m

11. 8921 m = _____ km

12. 8664 m = _____ km

13. 56.66 m = _____ km

14. 4.733 m = _____ km

15. 5666 m = _____ cm

16. 869 m = _____ cm

17. 477 cm = _____ m

18. 6.27 mm = _____ m

19. 6.88 m = _____ cm

20. 6.88 m = _____ dm

21. 1 mm = _____ cm

22. 1 cm = _____ km

23. 1 km = _____ cm

24. 2 km = _____ cm

25. 14.2 cm = _____ mm

26. 25.3 cm = _____ mm

27. 8.2 mm = _____ cm

28. 9.7 mm = _____ cm

29. 4500 mm = _____ cm

30. 8,000,000 m = _____ km

31. 0.024 mm = _____ m

32. 60,000 mm = _____ dam

33. 6.88 m = _____ dam

34. 7.44 m = _____ hm

35. 2.3 dam = _____ dm

36. 9 km = _____ hm

37. 392 dam = _____ km

38. 0.056 mm = _____ dm

Complete the following table.

	OBJECT	MILLIMETERS (mm)	CENTIMETERS (cm)	METERS (m)
39.	Length of a calculator		18	
40.	Width of a calculator	85		
41.	Length of a piece of typing paper			0.278
42.	Length of a football field			109.09
43.	Width of a football field		4844	
44.	Film size	33		
45.	Length of 4 meter sticks			4
46.	Length of 3 meter sticks		300	
47.	Thickness of an index card	0.27		
48.	Thickness of a piece of cardboard		0.23	
49.	Height of the Sears Tower			442
50.	Height of the CN Tower (Toronto)	553,000		

51. D_W Recall the guidelines for conversion: (1) "If the conversion is from a larger unit to a smaller unit, substitute." (2) "If the conversion is from a smaller unit to a larger unit, multiply by 1." Explain why each is the easier way to convert in that situation.

52. D_W Explain in your own words why metric units are easier to work with than American units.

Divide.   [4.4a]

53. $23.4 \div 100$

54. $23.4 \div 1000$

Multiply.

55. 3.14×4.41 [4.3a]

56. $4 \times 20\frac{1}{8}$ [3.6a]

Find decimal notation for the percent notation in the sentence.   [6.1b]

57. Blood is 90% water.

58. Of those accidents requiring medical attention, 10.8% of them occur on roads.

Each sentence is incorrect. Insert or alter a decimal point to make the sentence correct.

59. When my right arm is extended, the distance from my left shoulder to the end of my right hand is 10 m.

60. The height of the Empire State Building is 38.1 m.

61. A stack of ten quarters is 140 cm high.

62. The width of an adult's hand is 112 cm.

CHAPTER 8: Measurement

CONVERTING BETWEEN AMERICAN UNITS AND METRIC UNITS

Objective

a | Convert between American units of length and metric units of length.

a | Converting Units

We can make conversions between American and metric units by using the following table. These listings are rounded approximations. Again, we either make a substitution or multiply by 1 appropriately.

AMERICAN	METRIC
1 in.	2.540 cm
1 ft	0.305 m
1 yd	0.914 m
1 mi	1.609 km
0.621 mi	1 km
1.094 yd	1 m
3.281 ft	1 m
39.370 in.	1 m

THINK METRIC

1 Mile = 1.6 Kilometers

This table is set up to enable us to make all conversions by substitution.

EXAMPLE 1 Complete: 26.2 mi = _____ km.
(The length of the Olympic marathon)

$$26.2 \text{ mi} = 26.2 \times 1 \text{ mi}$$
$$\approx 26.2 \times 1.609 \text{ km} \qquad \text{Substituting 1.609 km for 1 mi}$$
$$= 42.1558 \text{ km}$$

EXAMPLE 2 Complete: 2.16 m = _____ in.
(The height of Shaquille O'Neal of the Los Angeles Lakers)

$$2.16 \text{ m} = 2.16 \times 1 \text{ m}$$
$$\approx 2.16 \times 39.37 \text{ in.} \quad \text{Substituting 39.37 in. for 1 m}$$
$$= 85.0392 \text{ in.}$$

In an application like this one, the answer would probably be rounded to the nearest one, as 85 in.

EXAMPLE 3 Complete: 100 m = _____ ft.
(The length of the 100-meter dash)

$$100 \text{ m} = 100 \times 1 \text{ m}$$
$$\approx 100 \times 3.281 \text{ ft} \qquad \text{Substituting 3.281 ft for 1 m}$$
$$= 328.1 \text{ ft}$$

EXAMPLE 4 Complete: 4544 km = _____ mi.
(The distance from New York to Los Angeles)

$$4544 = 4544 \times 1 \text{ km}$$
$$\approx 4544 \times 0.621 \text{ mi} \qquad \text{Substituting 0.621 mi for 1 km}$$
$$= 2821.824 \text{ mi}$$

In practical situations, we would probably round this answer to 2822 mi.

Complete.

1. 100 yd = _____ m
(The length of a football field)

2. 500 mi = _____ km
(The Indianapolis 500-mile race)

3. 2383 km = _____ mi
(The distance from St. Louis to Phoenix)

Answers on page A-20

507

Complete.

4. 568 mi = _____ km
(The distance from San
Francisco to Las Vegas)

5. The height of the John Hancock
Building in Chicago is 1127 ft.
Find the height in meters.

6. Complete:
0.125 in. = _____ mm.
(The thickness of a quarter)

Do Exercises 1–3 on the preceding page.

EXAMPLE 5 *Petronas Towers.* The height of the Petronas Towers in Kuala
Lumpur, Malaysia, is 1483 ft. Find the height in meters.
Source: *The New York Times Almanac*

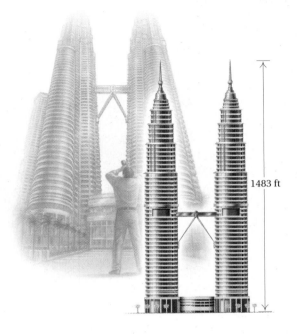

1483 ft

The height H, in meters, is given by

$$H = 1483 \text{ ft}$$
$$= 1483 \times 1 \text{ ft}$$
$$\approx 1483 \times 0.305 \text{ m} \qquad \text{Substituting 0.305 m for 1 ft}$$
$$= 452.315 \text{ m}.$$

Do Exercises 4 and 5.

EXAMPLE 6 Complete: 0.0041 in. = _____ mm.
(The thickness of a $1 bill)

In this case, we must make two substitutions since the chart on the
preceding page does not provide an easy way to convert from inches to
millimeters.

$$0.0041 \text{ in.} = 0.0041 \times 1 \text{ in.}$$
$$\approx 0.0041 \times 2.54 \text{ cm} \qquad \text{Substituting 2.54 cm for 1 in.}$$
$$= 0.0041 \times 2.54 \times 1 \text{ cm}$$
$$= 0.0041 \times 2.54 \times 10 \text{ mm} \qquad \text{Substituting 10 mm for 1 cm}$$
$$= 0.10414 \text{ mm}$$

Do Exercise 6.

Answers on page A-20

8.3 EXERCISE SET

a   Complete.

1. 330 ft = _____ m
(The length of most baseball foul lines)

2. 12 in. = _____ cm
(The length of a common ruler)

3. 1171.352 km = _____ mi
(The distance from Cleveland to Atlanta)

4. 2 m = _____ ft
(The length of a desk)

5. 65 mph = _____ km/h
(The common speed limit in the United States)

6. 100 km/h = _____ mph
(A common speed limit in Canada)

7. 180 mi = _____ km
(The distance from Indianapolis to Chicago)

8. 141,600,000 mi = _____ km
(The farthest distance of Mars from the sun)

9. 70 mph = _____ km/h
(An interstate speed limit in Arizona)

10. 60 km/h = _____ mph
(A city speed limit in Canada)

11. 10 yd = _____ m
(The length needed for a first down in football)

12. 450 ft = _____ m
(The length of a long home run in baseball)

13. 2.13 m = _____ in.
(The height of Tim Duncan of the San Antonio Spurs)

14. 87 in. = _____ m
(The height of Arvydas Sabonis of the Portland Trail Blazers)

15. 381 m = _____ ft
(The height of the Empire State Building)

16. 1127 ft = _____ m
(The height of the John Hancock Center)

17. 7.5 in. = _____ cm
(The length of a pencil)

18. 15.7 cm = _____ in.
(The length of a $1 bill)

19. 2216 km = _____ mi
(The distance from Chicago to Miami)

20. 1862 mi = _____ km
(The distance from Seattle to Kansas City)

21. 13 mm = _____ in.
(The thickness of a plastic case for a CD-ROM)

22. 0.25 in. = _____ mm
(The thickness of an eraser on a pencil)

Complete the following table.

	OBJECT	YARDS (yd)	CENTIMETERS (cm)	INCHES (in.)	METERS (m)	MILLIMETERS (mm)
23.	Length of a mousepad		23.8			
24.	Width of a mousepad		20.3			
25.	Width of a piece of typing paper			$8\frac{1}{2}$		
26.	Length of a football field	120 yd				
27.	Width of a football field		4844			
28.	Film size					33
29.	Length of 4 yard sticks	4				
30.	Length of 3 meter sticks		300			
31.	Thickness of an index card				0.00027	
32.	Thickness of a piece of cardboard		0.23			
33.	Height of the Sears Tower				442	
34.	Height of Central Plaza, Hong Kong	409				

35. D_W Do some research in a library or on the Internet about the metric system versus the American system. Why do you think the United States has not converted to the metric system?

36. D_W List all the memory devices you know for comparing metric and American units of measure.

SKILL MAINTENANCE

37. Convert to decimal notation: 56.1%   [6.1b]

38. Convert to percent notation: 0.6734.   [6.1b]

39. Convert to percent notation: $\frac{9}{8}$.   [6.2a]

Evaluate.   [1.9b]

40. 5^2

41. 10^2

42. 31^2

Convert the number in the sentence to standard notation.

43. *Oil Usage.*   In 2020, it is expected that the world will be using 113 million barrels of oil per day.   [4.3b]
Source: U.S. Geological Survey, Energy Information Administration

44. *Internet Spending.*   In 2002, teenagers spent $1.3 billion shopping on the Internet.   [4.3a]
Source: eMarketer

45. *Tax Reduction.*   In 2001, President George W. Bush presented a $1.6 trillion tax reduction proposal to Congress.   [4.3b]

46. *Health Insurance.*   In 1999, there were about 45 million Americans who lacked health insurance.   [4.3b]
Source: U.S. Bureau of the Census

SYNTHESIS

47. Develop a formula to convert from inches to millimeters.

48. Develop a formula to convert from millimeters to inches. How does it relate to the answer for Exercise 47?

49. Audio cassettes are generally played at a rate of $1\frac{7}{8}$ in. per second. How many meters of tape are used for a 60-min cassette? (*Note:* A 60-min cassette has 30 min of playing time on each side.)

50. In a recent year, the world record for the 100-m dash was 9.86 sec. How fast is this in miles per hour? Round to the nearest tenth of a mile per hour.

8.4

WEIGHT AND MASS; MEDICAL APPLICATIONS

Objectives

a   Convert from one American unit of weight to another.

b   Convert from one metric unit of mass to another.

c   Make conversions and solve applied problems concerning medical dosages.

There is a difference between **mass** and **weight,** but the terms are often used interchangeably. People sometimes use the word "weight" instead of "mass." Weight is related to the force of gravity. The farther you are from the center of the earth, the less you weigh. Your mass stays the same no matter where you are.

a   Weight: The American System

> **AMERICAN UNITS OF WEIGHT**
>
> 1 ton (T) = 2000 pounds (lb)   1 lb = 16 ounces (oz)

The term "ounce" used here for weight is different from the "ounce" we will use for capacity in Section 8.5. We convert units using the same techniques that we use with linear measure.

EXAMPLE 1   A well-known hamburger is called a "quarter-pounder." Find its name in ounces: a "_____ ouncer."

$$\frac{1}{4} \text{ lb} = \frac{1}{4} \cdot 1 \text{ lb} = \frac{1}{4} \cdot 16 \text{ oz}$$

$$= 4 \text{ oz}$$

Substituting 16 oz for 1 lb. Since we are converting from a larger unit to a smaller unit, we use substitution.

A "quarter-pounder" can also be called a "four-ouncer."

EXAMPLE 2   Complete: 15,360 lb = _____ T.

$$15{,}360 \text{ lb} = 15{,}360 \text{ lb} \times \frac{1 \text{ T}}{2000 \text{ lb}}$$

Multiplying by 1. Since we are converting from a smaller unit to a larger unit, we use multiplying by 1.

$$= \frac{15{,}360}{2000} \text{ T} = 7.68 \text{ T}$$

Do Exercises 1–3.

b   Mass: The Metric System

The basic unit of mass is the **gram** (g), which is the mass of 1 cubic centimeter (1 cm³) of water. Since a cubic centimeter is small, a gram is a small unit of mass.

1 g = 1 gram = the mass of 1 cm³ of water

Complete.

1. 5 lb = _____ oz

2. 8640 lb = _____ T

3. 1 T = _____ oz

Answers on page A-20

The following table lists the metric units of mass. The prefixes are the same as those for length.

METRIC UNITS OF MASS

1 metric ton (t) = 1000 kilograms (kg)

1 *kilo*gram (kg) = 1000 grams (g)

1 *hecto*gram (hg) = 100 grams (g)

1 *deka*gram (dag) = 10 grams (g)

1 gram (g)

1 *deci*gram (dg) = $\frac{1}{10}$ gram (g)

1 *centi*gram (cg) = $\frac{1}{100}$ gram (g)

1 *milli*gram (mg) = $\frac{1}{1000}$ gram (g)

THINKING METRIC

One gram is about the mass of 1 raisin or 1 paperclip or 1 package of "NutraSweet" sweetener. Since 1 kg is about 2.2 lb, 1000 kg is about 2200 lb, or 1 metric ton (t), which is just a little more than 1 American ton (T).

1 g

1 gram

1 kilogram

1 pound

Small masses, such as dosages of medicine and vitamins, may be measured in milligrams (mg). The gram (g) is used for objects ordinarily measured in ounces, such as the mass of a letter, a piece of candy, a coin, or a small package of food.

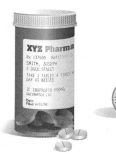

Each 2.5 mg

2 g

15 g

Ground beef
2 lb (0.9 kg)

90 kg

The kilogram (kg) is used for larger food packages, such as meat, or for human body mass. The metric ton (t) is used for very large masses, such as the mass of an automobile, a truckload of gravel, or an airplane.

Do Exercises 4–8.

CHANGING UNITS MENTALLY

As before, changing from one metric unit to another amounts to only the movement of a decimal point. We use this table.

1000 g	100 g	10 g	1 g	0.1 g	0.01 g	0.001 g
1 kg	1 hg	1 dag	1 g	1 dg	1 cg	1 mg

EXAMPLE 3   Complete: 8 kg = _____ g.

Think: To go from kg to g in the table is a move of three places to the right. Thus we move the decimal point three places to the right.

1000 g	100 g	10 g	1 g	0.1 g	0.01 g	0.001 g
1 kg	1 hg	1 dag	1 g	1 dg	1 cg	1 mg

3 places to the right

8.0   8.000.   8 kg = 8000 g

EXAMPLE 4   Complete: 4235 g = _____ kg.

Think: To go from g to kg in the table is a move of three places to the left. Thus we move the decimal point three places to the left.

1000 g	100 g	10 g	1 g	0.1 g	0.01 g	0.001 g
1 kg	1 hg	1 dag	1 g	1 dg	1 cg	1 mg

3 places to the left

4235.0   4.235.0   4235 g = 4.235 kg

Do Exercises 9 and 10.

Complete with mg, g, kg, or t.

4. A laptop computer has a mass of 6 _____ .

5. Eric has a body mass of 85.4 _____ .

6. This is a 3-_____ vitamin.

7. A pen has a mass of 12 _____ .

8. A minivan has a mass of 3 _____ .

Complete.

9. 6.2 kg = _____ g

10. 304.8 cg = _____ g

Answers on page A-20

Complete.

11. 7.7 cg = _____ mg

12. 2344 mg = _____ cg

13. 67 dg = _____ mg

14. Complete:

1 mcg = _____ mg.

Answers on page A-20

5̶1̶4̶

CHAPTER 8: Measurement

EXAMPLE 5 Complete: 6.98 cg = _____ mg.

Think: To go from cg to mg is a move of one place to the right. Thus we move the decimal point one place to the right.

1000 g	100 g	10 g	1 g	0.1 g	0.01 g	0.001 g
1 kg	1 hg	1 dag	1 g	1 dg	1 cg	1 mg

1 place to the right

6.98   6.9.8   6.98 cg = 69.8 mg

The most commonly used metric units of mass are kg, g, cg, and mg. We have purposely used those more than the others in the exercises.

EXAMPLE 6 Complete: 89.21 mg = _____ g.

Think: To go from mg to g is a move of three places to the left. Thus we move the decimal point three places to the left.

1000 g	100 g	10 g	1 g	0.1 g	0.01 g	0.001 g
1 kg	1 hg	1 dag	1 g	1 dg	1 cg	1 mg

3 places to the left

89.21   0.089.21   89.21 mg = 0.08921 g

Do Exercises 11–13.

C   Medical Applications

Another metric unit that is used in medicine is the microgram (mcg). It is defined as follows.

> **MICROGRAM**
>
> 1 microgram = 1 mcg = $\dfrac{1}{1,000,000}$ g = 0.000001 g
>
> 1,000,000 mcg = 1 g

One microgram is one-millionth of a gram and one million micrograms is one gram.

EXAMPLE 7 Complete: 1 mg = _____ mcg.

We convert to grams and then to micrograms:

$$1 \text{ mg} = 0.001 \text{ g}$$
$$= 0.001 \times 1 \text{ g}$$
$$= 0.001 \times 1,000,000 \text{ mcg} \qquad \text{Substituting 1,000,000 mcg for 1 g}$$
$$= 1000 \text{ mcg.}$$

Do Exercise 14.

EXAMPLE 8 *Medical Dosage.* Nitroglycerin sublingual tablets come in 0.4-mg tablets. How many micrograms are in each tablet?

Source: Steven R. Smith, M.D.

We are to complete: 0.4 mg = _____ mcg. Thus,

$$0.4 \text{ mg} = 0.4 \times 1 \text{ mg}$$

$$= 0.4 \times 1000 \text{ mcg} \quad \text{From Example 7, substituting 1000 mcg for 1 mg}$$

$$= 400 \text{ mcg}.$$

We can also do this problem in a manner similar to Example 7.

Do Exercise 15.

15. Medical Dosage. A physician prescribes 500 mcg of alprazolam, an antianxiety medication. How many milligrams is this dosage?

Source: Steven R. Smith, M.D.

Answer on page A-20

Study Tips

THE FOURTEEN BEST JOBS: HOW MATH STACKS UP

We list the Top 14 here and note whether math is an important aspect or requirement of the job. Note that math is significant in 11 of the 14 jobs and has at least some use in all the jobs!

	JOB	MATH EMPHASIS	MID-LEVEL SALARY
1.	Financial planner	Yes	$107,000
2.	Web site manager	Yes	68,000
3.	Computer systems analyst	Yes	54,000
4.	Actuary	Yes	71,000
4.	Computer programmer (tie)	Yes	57,000
6.	Software engineer	Yes	53,000
7.	Meteorologist	Some	57,000
8.	Biologist	Some	53,000
9.	Astronomer	Yes	74,000
10.	Paralegal assistant	Some	37,000
11.	Statistician	Yes	55,000
12.	Hospital administrator	Yes	64,000
13.	Dietician	Yes	39,000
14.	Mathematician	Yes	45,000

Source: Les Krantz, *Jobs Related Almanac.* New York: St. Martin's Press, 2000

515

8.4 EXERCISE SET

For Extra Help

Digital Video Tutor CD 5 Videotape 13 · InterAct Math · Math Tutor Center · MathXL · MyMathLab

a Complete.

1. 1 T = _____ lb

2. 1 lb = _____ oz

3. 6000 lb = _____ T

4. 8 T = _____ lb

5. 4 lb = _____ oz

6. 10 lb = _____ oz

7. 6.32 T = _____ lb

8. 8.07 T = _____ lb

9. 3200 oz = _____ T

10. 6400 oz = _____ T

11. 80 oz = _____ lb

12. 960 oz = _____ lb

13. *Excelsior.* Western Excelsior is a company that makes a packing material called excelsior. Excelsior is produced from the wood of aspen trees. The largest grove of aspen trees on record contained 13,000,000 tons of aspen. How many pounds of aspen were there?
Source: Western Excelsior Corporation, Mancos CO

14. *Excelsior.* Western Excelsior buys 44,800 tons of aspen each year to make excelsior. How many pounds of aspen does it buy each year?
Source: Western Excelsior Corporation, Mancos CO

b Complete.

15. 1 kg = _____ g

16. 1 hg = _____ g

17. 1 dag = _____ g

18. 1 dg = _____ g

19. 1 cg = _____ g

20. 1 mg = _____ g

21. 1 g = _____ mg

22. 1 g = _____ cg

23. 1 g = _____ dg

24. 25 kg = _____ g

25. 234 kg = _____ g

26. 9403 g = _____ kg

27. 5200 g = _____ kg

28. 1.506 kg = _____ g

29. 67 hg = _____ kg

30. 45 cg = _____ g

31. 0.502 dg = _____ g

32. 0.0025 cg = _____ mg

33. 8492 g = _____ kg

34. 9466 g = _____ kg

35. 585 mg = _____ cg

36. 96.1 mg = _____ cg

37. 8 kg = _____ cg

38. 0.06 kg = _____ mg

39. 1 t = _____ kg

40. 2 t = _____ kg

41. 3.4 cg = _____ dag

42. 115 mg = _____ g

C   Complete.

43. 1 mg = _____ mcg

44. 1 mcg = _____ mg

45. 325 mcg = _____ mg

46. 0.45 mg = _____ mcg

Medical Dosage.   Solve each of the following. (None of these medications should be taken without consulting your own physician.)

Source: Steven R. Smith, M.D.

47. Digoxin is a medication used to treat heart problems. A physician orders 0.125 mg of digoxin to be taken once daily. How many micrograms of digoxin are there in the daily dosage?

48. Digoxin is a medication used to treat heart problems. A physician orders 0.25 mg of digoxin to be taken once a day. How many micrograms of digoxin are there in the daily dosage?

49. Triazolam is a medication used for the short-term treatment of insomnia. A physician advises her patient to take one of the 0.125-mg tablets each night for 7 nights. How many milligrams of triazolam will the patient have ingested over that 7-day period? How many micrograms?

50. Clonidine is a medication used to treat high blood pressure. The usual starting dose of clonidine is one 0.1-mg tablet twice a day. If a patient is started on this dose by his physician, how many total milligrams of clonidine will the patient have taken before he returns to see his physician 14 days later? How many micrograms?

51. Cephalexin is an antibiotic that frequently is prescribed in a 500-mg tablet form. A physician prescribes 2 grams of cephalexin per day for a patient with a skin abscess. How many 500-mg tablets would have to be taken in order to achieve this daily dosage?

52. Quinidine gluconate is a liquid mixture, part medicine and part water, which is administered intravenously. There are 80 mg of quinidine gluconate in each cubic centimeter (cc) of the liquid mixture. A physician orders 900 mg of quinidine gluconate to be administered daily to a patient with malaria. How much of the solution would have to be administered in order to achieve the recommended daily dosage?

53. Amoxicillin is a common antibiotic prescribed for children. It is a liquid suspension composed of part amoxicillin and part water. In one formulation of amoxicillin suspension, there is 250 mg of amoxicillin in 5 cc of the liquid suspension. A physician prescribes 400 mg per day for a 2-year-old child with an ear infection. How much of the amoxicillin liquid suspension would the child's parent need to administer in order to achieve the recommended daily dosage of amoxicillin?

54. Albuterol is a medication used for the treatment of asthma. It comes in an inhaler that contains 17 mg albuterol mixed with a liquid. One actuation (inhalation) from the mouthpiece delivers a 90-mcg dose of albuterol.

a) A physician orders 2 inhalations 4 times per day. How many micrograms of albuterol does the patient inhale per day?

b) How many actuations/inhalations are contained in one inhaler?

c) Danielle is going away for 4 months of college and wants to take enough albuterol to last for that time. Her physician has prescribed 2 inhalations 4 times per day. Estimate how many inhalers Danielle will need to take with her for the 4-month period.

55. ᴰᴡ Give at least two reasons why someone might prefer the use of grams to the use of ounces.

56. ᴰᴡ Describe a situation in which one object weighs 70 kg, another weighs 3 g, and another weighs 125 mg.

Convert to fraction notation.   [6.2b]

57. 35%   **58.** 99%   **59.** 85.5%   **60.** 34.2%

61. $37\frac{1}{2}\%$ **62.** $66.\overline{6}\%$ **63.** $83.\overline{3}\%$ **64.** $16\frac{2}{3}\%$

Solve.   [1.8a]

65. A ream of paper contains 500 sheets. How many sheets are there in 15 reams?

66. A lab technician separates a vial containing 140 cc of blood into test tubes, each of which contains 3 cc of blood. How many test tubes can be filled? How much blood is left over?

67. A box of gelatin-mix packages weighs $15\frac{3}{4}$ lb. Each package weighs $1\frac{3}{4}$ oz. How many packages are in the box?

68. At \$0.90 a dozen, the cost of eggs is \$0.60 per pound. How much does an egg weigh?

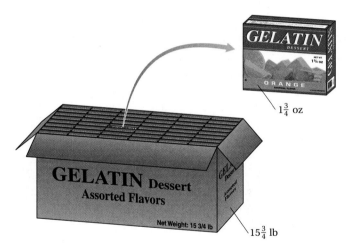

$1\frac{3}{4}$ oz

$15\frac{3}{4}$ lb

69. *Large Diamonds.*   A **carat** (also spelled **karat**) is a unit of weight for precious stones; 1 carat = 200 mg. The Golden Jubilee Diamond weighs 545.67 carats and is the largest cut diamond in the world. The Hope Diamond, located at the Smithsonian Institution Museum of Natural History, weighs 45.52 carats.
Source: *National Geographic*, February 2001

a) How many grams does the Golden Jubilee Diamond weigh?

b) How many grams does the Hope Diamond weigh?

c) ▦ Given that 1 lb = 453.6 g, how many ounces does each diamond weigh?

Objectives

a Convert from one unit of capacity to another.

b Solve applied problems concerning medical dosages.

Complete.

1. 5 gal = _____ pt

2. 80 qt = _____ gal

---

a Capacity

AMERICAN UNITS

To answer a question like "How much soda is in the can?" we need measures of **capacity.** American units of capacity are fluid ounces, cups, pints, quarts, and gallons. These units are related as follows.

AMERICAN UNITS OF CAPACITY	
1 gallon (gal) = 4 quarts (qt)	1 pt = 2 cups = 16 fluid ounces (fl oz)
1 qt = 2 pints (pt)	1 cup = 8 fluid oz

Fluid ounces, abbreviated fl oz, are often referred to as ounces, or oz.

EXAMPLE 1 Complete: 9 gal = _____ oz.

Since we are converting from a *larger* unit to a *smaller* unit, we use substitution.

$$9 \text{ gal} = 9 \cdot 1 \text{ gal} = 9 \cdot 4 \text{ qt} \qquad \text{Substituting 4 qt for 1 gal}$$
$$= 9 \cdot 4 \cdot 1 \text{ qt} = 9 \cdot 4 \cdot 2 \text{ pt} \qquad \text{Substituting 2 pt for 1 qt}$$
$$= 9 \cdot 4 \cdot 2 \cdot 1 \text{ pt} = 9 \cdot 4 \cdot 2 \cdot 16 \text{ oz} \qquad \text{Substituting 16 oz for 1 pt}$$
$$= 1152 \text{ oz}$$

EXAMPLE 2 Complete: 24 qt = _____ gal.

Since we are converting from a *smaller* unit to a *larger* unit, we multiply by 1 using 1 gal in the numerator and 4 qt in the denominator.

$$24 \text{ qt} = 24 \text{ qt} \cdot \frac{1 \text{ gal}}{4 \text{ qt}} = \frac{24}{4} \cdot 1 \text{ gal} = 6 \text{ gal}$$

Do Exercises 1 and 2.

METRIC UNITS

One unit of capacity in the metric system is a **liter.** A liter is just a bit more than a quart. It is defined as follows.

1 liter ≈ 1.06 quarts

1 liter   1 quart

METRIC UNITS OF CAPACITY

1 liter (L) = 1000 cubic centimeters (1000 cm³)

The script letter ℓ is also used for "liter."

The metric prefixes are also used with liters. The most common is **milli-**. The milliliter (mL) is, then, $\frac{1}{1000}$ liter. Thus,

$$1 \text{ L} = 1000 \text{ mL} = 1000 \text{ cm}^3;$$
$$0.001 \text{ L} = 1 \text{ mL} = 1 \text{ cm}^3.$$

Although the other metric prefixes are rarely used for capacity, we display them in the following table as we did for linear measure.

1000 L	100 L	10 L	1 L	0.1 L	0.01 L	0.001 L
1 kL	1 hL	1 daL	1 L	1 dL	1 cL	1 mL (cc)

A preferred unit for drug dosage is the milliliter (mL) or the cubic centimeter (cm³). The notation "cc" is also used for cubic centimeter, especially in medicine. The milliliter and the cubic centimeter represent the same measure of capacity. A milliliter is about $\frac{1}{5}$ of a teaspoon.

3 cm³

5 mL

$$1 \text{ mL} = 1 \text{ cm}^3 = 1 \text{ cc}$$

Volumes for which quarts and gallons are used are expressed in liters. Large volumes in business and industry are expressed using measures of cubic meters (m³).

Do Exercises 3–6.

EXAMPLE 3 Complete: 4.5 L = _____ mL.

$4.5 \text{ L} = 4.5 \times 1 \text{ L} = 4.5 \times 1000 \text{ mL}$ Substituting 1000 mL for 1 L

$= 4500 \text{ mL}$

1000 L	100 L	10 L	1 L	0.1 L	0.01 L	0.001 L
1 kL	1 hL	1 daL	1 L	1 dL	1 cL	1 mL (cc)

3 places to the right

Complete with mL or L.

3. The patient received an injection of 2 _____ of penicillin.

4. There are 250 _____ in a coffee cup.

5. The gas tank holds 80 _____.

6. Bring home 8 _____ of milk.

Answers on page A-21

Complete.

7. $0.97 \text{ L} = \underline{\hspace{2cm}} \text{ mL}$

8. $8990 \text{ mL} = \underline{\hspace{2cm}} \text{ L}$

9. Medical Dosage. A physician orders 2400 mL of 0.9% saline solution to be administered intravenously over a 24-hr period. How many liters were ordered?

10. Medical Dosage. A prescription calls for 2 oz of theophylline.
   a) For how many milliliters is the prescription?
   b) For how many liters is the prescription?

EXAMPLE 4   Complete: $280 \text{ mL} = \underline{\hspace{2cm}} \text{ L}$.

$$280 \text{ mL} = 280 \times 1 \text{ mL}$$
$$= 280 \times 0.001 \text{ L} \qquad \text{Substituting } 0.001 \text{ L for } 1 \text{ mL}$$
$$= 0.28 \text{ L}$$

1000 L	100 L	10 L	1 L	0.1 L	0.01 L	0.001 L
1 kL	1 hL	1 daL	1 L	1 dL	1 cL	1 mL (cc)

3 places to the left

We do find metric units of capacity in frequent use in the United States—for example, in sizes of soda bottles and automobile engines.

Do Exercises 7 and 8.

b   Medical Applications

The metric system is used extensively in medicine.

EXAMPLE 5   *Medical Dosage.*   A physician orders 3.5 L of 5% dextrose in water (abbrev. D5W) to be administered over a 24-hr period. How many milliliters were ordered?

We convert 3.5 L to milliliters:

$$3.5 \text{ L} = 3.5 \times 1 \text{ L} = 3.5 \times 1000 \text{ mL} = 3500 \text{ mL}.$$

The physician had ordered 3500 mL of D5W.

Do Exercise 9.

EXAMPLE 6   *Medical Dosage.*   Liquids at a pharmacy are often labeled in liters or milliliters. Thus if a physician's prescription is given in ounces, it must be converted. For conversion, a pharmacist knows that 1 oz ≈ 29.57 mL.* A prescription calls for 3 oz of theophylline. For how many milliliters is the prescription?

We convert as follows:

$$3 \text{ oz} = 3 \times 1 \text{ oz} \approx 3 \times 29.57 \text{ mL} = 88.71 \text{ mL}.$$

The prescription calls for 88.71 mL of theophylline.

Do Exercise 10.

*In practice, most physicians use 30 mL as an approximation to 1 oz.

For Extra Help

Digital Video
Tutor CD 5
Videotape 13

InterAct
Math

Math Tutor
Center

MathXL

MyMathLab

8.5 EXERCISE SET

a Complete.

1. 1 L = _____ mL = _____ cm³

2. _____ L = 1 mL = _____ cm³

3. 87 L = _____ mL

4. 806 L = _____ mL

5. 49 mL = _____ L

6. 19 mL = _____ L

7. 0.401 mL = _____ L

8. 0.816 mL = _____ L

9. 78.1 L = _____ cm³

10. 99.6 L = _____ cm³

11. 10 qt = _____ oz

12. 9.6 oz = _____ pt

13. 20 cups = _____ pt

14. 1 gal = _____ oz

15. 8 gal = _____ qt

16. 1 gal = _____ cups

17. 5 gal = _____ qt

18. 11 gal = _____ qt

19. 56 qt = _____ gal

20. 84 qt = _____ gal

21. 11 gal = _____ pt

22. 5 gal = _____ pt

Complete.

	OBJECT	GALLONS (gal)	QUARTS (qt)	PINTS (pt)	CUPS	OUNCES (oz)
23.	12-can package of 12-oz sodas					144
24.	6-bottle package of 16-oz sodas		3			
25.	Full tank of gasoline	16				
26.	Container of milk			8		
27.	Minute Maid Orange Juice				4	
28.	Scope Mouthwash					33
29.	Revlon Flex Shampoo					15
30.	Jhirmack Moisturizing Shampoo					20

Complete.

	OBJECT	LITERS (l)	MILLILITERS (ml)	CUBIC CENTIMETERS (cc)	CUBIC CENTIMETERS (cm³)
31.	2-L bottle of soda	2			
32.	Heinz Vinegar		3755		
33.	Full tank of gasoline in Europe	64			
34.	Old Spice Aftershave				125
35.	Revlon Flex Shampoo			443	
36.	Jhirmack Moisturizing Shampoo		591		

b *Medical Dosage.* Solve each of the following.
Source: Steven R. Smith, M.D.

37. An emergency-room physician orders 2.0 L of Ringer's lactate to be administered over 2 hr for a patient in shock. How many milliliters is this?

38. An emergency-room physician orders 2.5 L of 0.9% saline solution over 4 hr for a patient suffering from dehydration. How many milliliters is this?

39. A physician orders 320 mL of 5% dextrose in water (D5W) solution to be administered intravenously over 4 hr. How many liters of D5W is this?

40. A physician orders 40 mL of 5% dextrose in water (D5W) solution to be administered intravenously over 2 hr to an elderly patient. How many liters of D5W is this?

41. A physician orders 0.5 oz of magnesia and alumina oral suspension antacid 4 times per day for a patient with indigestion. How many milliliters of the antacid is the patient to ingest in a day?

42. A physician orders 0.25 oz of magnesia and alumina oral suspension antacid 3 times per day for a child with upper abdominal discomfort. How many milliliters of the antacid is the child to ingest in a day?

43. A physician orders 0.5 L of normal saline solution. How many milliliters are ordered?

44. A physician has ordered that his patient receive 60 mL per hour of normal saline solution intravenously. How many liters of the saline solution is the patient to receive in a 24-hr period?

45. A physician wants her patient to receive 3.0 L of normal saline intravenously over a 24-hr period. How many milliliters per hour must the nurse administer?

46. A physician tells a patient to purchase 0.5 L of hydrogen peroxide. Commercially, hydrogen peroxide is found on the shelf in bottles that hold 4 oz, 8 oz, and 16 oz. Which bottle comes closest to filling the prescription?

Medical Dosage. Because patients do not always have a working knowledge of the metric system, physicians often prescribe dosages in teaspoons (t or tsp) and tablespoons (T or tbs). The units are related to each other and to the metric system as follows:

$$5 \text{ mL} \approx 1 \text{ tsp}, \qquad 3 \text{ tsp} = 1 \text{ T}.$$

Complete.

47. 45 mL = _____ tsp

48. 3 T = _____ tsp

49. 1 mL = _____ tsp

50. 18.5 mL = _____ tsp

51. 2 T = _____ tsp

52. 8.5 tsp = _____ T

53. 1 T = _____ mL

54. 18.5 mL = _____ T

55. D_W What advantages does the use of metric units of capacity have over that of American units?

56. D_W Why do you think most liquid containers list both metric and American units of measure?

SKILL MAINTENANCE

Convert to percent notation. [6.1b], [6.2a]

57. 0.452

58. 0.999

59. $\dfrac{1}{3}$

60. $\dfrac{2}{3}$

61. $\dfrac{11}{20}$

62. $\dfrac{21}{20}$

63. $\dfrac{22}{25}$

64. $\dfrac{2}{25}$

65. *Tourist Spending.* Foreign tourists spend $13.1 billion in this country annually. The most money, $2.7 billion, is spent in Florida. What is the ratio of amount spent in Florida to total amount spent? What is the ratio of total amount spent to amount spent in Florida? [5.1a, b]

66. One person in four plays a musical instrument. In a given group of people, what is the ratio of those who play an instrument to total number of people? What is the ratio of those who do not play an instrument to total number of people? [5.1a]

SYNTHESIS

67. *Wasting Water.* Many people leave the water running while they are brushing their teeth. Suppose that 32 oz of water is wasted in such a way each day by one person. How much water, in gallons, is wasted in a week? in a month (30 days)? in a year? Assuming each of the 281 million people in this country wastes water in this way, estimate how much water is wasted in a day; in a year.

68. *Cost of Gasoline.* Suppose that premium gasoline is selling for about $1.69/gallon. Using the fact that 1 L = 1.057 qt, determine the price of the gasoline in liters.

69. *Bees and Honey.* The average bee produces only $\frac{1}{8}$ teaspoon of honey in its lifetime. It takes 60,000 honeybees to produce 100 lb of honey. How much does a teaspoon of honey weight?

TIME AND TEMPERATURE

Objectives

a Convert from one unit of time to another.

b Convert between Celsius and Fahrenheit temperatures using the formulas

$$F = \frac{9}{5} \cdot C + 32$$

and

$$C = \frac{5}{9} \cdot (F - 32).$$

a Time

A table of units of time is shown below. The metric system sometimes uses "h" for hour and "s" for second, but we will use the more familiar "hr" and "sec."

UNITS OF TIME	
1 day = 24 hours (hr)	1 year (yr) = $365\frac{1}{4}$ days
1 hr = 60 minutes (min)	1 week (wk) = 7 days
1 min = 60 seconds (sec)	

Since we cannot have $\frac{1}{4}$ day on the calendar, we give each year 365 days and every fourth year 366 days (a leap year), unless it is a year at the beginning of a century not divisible by 400.

EXAMPLE 1 Complete: 1 hr = _____ sec.

$$1 \text{ hr} = 60 \text{ min}$$
$$= 60 \cdot 1 \text{ min}$$
$$= 60 \cdot 60 \text{ sec} \qquad \text{Substituting 60 sec for 1 min}$$
$$= 3600 \text{ sec}$$

EXAMPLE 2 Complete: 5 yr = _____ days.

$$5 \text{ yr} = 5 \cdot 1 \text{ yr}$$
$$= 5 \cdot 365\frac{1}{4} \text{ days} \qquad \text{Substituting } 365\frac{1}{4} \text{ days for 1 yr}$$
$$= 1826\frac{1}{4} \text{ days}$$

EXAMPLE 3 Complete: 4320 min = _____ days.

$$4320 \text{ min} = 4320 \text{ min} \cdot \frac{1 \text{ hr}}{60 \text{ min}} \cdot \frac{1 \text{ day}}{24 \text{ hr}} = \frac{4320}{60 \cdot 24} \text{ days} = 3 \text{ days}$$

Do Exercises 1–4.

Complete.

1. 2 hr = _____ min

2. 4 yr = _____ days

3. 1 day = _____ min

4. 168 hr = _____ wk

b Temperature

Below are two temperature scales: **Fahrenheit** for American measure and **Celsius** for metric measure.

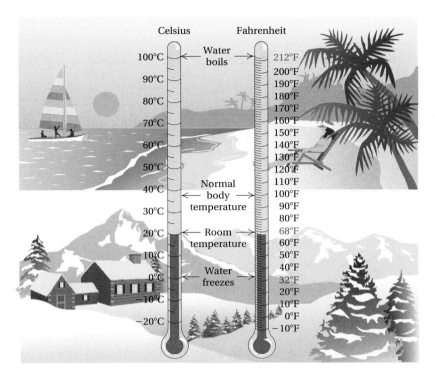

By laying a ruler or a piece of paper horizontally between the scales, we can make an approximate conversion from one measure of temperature to another and get an idea of how the temperature scales compare.

EXAMPLES Convert to Celsius (using the scales shown above). Approximate to the nearest ten degrees.

4. 212°F (Boiling point of water)	100°C	This is exact.
5. 32°F (Freezing point of water)	0°C	This is exact.
6. 105°F	40°C	This is approximate.

Do Exercises 5–7.

EXAMPLES Make an approximate conversion to Fahrenheit.

7. 44°C (Hot bath)	110°F	This is approximate.
8. 20°C (Room temperature)	68°F	This is exact.
9. 83°C	180°F	This is approximate.

Do Exercises 8–10.

Convert to Celsius. Approximate to the nearest ten degrees.

5. 180°F (Brewing coffee)

6. 25°F (Cold day)

7. −10°F (Miserably cold day)

Convert to Fahrenheit. Approximate to the nearest ten degrees.

8. 25°C (Warm day at the beach)

9. 40°C (Temperature of a patient with a high fever)

10. 10°C (A cold bath)

Answers on page A-21

Convert to Fahrenheit.

11. 80°C

12. 35°C

Convert to Celsius.

13. 95°F

14. 113°F

Answers on page A-21

The following formula allows us to make exact conversions from Celsius to Fahrenheit.

CELSIUS TO FAHRENHEIT

$$F = \frac{9}{5} \cdot C + 32, \quad \text{or} \quad F = 1.8 \cdot C + 32$$

$\left(\text{Multiply the Celsius temperature by } \dfrac{9}{5}, \text{ or } 1.8, \text{ and add } 32. \right)$

EXAMPLES Convert to Fahrenheit.

10. 0°C (Freezing point of water)

$$F = \frac{9}{5} \cdot C + 32 = \frac{9}{5} \cdot 0 + 32 = 0 + 32 = 32°$$

Thus, 0°C = 32°F.

11. 37°C (Normal body temperature)

$$F = 1.8 \cdot C + 32 = 1.8 \cdot 37 + 32 = 66.6 + 32 = 98.6°$$

Thus, 37°C = 98.6°F.

Check the answers to Examples 10 and 11 using the scales on p. 527.

Do Exercises 11 and 12.

The following formula allows us to make exact conversions from Fahrenheit to Celsius.

FAHRENHEIT TO CELSIUS

$$C = \frac{5}{9} \cdot (F - 32), \quad \text{or} \quad C = \frac{F - 32}{1.8}$$

$\left(\text{Subtract 32 from the Fahrenheit temperature and multiply by } \dfrac{5}{9} \text{ or divide by } 1.8. \right)$

EXAMPLES Convert to Celsius.

12. 212°F (Boiling point of water)

$$C = \frac{5}{9} \cdot (F - 32)$$
$$= \frac{5}{9} \cdot (212 - 32)$$
$$= \frac{5}{9} \cdot 180 = 100°$$

Thus, 212°F = 100°C.

13. 77°F

$$C = \frac{F - 32}{1.8}$$
$$= \frac{77 - 32}{1.8}$$
$$= \frac{45}{1.8} = 25°$$

Thus, 77°F = 25°C.

Check the answers to Examples 12 and 13 using the scales on p. 527.

Do Exercises 13 and 14.

8.6

EXERCISE SET

For Extra Help

Digital Video
Tutor CD 5
Videotape 13

InterAct
Math

Math Tutor
Center

MathXL

MyMathLab

a Complete.

1. 1 day = _____ hr

2. 1 hr = _____ min

3. 1 min = _____ sec

4. 1 wk = _____ days

5. 1 yr = _____ days

6. 2 yr = _____ days

7. 180 sec = _____ hr

8. 60 sec = _____ hr

9. 492 sec = _____ min
(The amount of time it takes for the rays of the sun to reach the earth)

10. 18,000 sec = _____ hr

11. 156 hr = _____ days

12. 444 hr = _____ days

13. 645 min = _____ hr

14. 375 min = _____ hr

15. 2 wk = _____ hr

16. 4 hr = _____ sec

17. 756 hr = _____ wk

18. 166,320 min = _____ wk

19. 2922 wk = _____ yr

20. 623 days = _____ wk

21. *Actual Time in a Day.* Although we round it to 24 hr, the actual length of a day is 23 hr, 56 min, and 4.2 sec. How many seconds are there in an actual day?
Source: *The Handy Geography Answer Book*

22. *Time Length.* What length of time is 86,400 sec? Is it 1 hr, 1 day, 1 week, or 1 month? (This was a question on the TV game show "Who Wants to Be a Millionaire?".)

b Convert to Fahrenheit. Use the formula $F = \frac{9}{5} \cdot C + 32$ or $F = 1.8 \cdot C + 32$.

23. 25°C

24. 85°C

25. 40°C

26. 90°C

27. 86°C

28. 93°C

29. 58°C

30. 35°C

31. 2°C

32. 78°C

33. 5°C

34. 15°C

35. 3000°C
(The melting point of iron)

36. 1000°C
(The melting point of gold)

Convert to Celsius. Use the formula $C = \dfrac{5}{9} \cdot (F - 32)$ or $C = \dfrac{F - 32}{1.8}$.

37. 86°F

38. 59°F

39. 131°F

40. 140°F

41. 178°F

42. 195°F

43. 140°F

44. 107°F

45. 68°F

46. 50°F

47. 44°F

48. 120°F

49. 98.6°F
(Normal body temperature)

50. 104°F
(High-fevered body temperature)

51. *Highest Temperatures.*   The highest temperature ever recorded in the world is 136°F in the desert of Libya in 1922. The highest temperature ever recorded in the United States is $56\frac{2}{3}$°C in California's Death Valley in 1913.
Source: *The Handy Geography Answer Book*

a) Convert each temperature to the other scale.
b) How much higher in degrees Fahrenheit was the world record than the U. S. record?

52. *Boiling Point and Altitude.*   The boiling point of water actually changes with altitude. The boiling point is 212°F at sea level, but lowers about 1°F for every 500 ft that the altitude increases above sea level.
Sources: *The Handy Geography Answer Book; The New York Times Almanac*

a) What is the boiling point at an elevation of 1500 ft above sea level?
b) The elevation of Tucson is 2564 ft above sea level and that of Phoenix is 1117 ft. What is the boiling point in each city?
c) How much lower is the boiling point in Denver, whose elevation is 5280 ft, than in Tucson?
d) What is the boiling point at the top of Mt. McKinley in Alaska, the highest point in the United States, at 20,320 ft?

530

53. D_W Write a report on the origin of the words *Fahrenheit* and *Celsius*. How is the word *centigrade* related to temperature?

54. D_W **a)** The temperature is 23°C. Would you want to play golf? Explain.
b) Your bathwater has a temperature of 10°C. Would you want to take a bath?
c) The nearby lake has a temperature of −10°C. Would it be safe to go ice skating?

SKILL MAINTENANCE

Convert to percent notation.   [6.1b]

55. 0.875

56. 0.58

57. $0.\overline{6}$

58. 0.4361

Convert to percent notation.   [6.2a]

59. $\dfrac{3}{8}$

60. $\dfrac{5}{8}$

61. $\dfrac{2}{3}$

62. $\dfrac{1}{5}$

Solve.

63. A state charges a meals tax of $4\frac{1}{2}\%$. What is the meals tax charged on a dinner party costing $540?   [6.6a]

64. The price of a cellular phone was reduced from $350 to $308. Find the percent of decrease in price.   [6.5b]

65. A country has a population that is increasing by 4% each year. This year the population is 180,000. What will it be next year?   [6.5b]

66. A college has a student body of 1850 students. Of these, 17.5% are seniors. How many students are seniors?   [6.5a]

SYNTHESIS

67. Estimate the number of years in one million seconds.

68. Estimate the number of years in one billion seconds.

69. Estimate the number of years in one trillion seconds.

Complete.

70. $88 \dfrac{\text{ft}}{\text{sec}} = \underline{\hspace{1cm}} \dfrac{\text{mi}}{\text{hr}}$

71. $0.9 \dfrac{\text{L}}{\text{hr}} = \underline{\hspace{1cm}} \dfrac{\text{mL}}{\text{sec}}$

Objectives

a Convert from one American unit of area to another.

b Convert from one metric unit of area to another.

Complete.

1. $1 \text{ yd}^2 =$ _____ ft^2

2. $5 \text{ yd}^2 =$ _____ ft^2

3. $20 \text{ ft}^2 =$ _____ in^2

Complete.

4. $360 \text{ in}^2 =$ _____ ft^2

5. $5 \text{ mi}^2 =$ _____ acres

Answers on page A-21

532

8.7 CONVERTING UNITS OF AREA

a American Units

Let's do some conversions from one American unit of area to another.

EXAMPLE 1 Complete: $1 \text{ ft}^2 =$ _____ in^2.

$$1 \text{ ft}^2 = 1 \cdot (12 \text{ in.})^2 \qquad \text{Substituting 12 in. for 1 ft}$$
$$= 12 \text{ in.} \cdot 12 \text{ in.} = 144 \text{ in}^2$$

EXAMPLE 2 Complete: $8 \text{ yd}^2 =$ _____ ft^2.

$$8 \text{ yd}^2 = 8 \cdot (3 \text{ ft})^2 \qquad \text{Substituting 3 ft for 1 yd}$$
$$= 8 \cdot 3 \text{ ft} \cdot 3 \text{ ft} = 8 \cdot 3 \cdot 3 \cdot \text{ft} \cdot \text{ft} = 72 \text{ ft}^2$$

Do Exercises 1–3.

AMERICAN UNITS OF AREA

1 square yard (yd^2) = 9 square feet (ft^2)
1 square foot (ft^2) = 144 square inches (in^2)
1 square mile (mi^2) = 640 acres
1 acre = 43,560 ft^2

EXAMPLE 3 Complete: $36 \text{ ft}^2 =$ _____ yd^2.

We are converting from "ft^2" to "yd^2". Thus we choose a symbol for 1 with yd^2 on top and ft^2 on the bottom.

$$36 \text{ ft}^2 = 36 \text{ ft}^2 \times \frac{1 \text{ yd}^2}{9 \text{ ft}^2} \qquad \text{Multiplying by 1 using } \frac{1 \text{ yd}^2}{9 \text{ ft}^2}$$

$$= \frac{36}{9} \times \frac{\text{ft}^2}{\text{ft}^2} \times 1 \text{ yd}^2 = 4 \text{ yd}^2$$

EXAMPLE 4 Complete: $7 \text{ mi}^2 =$ _____ acres.

$$7 \text{ mi}^2 = 7 \cdot 1 \text{ mi}^2$$
$$= 7 \cdot 640 \text{ acres} \qquad \text{Substituting 640 acres for 1 mi}^2$$
$$= 4480 \text{ acres}$$

Do Exercises 4 and 5.

b Metric Units

Let's now convert from one metric unit of area to another.

EXAMPLE 5 Complete: $1 \text{ km}^2 =$ _____ m^2.

$$1 \text{ km}^2 = 1 \cdot (1000 \text{ m})^2 \qquad \text{Substituting 1000 m for 1 km}$$
$$= 1000 \text{ m} \cdot 1000 \text{ m} = 1,000,000 \text{ m}^2$$

EXAMPLE 6 Complete: $10{,}000 \text{ cm}^2 = $ _____ m^2.

$$10{,}000 \text{ cm}^2 = 10{,}000 \text{ cm}^2 \cdot \frac{1 \text{ m}}{100 \text{ cm}} \cdot \frac{1 \text{ m}}{100 \text{ cm}}$$

$$= 10{,}000 \text{ cm}^2 \cdot \frac{1 \text{ m}^2}{10{,}000 \text{ cm}^2}$$

$$= 1 \text{ m}^2$$

Do Exercises 6 and 7.

MENTAL CONVERSION

To convert mentally, we first note that $10^2 = 100$, $100^2 = 10{,}000$, and $0.1^2 = 0.01$. We use the diagram as before and multiply the number of moves by 2 to determine the number of moves of the decimal point.

1000 m	100 m	10 m	1 m	0.1 m	0.01 m	0.001 m
1 km	1 hm	1 dam	1 m	1 dm	1 cm	1 mm

EXAMPLE 7 Complete: $3.48 \text{ km}^2 = $ _____ m^2.

Think: To go from km to m in the table is a move of 3 places to the right.

1000 m	100 m	10 m	1 m	0.1 m	0.01 m	0.001 m
1 km	1 hm	1 dam	1 m	1 dm	1 cm	1 mm

3 moves to the right

So we move the decimal point $2 \cdot 3$, or 6 places to the right.

$3.48 \qquad 3.480000. \qquad 3.48 \text{ km}^2 = 3{,}480{,}000 \text{ m}^2$

6 places to the right

EXAMPLE 8 Complete: $586.78 \text{ cm}^2 = $ _____ m^2.

Think: To go from cm to m in the table is a move of 2 places to the left.

1000 m	100 m	10 m	1 m	0.1 m	0.01 m	0.001 m
1 km	1 hm	1 dam	1 m	1 dm	1 cm	1 mm

2 moves to the left

So we move the decimal point $2 \cdot 2$, or 4 places to the left.

$586.78 \qquad 0.0586.78 \qquad 586.78 \text{ cm}^2 = 0.058678 \text{ m}^2$

4 places to the left

Do Exercises 8–10.

Complete.

6. $1 \text{ m}^2 = $ _____ mm^2

7. $100 \text{ mm}^2 = $ _____ cm^2

Complete.

8. $2.88 \text{ m}^2 = $ _____ cm^2

9. $4.3 \text{ mm}^2 = $ _____ cm^2

10. $678{,}000 \text{ m}^2 = $ _____ km^2

Answers on page A-21

a Complete.

1. $1 \text{ ft}^2 = $ _____ in^2

2. $1 \text{ yd}^2 = $ _____ ft^2

3. $1 \text{ mi}^2 = $ _____ acres

4. $1 \text{ acre} = $ _____ ft^2

5. $1 \text{ in}^2 = $ _____ ft^2

6. $1 \text{ ft}^2 = $ _____ yd^2

7. $22 \text{ yd}^2 = $ _____ ft^2

8. $40 \text{ ft}^2 = $ _____ in^2

9. $44 \text{ yd}^2 = $ _____ ft^2

10. $144 \text{ ft}^2 = $ _____ yd^2

11. $20 \text{ mi}^2 = $ _____ acres

12. $576 \text{ in}^2 = $ _____ ft^2

13. $1 \text{ mi}^2 = $ _____ ft^2

14. $1 \text{ mi}^2 = $ _____ yd^2

15. $720 \text{ in}^2 = $ _____ ft^2

16. $27 \text{ ft}^2 = $ _____ yd^2

17. $144 \text{ in}^2 = $ _____ ft^2

18. $72 \text{ in}^2 = $ _____ ft^2

19. $1 \text{ acre} = $ _____ mi^2

20. $4 \text{ acres} = $ _____ ft^2

b Complete.

21. $5.21 \text{ km}^2 = $ _____ m^2

22. $65 \text{ km}^2 = $ _____ m^2

23. $0.014 \text{ m}^2 = $ _____ cm^2

24. $0.028 \text{ m}^2 = $ _____ mm^2

25. 2345.6 mm^2 = _____ cm^2

26. 8.38 cm^2 = _____ mm^2

27. 852.14 cm^2 = _____ m^2

28. 125 mm^2 = _____ m^2

29. 250,000 mm^2 = _____ cm^2

30. 2400 mm^2 = _____ cm^2

31. 472,800 m^2 = _____ km^2

32. 1.37 cm^2 = _____ mm^2

33. ^{D}**w** Explain the difference between the way we move the decimal point for area conversion and the way we do so for length conversion.

34. ^{D}**w** Which is larger and why: one square meter or nine square feet?

━━━(SKILL MAINTENANCE)━━━

In Exercises 35 and 36, find the simple interest. [6.7a]

35. On $2000 at an interest rate of 8% for 1.5 yr

36. On $2000 at an interest rate of 5.3% for 2 yr

In each of Exercises 37–40, find (a) the amount of simple interest due and (b) the total amount that must be paid back. [6.7a]

37. A firm borrows $15,500 at 9.5% for 120 days.

38. A firm borrows $8500 at 10% for 90 days.

39. A firm borrows $6400 at 8.4% for 150 days.

40. A firm borrows $4200 at 11% for 30 days.

━━━(SYNTHESIS)━━━

Complete.

41. 1 m^2 = _____ ft^2

42. 1 in^2 = _____ cm^2

43. 2 yd^2 = _____ m^2

44. 1 acre = _____ m^2

45. *The White House.*   The president's family has about 20,175 ft^2 of living area in the White House. Estimate the amount of living area in square meters.

Summary and Review

The review that follows is meant to prepare you for a chapter exam. It consists of two parts. The first part is a checklist of some of the Study Tips referred to in this and preceding chapters, as well as a list of important properties and formulas. The second part is the Review Exercises. These provide practice exercises for the exam, together with references to section objectives so you can go back and review. Before beginning, stop and look back over the skills you have obtained. What skills in mathematics do you have now that you did not have before studying this chapter?

STUDY TIPS CHECKLIST

The foundation of all your study skills is TIME!

☐ Have you tried the suggestions for taking notes?

☐ Are you working together with a classmate?

☐ Are you working to improve your time-management skills?

☐ Have you established a learning relationship with your instructor?

☐ Are you keeping one section ahead in your studies?

IMPORTANT PROPERTIES AND FORMULAS

American Units of Length: 12 in. = 1 ft; 3 ft = 1 yd; 36 in. = 1 yd; 5280 ft = 1 mi

Metric Units of Length: 1 km = 1000 m; 1 hm = 100 m; 1 dam = 10 m; 1 dm = 0.1 m; 1 cm = 0.01 m; 1 mm = 0.001 m

American–Metric Conversion: 1 m = 39.370 in.; 1 m = 3.281 ft; 0.305 m = 1 ft; 2.540 cm = 1 in.; 1 km = 0.621 mi; 1.609 km = 1 mi

American System of Weights: 1 T = 2000 lb; 1 lb = 16 oz

Metric System of Mass: 1 t = 1000 kg; 1 kg = 1000 g; 1 hg = 100 g; 1 dag = 10 g; 1 dg = 0.1 g; 1 cg = 0.01 g; 1 mg = 0.001 g; 1 mcg = 0.000001 g

American Units of Capacity: 1 gal = 4 qt; 1 qt = 2 pt; 1 pt = 16 oz; 1 pt = 2 cups; 1 cup = 8 oz

Metric Units of Capacity: 1 L = 1000 mL = 1000 cm^3 = 1000 cc

American–Metric Conversion: 1 oz = 29.57 mL; 1 L = 1.057 qt

Units of Time: 1 min = 60 sec; 1 hr = 60 min; 1 day = 24 hr; 1 wk = 7 days; 1 yr = $365\frac{1}{4}$ days

Temperature Conversion: $F = \frac{9}{5} \cdot C + 32$, or $F = 1.8 \cdot C + 32$;

$C = \frac{5}{9} \cdot (F - 32)$, or $C = \frac{F - 32}{1.8}$

REVIEW EXERCISES

Complete.   [8.1a], [8.2a], [8.3a]

1. 8 ft = _____ yd

2. $\frac{5}{6}$ yd = _____ in.

3. 0.3 mm = _____ cm

4. 4 m = _____ km

5. 2 yd = _____ in.

6. 4 km = _____ cm

7. 14 in. = _____ ft

8. 15 cm = _____ m

9. 200 m = _____ yd

10. 20 mi = _____ km

Complete.   [8.4a, b], [8.5a], [8.6a]

13. 7 lb = _____ oz

14. 4 g = _____ kg

15. 16 min = _____ hr

16. 464 mL = _____ L

17. 3 min = _____ sec

18. 4.7 kg = _____ g

19. 8.07 T = _____ lb

20. 0.83 L = _____ mL

21. 6 hr = _____ days

22. 4 cg = _____ g

23. 0.2 g = _____ mg

24. 0.0003 kg = _____ cg

25. 0.7 mL = _____ L

26. 60 mL = _____ L

27. 0.8 T = _____ lb

28. 0.4 L = _____ mL

29. 20 oz = _____ lb

30. $\frac{5}{6}$ min = _____ sec

31. 20 gal = _____ pt

32. 960 oz = _____ gal

33. 54 qt = _____ gal

Complete the following table   [8.2a]

	OBJECT	MILLIMETERS (mm)	CENTIMETERS (cm)	METERS (m)
11.	Length of a key on a calculator		1	
12.	Height of Texas Commerce Center, Houston			305

Medical Dosage. **Solve.**

34. Amoxicillin is an antibiotic obtainable in a liquid suspension form, part medication and part water, and is frequently used to treat infections in infants. One formulation of the drug contains 125 mg of amoxicillin per 5 mL of liquid. A pediatrician orders 150 mg per day for a 4-month-old child with an ear infection. How much of the amoxicillin suspension would the parent need to administer to the infant in order to achieve the recommended daily dose?   [8.4c]

35. An emergency-room physician orders 3 L of Ringer's lactate to be administered over 4 hr for a patient suffering from shock and severe low blood pressure. How many milliliters is this?   [8.5b]

36. A physician prescribes 0.25 mg of alprazolam, an antianxiety medication. How many micrograms are in this dose?   [8.4c]

37. Convert 27°C to Fahrenheit.   [8.6b]

38. Convert 68°F to Celsius.   [8.6b]

Complete.   [8.7a, b]

39. $4 \text{ yd}^2 =$ _____ ft^2

40. $0.3 \text{ km}^2 =$ _____ m^2

41. $2070 \text{ in}^2 =$ _____ ft^2

42. $600 \text{ cm}^2 =$ _____ m^2

43. $^\text{D}_\text{W}$ Napoleon is credited with influencing the use of the metric system. Research this possibility and make a report.   [8.2a]

44. $^\text{D}_\text{W}$ It is known that 1 gal of water weighs 8.3453 lb. Which weighs more, an ounce of pennies or an ounce (as capacity) of water? Explain.   [8.4a], [8.5a]

SKILL MAINTENANCE

Certain objectives from two particular sections will be retested on the chapter test. The objectives are listed with the practice problems that follow.

45. Convert to percent notation: 0.47.   [6.1b]

46. Convert to percent notation: $\dfrac{23}{25}$.   [6.2a]

47. Convert to decimal notation: 56.7%.   [6.1b]

48. Convert to fraction notation: 73%.   [6.2b]

SYNTHESIS

49. *Running Record.* The world's record for running the 200-m dash is 19.32 sec, set by Michael Johnson of the United States in the 1996 Olympics in Atlanta. How should the record be changed if it were a 200-yd dash?   [8.3a]
Source: *The Guinness Book of Records*

Complete

1. 4 ft = _____ in.

2. 4 in. = _____ ft

3. 6 km = _____ m

4. 8.7 mm = _____ cm

5. 200 yd = _____ m

6. 2400 km = _____ mi

Complete the following table.

	OBJECT	MILLIMETERS (mm)	CENTIMETERS (cm)	METERS (m)
7.	Width of a key on a calculator		0.5	
8.	Height of your author			1.8542

Complete.

9. 3080 mL = _____ L

10. 0.24 L = _____ mL

11. 4 lb = _____ oz

12. 4.11 T = _____ lb

13. 3.8 kg = _____ g

14. 4.325 mg = _____ cg

15. 2200 mg = _____ g

16. 5 hr = _____ min

17. 15 days = _____ hr

18. 64 pt = _____ qt

19. 10 gal = _____ oz

20. 5 cups = _____ oz

21. 0.37 mg = _____ mcg

22. Convert 95°F to Celsius.

23. Convert 59°C to Fahrenheit.

Complete the following table.

	OBJECT	YARDS (yd)	CENTIMETERS (cm)	INCHES (in.)	METERS (m)	MILLIMETERS (mm)
24.	Length of a meter stick				1	
25.	Height of Xiamen Posts and Telecommunications Building, Xiamen, China	398				

Medical Dosage. **Solve each of the following.**

26. An emergency-room physician prescribes 2.5 L of normal saline intravenously over 8 hr for a patient who is severely dehydrated. How many milliliters is this?

27. A physician prescribes 0.5 mg of alprazolam to be taken 3 times a day by a patient suffering from anxiety. How many micrograms of alprazolam is the patient to ingest each day?

28. A prescription calls for 4 oz of dextromethorphan, a cough-suppressant medication. For how many milliliters is the prescription? (Use 1 oz = 29.57 mL.)

Complete.

29. $12 \text{ ft}^2 = $ _____ in^2

30. $3 \text{ cm}^2 = $ _____ m^2

31. Convert to percent notation: 0.93.

32. Convert to percent notation: $\dfrac{13}{16}$.

33. Convert to decimal notation: 93.2%.

34. Convert to fraction notation: $33\frac{1}{3}\%$

35. *Running Record.* The world's record for running the 400-m run is 43.18 sec, set by Michael Johnson of the United States in Seville, Spain, on August 26, 1999. How should the record be changed if it were a 400-yd run?
Source: *The Guinness Book of Records*

Solve.

1. *Population Growth.* The population of Los Angeles is projected to be 24.5 million by 2025. Find standard notation for 24.5 million.
   Sources: U.S. Bureau of the Census; bizjournals.com

2. *Gas Mileage.* A Ford Focus gets 312 miles of city driving on 12 gallons of gasoline. What is the gas mileage?
   Source: Ford Motor Company

Job Opportunities. The following table shows job opportunities, in thousands, in 1998 and then projected increases to 2008. Find the missing numbers. Round the percent of increase to the nearest tenth of a percent.

	OCCUPATION	NUMBER OF JOBS IN 1998 (in thousands)	NUMBER OF JOBS IN 2008 (in thousands)	CHANGE	PERCENT OF INCREASE
3.	Flight attendant	99	129		
4.	Short-order cook	677		124	
5.	Medical assistant	252	398		
6.	Loan and credit clerk		200	21	
7.	Manicurist	50			26%

Source: Handbook of U.S. Labor Statistics

Perform the indicated operation and simplify.

8. $46{,}231 \times 1100$

9. $\dfrac{1}{10} \cdot \dfrac{5}{6}$

10. $14.5 + \dfrac{4}{5} - 0.1$

11. $2\dfrac{3}{5} \div 3\dfrac{9}{10}$

12. $0.1\overline{)3.56}$

13. $3\dfrac{1}{2} - 2\dfrac{2}{3}$

14. Determine whether 1,298,032 is divisible by 8.

15. Determine whether 5,024,120 is divisible by 3.

16. Find the prime factorization of 99.

17. Find the LCM of 35 and 49.

18. Round $35.\overline{7}$ to the nearest tenth.

19. Write a word name for 103.064.

20. Find the average and the median of this set of numbers: 29, 21, 9, 13, 17, 18.

Find percent notation.

21. 0.08

22. $\dfrac{3}{5}$

Solve.

23. $0.07 \cdot x = 10.535$

24. $x + 12{,}843 = 32{,}091$

25. $\dfrac{2}{3} \cdot y = 5$

26. $\dfrac{4}{5} + y = \dfrac{6}{7}$

Complete.

27. 2 yd = _____ ft

28. 6 oz = _____ lb

29. 15°C = _____ F

30. 0.087 L = _____ mL

31. 9 sec = _____ min

32. 17 cm = _____ m

33. 2200 mi = _____ km

34. 6 qt = _____ L

35. 0.23 mg = _____ mcg

CD Sales. The following table shows the number of music CDs sold in recent years. Use it for Exercises 36–38.

YEAR	CD SALES (in millions)
1994	662
1995	723
1996	779
1997	759
1998	847
1999	939

Source: Recording Industry
Association of America

36. Make a bar graph of the data.

37. Make a line graph of the data.

38. Find the total number of CDs sold over this 6-yr period. Use it to make a circle graph of the data.

Solve.

39. *Seed Production.* The U. S. Department of Agriculture requires that 80% of the seeds that a company produces must sprout. To find out about the quality of the seeds it has produced, a company takes 500 seeds and plants them. It finds that 417 of the seeds sprout. Did the seeds pass government standards?
Source: U.S. Department of Agriculture

40. *Milk Production.* There are 11 million milk cows in America, each producing, on average, 15,000 lb of milk per year. How many pounds of milk are produced each year in America?
Source: U.S. Department of Agriculture

41. A driver bought gasoline when the odometer read 86,897.2. At the next gasoline purchase, the odometer read 87,153.0. How many miles had been driven? The tank was filled with 16 gal. What was the gas mileage?

42. A man on a diet loses $3\frac{1}{2}$ lb in 2 weeks. At this rate, how many pounds will he lose in 5 weeks?

43. A family has an annual income of $52,800. Of this, $\frac{1}{4}$ is spent for food. How much does the family spend for food?

44. A mechanic spent $\frac{1}{3}$ hr changing a car's oil, $\frac{1}{2}$ hr rotating the tires, $\frac{1}{10}$ hr changing the air filter, $\frac{1}{4}$ hr adjusting the idle speed, and $\frac{1}{15}$ hr checking the brake and transmission fluids. How many hours did the mechanic spend working on the car?

Sundae's Homemade Ice Cream & Coffee Co. With 4 outlets in the Indianapolis area, this company makes ice cream, sorbet, and frozen yogurt.

45. Ice cream is packaged in 15-lb tubs. How many ounces are in one tub?

46. While not a perfect process, Sundae's attempts to have about 4 oz in 1 dip of ice cream. How many dips are there in a tub of ice cream?

47. By weighing each tub, the owner can determine how many dips have been sold of that flavor. The weight of a tub changes from 15 lb to $8\frac{5}{8}$ lb over a busy weekend. How many dips of ice cream were served from that tub?

48. A 1-dip ice cream cone sells for $1.99. If the entire contents of a tub were used to make 1-dip cones, how much money is taken in from the sale of a tub of ice cream?

49. A 2-dip ice cream cone sells for $2.99. If the entire contents of a tub were used to make 2-dip cones, how much money is taken in from the sale of a tub of ice cream?

50. Each store offers 32 flavors in a case holding 32 tubs. Each tub starts out at 15 lb. If on any given day all 32 tubs are full, how many pounds of ice cream are in the store? How many ounces are in the store?

Assume that on a Monday all 32 tubs are full of ice cream. The table below shows how the ice cream sells for subsequent days. By Friday, 10 full ones must replace empties in the case at the end of day.

51. Complete the table and determine the income from Monday through Sunday, assuming only $1.99 one-dip cones were sold and then assuming only $2.99 two-dip cones were sold.

52. What do you think is a safe estimate of how much money was taken in from the ice cream for that week?

53. *Expenses.* Assume that it costs $14 per tub to make the ice cream and $6 per hour for one employee to work a 10-hr day for every day but Sunday. On Sunday, two employees are needed for the 10-hr shift. Rent is $800 per week and utilities are $225 per week. Assuming that paper products and cones raise the expenses per cone by 5¢, what are the total expenses to the store for the ice cream, assuming only 1-dip cones were sold? only 2-dip cones were sold?

54. Using the results of Exercises 51–53, what are the profits of the store each week from ice cream, assuming only 1-dip cones were sold? only 2-dip cones were sold?

DAY	TOTAL NUMBER OF POUNDS IN THE TUBS AT THE START OF THE DAY	NUMBER OF POUNDS SOLD	NUMBER OF OUNCES SOLD	NUMBER OF DIPS SOLD	AMOUNT OF REVENUE AT $1.99/CONE	AMOUNT OF REVENUE AT $2.99/CONE
Monday	480	93	1488			
Tuesday	387					
Wednesday	324					
Thursday	257					
Friday	302					
Saturday	224					
Sunday	138					
Monday	85					

Source: Sundae's Homemade Ice Cream & Coffee Co.

Geometry

Gateway to Chapter 9

Here we apply the concept of measure that we studied in Chapter 8 to finding the perimeters of polygons and the areas of squares, rectangles, triangles, parallelograms, trapezoids, and circles. We then study volume, properties of angles and triangles, and the Pythagorean theorem.

Real-World Application

Major league baseball underwent a rule change between the 2000 and 2001 seasons. Over the years before 2001, the strike zone evolved to something other than what was defined in the rule book. The zone in the rule book is described by rectangle *ABCD* in the illustration. The zone used before 2001 is described by the region *AQRST*. By what percent has the area of the strike zone been increased by the change?

Sources: The Cincinnati Enquirer; Major League Baseball; Gannett News Service; *The Sporting News Official Baseball Rules Book*

This problem appears as Example 10 in Section 9.2.

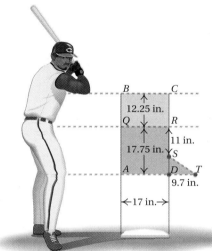

C H A P T E R

9

1. Find the perimeter.   [9.1a]

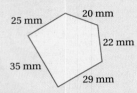

25 mm
20 mm
22 mm
35 mm
29 mm

2. Find the area of a square with sides of length 10 ft.   [9.2a]

3. Find the area of the shaded region.   [9.2c]

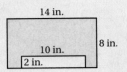

14 in.
10 in.
8 in.
2 in.

Find the area.   [9.2b]

4.

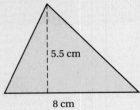

5.5 cm

8 cm

5.

$5\frac{1}{2}$ ft
5 ft
$7\frac{1}{2}$ ft

6.

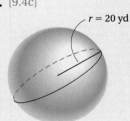

$1\frac{3}{5}$ m
$2\frac{1}{2}$ m

7. Find the length of a diameter of a circle with a radius of 4.8 m.   [9.3a]

8. Find the circumference of the circle in Question 7. Use 3.14 for π.   [9.3b]

9. Find the area of the circle in Question 7. Use 3.14 for π.   [9.3c]

Find the volume. Use 3.14 for π.

10. [9.4a]

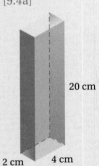

20 cm

2 cm   4 cm

11. [9.4b]

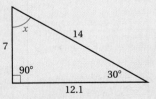

16 ft

5 ft

12. [9.4c]

$r = 20$ yd

13. [9.4d]

16 cm

3 cm

Use the triangle shown at right for Questions 14 and 15.

14. Find the missing angle measure.   [9.5e]

x
14
7
90°
30°
12.1

15. Classify the triangle as (a) equilateral, isosceles, or scalene and (b) as right, obtuse, or acute.   [9.5d]

16. Simplify: $\sqrt{81}$.   [9.6a]

17. Approximate to three decimal places: $\sqrt{97}$.   [9.6b]

18. In the right triangle, find the length of the side not given. Find an exact answer and an approximation to three decimal places.   [9.6c]

$c = 7$
$b = ?$
$a = 2$

9.1 PERIMETER

a Finding Perimeters

PERIMETER OF A POLYGON

A **polygon** is a geometric figure with three or more sides. The **perimeter** of a **polygon** is the distance around it, or the sum of the lengths of its sides.

EXAMPLE 1 Find the perimeter of this polygon.

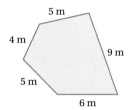

We add the lengths of the sides. Since all units are the same, we add the numbers, keeping meters (m) as the unit.

$$\text{Perimeter} = 6\,\text{m} + 5\,\text{m} + 4\,\text{m} + 5\,\text{m} + 9\,\text{m}$$
$$= (6 + 5 + 4 + 5 + 9)\,\text{m}$$
$$= 29\,\text{m}$$

Do Exercises 1 and 2.

A **rectangle** is a figure with four sides and four 90°-angles, like the one shown in Example 2.

EXAMPLE 2 Find the perimeter of a rectangle that is 3 cm by 4 cm.

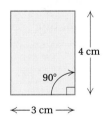

$$\text{Perimeter} = 3\,\text{cm} + 3\,\text{cm} + 4\,\text{cm} + 4\,\text{cm}$$
$$= (3 + 3 + 4 + 4)\,\text{cm}$$
$$= 14\,\text{cm}$$

Do Exercise 3.

Find the perimeter of the polygon.

1.

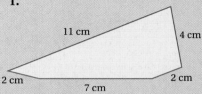

2.

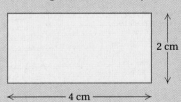

3. Find the perimeter of a rectangle that is 2 cm by 4 cm.

Answers on page A-23

4. Find the perimeter of a rectangle that is 5.25 yd by 3.5 yd.

PERIMETER OF A RECTANGLE

The **perimeter of a rectangle** is twice the sum of the length and the width, or 2 times the length plus 2 times the width:

$$P = 2 \cdot (l + w), \quad \text{or} \quad P = 2 \cdot l + 2 \cdot w.$$

EXAMPLE 3 Find the perimeter of a rectangle that is 4.3 ft by 7.8 ft.

$$
\begin{aligned}
P &= 2 \cdot (l + w) \\
&= 2 \cdot (4.3 \text{ ft} + 7.8 \text{ ft}) \\
&= 2 \cdot (12.1 \text{ ft}) \\
&= 24.2 \text{ ft}
\end{aligned}
$$

Do Exercises 4 and 5.

A square is a rectangle with all sides the same length.

EXAMPLE 4 Find the perimeter of a square whose sides are 9 mm long.

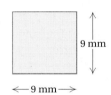

9 mm

←—9 mm—→

$$
\begin{aligned}
P &= 9 \text{ mm} + 9 \text{ mm} + 9 \text{ mm} + 9 \text{ mm} \\
&= (9 + 9 + 9 + 9) \text{ mm} \\
&= 36 \text{ mm}
\end{aligned}
$$

Do Exercise 6.

5. Find the perimeter of a rectangle that is $8\frac{1}{4}$ in. by $5\frac{2}{3}$ in.

PERIMETER OF A SQUARE

The **perimeter of a square** is four times the length of a side:

$$P = 4 \cdot s.$$

s

s ▢ *s*

s

6. Find the perimeter of a square with sides of length 10 km.

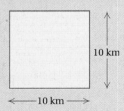

10 km

←—10 km—→

Answers on page A-23

EXAMPLE 5 Find the perimeter of a square whose sides are $20\frac{1}{8}$ in. long.

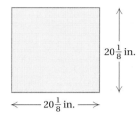

$$P = 4 \cdot s = 4 \cdot 20\frac{1}{8} \text{ in.}$$

$$= 4 \cdot \frac{161}{8} \text{ in.} = \frac{4 \cdot 161}{4 \cdot 2} \text{ in.}$$

$$= \frac{161}{2} \cdot \frac{4}{4} \text{ in.} = 80\frac{1}{2} \text{ in.}$$

Do Exercises 7 and 8.

b Solving Applied Problems

EXAMPLE 6 A vegetable garden is 20 ft by 15 ft. A fence is to be built around the garden. How many feet of fence will be needed? If fencing sells for $2.95 per foot, what will the fencing cost?

1. Familiarize. We make a drawing and let P = the perimeter.

15 ft

20 ft

2. Translate. The perimeter of the garden is given by

$$P = 2 \cdot (l + w) = 2 \cdot (20 \text{ ft} + 15 \text{ ft}).$$

3. Solve. We calculate the perimeter as follows:

$$P = 2 \cdot (20 \text{ ft} + 15 \text{ ft}) = 2 \cdot (35 \text{ ft}) = 70 \text{ ft}$$

Then we multiply by $2.95 to find the cost of the fencing:

$$\text{Cost} = \$2.95 \times \text{Perimeter} = \$2.95 \times 70 \text{ ft} = \$206.50.$$

4. Check. The check is left to the student.

5. State. The 70 ft of fencing that is needed will cost $206.50.

Do Exercise 9.

7. Find the perimeter of a square with sides of length $5\frac{1}{4}$ yd.

8. Find the perimeter of a square with sides of length 7.8 km.

9. A play area is 25 ft by 10 ft. A fence is to be built around the play area. How many feet of fencing will be needed? If fencing costs $4.95 per foot, what will the fencing cost?

Answers on page A-23

a   Find the perimeter of the polygon.

1.

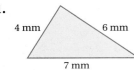

4 mm   6 mm
7 mm

2.

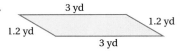

3 yd
1.2 yd
1.2 yd
3 yd

3.

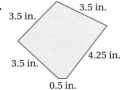

3.5 in.   3.5 in.
3.5 in.
4.25 in.
3.5 in.
0.5 in.

4.

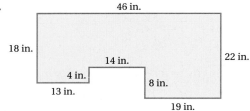

46 in.
18 in.
14 in.
4 in.
22 in.
13 in.
8 in.
19 in.

5.

3.4 km
5.6 km

6.

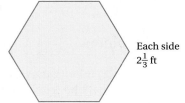

Each side $2\frac{1}{3}$ ft

Find the perimeter of the rectangle.

7. 5 ft by 10 ft

8. 2.5 m by 100 m

9. 34.67 cm by 4.9 cm

10. $3\frac{1}{2}$ yd by $4\frac{1}{2}$ yd

Find the perimeter of the square.

11. 22 ft on a side

12. 56.9 km on a side

13. 45.5 mm on a side

14. $3\frac{1}{8}$ yd on a side

b   Solve.

15. A security fence is to be built around a 173-m by 240-m rectangular field. What is the perimeter of the field? If fence wire costs $1.45 per meter, what will the fencing cost?

16. *Softball Diamond.* A standard-sized slow-pitch softball diamond is a square with sides of length 65 ft. What is the perimeter of this softball diamond? (This is the distance you would have to run if you hit a home run.)
Source: American Softball Association

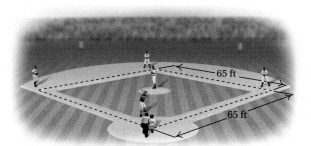

65 ft
65 ft

17. A piece of flooring tile is a square with sides of length 30.5 cm. What is the perimeter of a piece of tile?

18. A rectangular posterboard is 61.8 cm by 87.9 cm. What is the perimeter of the board?

19. A rain gutter is to be installed around the house shown in the figure.

    a) Find the perimeter of the house.
    b) If the gutter costs $4.59 per foot, what is the total cost of the gutter?

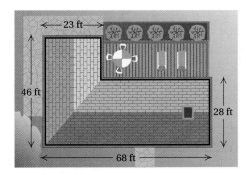

20. A carpenter is to build a fence around a 9-m by 12-m garden.

    a) The posts are 3 m apart. How many posts will be needed?
    b) The posts cost $2.40 each. How much will the posts cost?
    c) The fence will surround all but 3 m of the garden, which will be a gate. How long will the fence be?
    d) The fence costs $2.85 per meter. What will the cost of the fence be?
    e) The gate costs $9.95. What is the total cost of the materials?

21. $^\mathbf{D}\mathbf{w}$ Create for a fellow student a development of the formula

$$P = 2 \cdot (l + w) = 2 \cdot l + 2 \cdot w$$

for the perimeter of a rectangle.

22. $^\mathbf{D}\mathbf{w}$ Create for a fellow student a development of the formula

$$P = 4 \cdot s$$

for the perimeter of a square.

SKILL MAINTENANCE

23. Find the simple interest on $600 at 6.4% for $\frac{1}{2}$ yr. [6.7a]

24. Find the simple interest on $600 at 8% for 2 yr. [6.7a]

Evaluate. [1.9b]

25. 10^3

26. 11^3

27. 15^2

28. 22^2

29. 7^2

30. 4^3

Solve.

31. *Sales Tax.* In a certain state, a sales tax of $878 is collected on the purchase of a car for $17,560. What is the sales tax rate? [6.6a]

32. *Commission Rate.* Rich earns $1854.60 selling $16,860 worth of cellular phones. What is the commission rate? [6.6b]

SYNTHESIS

Find the perimeter, in feet, of the figure.

33.

18 in.

3 ft

34.

78 in.

5.5 yd

9.2 AREA

Objectives

a Find the area of a rectangle and a square.

b Find the area of a parallelogram, a triangle, and a trapezoid.

c Solve applied problems involving areas of rectangles, squares, parallelograms, triangles, and trapezoids.

a Rectangles and Squares

A polygon and its interior form a plane region. We can find the area of a *rectangular region*, or *rectangle*, by filling it in with square units. Two such units, a *square inch* and a *square centimeter*, are shown below.

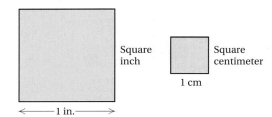

Square inch

Square centimeter
1 cm
←— 1 in. —→

1. What is the area of this region? Count the number of square centimeters.

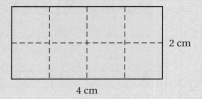

2 cm

4 cm

EXAMPLE 1 What is the area of this region?

We have a rectangular array. Since the region is filled with 12 square centimeters, its area is 12 square centimeters (sq cm), or 12 cm². The number of units is 3 × 4, or 12.

3 cm

4 cm

Do Exercise 1.

AREA OF A RECTANGLE

The **area of a rectangle** is the product of the length *l* and the width *w*:

$$A = l \cdot w.$$

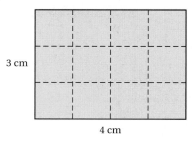

w

l

2. Find the area of a rectangle that is 7 km by 8 km.

EXAMPLE 2 Find the area of a rectangle that is 7 yd by 4 yd.

$$A = l \cdot w = 7 \text{ yd} \cdot 4 \text{ yd}$$
$$= 7 \cdot 4 \cdot \text{yd} \cdot \text{yd} = 28 \text{ yd}^2$$

We think of yd · yd as $(\text{yd})^2$ and denote it yd^2. Thus we read "28 yd²" as "28 square yards."

3. Find the area of a rectangle that is $5\frac{1}{4}$ yd by $3\frac{1}{2}$ yd.

Do Exercises 2 and 3.

Answers on page A-23

EXAMPLE 3 Find the area of a square with sides of length 9 mm.

$$A = (9 \text{ mm}) \cdot (9 \text{ mm})$$
$$= 9 \cdot 9 \cdot \text{mm} \cdot \text{mm}$$
$$= 81 \text{ mm}^2$$

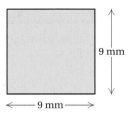
9 mm
9 mm

Do Exercise 4.

AREA OF A SQUARE

The **area of a square** is the square of the length of a side:

$$A = s \cdot s, \quad \text{or} \quad A = s^2.$$

s
s

EXAMPLE 4 Find the area of a square with sides of length 20.3 m.

$$A = s \cdot s = 20.3 \text{ m} \times 20.3 \text{ m} = 20.3 \times 20.3 \times \text{m} \times \text{m} = 412.09 \text{ m}^2$$

Do Exercises 5 and 6.

b   Finding Other Areas

PARALLELOGRAMS

A **parallelogram** is a four-sided figure with two pairs of parallel sides, as shown below.

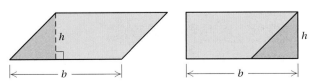

To find the area of a parallelogram, consider the one below.

If we cut off a piece and move it to the other end, we get a rectangle.

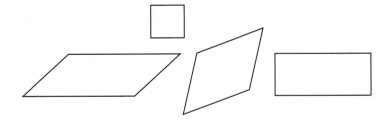
h
b
h
b

We can find the area by multiplying the length b, called a **base**, by h, called the **height.**

---

4. Find the area of a square with sides of length 12 km.

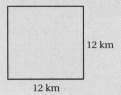
12 km
12 km

5. Find the area of a square with sides of length 10.9 m.

6. Find the area of a square with sides of length $3\frac{1}{2}$ yd.

Answers on page A-23

Study Tips

STUDYING THE ART PIECES

When you study a section of a mathematics text, read it slowly, observing all the details of the corresponding art pieces that are discussed in the paragraphs. Also note the precise color markings in the art that enhances the learning process. These tips apply especially to this chapter because geometry, by its nature, is quite visual.

Find the area.

7.

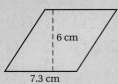

6 cm

7.3 cm

Answers on page A-23

CHAPTER 9: Geometry

8.

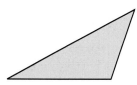

5.5 km

2.25 km

AREA OF A PARALLELOGRAM

The **area of a parallelogram** is the product of the length of a base b and the height h:

$$A = b \cdot h.$$

EXAMPLE 5   Find the area of this parallelogram.

$$A = b \cdot h$$
$$= 7 \text{ km} \cdot 5 \text{ km}$$
$$= 35 \text{ km}^2$$

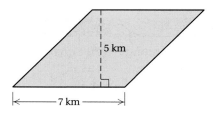

5 km

7 km

EXAMPLE 6   Find the area of this parallelogram.

$$A = b \cdot h$$
$$= 1.2 \text{ m} \times 6 \text{ m}$$
$$= 7.2 \text{ m}^2$$

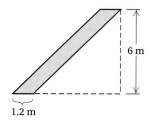

6 m

1.2 m

Do Exercises 7 and 8.

TRIANGLES

To find the area of a triangle like the one shown on the left below, think of cutting out another just like it and placing it as shown on the right below.

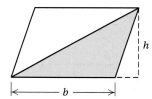

h

b

The resulting figure is a parallelogram whose area is

$$b \cdot h.$$

The triangle we started with has half the area of the parallelogram, or

$$\frac{1}{2} \cdot b \cdot h.$$

AREA OF A TRIANGLE

The **area of a triangle** is half the length of the base times the height:

$$A = \frac{1}{2} \cdot b \cdot h.$$

h

b

EXAMPLE 7 Find the area of this triangle.

$$A = \frac{1}{2} \cdot b \cdot h$$

$$= \frac{1}{2} \cdot 9 \text{ m} \cdot 6 \text{ m}$$

$$= \frac{9 \cdot 6}{2} \text{ m}^2$$

$$= 27 \text{ m}^2$$

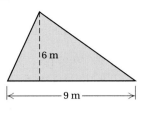

EXAMPLE 8 Find the area of this triangle.

$$A = \frac{1}{2} \cdot b \cdot h$$

$$= \frac{1}{2} \times 6.25 \text{ cm} \times 5.5 \text{ cm}$$

$$= 0.5 \times 6.25 \times 5.5 \text{ cm}^2$$

$$= 17.1875 \text{ cm}^2$$

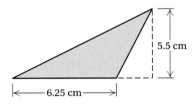

Do Exercises 9 and 10.

TRAPEZOIDS

A **trapezoid** is a polygon with four sides, two of which, the **bases,** are parallel to each other.

To find the area of a trapezoid, think of cutting out another just like it.

Then place the second one like this.

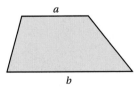

The resulting figure is a parallelogram whose area is

$$h \cdot (a + b). \qquad \text{The base is } a + b.$$

The trapezoid we started with has half the area of the parallelogram, or

$$\frac{1}{2} \cdot h \cdot (a + b).$$

Find the area.

9.

10.

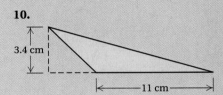

Answers on page A-23

Find the area.

11.

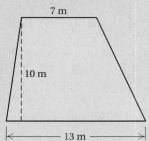

7 m

10 m

13 m

12.

6 cm

11 cm

10 cm

AREA OF A TRAPEZOID

The **area of a trapezoid** is half the product of the height and the sum of the lengths of the parallel sides (bases):

$$A = \frac{1}{2} \cdot h \cdot (a + b), \quad \text{or} \quad A = \frac{a + b}{2} \cdot h.$$

EXAMPLE 9   Find the area of this trapezoid.

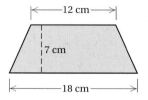

12 cm

7 cm

18 cm

$$A = \frac{1}{2} \cdot h \cdot (a + b)$$

$$= \frac{1}{2} \cdot 7 \text{ cm} \cdot (12 + 18) \text{ cm}$$

$$= \frac{7 \cdot 30}{2} \cdot \text{cm}^2 = \frac{7 \cdot 15 \cdot 2}{1 \cdot 2} \text{ cm}^2$$

$$= \frac{7 \cdot 15}{1} \cdot \frac{2}{2} \text{ cm}^2 = 105 \text{ cm}^2$$

Do Exercises 11 and 12.

C   Solving Applied Problems

EXAMPLE 10   *Baseball's Strike Zones.*   Major league baseball underwent a rule change between the 2000 and 2001 seasons. Over the years before 2001, the strike zone evolved to something other than what was defined in the rule book. The zone in the rule book is described by rectangle *ABCD* below. The zone used before 2001 is described by the region *AQRST*. The figure shown here represents the zones for a normal-sized player, but they vary depending on the height of the player.

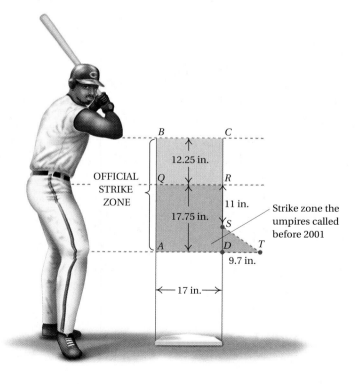

OFFICIAL STRIKE ZONE

B   C

12.25 in.

Q   R

11 in.

17.75 in.

S

Strike zone the umpires called before 2001

A   D   T

9.7 in.

17 in.

a) Find the area of the new zone.

b) Find the area of the former zone.

c) How much larger is the rule-book zone than the former zone?

d) By what percent has the area of the strike zone been increased by the change?

Sources: *The Cincinnati Inquirer*; Major League Baseball; Gannett News Service; *The Sporting News Official Baseball Rules Book*

This is a multistep problem that makes use of many of the skills we have learned in this book.

a) The area of rectangle *ABCD* (the rule-book zone), denoted A_1, is the length times width:

$A_1 = l \cdot w = (17.75 \text{ in.} + 12.25 \text{ in.}) \cdot (17 \text{ in.}) = (30 \text{ in.}) \cdot (17 \text{ in.}) = 510 \text{ in}^2.$

b) The area of the region *AQRST* (the former zone), denoted A_2, is shown shaded in the figure. To find that area, we add the area of rectangle *AQRD* to the area of triangle *SDT*. We first determine

Area of $AQRD = l \cdot w = (17.75 \text{ in.}) \cdot (17 \text{ in.}) = 301.75 \text{ in}^2.$

To find the area of triangle *SDT*, we first note that the length of the base is given as 9.7 in. To find the height that is the length of segment *SD*, we subtract 11 in. from 17.75 in.: 17.75 in. − 11 in. = 6.75 in. Then

Area of triangle $SDT = \frac{1}{2} \cdot b \cdot h = \frac{1}{2}(9.7 \text{ in.}) \cdot (6.75 \text{ in.}) = 32.7375 \text{ in}^2.$

The area of the former zone A_2 is the sum of the areas of the triangle *SDT* and the rectangle *AQRD*:

$A_2 = 301.75 \text{ in}^2 + 32.7375 \text{ in}^2 = 334.4875 \text{ in}^2 \approx 334.5 \text{ in}^2.$

c) To find the increase in the area, we subtract A_2 from A_1:

$510 \text{ in}^2 - 334.5 \text{ in}^2 = 175.5 \text{ in}^2.$

d) To determine the percent of increase, note that we are asking "What percent of the former area is the increase?" We translate this to an equation as follows:

$$\underbrace{\text{What percent}}_{p} \quad \underset{\downarrow}{\text{of}} \quad \underset{334.5}{\overset{\downarrow}{334.5}} \quad \underset{=}{\overset{\downarrow}{\text{is}}} \quad \underset{175.5}{\overset{\downarrow}{175.5?}}$$

We solve the equation:

$$p \cdot 334.5 = 175.5$$

$$\frac{p \cdot 334.5}{334.5} = \frac{175.5}{334.5}$$

$$p = \frac{175.5}{334.5} \approx 0.5247 \approx 52\%.$$

There was an increase of approximately 52% in the strike zone.

Do Exercise 13.

13. Find the area of this kite.

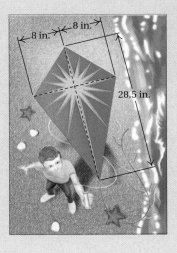

Answer on page A-23

9.2
EXERCISE SET

For Extra Help

Digital Video
Tutor CD 5
Videotape 14

InterAct
Math

Math Tutor
Center

MathXL

MyMathLab

a Find the area.

1.

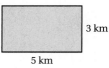

3 km

5 km

2.

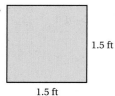

1.5 ft

1.5 ft

3.

2 in.

0.7 in.

4.

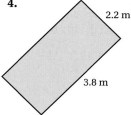

2.2 m

3.8 m

5.

$2\frac{1}{2}$ yd

$2\frac{1}{2}$ yd

6.

$3\frac{1}{2}$ mi

$3\frac{1}{2}$ mi

7.

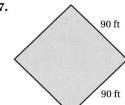

90 ft

90 ft

8.

65 ft

65 ft

Find the area of the rectangle.

9. 5 ft by 10 ft

10. 14 yd by 8 yd

11. 34.67 cm by 4.9 cm

12. 2.45 km by 100 km

13. $4\frac{2}{3}$ in. by $8\frac{5}{6}$ in.

14. $10\frac{1}{3}$ mi by $20\frac{2}{3}$ mi

Find the area of the square.

15. 22 ft on a side

16. 18 yd on a side

17. 56.9 km on a side

18. 45.5 m on a side

19. $5\frac{3}{8}$ yd on a side

20. $7\frac{2}{3}$ ft on a side

558

b Find the area.

21.

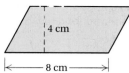

4 cm

|← 8 cm →|

22.

4 cm

|← 4 cm →|

23.

8 in.

|← 15 in. →|

24.

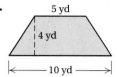

5 yd

4 yd

|← 10 yd →|

25.

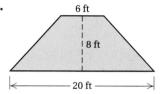

6 ft

8 ft

|← 20 ft →|

26.

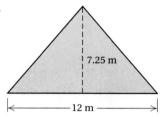

7.25 m

|← 12 m →|

27.

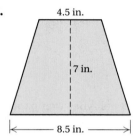

4.5 in.

7 in.

|← 8.5 in. →|

28.

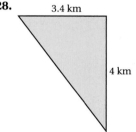

3.4 km

4 km

29.

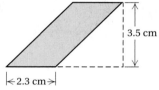

3.5 cm

|←2.3 cm→|

30.

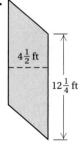

16 cm

35 cm

25 cm

31.

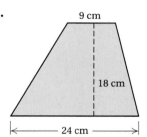

9 cm

18 cm

|← 24 cm →|

32.

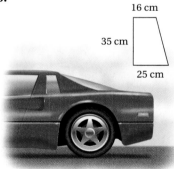

$4\frac{1}{2}$ ft

$12\frac{1}{4}$ ft

33.

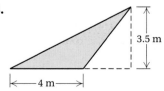

3.5 m

|← 4 m →|

34.

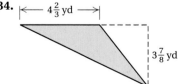

|← $4\frac{2}{3}$ yd →|

$3\frac{7}{8}$ yd

Solve.

35. *Area of a Lawn.*   A lot is 40 m by 36 m. A house 27 m by 9 m is built on the lot. How much area is left over for a lawn?

36. *Area of a Field.*   A field is 240.8 m by 450.2 m. Part of the field, 160.4 m by 90.6 m, is paved for a parking lot. How much area is unpaved?

37. *Mowing Expense.*   A square sandbox 4.5 ft on a side is placed on a 60-ft by $93\frac{2}{3}$-ft lawn.
   a) Find the area of the lawn.
   b) It costs $0.008 per square foot to have the lawn mowed. What is the total cost of the mowing?
   Source: Jackson's Lawn Care, Carmel IN

38. *Mowing Expense.*   A square flower bed 10.5 ft on a side is dug on a 90-ft by $67\frac{1}{4}$-ft lawn.
   a) Find the area of the lawn.
   b) It costs $0.03 per square foot to have the lawn mowed. What is the total cost of the mowing?
   Source: Jackson's Lawn Care, Carmel IN

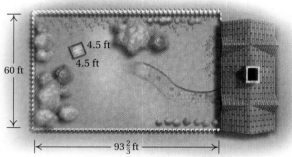

39. *Area of a Sidewalk.*   Franklin Construction Company builds a sidewalk around two sides of the Municipal Trust Bank building, as shown in the figure. What is the area of the sidewalk?

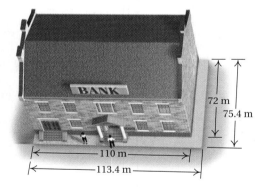

40. *Margin Area.*   A standard sheet of typewriter paper is $8\frac{1}{2}$ in. by 11 in. We generally type on a $7\frac{1}{2}$-in. by 9-in. area of the paper. What is the area of the margin?

41. *Painting Costs.*   A room is 15 ft by 20 ft. The ceiling is 8 ft above the floor. There are two windows in the room, each 3 ft by 4 ft. The door is $2\frac{1}{2}$ ft by $6\frac{1}{2}$ ft.
   a) What is the total area of the walls and the ceiling?
   b) A gallon of paint will cover 86.625 ft². How many gallons of paint are needed for the room, including the ceiling?
   c) Paint costs $17.95 a gallon. How much will it cost to paint the room?

42. *Carpeting Costs.*   A restaurant owner wants to carpet a 15-yd by 20-yd room.
   a) How many square yards of carpeting are needed?
   b) The carpeting she wants is $18.50 per square yard. How much will it cost to carpet the room?

Find the area of the shaded region.

43.

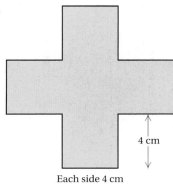

4 cm

Each side 4 cm

44.

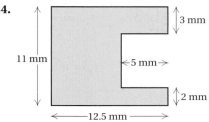

11 mm

3 mm

←5 mm→

2 mm

←—12.5 mm—→

45.

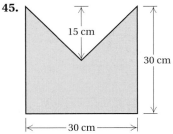

15 cm

30 cm

←—30 cm—→

46.

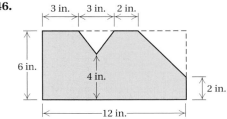

3 in.   3 in.   2 in.

6 in.

4 in.

2 in.

←——12 in.——→

47.

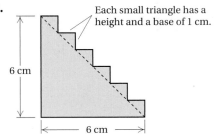

Each small triangle has a
height and a base of 1 cm.

6 cm

6 cm

48.

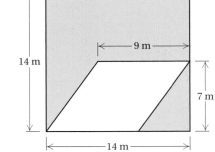

14 m

9 m

7 m

←——14 m——→

49. *Triangular Sail.* A rectangular piece of sailcloth is 36 ft by 24 ft. A triangular area with a height of 4.6 ft and a base of 5.2 ft is cut from the sailcloth. How much area is left over?

50. *Building Area.* Find the total area of the sides and ends of the building.

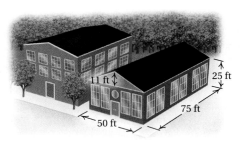

51. D_W The length and the width of one rectangle are each three times the length and the width of another rectangle. Is the area of the first rectangle three times the area of the other rectangle? Why or why not?

52. D_W Explain how the area of a triangle can be found by considering the area of a parallelogram.

SKILL MAINTENANCE

Complete.   [8.1a], [8.2a]

53. 23.4 cm = _____ mm

54. 0.23 km = _____ m

55. 28 ft = _____ in.

56. 72 ft = _____ yd

57. 72.4 cm = _____ m

58. 72.4 m = _____ km

59. 70 yd = _____ in.

60. 31,680 ft = _____ mi

61. 84 ft = _____ yd

62. $7\frac{1}{2}$ yd = _____ ft

63. 144 in. = _____ ft

64. 0.73 mi = _____ in.

SYNTHESIS

65. Find the area, in square inches, of the shaded region.

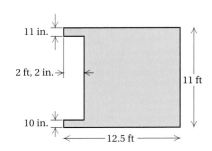

66. Find the area, in square feet, of the shaded region.

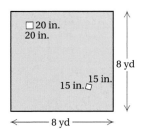

9.3 CIRCLES

Objectives

a Find the length of a radius of a circle given the length of a diameter, and find the length of a diameter given the length of a radius.

b Find the circumference of a circle given the length of a diameter or a radius.

c Find the area of a circle given the length of a radius.

d Solve applied problems involving circles.

a Radius and Diameter

Shown below is a circle with center O. Segment $\overline{AC}$ is a *diameter*. A **diameter** is a segment that passes through the center of the circle and has endpoints on the circle. Segment $\overline{OB}$ is called a *radius*. A **radius** is a segment with one endpoint on the center and the other endpoint on the circle.

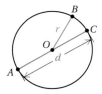

DIAMETER AND RADIUS

Suppose that d is the diameter of a circle and r is the radius. Then

$$d = 2 \cdot r \quad \text{and} \quad r = \frac{d}{2}.$$

EXAMPLE 1 Find the length of a radius of this circle.

$$r = \frac{d}{2}$$
$$= \frac{12 \text{ m}}{2}$$
$$= 6 \text{ m}$$

12 m

The radius is 6 m.

EXAMPLE 2 Find the length of a diameter of this circle.

$$d = 2 \cdot r$$
$$= 2 \cdot \frac{1}{4} \text{ ft}$$
$$= \frac{1}{2} \text{ ft}$$

$\frac{1}{4}$ ft

The diameter is $\frac{1}{2}$ ft.

Do Exercises 1 and 2.

1. Find the length of a radius.

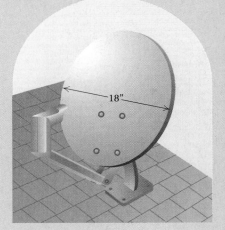

18"

2. Find the length of a diameter.

$2\frac{1}{2}$ ft

Answers on page A-23

3. Find the circumference of this circle. Use 3.14 for π.

20 m

b   Circumference

The **circumference** of a circle is the distance around it. Calculating circumference is similar to finding the perimeter of a polygon.

To find a formula for the circumference of any circle given its diameter, we first need to consider the ratio C/d. Take a 12-oz soda can and measure the circumference C with a tape measure. Also measure the diameter d. The results are shown in the figure. Then

$C \approx 7.8$ in.

$d \approx 2.5$ in.

$$\frac{C}{d} = \frac{7.8 \text{ in.}}{2.5 \text{ in.}} \approx 3.1.$$

Suppose we did this with cans and circles of several sizes. We would get a number close to 3.1. For any circle, if we divide the circumference C by the diameter d, we get the same number. We call this number π (pi).

CIRCUMFERENCE AND DIAMETER

The circumference C of a circle of diameter d is given by

$$C = \pi \cdot d.$$

The number π is about 3.14, or about $\frac{22}{7}$.

EXAMPLE 3   Find the circumference of this circle. Use 3.14 for π.

$C = \pi \cdot d$

$\approx 3.14 \times 6$ cm

$= 18.84$ cm

6 cm

The circumference is about 18.84 cm.

Do Exercise 3.

Answer on page A-23

Since $d = 2 \cdot r$, where r is the length of a radius, it follows that
$$C = \pi \cdot d = \pi \cdot (2 \cdot r).$$

4. Find the circumference of this circle. Use $\frac{22}{7}$ for π.

CIRCUMFERENCE AND RADIUS

The circumference C of a circle of radius r is given by
$$C = 2 \cdot \pi \cdot r.$$

EXAMPLE 4 Find the circumference of this circle. Use $\frac{22}{7}$ for π.

$C = 2 \cdot \pi \cdot r$

$\approx 2 \cdot \dfrac{22}{7} \cdot 70$ in.

$= 2 \cdot 22 \cdot \dfrac{70}{7}$ in.

$= 44 \cdot 10$ in.

$= 440$ in.

The circumference is about 440 in.

EXAMPLE 5 Find the perimeter of this figure. Use 3.14 for π.

We let $P = $ the perimeter. We see that we have half a circle attached to a square. Thus we add half the circumference of the circle to the lengths of the three line segments.

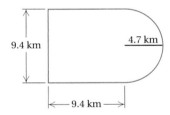

5. Find the perimeter of this figure. Use 3.14 for π.

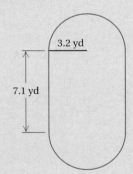

$\begin{aligned} P = \ &\text{Length of} &&\text{Half of the} \\ &\text{three sides} &+\ &\text{circumference} \\ &\text{of the square} &&\text{of the circle} \end{aligned}$

$= 3 \times 9.4 \text{ km} + \dfrac{1}{2} \times 2 \times \pi \times 4.7 \text{ km}$

$= 28.2 \text{ km} + 3.14 \times 4.7 \text{ km}$

$= 28.2 \text{ km} + 14.758 \text{ km}$

$= 42.958 \text{ km}$

The perimeter is about 42.958 km.

Do Exercises 4 and 5.

Answers on page A-23

6. Find the area of this circle. Use $\frac{22}{7}$ for π.

5 km

C   Area

Below is a circle of radius r.

r

Think of cutting half the circular region into small pieces and arranging them as shown below.

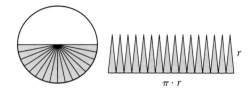

$\pi \cdot r$

r

Then imagine cutting the other half of the circular region and arranging the pieces in with the others as shown below.

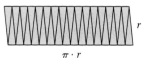

$\pi \cdot r$

r

This is almost a parallelogram. The base has length $\frac{1}{2} \cdot 2 \cdot \pi \cdot r$, or $\pi \cdot r$ (half the circumference) and the height is r. Thus the area is

$$(\pi \cdot r) \cdot r.$$

This is the area of a circle.

AREA OF A CIRCLE

The **area of a circle** with radius of length r is given by
$$A = \pi \cdot r \cdot r, \quad \text{or} \quad A = \pi \cdot r^2.$$

r

EXAMPLE 6   Find the area of this circle. Use $\frac{22}{7}$ for π.

$A = \pi \cdot r \cdot r$

$\approx \dfrac{22}{7} \cdot 14 \text{ cm} \cdot 14 \text{ cm}$

$= \dfrac{22}{7} \cdot 196 \text{ cm}^2$

$= 616 \text{ cm}^2$

14 cm

The area is about 616 cm².

Do Exercise 6.

Answer on page A-23

EXAMPLE 7 Find the area of this circle. Use 3.14 for π. Round to the nearest hundredth.

$$A = \pi \cdot r \cdot r$$
$$\approx 3.14 \times 2.1 \text{ m} \times 2.1 \text{ m}$$
$$= 3.14 \times 4.41 \text{ m}^2$$
$$= 13.8474 \text{ m}^2$$
$$\approx 13.85 \text{ m}^2$$

The area is about 13.85 m².

> **CAUTION!**
>
> Remember that circumference is always measured in linear units like ft, m, cm, yd, and so on. But area is measured in square units like ft², m², cm², yd², and so on.

Do Exercise 7.

d Solving Applied Problems

EXAMPLE 8 *Area of Pizza Pans.* Which is larger and by how much: a pizza made in a 16-in. square pizza pan or a pizza made in a 16-in. diameter circular pan?

First, we make a drawing of each.

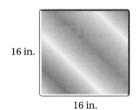  

Then we compute areas.
The area of the square is

$$A = s \cdot s$$
$$= 16 \text{ in.} \times 16 \text{ in.} = 256 \text{ in}^2.$$

The diameter of the circle is 16 in., so the radius is 16 in./2, or 8 in. The area of the circle is

$$A = \pi \cdot r \cdot r$$
$$\approx 3.14 \times 8 \text{ in.} \times 8 \text{ in.} = 200.96 \text{ in}^2.$$

We see that the square pizza is larger by about

$$256 \text{ in}^2 - 200.96 \text{ in}^2, \quad \text{or} \quad 55.04 \text{ in}^2.$$

Thus the pizza made in the square pan is larger, by about 55.04 in².

Do Exercise 8.

7. Find the area of this circle. Use 3.14 for π. Round to the nearest hundredth.

8. Which is larger and by how much: a 10-ft square flower bed or a 12-ft diameter flower bed?

Answers on page A-23

9.3
EXERCISE SET

For Extra Help

Digital Video
Tutor CD 5
Videotape 14

InterAct
Math

Math Tutor
Center

MathXL

MyMathLab

a, **b**, **c** For each circle, find the length of a diameter, the circumference, and the area. Use $\frac{22}{7}$ for π.

1.

7 cm

2.

8 m

3.

$\frac{3}{4}$ in.

4.

$8\frac{2}{3}$ mi

For each circle, find the length of a radius, the circumference, and the area. Use 3.14 for π.

5.

32 ft

6.

24 in.

7.

1.4 cm

8.

60.9 km

d Solve. Use 3.14 for π.

9. *Soda-Can Top.* The top of a soda can has a 6-cm diameter. What is its radius? its circumference? its area?

6 cm

10. *Penny.* A penny has a 1-cm radius. What is its diameter? its circumference? its area?

1 cm

11. *Trampoline.* The standard backyard trampoline has a diameter of 14 ft. What is its area?
Source: International Trampoline Industry Association, Inc.

14 ft

Frame height: 36 in.

12. *Area of Pizza Pans.* Which is larger and by how much: a pizza made in a 12-in square pizza pan or a pizza made in a 12-in diameter circular pan?

13. *Dimensions of a Quarter.* The circumference of a quarter is 7.85 cm. What is the diameter? the radius? the area?

14. *Dimensions of a Dime.* The circumference of a dime is 2.23 in. What is the diameter? the radius? the area?

15. *Gypsy-Moth Tape.* To protect an elm tree in your backyard, you need to attach gypsy moth caterpillar tape around the trunk. The tree has a 1.1-ft diameter. What length of tape is needed?

16. *Earth.* The diameter of the earth at the equator is 7926.41 mi. What is the circumference of the earth at the equator?
Source: *The Handy Geography Answer Book*

17. *Swimming-Pool Walk.* You want to install a 1-yd–wide walk around a circular swimming pool. The diameter of the pool is 20 yd. What is the area of the walk?

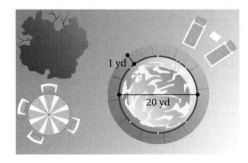

18. *Roller-Rink Floor.* A roller rink floor is shown below. What is its area? If hardwood flooring costs $10.50 per square meter, how much will the flooring cost?

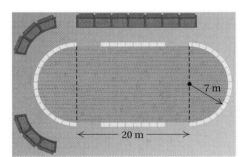

Find the perimeter. Use 3.14 for π.

19.

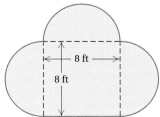

20.

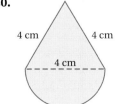

4 cm   4 cm

4 cm

21.

4 yd

4 yd

22.

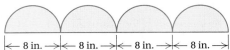

← 8 in. → ← 8 in. → ← 8 in. → ← 8 in. →

23.

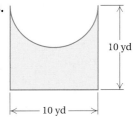

10 yd

← 10 yd →

24.

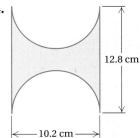

12.8 cm

← 10.2 cm →

Find the area of the shaded region. Use 3.14 for π.

25.

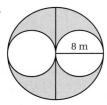

8 m

26.

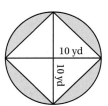

10 yd

10 yd

27.

← 2.8 cm →

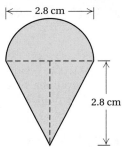

2.8 cm

28.

8 km

8 km

29.

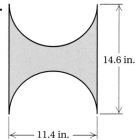

14.6 in.

← 11.4 in. →

30.

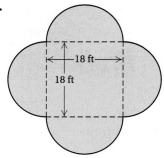

← 18 ft →

18 ft

31. ^{D}**w** Explain why a 16-in.–diameter pizza that costs $16.25 is a better buy than a 10-in.–diameter pizza that costs $7.85.

32. ^{D}**w** The radius of one circle is twice the length of another circle's radius. Is the area of the first circle twice the area of the other circle? Why or why not?

Evaluate. [1.9b]

33. 2^4

34. 17^2

35. 5^3

36. 8^2

Complete. [8.2a]

37. 5.43 m = _____ cm

38. 5.43 m = _____ km

39. Find the rate of discount. [6.6c]

Gabardine Slacks

100% Wool
Wrinkle-resistant and
inner waistband stretch
without binding.
Dry clean.

Reduced from $80

$39.90

40. Jack's commission is increased according to how much he sells. He receives a commission of 6% for the first $3000 and 10% on the amount over $3000. What is the total commission on sales of $8500? [6.6b]

41. If 2 cans of tomato paste cost $1.49, how many cans of tomato paste can you buy for $7.45? [5.4a]

42. The weight of a human brain is 2.5% of total body weight. A person weighs 200 lb. What does the brain weigh? [6.5a]

43. ▦ $\pi \approx \frac{3927}{1250}$ is another approximation for π. Find decimal notation using a calculator.

44. ▦ The distance from Kansas City to Indianapolis is 500 mi. A car was driven this distance using tires with a radius of 14 in. How many revolutions of each tire occurred on the trip? Use $\frac{22}{7}$ for π.

45. *Tennis Balls.* Tennis balls are generally packed vertically three in a can, one on top of another. Suppose the diameter of a tennis ball is d. Find the height of the stack of balls. Find the circumference of one ball. Which is greater? Explain.

Objectives

a Find the volume of a rectangular solid using the formula $V = l \cdot w \cdot h$.

b Given the radius and the height, find the volume of a circular cylinder.

c Given the radius, find the volume of a sphere.

d Given the radius and the height, find the volume of a circular cone.

e Solve applied problems involving volume of rectangular solids, circular cylinders, spheres, and cones.

1. Find the volume.

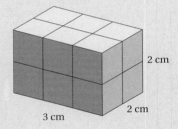

Answer on page A-23

a Rectangular Solids

The **volume** of a **rectangular solid** is the number of unit cubes needed to fill it.

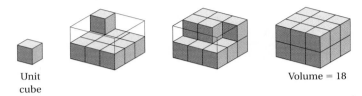

Unit cube   Volume = 18

Two other units are shown below.

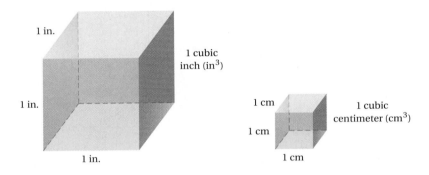

1 in. / 1 in. / 1 in.   1 cubic inch (in³)

1 cm / 1 cm / 1 cm   1 cubic centimeter (cm³)

EXAMPLE 1 Find the volume.

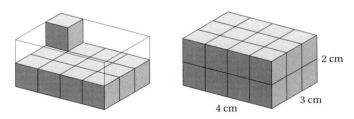

2 cm / 3 cm / 4 cm

The figure is made up of 2 layers of 12 cubes each, so its volume is 24 cubic centimeters (cm³).

Do Exercise 1.

VOLUME OF A RECTANGULAR SOLID

The **volume of a rectangular solid** is found by multiplying length by width by height:

$$V = l \cdot w \cdot h.$$

EXAMPLE 2 *Carry-on Luggage.* The largest piece of luggage that you can carry on an airplane measures 23 in. by 10 in. by 13 in. Find the volume of this solid.

$$V = l \cdot w \cdot h$$
$$= 23 \text{ in.} \cdot 10 \text{ in.} \cdot 13 \text{ in.}$$
$$= 230 \cdot 13 \text{ in}^3$$
$$= 2990 \text{ in}^3$$

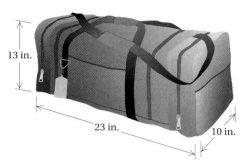

Do Exercises 2 and 3.

b Cylinders

A rectangular solid is shown below. Note that we can think of the volume as the product of the area of the base times the height:

$$V = l \cdot w \cdot h$$
$$= (l \cdot w) \cdot h$$
$$= (\text{Area of the base}) \cdot h$$
$$= B \cdot h,$$

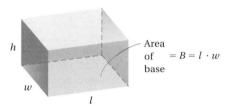

where B represents the area of the base.

Like rectangular solids, **circular cylinders** have bases of equal area that lie in parallel planes. The bases of circular cylinders are circular regions.

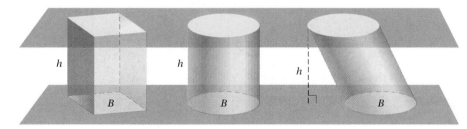

2. Popcorn. In a recent year, people in the United States bought enough unpopped popcorn to provide every person in the country with a bag of popped corn measuring 2 ft by 2 ft by 5 ft. Find the volume of such a bag.

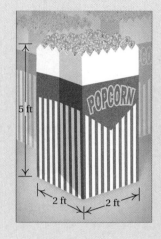

3. Cord of Wood. A cord of wood measures 4 ft by 4 ft by 8 ft. What is the volume of a cord of wood?

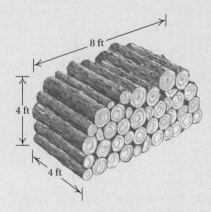

Answers on page A-23

573

9.4 Volume

4. Find the volume of the cylinder. Use 3.14 for π.

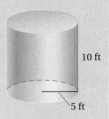

10 ft

5 ft

5. Find the volume of the cylinder. Use $\frac{22}{7}$ for π.

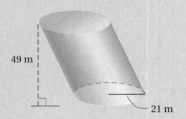

49 m

21 m

6. Find the volume of the sphere. Use $\frac{22}{7}$ for π.

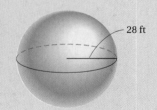

28 ft

7. The radius of a standard-sized golf ball is 2.1 cm. Find its volume. Use 3.14 for π.

Answers on page A-23

The volume of a circular cylinder is found in a manner similar to finding the volume of a rectangular solid. The volume is the product of the area of the base times the height. The height is always measured perpendicular to the base.

VOLUME OF A CIRCULAR CYLINDER

The **volume of a circular cylinder** is the product of the area of the base B and the height h:

$$V = B \cdot h, \quad \text{or} \quad V = \pi \cdot r^2 \cdot h.$$

EXAMPLE 3 Find the volume of this circular cylinder. Use 3.14 for π.

$$V = Bh = \pi \cdot r^2 \cdot h$$
$$\approx 3.14 \times 4 \text{ cm} \times 4 \text{ cm} \times 12 \text{ cm}$$
$$= 602.88 \text{ cm}^3$$

12 cm

4 cm

Do Exercises 4 and 5.

C Spheres

A **sphere** is the three-dimensional counterpart of a circle. It is the set of all points in space that are a given distance (the radius) from a given point (the center).

r

We find the volume of a sphere as follows.

VOLUME OF A SPHERE

The **volume of a sphere** of radius r is given by

$$V = \frac{4}{3} \cdot \pi \cdot r^3.$$

EXAMPLE 4 *Bowling Ball.* The radius of a standard-sized bowling ball is 4.2915 in. Find the volume of a standard-sized bowling ball. Round to the nearest hundredth of a cubic inch. Use 3.14 for π.

$$V = \frac{4}{3} \cdot \pi \cdot r^3 \approx \frac{4}{3} \times 3.14 \times (4.2915 \text{ in.})^3$$
$$\approx \frac{4 \times 3.14 \times 79.0364 \text{ in}^3}{3} \approx 330.90 \text{ in}^3 \qquad \text{Using a calculator}$$

Do Exercises 6 and 7.

d Cones

Consider a circle in a plane and choose any point P not in the plane. The circular region, together with the set of all segments connecting P to a point on the circle, is called a **circular cone.**

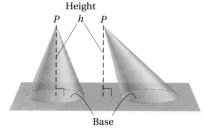

We find the volume of a cone as follows.

> ### VOLUME OF A CIRCULAR CONE
>
> The **volume of a circular cone** with base radius r is one-third the product of the base area and the height:
>
> $$V = \frac{1}{3} \cdot B \cdot h = \frac{1}{3} \pi \cdot r^2 \cdot h.$$

EXAMPLE 5   Find the volume of this circular cone. Use 3.14 for π.

$$V = \frac{1}{3} \pi \cdot r^2 \cdot h$$
$$\approx \frac{1}{3} \times 3.14 \times 3 \text{ cm} \times 3 \text{ cm} \times 7 \text{ cm}$$
$$= 65.94 \text{ cm}^3$$

Do Exercises 8 and 9.

8. Find the volume of this cone. Use 3.14 for π.

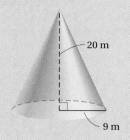

9. Find the volume of this cone. Use $\frac{22}{7}$ for π.

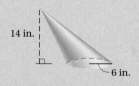

Answers on page A-23

CALCULATOR CORNER

The $\boxed{\pi}$ Key   Many calculators have a $\boxed{\pi}$ key that can be used to enter the value of π in a computation. It might be necessary to press a $\boxed{\text{2nd}}$ or $\boxed{\text{SHIFT}}$ key before pressing the $\boxed{\pi}$ key on some calculators. Since 3.14 is a rounded value for π, results obtained using the $\boxed{\pi}$ key might be slightly different from those obtained when 3.14 is used for the value of π in a computation.

To find the volume of the circular cylinder in Example 3, we press $\boxed{\text{2nd}}$ $\boxed{\pi}$ $\boxed{\times}$ $\boxed{4}$ $\boxed{\times}$ $\boxed{4}$ $\boxed{\times}$ $\boxed{1}$ $\boxed{2}$ $\boxed{=}$ or $\boxed{\text{SHIFT}}$ $\boxed{\pi}$ $\boxed{\times}$ $\boxed{4}$ $\boxed{\times}$ $\boxed{4}$ $\boxed{\times}$ $\boxed{1}$ $\boxed{2}$ $\boxed{=}$. The result is approximately 603.19. Note that this is slightly different from the result found using 3.14 for the value of π.

Exercises

1.   Use a calculator with a $\boxed{\pi}$ key to perform the computations in Examples 4 and 5.
2.   Use a calculator with a $\boxed{\pi}$ key to perform the computations in Margin Exercises 5–10.

10. Medicine Capsule. A cold capsule is 8 mm long and 4 mm in diameter. Find the volume of the capsule. Use 3.14 for π. (*Hint*: First find the length of the cylindrical section.)

Answer on page A-23

e ## Solving Applied Problems

EXAMPLE 6 *Propane Gas Tank.* A propane gas tank is shaped like a circular cylinder with half of a sphere at each end. Find the volume of the tank if the cylindrical section is 5 ft long with a 4-ft diameter. Use 3.14 for π.

1. **Familiarize.** We first make a drawing.

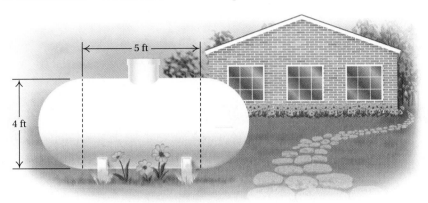

2. **Translate.** This is a two-step problem. We first find the volume of the cylindrical portion. Then we find the volume of the two ends and add. Note that the radius is 2 ft and that together the two ends make a sphere. We let

$$V = \pi \cdot r^2 \cdot h + \frac{4}{3} \cdot \pi \cdot r^3,$$

where V is the total volume. Then

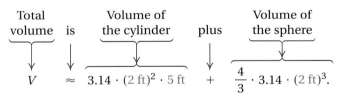

$$V \approx 3.14 \cdot (2\ \text{ft})^2 \cdot 5\ \text{ft} + \frac{4}{3} \cdot 3.14 \cdot (2\ \text{ft})^3.$$

3. **Solve.** The volume of the cylinder is approximately

$$3.14 \cdot (2\ \text{ft})^2 \cdot 5\ \text{ft} = 3.14 \cdot 2\ \text{ft} \cdot 2\ \text{ft} \cdot 5\ \text{ft}$$
$$= 62.8\ \text{ft}^3.$$

The volume of the two ends is approximately

$$\frac{4}{3} \cdot 3.14 \cdot (2\ \text{ft})^3 = \frac{4}{3} \cdot 3.14 \cdot 2\ \text{ft} \cdot 2\ \text{ft} \cdot 2\ \text{ft}$$
$$\approx 33.5\ \text{ft}^3.$$

The total volume is about

$$62.8\ \text{ft}^3 + 33.5\ \text{ft}^3 = 96.3\ \text{ft}^3.$$

4. **Check.** The check is left to the student.
5. **State.** The volume of the tank is about 96.3 ft³.

Do Exercise 10.

9.4

For Extra Help

Digital Video Tutor CD 5 Videotape 14 | InterAct Math | Math Tutor Center | MathXL | MyMathLab

a Find the volume.

1.

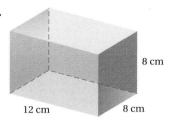

8 cm

12 cm   8 cm

2.

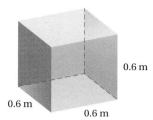

0.6 m

0.6 m   0.6 m

3.

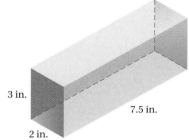

3 in.   7.5 in.

2 in.

4.

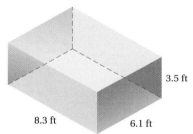

3.5 ft

8.3 ft   6.1 ft

5.

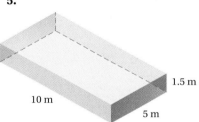

1.5 m

10 m   5 m

6.

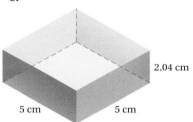

2.04 cm

5 cm   5 cm

7.

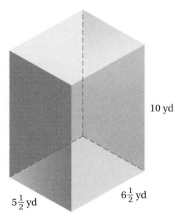

10 yd

$5\frac{1}{2}$ yd $6\frac{1}{2}$ yd

8.

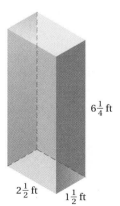

$6\frac{1}{4}$ ft

$2\frac{1}{2}$ ft $1\frac{1}{2}$ ft

b Find the volume of the circular cylinder. Use 3.14 for π in Exercises 9–12. Use $\frac{22}{7}$ for π in Exercises 13 and 14.

9.

4 in.
8 in.

10.

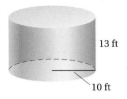
13 ft
10 ft

11.

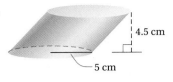
4.5 cm
5 cm

12.

40 cm
4 cm

13.

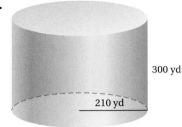
300 yd
210 yd

14.

28 km
4 km

c Find the volume of the sphere. Use 3.14 for π in Exercises 15–18. Use $\frac{22}{7}$ for π in Exercises 19 and 20.

15.

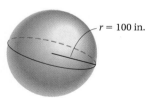
$r = 100$ in.

16.

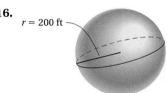
$r = 200$ ft

17.

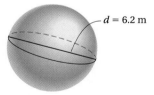
$d = 6.2$ m

18.

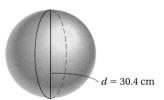
$d = 30.4$ cm

19.

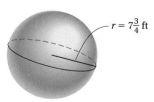
$r = 7\frac{3}{4}$ ft

20.

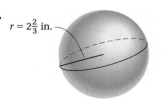
$r = 2\frac{2}{3}$ in.

 Find the volume of the circular cone. Use 3.14 for π in Exercises 21 and 22. Use $\frac{22}{7}$ for π in Exercises 23 and 24.

21.

100 ft

33 ft

22.

10 m

3 m

23.

12 cm

1.4 cm

24.

30 mm

35 mm

 Solve.

25. *Volume of a Trash Can.* The diameter of the base of a cylindrical trash can is 0.7 yd. The height is 1.1 yd. Find the volume. Use 3.14 for π.

26. *Ladder Rung.* A rung of a ladder is 2 in. in diameter and 16 in. long. Find the volume. Use 3.14 for π.

27. *Barn Silo.* A barn silo, excluding the top, is a circular cylinder. The silo is 6 m in diameter and the height is 13 m. Find the volume of the silo. Use 3.14 for π.

28. *Oak Log.* An oak log has a diameter of 12 cm and a length (height) of 42 cm. Find the volume. Use 3.14 for π.

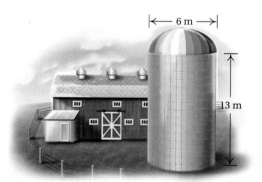

6 m

13 m

29. *Tennis Ball.* The diameter of a tennis ball is 6.5 cm. Find the volume. Use 3.14 for π.

30. *Spherical Gas Tank.* The diameter of a spherical gas tank is 6 m. Find the volume. Use 3.14 for π.

31. *Volume of Earth.* The diameter of the earth is about 3980 mi. Find the volume of the earth. Use 3.14 for π. Round to the nearest ten thousand cubic miles.

32. *Astronomy.* The radius of Pluto's moon is about 500 km. Find the volume of this satellite. Use $\frac{22}{7}$ for π.

33. *Tennis-Ball Packaging.* Tennis balls are generally packaged in circular cylinders that hold 3 balls each. The diameter of a tennis ball is 6.5 cm. Find the volume of a can of tennis balls. Use 3.14 for π.

34. *Golf-Ball Packaging.* The box shown is just big enough to hold 3 golf balls. If the radius of a golf ball is 2.1 cm, how much air surrounds the three balls? Use 3.14 for π.

35. *Water Storage.* A water storage tank is a right circular cylinder with a radius of 14 cm and a height of 100 cm. What is the tank's volume? Use $\frac{22}{7}$ for π.

36. *Oceanography.* A research submarine is capsule-shaped. Find the volume of the submarine if it has a length of 10 m and a diameter of 8 m. Use 3.14 for π. (*Hint*: First find the length of the cylindrical section.)

37. *Metallurgy.* If all the gold in the world could be gathered together, it would form a cube 18 yd on a side. Find the volume of the world's gold.

38. The volume of a ball is 36π cm^3. Find the dimensions of a rectangular box that is just large enough to hold the ball.

39. **D**w How could you use the volume formulas given in this section to help estimate the volume of an egg?

40. **D**w The design of a modern home includes a cylindrical tower that will be capped with either a 10-ft–high dome or a 10-ft–high cone. Which type of cap will be more energy-efficient and why?

Complete. [8.1a]

41. 11 yd = _____ in.

42. 15,840 ft = _____ mi

43. 42 ft = _____ yd

44. 48 mi = _____ ft

45. 144 in. = _____ ft

46. 5.3 mi = _____ in.

Complete. [8.5a]

47. 6 gal = _____ qt

48. 56 qt = _____ gal

49. 566 mL = _____ L

50. 13.4 L= _____ cm³

SYNTHESIS

51. ⊞ The width of a dollar bill is 2.3125 in., the length is 6.0625 in., and the thickness is 0.0041 in. Find the volume occupied by one million one-dollar bills.

© 1998 AL SATTERWHITE

52. ⊞ Audio-cassette cases are typically 7 cm by 10.75 cm by 1.5 cm and contain 90 min of music. Compact-disc cases are typically 12.4 cm by 14.1 cm by 1 cm and contain 50 min of music. Which container holds the most music per cubic centimeter?

53. ⊞ A 2-cm–wide stream of water passes through a 30-m–long garden hose. At the instant that the water is turned off, how many liters of water are in the hose? Use 3.141593 for π.

54. ⊞ The volume of a basketball is 2304π cm³. Find the volume of a cube-shaped box that is just large enough to hold a ball.

55. *Circumference of Earth.* The circumference of the earth at the equator is about 24,901.55 mi. Due to the irregular shape of the earth, the circumference of a circle of longitude wrapped around the earth between the north and south poles is about 24,859.82 mi. Describe and carry out a procedure for estimating the volume of the earth.
Source: *The Handy Geography Answer Book*

56. ⊞ A sphere with diameter 1 m is circumscribed by a cube. How much more volume is in the cube?

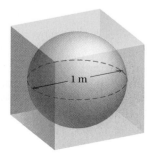

57. ⊞ A cube is circumscribed by a sphere with a 1-m diameter. How much more volume is in the sphere?

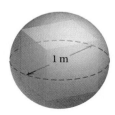

1 m

Objectives

a Name an angle in five different ways and given an angle, measure it with a protractor.

b Classify an angle as right, straight, acute, or obtuse.

c Identify complementary and supplementary angles and find the measure of a complement or a supplement of a given angle.

d Classify a triangle as equilateral, isosceles, or scalene, and as right, obtuse, or acute.

e Given two of the angle measures of a triangle, find the third.

Name the angle in five different ways.

1.

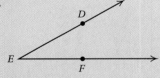

2.

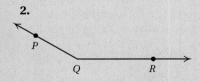

Answers on page A-23

a Measuring Angles

We see a real-world application of *angles* of various types in the spokes of these bicycles and the different back postures of the riders.

Style of Biking Determines Cycling Posture

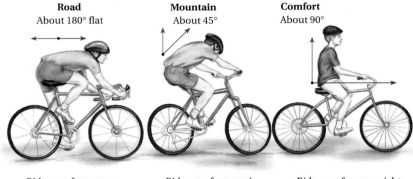

Road	Mountain	Comfort
About 180° flat	About 45°	About 90°
Riders prefer a more aerodynamic flat-back position.	Riders prefer a semi-upright position to help lift the front wheel over obstacles.	Riders prefer an upright position that lessens stress on the lower back and neck.

Source: USA TODAY research

An **angle** is a set of points consisting of two **rays,** or half-lines, with a common endpoint. The endpoint is called the **vertex.**

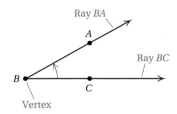

The rays are called the *sides*. The angle above can be named

angle *ABC*,   angle *CBA*, $\angle ABC$, $\angle CBA$,   or $\angle B$.

Note that the name of the vertex is either in the middle or, if no confusion results, listed by itself.

Do Exercises 1 and 2.

Measuring angles is similar to measuring segments. To measure angles, we start with some arbitrary angle and assign to it a measure of 1. We call it a *unit angle*. Suppose that $\angle U$, shown on the following page, is a unit angle. Let's measure $\angle DEF$. If we made 3 copies of $\angle U$, they would "fill up" $\angle DEF$. Thus the measure of $\angle DEF$ would be 3.

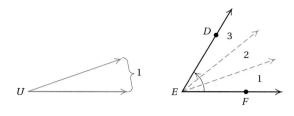

3. Use a protractor to measure this angle.

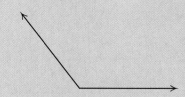

The unit most commonly used for angle measure is the degree. Below is such a unit. Its measure is 1 degree, or 1°.

A 1° angle:

Here are some other angles with their degree measures.

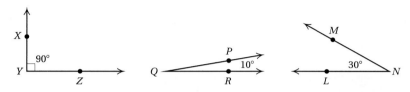

To indicate the *measure* of ∠XYZ, we write $m \angle XYZ = 90°$. The symbol ⌐ is sometimes drawn on a figure to indicate a 90° angle.

A device called a **protractor** is used to measure angles. Protractors have two scales. To measure an angle like ∠Q below, we place the protractor's ▲ at the vertex and line up one of the angle's sides at 0°. Then we check where the angle's other side crosses the scale. In the figure below, 0° is on the inside scale, so we check where the angle's other side crosses the inside scale. We see that $m \angle Q = 145°$. The notation $m \angle Q$ is read "the measure of angle Q."

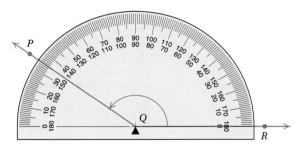

Do Exercise 3.

Answer on page A-23

4. Use a protractor to measure this angle.

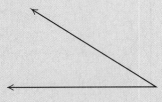

Classify the angle as right, straight, acute, or obtuse. Use a protractor if necessary.

5.

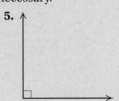

6.

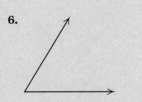

7.

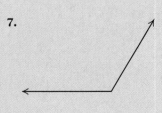

8.

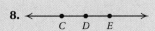

$C \quad D \quad E$

Answers on page A-23

Let's find the measure of $\angle ABC$. This time we will use the 0° on the outside scale. We see that $m\angle ABC = 42°$.

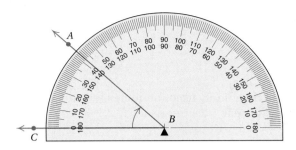

Do Exercise 4.

b Classifying Angles

The following are ways in which we classify angles.

> **TYPES OF ANGLES**
>
> **Right angle:** An angle whose measure is 90°.
> **Straight angle:** An angle whose measure is 180°.
> **Acute angle:** An angle whose measure is greater than 0° and less than 90°.
> **Obtuse angle:** An angle whose measure is greater than 90° and less than 180°.

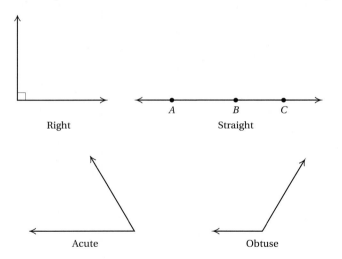

Right   Straight

Acute   Obtuse

Do Exercises 5–8.

C ‖ Complementary and Supplementary Angles

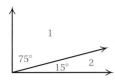

∠1 and ∠2 above are **complementary** angles.

$$m\angle 1 + m\angle 2 = 90°$$
$$75° \;+\; 15° \;= 90°$$

> ### COMPLEMENTARY ANGLES
>
> Two angles are **complementary** if the sum of their measures is 90°. Each angle is called a **complement** of the other.

If two angles are complementary, each is an acute angle. When complementary angles are adjacent to each other, they form a right angle.

EXAMPLE 1 Identify each pair of complementary angles.

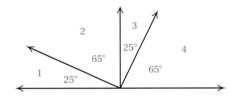

∠1 and ∠2	25° + 65° = 90°	∠2 and ∠3
∠1 and ∠4		∠3 and ∠4

EXAMPLE 2 Find the measure of a complement of an angle of 39°.

$$90° - 39° = 51°$$

The measure of a complement is 51°.

Do Exercises 9–12.

Next, consider ∠1 and ∠2 as shown below. Because the sum of their measures is 180°, ∠1 and ∠2 are said to be **supplementary.** Note that when supplementary angles are adjacent, they form a straight angle.

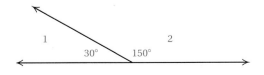

$$m\angle 1 + m\angle 2 = 180°$$
$$30° \;+\; 150° = 180°$$

---

9. Identify each pair of complementary angles.

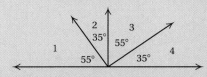

Find the measure of a complement of the angle.

10.

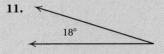

11.

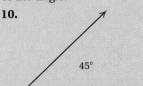

12.

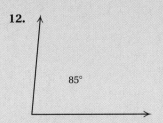

Answers on page A-23

13. Identify each pair of supplementary angles.

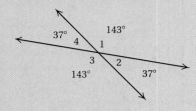

Find the measure of a supplement of an angle with the given measure.

14. 38°

15. 157°

16. 90°

Two angles are **supplementary** if the sum of their measures is 180°. Each angle is called a **supplement** of the other.

EXAMPLE 3 Identify each pair of supplementary angles.

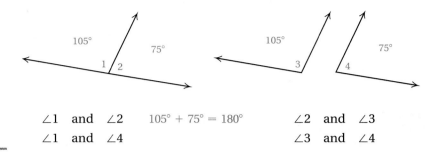

$\angle 1$ and $\angle 2$ $105° + 75° = 180°$ $\angle 2$ and $\angle 3$

$\angle 1$ and $\angle 4$ $\angle 3$ and $\angle 4$

EXAMPLE 4 Find the measure of a supplement of an angle of 112°.

$180° - 112° = 68°$

The measure of a supplement is 68°.

Do Exercises 13–16.

d   Triangles

A **triangle** is a polygon made up of three segments, or sides. Consider these triangles. The triangle with vertices *A*, *B*, and *C* can be named $\triangle ABC$.

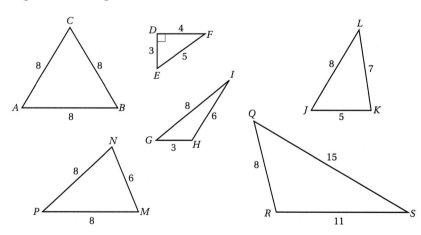

We can classify triangles according to sides and according to angles.

TYPES OF TRIANGLES

Equilateral triangle: All sides are the same length.
Isosceles triangle: Two or more sides are the same length.
Scalene triangle: All sides are of different lengths.
Right triangle: One angle is a right angle.
Obtuse triangle: One angle is an obtuse angle.
Acute triangle: All three angles are acute.

Do Exercises 17–20.

e Sum of the Angle Measures of a Triangle

The sum of the angle measures of a triangle is 180°. To see this, note that we can think of cutting apart a triangle as shown on the left below. If we reassemble the pieces, we see that a straight angle is formed.

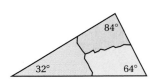
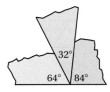

$$64° + 32° + 84° = 180°$$

SUM OF THE ANGLE MEASURES OF A TRIANGLE

In any triangle *ABC*, the sum of the measures of the angles is 180°:
$$m(\angle A) + m(\angle B) + m(\angle C) = 180°.$$

Do Exercise 21.

If we know the measures of two angles of a triangle, we can calculate the third.

EXAMPLE 5 Find the missing angle measure.

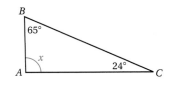

$$
\begin{aligned}
m(\angle A) + m(\angle B) + m(\angle C) &= 180° \\
x + 65° + 24° &= 180° \\
x + 89° &= 180° \\
x &= 180° - 89° \\
x &= 91°
\end{aligned}
$$

Do Exercise 22.

17. Which triangles on p. 586 are:
    a) equilateral?
    b) isosceles?
    c) scalene?

18. Are all equilateral triangles isosceles?

19. Are all isosceles triangles equilateral?

20. Which triangles on p. 586 are:
    a) right triangles?
    b) obtuse triangles?
    c) acute triangles?

21. Find $m(\angle P) + m(\angle Q) + m(\angle R)$.

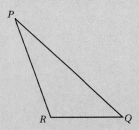

22. Find the missing angle measure.

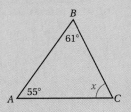

Answers on page A-23

9.5

EXERCISE SET

For Extra Help

Digital Video
Tutor CD 5
Videotape 14

InterAct
Math

Math Tutor
Center

MathXL

MyMathLab

a Name the angle in five different ways.

1.

2.

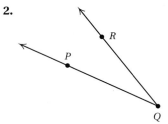

Use a protractor to measure the angle.

3.

4.

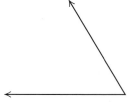

5.

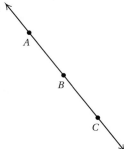

6.

7.

8.

b

9.–16. Classify each of the angles in Exercises 1–8 as right, straight, acute, or obtuse.

17.–20. Classify each of the angles in Margin Exercises 1–4 as right, straight, acute, or obtuse.

c Find the measure of a complement of an angle with the given measure.

21. 11°

22. 83°

23. 67°

24. 5°

25. 58°

26. 32°

27. 29°

28. 54°

Find the measure of a supplement of an angle with the given measure.

29. 3°

30. 54°

31. 139°

32. 13°

33. 85° **34.** 129° **35.** 102° **36.** 45°

---

d Classify the triangle as equilateral, isosceles, or scalene. Then classify it as right, obtuse, or acute.

37.

38.

39.
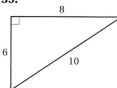

40.
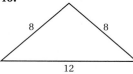

41.

42.
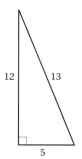

43.
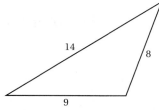

44.
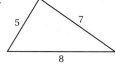

---

e Find the missing angle measure.

45.
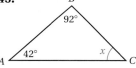

46.
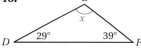

47.
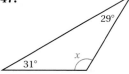

48.
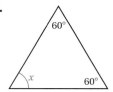

---

49. ^{D}W Explain how you might use triangles to find the sum of the angle measures of this figure.

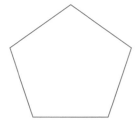

50. ^{D}W Explain a procedure that could be used to determine the measure of an angle's supplement from the measure of the angle's complement.

Find the simple interest.   [6.7a]

	PRINCIPAL	RATE OF INTEREST	TIME	SIMPLE INTEREST
51.	$2000	8%	1 year	
52.	$750	6%	$\frac{1}{2}$ year	
53.	$4000	7.4%	$\frac{1}{2}$ year	
54.	$200,000	6.7%	$\frac{1}{12}$ year	

Interest is compounded semiannually. Find the amount in the account after the given length of time. Round to the nearest cent.   [6.7b]

	PRINCIPAL	RATE OF INTEREST	TIME	AMOUNT IN THE ACCOUNT
55.	$25,000	6%	5 years	
56.	$150,000	$6\frac{7}{8}$%	15 years	
57.	$150,000	7.4%	20 years	
58.	$160,000	7.4%	20 years	

59. In the figure, $m\angle 1 = 79.8°$ and $m\angle 6 = 33.07°$. Find $m\angle 2$, $m\angle 3$, $m\angle 4$, and $m\angle 5$.

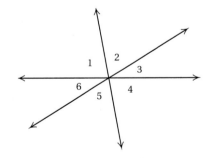

60. In the figure, $m\angle 2 = 42.17°$ and $m\angle 3 = 81.9°$. Find $m\angle 1$, $m\angle 4$, $m\angle 5$, and $m\angle 6$.

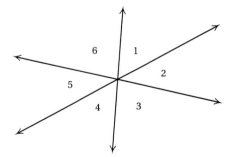

61. Find $m\angle ACB$, $m\angle CAB$, $m\angle EBC$, $m\angle EBA$, $m\angle AEB$, and $m\angle ADB$ in the rectangle shown below.

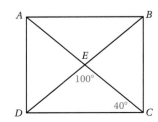

9.6 SQUARE ROOTS AND THE PYTHAGOREAN THEOREM

a Square Roots

Objectives

a	Simplify square roots of squares such as $\sqrt{25}$.
b	Approximate square roots.
c	Given the lengths of any two sides of a right triangle, find the length of the third side.
d	Solve applied problems involving right triangles.

SQUARE ROOT

If a number is a product of two identical factors, then either factor is called a **square root** of the number. (If $a = c^2$, then c is a square root of a.) The symbol $\sqrt{\ }$ (called a **radical sign**) is used in naming square roots.

For example, $\sqrt{36}$ is the square root of 36. It follows that

$$\sqrt{36} = \sqrt{6 \cdot 6} = 6 \qquad \text{The square root of 36 is 6.}$$

because $6^2 = 36$.

EXAMPLE 1 Simplify: $\sqrt{25}$.

$$\sqrt{25} = \sqrt{5 \cdot 5} = 5 \qquad \text{The square root of 25 is 5 because } 5^2 = 25.$$

EXAMPLE 2 Simplify: $\sqrt{144}$.

$$\sqrt{144} = \sqrt{12 \cdot 12} = 12 \qquad \text{The square root of 144 is 12 because } 12^2 = 144.$$

CAUTION!

It is common to confuse squares and square roots. A number squared is that number multiplied by itself. For example, $16^2 = 16 \cdot 16 = 256$. A square root of a number is a number that when multiplied by itself gives the original number. For example, $\sqrt{16} = 4$, because $4 \cdot 4 = 16$.

EXAMPLES Simplify.

3. $\sqrt{4} = 2$
4. $\sqrt{256} = 16$
5. $\sqrt{361} = 19$

Do Exercises 1–24.

Find the square. (See Section 1.9.)

1. 9^2 2. 10^2

3. 11^2 4. 12^2

> It would be helpful to memorize the squares of numbers from 1 to 25. We call these "perfect" squares.

5. 13^2 6. 14^2

7. 15^2 8. 16^2

9. 17^2 10. 18^2

11. 20^2 12. 25^2

Simplify. The results of Exercises 1–12 above may be helpful here.

13. $\sqrt{9}$ 14. $\sqrt{16}$

15. $\sqrt{121}$ 16. $\sqrt{100}$

17. $\sqrt{81}$ 18. $\sqrt{64}$

19. $\sqrt{324}$ 20. $\sqrt{400}$

21. $\sqrt{225}$ 22. $\sqrt{169}$

23. $\sqrt{1}$ 24. $\sqrt{0}$

Answers on page A-24

b. Approximating Square Roots

Many square roots can't be written as whole numbers or fractions. For example,

$$\sqrt{2}, \qquad \sqrt{3}, \qquad \sqrt{39}, \quad \text{and} \quad \sqrt{70}$$

cannot be precisely represented in decimal notation. To see this, consider the following decimal approximations for $\sqrt{2}$. Each gives a closer approximation, but none is exactly $\sqrt{2}$:

$$\sqrt{2} \approx 1.4 \qquad \text{because} \quad (1.4)^2 = 1.96;$$
$$\sqrt{2} \approx 1.41 \qquad \text{because} \quad (1.41)^2 = 1.9881;$$
$$\sqrt{2} \approx 1.414 \qquad \text{because} \quad (1.414)^2 = 1.999396;$$
$$\sqrt{2} \approx 1.4142 \quad \text{because} \quad (1.4142)^2 = 1.99996164.$$

Decimal approximations like these are commonly found by using a calculator.

CALCULATOR CORNER

Finding Square Roots Many calculators have a square root key, $\boxed{\sqrt{}}$. This is often the second function associated with the $\boxed{x^2}$ key and is accessed by pressing $\boxed{\text{2nd}}$ or $\boxed{\text{SHIFT}}$ followed by $\boxed{x^2}$. Often we enter the radicand first followed by the $\boxed{\sqrt{}}$ key. To find $\sqrt{30}$, for example, we press $\boxed{3}\,\boxed{0}\,\boxed{\text{2nd}}\,\boxed{\sqrt{}}$ or $\boxed{3}\,\boxed{0}\,\boxed{\text{SHIFT}}\,\boxed{\sqrt{}}$. We get 5.477225575.

It is always best to wait until calculations are complete before rounding. For example, to find $9 \cdot \sqrt{30}$ rounded to the nearest tenth, we do not first determine that $\sqrt{30} \approx 5.5$ and then multiply by 9 to get 49.5. Rather, we press $\boxed{9}\,\boxed{\times}\,\boxed{3}\,\boxed{0}\,\boxed{\text{2nd}}\,\boxed{\sqrt{}}\,\boxed{=}$ or $\boxed{9}\,\boxed{\times}\,\boxed{3}\,\boxed{0}\,\boxed{\text{SHIFT}}$ $\boxed{\sqrt{}}\,\boxed{=}$. The result is 49.29503018, so $9 \cdot \sqrt{30} \approx 49.3$.

Exercises: Use a calculator to find each of the following. Round to the nearest tenth.

1. $\sqrt{43}$
2. $\sqrt{94}$
3. $7 \cdot \sqrt{8}$
4. $5 \cdot \sqrt{12}$
5. $\sqrt{35} + 19$
6. $17 + \sqrt{57}$
7. $13\sqrt{68} + 14$
8. $24 \cdot \sqrt{31} - 18$
9. $5 \cdot \sqrt{30} - 3 \cdot \sqrt{14}$
10. $7 \cdot \sqrt{90} + 3 \cdot \sqrt{40}$

EXAMPLE 6 Approximate $\sqrt{3}$, $\sqrt{27}$, and $\sqrt{180}$ to three decimal places. Use a calculator.

We use a calculator to find each square root. Since more than three decimal places are given, we round back to three places.

$$\sqrt{3} \approx 1.732,$$
$$\sqrt{27} \approx 5.196,$$
$$\sqrt{180} \approx 13.416$$

As a check, note that $1 \cdot 1 = 1$ and $2 \cdot 2 = 4$, so we expect $\sqrt{3}$ to be between 1 and 2. Similarly, we expect $\sqrt{27}$ to be between 5 and 6 and $\sqrt{180}$ to be between 13 and 14.

Do Exercises 25–27.

C The Pythagorean Theorem

A **right triangle** is a triangle with a 90° angle, as shown here.

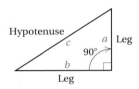

In a right triangle, the longest side is called the **hypotenuse.** It is also the side opposite the right angle. The other two sides are called **legs.** We generally use the letters a and b for the lengths of the legs and c for the length of the hypotenuse. They are related as follows.

THE PYTHAGOREAN THEOREM

In any right triangle, if a and b are the lengths of the legs and c is the length of the hypotenuse, then

$$a^2 + b^2 = c^2, \quad \text{or}$$
$$(\text{Leg})^2 + (\text{Other leg})^2 = (\text{Hypotenuse})^2.$$

The equation $a^2 + b^2 = c^2$ is called the **Pythagorean equation.***

---

*The *converse* of the Pythagorean theorem is also true. That is, if $a^2 + b^2 = c^2$, then the triangle is a right triangle.

Approximate to three decimal places.

25. $\sqrt{5}$

26. $\sqrt{78}$

27. $\sqrt{168}$

Answers on page A-24

28. Find the length of the hypotenuse of this right triangle.

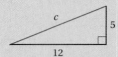

The Pythagorean theorem is named for the Greek mathematician Pythagoras (569?–500? B.C.). We can think of this relationship as adding areas.

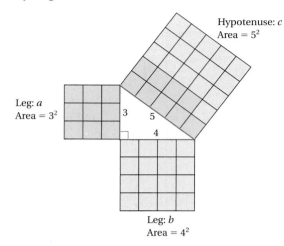

Hypotenuse: c
Area $= 5^2$

Leg: a
Area $= 3^2$

Leg: b
Area $= 4^2$

$$a^2 + b^2 = c^2$$
$$3^2 + 4^2 = 5^2$$
$$9 + 16 = 25$$

If we know the lengths of any two sides of a right triangle, we can use the Pythagorean equation to determine the length of the third side.

EXAMPLE 7   Find the length of the hypotenuse of this right triangle.

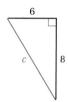

We substitute in the Pythagorean equation:

$$a^2 + b^2 = c^2$$
$$6^2 + 8^2 = c^2 \qquad \text{Substituting}$$
$$36 + 64 = c^2$$
$$100 = c^2.$$

The solution of this equation is the square root of 100, which is 10:

$$c = \sqrt{100} = 10.$$

Do Exercise 28.

EXAMPLE 8   Find the length b for the right triangle shown. Give an exact answer and an approximation to three decimal places.

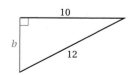

We substitute in the Pythagorean equation. Next, we solve for b^2 and then b, as follows:

$$a^2 + b^2 = c^2$$
$$10^2 + b^2 = 12^2 \qquad \text{Substituting}$$
$$100 + b^2 = 144.$$

Answer on page A-24

Then

$$100 + b^2 - 100 = 144 - 100 \qquad \text{Subtracting 100 on both sides}$$
$$b^2 = 144 - 100$$
$$b^2 = 44$$

Exact answer: $\quad b = \sqrt{44}$

Approximation: $\quad b \approx 6.633.$ $\qquad$ Using a calculator

Do Exercises 29–31.

d Applications

EXAMPLE 9 *Height of Ladder.* A 12-ft ladder leans against a building. The bottom of the ladder is 7 ft from the building. How high is the top of the ladder? Give an exact answer and an approximation to the nearest tenth of a foot.

1. **Familiarize.** We first make a drawing. In it we see a right triangle. We let $h =$ the unknown height.

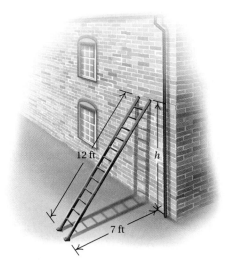

2. **Translate.** We substitute 7 for a, h for b, and 12 for c in the Pythagorean equation:

$$a^2 + b^2 = c^2 \qquad \text{Pythagorean equation}$$
$$7^2 + h^2 = 12^2.$$

3. **Solve.** We solve for h^2 and then h:

$$49 + h^2 = 144$$
$$49 + h^2 - 49 = 144 - 49$$
$$h^2 = 144 - 49$$
$$h^2 = 95$$

Exact answer: $\quad h = \sqrt{95}$

Approximation: $\quad h \approx 9.7$ ft.

4. **Check.** $7^2 + \left(\sqrt{95}\right)^2 = 49 + 95 = 144 = 12^2.$

5. **State.** The top of the ladder is $\sqrt{95}$, or about 9.7 ft from the ground.

Do Exercise 32.

Find the length of the leg of the right triangle. Give an exact answer and an approximation to three decimal places.

29.

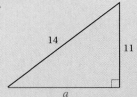

30.

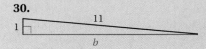

31.

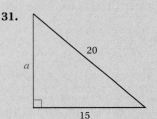

32. How long is a guy wire reaching from the top of an 18-ft pole to a point on the ground 10 ft from the pole? Give an exact answer and an approximation to the nearest tenth of a foot.

Answers on page A-24

9.6
EXERCISE SET

For Extra Help

Digital Video
Tutor CD 5
Videotape 14

InterAct
Math

Math Tutor
Center

MathXL

MyMathLab

a   Simplify.

1. $\sqrt{100}$

2. $\sqrt{25}$

3. $\sqrt{441}$

4. $\sqrt{225}$

5. $\sqrt{625}$

6. $\sqrt{576}$

7. $\sqrt{361}$

8. $\sqrt{484}$

9. $\sqrt{529}$

10. $\sqrt{169}$

11. $\sqrt{10,000}$

12. $\sqrt{4,000,000}$

b   Approximate to three decimal places.

13. $\sqrt{48}$

14. $\sqrt{17}$

15. $\sqrt{8}$

16. $\sqrt{3}$

17. $\sqrt{18}$

18. $\sqrt{7}$

19. $\sqrt{6}$

20. $\sqrt{61}$

21. $\sqrt{10}$

22. $\sqrt{21}$

23. $\sqrt{75}$

24. $\sqrt{220}$

25. $\sqrt{196}$

26. $\sqrt{123}$

27. $\sqrt{183}$

28. $\sqrt{300}$

c   Find the length of the third side of the right triangle. Give an exact answer and, where appropriate, an approximation to three decimal places.

29.

30.

31.

32.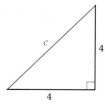

33.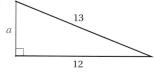

34.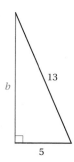

35.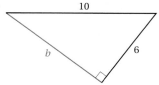

36.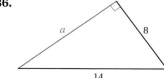

For each right triangle, find the length of the side not given. Assume that c represents the length of the hypotenuse. Give an exact answer and, when appropriate, an approximation to three decimal places.

37. $a = 10, b = 24$

38. $a = 5, b = 12$

39. $a = 9, c = 15$

40. $a = 18, c = 30$

41. $a = 1, c = 32$

42. $b = 1, c = 20$

43. $a = 4, b = 3$

44. $a = 1, c = 15$

d   In Exercises 45–54, give an exact answer and an approximation to the nearest tenth.

45. How long must a wire be in order to reach from the top of a 13-m telephone pole to a point on the ground 9 m from the base of the pole?

46. How long is a light cord reaching from the top of a 12-ft pole to a point on the ground 8 ft from the base of the pole?

47. *Softball Diamond.* A slow-pitch softball diamond is actually a square 65 ft on a side. How far is it from home plate to second base?

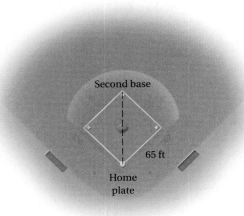

48. *Baseball Diamond.* A baseball diamond is actually a square 90 ft on a side. How far is it from home plate to second base?

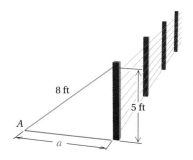

49. How tall is this tree?

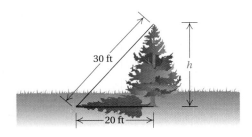

50. How far is the base of the fence post from point *A*?

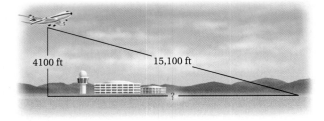

51. An airplane is flying at an altitude of 4100 ft. The slanted distance directly to the airport is 15,100 ft. How far is the airplane horizontally from the airport?

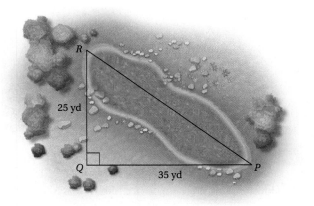

52. A surveyor had poles located at points *P*, *Q*, and *R* around a lake. The distances that the surveyor was able to measure are marked on the drawing. What is the approximate distance from *P* to *R* across the lake?

53. How long is a string of lights reaching from the top of a 24-ft pole to a point on the ground 16 ft from the base of the pole?

54. How long must a wire be in order to reach from the top of a 26-m telephone pole to a point on the ground 18 m from the base of the pole?

55. D_W Explain how the Pythagorean theorem can be used to prove that a triangle is a *right* triangle.

56. D_W Write a problem similar to Exercises 45–54 for a classmate to solve. Design the problem so that its solution involves the length $\sqrt{58}$ m.

SKILL MAINTENANCE

Evaluate.   [1.9b]

57. 10^3 **58.** 10^2 **59.** 10^5 **60.** 10^4

Find the simple interest.   [6.7a]

	PRINCIPAL	RATE OF INTEREST	TIME	SIMPLE INTEREST
61.	$2,600	8%	1 year	
62.	$750	6%	$\frac{1}{4}$ year	
63.	$20,600	6.7%	$\frac{1}{12}$ year	

Interest is compounded annually. Find the amount in the account after the given length of time. Round to the nearest cent.   [6.7b]

	PRINCIPAL	RATE OF INTEREST	TIME	AMOUNT IN THE ACCOUNT
64.	$200,000	$6\frac{5}{8}\%$	25 years	
65.	$150,000	8.4%	30 years	

SYNTHESIS

66. ▦ Find the area of the trapezoid shown. Round to the nearest hundredth.

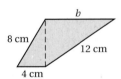

67. Which of the triangles below has the larger area?

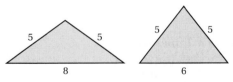

68. *Television Screen.*   A 19-in. television set has a rectangular screen that measures 19 in. diagonally. The ratio of length to width in a conventional television set is 4 to 3. Find the length and the width of the screen.

69. *Television Screen.*   A Philips 42-in. plasma television has a rectangular screen that measures 42 in. diagonally. The ratio of length to width is 16 to 9. Find the length and the width of the screen.

599

The review that follows is meant to prepare you for a chapter exam. It consists of two parts. The first part is a checklist of some of the Study Tips referred to in this and preceding chapters, as well as a list of important properties and formulas. The second part is the Review Exercises. These provide practice exercises for the exam, together with references to section objectives so you can go back and review. Before beginning, stop and look back over the skills you have obtained. What skills in mathematics do you have now that you did not have before studying this chapter?

STUDY TIPS CHECKLIST

The foundation of all your study skills is TIME!	☐ Have you started studying for the final exam?
	☐ Are you still working the margin exercises when directed to do so?
	☐ Are you working to improve your time-management skills?
	☐ Did you carefully analyze the art pieces as you studied this chapter?
	☐ Are you keeping one section ahead in your syllabus?

IMPORTANT PROPERTIES AND FORMULAS

Perimeter of a Rectangle:	$P = 2 \cdot (l + w)$, or $P = 2 \cdot l + 2 \cdot w$
Perimeter of a Square:	$P = 4 \cdot s$
Area of a Rectangle:	$A = l \cdot w$
Area of a Square:	$A = s \cdot s$, or $A = s^2$
Area of a Parallelogram:	$A = b \cdot h$
Area of a Triangle:	$A = \dfrac{1}{2} \cdot b \cdot h$
Area of a Trapezoid:	$A = \dfrac{1}{2} \cdot h \cdot (a + b)$
Radius and Diameter of a Circle:	$d = 2 \cdot r$, or $r = \dfrac{d}{2}$
Circumference of a Circle:	$C = \pi \cdot d$, or $C = 2 \cdot \pi \cdot r$
Area of a Circle:	$A = \pi \cdot r \cdot r$, or $A = \pi \cdot r^2$
Volume of a Rectangular Solid:	$V = l \cdot w \cdot h$
Volume of a Circular Cylinder:	$V = \pi \cdot r^2 \cdot h$
Volume of a Sphere:	$V = \frac{4}{3} \cdot \pi \cdot r^3$
Volume of a Cone:	$V = \frac{1}{3} \cdot \pi \cdot r^2 \cdot h$
Sum of Angle Measures of a Triangle:	$m(\angle A) + m(\angle B) + m(\angle C) = 180°$
Pythagorean Equation:	$a^2 + b^2 = c^2$

REVIEW EXERCISES

Find the perimeter. [9.1a]

1.

5 m

3 m

7 m

4 m

4 m

2. 0.5 m   1.9 m

0.8 m

1.2 m

3. *Tennis Court.* The dimensions of a standard-sized tennis court are 78 ft by 36 ft. Find the perimeter and the area of the tennis court. [9.1b], [9.2c]

36 ft

78 ft

Find the perimeter and the area. [9.1a], [9.2a]

4.

9 ft

9 ft

5.

1.8 cm

7 cm

Find the area. [9.2b]

6.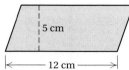

5 cm

12 cm

7.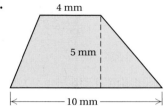

4 mm

5 mm

10 mm

8.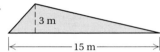

3 m

15 m

9.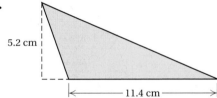

5.2 cm

11.4 cm

10.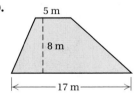

5 m

8 m

17 m

11.

$6\frac{2}{3}$ in.

$21\frac{5}{6}$ in.

601

12. *Seeded Area.* A grassy area is to be seeded around three sides of a building and has equal width on the three sides, as shown below. What is the seeded area?
[9.2c]

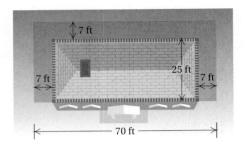

Find the length of a radius of the circle. [9.3a]

13.

16 m

14.

$\frac{28}{11}$ in.

Find the length of a diameter of the circle. [9.3a]

15.

7 ft

16.

10 cm

17. Find the circumference of the circle in Exercise 13. Use 3.14 for π. [9.3b]

18. Find the circumference of the circle in Exercise 14. Use $\frac{22}{7}$ for π. [9.3b]

19. Find the area of the circle in Exercise 13. Use 3.14 for π. [9.3c]

20. Find the area of the circle in Exercise 14. Use $\frac{22}{7}$ for π. [9.3c]

21. Find the area of the shaded region. Use 3.14 for π. [9.3d]

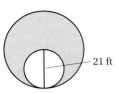

21 ft

Find the volume. [9.4a]

22.

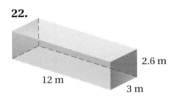

2.6 m

12 m

3 m

23.

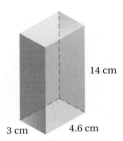

14 cm

3 cm   4.6 cm

Find the volume. Use 3.14 for π.

24. [9.4b]

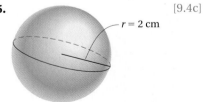

100 ft

20 ft

25. [9.4c]

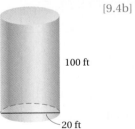

$r = 2$ cm

26. [9.4d]

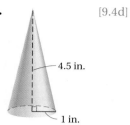

4.5 in.

1 in.

27. [9.4b]

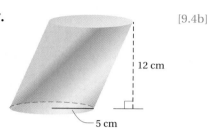

12 cm

5 cm

28. A "Norman" window is designed with dimensions as shown. Find its area and its perimeter. Use 3.14 for π. [9.3d]

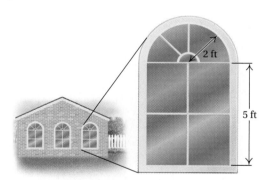

2 ft

5 ft

Use a protractor to measure each angle. [9.5a]

29.

30.

P

Q

R

31.

32.

33.–36. Classify each of the angles in Exercises 29–32 as right, straight, acute, or obtuse. [9.5b]

37. Find the measure of a complement of $\angle BAC$. [9.5c]

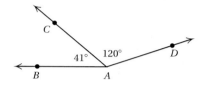

C

41°   120°   D

B   A

38. Find the measure of a supplement of a 44° angle. [9.5c]

Use the following triangle for Exercises 39–41.

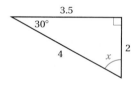

3.5

30°

4

x

2

39. Find the missing angle measure. [9.5e]

40. Classify the triangle as equilateral, isosceles, or scalene. [9.5d]

41. Classify the triangle as right, obtuse, or acute. [9.5d]

42. Simplify: $\sqrt{64}$. [9.6a]

43. Approximate to three decimal places: $\sqrt{83}$. [9.6b]

In a right triangle, find the length of the side not given. Give an exact answer and an approximation to three decimal places. [9.6c]

44. $a = 15, b = 25$

45. $a = 7, c = 10$

Find the length of the side not given. Give an exact answer and an approximation to three decimal places.   [9.6c]

46.

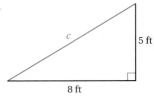

8 ft, 5 ft, c

47.

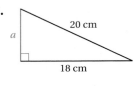

20 cm, 18 cm, a

Solve.   [9.6d]

48. How long is a wire reaching from the top of a 24-ft pole to a point on the ground 16 ft from the base of the pole?

49. How tall is this tree?

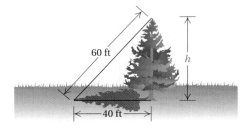

60 ft, 40 ft, h

50. Find the length of a diagonal from one corner to another of the tennis court in Exercise 3.

51. D_W List and describe all the volume formulas that you have learned in this chapter.   [9.4a, b, c, d]

52. D_W Which occupies more volume: two spheres, each with radius r, or one sphere with radius $2r$? Explain why.   [9.4c]

SKILL MAINTENANCE

Certain objectives from four particular sections will be retested on the chapter test. The objectives are listed with the practice problems that follow.

53. Find the simple interest on $5000 at 9.5% for 30 days.   [6.7a]

Evaluate.   [1.9b]

54. 3^3

55. $(4.7)^2$

56. 4.7^3

57. $\left(\dfrac{1}{2}\right)^4$

Complete.   [8.1a], [8.2a]

58. 2.5 mi = _____ ft

59. 144 in. = _____ yd

60. 4568 cm = _____ m

61. 4568 cm = _____ mm

SYNTHESIS

62. A square is cut in half so that the perimeter of the resulting rectangle is 30 ft. Find the area of the original square.   [9.1a], [9.2a]

63. Find the area, in square meters, of the shaded region.   [8.2a], [9.2c]

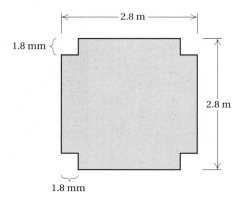

2.8 m, 1.8 mm, 2.8 m, 1.8 mm

64. Find the area, in square centimeters, of the shaded region.   [8.2a], [9.2c]

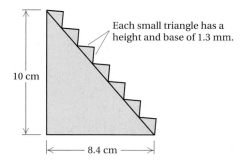

Each small triangle has a height and base of 1.3 mm.

10 cm, 8.4 cm

Chapter Test

Find the perimeter and the area.

1.

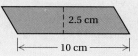

7.01 cm

9.4 cm

2.

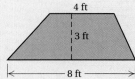

$4\frac{7}{8}$ in.

$4\frac{7}{8}$ in.

Find the area.

3.

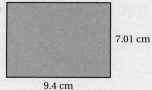

2.5 cm

10 cm

4.

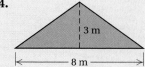

3 m

8 m

5.

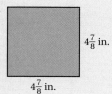

4 ft

3 ft

8 ft

6. Find the length of a diameter of this circle.

$\frac{1}{8}$ in.

7. Find the length of a radius of this circle.

18 cm

8. Find the circumference of the circle in Question 6. Use $\frac{22}{7}$ for π.

9. Find the area of the circle in Question 7. Use 3.14 for π.

10. Find the perimeter and the area of the shaded region. Use 3.14 for π.

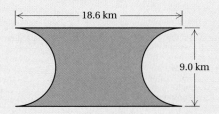

18.6 km

9.0 km

11. Find the volume.

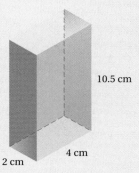

10.5 cm

4 cm

2 cm

12. A twelve-box carton of 12-oz juice boxes comes in a rectangular box $10\frac{1}{2}$ in. by 8 in. by 5 in. What is the volume of the carton?

Find the volume. Use 3.14 for π.

13.

15 ft

5 ft

14.

$d = 20$ yd

15.

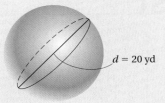

12 cm

3 cm

Use a protractor to measure each angle.

16.

17.

18.

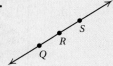

Q R S

19.

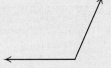

20.–23. Classify each of the angles in Questions 16–19 as right, straight, acute, or obtuse.

Use the following triangle for Questions 24–26.

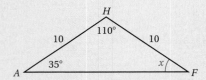

H

110°

10   10

35°   x

A   F

24. Find the missing angle measure.

25. Classify the triangle as equilateral, isosceles, or scalene.

26. Classify the triangle as right, obtuse, or acute.

27. Find the measure of a complement and a supplement of $\angle CAD$.

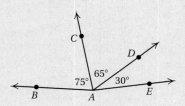

C

D

65°

75°   30°

B   A   E

28. Simplify: $\sqrt{225}$.

29. Approximate to three decimal places: $\sqrt{87}$

In a right triangle, find the length of the side not given. Give an exact answer and, where appropriate, an approximation to three decimal places.

30. $a = 24$, $b = 32$

31. $a = 2$, $c = 8$

32.

33.

34. How long must a wire be in order to reach from the top of a 13-m antenna to a point on the ground 9 m from the base of the antenna?

(SKILL MAINTENANCE)

35. Find the simple interest on $10,000 at 6.8% for 1 yr.

Evaluate.

36. 10^3

37. $\left(\dfrac{1}{4}\right)^2$

38. $(3.14)^2$

39. $(0.1)^5$

Complete.

40. 14 yd = _____ ft

41. 3000 in. = _____ ft

42. 2.3 km = _____ m

43. 34,000 mm = _____ cm

(SYNTHESIS)

Find the area of the shaded region. (Note that the figures are not drawn in perfect proportion.) Give the answer in square feet.

44.

45.

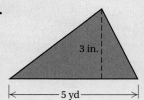

Find the volume of the solid. (Note that the solids are not drawn in perfect proportion.) Give the answer in cubic feet.

46.

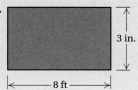

47.

48.

Solve.

1. *Quito, Ecuador.* In Quito, Ecuador, there are 1.5 million people living under threat of ash clouds and mudflows from a giant volcano, Mt. Antisana. Find standard notation for 1.5 million.
Source: *National Geographic Magazine, February 2001*

2. *Dead Sea.* The lowest point in the world is the Dead Sea on the border of Israel and Jordan. It is 1312 ft below sea level. Convert 1312 ft to yards; to meters.
Source: *The Handy Geography Answer Book*

Egg Consumption. As shown in the line graph below, egg consumption per person in the United States has been increasing in recent years. Use the graph for Exercises 3–8.

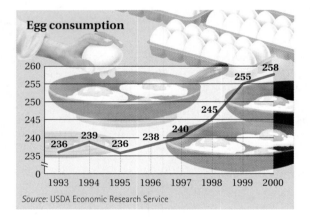

3. Find the lowest egg consumption and the year(s) in which it occurred.

4. Find the highest egg consumption and the year(s) in which it occurred.

5. Find the average, the median, and the mode of the egg consumptions.

6. Find the average egg consumption over the years 1997 to 2000.

7. Find the average egg consumption over the years 1993 to 1996. How does it compare to the answer to Exercise 6?

8. What was the percent of increase in egg consumption from 1996 to 2000?

9. *Medical Dosage.* Phenobarbital is used to control seizures. The dosage is based on the weight of the patient: 20 mg/kg. It comes in a 20-mg/5-mL solution. How many milliliters would be prescribed for someone who weighs 70 kg?

10. *Firefighting.* During a fire, the firefighters get a 1-ft layer of water on the 25-ft by 60-ft first floor of a 5-floor building. Water weighs $62\frac{1}{2}$ lb per cubic foot. What is the total weight of the water on the floor?

Calculate.

11. $1\frac{1}{2} + 2\frac{2}{3}$

12. $\left(\frac{1}{4}\right)^2 \div \left(\frac{1}{2}\right)^3 \times 2^4 + (10.3)(4)$

13. $120.5 - 32.98$

14. $22\overline{)27{,}148}$

15. $14 \div [33 \div 11 + 8 \times 2 - (15 - 3)]$

16. $8^3 + 45 \cdot 24 - 9^2 \div 3$

Find fraction notation.

17. 1.209

18. 17%

Use $<$, $>$, or $=$ for $\square$ to write a true sentence.

19. $\frac{5}{6} \square \frac{7}{8}$

20. $\frac{15}{18} \square \frac{10}{12}$

Complete.

21. $6 \text{ oz} = \underline{\hspace{1cm}} \text{ lb}$

22. $15°\text{C} = \underline{\hspace{1cm}} °\text{F}$

23. $0.087 \text{ L} = \underline{\hspace{1cm}} \text{ mL}$

24. $9 \text{ sec} = \underline{\hspace{1cm}} \text{ min}$

25. $3 \text{ yd}^2 = \underline{\hspace{1cm}} \text{ ft}^2$

26. $17 \text{ cm} = \underline{\hspace{1cm}} \text{ m}$

27. $2437 \text{ mcg} = \underline{\hspace{1cm}} \text{ mg}$

28. $29{,}082 \text{ ft} = \underline{\hspace{1cm}} \text{ m}$
(The height of Mt. Everest)

29. $9 \text{ L} = \underline{\hspace{1cm}} \text{ qt}$

Solve.

30. $\dfrac{12}{15} = \dfrac{x}{18}$

31. $\dfrac{3}{x} = \dfrac{7}{10}$

32. $25 \cdot x = 2835$

33. $x + \dfrac{3}{4} = \dfrac{7}{8}$

Find the perimeter and the area.

34.

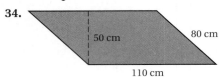

50 cm   80 cm

110 cm

35.

5.3 ft

6.8 ft   6.5 ft   8.1 ft

12.1 ft

36. Find the diameter and the area of this circle. Use $\frac{22}{7}$ for π.

35 in.

37. Find the volume of this sphere. Use $\frac{22}{7}$ for π.

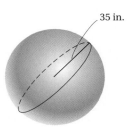

35 in.

Solve.

38. To get an A in math, a student must score an average of 90 on five tests. On the first four tests, the scores were 85, 92, 79, and 95. What is the lowest score that the student can get on the last test and still get an A?

39. Americans own 52 million dogs, 56 million cats, 45 million birds, 250 million fish, and 125 million other creatures as house pets. How many pets do Americans own altogether?

40. What is the simple interest on $8000 at 8.2% for $\frac{1}{4}$ year?

41. What is the amount in an account after 25 years if $8000 is invested at 8.2%, compounded annually?

42. How long must a rope be in order to reach from the top of an 8-m tree to a point on the ground 15 m from the bottom of the tree?

43. The sales tax on an office supply purchase of $5.50 is $0.33. What is the sales tax rate?

44. A bolt of fabric in a fabric store has $10\frac{3}{4}$ yd on it. A customer purchases $8\frac{5}{8}$ yd. How many yards remain on the bolt?

45. What is the cost, in dollars, of 15.6 gal of gasoline at 139.9¢ per gallon? Round to the nearest cent.

46. A box of powdered milk that makes 20 qt costs $4.99. A box that makes 8 qt costs $1.99. Which size has the lower unit price?

47. It is $\frac{7}{10}$ km from a student's dormitory to the library. Maria starts to walk there, changes her mind after going $\frac{1}{4}$ of the distance, and returns home. How far did she walk?

48. Find the missing angle measure.

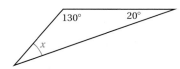

49. Classify the triangle as equilateral, isosceles, or scalene.

12 in. 12 in.

7 in.

50. Classify the triangle in Question 48 as right, obtuse, or acute.

51. *Matching.* Match each item in the first column with the appropriate item in the second column by drawing connecting lines. Some expressions in the second column will be used more than once. Some expressions will not be used.

Area of a circle of radius 4 ft

Area of a square of side 4 ft

Circumference of a circle of radius 4 ft

Volume of a cone with radius of the base 4 ft and height 8 ft

Area of a rectangle of length 8 ft and width 4 ft

Area of a triangle with base 4 ft and height 8 ft

Volume of a sphere of radius 4 ft

Volume of a right circular cylinder with radius of the base 4 ft and height 8 ft

Perimeter of a square of side 4 ft

Perimeter of a rectangle of length 8 ft and width 4 ft

24 ft

16 ft

$16 \cdot \pi \, \text{ft}^2$

$32 \, \text{ft}^2$

$\frac{4^4}{3} \cdot \pi \, \text{ft}^3$

$128 \cdot \pi \, \text{ft}^3$

$\left(21\frac{1}{3}\right) \cdot \pi \, \text{ft}^3$

$8 \cdot \pi \, \text{ft}$

64 ft

$128 \cdot \pi \, \text{ft}^2$

$16 \, \text{ft}^2$

$\frac{128}{3} \cdot \pi \, \text{ft}^3$

(SYNTHESIS)

52. Your house sits on a lot measuring 75 ft by 200 ft. The lot is at the intersection of two streets, so there are sidewalks on two sides of the lot. In the winter, you have to shovel the snow off the sidewalks. If the sidewalks are 3 ft wide and the snow is 4 in. deep, what volume of snow must you shovel?

Find the volume in cubic feet. Use 3.14 for π.

53.

100 yd

10 ft

54.

14 ft

3 in.   4.6 in.

55. J. C. Payne, a 71-year-old rancher at the time, recently asked *The Guinness Book of Records* to accept a world record for constructing a ball of string. The ball was 13.2 ft in diameter. What was the volume of the ball? Assuming the diameter of the string was 0.1 in., how long was the string in feet? in miles? Use the π key on your calculator.
Source: *The Guinness Book of Records*

Real Numbers

Gateway to Chapter 10

The preceding parts of this book have formed a foundation for algebra, which we begin to study in this chapter. Here we emphasize mainly manipulations involving real numbers and algebraic expressions, delaying equation solving until Chapter 11.

Real-World Application

The Viking 2 Lander spacecraft has determined that temperatures on Mars range from −125°C (Celsius) to 25°C. Find the difference between the highest value and the lowest value in this temperature range.

Source: The Lunar and Planetary Institute

This problem appears as Example 12 in Section 10.3.

Use either $<$ or $>$ for $\square$ to write a true sentence. [10.1d]

1. $0 \ \square \ -5$

2. $10 \ \square \ -5$

3. $-35 \ \square \ -45$

4. $-\dfrac{2}{3} \ \square \ \dfrac{4}{5}$

Find decimal notation. [10.1c]

5. $-\dfrac{5}{8}$

6. $-\dfrac{2}{3}$

7. $-\dfrac{10}{11}$

Find the absolute value. [10.1e]

8. $|-12|$

9. $|2.3|$

10. $|0|$

Find the opposite, or additive inverse. [10.2b]

11. 5.4

12. $-\dfrac{2}{3}$

Compute and simplify.

13. $-9 + (-8)$ [10.2a]

14. $20.2 - (-18.4)$ [10.3a]

15. $-\dfrac{5}{6} - \dfrac{3}{10}$ [10.3a]

16. $-11.5 + 6.5$ [10.2a]

17. $-9(-7)$ [10.4a]

18. $\dfrac{5}{8}\left(-\dfrac{2}{3}\right)$ [10.4a]

19. $-19.6 \div 0.2$ [10.5c]

20. $-56 \div (-7)$ [10.5a]

21. $12 - (-6) + 14 - 8$ [10.3a]

22. $20 - 10 \div 5 + 2^3$ [10.5e]

23. Temperature Extremes. In Churchill, Manitoba, Canada, the average daily low temperature in January is $-31°C$. The average daily low temperature in Key West, Florida, is $19°C$. How much higher is the average daily low temperature in Key West, Florida? [10.3b]

10.1

THE REAL NUMBERS

In this section, we introduce the *real numbers*. We begin with numbers called *integers* and build up to the real numbers. To describe integers, we start with the whole numbers, 0, 1, 2, 3, and so on. For each number 1, 2, 3, and so on, we obtain a new number to the left of zero on the number line:

> For the number 1, there will be an *opposite* number -1 (negative 1).
>
> For the number 2, there will be an *opposite* number -2 (negative 2).
>
> For the number 3, there will be an *opposite* number -3 (negative 3), and so on.

The **integers** consist of the whole numbers and these new numbers. We picture them on a number line as follows.

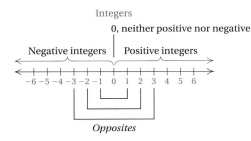

We call the numbers to the left of zero on the number line **negative integers.** The natural numbers are called **positive integers.** Zero is neither positive nor negative. We call -1 and 1 opposites of each other. Similarly, -2 and 2 are opposites, -3 and 3 are opposites, -100 and 100 are opposites, and 0 is its own opposite. Opposite pairs of numbers like -3 and 3 are the same distance from 0. The integers extend infinitely to the left and right of zero.

INTEGERS

The **integers:** $\ldots, -5, -4, -3, -2, -1, 0, 1, 2, 3, 4, 5, \ldots$

---

a | Integers and the Real World

Integers correspond to many real-world problems and situations. The following examples will help you get ready to translate problem situations to mathematical language.

Objectives

a Tell which integers correspond to a real-world situation.

b Graph rational numbers on a number line.

c Convert from fraction notation for a rational number to decimal notation.

d Determine which of two real numbers is greater and indicate which, using $<$ or $>$.

e Find the absolute value of a real number.

1. The halfback gained 8 yd on first down. The quarterback was sacked for a 5-yd loss on second down.

2. **Temperature High.** The highest human-made temperature on record is 950,000,000°F. It was created on May 27, 1994, at the Tokamak Fusion Test Reactor at the Princeton Plasma Physics Laboratory in New Jersey.
Source: *The Guinness Book of Records*

3. **Stock Decrease.** The stock of Sherwin Williams Co. decreased from $25 per share to $19 per share over a recent period.
Source: The New York Stock Exchange

4. At 10 sec before liftoff, ignition occurs. At 148 sec after liftoff, the first stage is detached from the rocket.

5. A student owes $137 to the bookstore. The student has $289 in a savings account.

Answers on page A-25

616

EXAMPLE 1 Tell which integer corresponds to this situation: The temperature is 3 degrees below zero.

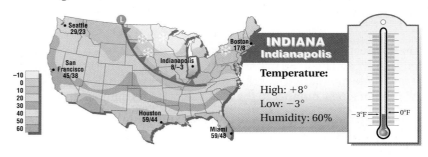

The integer −3 corresponds to the situation. The temperature is −3°.

EXAMPLE 2 *Elevation.* Tell which integer corresponds to this situation: The lowest point in New Orleans is 8 ft below sea level.

The integer −8 corresponds to the situation. The elevation is −8 ft.

EXAMPLE 3 *Stock Price Change.* Tell which integers correspond to the situation: The stock price of Pearson Education decreased from $24 per share to $17 per share over a recent time period. The stock of Sherwin Williams Co. increased from $21 per share to $25 per share over a recent period.
Source: The New York Stock Exchange

The integer −7 corresponds to the decrease in the value of the stock. The integer 4 represents the increase in the value of the stock.

Do Exercises 1–5.

b The Rational Numbers

To create a larger number system, called the **rational numbers,** we consider quotients of integers with nonzero divisors. The following are rational numbers:

$$\frac{2}{3}, \quad -\frac{2}{3}, \quad \frac{7}{1}, \quad 4, \quad -3, \quad 0, \quad \frac{23}{-8}, \quad 2.4, \quad -0.17, \quad 10\frac{1}{2}.$$

The number $-\frac{2}{3}$ (read "negative two-thirds") can also be named $\frac{2}{-3}$ or $\frac{-2}{3}$. The number 2.4 can be named $\frac{24}{10}$, or $\frac{12}{5}$, and −0.17 can be named $-\frac{17}{100}$.

Note that the rational numbers contain the natural numbers, the whole numbers, the integers, and the arithmetic numbers (also called the nonnegative rational numbers).

RATIONAL NUMBERS

The **rational numbers** consist of all numbers that can be named in the form $\frac{a}{b}$, where a and b are integers and b is not 0.

We picture the rational numbers on a number line, as follows.

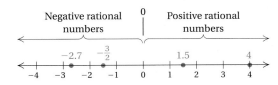

Negative rational numbers   0   Positive rational numbers

To **graph** a number means to find and mark its point on the number line. Some rational numbers are graphed in the preceding figure.

EXAMPLE 4   Graph: $\frac{5}{2}$.

The number $\frac{5}{2}$ can be named $2\frac{1}{2}$, or 2.5. Its graph is halfway between 2 and 3.

EXAMPLE 5   Graph: -3.2.

The graph of -3.2 is $\frac{2}{10}$ of the way from -3 to -4.

EXAMPLE 6   Graph: $\frac{13}{8}$.

The number $\frac{13}{8}$ can be named $1\frac{5}{8}$, or 1.625. The graph is about $\frac{6}{10}$ of the way from 1 to 2.

Do Exercises 6–8.

Graph on a number line.

6. $-\frac{7}{2}$

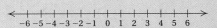

7. 1.4

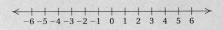

8. $-\frac{11}{4}$

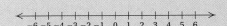

Answers on page A-25

Find decimal notation.

9. $-\dfrac{3}{8}$

10. $-\dfrac{6}{11}$

11. $\dfrac{4}{3}$

C Notation for Rational Numbers

Each rational number can be named using fraction or decimal notation.

EXAMPLE 7   Find decimal notation for $-\frac{5}{8}$.

We first find decimal notation for $\frac{5}{8}$. Since $\frac{5}{8}$ means $5 \div 8$, we divide,

$$
\begin{array}{r}
0.6\ 2\ 5 \\
8\ \overline{)\ 5.0\ 0\ 0} \\
4\ 8 \\
\hline
2\ 0 \\
1\ 6 \\
\hline
4\ 0 \\
4\ 0 \\
\hline
0
\end{array}
$$

Thus, $\frac{5}{8} = 0.625$, so $-\frac{5}{8} = -0.625$.

Decimal notation for $-\frac{5}{8}$ is -0.625. We consider -0.625 to be a **terminating decimal.** Decimal notation for some numbers repeats.

EXAMPLE 8   Find decimal notation for $-\frac{7}{11}$.

We divide to find decimal notation for $\frac{7}{11}$.

$$
\begin{array}{r}
0.6\ 3\ 6\ 3\ldots \\
1\ 1\ \overline{)\ 7.0\ 0\ 0\ 0} \\
6\ 6 \\
\hline
4\ 0 \\
3\ 3 \\
\hline
7\ 0 \\
6\ 6 \\
\hline
4\ 0 \\
3\ 3 \\
\hline
7
\end{array}
$$

Thus, $\frac{7}{11} = 0.6363\ldots$, so $-\frac{7}{11} = -0.6363\ldots$. Repeating decimal notation can be abbreviated by writing a bar over the repeating part; in this case, we write $-0.\overline{63}$.

The following are other examples to show how each rational number can be named using fraction or decimal notation:

$$
0 = \frac{0}{6}, \qquad \frac{27}{100} = 0.27, \qquad -8\frac{3}{4} = -8.75, \qquad -\frac{13}{6} = -2.1\overline{6}.
$$

Do Exercises 9–11.

d The Real Numbers and Order

The number line has a point for every rational number. However, there are some points on the line for which there are no rational numbers. These points correspond to what are called **irrational numbers.** Some examples of irrational numbers are π and $\sqrt{2}$.

Decimal notation for rational numbers *either* terminates *or* repeats. Decimal notation for irrational numbers *neither* terminates *nor* repeats. Some other examples of irrational numbers are $\sqrt{3}$, $-\sqrt{8}$, $\sqrt{11}$, and $0.121221222122221\ldots$. *Whenever we take the square root of a number that is not a perfect square (see Section 9.6), we will get an irrational number.*

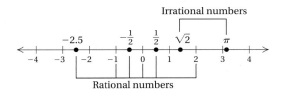

THE REAL-NUMBER SYSTEM

The rational numbers and the irrational numbers together correspond to all the points on a number line and make up what is called the **real-number system.**

The **real numbers** consist of the rational numbers and the irrational numbers. The following figure shows the relationship among various kinds of numbers.

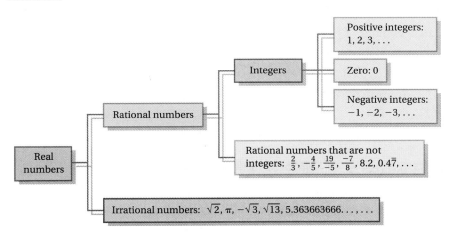

Order

Real numbers are named in order on the number line, with larger numbers named farther to the right. (See Section 1.4.) For any two numbers on the line, the one to the left is less than the one to the right.

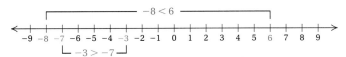

We use the symbol **<** to mean "**is less than.**" The sentence $-8 < 6$ means "-8 is less than 6." The symbol **>** means "**is greater than.**" The sentence $-3 > -7$ means "-3 is greater than -7."

◼ **EXAMPLES**   Use either $<$ or $>$ for $\square$ to write a true sentence.

9. $-7\ \square\ 3$ Since -7 is to the left of 3, we have $-7 < 3$.

10. $6\ \square\ -12$ Since 6 is to the right of -12, then $6 > -12$.

11. $-18\ \square\ -5$ Since -18 is to the left of -5, we have $-18 < -5$.

Use either $<$ or $>$ for $\square$ to write a true sentence.

12. $-3\ \square\ 7$

13. $-8\ \square\ -5$

14. $7\ \square\ -10$

15. $3.1\ \square\ -9.5$

16. $-\dfrac{2}{3}\ \square\ -1$

17. $-\dfrac{11}{8}\ \square\ \dfrac{23}{15}$

18. $-\dfrac{2}{3}\ \square\ -\dfrac{5}{9}$

19. $-4.78\ \square\ -5.01$

Answers on page A-25

Find the absolute value.

20. $|8|$

21. $|0|$

22. $|-9|$

23. $\left|-\dfrac{2}{3}\right|$

24. $|5.6|$

Answers on page A-25

12. $-2.7 \,\square\, -\dfrac{3}{2}$ The answer is $-2.7 < -\dfrac{3}{2}$.

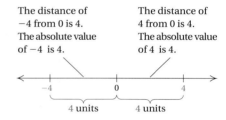

13. $1.5 \,\square\, -2.7$ The answer is $1.5 > -2.7$.

14. $-3.45 \,\square\, 1.32$ The answer is $-3.45 < 1.32$.

15. $\dfrac{5}{8} \,\square\, \dfrac{7}{11}$ We convert to decimal notation: $\dfrac{5}{8} = 0.625$, and $\dfrac{7}{11} = 0.6363\ldots$. Thus, $\dfrac{5}{8} < \dfrac{7}{11}$.

16. $-4 \,\square\, 0$ The answer is $-4 < 0$.

17. $5.8 \,\square\, 0$ The answer is $5.8 > 0$.

Do Exercises 12–19 on the preceding page.

e Absolute Value

From the number line, we see that numbers like 4 and -4 are the same distance from zero. Distance is always a nonnegative number. We call the distance of a number from zero the **absolute value** of the number.

The distance of -4 from 0 is 4. The absolute value of -4 is 4.

The distance of 4 from 0 is 4. The absolute value of 4 is 4.

4 units   4 units

ABSOLUTE VALUE

The **absolute value** of a number is its distance from zero on a number line. We use the symbol $|x|$ to represent the absolute value of a number x.

To find the absolute value of a number:

a) If a number is negative, make it positive.

b) If a number is positive or zero, leave it alone.

EXAMPLES   Find the absolute value.

18. $|-7|$ The distance of -7 from 0 is 7, so $|-7| = 7$.

19. $|12|$ The distance of 12 from 0 is 12, so $|12| = 12$.

20. $|0|$ The distance of 0 from 0 is 0, so $|0| = 0$.

21. $\left|\dfrac{3}{2}\right| = \dfrac{3}{2}$

22. $|-2.73| = 2.73$

Do Exercises 20–24.

EXERCISE SET

a Tell which integers correspond to the situation.

1. *Elevation.* The Dead Sea, between Jordan and Israel, is 1286 ft below sea level; Mt. Rainier in Washington State is 14,410 ft above sea level.

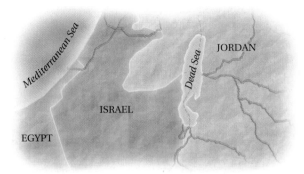

2. *Golf Score.* Tiger Woods' score in winning the 2000 PGA Championship was 18 below par.
Source: U.S. Golf Association

3. On Wednesday, the temperature was 24° above zero. On Thursday, it was 2° below zero.

4. A student deposited her tax refund of $750 in a savings account. Two weeks later, she withdrew $125 to pay sorority fees.

5. *U.S. Public Debt.* Recently, the total public debt of the United States was about $5,600,000,000,000.
Source: U.S. Department of the Treasury

6. *Birth and Death Rates.* Recently, the world birth rate was 27 per thousand. The death rate was 9.7 per thousand.
Source: United Nations Population Fund

7. In bowling, after the game, team A is 34 pins behind team B, and team B is 15 pins ahead of team C.

8. During a video game, Sara intercepted a missile worth 20 points, lost a starship worth 150 points, and captured a base worth 300 points.

b Graph the number on the number line.

9. $\frac{10}{3}$

10. $-\frac{17}{4}$

11. -5.2

12. 4.78

c Convert to decimal notation.

13. $-\frac{7}{8}$

14. $-\frac{1}{8}$

15. $\frac{5}{6}$

16. $\frac{5}{3}$

17. $\frac{7}{6}$

18. $\frac{5}{12}$

19. $\frac{2}{3}$

20. $\frac{1}{4}$

21. $-\frac{1}{2}$

22. $-\frac{5}{8}$

23. $-8\frac{7}{25}$

24. $-9\frac{5}{16}$

d Use either < or > for □ to write a true sentence.

25. 8 □ 0

26. 3 □ 0

27. −8 □ 3

28. 6 □ −6

29. −8 □ 8

30. 0 □ −9

31. −8 □ −5

32. −4 □ −3

33. −5 □ −11

34. −3 □ −4

35. −6 □ −5

36. −10 □ −14

37. 2.14 □ 1.24

38. −3.3 □ −2.2

39. −14.5 □ 0.011

40. 17.2 □ −1.67

41. $-12\frac{5}{8}$ □ $-6\frac{3}{8}$

42. $-7\frac{5}{16}$ □ $-3\frac{11}{16}$

43. $\frac{5}{12}$ □ $\frac{11}{25}$

44. $-\frac{13}{16}$ □ $-\frac{5}{9}$

e Find the absolute value.

45. |−3|

46. |−7|

47. |18|

48. |0|

49. |325|

50. |−4|

51. |−3.625|

52. $\left|-7\frac{4}{5}\right|$

53. $\left|-\frac{2}{3}\right|$

54. $\left|-\frac{10}{7}\right|$

55. $\left|\frac{0}{4}\right|$

56. |14.8|

57. $\mathbf{D_W}$ Give three examples of rational numbers that are not integers. Explain.

58. $\mathbf{D_W}$ Give three examples of irrational numbers. Explain the difference between an irrational number and a rational number.

SKILL MAINTENANCE

Find the prime factorization. [2.1d]

59. 54

60. 192

61. 102

62. 260

63. 864

64. 468

Find the LCM. [3.1a]

65. 6, 18

66. 18, 24

67. 6, 24, 32

68. 12, 24, 36

69. 48, 56, 64

70. 12, 36, 84

SYNTHESIS

Use either <, >, or = for □ to write a true sentence.

71. |−5| □ |−2|

72. |4| □ |−7|

73. |−8| □ |8|

List in order from the least to the greatest.

74. $-\frac{2}{3}, \frac{1}{2}, -\frac{3}{4}, -\frac{5}{6}, \frac{3}{8}, \frac{1}{6}$

75. $-8\frac{7}{8}, 7, -5, |-6|, 4, |3|, -8\frac{5}{8}, -100, 0, 1^7, \frac{14}{4}, -\frac{67}{8}$

10.2 ADDITION OF REAL NUMBERS

Objectives

a Add real numbers without using a number line.

b Find the opposite, or additive inverse, of a real number.

We now consider addition of real numbers. First, to gain an understanding, we add using a number line. Then we consider rules for addition.

Addition on a Number Line

Addition of numbers can be illustrated on a number line. To do the addition $a + b$, we start at 0. Then move to a and then move according to b.

a) If b is positive, we move to the right.

b) If b is negative, we move to the left.

c) If b is 0, we stay at a.

EXAMPLE 1 Add: $3 + (-5)$.

We start at 0 and move 3 units right since 3 is positive. Then we move 5 units left since -5 is negative.

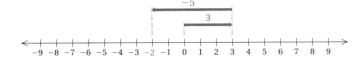

$$3 + (-5) = -2$$

EXAMPLE 2 Add: $-4 + (-3)$.

We start at 0 and move 4 units left since -4 is negative. Then we move 3 units further left since -3 is negative.

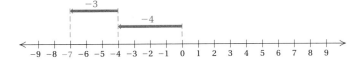

$$-4 + (-3) = -7$$

EXAMPLE 3 Add: $-4 + 9$.

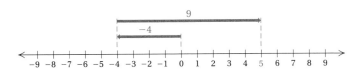

$$-4 + 9 = 5$$

EXAMPLE 4 Add: $-5.2 + 0$.

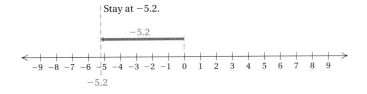

Add using a number line.

1. $0 + (-3)$

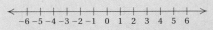

2. $1 + (-4)$

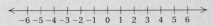

3. $-3 + (-2)$

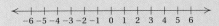

4. $-3 + 7$

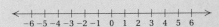

5. $-2.4 + 2.4$

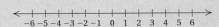

6. $-\dfrac{5}{2} + \dfrac{1}{2}$

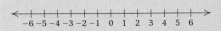

Answers on page A-25

623

Add without using a number line.

7. $-5 + (-6)$

8. $-9 + (-3)$

9. $-4 + 6$

10. $-7 + 3$

11. $5 + (-7)$

12. $-20 + 20$

13. $-11 + (-11)$

14. $10 + (-7)$

15. $-0.17 + 0.7$

16. $-6.4 + 8.7$

17. $-4.5 + (-3.2)$

18. $-8.6 + 2.4$

19. $\dfrac{5}{9} + \left(-\dfrac{7}{9}\right)$

20. $-\dfrac{1}{5} + \left(-\dfrac{3}{4}\right)$

Answers on page A-25

$-5.2 + 0 = -5.2$

Do Exercises 1–6 on the preceding page.

a   Addition Without a Number Line

You may have noticed some patterns in the preceding examples. These lead us to rules for adding without using a number line that are more efficient for adding larger or more complicated numbers.

> **RULES FOR ADDITION OF REAL NUMBERS**
>
> 1. *Positive numbers*: Add the same as arithmetic numbers. The answer is positive.
> 2. *Negative numbers*: Add absolute values. The answer is negative.
> 3. *A positive and a negative number*: Subtract the smaller absolute value from the larger. Then:
>   a) If the positive number has the greater absolute value, the answer is positive.
>   b) If the negative number has the greater absolute value, the answer is negative.
>   c) If the numbers have the same absolute value, the answer is 0.
> 4. *One number is zero*: The sum is the other number.

Rule 4 is known as the **identity property of 0.** It says that for any real number a, $a + 0 = a$.

EXAMPLES   Add without using a number line.

5. $-12 + (-7) = -19$ Two negatives. *Think*: Add the absolute values, 12 and 7, getting 19. Make the answer *negative*, -19.

6. $-1.4 + 8.5 = 7.1$ The absolute values are 1.4 and 8.5. The difference is 7.1. The positive number has the larger absolute value, so the answer is *positive*, 7.1.

7. $-36 + 21 = -15$ The absolute values are 36 and 21. The difference is 15. The negative number has the larger absolute value, so the answer is *negative*, -15.

8. $1.5 + (-1.5) = 0$ The numbers have the same absolute value. The sum is 0.

9. $-\dfrac{7}{8} + 0 = -\dfrac{7}{8}$ One number is zero. The sum is $-\dfrac{7}{8}$.

10. $-9.2 + 3.1 = -6.1$

11. $-\dfrac{3}{2} + \dfrac{9}{2} = \dfrac{6}{2} = 3$

12. $-\dfrac{2}{3} + \dfrac{5}{8} = -\dfrac{16}{24} + \dfrac{15}{24} = -\dfrac{1}{24}$

Do Exercises 7–20.

Suppose we wish to add several numbers, some positive and some negative, as follows. How can we proceed?

$$15 + (-2) + 7 + 14 + (-5) + (-12)$$

The commutative and associative laws hold for real numbers. Thus we can change grouping and order as we please when adding. For instance, we can group the positive numbers together and the negative numbers together and add them separately. Then we add the two results.

EXAMPLE 13   Add: $15 + (-2) + 7 + 14 + (-5) + (-12)$.

a) $15 + 7 + 14 = 36$ Adding the positive numbers

b) $-2 + (-5) + (-12) = -19$ Adding the negative numbers

c) $36 + (-19) = 17$ Adding the results of (a) and (b)

We can also add the numbers in any other order we wish, say, from left to right as follows:

$$
\begin{aligned}
15 + (-2) + 7 + 14 + (-5) + (-12) &= 13 + 7 + 14 + (-5) + (-12) \\
&= 20 + 14 + (-5) + (-12) \\
&= 34 + (-5) + (-12) \\
&= 29 + (-12) \\
&= 17
\end{aligned}
$$

Do Exercises 21–24.

b   Opposites, or Additive Inverses

Suppose we add two numbers that are **opposites,** such as 6 and -6. The result is 0. When opposites are added, the result is always 0. Such numbers are also called **additive inverses.** Every real number has an opposite, or additive inverse.

> **OPPOSITES, OR ADDITIVE INVERSES**
>
> Two numbers whose sum is 0 are called **opposites,** or **additive inverses,** of each other.

EXAMPLES   Find the opposite, or additive inverse, of each number.

14. 34   The opposite of 34 is -34 because $34 + (-34) = 0$.

15. -8 The opposite of -8 is 8 because $-8 + 8 = 0$.

16. 0   The opposite of 0 is 0 because $0 + 0 = 0$.

17. $-\dfrac{7}{8}$ The opposite of $-\dfrac{7}{8}$ is $\dfrac{7}{8}$ because $-\dfrac{7}{8} + \dfrac{7}{8} = 0$.

Do Exercises 25–30.

To name the opposite, we use the symbol $-$, as follows.

> **SYMBOLIZING OPPOSITES**
>
> The opposite, or additive inverse, of a number a can be named $-a$ (read "the opposite of a," or "the additive inverse of a").

Note that if we take a number, say 8, and find its opposite, -8, and then find the opposite of the result, we will have the original number, 8, again.

Add.

21. $(-15) + (-37) + 25 + 42 + (-59) + (-14)$

22. $42 + (-81) + (-28) + 24 + 18 + (-31)$

23. $-2.5 + (-10) + 6 + (-7.5)$

24. $\begin{aligned} &-35 \\ &17 \\ &14 \\ &-27 \\ &31 \\ &-12 \end{aligned}$

Find the opposite, or additive inverse, of each of the following.

25. -4

26. 8.7

27. -7.74

28. $-\dfrac{8}{9}$

29. 0

30. 12

Answers on page A-25

Evaluate $-x$ and $-(-x)$ when:

31. $x = 14$.

32. $x = 1$.

33. $x = -19$.

34. $x = -1.6$.

35. $x = \dfrac{2}{3}$.

36. $x = -\dfrac{9}{8}$.

Find the opposite. (Change the sign.)

37. -4

38. -13.4

39. 0

40. $\dfrac{1}{4}$

Answers on page A-25

CALCULATOR CORNER

Entering Negative Numbers On many calculators, we can enter negative numbers using the $\boxed{+/-}$ key. This allows us to perform calculations with real numbers. To enter -8, for example, we press $\boxed{8}$ $\boxed{+/-}$. To find the sum $-14 + (-9)$, we press $\boxed{1}\boxed{4}\boxed{+/-}$ $\boxed{+}\boxed{9}\boxed{+/-}\boxed{=}$. The result is -23. Note that it is not necessary to use parentheses when entering this expression.

Exercises: Add.

1. $-5 + 7$

2. $-4 + 17$

3. $-6 + (-9)$

4. $3 + (-11)$

5. $1.5 + (-4.8)$

6. $-2.8 + (-10.6)$

626

THE OPPOSITE OF THE OPPOSITE

The opposite of the opposite of a number is the number itself. (The additive inverse of the additive inverse of a number is the number itself.) That is, for any number a,

$$-(-a) = a.$$

EXAMPLE 18 Evaluate $-x$ and $-(-x)$ when $x = 16$.

We replace x in each case with 16.

a) If $x = 16$, then $-x = -16 = -16$. The opposite of 16 is -16.

b) If $x = 16$, then $-(-x) = -(-16) = 16$. The opposite of the opposite of 16 is 16.

EXAMPLE 19 Evaluate $-x$ and $-(-x)$ when $x = -3$.

We replace x in each case with -3.

a) If $x = -3$, then $-x = -(-3) = 3$.

b) If $x = -3$, then $-(-x) = -(-(-3)) = -(3) = -3$.

Note that in Example 19 we used an extra set of parentheses to show that we are substituting the negative number -3 for x. Symbolism like $--x$ is not considered meaningful.

Do Exercises 31–36.

A symbol such as -8 is usually read "negative 8." It could be read "the additive inverse of 8," because the additive inverse of 8 is negative 8. It could also be read "the opposite of 8," because the opposite of 8 is -8. Thus a symbol like -8 can be read in more than one way. A symbol like $-x$, which has a variable, should be read "the opposite of x" or "the additive inverse of x" and *not* "negative x," because we do not know whether x represents a positive number, a negative number, or 0. You can verify this by referring to the preceding examples.

We can use the symbolism $-a$ to restate the definition of opposite, or additive inverse.

THE SUM OF OPPOSITES

For any real number a, the opposite, or additive inverse, of a, expressed as $-a$, is such that

$$a + (-a) = (-a) + a = 0.$$

SIGNS OF NUMBERS

A negative number is sometimes said to have a "negative sign." A positive number is said to have a "positive sign." When we replace a number with its opposite, we can say that we have "changed its sign."

EXAMPLES Change the sign. (Find the opposite.)

20. -3 $-(-3) = 3$

21. -10 $-(-10) = 10$

22. 0 $-(0) = 0$

23. 14 $-(14) = -14$

Do Exercises 37–40.

a   Add. Do not use a number line except as a check.

1. $-9 + 2$

2. $-5 + 2$

3. $-10 + 6$

4. $4 + (-3)$

5. $-8 + 8$

6. $4 + (-4)$

7. $-3 + (-5)$

8. $-6 + (-8)$

9. $-7 + 0$

10. $-10 + 0$

11. $0 + (-27)$

12. $0 + (-36)$

13. $17 + (-17)$

14. $-20 + 20$

15. $-17 + (-25)$

16. $-23 + (-14)$

17. $18 + (-18)$

18. $-13 + 13$

19. $-18 + 18$

20. $11 + (-11)$

21. $8 + (-5)$

22. $-7 + 8$

23. $-4 + (-5)$

24. $10 + (-12)$

25. $13 + (-6)$

26. $-3 + 14$

27. $-25 + 25$

28. $40 + (-40)$

29. $63 + (-18)$

30. $85 + (-65)$

31. $-6.5 + 4.7$

32. $-3.6 + 1.9$

33. $-2.8 + (-5.3)$

34. $-7.9 + (-6.5)$

35. $-\dfrac{3}{5} + \dfrac{2}{5}$

36. $-\dfrac{4}{3} + \dfrac{2}{3}$

37. $-\dfrac{3}{7} + \left(-\dfrac{5}{7}\right)$

38. $-\dfrac{4}{9} + \left(-\dfrac{6}{9}\right)$

39. $-\dfrac{5}{8} + \dfrac{1}{4}$

40. $-\dfrac{5}{6} + \dfrac{2}{3}$

41. $-\dfrac{3}{7} + \left(-\dfrac{2}{5}\right)$

42. $-\dfrac{5}{8} + \left(-\dfrac{1}{3}\right)$

43. $-\dfrac{3}{5} + \left(-\dfrac{2}{15}\right)$

44. $-\dfrac{5}{9} + \left(-\dfrac{5}{18}\right)$

45. $-5.7 + (-7.2) + 6.6$

46. $-10.3 + (-7.5) + 3.1$

47. $-\dfrac{7}{16} + \dfrac{7}{8}$

48. $-\dfrac{3}{24} + \dfrac{7}{36}$

49. $75 + (-14) + (-17) + (-5)$

50. $28 + (-44) + 17 + 31 + (-94)$

51. $-44 + \left(-\dfrac{3}{8}\right) + 95 + \left(-\dfrac{5}{8}\right)$

52. $24 + 3.1 + (-44) + (-8.2) + 63$

53. $98 + (-54) + 113 + (-998) + 44 + (-612) + (-18) + 334$

54. $-455 + (-123) + 1026 + (-919) + 213 + 111 + (-874)$

b Find the opposite, or additive inverse.

55. 24

56. −84

57. −26.9

58. 27.4

Find −x when:

59. x = 9.

60. x = −26.

61. x = −$\frac{14}{3}$.

62. x = $\frac{1}{526}$.

Find −(−x) when:

63. x = −65.

64. x = 31.

65. x = $\frac{5}{3}$.

66. x = −7.8.

Change the sign. (Find the opposite.)

67. −14

68. −18.3

69. 10

70. −$\frac{5}{8}$

71. D_W Explain in your own words why the sum of two negative numbers is always negative.

72. D_W A student states that −93 is "bigger than" −47. What mistake is the student making?

SKILL MAINTENANCE

Find the area.

73. [9.2a]

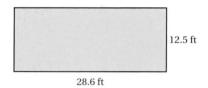

12.5 ft

28.6 ft

74. [9.2b]

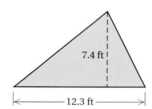

7.4 ft

12.3 ft

75. [9.2a]

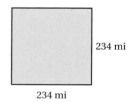

234 mi

234 mi

76. [9.2b]

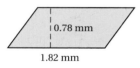

0.78 mm

1.82 mm

77. [9.2b]

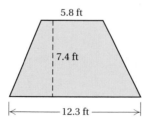

5.8 ft

7.4 ft

12.3 ft

78. [9.3.c]

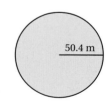

50.4 m

SYNTHESIS

79. For what numbers x is −x negative?

80. For what numbers x is −x positive?

Add.

81. 🖩 −345,882 + (−295,097)

82. 🖩 2706.835 + (−0.005684)

Tell whether the sum is positive, negative, or zero.

83. If n is positive and m is negative, then −n + m is _____ .

84. If n = m and n is negative, then −n + (−m) is _____ .

628

10.3 SUBTRACTION OF REAL NUMBERS

Objectives

a Subtract real numbers and simplify combinations of additions and subtractions.

b Solve applied problems involving addition and subtraction of real numbers.

a Subtraction

We now consider subtraction of real numbers.

SUBTRACTION

The difference $a - b$ is the number c for which $a = b + c$.

Consider, for example, $45 - 17$. *Think*: What number can we add to 17 to get 45? Since $45 = 17 + 28$, we know that $45 - 17 = 28$. Let's consider an example whose answer is a negative number.

EXAMPLE 1 Subtract: $3 - 7$.

Think: What number can we add to 7 to get 3? The number must be negative. Since $7 + (-4) = 3$, we know the number is -4: $3 - 7 = -4$. That is, $3 - 7 = -4$ because $7 + (-4) = 3$.

Do Exercises 1–3.

The definition above does not provide the most efficient way to do subtraction. From that definition, however, we can develop a faster way to subtract. As a rationale for the faster way, let's compare $3 + 7$ and $3 - 7$ on a number line.

To find $3 + 7$ on a number line, we move 3 units to the right from 0 since 3 is positive. Then we move 7 units farther to the right since 7 is positive.

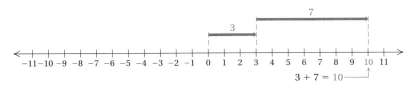

$$3 + 7 = 10$$

To find $3 - 7$, we do the "opposite" of adding 7: We move 7 units to the *left* to do the subtracting. This is the same as *adding* the opposite of 7, -7, to 3.

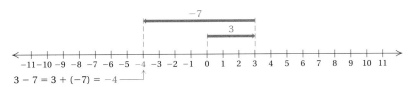

$$3 - 7 = 3 + (-7) = -4$$

Do Exercises 4–6.

Look for a pattern in the following examples.

SUBTRACTIONS	ADDING AN OPPOSITE
$5 - 8 = -3$	$5 + (-8) = -3$
$-6 - 4 = -10$	$-6 + (-4) = -10$
$-7 - (-10) = 3$	$-7 + 10 = 3$
$-7 - (-2) = -5$	$-7 + 2 = -5$

Subtract.

1. $-6 - 4$

Think: What number can be added to 4 to get -6:

$$\square + 4 = -6?$$

2. $-7 - (-10)$

Think: What number can be added to -10 to get -7:

$$\square + (-10) = -7?$$

3. $-7 - (-2)$

Think: What number can be added to -2 to get -7:

$$\square + (-2) = -7?$$

Subtract. Use a number line, doing the "opposite" of addition.

4. $-4 - (-3)$

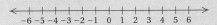

5. $-4 - (-6)$

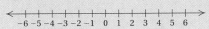

6. $5 - 9$

Answers on page A-25

Complete the addition and compare with the subtraction.

7. $4 - 6 = -2$;
$\quad 4 + (-6) = $ _____

8. $-3 - 8 = -11$;
$\quad -3 + (-8) = $ _____

9. $-5 - (-9) = 4$;
$\quad -5 + 9 = $ _____

10. $-5 - (-3) = -2$;
$\quad -5 + 3 = $ _____

Subtract.

11. $2 - 8$

12. $-6 - 10$

13. $12.4 - 5.3$

14. $-8 - (-11)$

15. $-8 - (-8)$

16. $\dfrac{2}{3} - \left(-\dfrac{5}{6}\right)$

Answers on page A-25

Do Exercises 7–10.

Perhaps you have noticed that we can subtract by adding the opposite of the number being subtracted. This can always be done.

> **SUBTRACTING BY ADDING THE OPPOSITE**
>
> For any real numbers a and b,
> $$a - b = a + (-b).$$
> (To subtract, add the opposite, or additive inverse, of the number being subtracted.)

This is the method generally used for quick subtraction of real numbers.

EXAMPLES Subtract.

2. $2 - 6 = 2 + (-6) = -4$

The opposite of 6 is -6. We change the subtraction to addition and add the opposite. *Check*: $-4 + 6 = 2$.

3. $4 - (-9) = 4 + 9 = 13$

The opposite of -9 is 9. We change the subtraction to addition and add the opposite. *Check*: $13 + (-9) = 4$.

4. $-4.2 - (-3.6) = -4.2 + 3.6 = -0.6$

Adding the opposite. *Check*: $-0.6 + (-3.6) = -4.2$.

5. $-\dfrac{1}{2} - \left(-\dfrac{3}{4}\right) = -\dfrac{1}{2} + \dfrac{3}{4} = \dfrac{1}{4}$

Adding the opposite. *Check*: $\dfrac{1}{4} + \left(-\dfrac{3}{4}\right) = -\dfrac{1}{2}$.

Do Exercises 11–16.

EXAMPLES Read each of the following. Then subtract by adding the opposite of the number being subtracted.

6. $3 - 5$ Read "three minus five is three plus the opposite of five"
$3 - 5 = 3 + (-5) = -2$

7. $\dfrac{1}{8} - \dfrac{7}{8}$ Read "one-eighth minus seven-eighths is one-eighth plus the opposite of seven-eighths"
$\dfrac{1}{8} - \dfrac{7}{8} = \dfrac{1}{8} + \left(-\dfrac{7}{8}\right) = -\dfrac{6}{8}$, or $-\dfrac{3}{4}$

8. $-4.6 - (-9.8)$ Read "negative four point six minus negative nine point eight is negative four point six plus the opposite of negative nine point eight"
$-4.6 - (-9.8) = -4.6 + 9.8 = 5.2$

9. $-\dfrac{3}{4} - \dfrac{7}{5}$ Read "negative three-fourths minus seven-fifths is negative three-fourths plus the opposite of seven-fifths"
$-\dfrac{3}{4} - \dfrac{7}{5} = -\dfrac{3}{4} + \left(-\dfrac{7}{5}\right) = -\dfrac{15}{20} + \left(-\dfrac{28}{20}\right) = -\dfrac{43}{20}$

Do Exercises 17–21 on the following page.

When several additions and subtractions occur together, we can make them all additions.

EXAMPLES Simplify.

10. $8 - (-4) - 2 - (-4) + 2 = 8 + 4 + (-2) + 4 + 2$ Adding the
$$= 16$$ opposite

11. $8.2 - (-6.1) + 2.3 - (-4) = 8.2 + 6.1 + 2.3 + 4 = 20.6$

Do Exercises 22–24.

b Applications and Problem Solving

Let's now see how we can use addition and subtraction of real numbers to solve applied problems.

EXAMPLE 12 *Temperatures on Mars.* The Viking 2 Lander spacecraft has determined that temperatures on Mars range from $-125°C$ (Celsius) to $25°C$. Find the difference between the highest value and the lowest value in this temperature range.
Source: The Lunar and Planetary Institute

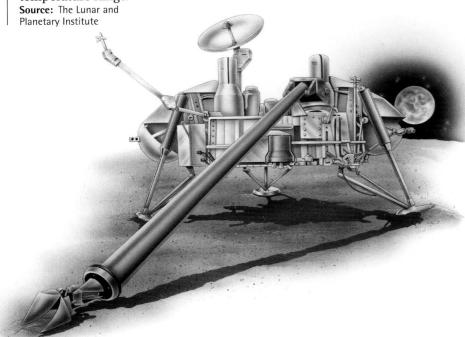

We let $D =$ the difference in the temperatures. Then the problem translates to the following subtraction:

Difference in temperature	is	Highest temperature	minus	Lowest temperature.
D	$=$	25	$-$	(-125)

We then solve the equation: $D = 25 - (-125) = 25 + 125 = 150$.

The difference in the temperatures is $150°C$.

Do Exercise 25.

Read each of the following. Then subtract by adding the opposite of the number being subtracted.

17. $3 - 11$

18. $12 - 5$

19. $-12 - (-9)$

20. $-12.4 - 10.9$

21. $-\dfrac{4}{5} - \left(-\dfrac{4}{5}\right)$

Simplify.
22. $-6 - (-2) - (-4) - 12 + 3$

23. $9 - (-6) + 7 - 11 - 14 - (-20)$

24. $-9.6 + 7.4 - (-3.9) - (-11)$

25. Temperature Extremes.
The highest temperature ever recorded in the United States was 134°F in Greenland Ranch, California, on July 10, 1913. The lowest temperature ever recorded was −80°F in Prospect Creek, Alaska, on January 23, 1971. How much higher was the temperature in Greenland Ranch than the temperature in Prospect Creek?
Source: National Oceanographic and Atmospheric Administration

Answers on page A-25

For Extra Help

Digital Video
Tutor CD 6
Videotape 15

InterAct
Math

Math Tutor
Center

MathXL

MyMathLab

a Subtract.

1. $3 - 7$

2. $5 - 10$

3. $0 - 7$

4. $0 - 8$

5. $-8 - (-2)$

6. $-6 - (-8)$

7. $-10 - (-10)$

8. $-8 - (-8)$

9. $12 - 16$

10. $14 - 19$

11. $20 - 27$

12. $26 - 7$

13. $-9 - (-3)$

14. $-6 - (-9)$

15. $-11 - (-11)$

16. $-14 - (-14)$

17. $8 - (-3)$

18. $-7 - 4$

19. $-6 - 8$

20. $6 - (-10)$

21. $-4 - (-9)$

22. $-14 - 2$

23. $2 - 9$

24. $2 - 8$

25. $0 - 5$

26. $0 - 10$

27. $-5 - (-2)$

28. $-3 - (-1)$

29. $2 - 25$

30. $18 - 63$

31. $-42 - 26$

32. $-18 - 63$

33. $-71 - 2$

34. $-49 - 3$

35. $24 - (-92)$

36. $48 - (-73)$

37. $-2.8 - 0$

38. $6.04 - 1.1$

39. $\dfrac{3}{8} - \dfrac{5}{8}$

40. $\dfrac{3}{9} - \dfrac{9}{9}$

41. $\dfrac{3}{4} - \dfrac{2}{3}$

42. $\dfrac{5}{8} - \dfrac{3}{4}$

43. $-\dfrac{3}{4} - \dfrac{2}{3}$

44. $-\dfrac{5}{8} - \dfrac{3}{4}$

45. $-\dfrac{5}{8} - \left(-\dfrac{3}{4}\right)$

46. $-\dfrac{3}{4} - \left(-\dfrac{2}{3}\right)$

47. $6.1 - (-13.8)$

48. $1.5 - (-3.5)$

49. $-3.2 - 5.8$

50. $-2.7 - 5.9$

51. $0.99 - 1$

52. $0.87 - 1$

53. $3 - 5.7$

54. $5.1 - 3.02$

55. $7 - 10.53$

56. $8 - (-9.3)$

57. $\dfrac{1}{6} - \dfrac{2}{3}$

58. $-\dfrac{3}{8} - \left(-\dfrac{1}{2}\right)$

59. $-\dfrac{4}{7} - \left(-\dfrac{10}{7}\right)$

60. $\dfrac{12}{5} - \dfrac{12}{5}$

61. $-\dfrac{7}{10} - \dfrac{10}{15}$

62. $-\dfrac{4}{18} - \left(-\dfrac{2}{9}\right)$

63. $\dfrac{1}{13} - \dfrac{1}{12}$

64. $-\dfrac{1}{7} - \left(-\dfrac{1}{6}\right)$

Simplify.

65. $18 - (-15) - 3 - (-5) + 2$

66. $22 - (-18) + 7 + (-42) - 27$

67. $-31 + (-28) - (-14) - 17$

68. $-43 - (-19) - (-21) + 25$

69. $-93 - (-84) - 41 - (-56)$

70. $84 + (-99) + 44 - (-18) - 43$

71. $-5 - (-30) + 30 + 40 - (-12)$

72. $14 - (-50) + 20 - (-32)$

73. $132 - (-21) + 45 - (-21)$

74. $81 - (-20) - 14 - (-50) + 53$

 Solve.

75. *Lake Level.* In the course of one four-month period, the water level of Lake Champlain went down 2 ft, up 1 ft, down 5 ft, and up 3 ft. How much had the lake level changed at the end of the four months?

76. *Credit Card Bills.* On August 1, Lyle's credit card bill shows that he owes $470. During August, he sends a check to the credit card company for $45, charges another $160 in merchandise, and then pays off another $500 of his bill. What is the new balance of Lyle's account at the end of August?

77. *Temperature Changes.* One day the temperature in Lawrence, Kansas, is 32° at 6:00 A.M. It rises 15° by noon, but falls 50° by midnight when a cold front moves in. What is the final temperature?

78. *Stock Price Changes.* On a recent day, the stock of Quaker Oats started at a value of $61.38. It rose $4.75, dropped $7.38, and rose $5.13. Find the value of the stock at the end of the day.
Source: The New York Stock Exchange

79. *Changes in Elevation.* The lowest elevation in Asia, the Dead Sea, is 1286 ft below sea level. The highest elevation in Asia, Mount Everest, is 29,028 ft. Find the difference in elevation between the highest point and the lowest point.

80. *Tallest Mountain.* The tallest mountain in the world, when measured from base to peak, is Mauna Kea (White Mountain) in Hawaii. From its base 19,684 ft below sea level in the Hawaiian Trough, it rises 33,480 ft. What is the elevation of the peak?
Source: The Guinness Book of Records

81. *Account Balance.* Leah has $460 in her checking account. She writes a check for $530, makes a deposit of $75, and then writes a check for $90. What is the balance in the account?

82. *Cell-Phone Bill.* Erika's cell-phone bill for July was $82. She made a payment of $50 and then made $37 worth of calls in August. How much did she then owe on her cell-phone bill?

83. *Temperature Records.* The greatest recorded temperature change in one day occurred in Browning, Montana, where the temperature fell from 44°F to -56°F. How much did the temperature drop that day?
Source: *The Guinness Book of Records*

84. *Low Points on Continents.* The lowest point in Africa is Lake Assal, which is 515 ft below sea level. The lowest point in South America is the Valdes Peninsula, which is 132 ft below sea level. How much lower is Lake Assal than the Valdes Peninsula?
Source: National Geographic Society

85. D_W If a negative number is subtracted from a positive number, will the result always be positive? Why or why not?

86. D_W Write a problem for a classmate to solve. Design the problem so that the solution is "The temperature dropped to $-9°$."

87. Find the area of a rectangle that is 8.4 cm by 11.5 cm. [9.2a]

88. Find the prime factorization of 750. [2.1d]

89. Find the LCM of 36 and 54. [3.1a]

90. Find the area of a square whose sides are of length 11.2 km. [9.2a]

Evaluate. [1.9b]

91. 4^3

92. 5^3

Solve. [1.8a]

93. How many 12-oz cans of soda can be filled with 96 oz of soda?

94. A case of soda contains 24 bottles. If each bottle contains 12 oz, how many ounces of soda are in the case?

Tell whether the statement is true or false for all integers m and n. If false, find a number that shows why.

95. $-n = 0 - n$

96. $n - 0 = 0 - n$

97. If $m \neq n$, then $m - n \neq 0$.

98. If $m = -n$, then $m + n = 0$.

99. If $m + n = 0$, then m and n are opposites.

100. If $m - n = 0$, then $m = -n$.

101. $m = -n$ if m and n are opposites.

102. If $m = -m$, then $m = 0$.

10.4

MULTIPLICATION OF REAL NUMBERS

a Multiplication

Multiplication of real numbers is very much like multiplication of arithmetic numbers. The only difference is that we must determine whether the answer is positive or negative.

MULTIPLICATION OF A POSITIVE NUMBER AND A NEGATIVE NUMBER

To see how to multiply a positive number and a negative number, consider the pattern of the following.

This number decreases by 1 each time.
$$
\begin{array}{rcr}
4 \cdot 5 = & 20 \\
3 \cdot 5 = & 15 \\
2 \cdot 5 = & 10 \\
1 \cdot 5 = & 5 \\
0 \cdot 5 = & 0 \\
-1 \cdot 5 = & -5 \\
-2 \cdot 5 = & -10 \\
-3 \cdot 5 = & -15
\end{array}
$$
This number decreases by 5 each time.

Do Exercise 1.

According to this pattern, it looks as though the product of a negative number and a positive number is negative. That is the case, and we have the first part of the rule for multiplying real numbers.

> **THE PRODUCT OF A POSITIVE NUMBER AND A NEGATIVE NUMBER**
>
> To multiply a positive number and a negative number, multiply their absolute values. The answer is negative.

EXAMPLES Multiply.

1. $8(-5) = -40$ **2.** $-\dfrac{1}{3} \cdot \dfrac{5}{7} = -\dfrac{5}{21}$ **3.** $(-7.2)5 = -36$

Do Exercises 2–7.

MULTIPLICATION OF TWO NEGATIVE NUMBERS

How do we multiply two negative numbers? Again we look for a pattern.

This number decreases by 1 each time.
$$
\begin{array}{rcr}
4 \cdot (-5) = & -20 \\
3 \cdot (-5) = & -15 \\
2 \cdot (-5) = & -10 \\
1 \cdot (-5) = & -5 \\
0 \cdot (-5) = & 0 \\
-1 \cdot (-5) = & 5 \\
-2 \cdot (-5) = & 10 \\
-3 \cdot (-5) = & 15
\end{array}
$$
This number increases by 5 each time.

Do Exercise 8.

Objective

| a | Multiply real numbers. |

1. Complete, as in the example.
$$
\begin{array}{rcl}
4 \cdot 10 & = & 40 \\
3 \cdot 10 & = & 30 \\
2 \cdot 10 & = & \\
1 \cdot 10 & = & \\
0 \cdot 10 & = & \\
-1 \cdot 10 & = & \\
-2 \cdot 10 & = & \\
-3 \cdot 10 & = &
\end{array}
$$

Multiply.

2. $-3 \cdot 6$

3. $20 \cdot (-5)$

4. $4 \cdot (-20)$

5. $-\dfrac{2}{3} \cdot \dfrac{5}{6}$

6. $-4.23(7.1)$

7. $\dfrac{7}{8}\left(-\dfrac{4}{5}\right)$

8. Complete, as in the example.
$$
\begin{array}{rcl}
3 \cdot (-10) & = & -30 \\
2 \cdot (-10) & = & -20 \\
1 \cdot (-10) & = & \\
0 \cdot (-10) & = & \\
-1 \cdot (-10) & = & \\
-2 \cdot (-10) & = & \\
-3 \cdot (-10) & = &
\end{array}
$$

Answers on page A-26

Multiply.

9. $-3 \cdot (-4)$

10. $-16 \cdot (-2)$

11. $-7 \cdot (-5)$

12. $-\frac{4}{7}\left(-\frac{5}{9}\right)$

13. $-\frac{3}{2}\left(-\frac{4}{9}\right)$

14. $-3.25(-4.14)$

Multiply.

15. $5(-6)$

16. $(-5)(-6)$

17. $(-3.2) \cdot 10$

18. $\left(-\frac{4}{5}\right)\left(\frac{10}{3}\right)$

Multiply.

19. $5 \cdot (-3) \cdot 2$

20. $-3 \times (-4.1) \times (-2.5)$

21. $-\frac{1}{2} \cdot \left(-\frac{4}{3}\right) \cdot \left(-\frac{5}{2}\right)$

22. $-2 \cdot (-5) \cdot (-4) \cdot (-3)$

23. $(-4)(-5)(-2)(-3)(-1)$

24. $(-1)(-1)(-2)(-3)(-1)(-1)$

Answers on page A-26

According to the pattern, it looks as though the product of two negative numbers is positive. That is actually so, and we have the second part of the rule for multiplying real numbers.

> ### THE PRODUCT OF TWO NEGATIVE NUMBERS
>
> To multiply two negative numbers, multiply their absolute values. The answer is positive.

Do Exercises 9–14.

> To multiply two nonzero real numbers:
>
> a) Multiply the absolute values.
> b) If the signs are the same, the answer is positive.
> c) If the signs are different, the answer is negative.

EXAMPLES Multiply.

4. $(-3)(-4) = 12$ **5.** $-1.6(2) = -3.2$ **6.** $\left(-\frac{5}{6}\right)\left(-\frac{1}{9}\right) = \frac{5}{54}$

Do Exercises 15–18.

MULTIPLYING MORE THAN TWO NUMBERS

When multiplying more than two real numbers, we can choose order and grouping as we please, using the commutative and associative laws.

EXAMPLES Multiply.

7. $-8 \cdot 2(-3) = -16(-3)$ Multiplying the first two numbers
$= 48$ Multiplying the results

8. $-8 \cdot 2(-3) = 24 \cdot 2$ Multiplying the negatives. Every
$= 48$ pair of negative numbers gives a positive product.

9. $-3(-2)(-5)(4) = 6(-5)(4)$ Multiplying the first two numbers
$= (-30)4 = -120$

10. $\left(-\frac{1}{2}\right)(8)\left(-\frac{2}{3}\right)(-6) = (-4)4$ Multiplying the first two numbers
$= -16$ and the last two numbers

11. $-5 \cdot (-2) \cdot (-3) \cdot (-6) = 10 \cdot 18 = 180$

12. $(-3)(-5)(-2)(-3)(-6) = (-30)(18) = -540$

Considering that the product of a pair of negative numbers is positive, we can see the following pattern in the results of Examples 11 and 12.

> The product of an even number of negative numbers is positive.
> The product of an odd number of negative numbers is negative.

Do Exercises 19–24.

a   Multiply.

1. $-8 \cdot 2$

2. $-3 \cdot 5$

3. $8 \cdot (-3)$

4. $-5 \cdot 2$

5. $-9 \cdot 8$

6. $-20 \cdot 3$

7. $-8 \cdot (-2)$

8. $-4 \cdot (-5)$

9. $-7 \cdot (-6)$

10. $-9 \cdot (-2)$

11. $15 \cdot (-8)$

12. $-11 \cdot (-10)$

13. $-14 \cdot 17$

14. $-13 \cdot (-15)$

15. $-25 \cdot (-48)$

16. $39 \cdot (-43)$

17. $-3.5 \cdot (-28)$

18. $97 \cdot (-2.1)$

19. $4 \cdot (-3.1)$

20. $3 \cdot (-2.2)$

21. $-6 \cdot (-4)$

22. $-5 \cdot (-6)$

23. $-7 \cdot (-3.1)$

24. $-4 \cdot (-3.2)$

25. $\dfrac{2}{3} \cdot \left(-\dfrac{3}{5}\right)$

26. $\dfrac{5}{7} \cdot \left(-\dfrac{2}{3}\right)$

27. $-\dfrac{3}{8} \cdot \left(-\dfrac{2}{9}\right)$

28. $-\dfrac{5}{8} \cdot \left(-\dfrac{2}{5}\right)$

29. -6.3×2.7

30. -6.2×8.5

31. $-\dfrac{5}{9} \cdot \dfrac{3}{4}$

32. $-\dfrac{8}{3} \cdot \dfrac{9}{4}$

33. $7 \cdot (-4) \cdot (-3) \cdot 5$

34. $9 \cdot (-2) \cdot (-6) \cdot 7$

35. $-\dfrac{2}{3} \cdot \dfrac{1}{2} \cdot \left(-\dfrac{6}{7}\right)$

36. $-\dfrac{1}{8} \cdot \left(-\dfrac{1}{4}\right) \cdot \left(-\dfrac{3}{5}\right)$

37. $-3 \cdot (-4) \cdot (-5)$

38. $-2 \cdot (-5) \cdot (-7)$

39. $-2 \cdot (-5) \cdot (-3) \cdot (-5)$

40. $-3 \cdot (-5) \cdot (-2) \cdot (-1)$

41. $\dfrac{1}{5}\left(-\dfrac{2}{9}\right)$

42. $-\dfrac{3}{5}\left(-\dfrac{2}{7}\right)$

43. $-7 \cdot (-21) \cdot 13$

44. $-14 \cdot 34 \cdot 12$

45. $-4 \cdot (-1.8) \cdot 7$

46. $-8 \cdot (-1.3) \cdot (-5)$

47. $-\dfrac{1}{9}\left(-\dfrac{2}{3}\right)\left(\dfrac{5}{7}\right)$

48. $-\dfrac{7}{2}\left(-\dfrac{5}{7}\right)\left(-\dfrac{2}{5}\right)$

49. $4 \cdot (-4) \cdot (-5) \cdot (-12)$

50. $-2 \cdot (-3) \cdot (-4) \cdot (-5)$

51. $0.07 \cdot (-7) \cdot 6 \cdot (-6)$

52. $80 \cdot (-0.8) \cdot (-90) \cdot (-0.09)$

53. $\left(-\dfrac{5}{6}\right)\left(\dfrac{1}{8}\right)\left(-\dfrac{3}{7}\right)\left(-\dfrac{1}{7}\right)$

54. $\left(\dfrac{4}{5}\right)\left(-\dfrac{2}{3}\right)\left(-\dfrac{15}{7}\right)\left(\dfrac{1}{2}\right)$

55. $(-14) \cdot (-27) \cdot (-2)$

56. $7 \cdot (-6) \cdot 5 \cdot (-4) \cdot 3 \cdot (-2) \cdot 1 \cdot (-1)$

57. $(-8)(-9)(-10)$

58. $(-7)(-8)(-9)(-10)$

59. $(-6)(-7)(-8)(-9)(-10)$

60. $(-5)(-6)(-7)(-8)(-9)(-10)$

61. $\mathbf{D_W}$ What rule have we developed that would tell you the sign of $(-7)^8$ and $(-7)^{11}$ without doing the computations? Explain.

62. $\mathbf{D_W}$ Which number is larger, $(-3)^{79}$ or $(-5)^{79}$? Why?

SKILL MAINTENANCE

63. Find the prime factorization of 4608.   [2.1d]

64. Find the LCM of 36 and 60.   [3.1a]

65. 23 is what percent of 69?   [6.3b], [6.4b]

66. What is 36% of 729?   [6.3b], [6.4b]

67. 40% of what number is 28.8?   [6.3b], [6.4b]

68. What percent of 72 is 28.8?   [6.3b], [6.4b]

Solve.

69. A rectangular rug measures 5 ft by 8 ft. What is the area of the rug?   [9.2c]

70. How many 12-egg cartons can be filled with 2880 eggs?   [1.8a]

SYNTHESIS

71. Jo wrote seven checks for $13 each. If she had a balance of $68 in her account before writing the checks, what was her balance after writing the checks?

72. After diving 62 m below the surface, a diver rises at a rate of 6 meters per minute for 7 min. What is the diver's new elevation?

73. What must be true of a and b if $-ab$ is to be (a) positive? (b) zero? (c) negative?

638

10.5 DIVISION OF REAL NUMBERS AND ORDER OF OPERATIONS

Objectives

a Divide integers.

b Find the reciprocal of a real number.

c Divide real numbers.

d Solve applied problems involving multiplication and division of real numbers.

e Simplify expressions using rules for order of operations.

We now consider division of real numbers. The definition of division results in rules for division that are the same as those for multiplication.

a Division of Integers

DIVISION

The quotient $a \div b$, or $\frac{a}{b}$, where $b \neq 0$, is that unique real number c for which $a = b \cdot c$.

EXAMPLES Divide, if possible. Check your answer.

1. $14 \div (-7) = -2$ *Think*: What number multiplied by -7 gives 14? That number is -2. *Check*: $(-2)(-7) = 14$.

2. $\dfrac{-32}{-4} = 8$ *Think*: What number multiplied by -4 gives -32? That number is 8. *Check*: $8(-4) = -32$.

3. $\dfrac{-10}{7} = -\dfrac{10}{7}$ *Think*: What number multiplied by 7 gives -10? That number is $-\frac{10}{7}$. *Check*: $-\frac{10}{7} \cdot 7 = -10$.

4. $\dfrac{-17}{0}$ is **not defined.**   *Think*: What number multiplied by 0 gives -17? There is no such number because the product of 0 and *any* number is 0.

Do Exercises 1–3.

The sign rules for division are the same as those for multiplication. We state them together.

> To multiply or divide two real numbers (where the divisor is nonzero):
> **a)** Multiply or divide the absolute values.
> **b)** If the signs are the same, the answer is positive.
> **c)** If the signs are different, the answer is negative.

Do Exercises 4–6.

EXCLUDING DIVISION BY ZERO

Example 4 shows why we cannot divide -17 by 0. We can use the same argument to show why we cannot divide any nonzero number b by 0. Consider $b \div 0$. We look for a number that when multiplied by 0 gives b. There is no such number because the product of 0 and any number is 0. Thus we cannot divide a nonzero number b by 0.

On the other hand, if we divide 0 by 0, we look for a number c such that $0 \cdot c = 0$. But $0 \cdot c = 0$ for any number c. Thus it appears that $0 \div 0$ could be any number we choose. Getting any answer we want when we divide 0 by 0 would be very confusing. Thus we agree that division by zero is not defined.

Divide.

1. $6 \div (-3)$

Think: What number multiplied by -3 gives 6?

2. $\dfrac{-15}{-3}$

Think: What number multiplied by -3 gives -15?

3. $-24 \div 8$

Think: What number multiplied by 8 gives -24?

4. $\dfrac{-72}{-8}$

5. $\dfrac{30}{-5}$

6. $\dfrac{30}{-7}$

Answers on page A-26

639

Divide, if possible.

7. $\dfrac{-5}{0}$

8. $\dfrac{0}{-3}$

Find the reciprocal.

9. $\dfrac{2}{3}$

10. $-\dfrac{5}{4}$

11. -3

12. $-\dfrac{1}{5}$

13. 5.78

14. $-\dfrac{2}{7}$

Answers on page A-26

> ### EXCLUDING DIVISION BY 0
>
> Division by 0 is not defined:
>
> $$a \div 0, \text{ or } \frac{a}{0}, \text{ is not defined for all real numbers } a.$$

DIVIDING 0 BY OTHER NUMBERS
Note that $0 \div 8 = 0$ because $0 = 0 \cdot 8$.

> ### DIVIDENDS OF 0
>
> Zero divided by any nonzero real number is 0:
>
> $$\frac{0}{a} = 0, \qquad a \neq 0.$$

EXAMPLES Divide.

5. $0 \div (-6) = 0$ **6.** $\dfrac{0}{12} = 0$ **7.** $\dfrac{-3}{0}$ is not defined.

Do Exercises 7 and 8.

b   Reciprocals

When two numbers like $\frac{7}{8}$ and $\frac{8}{7}$ are multiplied, the result is 1. Such numbers are called **reciprocals** of each other. Every nonzero real number has a reciprocal, also called a **multiplicative inverse.**

> ### RECIPROCALS
>
> Two numbers whose product is 1 are called **reciprocals** of each other.

EXAMPLES Find the reciprocal.

8. -5 The reciprocal of -5 is $-\frac{1}{5}$ because $-5\left(-\frac{1}{5}\right) = 1$.

9. $-\frac{1}{2}$ The reciprocal of $-\frac{1}{2}$ is -2 because $\left(-\frac{1}{2}\right)(-2) = 1$.

10. $-\frac{2}{3}$ The reciprocal of $-\frac{2}{3}$ is $-\frac{3}{2}$ because $\left(-\frac{2}{3}\right)\left(-\frac{3}{2}\right) = 1$.

> ### PROPERTIES OF RECIPROCALS
>
> For $a \neq 0$, the reciprocal of a can be named $\dfrac{1}{a}$ and the reciprocal of $\dfrac{1}{a}$ is a.
>
> The reciprocal of a nonzero number $\dfrac{a}{b}$ can be named $\dfrac{b}{a}$.
>
> The number 0 has no reciprocal.

Do Exercises 9–14.

The reciprocal of a positive number is also a positive number, because their product must be the positive number 1. The reciprocal of a negative number is also a negative number, because their product must be the positive number 1.

THE SIGN OF A RECIPROCAL

The reciprocal of a number has the same sign as the number itself.

It is important not to confuse *opposite* with *reciprocal*. Keep in mind that the opposite, or additive inverse, of a number is what we add to the number to get 0. A reciprocal, or multiplicative inverse, is what we multiply the number by to get 1. Compare the following.

NUMBER	OPPOSITE (Change the sign.)	RECIPROCAL (Invert but do not change the sign.)
$-\dfrac{3}{8}$	$\dfrac{3}{8}$	$-\dfrac{8}{3}$
19	-19	$\dfrac{1}{19}$
$\dfrac{18}{7}$	$-\dfrac{18}{7}$	$\dfrac{7}{18}$
-7.9	7.9	$-\dfrac{1}{7.9}$ or $-\dfrac{10}{79}$
0	0	Undefined

$$\left(-\frac{3}{8}\right)\left(-\frac{8}{3}\right) = 1$$

$$-\frac{3}{8} + \frac{3}{8} = 0$$

Do Exercise 15.

C Division of Real Numbers

We know that we can subtract by adding an opposite. Similarly, we can divide by multiplying by a reciprocal.

RECIPROCALS AND DIVISION

For any real numbers a and b, $b \neq 0$,

$$a \div b = \frac{a}{b} = a \cdot \frac{1}{b}.$$

(To divide, multiply by the reciprocal of the divisor.)

EXAMPLES Rewrite the division as a multiplication.

11. $-4 \div 3$ $-4 \div 3$ is the same as $-4 \cdot \dfrac{1}{3}$

12. $\dfrac{6}{-7}$ $\dfrac{6}{-7} = 6\left(-\dfrac{1}{7}\right)$

13. $\dfrac{3}{5} \div \left(-\dfrac{9}{7}\right)$ $\dfrac{3}{5} \div \left(-\dfrac{9}{7}\right) = \dfrac{3}{5}\left(-\dfrac{7}{9}\right)$

---

15. Complete the following table.

NUMBER	OPPOSITE	RECIPROCAL
$\dfrac{2}{3}$		
$-\dfrac{5}{4}$		
0		
1		
-4.5		

Rewrite the division as a multiplication.

16. $\dfrac{4}{7} \div \left(-\dfrac{3}{5}\right)$

17. $\dfrac{5}{-8}$

18. $\dfrac{-10}{7}$

19. $-\dfrac{2}{3} \div \dfrac{4}{7}$

20. $-5 \div 7$

Answers on page A-26

641

Divide by multiplying by the reciprocal of the divisor.

21. $\dfrac{4}{7} \div \left(-\dfrac{3}{5}\right)$

22. $-\dfrac{8}{5} \div \dfrac{2}{3}$

23. $-\dfrac{12}{7} \div \left(-\dfrac{3}{4}\right)$

24. Divide: $21.7 \div (-3.1)$.

Answers on page A-26

Do Exercises 16–20 on the preceding page.

When actually doing division calculations, we sometimes multiply by a reciprocal and we sometimes divide directly. With fraction notation, it is generally better to multiply by a reciprocal. With decimal notation, it is generally better to divide directly.

EXAMPLES Divide by multiplying by the reciprocal of the divisor.

14. $\dfrac{2}{3} \div \left(-\dfrac{5}{4}\right) = \dfrac{2}{3} \cdot \left(-\dfrac{4}{5}\right) = -\dfrac{8}{15}$

15. $-\dfrac{5}{6} \div \left(-\dfrac{3}{4}\right) = -\dfrac{5}{6} \cdot \left(-\dfrac{4}{3}\right) = \dfrac{20}{18} = \dfrac{10 \cdot 2}{9 \cdot 2} = \dfrac{10}{9} \cdot \dfrac{2}{2} = \dfrac{10}{9}$

> **CAUTION!** Be careful not to change the sign when taking a reciprocal!

16. $-\dfrac{3}{4} \div \dfrac{3}{10} = -\dfrac{3}{4} \cdot \left(\dfrac{10}{3}\right) = -\dfrac{30}{12} = -\dfrac{5}{2} \cdot \dfrac{6}{6} = -\dfrac{5}{2}$

With decimal notation, it is easier to carry out long division than to multiply by the reciprocal.

EXAMPLES Divide.

17. $-27.9 \div (-3) = \dfrac{-27.9}{-3} = 9.3$ Do the long division $3\overline{)27.9}$ → 9.3
The answer is positive.

18. $-6.3 \div 2.1 = -3$ Do the long division $2.1\overline{)6.3_{\wedge}0}$ → 3.0
The answer is negative.

Do Exercises 21–24.

d   Applications and Problem Solving

We can use multiplication and division to solve applied problems.

EXAMPLE 19 *Chemical Reaction.* During a chemical reaction, the temperature in the beaker decreased every minute by the same number of degrees. The temperature was 56°F at 10:10 A.M. By 10:42 A.M., the temperature had dropped to −12°F. By how many degrees did it change each minute?

We first determine by how many degrees d the temperature changed altogether. We subtract -12 from 56:

$$d = 56 - (-12) = 56 + 12 = 68.$$

The temperature changed a total of 68°. We can express this as $-68°$ since the temperature dropped.

The amount of time t that passed was $42 - 10$, or 32 min. Thus the number of degrees T that the temperature dropped each minute is given by

$$T = \frac{d}{t} = \frac{-68}{32} = -2.125.$$

The change was $-2.125°F$ per minute.

Do Exercise 25.

e ▌ Order of Operations

When several operations are to be done in a calculation or a problem, we apply the same rules that we did in Sections 1.9, 3.7, and 4.4. We repeat them here for review. If you did not study those sections before, you should do so before continuing.

> **RULES FOR ORDER OF OPERATIONS**
>
> 1. Do all calculations within grouping symbols before operations outside.
> 2. Evaluate all exponential expressions.
> 3. Do all multiplications and divisions in order from left to right.
> 4. Do all additions and subtractions in order from left to right.

EXAMPLE 20 Simplify: $-34 \cdot 56 - 17$.

There are no parentheses or powers so we start with the third step.

$$-34 \cdot 56 - 17 = -1904 - 17 \qquad \text{Carrying out all multiplications and divisions in order from left to right}$$

$$= -1921 \qquad \text{Carrying out all additions and subtractions in order from left to right}$$

EXAMPLE 21 Simplify: $2^4 + 51 \cdot 4 - (37 + 23 \cdot 2)$.

$$2^4 + 51 \cdot 4 - (37 + 23 \cdot 2)$$

$$= 2^4 + 51 \cdot 4 - (37 + 46) \qquad \text{Carrying out all operations inside parentheses first, multiplying 23 by 2, following the rules for order of operations within the parentheses}$$

$$= 2^4 + 51 \cdot 4 - 83 \qquad \text{Completing the addition inside parentheses}$$

$$= 16 + 51 \cdot 4 - 83 \qquad \text{Evaluating exponential expressions}$$

$$= 16 + 204 - 83 \qquad \text{Doing all multiplications}$$

$$= 220 - 83 \qquad \text{Doing all additions and subtractions in order from left to right}$$

$$= 137$$

25. Chemical Reaction. During a chemical reaction, the temperature in the beaker decreased every minute by the same number of degrees. The temperature was 71°F at 2:12 P.M. By 2:37 P.M., the temperature had changed to $-14°F$. By how many degrees did it change each minute?

Answer on page A-26

Simplify.

26. $23 - 42 \cdot 30$

27. $32 \div 8 \cdot 2$

28. $52 \cdot 5 + 5^3 - (4^2 - 48 \div 4)$

29. $\dfrac{5 - 10 - 5 \cdot 23}{2^3 + 3^2 - 7}$

Answers on page A-26

A fraction bar can play the role of a grouping symbol.

EXAMPLE 22   Simplify: $\dfrac{-64 \div (-16) \div (-2)}{2^3 - 3^2}$.

An equivalent expression with brackets as grouping symbols is

$$[-64 \div (-16) \div (-2)] \div [2^3 - 3^2].$$

This shows, in effect, that we can do the calculations in the numerator and then in the denominator, and divide the results:

$$\frac{-64 \div (-16) \div (-2)}{2^3 - 3^2} = \frac{4 \div (-2)}{8 - 9} = \frac{-2}{-1} = 2.$$

Do Exercises 26–29.

Study Tips

STUDYING FOR THE FINAL EXAM (PART 2)

The best scenario for preparing for a final exam is to do so over a period of at least two weeks. Work in a diligent, disciplined manner, doing some final-exam preparation *each* day. Here is a detailed plan that many find useful.

1. **Begin by browsing through each chapter, reviewing the highlighted or boxed information regarding important formulas in both the text and the Summary and Review.** There may be some formulas that you will need to memorize.
2. **Retake each chapter test that you took in class, assuming your instructor has returned it. Otherwise, use the chapter test in the book.** Restudy the objectives in the text that correspond to each question you missed.
3. **Then work the Cumulative Review that covers all chapters up to that point.** Be careful to avoid any questions corresponding to objectives not covered. Again, restudy the objectives in the text that correspond to each question you missed.
4. **If you are still missing questions, use the supplements for extra review.** For example, you might check out the video- or audiotapes, the *Student's Solutions Manual*, the InterAct Math Tutorial Software, or InterAct MathXL.
5. **For remaining difficulties, see your instructor, go to a tutoring session, or participate in a study group.**
6. **Check for former final exams that may be on file in the math department or a study center, or with students who have already taken the course.** Use them for practice, being alert to trouble spots.
7. **Take the Final Examination in the text during the last couple of days before the final.** Set aside the same amount of time that you will have for the final. See how much of the final exam you can complete under test-like conditions.

"Without some goal and effort to reach it, no man can live."

Fyodor Dostoyevsky, nineteenth-century Russian novelist

CHAPTER 10: Real Numbers

a   Divide, if possible. Check each answer.

1. $36 \div (-6)$

2. $\dfrac{42}{-7}$

3. $\dfrac{26}{-2}$

4. $24 \div (-12)$

5. $\dfrac{-16}{8}$

6. $-18 \div (-2)$

7. $\dfrac{-48}{-12}$

8. $-72 \div (-9)$

9. $\dfrac{-72}{9}$

10. $\dfrac{-50}{25}$

11. $-100 \div (-50)$

12. $\dfrac{-200}{8}$

13. $-108 \div 9$

14. $\dfrac{-64}{-7}$

15. $\dfrac{200}{-25}$

16. $-300 \div (-13)$

17. $\dfrac{75}{0}$

18. $\dfrac{0}{-5}$

19. $\dfrac{81}{-9}$

20. $\dfrac{-145}{-5}$

b   Find the reciprocal.

21. $-\dfrac{15}{7}$

22. $-\dfrac{5}{8}$

23. 13

24. -8

c   Divide.

25. $\dfrac{3}{4} \div \left(-\dfrac{2}{3}\right)$

26. $\dfrac{7}{8} \div \left(-\dfrac{1}{2}\right)$

27. $-\dfrac{5}{4} \div \left(-\dfrac{3}{4}\right)$

28. $-\dfrac{5}{9} \div \left(-\dfrac{5}{6}\right)$

29. $-\dfrac{2}{7} \div \left(-\dfrac{4}{9}\right)$

30. $-\dfrac{3}{5} \div \left(-\dfrac{5}{8}\right)$

31. $-\dfrac{3}{8} \div \left(-\dfrac{8}{3}\right)$

32. $-\dfrac{5}{8} \div \left(-\dfrac{6}{5}\right)$

33. $-6.6 \div 3.3$

34. $-44.1 \div (-6.3)$

35. $\dfrac{-11}{-13}$

36. $\dfrac{-1.7}{20}$

37. $\dfrac{48.6}{-3}$

38. $\dfrac{-17.8}{3.2}$

39. $\dfrac{-9}{17 - 17}$

40. $\dfrac{-8}{-5 + 5}$

d   Solve.

41. *Chemical Reaction.* The temperature of a chemical compound was at 0°C at 11:00 A.M. During a reaction, it dropped 3°C per minute until 11:18 A.M. What was the temperature at 11:18 A.M.?

42. *Chemical Reaction.* The temperature in a chemical compound was −5°C at 3:20 P.M. During a reaction, it increased 2°C per minute until 3:52 P.M. What was the temperature at 3:52 P.M.?

43. *Stock Price.* The price of ePDQ.com began the day at $23.75 per share and dropped $1.38 per hour for 8 hr. What was the price of the stock after 8 hr?

44. *Population Decrease.* The population of a rural town was 12,500. It decreased 380 each year for 4 yr. What was the population of the town after 4 yr?

45. *Diver's Position.* After diving 95 m below sea level, a diver rises at a rate of 7 meters per minute for 9 min. Where is the diver in relation to the surface?

46. *Debit Card Balance.* Karen had $234 in her checking account. After using her debit card to make seven purchases at $39 each, what was the balance in her checking account?

Percent of Increase or Decrease in Employment. A percent of increase is usually considered positive and a percent of decrease is considered negative. The following table lists estimates of the number of job opportunities for various occupations in 1998 and 2008. In Exercises 47–50, find the missing numbers.

	OCCUPATION	NUMBER OF JOBS IN 1998 (in thousands)	NUMBER OF JOBS IN 2008 (in thousands)	CHANGE	PERCENT OF INCREASE OR DECREASE
	Court clerk	100	112	12	12%
	Bank teller	560	529	−31	−5.5%
47.	Barber	54	50	−4	
48.	Child-care worker in private household	306	209	−97	
49.	Dental assistant	229	326	97	
50.	Cook (short-order and fast-food)	677	801	124	

Source: Handbook of U.S. Labor Statistics

e Simplify.

51. $8 - 2 \cdot 3 - 9$

52. $8 - (2 \cdot 3 - 9)$

53. $(8 - 2 \cdot 3) - 9$

54. $(8 - 2)(3 - 9)$

55. $16 \cdot (-24) + 50$

56. $10 \cdot 20 - 15 \cdot 24$

57. $2^4 + 2^3 - 10$

58. $40 - 3^2 - 2^3$

59. $5^3 + 26 \cdot 71 - (16 + 25 \cdot 3)$

60. $4^3 + 10 \cdot 20 + 8^2 - 23$

61. $4 \cdot 5 - 2 \cdot 6 + 4$

62. $4 \cdot (6 + 8)/(4 + 3)$

63. $4^3/8$

64. $5^3 - 7^2$

65. $8(-7) + 6(-5)$

66. $10(-5) + 1(-1)$

67. $19 - 5(-3) + 3$

68. $14 - 2(-6) + 7$

69. $9 \div (-3) + 16 \div 8$

70. $-32 - 8 \div 4 - (-2)$

71. $-4^2 + 6$

72. $-5^2 + 7$

73. $-8^2 - 3$

74. $-9^2 - 11$

75. $12 - 20^3$

76. $20 + 4^3 \div (-8)$

77. $2 \times 10^3 - 5000$

78. $-7(3^4) + 18$

79. $6[9 - (3 - 4)]$

80. $8[(6 - 13) - 11]$

81. $-1000 \div (-100) \div 10$

82. $256 \div (-32) \div (-4)$

83. $8 - (7 - 9)$

84. $(8 - 7) - 9$

85. $\dfrac{10 - 6^2}{9^2 + 3^2}$

86. $\dfrac{5^2 - 4^3 - 3}{9^2 - 2^2 - 1^5}$

87. $\dfrac{20(8 - 3) - 4(10 - 3)}{10(2 - 6) - 2(5 + 2)}$

88. $\dfrac{(3 - 5)^2 - (7 - 13)}{(12 - 9)^2 + (11 - 14)^2}$

89. ⌨ ^{D}W Jake keys 18/2 · 3 into his calculator and expects the result to be 3. What mistake is he making?

90. ^{D}W Explain how multiplication can be used to justify why the quotient of two negative integers is positive.

SKILL MAINTENANCE

Find the prime factorization. [2.1d]

91. 960

92. 1025

Find the LCM. [3.1a]

93. 5, 19, 35

94. 20, 40, 64

Solve. [6.3b], [6.4b]

95. What is 45% of 3800?

96. 344 is what percent of 8600?

97. *Sound Levels.* Use the information in the following table to make a horizontal bar graph illustrating the loudness of various sounds. (A decibel is a measure of the loudness of sound.) [7.3b]

SOUND	LOUDNESS (in decibels)
Light whisper	10
Normal conversation	30
Noisy office	60
Thunder	90
Moving car	80
Chain saw	100
Jet takeoff	140

Source: Barbara Ann Kipfer, *The Order of Things.*
New York: Random House, 1998

SYNTHESIS

Simplify.

98. ⌨ $\dfrac{19 - 17^2}{13^2 - 34}$

99. ⌨ $\dfrac{195 + (-15)^3}{195 - 7 \cdot 5^2}$

Determine the sign of the expression if m is negative and n is positive.

100. $\dfrac{-n}{m}$

101. $\dfrac{-n}{-m}$

102. $-\left(\dfrac{-n}{m}\right)$

103. $-\left(\dfrac{n}{-m}\right)$

104. $-\left(\dfrac{-n}{-m}\right)$

The review that follows is meant to prepare you for a chapter exam. It consists of two parts. The first part is a checklist of some of the Study Tips referred to in this and preceding chapters. The second part is the Review Exercises. These provide practice exercises for the exam, together with references to section objectives, so you can go back and review. Before beginning, stop and look back over the skills you have obtained. What skills in mathematics do you have now that you did not have before studying this chapter?

STUDY TIPS CHECKLIST

The foundation of all your study skills is TIME!	☐ Are you trying to approach your mathematics studies with an assertive, positive attitude?
	☐ Are you making use of the textbook supplements, such as the Math Tutor Center, the *Student's Solutions Manual*, and the videotapes?
	☐ Have you visited the learning resource centers on your campus, such as a math lab, tutor center, and your instructor's office?
	☐ Are you stopping to work the margin exercises when directed to do so?
	☐ Have you begun to prepare for the final exam?

REVIEW EXERCISES

1. Tell which integers correspond to this situation: [10.1a]

David has a debt of $45 and Joe has $72 in his savings account.

Find the absolute value. [10.1e]

2. $|-38|$ 　　　　　　　　**3.** $|7.3|$

4. $\left|\dfrac{5}{2}\right|$ 　　　　　　**5.** $-|-0.2|$

Find decimal notation. [10.1c]

6. $-\dfrac{5}{4}$ 　　　　　　　**7.** $-\dfrac{5}{6}$

8. $-\dfrac{5}{12}$ 　　　　　　**9.** $-\dfrac{3}{11}$

Graph the number on a number line. [10.1b]

10. -2.5

11. $\dfrac{8}{9}$

Use either $<$ or $>$ for ☐ to write a true sentence. [10.1d]

12. $-3 \ \square\ 10$ 　　　　　**13.** $-1 \ \square\ -6$

14. $0.126 \ \square\ -12.6$ 　　**15.** $-\dfrac{2}{3} \ \square\ -\dfrac{1}{10}$

Find the opposite, or additive inverse, of the number. [10.2b]

16. 3.8 　　　　　　　**17.** $-\dfrac{3}{4}$

18. Evaluate $-x$ when x is -34.   [10.2b]

19. Evaluate $-(-x)$ when x is 5.   [10.2b]

Find the reciprocal.   [10.5b]

20. $\dfrac{3}{8}$ **21.** -7

22. $-\dfrac{1}{10}$

Compute and simplify.

23. $4 + (-7)$ [10.2a]

24. $-\dfrac{2}{3} + \dfrac{1}{12}$ [10.2a]

25. $6 + (-9) + (-8) + 7$ [10.2a]

26. $-3.8 + 5.1 + (-12) + (-4.3) + 10$ [10.2a]

27. $-3 - (-7)$ [10.3a]

28. $-\dfrac{9}{10} - \dfrac{1}{2}$ [10.3a]

29. $-3.8 - 4.1$ [10.3a]

30. $-9 \cdot (-6)$ [10.4a]

31. $-2.7(3.4)$ [10.4a]

32. $\dfrac{2}{3} \cdot \left(-\dfrac{3}{7}\right)$ [10.4a]

33. $3 \cdot (-7) \cdot (-2) \cdot (-5)$ [10.4a]

34. $35 \div (-5)$ [10.5a]

35. $-5.1 \div 1.7$ [10.5c]

36. $-\dfrac{3}{11} \div \left(-\dfrac{4}{11}\right)$ [10.5c]

37. $(-3.4 - 12.2) - 8(-7)$ [10.5e]

Simplify.   [10.5e]

38. $[-12(-3) - 2^3] - (-9)(-10)$

39. $625 \div (-25) \div 5$

40. $-16 \div 4 - 30 \div (-5)$

41. $9[(7 - 14) - 13]$

Solve.

42. On the first, second, and third downs, a football team had these gains and losses: 5-yd gain, 12-yd loss, and 15-yd gain, respectively. Find the total gain (or loss). [10.3b]

43. Kaleb's total assets are $170. He borrows $300. What are his total assets now? [10.3b]

44. *Stock Price.* The price of EFX Corp. began the day at $17.68 per share and dropped by $1.63 per hour for 8 hr. What was the price of the stock after 8 hr? [10.5d]

45. *Checking Account Balance.* Yuri had $68 in his checking account. After writing checks to make seven purchases of DVDs at the same price for each, the balance in his account was −$64.65. What was the price of each DVD? [10.5d]

46. D_W Is it possible for a number to be its own reciprocal? Explain. [10.5b]

47. D_W Write as many arguments as you can to convince a fellow classmate that $-(-a) = a$ for all real numbers a. [10.2b]

48. Find the area of a rectangle of length 10.5 cm and width 20 cm. [8.4a]

49. Find the LCM of 15, 27, and 30. [3.1a]

50. Find the prime factorization of 648. [2.1d]

51. 2016 is what percent of 5600? [6.3b], [6.4b]

SYNTHESIS

52. The sum of two numbers is 800. The difference is 6. Find the numbers. [10.3b]

53. The sum of two numbers is 5. The product is −84. Find the numbers. [10.3b], [10.5d]

54. The following are examples of consecutive integers: 4, 5, 6, 7, 8: and −13, −12, −11, −10. [10.3b], [10.5d]
   a) Express the number 8 as the sum of 16 consecutive integers.
   b) Find the product of the 16 consecutive integers in part (a).

55. Describe how you might find the following product quickly: [10.4a]
$$\left(-\tfrac{1}{11}\right)\left(-\tfrac{1}{9}\right)\left(-\tfrac{1}{7}\right)\left(-\tfrac{1}{5}\right)\left(-\tfrac{1}{3}\right)(-1)(-3)(-5)(-7)(-9)(-11).$$

56. Simplify: $-\left|\dfrac{7}{8} - \left(-\dfrac{1}{2}\right) - \dfrac{3}{4}\right|.$ [10.1e], [10.3a]

57. Simplify: $\left(|2.7 - 3| + 3^2 - |-3|\right) \div (-3).$ [10.1e], [10.5e]

Use either < or > for ☐ to write a true sentence.

1. $-4 \ \square \ 0$

2. $-3 \ \square \ -8$

3. $-0.78 \ \square \ -0.87$

4. $-\dfrac{1}{8} \ \square \ \dfrac{1}{2}$

Find decimal notation.

5. $-\dfrac{1}{8}$

6. $-\dfrac{4}{9}$

7. $-\dfrac{2}{11}$

Find the absolute value.

8. $|-7|$

9. $\left| \dfrac{9}{4} \right|$

10. $-|-2.7|$

Find the opposite, or additive inverse.

11. $\dfrac{2}{3}$

12. -1.4

13. Evaluate $-x$ when x is -8.

Find the reciprocal.

14. -2

15. $\dfrac{4}{7}$

Compute and simplify.

16. $3.1 - (-4.7)$

17. $-8 + 4 + (-7) + 3$

18. $-\dfrac{1}{5} + \dfrac{3}{8}$

19. $2 - (-8)$

20. $3.2 - 5.7$

21. $\dfrac{1}{8} - \left(-\dfrac{3}{4} \right)$

22. $4 \cdot (-12)$

23. $-\dfrac{1}{2} \cdot \left(-\dfrac{3}{8}\right)$

24. $-45 \div 5$

25. $-\dfrac{3}{5} \div \left(-\dfrac{4}{5}\right)$

26. $4.864 \div (-0.5)$

27. $-2(16) - [2(-8) - 5^3]$

28. *Antarctica Highs and Lows.* The continent of Antarctica, which lies in the southern hemisphere, experiences winter in July. The average high temperature is $-67°F$ and the average low temperature is $-81°F$. How much higher is the average high than the average low?
Source: National Climatic Data Center

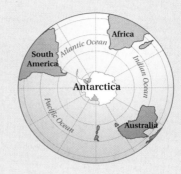

29. Maureen is a stockbroker. She kept track of the changes in the stock market over a period of 5 weeks. By how many points had the market risen or fallen over this time?

WEEK 1	WEEK 2	WEEK 3	WEEK 4	WEEK 5
Down 13 pts	Down 16 pts	Up 36 pts	Down 11 pts	Up 19 pts

30. *Population Decrease.* The population of a city was 18,600. It decreased by 420 each year for 6 yr. What was the population of the city after 6 yr?

31. *Chemical Reaction.* During a chemical reaction, the temperature in the beaker decreased every minute by the same number of degrees. The temperature was $16°C$ at 11:08 A.M. By 11:43 A.M., the temperature had dropped to $-17°C$. By how many degrees did it drop each minute?

SKILL MAINTENANCE

32. Find the area of a rectangle of length 12.4 ft and width 4.5 ft.

33. 24 is what percent of 50?

34. Find the prime factorization of 280.

35. Find the LCM of 16, 20, and 30.

SYNTHESIS

36. Simplify: $|-27 - 3(4)| - |-36| + |-12|$.

37. The deepest point in the Pacific Ocean is the Marianas Trench with a depth of 11,033 m. The deepest point in the Atlantic Ocean is the Puerto Rico Trench with a depth of 8648 m. How much higher is the Puerto Rico Trench than the Marianas Trench?
Source: Defense Mapping Agency, Hydrographic/Topographic Center

38. Find the next three numbers in each sequence.
   a) 6, 5, 3, 0, ___ , ___ , ___
   b) 14, 10, 6, 2, ___ , ___ , ___
   c) $-4, -6, -9, -13,$ ___ , ___ , ___
   d) 8, -4, 2, -1, 0.5, ___ , ___ , ___

CHAPTERS
1–10 Cumulative Review

Solve.

1. *Frozen Yogurt.* Frozen yogurt is 88% water. How much water, by weight, is in a 4-oz cone of frozen yogurt?

2. *Life on Earth.* A study of ancient rocks in South Africa indicates that life existed on Earth's land masses as far back as 2.6 billion years ago. Write standard notation for 2.6 billion.
Source: Yumiko Watanabe, et al., "Geochemical Evidence for Terrestrial Ecosystems 2.6 Billion Years Ago," *Nature* 408 (2000), 574–78

3. *Ticket Prices.* In 2001, the Indianapolis Colts raised their ticket prices from $125 to $149 per ticket for a seat between the 30-yd lines on the lower level. What was the percent of increase?
Source: The Indianapolis Colts

4. *Rick of Wood.* A *rick* of wood measures 2 ft by 4 ft by 8 ft. What is the volume of a rick of wood?

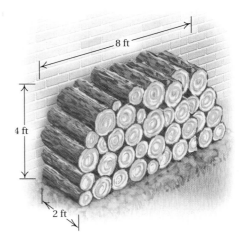

8 ft

4 ft

2 ft

Find decimal notation.

5. 26.3%

6. $-\dfrac{5}{11}$

Complete.

7. 83.4 cg = _____ mg

8. 2.75 mm^2 = _____ cm^2

9. Find the absolute value: $|-4.5|$.

10. Subtract: $2 - 13$.

11. What is the rate in meters per second?
    150 meters, 12 seconds

12. Find the radius, the circumference, and the area of a circle whose diameter is 70 mi. Use $\frac{22}{7}$ for π.

Sports Radio Stations. The data in the table below shows how the popularity of sports radio stations has risen over the past decade. Use the data for Exercises 13–17.

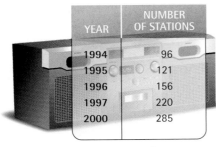

YEAR	NUMBER OF STATIONS
1994	96
1995	121
1996	156
1997	220
2000	285

Source: Interep

13. Make a bar graph of the data.

14. Make a line graph of the data.

15. Find the average, the median, and the mode of the data.

16. Find the percent of increase in the number of stations from 1996 to 2000.

17. Find the percent of increase in the number of stations from 1997 to 2000.

18. Simplify: $\sqrt{225}$.

19. Approximate to two decimal places: $\sqrt{69}$.

20. Multiply: $(-2)(5)$.

21. Divide: $\dfrac{-48}{-16}$.

22. Add: $-2 + 10$.

23. Subtract: $-2 - (-10)$.

Compute and simplify.

24. 14.85×0.001

25. $36 - (-3) + (-42)$

26. $\dfrac{5}{22} - \dfrac{4}{11}$

27. $\dfrac{2}{27} \cdot \left(-\dfrac{9}{16}\right)$

28. $4\dfrac{2}{9} - 2\dfrac{7}{18}$

29. $-\dfrac{3}{14} \div \dfrac{6}{7}$

30. $3(-4.5) + (2^2 - 3 \cdot 4^2)$

31. $12,854 \cdot 750,000$

32. $35.1 + (-2.61)$

33. $32 \div [(-2)(-8) - (15 - (-1))]$

Solve.

34. 7 is what percent of 8?

35. 4 is $12\frac{1}{2}\%$ of what number?

36. *Volume of a Can.* A can of fruit has a diameter of 7 cm and a height of 8 cm. Find the volume. Use 3.14 for π.

37. *Average Temperature.* The following temperatures were recorded every four hours on a certain day in Seattle: 42°, 40°, 45°, 52°, 50°, 40°. What was the average temperature for the day?

38. *Grade Reports.* Thirteen percent of a student body of 600 received all A's on their grade reports. How many students received all A's?

39. *Area of a Lot.* A lot is 125.5 m by 75 m. A house 60 m by 40.5 m and a rectangular swimming pool 10 m by 8 m are built on the lot. How much area is left?

Estimate each of the following as a whole number or as a mixed numeral where the fractional part is $\frac{1}{2}$.

40. $10\dfrac{8}{11}$

41. $7\dfrac{3}{10} + 4\dfrac{5}{6} - \dfrac{31}{29}$

42. $33\dfrac{14}{15} + 27\dfrac{4}{5} + 8\dfrac{27}{30} \cdot 8\dfrac{37}{76}$

For each of Exercises 43–46, choose the correct answer from the selections given.

43. Find the reciprocal of -0.25.

   a) 0.25   **b)** 4   **c)** -4
   d) $\frac{1}{4}$ **e)** None

44. Simplify: $-64 \div 16 \cdot (-8) \div 4$.

   a) 10   **b)** 18   **c)** -127.875
   d) -10 **e)** None

45. Complete: 13,829.19 kg = _____ t.

a) 1.38219   **b)** 0.1382919   **c)** 13.82919

d) 13,829,190   **e)** None

46. The area of a circle of radius r is _____ .

a) $2 \cdot \pi \cdot r$ **b)** $2 \cdot \pi \cdot r^2$ **c)** $\frac{1}{2} \cdot \pi \cdot r$

d) $\pi \cdot r^2$ **e)** None

47. *Truck Driving.* Truck drivers are mandated to keep records of their activity for an entire 24-hr day. The following is an example of a graph or chart that is used to keep a driver's records. The chart is arranged like a ruler with the smallest interval being $\frac{1}{4}$ hr, or 15 min. (Note that the total on the right must add to 24 hr.) In the example below, the driver spent 7 hr off duty. Then he was on duty for 30 min, probably making a delivery in Battle Creek. After that, he drove for $4\frac{1}{2}$ hr, and so on. Fill out the missing data in the second chart and check to make sure that it adds to 24 hr.

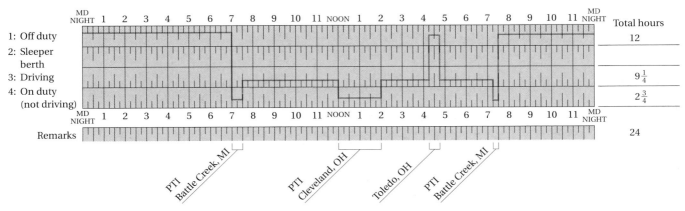

Source: Tony Way, *CON-WAY CENTRAL EXPRESS*

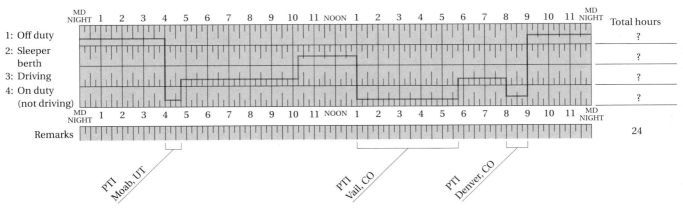

Source: Tony Way, *CON-WAY CENTRAL EXPRESS*

48. ▦ A large egg is about $5\frac{1}{2}$ cm tall with a diameter of 4 cm. Estimate the volume of such an egg by averaging the volumes of two spheres.

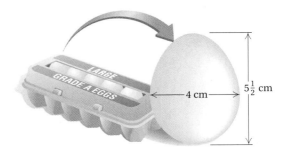

655

Algebra: Solving Equations and Problems

Gateway to Chapter 11

In this chapter, we continue our introduction to algebra. We first consider the manipulation of algebraic expressions. We then use the manipulations to do equation solving. Finally, we use manipulations and equation solving to do problem solving.

Real-World Application

In 1998, at age 79, Earl Shaffer became the oldest person to hike all 2100 miles of the Appalachian Trail—from Springer Mountain, Georgia, to Mount Katahdin, Maine. At one point, Shaffer stood atop Big Walker Mountain, Virginia, which is three times as far from the northern end as from the southern end. How far was Shaffer from each end of the trail?

Source: Appalachian Trail Conference

This problem appears as Example 6 in Section 11.5.

1. Evaluate $\dfrac{x}{2y}$ when $x = 5$ and $y = 8$. [11.1a]

2. Write an algebraic expression: [11.5a]
   Seventy-eight percent of some number.

Multiply. [11.1b]

3. $9(z - 2)$

4. $-2(2a + b - 5c)$

Factor. [11.1c]

5. $4x - 12$

6. $6y - 9z - 18$

Collect like terms. [11.1d]

7. $5x - 8x$

8. $6x - 9y - 4x + 11y + 18$

Solve.

9. $-7x = 49$ [11.3a]

10. $4y + 9 = 2y + 7$ [11.4b]

11. $6a - 2 = 10$ [11.4a]

12. $x + (x + 1) + (x + 2) = 12$ [11.4c]

Solve. [11.5b]

13. **Perimeter of Peach Orchard.** The perimeter of a rectangular peach orchard is 146 m. The width is 5 m less than the length. Find the dimensions of the orchard.

14. **Savings Investment.** Money is invested in a savings account at 6% simple interest. After 1 year, there is $826.80 in the account. How much was originally invested?

11.1 INTRODUCTION TO ALGEBRA

Objectives

a Evaluate an algebraic expression by substitution.

b Use the distributive laws to multiply expressions like 8 and $x - y$.

c Use the distributive laws to factor expressions like $4x - 12 + 24y$.

d Collect like terms.

Many types of problems require the use of equations in order to be solved efficiently. The study of algebra involves the use of equations to solve problems. Equations are constructed from algebraic expressions.

a   Algebraic Expressions

In arithmetic, you have worked with expressions such as

$$37 + 86, \quad 7 \times 8, \quad 19 - 7, \quad \text{and} \quad \frac{3}{8}.$$

In algebra, we use letters for numbers and work with *algebraic expressions* such as

$$x + 86, \quad 7 \times t, \quad 19 - y, \quad \text{and} \quad \frac{a}{b}.$$

Expressions like these should be familiar from the equation and problem solving that we have already done.

Sometimes a letter can stand for various numbers. In that case, we call the letter a **variable.** Let a = your age. Then a is a variable since a changes from year to year. Sometimes a letter can stand for just one number. In that case, we call the letter a **constant.** Let b = your date of birth. Then b is a constant.

An **algebraic expression** consists of variables, constants, numerals, and operation signs. When we replace a variable with a number, we say that we are **substituting** for the variable. This process is called **evaluating the expression.**

EXAMPLE 1   Evaluate $x + y$ for $x = 37$ and $y = 29$.

We substitute 37 for x and 29 for y and carry out the addition:

$$x + y = 37 + 29 = 66.$$

The number 66 is called the **value** of the expression.

Do Exercises 1 and 2.

Algebraic expressions involving multiplication can be written in several ways. For example, "8 times a" can be written as $8 \times a$, $8 \cdot a$, $8(a)$, or simply $8a$.

Two letters written together without an operation sign, such as ab, also indicates a multiplication.

EXAMPLE 2   Evaluate $3y$ for $y = 14$.

$$3y = 3(14) = 42$$

1. Evaluate $a + b$ when $a = 38$ and $b = 26$.

2. Evaluate $x - y$ when $x = 57$ and $y = 29$.

Answers on page A-27

3. Evaluate $4t$ when $t = 15$ and when $t = -6.8$.

4. Find the area of a rectangle when l is 24 ft and w is 8 ft.

Answers on page A-27

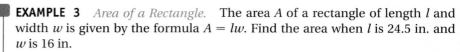

EXAMPLE 3 *Area of a Rectangle.* The area A of a rectangle of length l and width w is given by the formula $A = lw$. Find the area when l is 24.5 in. and w is 16 in.

We substitute 24.5 in. for l and 16 in. for w and carry out the multiplication:

$$A = lw = (24.5 \text{ in.})(16 \text{ in.})$$
$$= (24.5)(16)(\text{in.})(\text{in.})$$
$$= 392 \text{ in}^2, \text{ or } 392 \text{ square inches.}$$

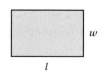

Do Exercises 3 and 4.

Algebraic expressions involving division can also be written in several ways. For example, "8 divided by t" can be written as $8 \div t$, $\dfrac{8}{t}$, $8/t$, or $8 \cdot \dfrac{1}{t}$, where the fraction bar is a division symbol.

EXAMPLE 4 Evaluate $\dfrac{a}{b}$ for $a = 63$ and $b = 9$.

We substitute 63 for a and 9 for b and carry out the division:

$$\frac{a}{b} = \frac{63}{9} = 7.$$

CALCULATOR CORNER

Evaluating Algebraic Expressions We can use a calculator to evaluate algebraic expressions. To evaluate $x - y$ when $x = 48$ and $y = 19$, for example, we press $\boxed{4}\boxed{8}\boxed{-}\boxed{1}\boxed{9}\boxed{=}$. The calculator displays the result, 29. To evaluate $3x$ when $x = -27$, we press $\boxed{3}\boxed{\times}\boxed{2}\boxed{7}\boxed{+/-}\boxed{=}$. The result is -81. When we evaluate an expression like $\dfrac{x + y}{4}$, we must enclose the numerator in parentheses. To evaluate this expression when $x = 29$ and $y = 15$, we press $\boxed{(}\boxed{2}\boxed{9}\boxed{+}\boxed{1}\boxed{5}\boxed{)}\boxed{\div}\boxed{4}\boxed{=}$. The calculator displays the result, 11.

A calculator can also be used to demonstrate the distributive laws. We can evaluate $5(x + y)$ and $5x + 5y$ when $x = 3$ and $y = 9$, for example, and see that the results are the same. To evaluate $5(x + y)$ when $x = 3$ and $y = 9$, we press $\boxed{5}\boxed{\times}\boxed{(}\boxed{3}\boxed{+}\boxed{9}\boxed{)}\boxed{=}$. The calculator displays the result, 60. Next, we evaluate $5x + 5y$ when $x = 3$ and $y = 9$. We press $\boxed{5}\boxed{\times}\boxed{3}\boxed{+}\boxed{5}\boxed{\times}\boxed{9}\boxed{=}$. Again the calculator displays 60. This verifies that $5(3 + 9) = 5 \cdot 3 + 5 \cdot 9$.

Exercises: Evaluate.

1. $x + y$, when $x = 35$ and $y = 16$

2. $7t$, when $t = 14$

3. $\dfrac{a}{b}$, when $a = 54$ and $b = -9$

4. $\dfrac{2m}{n}$, when $m = 38$ and $n = -4$

5. $\dfrac{x - y}{7}$, when $x = 94$ and $y = 31$

6. $\dfrac{p + q}{12}$, when $p = 47$ and $q = 97$

7. $4(x + y)$ and $4x + 4y$, when $x = 6$ and $y = 7$

8. $3(x + y)$ and $3x + 3y$, when $x = 34$ and $y = 18$

9. $6(a - b)$ and $6a - 6b$, when $a = 67$ and $b = 29$

10. $7(a - b)$ and $7a - 7b$, when $a = 18$ and $b = 6$

EXAMPLE 5 Evaluate $\dfrac{12m}{n}$ for $m = 8$ and $n = 16$.

$$\frac{12m}{n} = \frac{12 \cdot 8}{16} = \frac{96}{16} = 6$$

Do Exercises 5 and 6.

b   Equivalent Expressions and the Distributive Laws

In solving and doing other kinds of work in algebra, we manipulate expressions in various ways. To see how to do this, we consider some examples in which we evaluate expressions.

EXAMPLE 6   Evaluate $1 \cdot x$ when $x = 5$ and $x = -8$ and compare the results to x.

We substitute 5 for x:

$$1 \cdot x = 1 \cdot 5 = 5.$$

Then we substitute -8 for x:

$$1 \cdot x = 1 \cdot (-8) = -8.$$

We see that $1 \cdot x$ and x represent the same number.

Do Exercises 7 and 8.

We see in Example 6 and Margin Exercise 7 that the expressions represent the same number for any allowable replacement of x. In that sense, the expressions $1 \cdot x$ and x are **equivalent.**

> ### EQUIVALENT EXPRESSIONS
>
> Two expressions that have the same value for all allowable replacements are called **equivalent.**

In the expression $3/x$, the number 0 is not allowable because $3/0$ is not defined. Even so, the expressions $6/2x$ and $3/x$ are *equivalent* because they represent the same number for any allowable (not 0) replacement of x. For example, when $x = 5$,

$$\frac{6}{2x} = \frac{6}{2 \cdot 5} = \frac{6}{10} = \frac{3}{5} \quad \text{and} \quad \frac{3}{x} = \frac{3}{5}.$$

We see in Margin Exercise 8 that the expressions $2x$ and $5x$ are *not* equivalent.

The fact that $1 \cdot x$ and x are equivalent is a law of real numbers. It is called the **identity property of 1.** We often refer to the use of the identity property of 1 as "multiplying by 1." We have used multiplying by 1 for understanding many times in this text.

> ### THE IDENTITY PROPERTY OF 1 (MULTIPLICATIVE IDENTITY)
>
> For any real number a,
> $$a \cdot 1 = 1 \cdot a = a.$$

---

5. Evaluate $\dfrac{a}{b}$ when $a = -200$ and $b = 8$.

6. Evaluate $\dfrac{10p}{q}$ when $p = 40$ and $q = 25$.

Complete the table by evaluating each expression for the given values.

7.

VALUE OF x	$1 \cdot x$	x
$x = 3$		
$x = -6$		
$x = 4.8$		

8.

VALUE OF x	$2x$	$5x$
$x = 2$		
$x = -6$		
$x = 4.8$		

Answers on page A-27

9. Complete this table.

VALUES OF x AND y	$5(x + y)$	$5x + 5y$
$x = 2, y = 8$		
$x = -3, y = 4$		
$x = -10, y = 5$		

10. Evaluate $6x + 6y$ and $6(x + y)$ when $x = 10$ and $y = 5$.

11. Evaluate $4(x + y)$ and $4x + 4y$ when $x = 11$ and $y = 5$.

12. Evaluate $7(x - y)$ and $7x - 7y$ when $x = 9$ and $y = 7$.

13. Evaluate $6x - 6y$ and $6(x - y)$ when $x = 10$ and $y = 5$.

14. Evaluate $2(x - y)$ and $2x - 2y$ when $x = 11$ and $y = 5$.

Answers on page A-27

We now consider two other laws of real numbers called the **distributive laws.** They are the basis of many procedures in both arithmetic and algebra and are probably the most important laws that we use to manipulate algebraic expressions. The first distributive law involves two operations: addition and multiplication.

Let's begin by considering a multiplication problem from arithmetic:

$$\begin{array}{r} 4\ 5 \\ \times\quad 7 \\ \hline 3\ 5 \\ 2\ 8\ 0 \\ \hline 3\ 1\ 5 \end{array}$$

3 5 ← This is $7 \cdot 5$.
2 8 0 ← This is $7 \cdot 40$.
3 1 5 ← This is the sum $7 \cdot 40 + 7 \cdot 5$.

To carry out the multiplication, we actually added two products. That is,

$$7 \cdot 45 = 7(40 + 5) = 7 \cdot 40 + 7 \cdot 5.$$

Let's examine this further. If we wish to multiply a sum of several numbers by a factor, we can either add and then multiply or multiply and then add.

EXAMPLE 7   Evaluate $5(x + y)$ and $5x + 5y$ when $x = 2$ and $y = 8$ and compare the results.

We substitute 2 for x and 8 for y in each expression. Then we use the rules for order of operations to calculate.

a) $5(x + y) = 5(2 + 8)$
$\qquad\qquad = 5(10)$ Adding within parentheses first, and then multiplying
$\qquad\qquad = 50$

b) $5x + 5y = 5 \cdot 2 + 5 \cdot 8$
$\qquad\qquad = 10 + 40$ Multiplying first and then adding
$\qquad\qquad = 50$

The results of (a) and (b) are the same.

Do Exercises 9–11.

The expressions $5(x + y)$ and $5x + 5y$, in Example 7 and Margin Exercise 9, are equivalent. They illustrate the distributive law of multiplication over addition.

> ### THE DISTRIBUTIVE LAW OF MULTIPLICATION OVER ADDITION
>
> For any numbers a, b, and c,
> $$a(b + c) = ab + ac.$$

Margin Exercises 10 and 11 also illustrate the distributive law.

In the statement of the distributive law, we know that in an expression such as $ab + ac$, the multiplications are to be done first according to the rules for order of operations. So, instead of writing $(4 \cdot 5) + (4 \cdot 7)$, we can write $4 \cdot 5 + 4 \cdot 7$. However, in $a(b + c)$, we cannot omit the parentheses. If we did, we would have $ab + c$, which means $(ab) + c$. For example, $3(4 + 2) = 18$, but $3 \cdot 4 + 2 = 14$.

The second distributive law relates multiplication and subtraction. This law says that to multiply by a difference, we can either subtract and then multiply or multiply and then subtract.

> ### THE DISTRIBUTIVE LAW OF MULTIPLICATION OVER SUBTRACTION
>
> For any numbers a, b, and c,
> $$a(b - c) = ab - ac.$$

We often refer to "*the* distributive law" when we mean *either or both* of these laws.

Do Exercises 12–14 on the preceding page.

What do we mean by the *terms* of an expression? **Terms** are separated by addition signs. If there are subtraction signs, we can find an equivalent expression that uses addition signs.

EXAMPLE 8 What are the terms of $3x - 4y + 2z$?

$$3x - 4y + 2z = 3x + (-4y) + 2z \qquad \text{Separating parts with + signs}$$

The terms are $3x$, $-4y$, and $2z$.

Do Exercises 15 and 16.

The distributive laws are the basis for a procedure in algebra called **multiplying.** In an expression such as $8(a + 2b - 7)$, we multiply each term inside the parentheses by 8:

$$8(a + 2b - 7) = 8 \cdot a + 8 \cdot 2b - 8 \cdot 7 = 8a + 16b - 56.$$

EXAMPLES Multiply.

9. $9(x - 5) = 9x - 9(5) \qquad$ Using the distributive law of multiplication over subtraction

$$= 9x - 45$$

10. $\dfrac{2}{3}(w + 1) = \dfrac{2}{3} \cdot w + \dfrac{2}{3} \cdot 1 \qquad$ Using the distributive law of multiplication over addition

$$= \dfrac{2}{3}w + \dfrac{2}{3}$$

Multiply.

17. $3(x - 5)$

18. $5(x + 1)$

19. $\dfrac{5}{4}(x - y + 4)$

20. $-2(x - 3)$

21. $-5(x - 2y + 4z)$

Answers on page A-27

663

Factor.

22. $6z - 12$

23. $3x - 6y + 9$

24. $16a - 36b + 42$

25. $-12x + 32y - 16z$

EXAMPLE 11 Multiply: $-4(x - 2y + 3z)$.

$$-4(x - 2y + 3z) = -4 \cdot x - (-4)(2y) + (-4)(3z) \qquad \text{Using both distributive laws}$$

$$= -4x - (-8y) + (-12z) \qquad \text{Multiplying}$$

$$= -4x + 8y - 12z.$$

We can also do this problem by first finding an equivalent expression with all plus signs and then multiplying:

$$-4(x - 2y + 3z) = -4[x + (-2y) + 3z]$$

$$= -4 \cdot x + (-4)(-2y) + (-4)(3z) = -4x + 8y - 12z.$$

Do Exercises 17–21 on the preceding page.

C Factoring

Factoring is the reverse of multiplying. To factor, we can use the distributive laws in reverse:

$$ab + ac = a(b + c) \quad \text{and} \quad ab - ac = a(b - c).$$

> **FACTOR**
>
> To **factor** an expression is to find an equivalent expression that is a product.

Look at Example 9. To *factor* $9x - 45$, we find an equivalent expression that is a product, $9(x - 5)$. When all the terms of an expression have a factor in common, we can "factor it out" using the distributive laws. Note the following.

$9x$ has the factors $9, -9, 3, -3, 1, -1, x, -x, 3x, -3x, 9x, -9x$;

-45 has the factors $1, -1, 3, -3, 5, -5, 9, -9, 15, -15, 45, -45$.

We remove the largest common factor. In this case, that factor is 9. Thus,

$$9x - 45 = 9 \cdot x - 9 \cdot 5 = 9(x - 5).$$

Remember that an expression is factored when we find an equivalent expression that is a product.

EXAMPLES Factor.

12. $5x - 10 = 5 \cdot x - 5 \cdot 2 \qquad$ Try to do this step mentally.

$\qquad\qquad = 5(x - 2) \qquad \longleftarrow$ You can check by multiplying.

13. $9x + 27y - 9 = 9 \cdot x + 9 \cdot 3y - 9 \cdot 1 = 9(x + 3y - 1)$

> **CAUTION!**
>
> Note that although $3(3x + 9y - 3)$ is also equivalent to $9x + 27y - 9$, it is *not* the desired form. However, we can complete the process by factoring out another factor of 3:
>
> $$9x + 27y - 9 = 3(3x + 9y - 3) = 3 \cdot 3(x + 3y - 1) = 9(x + 3y - 1).$$
>
> Remember to factor out the *largest common factor*.

EXAMPLES Factor. Try to write just the answer, if you can.

14. $5x - 5y = 5(x - y)$

15. $-3x + 6y - 9z = -3 \cdot x - 3(-2y) - 3(3z) = -3(x - 2y + 3z)$

We usually factor out a negative factor when the first term is negative. The way we factor can depend on the situation in which we are working. We might also factor the expression in Example 15 as follows:

$$-3x + 6y - 9z = 3(-x + 2y - 3z).$$

16. $18z - 12x - 24 = 6(3z - 2x - 4)$

Remember that you can always check such factoring by multiplying. Keep in mind that an expression is factored when it is written as a product.

Do Exercises 22–25 on the preceding page.

d   Collecting Like Terms

Terms such as $5x$ and $-4x$, whose variable factors are exactly the same, are called **like terms.** Similarly, numbers, such as -7 and 13, are like terms. Also, $3y^2$ and $9y^2$ are like terms because the variables are raised to the same power. Terms such as $4y$ and $5y^2$ are not like terms, and $7x$ and $2y$ are not like terms.

The process of **collecting like terms** is based on the distributive laws. We can also apply the distributive law when a factor is on the right.

EXAMPLES Collect like terms. Try to write just the answer, if you can.

17. $4x + 2x = (4 + 2)x = 6x$ Factoring out the x using a distributive law

18. $2x + 3y - 5x - 2y = 2x - 5x + 3y - 2y$

$$= (2 - 5)x + (3 - 2)y = -3x + y$$

19. $3x - x = 3x - 1x = (3 - 1)x = 2x$

20. $x - 0.24x = 1 \cdot x - 0.24x = (1 - 0.24)x = 0.76x$

21. $x - 6x = 1 \cdot x - 6 \cdot x = (1 - 6)x = -5x$

22. $4x - 7y + 9x - 5 + 3y - 8 = 13x - 4y - 13$

23. $\dfrac{2}{3}a - b + \dfrac{4}{5}a + \dfrac{1}{4}b - 10 = \dfrac{2}{3}a - 1 \cdot b + \dfrac{4}{5}a + \dfrac{1}{4}b - 10$

$$= \left(\dfrac{2}{3} + \dfrac{4}{5}\right)a + \left(-1 + \dfrac{1}{4}\right)b - 10$$

$$= \left(\dfrac{10}{15} + \dfrac{12}{15}\right)a + \left(-\dfrac{4}{4} + \dfrac{1}{4}\right)b - 10$$

$$= \dfrac{22}{15}a - \dfrac{3}{4}b - 10$$

Do Exercises 26–32.

Collect like terms.

26. $6x - 3x$

27. $7x - x$

28. $x - 9x$

29. $x - 0.41x$

30. $5x + 4y - 2x - y$

31. $3x - 7x - 11 + 8y + 4 - 13y$

32. $-\dfrac{2}{3} - \dfrac{3}{5}x + y + \dfrac{7}{10}x - \dfrac{2}{9}y$

Answers on page A-27

11.1
EXERCISE SET

For Extra Help

Digital Video
Tutor CD 6
Videotape 16

InterAct
Math

Math Tutor
Center

MathXL

MyMathLab

a Evaluate.

1. $6x$, when $x = 7$

2. $9t$, when $t = 8$

3. $\dfrac{x}{y}$, when $x = 9$ and $y = 3$

4. $\dfrac{m}{n}$, when $m = 18$ and $n = 3$

5. $\dfrac{3p}{q}$, when $p = -2$ and $q = 6$

6. $\dfrac{5y}{z}$, when $y = -15$ and $z = -25$

7. $\dfrac{x + y}{5}$, when $x = 10$ and $y = 20$

8. $\dfrac{p - q}{2}$, when $p = 17$ and $q = 3$

9. ab, when $a = -5$ and $b = 4$

10. ba, when $a = -5$ and $b = 4$

b Evaluate.

11. $10(x + y)$ and $10x + 10y$, when $x = 20$ and $y = 4$

12. $5(a + b)$ and $5a + 5b$, when $a = 16$ and $b = 6$

13. $10(x - y)$ and $10x - 10y$, when $x = 20$ and $y = 4$

14. $5(a - b)$ and $5a - 5b$, when $a = 16$ and $b = 6$

Multiply.

15. $2(b + 5)$

16. $4(x + 3)$

17. $7(1 - t)$

18. $4(1 - y)$

19. $6(5x + 2)$

20. $9(6m + 7)$

21. $7(x + 4 + 6y)$

22. $4(5x + 8 + 3p)$

23. $-7(y - 2)$

24. $-9(y - 7)$

25. $-9(-5x - 6y + 8)$

26. $-7(-2x - 5y + 9)$

27. $\dfrac{3}{4}(x - 3y - 2z)$

28. $\dfrac{2}{5}(2x - 5y - 8z)$

29. $3.1(-1.2x + 3.2y - 1.1)$

30. $-2.1(-4.2x - 4.3y - 2.2)$

c Factor. Check by multiplying.

31. $2x + 4$

32. $5y + 20$

33. $30 + 5y$

34. $7x + 28$

35. $14x + 21y$

36. $18a + 24b$

37. $5x + 10 + 15y$

38. $9a + 27b + 81$

39. $8x - 24$

40. $10x - 50$

41. $32 - 4y$

42. $24 - 6m$

43. $8x + 10y - 22$ **44.** $9a + 6b - 15$ **45.** $-18x - 12y + 6$ **46.** $-14x + 21y + 7$

d Collect like terms.

47. $9a + 10a$

48. $14x + 3x$

49. $10a - a$

50. $-10x + x$

51. $2x + 9z + 6x$

52. $3a - 5b + 4a$

53. $41a + 90 - 60a - 2$

54. $42x - 6 - 4x + 20$

55. $23 + 5t + 7y - t - y - 27$

56. $95 - 90d - 87 - 9d + 3 + 7d$

57. $11x - 3x$

58. $9t - 13t$

59. $6n - n$

60. $10t - t$

61. $y - 17y$

62. $5m - 8m + 4$

63. $-8 + 11a - 5b + 6a - 7b + 7$

64. $8x - 5x + 6 + 3y - 2y - 4$

65. $9x + 2y - 5x$

66. $8y - 3z + 4y$

67. $\frac{11}{4}x + \frac{2}{3}y - \frac{4}{5}x - \frac{1}{6}y + 12$

68. $\frac{13}{2}a + \frac{9}{5}b - \frac{2}{3}a - \frac{3}{10}b - 42$

69. $2.7x + 2.3y - 1.9x - 1.8y$

70. $6.7a + 4.3b - 4.1a - 2.9b$

71. D_W Determine whether $(a + b)^2$ and $a^2 + b^2$ are equivalent for all real numbers. Explain.

72. D_W The distributive law is introduced before the material on collecting like terms. Why do you think this is?

SKILL MAINTENANCE

For a circle with the given radius, find the diameter, the circumference, and the area. Use 3.14 for π.
[9.3a, b, c]

73. $r = 15$ yd **74.** $r = 8.2$ m **75.** $r = 9\frac{1}{2}$ mi **76.** $r = 2400$ cm

For a circle with the given diameter, find the radius, the circumference, and the area. Use 3.14 for π.
[9.3a, b, c]

77. $d = 20$ mm **78.** $d = 264$ km **79.** $d = 4.6$ ft **80.** $d = 10.3$ m

---

SYNTHESIS

Collect like terms, if possible, and factor the result.

81. $q + qr + qrs + qrst$

82. $21x + 44xy + 15y - 16x - 8y - 38xy + 2y + xy$

Objective

Solve equations using the addition principle.

1. Solve $x + 2 = 11$ using the addition principle.

   a) First, complete this sentence:
   $$2 + \square = 0.$$

   b) Then solve the equation.

11.2 SOLVING EQUATIONS: THE ADDITION PRINCIPLE

a Using the Addition Principle

Consider the equation $x = 7$. We can easily "see" that the solution of this equation is 7. If we replace x with 7, we get $7 = 7$, which is true. Now consider the equation $x + 6 = 13$. The solution of this equation is also 7, but the fact that 7 is the solution is not as obvious. We now begin to consider principles that allow us to start with an equation like $x + 6 = 13$ and end up with an equation like $x = 7$, in which the variable is alone on one side and for which the solution is easier to find. The equations $x + 6 = 13$ and $x = 7$ are **equivalent**.

> ### EQUIVALENT EQUATIONS
>
> Equations with the same solutions are called **equivalent equations**.

One principle that we use to solve equations involves the addition principle, which we have used throughout this text.

> ### THE ADDITION PRINCIPLE
>
> For any real numbers a, b, and c,
> $$a = b \quad \text{is equivalent to} \quad a + c = b + c.$$

Let's solve $x + 6 = 13$ using the addition principle. We want to get x alone on one side. To do so, we use the addition principle, choosing to add -6 on both sides because $6 + (-6) = 0$:

$$x + 6 = 13$$
$$x + 6 + (-6) = 13 + (-6) \qquad \text{Using the addition principle; adding } -6 \text{ on both sides}$$
$$x + 0 = 7 \qquad \text{Simplifying}$$
$$x = 7. \qquad \text{Identity property of 0}$$

The solution of $x + 6 = 13$ is 7.

Do Exercise 1.

To visualize the addition principle, think of a jeweler's balance. When both sides of the balance hold equal amounts of weight, the balance is level. If weight is added or removed, equally, on both sides, the balance remains level.

Answers on page A-28

CHAPTER 11: Algebra: Solving Equations and Problems

When we use the addition principle, we sometimes say that we "add the same number on both sides of the equation." This is also true for subtraction, since we can express every subtraction as an addition. That is, since

$$a - c = b - c \quad \text{is equivalent to} \quad a + (-c) = b + (-c),$$

the addition principle tells us that we can "subtract the same number on both sides of an equation."

EXAMPLE 1 Solve: $x + 5 = -7$.

We have

$$x + 5 = -7$$
$$x + 5 - 5 = -7 - 5 \qquad \text{Using the addition principle: adding } -5 \text{ on both sides or subtracting 5 on both sides}$$
$$x + 0 = -12 \qquad \text{Simplifying}$$
$$x = -12. \qquad \text{Identity property of 0}$$

We can see that the solution of $x = -12$ is the number -12. To check the answer, we substitute -12 in the original equation.

CHECK:
$$\frac{x + 5 = -7}{-12 + 5 \;?\; -7}$$
$$-7 \;|\; \qquad \textbf{TRUE}$$

The solution of the original equation is -12.

In Example 1, to get x alone, we used the addition principle and subtracted 5 on both sides. This eliminated the 5 on the left. We started with $x + 5 = -7$, and using the addition principle we found a simpler equation $x = -12$, for which it was easier to "*see*" the solution. The equations $x + 5 = -7$ and $x = -12$ are *equivalent*.

Do Exercise 2.

Now we solve an equation that involves a subtraction by using the addition principle.

EXAMPLE 2 Solve: $-6.5 = y - 8.4$.

We have

$$-6.5 = y - 8.4$$
$$-6.5 + 8.4 = y - 8.4 + 8.4 \qquad \text{Using the addition principle: adding 8.4 to eliminate } -8.4 \text{ on the right}$$
$$1.9 = y.$$

CHECK:
$$\frac{-6.5 = y - 8.4}{-6.5 \;?\; 1.9 - 8.4}$$
$$|\; -6.5 \qquad \textbf{TRUE}$$

The solution is 1.9.

Note that equations are reversible. That is, if $a = b$ is true, then $b = a$ is true. Thus, when we solve $-6.5 = y - 8.4$, we can reverse it and solve $y - 8.4 = -6.5$ if we wish.

Do Exercises 3 and 4.

---

2. Solve using the addition principle:

$$x + 7 = 2.$$

Solve.

3. $8.7 = n - 4.5$

4. $y + 17.4 = 10.9$

Answers on page A-28

11.2 Solving Equations: The Addition Principle

EXERCISE SET

a Solve using the addition principle. Don't forget to check!

1. $x + 2 = 6$

CHECK: $x + 2 = 6$
?

2. $y + 4 = 11$

CHECK: $y + 4 = 11$
?

3. $x + 15 = -5$

CHECK: $x + 15 = -5$
?

4. $t + 10 = 44$

CHECK: $t + 10 = 44$
?

5. $x + 6 = -8$

6. $y + 8 = 37$

7. $x + 5 = 12$

8. $x + 3 = 7$

9. $-22 = t + 4$

10. $t + 8 = -14$

11. $x + 16 = -2$

12. $y + 34 = -8$

13. $x - 9 = 6$

14. $x - 9 = 2$

15. $x - 7 = -21$

16. $x - 5 = -16$

17. $5 + t = 7$

18. $6 + y = 22$

19. $-7 + y = 13$

20. $-8 + z = 16$

21. $-3 + t = -9$

22. $-8 + y = -23$

23. $r + \dfrac{1}{3} = \dfrac{8}{3}$

24. $t + \dfrac{3}{8} = \dfrac{5}{8}$

25. $m + \dfrac{5}{6} = -\dfrac{11}{12}$

26. $x + \dfrac{2}{3} = -\dfrac{5}{6}$

27. $x - \dfrac{5}{6} = \dfrac{7}{8}$

28. $y - \dfrac{3}{4} = \dfrac{5}{6}$

29. $-\dfrac{1}{5} + z = -\dfrac{1}{4}$

30. $-\dfrac{1}{8} + y = -\dfrac{3}{4}$

31. $7.4 = x + 2.3$

32. $9.3 = 4.6 + x$

33. $7.6 = x - 4.8$ **34.** $9.5 = y - 8.3$ **35.** $-9.7 = -4.7 + y$ **36.** $-7.8 = 2.8 + x$

37. $5\frac{1}{6} + x = 7$ **38.** $5\frac{1}{4} = 4\frac{2}{3} + x$ **39.** $q + \frac{1}{3} = -\frac{1}{7}$ **40.** $47\frac{1}{8} = -76 + z$

41. ^{D}W Explain the following mistake made by a fellow student.

$$x + \frac{1}{3} = -\frac{5}{3}$$
$$x = -\frac{4}{3}$$

42. ^{D}W Explain the role of the opposite of a number when using the addition principle.

SKILL MAINTENANCE

Add.   [10.2a]

43. $-3 + (-8)$ **44.** $-\frac{2}{3} + \frac{5}{8}$ **45.** $-14.3 + (-19.8)$ **46.** $3.2 + (-4.9)$

Subtract.   [10.3a]

47. $-3 - (-8)$ **48.** $-\frac{2}{3} - \frac{5}{8}$ **49.** $-14.3 - (-19.8)$ **50.** $3.2 - (-4.9)$

Multiply.   [10.4a]

51. $-3(-8)$ **52.** $-\frac{2}{3} \cdot \frac{5}{8}$ **53.** $-14.3 \times (-19.8)$ **54.** $3.2(-4.9)$

Divide.   [10.5c]

55. $\dfrac{-24}{-3}$ **56.** $-\frac{2}{3} \div \frac{5}{8}$ **57.** $\dfrac{283.14}{-19.8}$ **58.** $\dfrac{-15.68}{3.2}$

SYNTHESIS

Solve.

59. $\boxed{}$ $-356.788 = -699.034 + t$

60. $-\frac{4}{5} + \frac{7}{10} = x - \frac{3}{4}$

61. $x + \frac{4}{5} = -\frac{2}{3} - \frac{4}{15}$

62. $8 - 25 = 8 + x - 21$

63. $16 + x - 22 = -16$

64. $x + x = x$

65. $-\frac{3}{2} + x = -\frac{5}{17} - \frac{3}{2}$

66. $|x| = 5$

671

Objective

a Solve equations using the multiplication principle.

1. Solve. Multiply on both sides.

$$6x = 90$$

2. Solve. Divide on both sides.

$$4x = -7$$

Answers on page A-28

11.3 SOLVING EQUATIONS: THE MULTIPLICATION PRINCIPLE

a Using the Multiplication Principle

Suppose that $a = b$ is true, and we multiply a by some number c. We get the same answer if we multiply b by c, because a and b are the same number.

THE MULTIPLICATION PRINCIPLE

For any real numbers a, b, and c, $c \neq 0$,

$$a = b \quad \text{is equivalent to} \quad a \cdot c = b \cdot c.$$

When using the multiplication principle, we sometimes say that we "multiply on both sides of the equation by the same number."

EXAMPLE 1 Solve: $5x = 70$.

To get x alone, we multiply by the *multiplicative inverse,* or *reciprocal,* of 5. Then we get the *multiplicative identity* 1 times x, or $1 \cdot x$, which simplifies to x. This allows us to eliminate 5 on the left.

$$5x = 70 \qquad \text{The reciprocal of 5 is } \tfrac{1}{5}.$$

$$\frac{1}{5} \cdot 5x = \frac{1}{5} \cdot 70 \qquad \begin{array}{l}\text{Multiplying by } \tfrac{1}{5} \text{ to get } 1 \cdot x \text{ and} \\ \text{eliminate 5 on the left}\end{array}$$

$$1 \cdot x = 14 \qquad \text{Simplifying}$$

$$x = 14 \qquad \text{Identity property of 1: } 1 \cdot x = x$$

CHECK:
$$\frac{5x = 70}{5 \cdot 14 \;?\; 70}$$
$$70 \;\big|\; \qquad \textbf{TRUE}$$

The solution is 14.

The multiplication principle also tells us that we can "divide on both sides of the equation by a nonzero number." This is because division is the same as multiplying by a reciprocal. That is,

$$\frac{a}{c} = \frac{b}{c} \quad \text{is equivalent to} \quad a \cdot \frac{1}{c} = b \cdot \frac{1}{c}, \quad \text{when } c \neq 0.$$

In an expression like $5x$ in Example 1, the number 5 is called the **coefficient.** Example 1 could be done as follows, dividing by 5, the coefficient of x, on both sides.

EXAMPLE 2 Solve: $5x = 70$.

$$5x = 70$$

$$\frac{5x}{5} = \frac{70}{5} \qquad \text{Dividing by 5 on both sides}$$

$$1 \cdot x = 14 \qquad \text{Simplifying}$$

$$x = 14 \qquad \text{Identity property of 1}$$

Do Exercises 1 and 2.

EXAMPLE 3 Solve: $-4x = 92$.

We have

$$-4x = 92$$

$$\frac{-4x}{-4} = \frac{92}{-4}$$ Using the multiplication principle. Dividing by -4 on both sides is the same as multiplying by $-\frac{1}{4}$.

$$1 \cdot x = -23$$ Simplifying

$$x = -23.$$ Identity property of 1

CHECK: $$\frac{-4x = 92}{-4(-23) \;?\; 92}$$
$$92 \;|\; \quad \textbf{TRUE}$$

The solution is -23.

Do Exercise 3.

EXAMPLE 4 Solve: $-x = 9$.

We have

$$-x = 9$$

$$-1(-x) = -1 \cdot 9$$ Multiplying by -1 on both sides

$$-1 \cdot (-1) \cdot x = -9$$

$$1 \cdot x = -9$$

$$x = -9.$$

CHECK: $$\frac{-x = 9}{-(-9) \;?\; 9}$$
$$9 \;|\; \quad \textbf{TRUE}$$

The solution is -9.

Do Exercise 4.

In practice, it is generally more convenient to "divide" on both sides of the equation if the coefficient of the variable is in decimal notation or is an integer. If the coefficient is in fraction notation, it is more convenient to "multiply" by a reciprocal.

EXAMPLE 5 Solve: $\dfrac{3}{8} = -\dfrac{5}{4}x$.

$$\frac{3}{8} = -\frac{5}{4}x$$

The reciprocal of $-\frac{5}{4}$ is $-\frac{4}{5}$. There is no sign change.

$$-\frac{4}{5} \cdot \frac{3}{8} = -\frac{4}{5} \cdot \left(-\frac{5}{4}x\right)$$ Multiplying by $-\frac{4}{5}$ to get $1 \cdot x$ and eliminate $-\frac{5}{4}$ on the right

$$-\frac{12}{40} = 1 \cdot x$$

$$-\frac{3}{10} = 1 \cdot x$$ Simplifying

$$-\frac{3}{10} = x$$ Identity property of 1

3. Solve: $-6x = 108$.

4. Solve: $-x = -10$.

Answers on page A-28

673

5. Solve: $\dfrac{2}{3} = -\dfrac{5}{6}y$.

Solve.

6. $1.12x = 8736$

7. $6.3 = -2.1y$

8. Solve: $-14 = \dfrac{-y}{2}$.

Answers on page A-28

CHECK:
$$\frac{3}{8} = -\frac{5}{4}x$$

$$\frac{3}{8} \;\Big|\;?\; -\frac{5}{4}\left(-\frac{3}{10}\right)$$

$$\frac{3}{8} \qquad \text{TRUE}$$

The solution is $-\dfrac{3}{10}$.

If $a = b$ is true, then $b = a$ is true. Thus when we solve $\frac{3}{8} = -\frac{5}{4}x$, we can reverse it and solve $-\frac{5}{4}x = \frac{3}{8}$ if we wish.

Do Exercise 5.

EXAMPLE 6 Solve: $1.16y = 9744$.

$$\frac{1.16y}{1.16} = \frac{9744}{1.16} \qquad \text{Dividing by 1.16 on both sides}$$

$$y = \frac{9744}{1.16}$$

$$y = 8400 \qquad \text{Using a calculator to divide}$$

CHECK:
$$\frac{1.16y = 9744}{1.16(8400) \;\;?\;\; 9744}$$
$$9744 \;\Big|\qquad \text{TRUE}$$

The solution is 8400.

Do Exercises 6 and 7.

Now we solve an equation by using the multiplication principle.

EXAMPLE 7 Solve: $\dfrac{-y}{9} = 14$.

$$9 \cdot \frac{-y}{9} = 9 \cdot 14 \qquad \text{Multiplying by 9 on both sides}$$

$$-y = 126$$

$$-1 \cdot (-y) = -1 \cdot 126 \qquad \text{Multiplying by } -1 \text{ on both sides}$$

$$y = -126$$

CHECK:
$$\frac{-y}{9} = 14$$

$$\frac{-(-126)}{9} \;\Big|\;?\; 14$$

$$\frac{126}{9}$$

$$14 \;\Big|\qquad \text{TRUE}$$

The solution is -126.

Do Exercise 8.

EXERCISE SET

For Extra Help

Digital Video
Tutor CD 6
Videotape 16

InterAct
Math

Math Tutor
Center

MathXL

MyMathLab

a Solve using the multiplication principle. Don't forget to check!

1. $6x = 36$

2. $4x = 52$

3. $5x = 45$

4. $8x = 56$

5. $84 = 7x$

6. $63 = 7x$

7. $-x = 40$

8. $50 = -x$

9. $-2x = -10$

10. $-78 = -39p$

11. $7x = -49$

12. $9x = -54$

13. $-12x = 72$

14. $-15x = 105$

15. $-21x = -126$

16. $-13x = -104$

17. $\frac{1}{7}t = -9$

18. $-\frac{1}{8}y = 11$

19. $\frac{3}{4}x = 27$

20. $\frac{4}{5}x = 16$

21. $-\frac{1}{3}t = 7$

22. $-\frac{1}{6}x = 9$

23. $-\frac{1}{3}m = \frac{1}{5}$

24. $\frac{1}{5} = -\frac{1}{8}z$

25. $-\frac{3}{5}r = \frac{9}{10}$

26. $\frac{2}{5}y = -\frac{4}{15}$

27. $-\frac{3}{2}r = -\frac{27}{4}$

28. $-\frac{5}{7}x = -\frac{10}{14}$

29. $6.3x = 44.1$

30. $2.7y = 54$

31. $-3.1y = 21.7$

32. $-3.3y = 6.6$

33. $38.7m = 309.6$

34. $29.4m = 235.2$

35. $-\dfrac{2}{3}y = -10.6$

36. $-\dfrac{9}{7}y = 12.06$

37. $\dfrac{-x}{5} = 10$

38. $\dfrac{-x}{8} = -16$

39. $\dfrac{t}{-2} = 7$

40. $\dfrac{m}{-3} = 10$

41. ^{D}W Explain the following mistake made by a fellow student.

$$\frac{2}{3}x = -\frac{5}{3}$$

$$x = \frac{10}{9}$$

42. ^{D}W Explain the role of the reciprocal of a number when using the multiplication principle.

SKILL MAINTENANCE

43. Find the circumference, the diameter, and the area of a circle painted on a driveway. The radius is 10 ft. Use 3.14 for π.   [9.3a, b, c]

44. Find the circumference, the radius, and the area of a circle whose diameter is 24 cm. Use 3.14 for π. [9.3a, b, c]

45. Find the volume of a rectangular block of granite of length 25 ft, width 10 ft, and height 32 ft.   [9.4a]

46. Find the volume of a rectangular solid of length 1.3 cm, width 10 cm, and height 2.4 cm.   [9.4a]

Find the area of the figure.   [9.2b]

47.

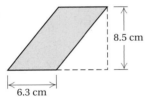

8.5 cm

6.3 cm

48.

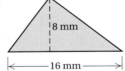

8 mm

16 mm

49.

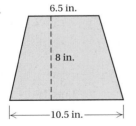

6.5 in.

8 in.

10.5 in.

50.

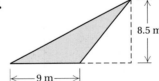

8.5 m

9 m

SYNTHESIS

Solve.

51. ▦ $-0.2344m = 2028.732$

52. $0 \cdot x = 0$

53. $0 \cdot x = 9$

54. $4|x| = 48$

55. $2|x| = -12$

56. A student makes a calculation and gets an answer of 22.5. On the last step, the student multiplies by 0.3 when a division by 0.3 should have been done. What should the correct answer be?

11.4

USING THE PRINCIPLES TOGETHER

Objectives

a Solve equations using both the addition and the multiplication principles.

b Solve equations in which like terms may need to be collected.

c Solve equations by first removing parentheses and collecting like terms.

a Applying Both Principles

Consider the equation $3x + 4 = 13$. It is more complicated than those in the preceding two sections. In order to solve such an equation, we first isolate the x-term, $3x$, using the addition principle. Then we apply the multiplication principle to get x by itself.

EXAMPLE 1 Solve: $3x + 4 = 13$.

$$3x + 4 = 13$$

$$3x + 4 - 4 = 13 - 4 \qquad \text{Using the addition principle: adding } -4 \text{ or subtracting 4 on both sides}$$

$$3x = 9 \qquad \text{Simplifying}$$

$$\frac{3x}{3} = \frac{9}{3} \qquad \text{Using the multiplication principle: multiplying by } \frac{1}{3} \text{ or dividing by 3 on both sides}$$

$$x = 3 \qquad \text{Simplifying}$$

CHECK:
$$\frac{3x + 4 = 13}{3 \cdot 3 + 4 \; ? \; 13}$$
$$9 + 4 \;\bigg|$$
$$13 \;\bigg| \qquad \textbf{TRUE}$$

The solution is 3.

Do Exercise 1.

1. Solve: $9x + 6 = 51$.

EXAMPLE 2 Solve: $-5x - 6 = 16$.

$$-5x - 6 = 16$$

$$-5x - 6 + 6 = 16 + 6 \qquad \text{Adding 6 on both sides}$$

$$-5x = 22$$

$$\frac{-5x}{-5} = \frac{22}{-5} \qquad \text{Dividing by } -5 \text{ on both sides}$$

$$x = -\frac{22}{5}, \text{ or } -4\frac{2}{5} \qquad \text{Simplifying}$$

CHECK:
$$\frac{-5x - 6 = 16}{-5\left(-\frac{22}{5}\right) - 6 \; ? \; 16}$$
$$22 - 6 \;\bigg|$$
$$16 \;\bigg| \qquad \textbf{TRUE}$$

The solution is $-\frac{22}{5}$.

Do Exercises 2 and 3.

Solve.

2. $8x - 4 = 28$

3. $-\dfrac{1}{2}x + 3 = 1$

Answers on page A-28

4. Solve: $-18 - x = -57$.

Solve.

5. $-4 - 8x = 8$

6. $41.68 = 4.7 - 8.6y$

Solve.

7. $4x + 3x = -21$

8. $x - 0.09x = 728$

Answers on page A-28

678

EXAMPLE 3   Solve: $45 - x = 13$.

$$45 - x = 13$$
$$-45 + 45 - x = -45 + 13 \qquad \text{Adding } -45 \text{ on both sides}$$
$$-x = -32$$
$$-1 \cdot (-x) = -1 \cdot (-32) \qquad \text{Multiplying by } -1 \text{ on both sides}$$
$$-1 \cdot (-1) \cdot x = 32$$
$$x = 32$$

CHECK:
$$\frac{45 - x = 13}{45 - 32 \ ? \ 13}$$
$$13 \ | \qquad \textbf{TRUE}$$

The solution is 32.

Do Exercise 4.

EXAMPLE 4   Solve: $16.3 - 7.2y = -8.18$.

$$16.3 - 7.2y = -8.18$$
$$-16.3 + 16.3 - 7.2y = -16.3 + (-8.18) \qquad \text{Adding } -16.3 \text{ on both sides}$$
$$-7.2y = -24.48$$
$$\frac{-7.2y}{-7.2} = \frac{-24.48}{-7.2} \qquad \text{Dividing by } -7.2 \text{ on both sides}$$
$$y = 3.4$$

CHECK:
$$\frac{16.3 - 7.2y = -8.18}{16.3 - 7.2(3.4) \ ? \ -8.18}$$
$$16.3 - 24.48 \ |$$
$$-8.18 \ | \qquad \textbf{TRUE}$$

The solution is 3.4.

Do Exercises 5 and 6.

b   Collecting Like Terms

If there are like terms on one side of the equation, we collect them before using the addition or multiplication principle.

EXAMPLE 5   Solve: $3x + 4x = -14$.

$$3x + 4x = -14$$
$$7x = -14 \qquad \text{Collecting like terms}$$
$$\frac{7x}{7} = \frac{-14}{7} \qquad \text{Dividing by 7 on both sides}$$
$$x = -2.$$

The number -2 checks, so the solution is -2.

Do Exercises 7 and 8.

If there are like terms on opposite sides of the equation, we get them on the same side by using the addition principle. Then we collect them. In other words, we get all terms with a variable on one side and all numbers on the other.

EXAMPLE 6   Solve: $2x - 2 = -3x + 3$.

$$2x - 2 = -3x + 3$$
$$2x - 2 + 2 = -3x + 3 + 2 \qquad \text{Adding 2}$$
$$2x = -3x + 5 \qquad \text{Collecting like terms}$$
$$2x + 3x = -3x + 5 + 3x \qquad \text{Adding } 3x$$
$$5x = 5 \qquad \text{Simplifying}$$
$$\frac{5x}{5} = \frac{5}{5} \qquad \text{Dividing by 5}$$
$$x = 1 \qquad \text{Simplifying}$$

CHECK:
$$\frac{2x - 2 = -3x + 3}{2 \cdot 1 - 2 \; ? \; -3 \cdot 1 + 3}$$
$$2 - 2 \;\bigg|\; -3 + 3$$
$$0 \;\bigg|\; 0 \qquad \textbf{TRUE}$$

The solution is 1.

Do Exercises 9 and 10.

In Example 6, we used the addition principle to get all terms with a variable on one side and all numbers on the other side. Then we collected like terms and proceeded as before. If there are like terms on one side at the outset, they should be collected first.

EXAMPLE 7   Solve: $6x + 5 - 7x = 10 - 4x + 3$.

$$6x + 5 - 7x = 10 - 4x + 3$$
$$-x + 5 = 13 - 4x \qquad \text{Collecting like terms}$$
$$4x - x + 5 = 13 - 4x + 4x \qquad \text{Adding } 4x$$
$$3x + 5 = 13 \qquad \text{Simplifying}$$
$$3x + 5 - 5 = 13 - 5 \qquad \text{Subtracting 5}$$
$$3x = 8 \qquad \text{Simplifying}$$
$$\frac{3x}{3} = \frac{8}{3} \qquad \text{Dividing by 3}$$
$$x = \frac{8}{3} \qquad \text{Simplifying}$$

The number $\frac{8}{3}$ checks, so $\frac{8}{3}$ is the solution.

Do Exercises 11 and 12.

Solve.

9. $7y + 5 = 2y + 10$

10. $5 - 2y = 3y - 5$

Solve.

11. $7x - 17 + 2x = 2 - 8x + 15$

12. $3x - 15 = 5x + 2 - 4x$

Answers on page A-28

BEGINNING TO STUDY FOR THE FINAL EXAM (PART 3): THREE DAYS TO TWO WEEKS OF STUDY TIME

1. Begin by browsing through each chapter, reviewing the highlighted or boxed information regarding important formulas in both the text and the Summary and Review. There may be some formulas that you will need to memorize.

2. Retake each chapter test that you took in class, assuming your instructor has returned it. Otherwise, use the chapter test in the book. Restudy the objectives in the text that correspond to each question you missed.

3. Work the Cumulative Review/Final Examination during the last couple of days before the final. Set aside the same amount of time that you will have for the final. See how much of the final exam you can complete under test-like conditions. Be careful to avoid any questions corresponding to objectives not covered. Again, restudy the objectives in the text that correspond to each question you missed.

4. For remaining difficulties, see your instructor, go to a tutoring session, or participate in a study group.

"It is a great piece of skill to know how to guide your luck, even while waiting for it."

Baltasar Gracian, seventeenth-century Spanish philosopher and writer

CLEARING FRACTIONS AND DECIMALS

For the equations considered thus far, we generally use the addition principle first. There are, however, some situations in which it is to our advantage to use the multiplication principle first. Consider, for example,

$$\frac{1}{2}x = \frac{3}{4}.$$

The LCM of the denominators is 4. If we multiply by 4 on both sides, we get $2x = 3$, which has no fractions. We have "cleared fractions." Now consider

$$2.3x = 4.78.$$

If we multiply by 100 on both sides, we get $230x = 478$, which has no decimal points. We have "cleared decimals." The equations are then easier to solve. It is your choice whether to clear fractions or decimals, but doing so often eases computations.

In what follows, we use the multiplication principle first to "clear," or "eliminate," fractions or decimals. For fractions, the number by which we multiply is the **least common multiple of all the denominators.**

EXAMPLE 8 Solve:

$$\frac{2}{3}x - \frac{1}{6} + \frac{1}{2}x = \frac{7}{6} + 2x.$$

The number 6 is the least common multiple of all the denominators. We multiply by 6 on both sides:

$$6\left(\frac{2}{3}x - \frac{1}{6} + \frac{1}{2}x\right) = 6\left(\frac{7}{6} + 2x\right) \quad \text{Multiplying by 6 on both sides}$$

$$6 \cdot \frac{2}{3}x - 6 \cdot \frac{1}{6} + 6 \cdot \frac{1}{2}x = 6 \cdot \frac{7}{6} + 6 \cdot 2x \quad \text{Using the distributive laws. (Caution! Be sure to multiply all terms by 6.)}$$

$$4x - 1 + 3x = 7 + 12x \quad \text{Simplifying. Note that the fractions are cleared.}$$

$$7x - 1 = 7 + 12x \quad \text{Collecting like terms}$$

$$7x - 1 - 12x = 7 + 12x - 12x \quad \text{Subtracting } 12x$$

$$-5x - 1 = 7 \quad \text{Simplifying}$$

$$-5x - 1 + 1 = 7 + 1 \quad \text{Adding 1}$$

$$-5x = 8 \quad \text{Collecting like terms}$$

$$\frac{-5x}{-5} = \frac{8}{-5} \quad \text{Dividing by } -5$$

$$x = -\frac{8}{5}.$$

CHECK:

$$\frac{2}{3}x - \frac{1}{6} + \frac{1}{2}x = \frac{7}{6} + 2x$$

$$\frac{2}{3}\left(-\frac{8}{5}\right) - \frac{1}{6} + \frac{1}{2}\left(-\frac{8}{5}\right) \;?\; \frac{7}{6} + 2\left(-\frac{8}{5}\right)$$

$$-\frac{16}{15} - \frac{1}{6} - \frac{8}{10} \;\bigg|\; \frac{7}{6} - \frac{16}{5}$$

$$-\frac{32}{30} - \frac{5}{30} - \frac{24}{30} \;\bigg|\; \frac{35}{30} - \frac{96}{30}$$

$$\frac{-32 - 5 - 24}{30} \;\bigg|\; -\frac{61}{30}$$

$$-\frac{61}{30}$$ **TRUE**

The solution is $-\frac{8}{5}$.

Do Exercise 13.

To illustrate clearing decimals, we repeat Example 4, but this time we clear the decimals first.

EXAMPLE 9   Solve: $16.3 - 7.2y = -8.18$.

The greatest number of decimal places in any one number is *two*. Multiplying by 100, which has *two* 0's, will clear the decimals.

$$100(16.3 - 7.2y) = 100(-8.18)$$ Multiplying by 100 on both sides

$$100(16.3) - 100(7.2y) = 100(-8.18)$$ Using a distributive law

$$1630 - 720y = -818$$ Simplifying

$$1630 - 720y - 1630 = -818 - 1630$$ Subtracting 1630 on both sides

$$-720y = -2448$$ Collecting like terms

$$\frac{-720y}{-720} = \frac{-2448}{-720}$$ Dividing by -720 on both sides

$$y = \frac{17}{5}, \text{ or } 3.4$$

The number $\frac{17}{5}$, or 3.4, checks and is the solution.

Do Exercise 14.

13. Solve: $\frac{7}{8}x - \frac{1}{4} + \frac{1}{2}x = \frac{3}{4} + x$.

14. Solve: $41.68 = 4.7 - 8.6y$.

Answers on page A-28

Solve.

15. $2(2y + 3) = 14$

C Equations Containing Parentheses

To solve certain kinds of equations that contain parentheses, we first use the distributive laws to remove the parentheses. Then we proceed as before.

EXAMPLE 10 Solve: $4x = 2(12 - 2x)$.

$$4x = 2(12 - 2x)$$

$$4x = 24 - 4x \qquad \text{Using the distributive law to multiply and remove parentheses}$$

$$4x + 4x = 24 - 4x + 4x \qquad \text{Adding } 4x \text{ to get all } x\text{-terms on one side}$$

$$8x = 24 \qquad \text{Collecting like terms}$$

$$\frac{8x}{8} = \frac{24}{8} \qquad \text{Dividing by 8}$$

$$x = 3$$

CHECK:

$$\frac{4x = 2(12 - 2x)}{4 \cdot 3 \ ? \ 2(12 - 2 \cdot 3)}$$

$$12 \ \bigg| \ 2(12 - 6) \qquad \text{We use the rules for order of operations to}$$

$$2 \cdot 6 \qquad \text{carry out the calculations on each side of}$$

$$12 \qquad \textbf{TRUE} \qquad \text{the equation.}$$

The solution is 3.

Do Exercises 15 and 16.

16. $5(3x - 2) = 35$

Here is a procedure for solving the types of equation discussed in this section.

> ### AN EQUATION-SOLVING PROCEDURE
>
> 1. Multiply on both sides to clear the equation of fractions or decimals. (This is optional, but it can ease computations.)
> 2. If parentheses occur, multiply using the *distributive laws* to remove them.
> 3. Collect like terms on each side, if necessary.
> 4. Get all terms with variables on one side and all constant terms on the other side, using the *addition principle*.
> 5. Collect like terms again, if necessary.
> 6. Multiply or divide to solve for the variable, using the *multiplication principle*.
> 7. Check all possible solutions in the original equation.

Answers on page A-28

EXAMPLE 11 Solve: $2 - 5(x + 5) = 3(x - 2) - 1$.

$$2 - 5(x + 5) = 3(x - 2) - 1$$

$2 - 5x - 25 = 3x - 6 - 1$	Using the distributive laws to multiply and remove parentheses
$-5x - 23 = 3x - 7$	Collecting like terms
$-5x - 23 + 5x = 3x - 7 + 5x$	Adding $5x$
$-23 = 8x - 7$	Collecting like terms
$-23 + 7 = 8x - 7 + 7$	Adding 7
$-16 = 8x$	Collecting like terms
$\dfrac{-16}{8} = \dfrac{8x}{8}$	Dividing by 8
$-2 = x$	

CHECK:

$$\begin{array}{c|c} \multicolumn{2}{c}{2 - 5(x + 5) = 3(x - 2) - 1} \\ \hline 2 - 5(-2 + 5) \;?\; & 3(-2 - 2) - 1 \\ 2 - 5(3) & 3(-4) - 1 \\ 2 - 15 & -12 - 1 \\ -13 & -13 \quad \textbf{TRUE} \end{array}$$

The solution is -2.

Note that the solution of $-2 = x$ is -2, which is also the solution of $x = -2$.

Do Exercises 17 and 18.

Solve.

17. $3(7 + 2x) = 30 + 7(x - 1)$

18. $4(3 + 5x) - 4 = 3 + 2(x - 2)$

Answers on page A-28

CALCULATOR CORNER

Checking Solutions of Equations   We can use a calculator to check solutions of equations by substituting the possible solution for each occurrence of the variable on the left side of the equation and then on the right side of the equation. If both sides have the same value, then the number that was substituted for the variable is the solution of the equation. In Example 5, for instance, to check if -2 is the solution of the equation $3x + 4x = -14$, we substitute -2 for x on the left side of the equation. To do this, we press $\boxed{3}\;\boxed{\times}\;\boxed{2}\;\boxed{+/-}\;\boxed{+}\;\boxed{4}\;\boxed{\times}\;\boxed{2}\;\boxed{+/-}\;\boxed{=}$. The result is -14. Note that the right side of the equation does not contain a variable. It is -14, the value that was obtained when -2 was substituted for x on the left side. Thus, the number -2 is the solution of the equation.

In Example 6, we can check if 1 is the solution of the equation $2x - 2 = -3x + 3$. First, we substitute 1 for x on the left side of the equation by pressing $\boxed{2}\;\boxed{\times}\;\boxed{1}\;\boxed{-}\;\boxed{2}\;\boxed{=}$. We get 0. Next, we substitute 1 for x on the right side of the equation by pressing $\boxed{3}\;\boxed{+/-}\;\boxed{\times}\;\boxed{1}\;\boxed{+}\;\boxed{3}\;\boxed{=}$. We get 0 again. Since the same value was obtained on both sides of the equation, the number 1 is the solution.

Exercises:

1. Use a calculator to check the solutions of the equations in Examples 1–4 and 7–11.
2. Use a calculator to check the solutions of the equations in Margin Exercises 1–18.

11.4

EXERCISE SET

For Extra Help

Digital Video
Tutor CD 6
Videotape 16

InterAct
Math

Math Tutor
Center

MathXL

MyMathLab

a   Solve. Don't forget to check!

1. $5x + 6 = 31$ **2.** $8x + 6 = 30$ **3.** $8x + 4 = 68$ **4.** $8z + 7 = 79$

5. $4x - 6 = 34$ **6.** $4x - 11 = 21$ **7.** $3x - 9 = 33$ **8.** $6x - 9 = 57$

9. $7x + 2 = -54$ **10.** $5x + 4 = -41$ **11.** $-45 = 3 + 6y$ **12.** $-91 = 9t + 8$

13. $-4x + 7 = 35$ **14.** $-5x - 7 = 108$ **15.** $-7x - 24 = -129$ **16.** $-6z - 18 = -132$

b   Solve.

17. $5x + 7x = 72$ **18.** $4x + 5x = 45$ **19.** $8x + 7x = 60$

20. $3x + 9x = 96$ **21.** $4x + 3x = 42$ **22.** $6x + 19x = 100$

23. $-6y - 3y = 27$ **24.** $-4y - 8y = 48$ **25.** $-7y - 8y = -15$

26. $-10y - 3y = -39$ **27.** $10.2y - 7.3y = -58$ **28.** $6.8y - 2.4y = -88$

29. $x + \dfrac{1}{3}x = 8$

30. $x + \dfrac{1}{4}x = 10$

31. $8y - 35 = 3y$

32. $4x - 6 = 6x$

33. $8x - 1 = 23 - 4x$

34. $5y - 2 = 28 - y$

35. $2x - 1 = 4 + x$

36. $5x - 2 = 6 + x$

37. $6x + 3 = 2x + 11$

38. $5y + 3 = 2y + 15$

39. $5 - 2x = 3x - 7x + 25$

40. $10 - 3x = 2x - 8x + 40$

41. $4 + 3x - 6 = 3x + 2 - x$

42. $5 + 4x - 7 = 4x - 2 - x$

43. $4y - 4 + y + 24 = 6y + 20 - 4y$

44. $5y - 7 + y = 7y + 21 - 5y$

Solve. Clear fractions or decimals first.

45. $\dfrac{7}{2}x + \dfrac{1}{2}x = 3x + \dfrac{3}{2} + \dfrac{5}{2}x$

46. $\dfrac{7}{8}x - \dfrac{1}{4} + \dfrac{3}{4}x = \dfrac{1}{16} + x$

47. $\dfrac{2}{3} + \dfrac{1}{4}t = \dfrac{1}{3}$

48. $-\dfrac{3}{2} + x = -\dfrac{5}{6} - \dfrac{4}{3}$

49. $\dfrac{2}{3} + 3y = 5y - \dfrac{2}{15}$

50. $\dfrac{1}{2} + 4m = 3m - \dfrac{5}{2}$

51. $\dfrac{5}{3} + \dfrac{2}{3}x = \dfrac{25}{12} + \dfrac{5}{4}x + \dfrac{3}{4}$

52. $1 - \dfrac{2}{3}y = \dfrac{9}{5} - \dfrac{y}{5} + \dfrac{3}{5}$

53. $2.1x + 45.2 = 3.2 - 8.4x$

54. $0.96y - 0.79 = 0.21y + 0.46$

55. $1.03 - 0.62x = 0.71 - 0.22x$

56. $1.7t + 8 - 1.62t = 0.4t - 0.32 + 8$

57. $\dfrac{2}{7}x - \dfrac{1}{2}x = \dfrac{3}{4}x + 1$

58. $\dfrac{5}{16}y + \dfrac{3}{8}y = 2 + \dfrac{1}{4}y$

C Solve.

59. $3(2y - 3) = 27$

60. $4(2y - 3) = 28$

61. $40 = 5(3x + 2)$

62. $9 = 3(5x - 2)$

63. $2(3 + 4m) - 9 = 45$

64. $3(5 + 3m) - 8 = 88$

65. $5r - (2r + 8) = 16$

66. $6b - (3b + 8) = 16$

67. $6 - 2(3x - 1) = 2$

68. $10 - 3(2x - 1) = 1$

69. $5(d + 4) = 7(d - 2)$

70. $3(t - 2) = 9(t + 2)$

71. $8(2t + 1) = 4(7t + 7)$

72. $7(5x - 2) = 6(6x - 1)$

73. $3(r - 6) + 2 = 4(r + 2) - 21$

74. $5(t + 3) + 9 = 3(t - 2) + 6$

CHAPTER 11: Algebra: Solving Equations
and Problems

75. $19 - (2x + 3) = 2(x + 3) + x$

76. $13 - (2c + 2) = 2(c + 2) + 3c$

77. $0.7(3x + 6) = 1.1 - (x + 2)$

78. $0.9(2x + 8) = 20 - (x + 5)$

79. $a + (a - 3) = (a + 2) - (a + 1)$

80. $0.8 - 4(b - 1) = 0.2 + 3(4 - b)$

81. $\mathbf{D_W}$ A student begins solving the equation $\frac{2}{3}x + 1 = \frac{5}{6}$ by multiplying by 6 on both sides. Is this a wise thing to do? Why or why not?

82. $\mathbf{D_W}$ Describe a procedure that a classmate could use to solve the equation $ax + b = c$ for x.

(**SKILL MAINTENANCE**)

Divide. [10.5c]

83. $22.1 \div 3.4$

84. $-22.1 \div (-3.4)$

85. $22.1 \div (-3.4)$

86. $-22.1 + 3.4$

Factor. [11.1c]

87. $7x - 21 - 14y$

88. $25a - 625b + 75$

89. $42t + 14m - 56$

90. $16a - 64b + 224 - 32q$

91. Evaluate $-(-x)$ when $x = -8$. [10.2b]

92. Use $<$ or $>$ for $\square$ to write a true sentence: [10.1d]
$-15 \,\square\, -13$.

(**SYNTHESIS**)

Solve.

93. $\dfrac{y - 2}{3} = \dfrac{2 - y}{5}$

94. $3x = 4x$

95. $\dfrac{5 + 2y}{3} = \dfrac{25}{12} + \dfrac{5y + 3}{4}$

96. $\boxed{\blacksquare}$ $0.05y - 1.82 = 0.708y - 0.504$

97. $\dfrac{2}{3}(2x - 1) = 10$

98. $\dfrac{2}{3}\left(\dfrac{7}{8} - 4x\right) - \dfrac{5}{8} = \dfrac{3}{8}$

99. The perimeter of the figure shown is 15 cm. Solve for x.

11.5 APPLICATIONS AND PROBLEM SOLVING

a Translating to Algebraic Expressions

In algebra, we translate problems to equations. The different parts of an equation are translations of word phrases to algebraic expressions. To translate, it helps to learn which words translate to certain operation symbols.

KEY WORDS

ADDITION (+)	SUBTRACTION (−)	MULTIPLICATION (·)	DIVISION (÷)
add	subtract	multiply	divide
added to	subtracted from	multiplied by	quotient
sum	difference	product	divided by
total	minus	times	
plus	less than	of	
more than	decreased by		
increased by	take away		

EXAMPLE 1 Translate to an algebraic expression:

Twice (or two times) some number.

Think of some number—say, 8. What number is twice 8? It is 16. How did you get 16? You multiplied by 2. Do the same thing using a variable. We can use any variable we wish, such as x, y, m, or n. Let's use y to stand for some number. If we multiply by 2, we get an expression

$$y \times 2, \quad 2 \times y, \quad 2 \cdot y, \quad \text{or} \quad 2y.$$

In algebra, $2y$ is the expression used most often.

EXAMPLE 2 Translate to an algebraic expression:

Seven less than some number.

We let

$$x = \text{the number.}$$

If the number were 23, then the translation would be "7 subtracted from 23," or $23 - 7$. If we knew the number to be 345, then the translation would be $345 - 7$. Thus if the number is x, then the translation is

$$x - 7.$$

Note that $7 - x$ is *not* a correct translation of the expression in Example 2. The expression $7 - x$ is a translation of "seven minus some number" or "some number less than seven."

EXAMPLE 3   Translate to an algebraic expression:

Eighteen more than a number.

We let

t = the number.

Now if the number were 26, then the translation would be $18 + 26$, or $26 + 18$. If we knew the number to be 174, then the translation would be $18 + 174$, or $174 + 18$. The translation is

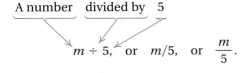

Eighteen   more than   a number

$18 + t,$ or $t + 18.$

EXAMPLE 4   Translate to an algebraic expression:

A number divided by 5.

We let

m = the number.

If the number were 76, then the translation would be $76 \div 5$, or $76/5$, or $\frac{76}{5}$. If the number were 213, then the translation would be $213 \div 5$, or $213/5$, or $\frac{213}{5}$. The translation is found as follows:

A number   divided by   5

$m \div 5,$ or $m/5,$ or $\dfrac{m}{5}.$

EXAMPLE 5   Translate to an algebraic expression.

PHRASE	ALGEBRAIC EXPRESSION
Five more than some number	$5 + n$, or $n + 5$
Half of a number	$\frac{1}{2}t$, or $\frac{t}{2}$
Five more than three times some number	$5 + 3p$, or $3p + 5$
The difference of two numbers	$x - y$
Six less than the product of two numbers	$mn - 6$
Seventy-six percent of some number	$76\%z$, or $0.76z$

Do Exercises 1–9.

Translate to an algebraic expression.

1. Twelve less than some number

2. Twelve more than some number

3. Four less than some number

4. Half of some number

5. Six more than eight times some number

6. The difference of two numbers

7. Fifty-nine percent of some number

8. Two hundred less than the product of two numbers

9. The sum of two numbers

Answers on page A-28

b   Five Steps for Solving Problems

We have studied many new equation-solving tools in this chapter. We now apply them to problem solving. We have purposely used the following strategy throughout this text in order to introduce you to algebra.

> ### FIVE STEPS FOR PROBLEM SOLVING IN ALGEBRA
>
> 1. *Familiarize* yourself with the problem situation.
> 2. *Translate* to an equation.
> 3. *Solve* the equation.
> 4. *Check* your possible answer in the original problem.
> 5. *State* the answer clearly.

Of the five steps, the most important is probably the first one: becoming familiar with the problem situation. The table below lists some hints for familiarization.

> To familiarize yourself with a problem:
> - If a problem is given in words, read it carefully. Reread the problem, perhaps aloud. Try to verbalize the problem as if you were explaining it to someone else.
> - Choose a variable (or variables) to represent the unknown and clearly state what the variable represents. Be descriptive! For example, let L = the length, d = the distance, and so on.
> - Make a drawing and label it with known information, using specific units if given. Also, indicate unknown information.
> - Find further information. Look up formulas or definitions with which you are not familiar. (Geometric formulas appear on the inside front cover of this text.) Consult a reference librarian or an expert in the field.
> - Create a table that lists all the information you have available. Look for patterns that may help in the translation to an equation.
> - Think of a possible answer and check the guess. Observe the manner in which the guess is checked.

EXAMPLE 6   *Hiking.*   In 1998, at age 79, Earl Shaffer became the oldest person to hike all 2100 miles of the Appalachian Trail—from Springer Mountain, Georgia, to Mount Katahdin, Maine. At one point, Shaffer stood atop Big Walker Mountain, Virginia, which is three times as far from the northern end as from the southern end. How far was Shaffer from each end of the trail?
Source: Appalachian Trail Conference

1. Familiarize. Let's consider a drawing.

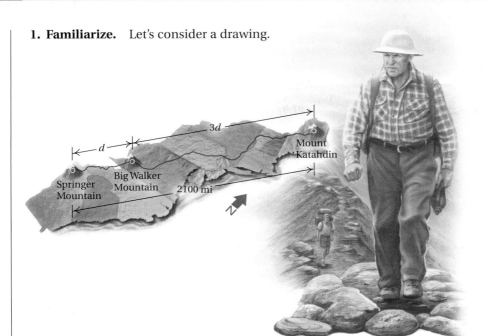

To become familiar with the problem, let's guess a possible distance that Shaffer stood from Springer Mountain—say, 600 mi. Three times 600 mi is 1800 mi. Since 600 mi + 1800 mi = 2400 mi and 2400 mi is greater than 2100 mi, we see that our guess is too large. Rather than guess again, let's use the skills we have obtained in the ability to solve equations. We let

d = the distance, in miles, to the southern end,

and

$3d$ = the distance, in miles, to the northern end.

(We could also let x = the distance to the northern end and $\frac{1}{3}x$ = the distance to the southern end.)

2. Translate. From the drawing, we see that the lengths of the two parts of the trail must add up to 2100 mi. This leads to our translation.

$$
\underbrace{\text{Distance to southern end}}_{d} \underset{+}{\text{ plus }} \underbrace{\text{Distance to northern end}}_{3d} \underset{=}{\text{ is }} \underset{2100}{2100 \text{ mi}}
$$

3. Solve. We solve the equation:

$d + 3d = 2100$

$4d = 2100$ Collecting like terms

$\dfrac{4d}{4} = \dfrac{2100}{4}$ Dividing by 4 on both sides

$d = 525.$

4. Check. As expected, d is less than 600 mi. If $d = 525$ mi, then $3d = 1575$ mi. Since 525 mi + 1575 mi = 2100 mi, we have a check.

5. State. Atop Big Walker Mountain, Shaffer stood 525 mi from Springer Mountain and 1575 mi from Mount Katahdin.

Do Exercise 10.

10. Running. In 1997, Yiannis Kouros of Australia set the record for the greatest distance run in 24 hr by running 188 mi. After 8 hr, he was approximately twice as far from the finish line as he was from the start. How far had he run?
Source: *Guinness World Records 2000 Millennium Edition*

Answer on page A-28

11. Board Cutting. A 52-in. board is cut into three pieces. The first piece is one-half the length of the third piece. The second piece is one-eighth the length of the third piece. Find the length of each piece.

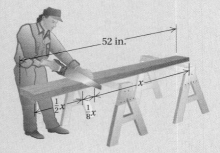

52 in.

x

$\frac{1}{2}x$ $\frac{1}{8}x$

Answer on page A-28

EXAMPLE 7 *Rocket Sections.* A rocket is divided into three sections: the payload and navigation at the top, the fuel in the middle, and the rocket engine at the bottom. The top section is one-sixth the length of the bottom section. The middle section is one-half the length of the bottom section. The total length of all three is 240 ft. Find the length of each section.

1. **Familiarize.** We first make a drawing. Noting that the lengths of the top and middle sections are expressed in terms of the bottom section, we let

$$x = \text{the length of the bottom section.}$$

Then $\frac{1}{2}x = $ the length of the middle section

and $\frac{1}{6}x = $ the length of the top section.

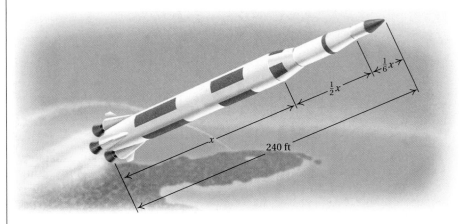

2. **Translate.** From the statement of the problem and the drawing, we see that the lengths add up to 240 ft. This gives us our translation.

Length of top	plus	Length of middle	plus	Length of bottom	is	Total length
$\frac{1}{6}x$	$+$	$\frac{1}{2}x$	$+$	x	$=$	240

3. **Solve.** We solve the equation by clearing fractions, as follows:

$$\frac{1}{6}x + \frac{1}{2}x + x = 240 \qquad \text{The LCM of all the denominators is 6.}$$

$$6 \cdot \left(\frac{1}{6}x + \frac{1}{2}x + x\right) = 6 \cdot 240 \qquad \text{Multiplying by the LCM, 6}$$

$$6 \cdot \frac{1}{6}x + 6 \cdot \frac{1}{2}x + 6 \cdot x = 6 \cdot 240 \qquad \text{Using the distributive law}$$

$$x + 3x + 6x = 1440 \qquad \text{Simplifying}$$

$$10x = 1440 \qquad \text{Adding}$$

$$\frac{10x}{10} = \frac{1440}{10} \qquad \text{Dividing by 10}$$

$$x = 144.$$

4. **Check.** Do we have an answer *to the problem*? If the rocket engine (bottom section) is 144 ft, then the middle section is $\frac{1}{2} \cdot 144$, or 72 ft, and the top section is $\frac{1}{6} \cdot 144$, or 24 ft. Since these lengths total 240 ft, our answer checks.

5. **State.** The top section is 24 ft, the middle 72 ft, and the bottom 144 ft.

Do Exercise 11.

EXAMPLE 8 *Truck Rental.* Truck-Rite Rentals rents trucks at a daily rate of $59.95 plus 49¢ per mile. Concert Productions has budgeted $400 per day for renting a truck to haul equipment to an upcoming concert. How many miles can a rental truck be driven on a $400 daily budget?

1. **Familiarize.** Suppose that Concert Productions drives 75 mi. Then the cost is

Daily charge   plus   Mileage charge

($59.95)   plus   (Cost per mile)   times   (Number of miles driven)
$59.95   + $0.49   ·   75,

which is $59.95 + $36.75, or $96.70. This familiarizes us with the way in which a calculation is made. Note that we convert 49 cents to $0.49 so that we are using the same units, dollars, throughout the problem.
   Let m = the number of miles that can be driven for $400.

2. **Translate.** We reword the problem and translate as follows.

Daily rate   plus   Cost per mile   times   Number of miles driven   is   Total cost

$59.95   + $0.49   · m = $400

3. **Solve.** We solve the equation:

$$\$59.95 + 0.49m = 400$$
$$100(59.95 + 0.49m) = 100(400) \quad \text{Multiplying by 100 on both sides to clear decimals}$$
$$100(59.95) + 100(0.49m) = 40,000 \quad \text{Using the distributive law}$$
$$5995 + 49m = 40,000$$
$$49m = 34,005 \quad \text{Subtracting 5995}$$
$$\frac{49m}{49} = \frac{34,005}{49} \quad \text{Dividing by 49}$$
$$m \approx 694.0. \quad \text{Rounding to the nearest tenth}$$

4. **Check.** We check in the original problem. We multiply 694 by $0.49, getting $340.06. Then we add $340.06 to $59.95 and get $400.01, which is just about the $400 allotted.

5. **State.** The truck can be driven about 694 mi on the rental allotment of $400.

Do Exercise 12.

---

12. **Van Rental.** Truck-Rite also rents vans at a daily rate of $64.95 plus 57 cents per mile. What mileage will allow a salesperson to stay within a daily budget of $400?

Answer on page A-28

Study Tips

BEGINNING TO STUDY FOR THE FINAL EXAM (PART 4): ONE OR TWO DAYS OF STUDY TIME

1. **Begin by browsing through each chapter, reviewing the highlighted or boxed information regarding important formulas in both the text and the Summary and Review.** There may be some formulas that you will need to memorize.

2. **Take the Final Examination in the text during the last couple of days before the final.** Set aside the same amount of time that you will have for the final. See how much of the final exam you can complete under test-like conditions. Be careful to avoid any questions corresponding to objectives not covered. Restudy the objectives in the text that correspond to each question you missed.

3. **Attend a final-exam review session if one is available.**

"Great is the art of beginning, but greater is the art of ending."

Henry Wadsworth Longfellow,
nineteenth-century
American poet

13. Angles of a Triangle. The second angle of a triangle is three times as large as the first. The third angle measures 30° more than the first angle. Find the measures of the angles.

EXAMPLE 9 *Angles of a Triangle.* The second angle of a triangle is twice as large as the first. The measure of the third angle is 20° greater than that of the first angle. How large are the angles?

1. Familiarize. We first make a drawing, letting

the measure of the first angle $= x$.

Then the measure of the second angle $= 2x$

and the measure of the third angle $= x + 20$.

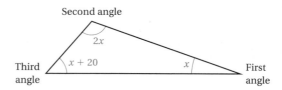

Second angle

$2x$

Third angle $x + 20$ x First angle

2. Translate. To translate, we recall from Section 9.5 that the measures of the angles of any triangle add up to 180°.

$$\underbrace{\text{Measure of first angle}}_{x} + \underbrace{\text{Measure of second angle}}_{2x} + \underbrace{\text{Measure of third angle}}_{(x + 20)} = 180°$$

$$x + 2x + (x + 20) = 180$$

3. Solve. We solve the equation:

$$x + 2x + (x + 20) = 180$$
$$4x + 20 = 180$$
$$4x + 20 - 20 = 180 - 20$$
$$4x = 160$$
$$\frac{4x}{4} = \frac{160}{4}$$
$$x = 40.$$

Possible measures for the angles are as follows:

First angle: $x = 40°$;
Second angle: $2x = 2(40) = 80°$;
Third angle: $x + 20 = 40 + 20 = 60°$.

4. Check. Consider 40°, 80°, and 60°. The second is twice the first, and the third is 20° greater than the first. The sum is 180°. These numbers check.

5. State. The measures of the angles are 40°, 80°, and 60°.

CAUTION!

Units are important in answers. Remember to include them, where appropriate.

Do Exercise 13.

Answer on page A-28

CHAPTER 11: Algebra: Solving Equations and Problems

EXAMPLE 10 *Area of Colorado.* The state of Colorado is roughly in the shape of a rectangle whose perimeter is 1300 mi. The length is 110 mi more than the width. Find the dimensions.

1. **Familiarize.** We first make a drawing, letting

$$w = \text{the width of the rectangle.}$$

Then $w + 110 =$ the length.

(We can also let $l =$ the length and $l - 110 =$ the width.)

The perimeter P of a rectangle is the distance around it and is given by the formula $2l + 2w = P$, where $l =$ the length and $w =$ the width.

2. **Translate.** To translate the problem, we substitute $w + 110$ for l and 1300 for P, as follows:

$$2l + 2w = P$$
$$2(w + 110) + 2w = 1300.$$

> **CAUTION!**
> Parentheses are important here.

3. **Solve.** We solve the equation:

$$2(w + 110) + 2w = 1300$$
$$2w + 220 + 2w = 1300 \qquad \text{Multiplying using the distributive law}$$
$$4w + 220 = 1300$$
$$4w = 1080$$
$$w = 270.$$

Possible dimensions are $w = 270$ mi and $l = w + 110 = 380$ mi.

4. **Check.** If the width is 270 mi and the length is 380 mi, the perimeter is 2(380 mi) + 2(270 mi), or 1300 mi. This checks.

5. **State.** The width is 270 mi, and the length is 380 mi.

Do Exercise 14.

> **CAUTION!**
>
> Always be sure to answer the original problem completely. For instance, in Example 6 we need to find *two* numbers: the distances from *each* end of the trail to the hiker. Similarly, in Example 10, we need to find two dimensions, not just the width.

14. Dimensions of Rug. A hooked rug has a perimeter of 42 ft. The length is 3 ft more than the width. Find the dimensions of the rug.

Answer on page A-28

15. Savings Investment. An investment is made at 7% simple interest for 1 year. It grows to $8988. How much was originally invested (the principal)?

EXAMPLE 11 *Savings Investment.* An investment is made at 6% simple interest for 1 year. It grows to $768.50. How much was originally invested (the principal)?

1. **Familiarize.** Suppose that $100 was invested. Recalling the formula for simple interest, $I = Prt$, we know that the interest for 1 year on $100 at 6% simple interest is given by $I = \$100 \cdot 0.06 \cdot 1 = \$6$. Then, at the end of the year, the amount in the account is found by adding the principal and the interest:

$$
\begin{array}{ccccc}
\text{Principal} & + & \text{Interest} & = & \text{Amount} \\
\downarrow & \downarrow & \downarrow & \downarrow & \downarrow \\
\$100 & + & \$6 & = & \$106.
\end{array}
$$

In this problem, we are working backward. We are trying to find the principal, which is the original investment. We let $x =$ the principal.

2. **Translate.** We reword the problem and then translate.

$$
\begin{array}{ccccc}
\text{Principal} & + & \text{Interest} & = & \text{Amount} \\
\downarrow & \downarrow & \downarrow & \downarrow & \downarrow \\
x & + & 6\%x & = & 768.50
\end{array}
$$

Interest is 6% of the principal.

3. **Solve.** We solve the equation:

$$
\begin{aligned}
x + 6\%x &= 768.50 \\
x + 0.06x &= 768.50 \qquad \text{Converting to decimal notation} \\
1x + 0.06x &= 768.50 \qquad \text{Identity property of 1} \\
1.06x &= 768.50 \qquad \text{Collecting like terms} \\
\frac{1.06x}{1.06} &= \frac{768.50}{1.06} \qquad \text{Dividing by 1.06} \\
x &= 725.
\end{aligned}
$$

4. **Check.** We check by taking 6% of $725 and adding it to $725:

$$6\% \times \$725 = 0.06 \times 725 = \$43.50.$$

Then $725 + \$43.50 = \$768.50$, so $725 checks.

5. **State.** The original investment was $725.

Do Exercise 15.

Answer on page A-28

Study Tips

FINAL STUDY TIP

You are arriving at the end of your course in Basic Mathematics. If you have not begun to prepare for the final examination, be sure to read the comments in the Study Tips on pp. 576, 644, 680, and 693.

"We make a living by what we get, but we make a life by what we give."

Winston Churchill

696

CHAPTER 11: Algebra: Solving Equations and Problems

For Extra Help

Digital Video Tutor CD 6 Videotape 16   InterAct Math   Math Tutor Center   MathXL   MyMathLab

a   Translate to an algebraic expression.

1. Three less than twice a number

2. Three times a number divided by a

3. The product of 97% and some number

4. 43% of some number

5. Four more than five times some number

6. Seventy-five less than eight times a number

7. A 240-in. pipe is cut into two pieces. One piece is three times the length of the other. Let $x =$ the length of the longer piece. Write an expression for the length of the shorter piece.

8. The price of a book is decreased by 30% during a sale. Let $b =$ the price of the book before the reduction. Write an expression for the sale price.

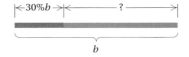

b   Solve.

9. What number added to 85 is 117?

10. Eight times what number is 2552?

11. *Statue of Liberty.* The height of the Eiffel Tower is 974 ft, which is about 669 ft higher than the Statue of Liberty. What is the height of the Statue of Liberty?

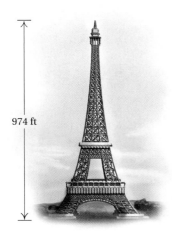

12. *Area of Lake Ontario.* The area of Lake Superior is about four times the area of Lake Ontario. The area of Lake Superior is 30,172 mi². What is the area of Lake Ontario?

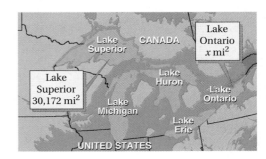

13. *Wheaties.* Recently, the cost of four 18-oz boxes of Wheaties cereal was $14.68. What was the cost of one box?
Source: General Mills

14. *Women's Dresses.* In a recent year, the total amount spent on women's blouses was $6.5 billion. This was $0.2 billion more than what was spent on women's dresses. How much was spent on women's dresses?

15. When 17 is subtracted from four times a certain number, the result is 211. What is the number?

16. When 36 is subtracted from five times a certain number, the result is 374. What is the number?

17. If you double a number and then add 16, you get $\frac{2}{3}$ of the original number. What is the original number?

18. If you double a number and then add 85, you get $\frac{3}{4}$ of the original number. What is the original number?

19. *Iditarod Race.* The Iditarod sled-dog race extends for 1049 mi from Anchorage to Nome. If a musher is twice as far from Anchorage as from Nome, how many miles has the musher traveled?
Source: Iditarod Trail Commission

20. *Home Remodeling.* In a recent year, Americans spent a total of approximately $32.8 billion to remodel bathrooms and kitchens. Twice as much was spent on kitchens as on bathrooms. How much was spent on each?
Source: *Remodeling Magazine*

698

21. *Pipe Cutting.* A 480-m pipe is cut into three pieces. The second piece is three times as long as the first. The third piece is four times as long as the second. How long is each piece?

22. *Rope Cutting.* A 180-ft rope is cut into three pieces. The second piece is twice as long as the first. The third piece is three times as long as the second. How long is each piece of rope?

23. *Perimeter of NBA Court.* The perimeter of an NBA basketball court is 288 ft. The length is 44 ft longer than the width. Find the dimensions of the court.
Source: National Basketball Association

24. *Perimeter of High School Court.* The perimeter of a standard high school basketball court is 268 ft. The length is 34 ft longer than the width. Find the dimensions of the court.
Source: Indiana High School Athletic Association

25. *Hancock Building Dimensions.* The ground floor of the John Hancock Building in Chicago is in the shape of a rectangle whose length is 100 ft more than the width. The perimeter is 860 ft. Find the length, the width, and the area of the ground floor.

26. *Hancock Building Dimensions.* The top floor of the John Hancock Building in Chicago is in the shape of a rectangle whose length is 60 ft more than the width. The perimeter is 520 ft. Find the length, the width, and the area of the top floor.

27. *Car Rental.* Value Rent-A-Car rents a family-sized car at a daily rate of $74.95 plus 40 cents per mile. Rick Moneycutt is allotted a daily budget of $240. How many miles can he drive per day and stay within his budget?

28. *Van Rental.* Value Rent-A-Car rents a van at a daily rate of $84.95 plus 60 cents per mile. Molly Rickert rents a van to deliver electrical parts to her customers. She is allotted a daily budget of $320. How many miles can she drive per day and stay within her budget?

29. *Angles of a Triangle.* The second angle of a triangular parking lot is four times as large as the first. The third angle is 45° less than the sum of the other two angles. How large are the angles?

30. *Angles of a Triangle.* The second angle of a triangular field is three times as large as the first. The third angle is 40° greater than the first. How large are the angles?

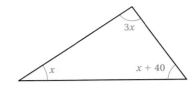

31. *Stock Prices.* Sarah's investment in AOL/Time Warner stock grew 28% to $448. How much did she invest?

32. *Savings Interest.* Sharon invested money in a savings account at a rate of 6% simple interest. After 1 yr, she has $6996 in the account. How much did Sharon originally invest?

33. *Credit Cards.* The balance in Will's Mastercard® account grew 2%, to $870, in one month. What was his balance at the beginning of the month?

34. *Loan Interest.* Alvin borrowed money from a cousin at a rate of 10% simple interest. After 1 yr, $7194 paid off the loan. How much did Alvin borrow?

35. *Taxi Fares.* In Beniford, taxis charge $3 plus 75¢ per mile for an airport pickup. How far from the airport can Courtney travel for $12?

36. *Taxi Fares.* In Cranston, taxis charge $4 plus 90¢ per mile for an airport pickup. How far from the airport can Ralph travel for $17.50?

37. *Tipping.* Leon left a 15% tip for a meal. The total cost of the meal, including the tip, was $41.40. What was the cost of the meal before the tip was added?

38. *Tipping.* Selena left an 18% tip for a meal. The total cost of the meal, including the tip, was $40.71. What was the cost of the meal before the tip was added?

39. *Standard Billboard Sign.* A standard rectangular highway billboard sign has a perimeter of 124 ft. The length is 6 ft more than three times the width. Find the dimensions.

40. *Two-by-Four.* The perimeter of a cross section of a "two-by-four" piece of lumber is $10\frac{1}{2}$ in. The length is twice the width. Find the actual dimensions of the cross section of a two-by-four.

Two-by-four $P = 10\frac{1}{2}$ in.

41. $^{\text{D}}\text{W}$ A fellow student claims to be able to solve most of the problems in this section by guessing. Is there anything wrong with this approach? Why or why not?

42. $^{\text{D}}\text{W}$ Write a problem for a classmate to solve so that it can be translated to the equation
$$\tfrac{2}{3}x + (x + 5) + x = 375.$$

SKILL MAINTENANCE

Calculate.

43. $-\dfrac{4}{5} - \left(\dfrac{3}{8}\right)$ [10.3a]

44. $-\dfrac{4}{5} + \dfrac{3}{8}$ [10.2a]

45. $-\dfrac{4}{5} \cdot \dfrac{3}{8}$ [10.4a]

46. $-\dfrac{4}{5} \div \left(\dfrac{3}{8}\right)$ [10.5c]

47. $-25.6 \div (-16)$
[10.5c]

48. $-25.6(-16)$ [10.4a]

49. $-25.6 - (-16)$
[10.3a]

50. $-25.6 + (-16)$
[10.2a]

SYNTHESIS

51. Abraham Lincoln's 1863 Gettysburg Address refers to the year 1776 as "Four *score* and seven years ago." Write an equation to find out what a score is.

52. A student scored 78 on a test that had 4 seven-point fill-ins and 24 three-point multiple-choice questions. The student had 1 fill-in wrong. How many multiple-choice questions did the student get right?

53. The width of a rectangle is $\frac{3}{4}$ of the length. The perimeter of the rectangle becomes 50 cm when the length and the width are each increased by 2 cm. Find the length and the width.

54. Cookies are set out on a tray for six people to take home. One-third, one-fourth, one-eighth, and one-fifth are given to four people, respectively. The fifth person eats ten cookies on the spot, with one cookie remaining for the sixth person. Find the original number of cookies on the tray.

55. A storekeeper goes to the bank to get $10 worth of change. She requests twice as many quarters as half dollars, twice as many dimes as quarters, three times as many nickels as dimes, and no pennies or dollars. How many of each coin did the storekeeper receive?

56. A student has an average score of 82 on three tests. His average score on the first two tests is 85. What was the score on the third test?

11 Summary and Review

The review that follows is meant to prepare you for a chapter exam. It consists of two parts. The first part is a checklist of some of the Study Tips referred to in this and preceding chapters, as well as a list of important properties and formulas. The second part is the Review Exercises. These provide practice exercises for the exam, together with references to section objectives so you can go back and review. Before beginning, stop and look back over the skills you have obtained. What skills in mathematics do you have now that you did not have before studying this chapter?

STUDY TIPS CHECKLIST

The foundation of all your study skills is TIME!	☐ Have you begun to prepare for the final exam?
	☐ Have you made use of the many supplements available with this text?
	☐ Are you practicing the five-step problem-solving strategy?
	☐ Are you using the tutoring resources on campus?
	☐ Have you found someone with whom to study for the final exam?

IMPORTANT PROPERTIES AND FORMULAS

The Addition Principle: For any real numbers a, b, and c, $a = b$ is equivalent to $a + c = b + c$.
The Multiplication Principle: For any real numbers a, b, and c, $c \neq 0$, $a = b$ is equivalent to $a \cdot c = b \cdot c$.
The Distributive Laws: $a(b + c) = ab + ac$, $a(b - c) = ab - ac$

REVIEW EXERCISES

1. Evaluate $\dfrac{x - y}{3}$ when $x = 17$ and $y = 5$.   [11.1a]

Multiply.   [11.1b]

2. $5(3x - 7)$

3. $-2(4x - 5)$

4. $10(0.4x + 1.5)$

5. $-8(3 - 6x)$

Factor.   [11.1c]

6. $2x - 14$

7. $6x - 6$

8. $5x + 10$

9. $12 - 3x$

Collect like terms.   [11.1d]

10. $11a + 2b - 4a - 5b$

11. $7x - 3y - 9x + 8y$

12. $6x + 3y - x - 4y$

13. $-3a + 9b + 2a - b$

Solve.   [11.2a], [11.3a]

14. $x + 5 = -17$

15. $-8x = -56$

16. $-\dfrac{x}{4} = 48$

17. $n - 7 = -6$

18. $15x = -35$

19. $x - 11 = 14$

20. $-\dfrac{2}{3} + x = -\dfrac{1}{6}$

21. $\dfrac{4}{5}\,y = -\dfrac{3}{16}$

22. $y - 0.9 = 9.09$

23. $5 - x = 13$

Solve.   [11.4a, b, c]

24. $5t + 9 = 3t - 1$

25. $7x - 6 = 25x$

26. $\dfrac{1}{4}x - \dfrac{5}{8} = \dfrac{3}{8}$

27. $14y = 23y - 17 - 10$

28. $0.22y - 0.6 = 0.12y + 3 - 0.8y$

29. $\dfrac{1}{4}x - \dfrac{1}{8}x = 3 - \dfrac{1}{16}x$

30. $4(x + 3) = 36$

31. $3(5x - 7) = -66$

32. $8(x - 2) - 5(x + 4) = 20x + x$

33. $-5x + 3(x + 8) = 16$

34. Translate to an algebraic expression: [11.5a]
Nineteen percent of some number.

Solve.   [11.5b]

35. *Dimensions of Wyoming.*   The state of Wyoming is roughly in the shape of a rectangle whose perimeter is 1280 mi. The length is 90 mi more than the width. Find the dimensions.

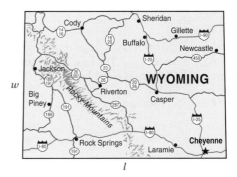

36. An entertainment center sold for $2449 in June. This was $332 more than the cost in February. Find the cost in February.

37. Ty is paid a commission of $4 for each appliance he sells. One week, he received $108 in commissions. How many appliances did he sell?

38. The measure of the second angle of a triangle is 50° more than that of the first angle. The measure of the third angle is 10° less than twice the first angle. Find the measures of the angles.

39. After a 30% reduction, a bread maker is on sale for $154. What was the marked price (the price before the reduction)?

40. A hotel manager's salary is $30,000, which is a 15% increase over the previous year's salary. What was the previous salary (to the nearest dollar)?

41. A tax-exempt charity received a bill of $145.90 for a sump pump. The bill incorrectly included sales tax of 5%. How much does the charity actually owe?

42. *Nile and Amazon Rivers.* The total length of the Nile and Amazon Rivers is 13,108 km. If the Amazon were 234 km longer, it would be as long as the Nile. Find the length of each river.
Source: *The Handy Geography Answer Book*

43. *HDTV Costs.* An HDTV television sold for $829 in May. This was $38 more than the cost in January. Find the cost in January.

44. *Writing Pad.* The perimeter of a rectangular writing pad is 56 cm. The width is 6 cm less than the length. Find the width and the length.

45. D_W Explain at least three uses of the distributive laws considered in this chapter. [11.1b, c, d]

46. D_W Explain all the errors in each of the following. [11.4a, c]

a) $4 - 3x = 9$
$3x = 9$
$x = 3$

b) $2(x - 5) = 7$
$2x - 5 = 7$
$2x = 12$
$x = 6$

SKILL MAINTENANCE

Certain objectives from four particular sections will be retested on the chapter test. The objectives are listed with the practice problems that follow.

47. Find the diameter, the circumference, and the area of a circle when $r = 20$ ft. Use 3.14 for π. [9.3a, b, c]

48. Find the volume of a rectangular solid when the length is 20 cm, the width is 18.5 cm, and the height is 4.6 cm. [9.4a]

49. Add: $-12 + 10 + (-19) + (-24)$. [10.2a]

50. Multiply: $(-2) \cdot (-3) \cdot (-5) \cdot (-2) \cdot (-1)$. [10.4a]

SYNTHESIS

Solve. [10.1e], [11.4a, b]

51. $2|n| + 4 = 50$

52. $|3n| = 60$

Chapter Test

1. Evaluate $\dfrac{3x}{y}$ when $x = 10$ and $y = 5$.

Multiply.

2. $3(6 - x)$

3. $-5(y - 1)$

Factor.

4. $12 - 22x$

5. $7x + 21 + 14y$

Collect like terms.

6. $9x - 2y - 14x + y$

7. $-a + 6b + 5a - b$

Solve.

8. $x + 7 = 15$

9. $t - 9 = 17$

10. $3x = -18$

11. $-\dfrac{4}{7}x = -28$

12. $3t + 7 = 2t - 5$

13. $\dfrac{1}{2}x - \dfrac{3}{5} = \dfrac{2}{5}$

14. $8 - y = 16$

15. $-\dfrac{2}{5} + x = -\dfrac{3}{4}$

16. $0.4p + 0.2 = 4.2p - 7.8 - 0.6p$

17. $3(x + 2) = 27$

18. $-3x - 6(x - 4) = 9$

19. Translate to an algebraic expression:
Nine less than some number.

Solve.

20. *Perimeter of a Photograph.* The perimeter of a rectangular photograph is 36 cm. The length is 4 cm greater than the width. Find the width and the length.

21. *Charitable Contributions.* About 53% of all charitable contributions are made to religious organizations. In 1997, $75 billion was given to religious organizations. How much was given to charities in general?
Source: *Statistical Abstract of the United States*, 1999

22. *Board Cutting.* An 8-m board is cut into two pieces. One piece is 2 m longer than the other. How long are the pieces?

23. *Savings Account.* Money is invested in a savings account at 5% simple interest. After 1 year, there is $924 in the account. How much was originally invested?

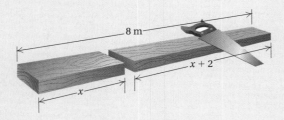

24. If you triple a number and then subtract 14, you get $\frac{2}{3}$ of the original number. What is the original number?

25. The second angle of a triangle is three times as large as the first. The third angle is 25° less than the sum of the other two angles. Find the measure of the first angle.

26. Multiply:
$$(-9) \cdot (-2) \cdot (-2) \cdot (-5).$$

27. Add:
$$\frac{2}{3} + \left(-\frac{8}{9}\right).$$

28. Find the diameter, the circumference, and the area of a circle when the radius is 70 yd. Use 3.14 for π.

29. Find the volume of a rectangular solid when the length is 22 ft, the width is 10 ft, and the height is 6 ft.

30. Solve: $3|w| - 8 = 37$.

31. A movie theater had a certain number of tickets to give away. Five people got the tickets. The first got $\frac{1}{3}$ of the tickets, the second got $\frac{1}{4}$ of the tickets, and the third got $\frac{1}{5}$ of the tickets. The fourth person got 8 tickets, and there were 5 tickets left for the fifth person. Find the total number of tickets given away.

Cumulative Review/ Final Examination

This cumulative review also serves as a final examination for the entire book. A question that may occur at this point is what notation to use for a particular problem or exercise. Although there is no particular rule, especially as you use mathematics outside the classroom, here is the guideline that we follow: Use the notation given in the problem. That is, if the problem is given using mixed numerals, give the answer in mixed numerals. If the problem is given in decimal notation, give the answer in decimal notation.

1. *Pears.* Find the unit price of each brand. Then determine which package has the lower unit price.

BRAND	PACKAGE	PRICE	UNIT PRICE
A	$8\frac{1}{2}$ oz	$0.65	
B	15 oz	$1.07	
C	$15\frac{1}{4}$ oz	$1.07	
D	29 oz	$1.69	
E	29 oz	$1.32	

2. Write a word name for 7.463.

3. *Pluto.* The planet Pluto has a diameter of 1400 mi. Use $\frac{22}{7}$ for π.

a) Find the circumference of Pluto.
b) Find the volume of Pluto.

Kitchen Costs. The average cost of a remodeled kitchen is $26,888, according to the National Kitchen and Bath Association. Complete the table below, which relates percents and costs of certain items.

	ITEM	PERCENT OF COST	COST
4.	Cabinets	50%	
5.	Countertops		$4033.20
6.	Appliances	8%	
7.	Fixtures		$8066.40
8.	Flooring	2%	

Source: National Kitchen and Bath Association

9. In 47,201, what digit tells the number of thousands?

10. Write expanded notation for 7405.

Add and simplify, if appropriate.

11.
$$\begin{array}{r} 7\ 4\ 1 \\ +\ 2\ 7\ 1 \\ \hline \end{array}$$

12.
$$\begin{array}{r} 4\ 9\ 0\ 3 \\ 5\ 2\ 7\ 8 \\ 6\ 3\ 9\ 1 \\ +\ 4\ 5\ 1\ 3 \\ \hline \end{array}$$

13. $\dfrac{2}{13} + \dfrac{1}{26}$

14.
$$\begin{array}{r} 2\dfrac{4}{9} \\ +\ 3\dfrac{1}{3} \\ \hline \end{array}$$

15.
$$\begin{array}{r} 2.0\ 4\ 8 \\ 6\ 3.9\ 1\ 4 \\ +\ 4\ 2\ 8.0\ 0\ 9 \\ \hline \end{array}$$

16. $34.56 + 2.783 + 0.433 + 765.1$

Subtract and simplify, if possible.

17.
$$\begin{array}{r} 6\ 7\ 4 \\ -\ 5\ 2\ 2 \\ \hline \end{array}$$

18.
$$\begin{array}{r} 9\ 4\ 6\ 5 \\ -\ 8\ 7\ 9\ 1 \\ \hline \end{array}$$

19. $\dfrac{7}{8} - \dfrac{2}{3}$

20.
$$\begin{array}{r} 4\dfrac{1}{3} \\ -\ 1\dfrac{5}{8} \\ \hline \end{array}$$

21.
$$\begin{array}{r} 2\ 0.0 \\ -\ \ \ 0.0\ 0\ 2\ 7 \\ \hline \end{array}$$

22. $40.03 - 5.789$

Multiply and simplify, if possible.

23.
$$\begin{array}{r} 2\ 9\ 7 \\ \times\ \ \ 1\ 6 \\ \hline \end{array}$$

24.
$$\begin{array}{r} 3\ 4\ 9 \\ \times\ 7\ 6\ 3 \\ \hline \end{array}$$

25. $1\dfrac{3}{4} \cdot 2\dfrac{1}{3}$

26. $\dfrac{9}{7} \cdot \dfrac{14}{15}$

27. $12 \cdot \dfrac{5}{6}$

28.
$$\begin{array}{r} 3\ 4.0\ 9 \\ \times\ \ \ \ \ 7.6 \\ \hline \end{array}$$

Divide and simplify. State the answer using a whole-number quotient and remainder.

29. $6\ )\overline{\ 3\ 4\ 3\ 8\ }$

30. $3\ 4\ )\overline{\ 1\ 9\ 1\ 4\ }$

31. Give a mixed numeral for the quotient in Question 30.

Divide and simplify, if possible.

32. $\dfrac{4}{5} \div \dfrac{8}{15}$

33. $2\dfrac{1}{3} \div 30$

34. $2.7\ )\overline{\ 1\ 0\ 5.3\ }$

35. Round 68,489 to the nearest thousand.

36. Round 0.4275 to the nearest thousandth.

37. Round $21.\overline{83}$ to the nearest hundredth.

38. Determine whether 1368 is divisible by 8.

39. Find all the factors of 15.

40. Find the LCM of 16, 25, and 32.

Simplify.

41. $\dfrac{21}{30}$

42. $\dfrac{275}{5}$

43. Convert to a mixed numeral: $\dfrac{18}{5}$.

44. Use = or ≠ for ☐ to write a true sentence:
$$\dfrac{4}{7} \,\square\, \dfrac{3}{5}.$$

45. Use < or > for ☐ to write a true sentence:
$$\dfrac{4}{7} \,\square\, \dfrac{3}{5}.$$

46. Which number is greater, 1.001 or 0.9976?

47. Use < or > for ☐ to write a true sentence: 987 ☐ 879.

48. What part is shaded?

Convert to decimal notation.

49. $\dfrac{37}{1000}$

50. $\dfrac{13}{25}$

51. $\dfrac{8}{9}$

52. 7%

Convert to fraction notation.

53. 4.63

54. $7\dfrac{1}{4}$

55. 40%

Convert to percent notation.

56. $\dfrac{17}{20}$

57. 1.5

Solve.

58. $234 + y = 789$

59. $3.9 \times y = 249.6$

60. $\dfrac{2}{3} \cdot t = \dfrac{5}{6}$

61. $\dfrac{8}{17} = \dfrac{36}{x}$

Golf Rounds.   The data in the following table show the percent of golfers who play a certain number of rounds of golf per year.

NUMBER OF ROUNDS PER YEAR	PERCENT
1–7	2.2
8–24	14.9
25–49	25.3
50–99	39.0
100 or more	18.6

Source: U.S. Golf Foundation

62. Make a circle graph of the data.

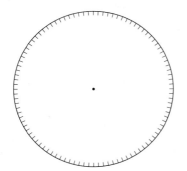

63. Make a bar graph of the data.

64. Find the missing angle measure.

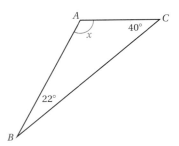

65. Classify the triangle in Exercise 64 as right, obtuse, or acute.

Solve.

66. *Donations.* Lorenzo makes donations of $627 and $48. What was the total donation?

67. *Candy Bars.* A machine wraps 134 candy bars per minute. How long does it take this machine to wrap 8710 bars?

68. *Stock Prices.* A share of stock bought for $29.63 dropped $3.88 before it was resold. What was the price when it was resold?

69. *Length of Trip.* At the start of a trip, a car's odometer read 27,428.6 mi, and at the end of the trip, the reading was 27,914.5 mi. How long was the trip?

70. *Taxes.* From an income of $12,000, amounts of $2300 and $1600 are paid for federal and state taxes. How much remains after these taxes have been paid?

71. *Teacher Salary.* A substitute teacher is paid $47 a day for 9 days. How much was received altogether?

72. *Walking Distance.* A person walks $\frac{3}{5}$ km per hour. At this rate, how far would the person walk in $\frac{1}{2}$ hr?

73. *Sweater Costs.* Eight identical sweaters cost a total of $679.68. What is the cost of each sweater?

74. *Paint Needs.* Eight gallons of exterior paint covers 400 ft^2. How much paint is needed to cover 650 ft^2?

75. *Simple Interest.* What is the simple interest on $4000 principal at 5% for $\frac{3}{4}$ year?

76. *Commission Rate.* A real estate agent received $5880 commission on the sale of an $84,000 home. What was the rate of commission?

77. *Population Growth.* The population of a city is 29,000 this year and is increasing at 4% per year. What will the population be next year?

78. *Student Ages.* The ages of students in a math class at a community college are as follows:

18, 21, 26, 31, 32, 18, 50.

Find the average, the median, and the mode of their ages.

Evaluate.

79. 18^2

80. 20^2

Simplify.

81. $\sqrt{9}$

82. $\sqrt{121}$

83. Approximate to three decimal places: $\sqrt{20}$.

Complete.

84. $\dfrac{1}{3}$ yd = _____ in.

85. 4280 mm = _____ cm

86. 3 days = _____ hr

87. 20,000 g = _____ kg

88. 5 lb = _____ oz

89. 0.008 cg = _____ mg

90. 8190 mL = _____ L

91. 20 qt = _____ gal

92. Find the length of the third side of this right triangle. Give an exact answer and an approximation to three decimal places.

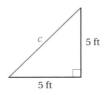

93. Find the perimeter and the area.

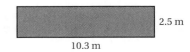

Find the area.

94.

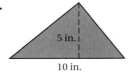

95.

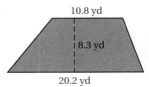

96.

97. Find the diameter, the circumference, and the area of this circle. Use 3.14 for π.

98. Find the volume.

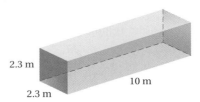

Find the volume. Use 3.14 for π.

99.

100.

Compute and simplify.

101. $12 \times 20 - 10 \div 5$

102. $4^3 - 5^2 + (16 \cdot 4 + 23 \cdot 3)$

103. $|(-1) \cdot 3|$

104. $17 + (-3)$

105. $\left(-\dfrac{1}{3}\right) - \left(-\dfrac{2}{3}\right)$

106. $(-6) \cdot (-5)$

107. $-\dfrac{5}{7} \cdot \dfrac{14}{35}$

108. $\dfrac{48}{-6}$

Solve.

109. $7 - x = 12$

110. $-4.3x = -17.2$

111. $5x + 7 = 3x - 9$

112. $5(x - 2) - 8(x - 4) = 20$

Translate to an algebraic expression.

113. 17 more than some number

114. 38 percent of some number

Solve.

115. *Rollerblade Costs.*   Melinda and Susan purchased rollerblades for a total of $107. Melinda paid $17 more for her rollerblades than Susan did. What did Susan pay?

116. *Savings Investment.*   Money is invested in a savings account at 8% simple interest. After 1 year, there is $1134 in the account. How much was originally invested?

117. *Wire Cutting.*   A 143-m wire is cut into three pieces. The second piece is 3 m longer than the first. The third is four-fifths as long as the first. How long is each piece?

Solve.

118. $\dfrac{2}{3}x + \dfrac{1}{6} - \dfrac{1}{2}x = \dfrac{1}{6} - 3x$

119. $29.966 - 8.673y = -8.18 + 10.4y$

For each of Questions 120–123, choose the correct answer from the selection given.

120. Collect like terms: $\dfrac{1}{4}x - \dfrac{3}{4}y + \dfrac{1}{4}x - \dfrac{3}{4}y$.

   a) 0   **b)** $-\dfrac{2}{3}y$ **c)** $\dfrac{1}{2}x - \dfrac{3}{2}y$

   d) $\dfrac{1}{2}x$ **e)** None

121. Factor: $8x + 4y - 12z$.

   a) $2x + y - 3z$ **b)** $4x + 2y - 6z$ **c)** $2x - 3z$
   d) $2x + 4y - 3z$ **e)** None

122. Divide: $-\dfrac{13}{25} \div \left(-\dfrac{13}{5}\right)$.

   a) $\dfrac{169}{125}$ **b)** 5   **c)** $\dfrac{125}{169}$

   d) $\dfrac{1}{5}$ **e)** None

123. Add: $-27 + (-11)$.

   a) -38 **b)** -16 **c)** 16
   d) 38   **e)** None

Answers

Chapter 1

Pretest: Chapter 1, p. 2

1. Three million, seventy-eight thousand, fifty-nine
2. 6 thousands + 9 hundreds + 8 tens + 7 ones
3. 2,047,398,589   4. 6 ten thousands   5. 956,000
6. 60,000   7. 10,216   8. 4108   9. 22,976
10. 503 R 11   11. <   12. >   13. 5542   14. 22
15. 34   16. 25   17. 4500 sheets   18. $299   19. $369
20. $30,793   21. 25   22. 64   23. 0   24. 3

Margin Exercises, Section 1.1, pp. 3–6

1. 2 ten thousands   2. 2 hundred thousands
3. 2 millions   4. 2 ten millions
5. 4 hundred thousands; 8 ten thousands; 6 thousands;
5 hundreds; 7 tens; 5 ones
6. 1 thousand + 8 hundreds + 9 tens + 5 ones
7. 2 ten thousands + 2 thousands +
1 hundred + 3 tens + 2 ones
8. 3 thousands + 0 hundreds + 3 tens + 1 one, or
3 thousands + 3 tens + 1 one
9. 4 thousands + 1 hundred + 0 tens + 0 ones, or
4 thousands + 1 hundred
10. 3 thousands + 8 hundreds + 6 tens + 0 ones, or
3 thousands + 8 hundreds + 6 tens   11. 5689
12. 87,128   13. 9003   14. Eighty-eight   15. Sixteen
16. Thirty-two   17. Two hundred four
18. Forty-three thousand, seven hundred eighty-two
19. One million, eight hundred seventy-nine thousand,
two hundred four   20. Six billion, two hundred fifty-nine
million, six hundred thousand   21. 213,105,329

Exercise Set 1.1, p. 8

1. 5 thousands   3. 5 hundreds   5. The last 0   7. 1
9. 5 thousands + 7 hundreds + 0 tens + 2 ones, or
5 thousands + 7 hundreds + 2 ones
11. 9 ten thousands + 3 thousands + 9 hundreds +
8 tens + 6 ones

13. 2 thousands + 0 hundreds + 5 tens + 8 ones, or
2 thousands + 5 tens + 8 ones
15. 1 thousand + 2 hundreds + 6 tens + 8 ones
17. 2475   19. 68,939   21. 7304   23. 1009
25. Eighty-five   27. Eighty-eight thousand
29. One hundred twenty-three thousand, seven hundred
sixty-five   31. Seven billion, seven hundred fifty-four
million, two hundred eleven thousand, five hundred
seventy-seven   33. 2,233,812   35. 8,000,000,000
37. Five hundred sixty-six thousand, two hundred eighty
39. Four hundred sixty-seven million, three hundred
twenty-two thousand, three hundred eighty-eight
41. 9,460,000,000,000   43. 64,186,000   45. D_W
47. 138

Margin Exercises, Section 1.2, pp. 12–15

1. 97   2. 9745   3. 13,465   4. 16,182   5. 27   6. 34
7. 27   8. 38   9. 47   10. 61   11. 27,474
12. 16 in.; 26 in.   13. 29 in.   14. 22 ft

Calculator Corner, p. 16

1. 55   2. 121   3. 1602   4. 734   5. 1932   6. 864

Exercise Set 1.2, p. 17

1. 387   3. 5198   5. 164   7. 100   9. 8503   11. 5266
13. 4466   15. 8310   17. 6608   19. 16,784
21. 34,432   23. 101,310   25. 100,111   27. 28
29. 26   31. 67   33. 230   35. 1349   37. 18,424
39. 2320   41. 31,685   43. 114 mi   45. 570 ft
47. D_W 49. 8 ten thousands   50. One hundred
fourteen million, three hundred thirty-six thousand, six
hundred ten
51. $1 + 99 = 100, 2 + 98 = 100, \ldots, 49 + 51 = 100$. Then
$49 \cdot 100 = 4900$ and $4900 + 50 + 100 = 5050$.

Margin Exercises, Section 1.3, pp. 21–24

1. $7 = 2 + 5$, or $7 = 5 + 2$ 2. $17 = 9 + 8$, or $17 = 8 + 9$
3. $5 = 13 - 8; 8 = 13 - 5$ 4. $11 = 14 - 3; 3 = 14 - 11$

5. 3801   **6.** 6328   **7.** 4747   **8.** 56   **9.** 205
10. 658   **11.** 2851   **12.** 1546

Calculator Corner, p. 25

1. 28   **2.** 47   **3.** 67   **4.** 119   **5.** 2128   **6.** 2593

Exercise Set 1.3, p. 26

1. $7 = 3 + 4$, or $7 = 4 + 3$ **3.** $13 = 5 + 8$, or $13 = 8 + 5$
5. $23 = 14 + 9$, or $23 = 9 + 14$ **7.** $43 = 27 + 16$, or
$43 = 16 + 27$ **9.** $6 = 15 - 9$; $9 = 15 - 6$
11. $8 = 15 - 7$; $7 = 15 - 8$ **13.** $17 = 23 - 6$; $6 = 23 - 17$
15. $23 = 32 - 9$; $9 = 32 - 23$ **17.** 12   **19.** 44
21. 533   **23.** 1126   **25.** 39   **27.** 298   **29.** 226
31. 234   **33.** 5382   **35.** 1493   **37.** 2187   **39.** 3831
41. 7748   **43.** 33,794   **45.** 2168   **47.** 43,028   **49.** 56
51. 36   **53.** 84   **55.** 454   **57.** 771   **59.** 2191
61. 3749   **63.** 7019   **65.** 5745   **67.** 95,974   **69.** 9989
71. 83,818   **73.** 4206   **75.** 10,305   **77.** D_W **79.** 1024
80. 12,732   **81.** 90,283   **82.** 29,364   **83.** 1345
84. 924   **85.** 22,692   **86.** 10,920   **87.** Six million,
three hundred seventy-five thousand, six hundred two
88. 7 ten thousands   **89.** 3; 4

Margin Exercises, Section 1.4, pp. 29–34

1. 40   **2.** 50   **3.** 70   **4.** 100   **5.** 40   **6.** 80   **7.** 90
8. 140   **9.** 470   **10.** 240   **11.** 290   **12.** 600   **13.** 800
14. 800   **15.** 9300   **16.** 8000   **17.** 8000   **18.** 19,000
19. 69,000   **20.** 48,970; 49,000; 49,000   **21.** 269,580;
269,600; 270,000   **22. (a)** $19,500; **(b)** yes   **23.** Eliminate
driver's seat with 6-way power adjustment, sunroof, feature
package, and sport package. Answers may vary.   **24.** 200
25. 1800   **26.** 2600   **27.** 11,000   **28.** <   **29.** >
30. >   **31.** <   **32.** <   **33.** >

Exercise Set 1.4, p. 35

1. 50   **3.** 470   **5.** 730   **7.** 900   **9.** 100   **11.** 1000
13. 9100   **15.** 32,900   **17.** 6000   **19.** 8000
21. 45,000   **23.** 373,000   **25.** 180   **27.** 5720
29. 220; incorrect   **31.** 890; incorrect   **33.** 16,500
35. 5200   **37.** $1500   **39.** $1300, yes   **41.** Answers will
vary depending on the options chosen.   **43.** 1600
45. 1500   **47.** 31,000   **49.** 69,000   **51.** <   **53.** >
55. <   **57.** >   **59.** >   **61.** >   **63.** Thrust SCC;
54 ft > 47 ft   **65.** 87 yr > 81 yr, or 81 yr < 87 yr   **67.** D_W
69. 7992   **70.** 23,000,000   **71.** Two hundred forty-six
billion, six hundred five million, four thousand, thirty-two
72. 8 thousands + 0 hundreds + 1 ten + 7 ones, or
8 thousands + 1 ten + 7 ones   **73.** 86,754   **74.** 13,589
75. 48,824   **76.** 4415   **77.** Left to the student
79. Left to the student

Margin Exercises, Section 1.5, pp. 40–44

1. 116   **2.** 148   **3.** 4938   **4.** 6740   **5.** 1035   **6.** 3024

7. 46,252   **8.** 205,065   **9.** 144,432   **10.** 287,232
11. 14,075,720   **12.** 391,760   **13.** 17,345,600
14. 56,200   **15.** 562,000   **16. (a)** 1081; **(b)** 1081; **(c)** same
17. 40   **18.** 15   **19.** $400   **20.** 210,000; 160,000
21. 45 sq ft

Calculator Corner, p. 42

1. 448   **2.** 21,970   **3.** 6380   **4.** 39,564   **5.** 180,480
6. 2,363,754

Exercise Set 1.5, p. 45

1. 870   **3.** 2,340,000   **5.** 520   **7.** 564   **9.** 1527
11. 64,603   **13.** 4770   **15.** 3995   **17.** 46,080
19. 14,652   **21.** 207,672   **23.** 798,408   **25.** 20,723,872
27. 362,128   **29.** 20,064,048   **31.** 25,236,000
33. 302,220   **35.** 49,101,136   **37.** $50 \cdot 70 = 3500$
39. $30 \cdot 30 = 900$ **41.** $900 \cdot 300 = 270,000$
43. $400 \cdot 200 = 80,000$ **45. (a)** $5,260,000; **(b)** $5,400,000
47. 529,984 sq mi   **49.** 8100 sq ft   **51.** D_W **53.** 12,685
54. 10,834   **55.** 427,477   **56.** 111,110   **57.** 1241
58. 8889   **59.** 254,119   **60.** 66,444   **61.** 6,376,000
62. 6,375,600   **63.** 247,464 sq ft

Margin Exercises, Section 1.6, pp. 48–55

1. $\frac{54}{6} = 9$, $6\overline{)54}$ with 9   **2.** $15 = 5 \cdot 3$, or $15 = 3 \cdot 5$
3. $72 = 9 \cdot 8$, or $72 = 8 \cdot 9$ **4.** $6 = 12 \div 2$; $2 = 12 \div 6$
5. $6 = 42 \div 7$; $7 = 42 \div 6$ **6.** 6; $6 \cdot 9 = 54$ **7.** 6 R 7;
$6 \cdot 9 = 54$, $54 + 7 = 61$ **8.** 4 R 5; $4 \cdot 12 = 48$, $48 + 5 = 53$
9. 6 R 13; $6 \cdot 24 = 144$, $144 + 13 = 157$ **10.** 59 R 3
11. 1475 R 5   **12.** 1015   **13.** 134   **14.** 63 R 12
15. 807 R 4   **16.** 1088   **17.** 360 R 4   **18.** 800 R 47

Calculator Corner, p. 56

1. 3 R 11   **2.** 28   **3.** 124 R 2   **4.** 131 R 18   **5.** 283 R 57
6. 843 R 187

Exercise Set 1.6, p. 57

1. $18 = 3 \cdot 6$, or $18 = 6 \cdot 3$ **3.** $22 = 22 \cdot 1$, or $22 = 1 \cdot 22$
5. $54 = 6 \cdot 9$, or $54 = 9 \cdot 6$ **7.** $37 = 1 \cdot 37$, or $37 = 37 \cdot 1$
9. $9 = 45 \div 5$; $5 = 45 \div 9$ **11.** $37 = 37 \div 1$; $1 = 37 \div 37$
13. $8 = 64 \div 8$ **15.** $11 = 66 \div 6$; $6 = 66 \div 11$ **17.** 12
19. 1   **21.** 22   **23.** Not defined   **25.** 55 R 2   **27.** 108
29. 307   **31.** 753 R 3   **33.** 74 R 1   **35.** 92 R 2
37. 1703   **39.** 987 R 5   **41.** 12,700   **43.** 127
45. 52 R 52   **47.** 29 R 5   **49.** 40 R 12   **51.** 90 R 22
53. 29   **55.** 105 R 3   **57.** 1609 R 2   **59.** 1007 R 1
61. 23   **63.** 107 R 1   **65.** 370   **67.** 609 R 15   **69.** 304
71. 3508 R 219   **73.** 8070   **75.** D_W
77. 7 thousands + 8 hundreds + 8 tens + 2 ones   **78.** >
79. $21 = 16 + 5$, or $21 = 5 + 16$ **80.** $56 = 14 + 42$, or
$56 = 42 + 14$ **81.** $47 = 56 - 9$; $9 = 56 - 47$
82. $350 = 414 - 64$; $64 = 414 - 350$ **83.** 359
84. 21,300   **85.** 209   **86.** 3534   **87.** 54, 122; 33, 2772;
4, 8; 16, 19   **89.** 30 buses

Margin Exercises, Section 1.7, pp. 61–65

1. 7 **2.** 5 **3.** No **4.** Yes **5.** 5 **6.** 10 **7.** 5
8. 22 **9.** 22,490 **10.** 9022 **11.** 570 **12.** 3661
13. 8 **14.** 45 **15.** 77 **16.** 3311 **17.** 6114 **18.** 8
19. 16 **20.** 644 **21.** 96 **22.** 94

Exercise Set 1.7, p. 66

1. 14 **3.** 0 **5.** 29 **7.** 0 **9.** 8 **11.** 14 **13.** 1035
15. 25 **17.** 450 **19.** 90,900 **21.** 32 **23.** 143
25. 79 **27.** 45 **29.** 324 **31.** 743 **33.** 37 **35.** 66
37. 15 **39.** 48 **41.** 175 **43.** 335 **45.** 104 **47.** 45
49. 4056 **51.** 17,603 **53.** 18,252 **55.** 205 **57.** D_W
59. $7 = 15 - 8$; $8 = 15 - 7$ **60.** $6 = 48 \div 8$; $8 = 48 \div 6$
61. < **62.** > **63.** > **64.** < **65.** 142 R 5
66. 142 **67.** 334 **68.** 334 R 11 **69.** 347

Margin Exercises, Section 1.8, pp. 69–76

1. 277 home runs **2.** 249 home runs **3.** Mark McGwire
4. $1874 **5.** $223 **6.** $34,776 **7.** 9180 sq in.
8. 378 cartons with 1 can left over **9.** 34 gal
10. 181 seats

Exercise Set 1.8, p. 77

1. 178,668 million, or 178,668,000,000 **3.** 12,113 million,
or 12,113,000,000 **5.** 2000 **7.** 66,672 sq mi
9. $459,860 **11.** 211,000 people **13.** 240 mi
15. 256 gal **17.** 480,000 pixels **19.** $11,976
21. 216,000,000 CDs **23.** 2,545,000,000 CDs
25. 35 weeks; 2 episodes left over **27.** 168 hr **29.** $247
31. 2,959,500 **33.** 21 columns **35.** 563 packages;
7 bars left over **37.** (a) 4200 sq ft; (b) 268 ft
39. 35 cartons **41.** 56 full cartons; 11 books left over. If
1355 books are shipped, it will take 57 cartons.
43. 384 mi; 27 in. **45.** 18 rows **47.** 32 $10 bills
49. $400 **51.** 280 min, or 4 hr 40 min **53.** 525 min, or
8 hr 45 min **55.** 106 bones **57.** $704 **59.** D_W
61. 234,600 **62.** 234,560 **63.** 235,000 **64.** 22,000
65. 16,000 **66.** 8000 **67.** 4000 **68.** 320,000
69. 720,000 **70.** 46,800,000 **71.** 792,000 mi;
1,386,000 mi

Margin Exercises, Section 1.9, pp. 84–89

1. 5^4 **2.** 5^5 **3.** 10^2 **4.** 10^4 **5.** 10,000 **6.** 100
7. 512 **8.** 32 **9.** 51 **10.** 30 **11.** 584 **12.** 84
13. 4; 1 **14.** 52; 52 **15.** 29 **16.** 1880 **17.** 253
18. 93 **19.** 1880 **20.** 305 **21.** 75 **22.** 4
23. 86 in. **24.** 46 **25.** 4

Calculator Corner, p. 85

1. 243 **2.** 15,625 **3.** 20,736 **4.** 2048

Calculator Corner, p. 87

1. 49 **2.** 85 **3.** 36 **4.** 0 **5.** 73 **6.** 49

Exercise Set 1.9, p. 90

1. 3^4 **3.** 5^2 **5.** 7^5 **7.** 10^3 **9.** 49 **11.** 729
13. 20,736 **15.** 121 **17.** 22 **19.** 20 **21.** 100
23. 1 **25.** 49 **27.** 5 **29.** 434 **31.** 41 **33.** 88
35. 4 **37.** 303 **39.** 20 **41.** 70 **43.** 295 **45.** 32
47. 906 **49.** 62 **51.** 102 **53.** 256 **55.** $94
57. 110 **59.** 7 **61.** 544 **63.** 708 **65.** 27 **67.** D_W
69. 452 **70.** 835 **71.** 13 **72.** 37 **73.** 2342
74. 4898 **75.** 25 **76.** 100 **77.** 102,600 mi^2
78. 98 gal **79.** 24; $1 + 5 \cdot (4 + 3) = 36$
81. 7; $12 \div (4 + 2) \cdot 3 - 2 = 4$

Summary and Review: Chapter 1, p. 93

1. 2 thousands + 7 hundreds + 9 tens + 3 ones
2. 5 ten thousands + 6 thousands + 0 hundreds +
7 tens + 8 ones, or 5 ten thousands + 6 thousands +
7 tens + 8 ones **3.** 8669 **4.** 90,844 **5.** Sixty-seven
thousand, eight hundred nineteen **6.** Two million, seven
hundred eighty-one thousand, four hundred twenty-seven
7. 476,588 **8.** 2,000,400,000 **9.** 8 thousands **10.** 3
11. 14,272 **12.** 66,024 **13.** 22,098 **14.** 98,921
15. $10 = 6 + 4$, or $10 = 4 + 6$ **16.** $8 = 11 - 3$;
$3 = 11 - 8$ **17.** 5148 **18.** 1153 **19.** 2274
20. 17,757 **21.** 345,800 **22.** 345,760 **23.** 346,000
24. $41,300 + 19,700 = 61,000$
25. $38,700 - 24,500 = 14,200$ **26.** $400 \cdot 700 = 280,000$
27. > **28.** < **29.** 420,000 **30.** 6,276,800
31. 506,748 **32.** 27,589 **33.** 5,331,810
34. $56 = 8 \cdot 7$, or $56 = 7 \cdot 8$ **35.** $4 = 52 \div 13$;
$13 = 52 \div 4$ **36.** 12 R 3 **37.** 5 **38.** 913 R 3
39. 384 R 1 **40.** 4 R 46 **41.** 54 **42.** 452 **43.** 5008
44. 4389 **45.** 8 **46.** 45 **47.** 546 **48.** 4^3
49. 10,000 **50.** 36 **51.** 65 **52.** 233 **53.** 56
54. 32 **55.** 260 **56.** 165 **57.** $502 **58.** $484
59. 1982 **60.** 37 cartons **61.** 14 beehives **62.** $2413
63. $19,748 **64.** 137 beakers filled; 13 mL left over
65. 98 ft^2; 42 ft **66.** D_W A vat contains 1152 oz of hot
sauce. If 144 bottles are to be filled equally, how much will
each bottle contain? Answers may vary.
67. D_W No; if subtraction were associative, then
$a - (b - c) = (a - b) - c$ for any a, b, and c. But, for
example,
$$12 - (8 - 4) = 12 - 4 = 8,$$
whereas
$$(12 - 8) - 4 = 4 - 4 = 0.$$
Since $8 \neq 0$, this example shows that subtraction is not
associative. **68.** $d = 8$ **69.** $a = 8$, $b = 4$ **70.** 6 days

Test: Chapter 1, p. 96

1. [1.1b] 8 thousands + 8 hundreds + 4 tens + 3 ones
2. [1.1c] Thirty-eight million, four hundred three thousand,
two hundred seventy-seven **3.** [1.1a] 5 **4.** [1.2a] 9989
5. [1.2a] 63,791 **6.** [1.2a] 34 **7.** [1.2a] 10,515
8. [1.3b] 3630 **9.** [1.3b] 1039 **10.** [1.3b] 6848

11. [1.3b] 5175   **12.** [1.5a] 41,112   **13.** [1.5a] 5,325,600
14. [1.5a] 2405   **15.** [1.5a] 534,264   **16.** [1.6b] 3 R 3
17. [1.6b] 70   **18.** [1.6b] 97   **19.** [1.6b] 805 R 8
20. [1.8a] 1852 12-packs; 7 cakes left over
21. [1.8a] 1,285,156 sq mi   **22.** (a) [1.2b], [1.5c] 300 in.,
5000 in²; 264 in., 3872 in²; 228 in., 2888 in²;
(b) [1.8a] 2112 in²   **23.** [1.8a] 99,076 patents
24. [1.8a] 1808 lb   **25.** [1.8a] 20 staplers   **26.** [1.7b] 46
27. [1.7b] 13   **28.** [1.7b] 14   **29.** [1.4a] 35,000
30. [1.4a] 34,580   **31.** [1.4a] 34,600
32. [1.4b] 23,600 + 54,700 = 78,300
33. [1.4b] 54,800 − 23,600 = 31,200
34. [1.5b] 800 · 500 = 400,000   **35.** [1.4c] >
36. [1.4c] <   **37.** [1.9a] 12^4 **38.** [1.9b] 343
39. [1.9b] 8   **40.** [1.9c] 64   **41.** [1.9c] 96
42. [1.9c] 2   **43.** [1.9d] 216   **44.** [1.9c] 18
45. [1.9c] 92   **46.** [1.5c], [1.8a] 336 in²
47. [1.8a] 80 payments   **48.** [1.9c] 83   **49.** [1.9c] 9

Chapter 2

Pretest: Chapter 2, p. 100

1. Prime   **2.** 2 · 2 · 3 · 5 · 7, or $2^2 \cdot 3 \cdot 5 \cdot 7$ **3.** Yes
4. Yes   **5.** 1   **6.** 68   **7.** 0   **8.** $\frac{1}{4}$ **9.** $\frac{6}{5}$ **10.** 20
11. $\frac{5}{4}$ **12.** $\frac{8}{7}$ **13.** $\frac{1}{11}$ **14.** 24   **15.** $\frac{3}{4}$ **16.** 30   **17.** $\neq$
18. $72   **19.** $\frac{1}{24}$ m

Margin Exercises, Section 2.1, pp. 102–106

1. Yes   **2.** No   **3.** 1, 2, 5   **4.** 1, 3, 5, 9, 15, 45
5. 1, 2, 31, 62   **6.** 1, 2, 3, 4, 6, 8, 12, 24   **7.** 5 = 1 · 5;
45 = 9 · 5; 100 = 20 · 5   **8.** 10 = 1 · 10; 60 = 6 · 10;
110 = 11 · 10   **9.** 5, 10, 15, 20, 25, 30, 35, 40, 45, 50
10. Yes   **11.** Yes   **12.** No   **13.** 2, 13, 19, 41, 73 are
prime; 6, 12, 65, 99 are composite; 1 is neither   **14.** 2 · 3
15. 2 · 2 · 3   **16.** 3 · 3 · 5   **17.** 2 · 7 · 7
18. 2 · 3 · 3 · 7   **19.** 2 · 2 · 2 · 2 · 3 · 3

Calculator Corner, p. 103

1. Yes   **2.** No   **3.** No   **4.** Yes   **5.** No   **6.** Yes
7. Yes   **8.** No   **9.** Yes   **10.** No

Exercise Set 2.1, p. 107

1. No   **3.** Yes   **5.** 1, 2, 3, 6, 9, 18
7. 1, 2, 3, 6, 9, 18, 27, 54   **9.** 1, 2, 4   **11.** 1, 7   **13.** 1
15. 1, 2, 7, 14, 49, 98   **17.** 4, 8, 12, 16, 20, 24, 28, 32, 36, 40
19. 20, 40, 60, 80, 100, 120, 140, 160, 180, 200
21. 3, 6, 9, 12, 15, 18, 21, 24, 27, 30
23. 12, 24, 36, 48, 60, 72, 84, 96, 108, 120
25. 10, 20, 30, 40, 50, 60, 70, 80, 90, 100
27. 9, 18, 27, 36, 45, 54, 63, 72, 81, 90   **29.** No   **31.** Yes
33. Yes   **35.** No   **37.** No   **39.** Neither
41. Composite   **43.** Prime   **45.** Prime   **47.** 2 · 2 · 2
49. 2 · 7   **51.** 2 · 3 · 7   **53.** 5 · 5   **55.** 2 · 5 · 5

57. 13 · 13   **59.** 2 · 2 · 5 · 5   **61.** 5 · 7
63. 2 · 2 · 2 · 3 · 3   **65.** 7 · 11   **67.** 2 · 2 · 7 · 103
69. 3 · 17   **71.** $\mathbf{D_W}$ **73.** 26   **74.** 256   **75.** 425
76. 4200   **77.** 0   **78.** 22   **79.** 1   **80.** 3   **81.** $612
82. 201 min, or 3 hr 21 min   **83.** Row 1: 48, 90, 432, 63;
row 2: 7, 2, 2, 10, 8, 6, 21, 10; row 3: 9, 18, 36, 14, 12, 11,
21; row 4: 29, 19, 42

Margin Exercises, Section 2.2, pp. 109–112

1. Yes   **2.** No   **3.** Yes   **4.** No   **5.** Yes   **6.** No
7. Yes   **8.** No   **9.** Yes   **10.** No   **11.** No   **12.** Yes
13. No   **14.** Yes   **15.** No   **16.** Yes   **17.** No   **18.** Yes
19. No   **20.** Yes   **21.** Yes   **22.** No   **23.** No   **24.** Yes
25. Yes   **26.** No   **27.** No   **28.** Yes   **29.** No   **30.** Yes
31. Yes   **32.** No

Exercise Set 2.2, p. 113

1. 46, 224, 300, 36, 45,270, 4444, 256, 8064, 21,568
3. 224, 300, 36, 4444, 256, 8064, 21,568
5. 300, 36, 45,270, 8064   **7.** 36, 45,270, 711, 8064
9. 324, 42, 501, 3009, 75, 2001, 402, 111,111, 1005
11. 55,555, 200, 75, 2345, 35, 1005   **13.** 324   **15.** 200
17. $\mathbf{D_W}$ **19.** 138   **20.** 139   **21.** 874   **22.** 56
23. 26   **24.** 13   **25.** 234   **26.** 4003   **27.** 45 gal
28. 4320 min   **29.** 2 · 2 · 2 · 3 · 5 · 5 · 13
31. 2 · 2 · 3 · 3 · 7 · 11   **33.** 95,238

Margin Exercises, Section 2.3, pp. 115–120

1. 1, numerator; 6, denominator   **2.** 5, numerator; 7,
denominator   **3.** 22, numerator; 3, denominator
4. $\frac{1}{2}$ **5.** $\frac{1}{3}$ **6.** $\frac{1}{3}$ **7.** $\frac{2}{3}$ **8.** $\frac{3}{4}$ **9.** $\frac{15}{16}$ **10.** $\frac{5}{4}$ **11.** $\frac{7}{4}$
12. $\frac{3}{5}; \frac{2}{5}$ **13.** $\frac{86}{76}; \frac{86}{162}; \frac{76}{162}$ **14.** 1   **15.** 1   **16.** 1   **17.** 1
18. 1   **19.** 1   **20.** 0   **21.** 0   **22.** 0   **23.** 0
24. Not defined   **25.** Not defined   **26.** 8   **27.** 10
28. 346   **29.** 23

Exercise Set 2.3, p. 121

1. 3, numerator; 4, denominator   **3.** 11, numerator; 20,
denominator   **5.** $\frac{2}{4}$ **7.** $\frac{1}{8}$ **9.** $\frac{4}{3}$ **11.** $\frac{12}{16}$ **13.** $\frac{38}{16}$
15. $\frac{3}{4}$ **17.** $\frac{4}{8}$ **19.** $\frac{6}{12}$ **21.** $\frac{5}{8}$ **23.** $\frac{4}{7}$ **25.** (a) $\frac{2}{8}$; (b) $\frac{6}{8}$
27. (a) $\frac{3}{8}$; (b) $\frac{5}{8}$ **29.** (a) $\frac{3}{7}$; (b) $\frac{3}{4}$; (c) $\frac{4}{7}$; (d) $\frac{4}{3}$ **31.** $\frac{67}{10,000}$
33. (a) $\frac{4}{15}$; (b) $\frac{4}{11}$; (c) $\frac{11}{15}$ **35.** 0   **37.** 7   **39.** 1   **41.** 1
43. 0   **45.** 1   **47.** 1   **49.** 1   **51.** 1   **53.** 18
55. Not defined   **57.** Not defined   **59.** $\mathbf{D_W}$
61. 34,560   **62.** 34,600   **63.** 35,000   **64.** 30,000
65. $27,729   **66.** 130 gal   **67.** 2203
68. 848   **69.** 37,239   **70.** 11,851   **71.** $\frac{1}{6}$ **73.** $\frac{2}{16}$, or $\frac{1}{8}$
75. ⬡⬡⬢⬢⬢   **77.**

Margin Exercises, Section 2.4, pp. 126–130

1. $\frac{2}{3}$ **2.** $\frac{5}{8}$ **3.** $\frac{10}{3}$ **4.** $\frac{33}{8}$ **5.** $\frac{46}{5}$
6. 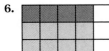 **7.** $\frac{15}{56}$ **8.** $\frac{32}{15}$ **9.** $\frac{3}{100}$ **10.** $\frac{14}{3}$
11. $\frac{3}{8}$ **12.** $\frac{63}{100}$ cm^2 **13.** $\frac{3}{40}$

Exercise Set 2.4, p. 131

1. $\frac{3}{5}$ **3.** $\frac{5}{8}$ **5.** $\frac{8}{11}$ **7.** $\frac{70}{9}$ **9.** $\frac{2}{5}$ **11.** $\frac{6}{5}$ **13.** $\frac{21}{4}$
15. $\frac{85}{6}$ **17.** $\frac{1}{6}$ **19.** $\frac{1}{6}$ **21.** $\frac{2}{15}$ **23.** $\frac{4}{15}$ **25.** $\frac{9}{16}$ **27.** $\frac{14}{39}$
29. $\frac{7}{100}$ **31.** $\frac{49}{64}$ **33.** $\frac{1}{1000}$ **35.** $\frac{182}{285}$ **37.** $\frac{12}{25}$ m^2 **39.** $\frac{1}{1521}$
41. $\frac{3}{8}$ cup **43.** $\frac{9}{20}$ **45.** D_W **47.** 204 **48.** 700
49. 3001 **50.** 204 R 8 **51.** 8 thousands
52. 8 millions **53.** 8 ones **54.** 8 hundreds **55.** 3
56. 81 **57.** 50 **58.** 6399 **59.** $\frac{71,269}{180,433}$ **61.** $\frac{56}{1125}$

Margin Exercises, Section 2.5, pp. 133–137

1. $\frac{8}{16}$ **2.** $\frac{30}{50}$ **3.** $\frac{52}{100}$ **4.** $\frac{200}{75}$ **5.** $\frac{12}{9}$ **6.** $\frac{18}{24}$ **7.** $\frac{90}{100}$
8. $\frac{9}{45}$ **9.** $\frac{56}{49}$ **10.** $\frac{1}{4}$ **11.** $\frac{5}{6}$ **12.** 5 **13.** $\frac{4}{3}$ **14.** $\frac{7}{8}$
15. $\frac{89}{78}$ **16.** $\frac{8}{7}$ **17.** $\frac{1}{4}$
18. $\frac{2}{100} = \frac{1}{50}; \frac{4}{100} = \frac{1}{25}; \frac{32}{100} = \frac{8}{25}; \frac{44}{100} = \frac{11}{25}; \frac{18}{100} = \frac{9}{50}$ **19.** =
20. ≠

Calculator Corner, p. 136

1. $\frac{14}{15}$ **2.** $\frac{7}{8}$ **3.** $\frac{138}{167}$ **4.** $\frac{7}{25}$

Exercise Set 2.5, p. 138

1. $\frac{5}{10}$ **3.** $\frac{20}{32}$ **5.** $\frac{27}{30}$ **7.** $\frac{28}{32}$ **9.** $\frac{20}{48}$ **11.** $\frac{51}{54}$ **13.** $\frac{75}{45}$
15. $\frac{42}{132}$ **17.** $\frac{1}{2}$ **19.** $\frac{3}{4}$ **21.** $\frac{1}{5}$ **23.** 3 **25.** $\frac{3}{4}$ **27.** $\frac{7}{8}$
29. $\frac{6}{5}$ **31.** $\frac{1}{3}$ **33.** 6 **35.** $\frac{1}{3}$ **37.** = **39.** ≠ **41.** =
43. ≠ **45.** = **47.** ≠ **49.** = **51.** ≠ **53.** D_W
55. 4992 ft^2; 284 ft **56.** $928 **57.** 11 **58.** 32
59. 186 **60.** 2737 **61.** 5 **62.** 89 **63.** 3520
64. 9001 **65.** $\frac{137}{149}$ **67.** $\frac{2}{5}; \frac{3}{5}$ **69.** No. $\frac{216}{580} \neq \frac{197}{529}$ because
$216 \cdot 529 \neq 580 \cdot 197$.

Margin Exercises, Section 2.6, p. 141

1. $\frac{7}{12}$ **2.** $\frac{1}{3}$ **3.** 6 **4.** $\frac{5}{2}$ **5.** 14 lb

Exercise Set 2.6, p. 142

1. $\frac{1}{3}$ **3.** $\frac{1}{8}$ **5.** $\frac{1}{10}$ **7.** $\frac{1}{6}$ **9.** $\frac{27}{10}$ **11.** $\frac{14}{9}$ **13.** 1
15. 1 **17.** 1 **19.** 1 **21.** 2 **23.** 4 **25.** 9 **27.** 9
29. $\frac{26}{5}$ **31.** $\frac{98}{5}$ **33.** 60 **35.** 30 **37.** $\frac{1}{5}$ **39.** $\frac{9}{25}$
41. $\frac{11}{40}$ **43.** $\frac{5}{14}$ **45.** $\frac{5}{8}$ in. **47.** 18 mph
49. 625 addresses **51.** $\frac{1}{3}$ cup **53.** $115,500
55. 160 mi **57.** Food: $9000; housing: $7200; clothing:
$3600; savings: $4000; taxes: $9000; other expenses: $3200
59. D_W **61.** 35 **62.** 85 **63.** 125 **64.** 120

65. 4989 **66.** 8546 **67.** 6498 **68.** 6407 **69.** 4673
70. 5338 **71.** $\frac{129}{485}$ **73.** $\frac{1}{12}$ **75.** $\frac{1}{168}$

Margin Exercises, Section 2.7, pp. 146–150

1. $\frac{5}{2}$ **2.** $\frac{7}{10}$ **3.** $\frac{1}{9}$ **4.** 5 **5.** $\frac{8}{7}$ **6.** $\frac{8}{3}$ **7.** $\frac{1}{10}$ **8.** 100
9. 1 **10.** $\frac{14}{15}$ **11.** $\frac{4}{5}$ **12.** 32 **13.** 320 loops
14. 200 gal

Exercise Set 2.7, p. 151

1. $\frac{6}{5}$ **3.** $\frac{1}{6}$ **5.** 6 **7.** $\frac{3}{10}$ **9.** $\frac{4}{5}$ **11.** $\frac{4}{15}$ **13.** 4 **15.** 2
17. $\frac{1}{8}$ **19.** $\frac{3}{7}$ **21.** 8 **23.** 35 **25.** 1 **27.** $\frac{2}{3}$ **29.** $\frac{9}{4}$
31. 144 **33.** 75 **35.** 2 **37.** $\frac{3}{5}$ **39.** 315
41. 75 times **43.** 32 pairs **45.** 24 bowls **47.** 16 L
49. 288 km; 108 km **51.** $\frac{1}{16}$ in. **53.** D_W **55.** 67
56. 33 R 4 **57.** 285 R 2 **58.** 103 R 10 **59.** 37
60. 27 **61.** 17 **62.** 56 **63.** 67 **64.** 264 **65.** 8499
66. 4368 **67.** $\frac{9}{19}$ **69.** 36 **71.** $\frac{3}{8}$

Summary and Review: Chapter 2, p. 154

1. $2 \cdot 5 \cdot 7$ **2.** $2 \cdot 3 \cdot 5$ **3.** $3 \cdot 3 \cdot 5$ **4.** $2 \cdot 3 \cdot 5 \cdot 5$
5. No **6.** No **7.** No **8.** Yes **9.** Prime
10. 2, numerator; 7, denominator **11.** $\frac{3}{5}$
12. $\frac{3}{100}; \frac{8}{100} = \frac{2}{25}; \frac{10}{100} = \frac{1}{10}; \frac{15}{100} = \frac{3}{20}; \frac{21}{100}; \frac{43}{100}$ **13.** 0 **14.** 1
15. 48 **16.** 6 **17.** $\frac{2}{3}$ **18.** $\frac{1}{4}$ **19.** 1 **20.** 0 **21.** $\frac{2}{5}$
22. 18 **23.** 4 **24.** $\frac{1}{3}$ **25.** Not defined
26. Not defined **27.** ≠ **28.** = **29.** ≠ **30.** =
31. $\frac{3}{2}$ **32.** 56 **33.** $\frac{5}{2}$ **34.** 24 **35.** $\frac{2}{3}$ **36.** $\frac{1}{14}$ **37.** $\frac{2}{3}$
38. $\frac{1}{22}$ **39.** $\frac{5}{4}$ **40.** $\frac{1}{3}$ **41.** 9 **42.** $\frac{36}{47}$ **43.** $\frac{9}{2}$ **44.** 2
45. $\frac{11}{6}$ **46.** $\frac{1}{4}$ **47.** $\frac{9}{4}$ **48.** 300 **49.** 1 **50.** $\frac{4}{9}$ **51.** $\frac{3}{10}$
52. 240 **53.** 9 days **54.** 100 km **55.** $\frac{1}{3}$ cup; 2 cups
56. $15 **57.** D_W To simplify fraction notation, first factor
the numerator and the denominator into prime numbers.
Examine the factorizations for factors common to both the
numerator and the denominator. Factor the fraction, with
each pair of like factors forming a factor of 1. Remove the
factors of 1, and multiply the remaining factors in the
numerator and in the denominator, if necessary.
58. D_W Taking $\frac{1}{2}$ of a number is equivalent to multiplying
the number by $\frac{1}{2}$. Dividing by $\frac{1}{2}$ is equivalent to multiplying
by the reciprocal of $\frac{1}{2}$, or 2. Thus taking $\frac{1}{2}$ of a number is not
the same as dividing by $\frac{1}{2}$. **59.** 24 **60.** 469 **61.** $912
62. 774 mi **63.** 408 R 9 **64.** 3607
65. $a = 11{,}176$; $b = 9887$ **66.** 13, 11, 101, 37

Test: Chapter 2, p. 157

1. [2.1d] $2 \cdot 3 \cdot 3$ **2.** [2.1d] $2 \cdot 2 \cdot 3 \cdot 5$ **3.** [2.2a] Yes
4. [2.2a] No **5.** [2.3a] 4, numerator; 9, denominator
6. [2.3a] $\frac{3}{4}$ **7.** [2.3a] **(a)** $\frac{286}{473}$; **(b)** $\frac{187}{473}$ **8.** [2.3b] 26
9. [2.3b] 1 **10.** [2.3b] 0 **11.** [2.5b] $\frac{1}{2}$ **12.** [2.5b] 6
13. [2.5b] $\frac{1}{14}$ **14.** [2.3b] Not defined
15. [2.3b] Not defined **16.** [2.5c] = **17.** [2.5c] ≠
18. [2.6a] 32 **19.** [2.6a] $\frac{3}{2}$ **20.** [2.6a] $\frac{5}{2}$ **21.** [2.6a] $\frac{1}{10}$
22. [2.7a] $\frac{8}{5}$ **23.** [2.7a] 4 **24.** [2.7a] $\frac{1}{18}$ **25.** [2.7b] $\frac{3}{10}$

26. [2.7b] $\frac{8}{5}$ **27.** [2.7b] 18 **28.** [2.7c] 64 **29.** [2.7c] $\frac{7}{4}$
30. [2.6b] 4375 students **31.** [2.7d] $\frac{3}{40}$ m
32. [1.7b] 1805 **33.** [1.7b] 101 **34.** [1.8a] 3635 mi
35. [1.6b] 380 R 7 **36.** [1.3b] 4434 **37.** [2.6b] $\frac{15}{8}$ tsp
38. [2.6b] $\frac{7}{48}$ acre **39.** [2.6a], [2.7b] $\frac{7}{960}$ **40.** [2.7c] $\frac{7}{5}$

Cumulative Review: Chapters 1–2, p. 159

1. [1.1c] 584,017,800 **2.** [1.1c] Five million, three hundred eighty thousand, six hundred twenty-one **3.** [1.1a] 0
4. [2.3a] $\frac{25}{16}$ **5.** [2.3a] $\frac{3}{4}$ **6.** [1.2a] 17,797 **7.** [1.2a] 8866
8. [1.3b] 4946 **9.** [1.3b] 1425 **10.** [1.5a] 16,767
11. [1.5a] 8,266,500 **12.** [2.6a] $\frac{3}{20}$ **13.** [2.6a] $\frac{1}{6}$
14. [1.6b] 241 R 1 **15.** [1.6b] 62 **16.** [2.7b] $\frac{3}{50}$
17. [2.7b] $\frac{16}{45}$ **18.** [1.4a] 428,000 **19.** [1.4a] 5300
20. [1.4b] 749,600 + 301,400 = 1,051,000
21. [1.5b] 700 · 500 = 350,000 **22.** [1.4c] >
23. [2.5c] $\neq$ **24.** [1.9b] 81 **25.** [1.9b] 100
26. [1.9b] 32 **27.** [1.9b] 1000 **28.** [1.9c] 36
29. [1.9d] 2 **30.** [2.1a] 1, 2, 4, 7, 14, 28
31. [2.1d] 2 · 2 · 7 **32.** [2.1c] Composite **33.** [2.2a] Yes
34. [2.2a] No **35.** [2.3b] 35 **36.** [2.5b] 7 **37.** [2.5b] $\frac{2}{7}$
38. [2.3b] 0 **39.** [1.7b] 37 **40.** [2.7c] $\frac{3}{2}$ **41.** [1.7b] 3
42. [1.7b] 24 **43.** [1.8a] $12,535
44. [1.8a] 480 rows **45.** [1.8a] (a) 2062; (b) 2138
46. [2.6b], [1.8a] (a) $\frac{3}{2}$ cups; (b) 1968 calories
47. [2.7d] 8 days **48.** [1.8a] $920
49. [2.1a, b, c, d], [2.2a]
Factors of 68: 1, 2, 4, 17, 34, 68
Factorization of 68: 2 · 34, or 2 · 2 · 17
Prime factorization of 68: 2 · 2 · 17
Numbers divisible by 6: 12, 42, 54, 300
Numbers divisible by 8: 8, 16, 24, 32, 40, 48, 64, 864
Numbers divisible by 5: 70, 95, 215
Prime numbers: 2, 3, 17, 19, 23, 31, 47, 101
50. [2.6b] (d) **51.** [2.6a] (d) **52.** [2.7b] (c)
53. [2.7d] (a) **54.** [1.1a] 36,542, 63,542, 18,542, or 81,542
55. [1.9b] $\frac{1}{1,073,741,824}$

Chapter 3

Pretest: Chapter 3, p. 164

1. 120 **2.** < **3.** $\frac{11}{12}$ **4.** $\frac{29}{18}$ **5.** $\frac{1}{2}$ **6.** $\frac{1}{40}$ **7.** $\frac{61}{8}$
8. $5\frac{1}{2}$ **9.** $399\frac{1}{12}$ **10.** $11\frac{31}{60}$ **11.** $6\frac{1}{6}$ **12.** $13\frac{3}{5}$ **13.** $21\frac{2}{3}$
14. 6 **15.** $1\frac{2}{3}$ **16.** $\frac{2}{9}$ **17.** $21\frac{1}{4}$ lb **18.** $4\frac{1}{4}$ cu ft
19. $351\frac{1}{5}$ mi **20.** $22\frac{1}{2}$ cups **21.** 1 **22.** 0 **23.** 2
24. 58

Margin Exercises, Section 3.1, pp. 165–172

1. 45 **2.** 40 **3.** 30 **4.** 24 **5.** 10 **6.** 80 **7.** 40
8. 360 **9.** 864 **10.** 2520 **11.** $2^3 \cdot 3^2 \cdot 5 \cdot 7$, or 2520
12. $2^3 \cdot 5$, or 40; $2^3 \cdot 3^2 \cdot 5$, or 360; $2^5 \cdot 3^3$, or 864 **13.** 18
14. 24 **15.** 36 **16.** 210 **17.** 600 **18.** 3780
19. 600

Exercise Set 3.1, p. 171

1. 4 **3.** 50 **5.** 40 **7.** 54 **9.** 150 **11.** 120
13. 72 **15.** 420 **17.** 144 **19.** 288 **21.** 30
23. 105 **25.** 72 **27.** 60 **29.** 36 **31.** 900 **33.** 48
35. 50 **37.** 143 **39.** 420 **41.** 378 **43.** 810
45. Every 60 yr **47.** Every 420 yr **49.** $\mathbf{D_W}$
51. 90 days **52.** $539 **53.** 33,135 **54.** 6939 **55.** $\frac{2}{3}$
56. $\frac{8}{7}$ **57.** 18,900 **59.** 5 in. by 24 in.

Margin Exercises, Section 3.2, pp. 173–176

1. $\frac{4}{5}$ **2.** 1 **3.** $\frac{1}{2}$ **4.** $\frac{3}{4}$ **5.** $\frac{5}{6}$ **6.** $\frac{29}{24}$ **7.** $\frac{5}{9}$ **8.** $\frac{413}{1000}$
9. $\frac{197}{210}$ **10.** $\frac{65}{72}$ **11.** $\frac{9}{10}$ mi

Exercise Set 3.2, p. 177

1. 1 **3.** $\frac{3}{4}$ **5.** $\frac{3}{2}$ **7.** $\frac{7}{24}$ **9.** $\frac{3}{2}$ **11.** $\frac{19}{24}$ **13.** $\frac{9}{10}$
15. $\frac{29}{18}$ **17.** $\frac{31}{100}$ **19.** $\frac{41}{60}$ **21.** $\frac{189}{100}$ **23.** $\frac{7}{8}$ **25.** $\frac{13}{24}$
27. $\frac{17}{24}$ **29.** $\frac{3}{4}$ **31.** $\frac{437}{500}$ **33.** $\frac{53}{40}$ **35.** $\frac{391}{144}$ **37.** $\frac{5}{6}$ lb
39. $\frac{23}{12}$ mi **41.** 690 kg; $\frac{14}{23}$ cement; $\frac{5}{23}$ stone; $\frac{4}{23}$ sand; 1
43. $\frac{51}{32}$ in. **45.** $\mathbf{D_W}$ **47.** 210,528 **48.** 4,194,000
49. 3,387,807 **50.** 352,350 **51.** $73 **52.** $11
53. $3 **54.** $9 **55.** $1558 **56.** $1684 **57.** 12 tickets
58. $\frac{3}{64}$ acre **59.** $\frac{4}{15}$; $320

Margin Exercises, Section 3.3, pp. 180–183

1. $\frac{1}{2}$ **2.** $\frac{3}{8}$ **3.** $\frac{1}{2}$ **4.** $\frac{1}{12}$ **5.** $\frac{13}{18}$ **6.** $\frac{1}{2}$ **7.** $\frac{9}{112}$ **8.** <
9. > **10.** > **11.** > **12.** < **13.** $\frac{1}{6}$ **14.** $\frac{11}{40}$
15. $\frac{5}{24}$ mi

Exercise Set 3.3, p. 184

1. $\frac{2}{3}$ **3.** $\frac{3}{4}$ **5.** $\frac{5}{8}$ **7.** $\frac{1}{24}$ **9.** $\frac{1}{2}$ **11.** $\frac{9}{14}$ **13.** $\frac{3}{5}$ **15.** $\frac{7}{10}$
17. $\frac{17}{60}$ **19.** $\frac{53}{100}$ **21.** $\frac{26}{75}$ **23.** $\frac{9}{100}$ **25.** $\frac{13}{24}$ **27.** $\frac{1}{10}$
29. $\frac{1}{24}$ **31.** $\frac{13}{16}$ **33.** $\frac{31}{75}$ **35.** $\frac{13}{75}$ **37.** < **39.** >
41. < **43.** < **45.** > **47.** > **49.** < **51.** $\frac{1}{15}$
53. $\frac{2}{15}$ **55.** $\frac{1}{2}$ **57.** $\frac{5}{12}$ hr **59.** $\frac{1}{32}$ in.
61. $\frac{1}{4}$ of the business **63.** $\frac{2}{3}$ cup **65.** $\mathbf{D_W}$ **67.** 1
68. Not defined **69.** Not defined **70.** 4 **71.** $\frac{4}{21}$
72. $\frac{3}{2}$ **73.** 21 **74.** $\frac{1}{32}$ **75.** 6 lb **76.** 9 cups **77.** $\frac{14}{3553}$
79. $\frac{227}{420}$ km **81.** $\frac{19}{24}$ **83.** $\frac{145}{144}$ **85.** $\frac{21}{40}$ km **87.** >

Margin Exercises, Section 3.4, pp. 187–190

1. $1\frac{2}{3}$ **2.** $2\frac{3}{4}$ **3.** $8\frac{3}{4}$ **4.** $12\frac{2}{3}$ **5.** $\frac{22}{5}$ **6.** $\frac{61}{10}$ **7.** $\frac{29}{6}$
8. $\frac{37}{12}$ **9.** $\frac{62}{3}$ **10.** $2\frac{1}{3}$ **11.** $1\frac{1}{10}$ **12.** $18\frac{1}{3}$ **13.** $807\frac{2}{3}$
14. $134\frac{23}{45}$

Calculator Corner, p. 190

1. $1476\frac{1}{6}$ **2.** $676\frac{4}{9}$ **3.** $800\frac{51}{56}$ **4.** $13,031\frac{1}{2}$ **5.** $51,626\frac{9}{11}$
6. $7330\frac{7}{32}$ **7.** $134\frac{1}{15}$ **8.** $2666\frac{130}{213}$ **9.** $3571\frac{51}{112}$
10. $12\frac{169}{454}$

Exercise Set 3.4, p. 191

1. $\frac{29}{8}$; $\frac{11}{4}$ **3.** $\frac{17}{3}$ **5.** $\frac{13}{4}$ **7.** $\frac{81}{8}$ **9.** $\frac{51}{10}$ **11.** $\frac{103}{5}$ **13.** $\frac{59}{6}$
15. $\frac{73}{10}$ **17.** $\frac{13}{8}$ **19.** $\frac{51}{4}$ **21.** $\frac{43}{10}$ **23.** $\frac{203}{100}$ **25.** $\frac{200}{3}$
27. $\frac{279}{50}$ **29.** $3\frac{3}{5}$ **31.** $4\frac{2}{3}$ **33.** $4\frac{1}{2}$ **35.** $5\frac{7}{10}$ **37.** $7\frac{4}{7}$
39. $7\frac{1}{2}$ **41.** $11\frac{1}{2}$ **43.** $1\frac{1}{2}$ **45.** $7\frac{57}{100}$ **47.** $43\frac{1}{8}$
49. $108\frac{5}{8}$ **51.** $618\frac{1}{5}$ **53.** $40\frac{4}{7}$ **55.** $55\frac{1}{51}$ **57.** D_W
59. 18 **60.** $\frac{5}{2}$ **61.** $\frac{1}{4}$ **62.** $\frac{2}{5}$ **63.** 24 **64.** 49
65. $\frac{2560}{3}$ **66.** $\frac{4}{3}$ **67.** $237\frac{19}{541}$ **69.** $8\frac{2}{3}$ **71.** $52\frac{2}{7}$

Margin Exercises, Section 3.5, pp. 194–198

1. $7\frac{2}{5}$ **2.** $12\frac{1}{10}$ **3.** $13\frac{7}{12}$ **4.** $1\frac{1}{2}$ **5.** $3\frac{1}{6}$ **6.** $3\frac{5}{18}$
7. $3\frac{2}{3}$ **8.** $17\frac{1}{12}$ yd **9.** $\frac{3}{8}$ in. **10.** $23\frac{1}{4}$ gal

Exercise Set 3.5, p. 199

1. $28\frac{3}{4}$ **3.** $185\frac{7}{8}$ **5.** $6\frac{1}{2}$ **7.** $2\frac{11}{12}$ **9.** $14\frac{7}{12}$ **11.** $12\frac{1}{10}$
13. $16\frac{5}{24}$ **15.** $21\frac{1}{2}$ **17.** $27\frac{7}{8}$ **19.** $27\frac{13}{24}$ **21.** $1\frac{3}{5}$
23. $4\frac{1}{10}$ **25.** $21\frac{17}{24}$ **27.** $12\frac{1}{4}$ **29.** $15\frac{3}{8}$ **31.** $7\frac{5}{12}$
33. $13\frac{3}{8}$ **35.** $11\frac{5}{18}$ **37.** $5\frac{3}{8}$ yd **39.** $7\frac{5}{12}$ lb **41.** $6\frac{5}{12}$ in.
43. $19\frac{1}{16}$ in. **45.** $95\frac{1}{5}$ mi **47.** $36\frac{1}{2}$ in. **49.** $20\frac{1}{8}$ in.
51. $78\frac{1}{12}$ in. **53.** $3\frac{4}{5}$ hr **55.** $28\frac{3}{4}$ yd **57.** $7\frac{3}{8}$ ft
59. $1\frac{9}{16}$ in. **61.** D_W **63.** 16 packages
64. 286 cartons; 2 oz left over **65.** Yes **66.** No
67. No **68.** Yes **69.** No **70.** Yes **71.** Yes
72. Yes **73.** $\frac{10}{13}$ **74.** $\frac{1}{10}$ **75.** $8568\frac{786}{1189}$ **77.** $5\frac{3}{4}$ ft

Margin Exercises, Section 3.6, pp. 204–207

1. 20 **2.** $1\frac{7}{8}$ **3.** $12\frac{4}{5}$ **4.** $8\frac{1}{3}$ **5.** 16 **6.** $7\frac{3}{7}$ **7.** $1\frac{7}{8}$
8. $\frac{7}{10}$ **9.** $227\frac{1}{2}$ mi **10.** 20 mpg **11.** $240\frac{3}{4}$ ft²

Calculator Corner, p. 208

1. $\frac{7}{12}$ **2.** $\frac{11}{10}$ **3.** $\frac{35}{16}$ **4.** $\frac{3}{10}$ **5.** $10\frac{2}{15}$ **6.** $1\frac{1}{28}$ **7.** $10\frac{11}{15}$
8. $2\frac{91}{115}$

Exercise Set 3.6, p. 209

1. $22\frac{2}{3}$ **3.** $2\frac{5}{12}$ **5.** $8\frac{1}{6}$ **7.** $9\frac{31}{40}$ **9.** $24\frac{91}{100}$ **11.** $975\frac{4}{5}$
13. $6\frac{1}{4}$ **15.** $1\frac{1}{5}$ **17.** $3\frac{9}{16}$ **19.** $1\frac{1}{8}$ **21.** $1\frac{8}{43}$ **23.** $\frac{9}{40}$
25. 24 cassettes **27.** $13\frac{1}{3}$ tsp **29.** $343\frac{3}{4}$ lb
31. $\frac{1}{2}$ **recipe:** $\frac{3}{4}$ cup sugar, 1 egg, $\frac{1}{2}$ can fruit cocktail, $\frac{1}{2}$ teaspoon vanilla, $1\frac{1}{8}$ cups flour, $\frac{3}{4}$ teaspoon soda, $\frac{1}{2}$ teaspoon salt, $\frac{2}{3}$ cup coconut, $\frac{1}{2}$ cup walnuts; (glaze) $\frac{1}{4}$ cup sugar, $\frac{1}{8}$ cup butter or margarine, 1 tablespoon milk, $\frac{1}{8}$ teaspoon vanilla; **3 recipes:** $4\frac{1}{2}$ cups sugar, 6 eggs, 3 cans fruit cocktail, 3 teaspoons vanilla, $6\frac{3}{4}$ cups flour, $4\frac{1}{2}$ teaspoons soda, 3 teaspoons salt, 4 cups coconut, 3 cups walnuts; (glaze) $1\frac{1}{2}$ cups sugar, $\frac{3}{4}$ cup butter or margarine, 6 tablespoons milk, $\frac{3}{4}$ teaspoon vanilla
33. 68°F **35.** $82\frac{1}{2}$ in. **37.** 15 mpg **39.** 4 cu ft
41. $16\frac{1}{2}$ servings **43.** $35\frac{115}{256}$ sq in. **45.** $59{,}538\frac{1}{8}$ sq ft

47. D_W **49.** 1,429,017 **50.** 45,800 **51.** 588 **52.** $\frac{2}{3}$
53. $\frac{1}{6}$ **54.** 45,770 **55.** $\frac{8}{15}$ **56.** $\frac{21}{25}$ **57.** $\frac{16}{27}$ **58.** $\frac{583}{669}$
59. $360\frac{60}{473}$ **61.** $35\frac{57}{64}$ **63.** $\frac{4}{9}$ **65.** $1\frac{4}{5}$

Margin Exercises, Section 3.7, pp. 214–217

1. $\frac{1}{2}$ **2.** $\frac{3}{10}$ **3.** $20\frac{2}{3}$, or $\frac{62}{3}$ **4.** $9\frac{7}{8}$ in. **5.** $\frac{5}{9}$ **6.** $\frac{31}{40}$
7. $\frac{27}{56}$ **8.** 0 **9.** 1 **10.** $\frac{1}{2}$ **11.** 1
12. 12; answers may vary **13.** 32; answers may vary
14. 27; answers may vary **15.** 15; answers may vary
16. 1; answers may vary **17.** 1,000,000; answers may vary
18. $22\frac{1}{2}$ **19.** 132 **20.** 37

Exercise Set 3.7, p. 218

1. $\frac{1}{24}$ **3.** $\frac{2}{5}$ **5.** $\frac{4}{7}$ **7.** $\frac{59}{30}$, or $1\frac{29}{30}$ **9.** $\frac{3}{20}$ **11.** $\frac{211}{8}$, or $26\frac{3}{8}$
13. $\frac{7}{16}$ **15.** $\frac{1}{36}$ **17.** $\frac{3}{8}$ **19.** $\frac{37}{48}$ **21.** $\frac{25}{72}$ **23.** $\frac{103}{16}$, or $6\frac{7}{16}$
25. $2\frac{41}{128}$ lb **27.** $18\frac{3}{5}$ days **29.** $\frac{17}{6}$, or $2\frac{5}{6}$
31. $\frac{8395}{84}$, or $99\frac{79}{84}$ **33.** 0 **35.** 0 **37.** $\frac{1}{2}$ **39.** $\frac{1}{2}$ **41.** 0
43. 1 **45.** 6 **47.** 12 **49.** 19 **51.** 6 **53.** 12
55. 16 **57.** 3 **59.** 13 **61.** 2 **63.** 2 **65.** $\frac{1}{2}$
67. $271\frac{1}{2}$ **69.** 3 **71.** 100 **73.** $29\frac{1}{2}$ **75.** D_W
77. 3402 **78.** 1,038,180 **79.** 59 R 77 **80.** 348
81. 783 **82.** $\frac{8}{3}$ **83.** $\frac{3}{8}$
84. Prime: 5, 7, 23, 43; composite: 9, 14; neither: 1
85. 16 people **86.** 43 mg
87. (a) $13 \cdot 9\frac{1}{4} + 8\frac{1}{4} \cdot 7\frac{1}{4}$; (b) $\frac{2881}{16}$, or $180\frac{1}{16}$ in²;
(c) Multiply before adding. **89.** $a = 2$, $b = 8$
91. The largest is $\frac{4}{3} + \frac{5}{2} = \frac{23}{6}$.

Summary and Review: Chapter 3, p. 222

1. 36 **2.** 90 **3.** 30 **4.** 1404 **5.** $\frac{63}{40}$ **6.** $\frac{19}{48}$ **7.** $\frac{29}{15}$
8. $\frac{7}{16}$ **9.** $\frac{1}{3}$ **10.** $\frac{1}{8}$ **11.** $\frac{5}{27}$ **12.** $\frac{11}{18}$ **13.** > **14.** >
15. $\frac{19}{40}$ **16.** $\frac{2}{5}$ **17.** $\frac{15}{2}$ **18.** $\frac{67}{8}$ **19.** $\frac{13}{3}$ **20.** $\frac{75}{7}$
21. $2\frac{1}{3}$ **22.** $6\frac{3}{4}$ **23.** $12\frac{3}{5}$ **24.** $3\frac{1}{2}$ **25.** $877\frac{1}{3}$
26. $456\frac{5}{23}$ **27.** $10\frac{2}{5}$ **28.** $11\frac{11}{15}$ **29.** $10\frac{2}{3}$ **30.** $8\frac{1}{4}$
31. $7\frac{7}{9}$ **32.** $4\frac{11}{15}$ **33.** $4\frac{3}{20}$ **34.** $13\frac{3}{8}$ **35.** 16 **36.** $3\frac{1}{2}$
37. $2\frac{21}{50}$ **38.** 6 **39.** 12 **40.** $1\frac{7}{17}$ **41.** $\frac{1}{8}$ **42.** $\frac{9}{10}$
43. $4\frac{1}{4}$ yd **44.** 24 lb **45.** About $69\frac{3}{8}$ kg **46.** $1\frac{73}{100}$ in.
47. $177\frac{3}{4}$ in² **48.** $50\frac{1}{4}$ in² **49.** $8\frac{3}{8}$ cups
50. $63\frac{2}{3}$ pies; $19\frac{1}{3}$ pies **51.** 1 **52.** $\frac{77}{240}$ **53.** $\frac{1}{2}$ **54.** 0
55. 1 **56.** 7 **57.** 10 **58.** 2 **59.** $28\frac{1}{2}$
60. D_W It might be necessary to find the least common denominator before adding or subtracting. The least common denominator is the least common multiple of the denominators.
61. D_W Suppose that a room has dimensions $15\frac{3}{4}$ ft by $28\frac{5}{8}$ ft. The equation $2 \cdot 15\frac{3}{4} + 2 \cdot 28\frac{5}{8} = 88\frac{3}{4}$ gives the perimeter of the room, in feet. Answers may vary.
62. $\frac{6}{5}$ **63.** $\frac{3}{2}$ **64.** 708,048 **65.** 17 days **66.** 12 min
67. $\frac{6}{3} + \frac{5}{4} = 3\frac{1}{4}$

Test: Chapter 3, p. 225

1. [3.1a] 48   2. [3.2a] 3   3. [3.2a] $\frac{37}{24}$ 4. [3.2a] $\frac{79}{100}$
5. [3.3a] $\frac{1}{3}$ 6. [3.3a] $\frac{1}{12}$ 7. [3.3a] $\frac{1}{12}$ 8. [3.3b] $>$
9. [3.3c] $\frac{1}{4}$ 10. [3.4a] $\frac{7}{2}$ 11. [3.4a] $\frac{79}{8}$ 12. [3.4a] $4\frac{1}{2}$
13. [3.4a] $8\frac{2}{9}$ 14. [3.4b] $162\frac{7}{11}$ 15. [3.5a] $14\frac{1}{5}$
16. [3.5a] $14\frac{5}{12}$ 17. [3.5b] $4\frac{7}{24}$ 18. [3.5b] $6\frac{1}{6}$
19. [3.6a] 39   20. [3.6a] $4\frac{1}{2}$ 21. [3.6b] 2   22. [3.6b] $\frac{1}{36}$
23. [3.6c] 105 kg   24. [3.6c] 80 books
25. [3.5c] (a) 3 in.; (b) $4\frac{1}{2}$ in.   26. [3.3d] $\frac{1}{16}$ in.
27. [3.7a] $6\frac{11}{36}$ ft   28. [3.7a] $3\frac{1}{2}$ 29. [3.7b] 0
30. [3.7b] 1   31. [3.7b] 4   32. [3.7b] $18\frac{1}{2}$ 33. [3.7b] 16
34. [3.7b] $1214\frac{1}{2}$ 35. [1.5a] 346,636   36. [2.7b] $\frac{8}{5}$
37. [2.6a] $\frac{10}{9}$ 38. [1.8a] 535 bottles; 10 oz
39. [3.1a] (a) 24, 48, 72; (b) 24
40. [3.3b], [3.5c] Dolores runs $\frac{17}{56}$ mi farther.

Cumulative Review: Chapters 1–3, p. 227

1. [3.3d] (a) $\frac{1}{48}$ in.; (b) $\frac{1}{12}$ in.   2. [2.7d] 61
3. [1.8a], [1.9c], [3.6c] (a) \$517; (b) $172\frac{1}{3}$
4. [3.6c], [3.5c] (a) $142\frac{1}{4}$ ft^2; (b) 54 ft   5. [1.8a] 31 people
6. [1.8a] \$108   7. [2.6b] $\frac{2}{5}$ tsp; 4 tsp   8. [3.6c] 39 lb
9. [3.6c] 16 pieces   10. [3.2b] $\frac{33}{20}$ mi   11. [1.1a] 5
12. [1.1b] 6 thousands + 7 tens + 5 ones
13. [1.1c] Twenty-nine thousand, five hundred
14. [2.3a] $\frac{5}{16}$ 15. [1.2a] 899   16. [1.2a] 8982
17. [3.2a] $\frac{5}{12}$ 18. [3.5a] $8\frac{1}{4}$ 19. [1.3b] 5124
20. [1.3b] 4518   21. [3.3a] $\frac{5}{12}$ 22. [3.5b] $1\frac{1}{6}$
23. [1.5a] 5004   24. [1.5a] 293,232   25. [2.6a] $\frac{3}{2}$
26. [2.6a] 15   27. [3.6a] $7\frac{1}{3}$ 28. [1.6b] 715
29. [1.6b] 56 R 11   30. [3.4a] $56\frac{11}{45}$ 31. [2.7b] $\frac{4}{7}$
32. [3.6b] $7\frac{1}{3}$ 33. [1.4a] 38,500   34. [3.1a] 72
35. [2.2a] No   36. [2.1a] 1, 2, 4, 8, 16   37. [3.3b] $>$
38. [3.3b] $<$ 39. [2.5b] $\frac{4}{5}$ 40. [2.5b] 32   41. [3.4a] $\frac{37}{8}$
42. [3.4a] $5\frac{2}{3}$ 43. [1.7b] 93   44. [3.3c] $\frac{5}{9}$ 45. [2.7c] $\frac{12}{7}$
46. [1.7b] 905   47. [3.7b] 1   48. [3.7b] $\frac{1}{2}$ 49. [3.7b] 0
50. [3.7b] 30   51. [3.7b] 1   52. [3.7b] 42
53. [2.5b], [2.7a], [Chapter 3] $\frac{3}{5} \cdot 4\frac{1}{6} = 2\frac{1}{2}$; $\frac{72}{48} = \frac{3}{2}$;
$1\frac{1}{8} + 1\frac{3}{16} = \frac{37}{16}$; $1\frac{1}{2} \div \frac{2}{3} = \frac{9}{4}$; the reciprocal of $\frac{4}{9} = \frac{9}{4}$; $\frac{108}{54} = 2$;
$\frac{24}{54} = \frac{4}{9}$; $14\frac{1}{8} - 11\frac{5}{8} = 2\frac{1}{2}$; $3^2 \div 2^2 = \frac{9}{4}$; $\frac{216}{96} = \frac{9}{4}$
54. [3.6c] (a)   55. [3.6c] (a)   56. [3.2a] (a) $\frac{1}{2}, \frac{2}{3}, \frac{3}{4}, \frac{4}{5}$;
(b) $\frac{9}{10}$ 57. [2.1c] 2003

Chapter 4

Pretest: Chapter 4, p. 232

1. Two and three hundred forty-seven thousandths
2. Three thousand, two hundred sixty-four and $\frac{78}{100}$ dollars
3. $\frac{21}{100}$ 4. $\frac{5408}{1000}$ 5. 0.379   6. 28.439   7. 3.2   8. 0.099
9. 21.0   10. 21.045   11. 607.219   12. 39.0901

13. 0.6179   14. 0.32456   15. 30.4   16. 0.57698
17. 84.26   18. 6345.157   19. 1081.6 mi   20. \$285.95
21. \$159.84   22. \$3397.71   23. 224   24. 1.4
25. 0.925   26. 2.75   27. $4.\overline{142857}$ 28. 4.6   29. 4.62
30. 4.616   31. \$9.49   32. 490,000,000,000,000
33. 1548.8836   34. 58.17

Margin Exercises, Section 4.1, pp. 234–239

1. Fifteen and three tenths   2. Two and five thousand three hundred thirty-three hundred-thousandths
3. Two hundred forty-five and eighty-nine hundredths
4. Thirty-four and sixty-four ten-thousandths
5. Thirty-one thousand, seventy-nine and seven hundred sixty-four thousandths   6. Four thousand, two hundred seventeen and $\frac{56}{100}$ dollars   7. Thirteen and $\frac{98}{100}$ dollars
8. $\frac{896}{1000}$ 9. $\frac{2378}{100}$ 10. $\frac{56,789}{10,000}$ 11. $\frac{19}{10}$ 12. 7.43
13. 0.406   14. 6.7089   15. 0.9   16. 0.057   17. 0.083
18. 4.3   19. 283.71   20. 456.013   21. 2.04   22. 0.06
23. 0.58   24. 1   25. 0.8989   26. 21.05   27. 2.8
28. 13.9   29. 234.4   30. 7.0   31. 0.64   32. 7.83
33. 34.68   34. 0.03   35. 0.943   36. 8.004
37. 43.112   38. 37.401   39. 7459.355   40. 7459.35
41. 7459.4   42. 7459   43. 7460   44. 7500   45. 7000

Exercise Set 4.1, p. 240

1. Two hundred forty-nine and ninety-four hundredths
3. Ninety-six and four thousand three hundred seventy-five ten-thousandths   5. Thirty-four and eight hundred ninety-one thousandths   7. Three hundred twenty-six and $\frac{48}{100}$ dollars   9. Thirty-six and $\frac{72}{100}$ dollars
11. $\frac{83}{10}$ 13. $\frac{356}{100}$ 15. $\frac{4603}{100}$ 17. $\frac{13}{100,000}$ 19. $\frac{10,008}{10,000}$
21. $\frac{20,003}{1000}$ 23. 0.8   25. 8.89   27. 3.798   29. 0.0078
31. 0.00019   33. 0.376193   35. 99.44   37. 3.798
39. 2.1739   41. 8.953073   43. 0.58   45. 0.91
47. 0.001   49. 235.07   51. $\frac{4}{100}$ 53. 0.4325   55. 0.1
57. 0.5   59. 2.7   61. 123.7   63. 0.89   65. 0.67
67. 1.00   69. 0.09   71. 0.325   73. 17.002
75. 10.101   77. 9.999   79. 800   81. 809.473
83. 809   85. 34.5439   87. 34.54   89. 35   91. $\mathbf{D_W}$
93. 6170   94. 6200   95. 6000   96. 830
97. $\frac{830}{1000}$, or $\frac{83}{100}$ 98. 182   99. $\frac{182}{100}$, or $\frac{91}{50}$
100. $2 \cdot 2 \cdot 2 \cdot 2 \cdot 5 \cdot 5 \cdot 5$, or $2^4 \cdot 5^3$
101. $2 \cdot 3 \cdot 3 \cdot 5 \cdot 17$, or $2 \cdot 3^2 \cdot 5 \cdot 17$ 102. $2 \cdot 7 \cdot 11 \cdot 13$
103. $2 \cdot 2 \cdot 2 \cdot 7 \cdot 7 \cdot 11$, or $2^3 \cdot 7^2 \cdot 11$
105. 2.000001, 2.0119, 2.018, 2.0302, 2.1, 2.108, 2.109
107. 6.78346   109. 0.03030

Margin Exercises, Section 4.2, pp. 243–247

1. 10.917   2. 34.2079   3. 4.969   4. 3.5617
5. 9.40544   6. 912.67   7. 2514.773   8. 10.754
9. 0.339   10. 0.5345   11. 0.5172   12. 7.36992
13. 1194.22   14. 4.9911   15. 38.534   16. 14.164
17. 2133.5

18. The "balance forward" column should read:

$3078.92
2738.23
2659.67
2890.47
2877.33
2829.33
2868.91
2766.04
2697.45
2597.45

Calculator Corner, p. 245

1. 317.645   **2.** 506.553   **3.** 17.15   **4.** 49.08   **5.** 4.4
6. 33.83   **7.** 454.74   **8.** 0.99

Exercise Set 4.2, p. 248

1. 334.37   **3.** 1576.215   **5.** 132.560   **7.** 84.417
9. 50.0248   **11.** 40.007   **13.** 771.967   **15.** 20.8649
17. 227.4680   **19.** 8754.8221   **21.** 1.3   **23.** 49.02
25. 45.61   **27.** 85.921   **29.** 2.4975   **31.** 3.397
33. 8.85   **35.** 3.37   **37.** 1.045   **39.** 3.703   **41.** 0.9902
43. 99.66   **45.** 4.88   **47.** 0.994   **49.** 17.802
51. 51.13   **53.** 2.491   **55.** 32.7386   **57.** 1.6666
59. 2344.90886   **61.** 11.65   **63.** 19.251   **65.** 384.68
67. 582.97   **69.** 15,335.3
71. The balance forward should read:

$ 9704.56
9677.12
10677.12
10553.17
10429.15
10416.72
12916.72
12778.94
12797.82
9997.82

73. $\mathbf{D_W}$ **75.** 35,000   **76.** 34,000   **77.** $\frac{1}{6}$ **78.** $\frac{34}{45}$
79. 6166   **80.** 5366   **81.** $16\frac{1}{2}$ servings   **82.** $60\frac{1}{5}$ mi
83. 345.8

Margin Exercises, Section 4.3, pp. 253–256

1. 529.48   **2.** 5.0594   **3.** 34.2906   **4.** 0.348   **5.** 0.0348
6. 0.00348   **7.** 0.000348   **8.** 34.8   **9.** 348   **10.** 3480
11. 34,800   **12.** 4,400,000,000   **13.** 3,700,000
14. 1569¢   **15.** 17¢   **16.** $0.35   **17.** $5.77

Calculator Corner, p. 255

1. 48.6   **2.** 6930.5   **3.** 142.803   **4.** 0.5076   **5.** 7916.4
6. 20.4153

Exercise Set 4.3, p. 257

1. 60.2   **3.** 6.72   **5.** 0.252   **7.** 0.522   **9.** 237.6
11. 583,686.852   **13.** 780   **15.** 8.923   **17.** 0.09768
19. 0.782   **21.** 521.6   **23.** 3.2472   **25.** 897.6
27. 322.07   **29.** 55.68   **31.** 3487.5   **33.** 50.0004
35. 114.42902   **37.** 13.284   **39.** 90.72   **41.** 0.0028728
43. 0.72523   **45.** 1.872115   **47.** 45,678   **49.** 2888¢
51. 66¢   **53.** $0.34   **55.** $34.45   **57.** 93,000,000
59. 7,200,000,000   **61.** $\mathbf{D_W}$ **63.** $11\frac{1}{5}$ **64.** $\frac{35}{72}$ **65.** $2\frac{7}{15}$
66. $7\frac{2}{15}$ **67.** 342   **68.** 87   **69.** 4566   **70.** 1257
71. 87   **72.** 1176 R 14   **73.** $10^{21} = 1$ sextillion
75. $10^{24} = 1$ septillion

Margin Exercises, Section 4.4, pp. 260–266

1. 0.6   **2.** 1.5   **3.** 0.47   **4.** 0.32   **5.** 3.75   **6.** 0.25
7. (a) 375; (b) 15   **8.** 4.9   **9.** 12.8   **10.** 15.625
11. 12.78   **12.** 0.001278   **13.** 0.09847   **14.** 67.832
15. 0.78314   **16.** 1105.6   **17.** 0.04   **18.** 0.2426
19. 593.44   **20.** 5967.5 m

Calculator Corner, p. 263

1. 14.3   **2.** 2.56   **3.** 200   **4.** 0.75   **5.** 20   **6.** 0.064
7. 15.7   **8.** 75.8

Exercise Set 4.4, p. 267

1. 2.99   **3.** 23.78   **5.** 7.48   **7.** 7.2   **9.** 1.143
11. 4.041   **13.** 0.07   **15.** 70   **17.** 20   **19.** 0.4
21. 0.41   **23.** 8.5   **25.** 9.3   **27.** 0.625   **29.** 0.26
31. 15.625   **33.** 2.34   **35.** 0.47   **37.** 0.2134567
39. 21.34567   **41.** 1023.7   **43.** 9.3   **45.** 0.0090678
47. 45.6   **49.** 2107   **51.** 303.003   **53.** 446.208
55. 24.14   **57.** 13.0072   **59.** 19.3204   **61.** 473.188278
63. 10.49   **65.** 911.13   **67.** 205   **69.** $1288.36
71. 16,249.6   **73.** $\mathbf{D_W}$ **75.** $\frac{6}{7}$ **76.** $\frac{7}{8}$ **77.** $\frac{19}{73}$ **78.** $\frac{19}{73}$
79. $2 \cdot 2 \cdot 3 \cdot 3 \cdot 19$, or $2^2 \cdot 3^2 \cdot 19$
80. $2 \cdot 3 \cdot 3 \cdot 3 \cdot 3$, or $2 \cdot 3^4$ **81.** $3 \cdot 3 \cdot 223$, or $3^2 \cdot 223$
82. $5 \cdot 401$ **83.** $15\frac{1}{8}$ **84.** $5\frac{7}{8}$ **85.** 6.254194585
87. 1000   **89.** 100

Margin Exercises, Section 4.5, pp. 271–275

1. 0.8   **2.** 0.45   **3.** 0.275   **4.** 1.32   **5.** 0.4   **6.** 0.375
7. $0.1\overline{6}$ **8.** $0.\overline{6}$ **9.** $0.\overline{45}$ **10.** $1.\overline{09}$ **11.** $0.\overline{428571}$
12. 0.7; 0.67; 0.667   **13.** 0.8; 0.81; 0.808   **14.** 6.2; 6.25;
6.245   **15.** 0.510   **16.** 24.2 mpg   **17.** 42.1 models
18. 0.72   **19.** 0.552   **20.** 9.6575

Exercise Set 4.5, p. 276

1. 0.23   **3.** 0.6   **5.** 0.325   **7.** 0.2   **9.** 0.85   **11.** 0.375
13. 0.975   **15.** 0.52   **17.** 20.016   **19.** 0.25   **21.** $1.1\overline{6}$
23. 1.1875   **25.** $0.2\overline{6}$ **27.** $0.\overline{3}$ **29.** $1.\overline{3}$ **31.** $1.1\overline{6}$
33. $0.\overline{571428}$ **35.** $0.91\overline{6}$ **37.** 0.3; 0.27; 0.267
39. 0.3; 0.33; 0.333   **41.** 1.3; 1.33; 1.333   **43.** 1.2; 1.17;
1.167   **45.** 0.6; 0.57; 0.571   **47.** 0.9; 0.92; 0.917

49. 0.2; 0.18; 0.182   **51.** 0.3; 0.28; 0.278   **53.** (a) 0.429;
(b) 0.75; (c) 0.571; (d) 1.333   **55.** 15.8 mpg
57. 17.8 mpg   **59.** 15.2 mph   **61.** $41.6875; $41.69
63. $25.875; $25.88   **65.** $19.046875; $19.05   **67.** 11.06
69. 8.4   **71.** $417.51\overline{6}$ **73.** 0   **75.** 2.8125   **77.** 0.20425
79. 317.14   **81.** 0.1825   **83.** 18   **85.** 2.736
87. $\mathbf{D_W}$ **89.** 21   **90.** $238\frac{7}{8}$ **91.** 10   **92.** $\frac{43}{52}$
93. $50\frac{5}{24}$ **94.** $30\frac{7}{10}$ **95.** $1\frac{1}{2}$ **96.** $14\frac{13}{24}$ **97.** $1\frac{1}{24}$ cups
98. $1\frac{33}{100}$ in.   **99.** $0.\overline{142857}$ **101.** $0.\overline{428571}$
103. $0.\overline{714285}$ **105.** $0.\overline{1}$ **107.** $0.\overline{001}$

Margin Exercises, Section 4.6, pp. 280–282

1. (b)   **2.** (a)   **3.** (d)   **4.** (b)   **5.** (a)   **6.** (d)   **7.** (b)
8. (c)   **9.** (b)   **10.** (b)   **11.** (c)   **12.** (a)   **13.** (c)
14. (c)

Exercise Set 4.6, p. 283

1. (d)   **3.** (c)   **5.** (a)   **7.** (c)   **9.** 1.6   **11.** 6   **13.** 60
15. 2.3   **17.** 180   **19.** (a)   **21.** (c)   **23.** (b)   **25.** (b)
27. $1500 \div 0.5 = 3000$; answers may vary   **29.** $\mathbf{D_W}$
31. $2 \cdot 2 \cdot 3 \cdot 3 \cdot 3$, or $2^2 \cdot 3^3$
32. $2 \cdot 2 \cdot 2 \cdot 2 \cdot 5 \cdot 5$, or $2^4 \cdot 5^2$ **33.** $5 \cdot 5 \cdot 13$, or $5^2 \cdot 13$
34. $2 \cdot 3 \cdot 3 \cdot 37$, or $2 \cdot 3^2 \cdot 37$
35. $2 \cdot 2 \cdot 2 \cdot 2 \cdot 2 \cdot 3 \cdot 3 \cdot 3$, or $2^6 \cdot 3^3$ **36.** $\frac{5}{16}$ **37.** $\frac{129}{251}$
38. $\frac{8}{9}$ **39.** $\frac{13}{25}$ **40.** $\frac{25}{19}$ **41.** Yes   **43.** No
45. (a) +, ×; (b) +, ×, −

Margin Exercises, Section 4.7, pp. 286–293

1. 8.4°   **2.** 148.1 gal   **3.** $51.26   **4.** $368.75
5. 96.52 cm²   **6.** $1.33   **7.** 28.6 mpg   **8.** $289,683
9. $582,278

Exercise Set 4.7, p. 294

1. $10.50   **3.** $3.87   **5.** 102.8°F   **7.** $21,219.17
9. Area: 8.125 cm²; perimeter: 11.5 cm   **11.** 22,691.5 mi
13. 20.2 mpg   **15.** $10   **17.** 11.9752 ft³   **19.** 78.1 cm
21. 28.5 cm   **23.** 2.31 cm   **25.** 876 calories
27. $1171.74   **29.** 227.75 ft²   **31.** 0.372   **33.** $69.24
35. $906.50   **37.** 5.8¢, or $0.058   **39.** 2152.56 yd²
41. 4.229 billion   **43.** 1.4°F   **45.** $142,989
47. $112,762   **49.** $75,097   **51.** $\mathbf{D_W}$ **53.** 6335
54. $\frac{31}{24}$ **55.** $6\frac{5}{6}$ **56.** $\frac{23}{15}$ **57.** 13,766   **58.** 2803   **59.** $\frac{1}{24}$
60. $1\frac{5}{6}$ **61.** $\frac{2}{15}$ **62.** 2432   **63.** 28 min   **64.** $7\frac{1}{5}$ min
65. $17.28

Summary and Review: Chapter 4, p. 301

1. 6,590,000   **2.** 6,900,000   **3.** Three and
forty-seven hundredths   **4.** Thirty-one thousandths
5. Five hundred ninety-seven and $\frac{25}{100}$ dollars   **6.** Zero and
$\frac{96}{100}$ dollars   **7.** $\frac{9}{100}$ **8.** $\frac{4561}{1000}$ **9.** $\frac{89}{1000}$ **10.** $\frac{30,227}{10,000}$
11. 0.034   **12.** 4.2603   **13.** 27.91   **14.** 867.006
15. 0.034   **16.** 0.91   **17.** 0.741   **18.** 1.041   **19.** 17.4
20. 17.43   **21.** 17.429   **22.** 17   **23.** 574.519

24. 0.6838   **25.** 229.1   **26.** 45.551   **27.** 29.2092
28. 790.29   **29.** 29.148   **30.** 70.7891   **31.** 12.96
32. 0.14442   **33.** 4.3   **34.** 0.02468   **35.** 7.5   **36.** 0.45
37. 45.2   **38.** 1.022   **39.** 0.2763   **40.** 1389.2
41. 496.2795   **42.** 6.95   **43.** 42.54   **44.** 4.9911
45. 24.36 cups; 104.4 cups   **46.** $15.52   **47.** $5788.56
48. $224.99   **49.** 14.5 mpg   **50.** (a) 54.6 lb; (b) 13.65 lb
51. 272   **52.** 216   **53.** 4   **54.** $125   **55.** 2.6
56. 1.28   **57.** 2.75   **58.** 3.25   **59.** $1.1\overline{6}$ **60.** $1.\overline{54}$
61. 1.5   **62.** 1.55   **63.** 1.545   **64.** $82.73   **65.** $4.87
66. 2493¢   **67.** 986¢   **68.** 1.8045   **69.** 57.1449
70. 15.6375   **71.** $41.537\overline{3}$
72. $\mathbf{D_W}$ Multiply by 1 to get a denominator that is a power
of 10:
$$\frac{44}{125} = \frac{44}{125} \cdot \frac{8}{8} = \frac{352}{1000} = 0.352.$$
We can also divide to find that $\frac{44}{125} = 0.352$.
73. $\mathbf{D_W}$ Each decimal place in the decimal notation
corresponds to one zero in the power of ten in the fraction
notation. When the fractions are multiplied, the number of
zeros in the denominator of the product is the sum of the
number of zeros in the denominators of the factors. So the
number of decimal places in the product is the sum of
the number of decimal places in the factors.
74. $43\frac{3}{4}$ **75.** $3\frac{3}{4}$ **76.** $19\frac{4}{5}$ **77.** $6\frac{3}{5}$ **78.** $\frac{1}{2}$
79. $2 \cdot 2 \cdot 2 \cdot 2 \cdot 2 \cdot 2 \cdot 3$, or $2^6 \cdot 3$
80. (a) $2.56 \times 6.4 \div 51.2 - 17.4 + 89.7 = 72.62$;
(b) $(11.12 - 0.29) \times 3^4 = 877.23$
81. $\frac{1}{3} + \frac{2}{3} = 0.33333333\ldots + 0.66666666\ldots$
  $= 0.99999999\ldots$.
Therefore, $1 = 0.99999999\ldots$ because $\frac{1}{3} + \frac{2}{3} = 1$.
82. $2 = 1.\overline{9}$

Test: Chapter 4, p. 304

1. [4.3b] 8,900,000,000   **2.** [4.3b] 3,756,000
3. [4.1a] Two and thirty-four hundredths   **4.** [4.1a] One
thousand, two hundred thirty-four and $\frac{78}{100}$ dollars
5. [4.1b] $\frac{91}{100}$ **6.** [4.1b] $\frac{2769}{1000}$ **7.** [4.1b] 0.074
8. [4.1b] 3.7047   **9.** [4.1b] 756.09   **10.** [4.1b] 91.703
11. [4.1c] 0.162   **12.** [4.1c] 0.078   **13.** [4.1c] 0.9
14. [4.1d] 6   **15.** [4.1d] 5.68   **16.** [4.1d] 5.678
17. [4.1d] 5.7   **18.** [4.2a] 0.7902   **19.** [4.2a] 186.5
20. [4.2a] 1033.23   **21.** [4.2b] 48.357   **22.** [4.2b] 19.0901
23. [4.2b] 152.8934   **24.** [4.3a] 0.03   **25.** [4.3a] 0.21345
26. [4.3a] 73,962   **27.** [4.4a] 4.75   **28.** [4.4a] 30.4
29. [4.4a] 0.19   **30.** [4.4a] 0.34689   **31.** [4.4a] 34,689
32. [4.4b] 84.26   **33.** [4.2c] 8.982   **34.** [4.7a] $120.49
35. [4.7a] 28.3 mpg   **36.** [4.7a] $6572.45
37. [4.7a] $1199.94   **38.** [4.7a] 53.9 million passengers
39. [4.6a] 198   **40.** [4.6a] 4   **41.** [4.5a] 1.6
42. [4.5a] 0.88   **43.** [4.5a] 5.25   **44.** [4.5a] 0.75
45. [4.5a] $1.\overline{2}$ **46.** [4.5a] $2.\overline{142857}$ **47.** [4.5b] 2.1
48. [4.5b] 2.14   **49.** [4.5b] 2.143   **50.** [4.3b] $9.49

51. [4.4c] 40.0065　**52.** [4.4c] 384.8464　**53.** [4.5c] 302.4
54. [4.5c] 52.3394̄　**55.** [3.6a] 14　**56.** [3.5a] $2\frac{11}{16}$
57. [3.5b] $26\frac{1}{2}$　**58.** [3.6b] $1\frac{1}{8}$　**59.** [2.5b] $\frac{11}{18}$
60. [2.1d] $2 \cdot 2 \cdot 2 \cdot 3 \cdot 3 \cdot 5$, or $2^3 \cdot 3^2 \cdot 5$　**61.** [4.7a] \$35
62. [4.1b, c] $\frac{2}{3}, \frac{5}{7}, \frac{15}{19}, \frac{11}{13}, \frac{17}{20}, \frac{13}{15}$

Cumulative Review: Chapters 1–4, p. 307

1. [3.4a] $\frac{20}{9}$　**2.** [4.1b] $\frac{3052}{1000}$　**3.** [4.5a] 1.4　**4.** [4.5a] $0.\overline{54}$
5. [2.1c] Prime　**6.** [2.2a] Yes　**7.** [1.9c] 1754
8. [4.4c] 4.364　**9.** [4.1d] 584.90　**10.** [4.5b] 218.56
11. [4.6a] 160　**12.** [4.6a] 4　**13.** [1.5b] 12,800,000
14. [4.6a] 6　**15.** [4.5b] 59.55°F　**16.** [4.5b] 59.82°F
17. [1.8a] 14,274 transplants　**18.** [4.7a] \$23.31
19. (a) [3.5c], [3.6c] Area: $194\frac{2}{3}$ ft^2, perimeter: $56\frac{1}{3}$ ft;
(b) [1.8a] \$1700
20. [2.7d] \$500　**21.** [3.5c] $6\frac{1}{2}$ lb　**22.** [3.2b] 2 lb
23. [4.4a] 2.122　**24.** [1.6b] 1843　**25.** [4.4a] 13,862.1
26. [2.7b] $\frac{5}{6}$　**27.** [4.2c] 0.78　**28.** [1.7b] 28
29. [4.4b] 8.62　**30.** [1.7b] 367,251　**31.** [3.3c] $\frac{1}{18}$
32. [2.7c] $\frac{1}{2}$　**33.** [3.5a] $6\frac{1}{20}$　**34.** [1.2a] 139,116
35. [3.2a] $\frac{31}{18}$　**36.** [4.2a] 145.953　**37.** [1.3b] 710,137
38. [4.2b] 13.097　**39.** [3.5b] $\frac{5}{7}$　**40.** [3.3a] $\frac{1}{110}$
41. [2.6a] $\frac{1}{6}$　**42.** [1.5a] 5,317,200　**43.** [4.3a] 4.78
44. [4.3a] 0.0279431　**45.** [3.7a] $\frac{9}{32}$
46. [4.2a, b] \$66.70, \$77.82, \$88.94, \$100.06, \$111.18,
\$122.30, \$133.42　**47.** [4.2a], [4.2b] \$2029.66, \$1950.88,
\$1872.10, \$1793.32
48. [4.2a], [4.2b] First row: 4.55, 1.3, 1.95; second row: 0, 2.6,
5.2; third row: 3.25, 3.9, 0.65. Magic sum = 7.8.
49. [4.2a], [4.2b] First row: 2.16, 0.81, 1.08; second row: 0.27,
1.35, 2.43; third row: 1.62, 1.89, 0.54.
50. [4.2a], [4.2b] First row: 6.16, 43.12, 34.72, 16.24;
second row: 38.64, 12.32, 9.52, 39.76; third row: 15.12, 34.16,
44.24, 6.72; fourth row: 40.32, 10.64, 11.76, 37.52
51. [1.3a] (c)　**52.** [4.5a] (c)　**53.** [4.7a] (e)
54. [4.1b] (b)　**55.** [4.7a] \$0.91

Chapter 5

Pretest: Chapter 5, p. 312

1. $\frac{35}{43}$　**2.** $\frac{0.079}{1.043}$　**3.** 22.5　**4.** 0.75　**5.** $\frac{117}{14}$, or $8\frac{5}{14}$
6. 25.5 mpg　**7.** $\frac{2}{9}$ qt/min　**8.** 20.17 cents/oz;
16.43 cents/oz; the 34.5-oz package has the lower unit price
9. 1944 km　**10.** 22 packs　**11.** 12 min　**12.** 393.75 mi
13. $x = 15$; $y = 9$　**14.** $\frac{2}{3}$

Margin Exercises, Section 5.1, pp. 313–315

1. $\frac{5}{11}$, or 5:11　**2.** $\frac{57.3}{86.1}$, or 57.3:86.1　**3.** $\frac{6\frac{3}{4}}{7\frac{2}{5}}$, or $6\frac{3}{4}:7\frac{2}{5}$

4. $\frac{739}{12}$　**5.** $\frac{12}{14}$　**6.** $\frac{73}{248\frac{2}{3}}$; $\frac{248\frac{2}{3}}{73}$　**7.** $\frac{38.2}{56.1}$
8. 18 is to 27 as 2 is to 3　**9.** 3.6 is to 12 as 3 is to 10
10. 1.2 is to 1.5 as 4 is to 5　**11.** $\frac{3}{4}$

Exercise Set 5.1, p. 316

1. $\frac{4}{5}$　**3.** $\frac{178}{572}$　**5.** $\frac{0.4}{12}$　**7.** $\frac{3.8}{7.4}$　**9.** $\frac{56.78}{98.35}$　**11.** $\frac{8\frac{3}{4}}{9\frac{5}{6}}$　**13.** $\frac{4}{1}$
15. $\frac{201}{593}$; $\frac{593}{201}$　**17.** $\frac{93.2}{1000}$　**19.** $\frac{28}{100}$; $\frac{100}{28}$　**21.** $\frac{60}{100}$; $\frac{100}{60}$　**23.** $\frac{2}{3}$
25. $\frac{3}{4}$　**27.** $\frac{12}{25}$　**29.** $\frac{7}{9}$　**31.** $\frac{2}{3}$　**33.** $\frac{14}{25}$　**35.** $\frac{1}{2}$　**37.** $\frac{3}{4}$
39. $\frac{478}{213}$; $\frac{213}{478}$　**41.** D$_\text{W}$　**43.** $=$　**44.** $\neq$　**45.** $\neq$　**46.** $=$
47. 50　**48.** 9.5　**49.** 14.5　**50.** 152　**51.** $6\frac{7}{20}$ cm
52. $17\frac{11}{20}$ cm　**53.** $\frac{30}{47}$　**55.** 1:2:3

Margin Exercises, Section 5.2, pp. 320–321

1. 5 mi/hr, or 5 mph　**2.** 12 mi/hr, or 12 mph
3. $\frac{89}{13}$ km/h, or 6.85 km/h　**4.** 1100 ft/sec　**5.** 4 ft/sec
6. $\frac{121}{8}$ ft/sec, or 15.125 ft/sec　**7.** $\frac{33 \text{ home runs}}{107 \text{ strikeouts}} \approx$
0.308 home run per strikeout　**8.** $\frac{714 \text{ home runs}}{1330 \text{ strikeouts}} \approx$
0.537 home run per strikeout; Babe Ruth's rate is
approximately 0.229 higher　**9.** Approximately 24.174 ¢/oz
10. 24.143 ¢/oz; 25.9 ¢/oz; 24.174 ¢/oz; the 7-oz package
has the lowest unit price

Exercise Set 5.2, p. 322

1. 40 km/h　**3.** 7.48 mi/sec　**5.** 18 mpg　**7.** 28 mpg
9. About 3.2 servings/lb　**11.** 25 mph; 0.04 hr/mi
13. About 26.8 points/game　**15.** 0.623 gal/ft^2
17. 186,000 mi/sec　**19.** 124 km/h　**21.** 25 beats/min
23. 12.030 ¢/oz; 11.56 ¢/oz; 50 fl oz　**25.** 47.903 ¢/oz;
43.375 ¢/oz; 8.0 oz　**27.** 18.174 ¢/oz; 15.275 ¢/oz; 34.5 oz
29. 1.583 ¢/sq ft; 3.311 ¢/sq ft; 1.537 ¢/sq ft; 2.419 ¢/sq ft;
1.544 ¢/sq ft; 1.904 ¢/sq ft; 61.8 sq ft　**31.** 7.94 ¢/oz;
4.99 ¢/oz; 6.12 ¢/oz; 5.99 ¢/oz; 100 fl oz　**33.** D$_\text{W}$
35. 1.7 million　**36.** $37\frac{1}{2}$ servings　**37.** 30 tests
38. 8.9 billion　**39.** 109.608　**40.** 67,819　**41.** 5833.56
42. 466,190.4　**43.** 6-oz: 10.83 ¢/oz; 5.5-oz: 10.91 ¢/oz

Margin Exercises, Section 5.3, pp. 327–330

1. Yes　**2.** No　**3.** No　**4.** Yes　**5.** No　**6.** Yes
7. 14　**8.** $11\frac{1}{4}$　**9.** 10.5　**10.** 2.64　**11.** 10.8
12. $\frac{125}{42}$, or $2\frac{41}{42}$

Calculator Corner, p. 330

1. Left to the student   **2.** Left to the student   **3.** 27.5625
4. 25.6   **5.** 15.140625   **6.** 40.03952941
7. 39.74857143   **8.** 119

Exercise Set 5.3, p. 331

1. No   **3.** Yes   **5.** Yes   **7.** No   **9.** 0.59; 0.59; 0.64;
0.60; the completion rates (*rounded* to the nearest
hundredth) are the same for Collins and Dilfer   **11.** 45
13. 12   **15.** 10   **17.** 20   **19.** 5   **21.** 18   **23.** 22
25. 28   **27.** $9\frac{1}{3}$ **29.** $2\frac{8}{9}$ **31.** 0.06   **33.** 5   **35.** 1
37. 1   **39.** 14   **41.** $2\frac{3}{16}$ **43.** $\frac{51}{16}$, or $3\frac{3}{16}$ **45.** 12.5725
47. $\frac{1748}{249}$, or $7\frac{5}{249}$ **49.** $\mathbf{D_W}$ **51.** $\neq$ **52.** $=$ **53.** $\neq$
54. $\neq$ **55.** 65   **56.** 39.5   **57.** 290.5   **58.** $1523.\overline{1}$
59. 0.588   **60.** 0.593   **61.** 0.643   **62.** 0.600
63. Approximately 2731.4   **65. (a)** Ruth: 1.863 strikeouts
per home run; Schmidt: 3.436 strikeouts per home run;
(b) Schmidt

Margin Exercises, Section 5.4, pp. 334–338

1. 445 calories   **2.** 15 gal   **3.** 8 shirts   **4.** 38 in. or less
5. 9.5 in.   **6.** 2074 deer

Exercise Set 5.4, p. 339

1. 11.04 hr   **3.** 600 calories   **5.** 168.6 million, or
168,600,000   **7.** 9.75 gal   **9.** 175 bulbs   **11.** 2975 ft^2
13. 450 pages   **15. (a)** 450 Australian dollars; **(b)** $27.78
17. (a) About 96 gal; **(b)** 3920 mi   **19.** 60 students
21. 13,500 mi   **23.** 120 lb   **25.** 64 gal   **27.** 100 oz
29. 954 deer   **31.** 58.1 mi   **33. (a)** 56 games; **(b)** about
2197 points   **35.** $\mathbf{D_W}$ **37.** Neither   **38.** Composite
39. Prime   **40.** Composite   **41.** Prime
42. $2 \cdot 2 \cdot 2 \cdot 101$, or $2^3 \cdot 101$ **43.** $2 \cdot 2 \cdot 7$, or $2^2 \cdot 7$
44. $2 \cdot 433$ **45.** $3 \cdot 31$ **46.** $2 \cdot 2 \cdot 5 \cdot 101$, or $2^2 \cdot 5 \cdot 101$
47. 17 positions   **49.** 2150 earned runs   **51.** CD player:
$150; receiver: $450; speakers: $300

Margin Exercises, Section 5.5, pp. 344–347

1. 15   **2.** 24.75 ft   **3.** 34.9 ft   **4.** 21 cm   **5.** 29 ft

Exercise Set 5.5, p. 348

1. 25   **3.** $\frac{4}{3}$, or $1\frac{1}{3}$ **5.** $x = \frac{27}{4}$, or $6\frac{3}{4}$; $y = 9$ **7.** $x = 7.5$;
$y = 7.2$ **9.** 1.25 m   **11.** 36 ft   **13.** 7 ft   **15.** 100 ft
17. 4   **19.** $10\frac{1}{2}$ **21.** $x = 6$; $y = 5.25$; $z = 3$ **23.** $x = 5\frac{1}{3}$,
or $5.\overline{3}$; $y = 4\frac{2}{3}$, or $4.\overline{6}$; $z = 5\frac{1}{3}$, or $5.\overline{3}$ **25.** 20 ft   **27.** 152 ft
29. $\mathbf{D_W}$ **31.** $59.81   **32.** 9.63   **33.** 679.4928
34. 2.74568   **35.** 27,456.8   **36.** 0.549136   **37.** 0.85
38. 1.825   **39.** 0.909   **40.** 0.843   **41.** 13.75 ft
43. 1.034 cm   **45.** 3681.437   **47.** $x = 0.4$; $y \approx 0.35$

Summary and Review: Chapter 5, p. 352

1. $\frac{47}{84}$ **2.** $\frac{46}{1.27}$ **3.** $\frac{83}{100}$ **4.** $\frac{0.72}{197}$ **5. (a)** $\frac{12,480}{16,640}$, or $\frac{3}{4}$;
(b) $\frac{16,640}{29,120}$, or $\frac{4}{7}$ **6.** $\frac{3}{4}$ **7.** $\frac{9}{16}$ **8.** 26 mpg
9. 6300 rpm   **10.** 0.638 gal/ft^2 **11.** 0.72 serving/lb
12. 4.33 ¢/tablet   **13.** 14.173 ¢/oz   **14.** 5.4 ¢/oz;
6.46 ¢/oz; 100 oz   **15.** 8.063 ¢/oz; 5.406 ¢/oz; 4.146 ¢/oz;
4.125 ¢/oz; 4.414 ¢/oz; 64 oz   **16.** No   **17.** No   **18.** 32
19. 7   **20.** $\frac{1}{40}$ **21.** 24   **22.** $4.45   **23.** 351 circuits
24. (a) 270 Euros; **(b)** $46.30   **25.** 832 mi   **26.** 27 acres
27. Approximately 3,173,732 kg   **28.** 6 in.
29. Approximately 9906 lawyers   **30.** $x = \frac{14}{3}$, or $4\frac{2}{3}$
31. $x = \frac{56}{5}$, or $11\frac{1}{5}$; $y = \frac{63}{5}$, or $12\frac{3}{5}$ **32.** 40 ft
33. $x = 3$; $y = 9$; $z = \frac{15}{2}$, or $7\frac{1}{2}$
34. $\mathbf{D_W}$ In terms of cost, a low faculty-to-student ratio is
less expensive than a high faculty-to-student ratio. In terms
of quality of education and student satisfaction, a high
faculty-to-student ratio is more desirable. A college
president must balance the cost and quality issues.
35. $\mathbf{D_W}$ Leslie used 4 gal of gasoline to drive 92 mi. At the
same rate, how many gallons would be needed to travel
368 mi?   **36.** $1630.40   **37.** $2799.60   **38.** $=$ **39.** $\neq$
40. 10,672.74   **41.** 45.5   **42.** 105 min, or 1 hr 45 min
43. $x = 4258.5$; $z \approx 10,094.3$ **44.** Finishing paint: 11 gal;
primer: 16.5 gal

Test: Chapter 5, p. 355

1. [5.1a] $\frac{85}{97}$ **2.** [5.1a] $\frac{0.34}{124}$ **3.** [5.1b] $\frac{9}{10}$ **4.** [5.1b] $\frac{25}{32}$
5. [5.2a] 0.625 ft/sec   **6.** [5.2a] $1\frac{1}{3}$ servings/lb
7. [5.2a] About 23.5 mpg   **8.** [5.2b] 7.765 ¢/oz
9. [5.2b] 11.182 ¢/oz; 7.149 ¢/oz; 8.389 ¢/oz; 6.840 ¢/oz;
263 oz   **10.** [5.3a] Yes   **11.** [5.3a] No   **12.** [5.3b] 12
13. [5.3b] 360   **14.** [5.3b] 42.1875   **15.** [5.3b] 100
16. [5.4a] 1512 km   **17.** [5.4a] 4.8 min   **18.** [5.4a] 525 mi
19. [5.5a] 66 m   **20.** [5.4a] **(a)** 684 Canadian dollars;
(b) $368.42   **21.** [5.4a] About 86,151 arrests
22. [5.5a] $x = 8$; $y = 8.8$ **23.** [5.5b] $x = 8$; $y = 8$; $z = 12$
24. [4.7a] 25.8 million lb   **25.** [2.5c] $\neq$
26. [4.3a] 17,324.14   **27.** [4.4a] 0.9944   **28.** [5.4a] 5888

Cumulative Review: Chapters 1–5, p. 357

1. [1.6b], [4.3b], [4.4a] **(a)** $252,000,000; **(b)** $0.252 billion;
(c) $25,200,000; **(d)** $45,487.36   **2.** [5.2a] 24 mpg
3. [4.2a] 513.996   **4.** [3.5a] $6\frac{3}{4}$ **5.** [3.2a] $\frac{7}{20}$
6. [4.2b] 30.491   **7.** [4.2b] 72.912   **8.** [3.3a] $\frac{7}{60}$
9. [4.3a] 222.076   **10.** [4.3a] 567.8   **11.** [3.6a] 3
12. [4.4a] 43   **13.** [1.6b] 899   **14.** [2.7b] $\frac{3}{2}$
15. [1.1b] 3 ten thousands + 7 tens + 4 ones
16. [4.1a] One hundred twenty and seven hundredths
17. [4.1c] 0.7   **18.** [4.1c] 0.8
19. [2.1d] $2 \cdot 2 \cdot 2 \cdot 2 \cdot 3 \cdot 3$, or $2^4 \cdot 3^2$ **20.** [3.1a] 140
21. [2.3a] $\frac{5}{8}$ **22.** [2.5b] $\frac{5}{8}$ **23.** [4.5c] 5.718

24. [4.5c] 0.179   **25.** [5.1a] $\frac{0.3}{15}$ **26.** [5.3a] Yes
27. [5.2a] 55 m/sec   **28.** [5.2b] 11.769 ¢/oz; 9.344 ¢/oz; 9.728 ¢/oz; 6.357 ¢/oz; 7.002 ¢/oz; 28-oz Joy
29. [5.3b] 30.24   **30.** [4.4b] 26.4375   **31.** [2.7c] $\frac{8}{9}$
32. [5.3b] 128   **33.** [4.2c] 33.34   **34.** [3.3c] $\frac{76}{175}$
35. [2.6b] 390 cal   **36.** [5.4a] **(a)** 39,462.5 Yen; **(b)** $354.77
37. [4.7a] 976.9 mi   **38.** [5.4a] 7 min   **39.** [3.5c] $2\frac{1}{4}$ cups
40. [2.7d] 12 doors   **41.** [4.2b], [2.7d] **(a)** 40,200 yr;
(b) answers may vary   **42.** [1.8a] **(a)** $360,000;
(b) $144,000,000;   **(c)** $1,728,000,000
43. [4.7a], [5.2a] 42.2025 mi   **44.** [4.7a] 132 orbits
45. [2.1c] (d)   **46.** [5.2a] (b)   **47.** [1.5c] (b)
48. [5.5a] $10\frac{1}{2}$ ft

Chapter 6

Pretest: Chapter 6, p. 362

1. 0.133   **2.** 50.4%   **3.** 4%   **4.** $\frac{19}{100}$
5. $x = 60\% \times 75$; 45   **6.** $\frac{n}{100} = \frac{35}{50}$; 70%   **7.** 90 lb
8. 19.2%   **9.** $14.30; $300.30   **10.** $5152
11. $1112.50; $3337.50   **12.** $99.60   **13.** $20
14. $7128.60   **15.** $10,116.38

Margin Exercises, Section 6.1, pp. 363–366

1. $\frac{70}{100}$; $70 \times \frac{1}{100}$; 70×0.01 **2.** $\frac{23.4}{100}$; $23.4 \times \frac{1}{100}$; 23.4×0.01
3. $\frac{100}{100}$; $100 \times \frac{1}{100}$; 100×0.01 **4.** 0.34   **5.** 0.789
6. 0.06625   **7.** 0.63   **8.** 0.0008   **9.** 24%   **10.** 347%
11. 100%   **12.** 60%   **13.** 25.3%

Calculator Corner, p. 364

1. 0.14   **2.** 0.00069   **3.** 0.438   **4.** 1.25

Exercise Set 6.1, p. 367

1. $\frac{90}{100}$; $90 \times \frac{1}{100}$; 90×0.01 **3.** $\frac{12.5}{100}$; $12.5 \times \frac{1}{100}$; 12.5×0.01
5. 0.67   **7.** 0.456   **9.** 0.5901   **11.** 0.1   **13.** 0.01
15. 2   **17.** 0.001   **19.** 0.0009   **21.** 0.0018   **23.** 0.2319
25. 0.14875   **27.** 0.565   **29.** 0.4   **31.** 0.186   **33.** 0.29
35. 47%   **37.** 3%   **39.** 870%   **41.** 33.4%   **43.** 75%
45. 40%   **47.** 0.6%   **49.** 1.7%   **51.** 27.18%
53. 2.39%   **55.** 52.6%   **57.** 17%   **59.** 41.1%   **61.** $\mathbf{D_W}$
63. $33\frac{1}{3}$ **64.** $37\frac{1}{2}$ **65.** $9\frac{3}{8}$ **66.** $18\frac{9}{16}$ **67.** $5\frac{11}{14}$
68. $111\frac{2}{3}$ **69.** $0.\overline{6}$ **70.** $0.\overline{3}$ **71.** $0.8\overline{3}$ **72.** $1.41\overline{6}$
73. $2.\overline{6}$ **74.** 0.9375

Margin Exercises, Section 6.2, pp. 371–373

1. 25%   **2.** 62.5%, or $62\frac{1}{2}\%$ **3.** $66.\overline{6}\%$, or $66\frac{2}{3}\%$
4. $83.\overline{3}\%$, or $83\frac{1}{3}\%$ **5.** 57%   **6.** 76%   **7.** $\frac{3}{5}$ **8.** $\frac{13}{400}$
9. $\frac{2}{3}$

10.

Fraction Notation	$\frac{1}{5}$	$\frac{5}{6}$	$\frac{3}{8}$
Decimal Notation	0.2	$0.8\overline{3}$	0.375
Percent Notation	20%	$83.\overline{3}\%$, or $83\frac{1}{3}\%$	$37\frac{1}{2}\%$

Calculator Corner, p. 371

1. 52%   **2.** 38.46%   **3.** 110.26%   **4.** 171.43%
5. 59.62%   **6.** 28.31%

Calculator Corner, p. 374

1. 30.54; 1.31%   **2.** 32.05; 1.20%   **3.** 34.47; 1.19%
4. 26.47; 1.00%   **5.** 11.98; 4.32%   **6.** 17.52; 0.89%

Exercise Set 6.2, p. 375

1. 41%   **3.** 5%   **5.** 20%   **7.** 30%   **9.** 50%
11. 87.5%, or $87\frac{1}{2}\%$ **13.** 80%   **15.** $66.\overline{6}\%$, or $66\frac{2}{3}\%$
17. $16.\overline{6}\%$, or $16\frac{2}{3}\%$ **19.** 18.75%, or $18\frac{3}{4}\%$
21. 81.25%, or $81\frac{1}{4}\%$ **23.** 16%   **25.** 5%   **27.** 34%
29. 8%   **31.** 21%   **33.** 24%   **35.** $\frac{17}{20}$ **37.** $\frac{5}{8}$ **39.** $\frac{1}{3}$
41. $\frac{1}{6}$ **43.** $\frac{29}{400}$ **45.** $\frac{1}{125}$ **47.** $\frac{203}{800}$ **49.** $\frac{176}{225}$ **51.** $\frac{711}{1100}$
53. $\frac{3}{2}$ **55.** $\frac{13}{40,000}$ **57.** $\frac{1}{3}$ **59.** $\frac{13}{50}$ **61.** $\frac{1}{20}$ **63.** $\frac{3}{50}$
65. $\frac{9}{20}$ **67.** $\frac{47}{100}$
69.

Fraction Notation	Decimal Notation	Percent Notation
$\frac{1}{8}$	0.125	12.5%, or $12\frac{1}{2}\%$
$\frac{1}{6}$	$0.1\overline{6}$	$16.\overline{6}\%$, or $16\frac{2}{3}\%$
$\frac{1}{5}$	0.2	20%
$\frac{1}{4}$	0.25	25%
$\frac{1}{3}$	$0.\overline{3}$	$33.\overline{3}\%$, or $33\frac{1}{3}\%$
$\frac{3}{8}$	0.375	37.5%, or $37\frac{1}{2}\%$
$\frac{2}{5}$	0.4	40%
$\frac{1}{2}$	0.5	50%

71.

Fraction Notation	Decimal Notation	Percent Notation
$\frac{1}{2}$	0.5	50%
$\frac{1}{3}$	$0.\overline{3}$	$33.\overline{3}\%$, or $33\frac{1}{3}\%$
$\frac{1}{4}$	0.25	25%
$\frac{1}{6}$	$0.1\overline{6}$	$16.\overline{6}\%$, or $16\frac{2}{3}\%$
$\frac{1}{8}$	0.125	12.5%, or $12\frac{1}{2}\%$
$\frac{3}{4}$	0.75	75%
$\frac{5}{6}$	$0.8\overline{3}$	$83.\overline{3}\%$, or $83\frac{1}{3}\%$
$\frac{3}{8}$	0.375	37.5%, or $37\frac{1}{2}\%$

73. $\mathbf{D_W}$ **75.** 70 **76.** 5 **77.** 400 **78.** 18.75
79. 23.125 **80.** 25.5 **81.** $33\frac{1}{3}$ **82.** $37\frac{1}{2}$ **83.** $83\frac{1}{3}$
84. $20\frac{1}{2}$ **85.** $43\frac{1}{8}$ **86.** $62\frac{1}{6}$ **87.** $18\frac{3}{4}$ **88.** $7\frac{4}{9}$
89. $11.\overline{1}\%$ **91.** $257.\overline{46317}\%$ **93.** $0.01\overline{5}$
95. $1.04\overline{142857}$

Margin Exercises, Section 6.3, pp. 379–382

1. $12\% \times 50 = a$ **2.** $a = 40\% \times 60$ **3.** $45 = 20\% \times t$
4. $120\% \times y = 60$ **5.** $16 = n \times 40$ **6.** $b \times 84 = 10.5$
7. 6 **8.** $35.20 **9.** 225 **10.** $50 **11.** 40%
12. 12.5%

Calculator Corner, p. 383

1. 1.2 **2.** $5.04 **3.** 48.64 **4.** $22.40 **5.** 0.0112
6. $29.70 **7.** Left to the student **8.** Left to the student

Exercise Set 6.3, p. 384

1. $y = 32\% \times 78$ **3.** $89 = a \times 99$ **5.** $13 = 25\% \times y$
7. 234.6 **9.** 45 **11.** $18 **13.** 1.9 **15.** 78%
17. 200% **19.** 50% **21.** 125% **23.** 40 **25.** $40
27. 88 **29.** 20 **31.** 6.25 **33.** $846.60 **35.** $\mathbf{D_W}$
37. $\frac{9}{100}$ **38.** $\frac{179}{100}$ **39.** $\frac{875}{1000}$, or $\frac{7}{8}$ **40.** $\frac{125}{1000}$, or $\frac{1}{8}$
41. $\frac{9375}{10,000}$, or $\frac{15}{16}$ **42.** $\frac{6875}{10,000}$, or $\frac{11}{16}$ **43.** 0.89 **44.** 0.07

45. 0.3 **46.** 0.017 **47.** $800 (can vary); $843.20
49. $10,000 (can vary); $10,400 **51.** $1875

Margin Exercises, Section 6.4, pp. 387–389

1. $\frac{12}{100} = \frac{a}{50}$ **2.** $\frac{40}{100} = \frac{a}{60}$ **3.** $\frac{130}{100} = \frac{a}{72}$ **4.** $\frac{20}{100} = \frac{45}{b}$
5. $\frac{120}{100} = \frac{60}{b}$ **6.** $\frac{P}{100} = \frac{16}{40}$ **7.** $\frac{P}{100} = \frac{10.5}{84}$ **8.** $225

9. 35.2 **10.** 6 **11.** 50 **12.** 30% **13.** 12.5%

Exercise Set 6.4, p. 390

1. $\frac{37}{100} = \frac{a}{74}$ **3.** $\frac{P}{100} = \frac{4.3}{5.9}$ **5.** $\frac{25}{100} = \frac{14}{b}$ **7.** 68.4
9. 462 **11.** 40 **13.** 2.88 **15.** 25% **17.** 102%
19. 25% **21.** 93.75% **23.** $72 **25.** 90 **27.** 88
29. 20 **31.** 25 **33.** $780.20 **35.** $\mathbf{D_W}$ **37.** 8
38. 4000 **39.** 8 **40.** 2074 **41.** 100 **42.** 15
43. $8.0\overline{4}$ **44.** $\frac{3}{16}$, or 0.1875 **45.** $\frac{43}{48}$ qt **46.** $\frac{1}{8}$ T
47. $1134 (can vary); $1118.64

Margin Exercises, Section 6.5, pp. 393–398

1. About 9.3% **2.** 750 mL **3.** (a) $1475; (b) $38,350
4. (a) $9218.75; (b) $27,656.25 **5.** About 30%
6. About 16.2%

Exercise Set 6.5, p. 399

1. (a) About 67.7%; (b) 90% **3.** 134 passes
5. Overweight: 168.6 million; obese: 70.25 million
7. 20.4 mL; 659.6 mL **9.** 637 field goals **11.** 95%; 5%
13. 36.4 correct; 3.6 incorrect **15.** 95 items **17.** 25%
19. 166; 156; 146; 140; 122 **21.** 8% **23.** 20%
25. About 30.6% **27.** $30,030 **29.** $16,174.50;
$12,130.88 **31.** About 25% **33.** 34.375%, or $34\frac{3}{8}\%$
35. 5.7%; 539,452 **37.** 71% **39.** $1560 **41.** 80%
43. 46,069; 8.2% **45.** 1,027,429; 21.1% **47.** 4,891,769;
9.6% **49.** $36,400 **51.** 40% **53.** $\mathbf{D_W}$ **55.** $2.\overline{27}$
56. 0.44 **57.** 3.375 **58.** $4.\overline{7}$ **59.** 0.92 **60.** $0.8\overline{3}$
61. 0.4375 **62.** 2.317 **63.** 3.4809 **64.** 0.675
65. About 5 ft 6 in. **67.** $83\frac{1}{3}\%$

Margin Exercises, Section 6.6, pp. 405–409

1. $53.52; $722.47 **2.** $9.43; $144.18 **3.** 6% **4.** $420
5. $5628 **6.** 12.5%, or $12\frac{1}{2}\%$ **7.** $1675 **8.** $180; $360
9. 20%

Exercise Set 6.6, p. 410

1. $10.95 **3.** $2.14 **5.** $18.55; $283.55 **7.** 5%
9. 4% **11.** $2000 **13.** $800 **15.** $719.86 **17.** 5.6%
19. $2700 **21.** 5% **23.** $980 **25.** $5880 **27.** 12%
29. $420 **31.** $30; $270 **33.** $2.55; $14.45
35. $125; $112.50 **37.** 40%; $360 **39.** $387; 30.4%
41. $460; 18.043% **43.** $\mathbf{D_W}$ **45.** $\mathbf{D_W}$ **47.** 18 **48.** $\frac{22}{7}$

49. 265.625 **50.** 1.113 **51.** $0.\overline{5}$ **52.** $2.\overline{09}$ **53.** $0.91\overline{6}$
54. $1.\overline{857142}$ **55.** $2.\overline{142857}$ **56.** $1.58\overline{3}$
57. 4,030,000,000,000 **58.** 5,800,000 **59.** 42,700,000
60. 6,090,000,000,000 **61.** $2.69
63. He bought the plaques for $166\frac{2}{3}$ + $250, or $416\frac{2}{3}$, and sold them for $400, so he lost money.

Margin Exercises, Section 6.7, pp. 414–417

1. $602 **2.** $451.50 **3.** $37.48; $4837.48 **4.** $2464.20
5. $8146.86

Calculator Corner, p. 417

1. $16,357.18 **2.** $12,764.72

Exercise Set 6.7, p. 418

1. $16 **3.** $84 **5.** $113.52 **7.** $1525 **9.** $671.88
11. (a) $147.95; **(b)** $10,147.95 **13. (a)** $128.22;
(b) $6628.22 **15. (a)** $46.03; **(b)** $5646.03 **17.** $484
19. $2802.50 **21.** $7853.38 **23.** $125,562.26
25. $4284.90 **27.** $28,225.00 **29.** $9270.87
31. $129,871.09 **33.** $4101.01 **35.** $1324.58 **37.** $\mathbf{D_W}$
39. 4.5 **40.** $8\frac{3}{4}$ **41.** 8 **42.** 87.5 **43.** $33\frac{1}{3}$ **44.** $3\frac{13}{17}$
45. $12\frac{2}{3}$ **46.** $3\frac{5}{11}$ **47.** $\frac{18}{17}$ **48.** $\frac{209}{10}$ **49.** $\frac{203}{2}$ **50.** $\frac{259}{8}$
51. 9.38%

Margin Exercises, Section 6.8, pp. 422–426

1. (a) $97; **(b)** interest: $86.40; amount applied to principal: $10.60; **(c)** interest: $55.17; amount applied to principal: $41.83; **(d)** At 13.6%, the principal was decreased by $31.23 more than at the 21.3% rate. The interest at 13.6% is $31.23 less than at 21.3%. **2. (a)** Interest: $197.44; amount applied to principal: $186.56; **(b)** $34.04; **(c)** $2520
3. Interest: $909.18; amount applied to principal: $69.46
4. (a) Interest: $909.69; amount applied to principal: $328.73 **(b)** $99,915.60; **(c)** The Sawyers will pay $129,394.80 less in interest with the 15-yr loan than with the 30-yr loan.
5. (a) Interest: $666.25; amount applied to principal: $405.21 **(b)** $69,862.80; **(c)** The Sawyers will pay $30,052.80 less in interest with the 15-yr loan at $6\frac{1}{2}$% than with the 15-yr loan at $8\frac{7}{8}$%.

Exercise Set 6.8, p. 427

1. (a) $98; **(b)** interest: $86.56; amount applied to principal: $11.44 **(c)** interest: $51.20; amount applied to principal: $46.80 **(d)** At 12.6%, the principal is decreased by $35.36 more than at the 21.3% rate. The interest at 12.6% is $35.36 less than at 21.3%. **3. (a)** Interest: $241.37; amount applied to principal: $264.60; **(b)** $74.26; **(c)** $16,156.40, $21,737.60, $5581.20 **5. (a)** Interest: $872.50; amount applied to principal: $123.44; **(b)** $208,538.40; **(c)** new principal: $149,876.56; interest: $871.78; amount applied to principal: $124.16 **7. (a)** Interest: $872.50; amount

applied to principal: $474.07; **(b)** $92,382.60; **(c)** The Martinez family will pay $116,155.80 less in interest with the 15-yr loan than with the 30-yr loan. **9.** $99,917.71;
$99,834.94 **11.** $99,712.04; $99,422.15 **13.** $149,882.75;
$149,764.79 **15.** $199,382.07; $198,760.41
17. (a) Interest: $112.38; amount applied to principal: $260.82; **(b)** new principal: $14,739.18; $1.96 less interest in second payment; **(c)** $2913.60 **19. (a)** $790; $7110;
(b) interest: $74; amount applied to principal: $163.82;
(c) $1451.52 **21.** $\mathbf{D_W}$ **23.** $\mathbf{D_W}$ **25.** 17.5, or $\frac{35}{2}$
26. 28 **27.** 100 **28.** 56.25, or $\frac{225}{4}$ **29.** $\frac{66,875}{36}$, or approximately 1857.64 **30.** 787.69 **31.** 71.81
32. 17.5, or $\frac{35}{2}$ **33.** 56.25, or $\frac{225}{4}$ **34.** $\frac{66,875}{36}$, or approximately 1857.64

Summary and Review: Chapter 6, p. 430

1. 1.7% **2.** 56% **3.** 37.5% **4.** $33.\overline{3}$%, or $33\frac{1}{3}$%
5. 0.735 **6.** 0.065 **7.** $\frac{6}{25}$ **8.** $\frac{63}{1000}$
9. $30.6 = p \times 90$; 34% **10.** $63 = 84\% \times n$; 75
11. $y = 38\frac{1}{2}\% \times 168$; 64.68 **12.** $\frac{24}{100} = \frac{16.8}{b}$; 70
13. $\frac{42}{30} = \frac{P}{100}$; 140% **14.** $\frac{10.5}{100} = \frac{a}{84}$; 8.82
15. 223 students; 105 students **16.** 44% **17.** 2500 mL
18. 12% **19.** 93.15 **20.** $14.40 **21.** 5% **22.** 11%
23. $42; $308 **24.** $42.70; $262.30 **25.** $2940
26. Approximately 25% **27.** $36 **28. (a)** $394.52;
(b) $24,394.52 **29.** $121 **30.** $7727.26 **31.** $9504.80
32. (a) $129; **(b)** interest: $100.18; amount applied to principal: $28.82; **(c)** interest: $70.72; amount applied to principal: $58.28; **(d)** At 13.2%, the principal is decreased by $29.46 more than at the 18.7% rate. The interest at 13.2% is $29.46 less than at 18.7%. **33.** $\frac{56}{3}$ **34.** 42 **35.** 64
36. 7.6123 **37.** $3.\overline{6}$ **38.** $1.\overline{571428}$ **39.** $3\frac{2}{3}$ **40.** $17\frac{2}{7}$
41. $\mathbf{D_W}$ No; the 10% discount was based on the original price rather than on the sale price.
42. $\mathbf{D_W}$ A 40% discount is better. When successive discounts are taken, each is based on the previous discounted price rather than on the original price. A 20% discount followed by a 22% discount is the same as a 37.6% discount off the original price.
43. 19.5% increase **44.** $66\frac{2}{3}$% **45.** $168

Test: Chapter 6, p. 433

1. [6.1b] 90.5% **2.** [6.1b] 0.2 **3.** [6.2a] 137.5%
4. [6.2b] $\frac{13}{20}$ **5.** [6.3a, b] $a = 40\% \cdot 55$; 22
6. [6.4a, b] $\frac{P}{100} = \frac{65}{80}$; 81.25% **7.** [6.5a] 400; 575
8. [6.5a] About 539 at-bats **9.** [6.5b] $50.\overline{90}$%
10. [6.5a] 5.5% **11.** [6.6a] $16.20; $340.20
12. [6.6b] $630 **13.** [6.6c] $40; $160 **14.** [6.7a] $8.52
15. [6.7a] $5356 **16.** [6.7b] $1110.39
17. [6.7b] $13,086.45 **18.** [6.5b] Dental assistant: 326; 42.4%; Nurse/psychiatric aide: 333, 22.8%; Child-care

worker: 905, 26.1%; Hairdresser/hairstylist/cosmetologist: 608, 62   **19.** [6.6c] $131.95; 52.8%
20. [6.8a] $119,909.14; $119,817.72   **21.** [4.4b] 222
22. [5.3b] 16   **23.** [4.5a] $1.41\overline{6}$ **24.** [3.4a] $3\frac{21}{44}$
25. [6.6b] $194,600   **26.** [6.6b], [6.7b] $2546.16

Cumulative Review: Chapters 1–6, p. 435

1. [6.1b] 0.53   **2.** [6.1b] 26.9%   **3.** [6.2a] 112.5%
4. [4.5a] $2.1\overline{6}$ **5.** [5.1a] $\frac{10}{1}$ **6.** [5.2a] $23\frac{1}{3}$ km/h
7. [3.3b] $<$ **8.** [3.3b] $<$ **9.** [4.6a] 296,200
10. [1.4b] 50,000   **11.** [1.9c, d] 13   **12.** [4.4c] 1.5
13. [3.5a] $3\frac{1}{30}$ **14.** [4.2a] 49.74   **15.** [1.2a] 515,150
16. [4.2b] 0.02   **17.** [3.5b] $\frac{2}{3}$ **18.** [3.3a] $\frac{2}{63}$
19. [2.4b] $\frac{1}{6}$ **20.** [1.5a] 853,142,400   **21.** [4.3a] 1.38036
22. [3.6b] $1\frac{1}{2}$ **23.** [4.4a] 12.25   **24.** [3.4b] $123\frac{1}{3}$
25. [1.7b] 95   **26.** [4.2c] 8.13   **27.** [2.7c] 9   **28.** [3.3c] $\frac{1}{12}$
29. [5.3b] 40   **30.** [5.3b] $8\frac{8}{21}$
31. [4.3b], [4.7a], [6.5b] **(a)** $15,100,000; $16,400,000; $9,100,000; $6,600,000; $3,300,000; **(b)** $50.5 million; **(c)** $10.1 million; **(d)** 8.6%; **(e)** 44.5%
32. [4.7a], [6.5b] **(a)** $7.31 billion; **(b)** 1.9%
33. [4.7a] About 47 messages   **34.** [5.4a] About 53 games
35. [4.7a] $42.50   **36.** [5.2b] 4.995 ¢/oz   **37.** [3.2b] $1\frac{1}{2}$ mi
38. [5.4a] 60 mi   **39.** [6.7b] $12,663.69
40. [3.6c] 5 pieces   **41.** [6.5b] Office manager: 1924, 19.4%; Office clerk: 463, 15.3%; Teacher assistant: 1192, 31.5%; Host/hostess: 297, 54   **42.** [6.5b] (b)
43. [4.5a] (e)   **44.** [4.7a] (d)   **45.** [3.3a] (c)
46. (a) [5.4a] Calories 120; 160

Calories from fat $6.\overline{6}$; $6.\overline{6}$		
Total fat $0.\overline{6}$ g	1.3%	1.3%
Saturated fat 0 g	0%	0%
Polyunsaturated fat 0 g		
Monounsaturated fat 0 g		
Cholesterol 0 mg	0%	0%
Sodium 280 mg	12%	14%
Potassium $226.\overline{6}$ mg	$6.\overline{6}$%	$12.\overline{6}$%
Total carbohydrate: $30.\overline{6}$ g	$10.\overline{6}$%	$12.\overline{6}$%
Dietary fiber $6.\overline{6}$ g	$26.\overline{6}$%	$26.\overline{6}$%
Soluble fiber $1.\overline{3}$ g		
Insoluble fiber $5.\overline{3}$ g		
Sugars $6.\overline{6}$ g		
Other carbohydrate $17.\overline{3}$ g		
Protein 4 g;		

(b) Complete; **(c)** Wheaties; **(d)** Complete; **(e)** Complete

Chapter 7

Pretest: Chapter 7, p. 440

1. (a) 51; **(b)** 51.5; **(c)** no mode exists   **2. (a)** 4.4; **(b)** 3; **(c)** 8   **3. (a)** 50; **(b)** 50; **(c)** 50   **4.** 55 mph   **5.** 76

6.
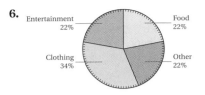

7. (a) $172; **(b)** $134
8.
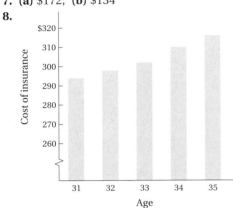

9.
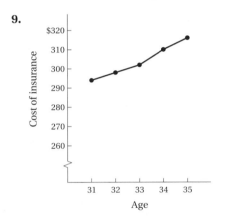

10. 260   **11.** 160 occurrences per 1000
12.

Margin Exercises, Section 7.1, pp. 443–448

1. 75   **2.** 54.9   **3.** 81   **4.** 19.4   **5.** 36.9   **6.** 68 mpg
7. 2.5   **8.** 94   **9.** 17   **10.** 17   **11.** 91   **12.** $3700
13. 67.5   **14.** 45   **15.** 34, 67   **16.** No mode exists.

17. **(a)** 17 g; **(b)** 18 g; **(c)** 19 g
18. Wheat A: average stalk height ≈ 25.21 in.; wheat B: average stalk height ≈ 22.54 in.; wheat B is better.

Calculator Corner, p. 445

1. 285.5   **2.** 75; 54.9; 81; 19.4   **3.** $202.\overline{3}$

Exercise Set 7.1, p. 449

1. Average: 21; median: 18.5; mode: 29
3. Average: 21; median: 20; mode: 5, 20
5. Average: 5.2; median: 5.7; mode: 7.4
7. Average: 239.5; median: 234; mode: 234
9. Average: $23.\overline{8}$; median: 15; mode: 1   **11.** 38 mpg
13. 2.7   **15.** Average: $8.19; median: $8.49; mode: $6.99
17. 90   **19.** 263 days
21. Bulb A: average time = 1171.25 hr; bulb B: average time ≈ 1251.58 hr; bulb B is better   **23.** **D**$_W$ **25.** 196
26. $\frac{4}{9}$ **27.** 1.96   **28.** 1.999396   **29.** 225.05
30. 126.0516   **31.** $\frac{3}{35}$ **32.** $\frac{14}{15}$ **33.** 182
35. 10 home runs   **37.** 58

Margin Exercises, Section 7.2, pp. 452–455

1. Cinnamon Life   **2.** Wheaties   **3.** Wheaties   **4.** 1.2 g
5. 224.66 mg   **6.** 213.3 mg   **7.** Average: 134; median: 120; mode: 120, 160   **8.** 20 calories   **9.** Yes
10. 560 mg   **11.** 24%   **12.** 14 g   **13.** 60,000 elephants
14. Two and one half as many in Zimbabwe as in Cameroon
15. 55,000 elephants   **16.** 795 cups; answers may vary
17. 750 cups; answers may vary
18. 1830 cups; answers may vary

Exercise Set 7.2, p. 456

1. 483,612,200 mi   **3.** Neptune   **5.** All   **7.** 11
9. Average: $27,884.\overline{1}$ mi; median: 7926 mi; no mode exists
11. 92°   **13.** 108°   **15.** 3   **17.** 90° and higher
19. 30% and higher   **21.** 50%   **23.** 1997: 59.74°; 1998: 60.26°; 0.87%   **25.** 59.545°; 59.937°; 0.392°
27. 1.0 billion   **29.** 2070   **31.** 1650 and 1850
33. 3 billion; 75%   **35.** Africa   **37.** 475,000 gal
39. 325,000 gal   **41.** **D**$_W$
43. Cabinets: $13,444; countertops: $4033.20; appliances: $2151.04; fixtures: $806.64
44. Cabinets: $3597.55; labor: $2901.25; countertops: $1276.55; flooring: $696.30   **45.** $\frac{6}{25}$ **46.** $\frac{9}{20}$
47. $\frac{6}{125}$ **48.** $\frac{8}{125}$ **49.** $\frac{531}{1000}$ **50.** $\frac{873}{1000}$ **51.** 1   **52.** $\frac{1}{50}$

53.

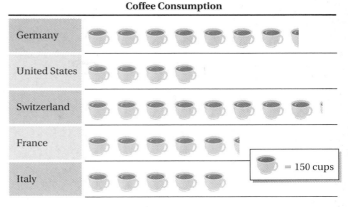

Coffee Consumption

Margin Exercises, Section 7.3, pp. 461–466

1. 18 g   **2.** Big Bacon Classic   **3.** Spicy Chicken, Plain Single, Breaded Chicken, Chicken Club, Single with Everything, Big Bacon Classic   **4.** 60   **5.** 85+
6. 60–64   **7.** Yes
8.

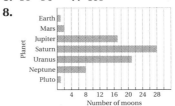

9. June   **10.** December and January, March and April, May and June, October and November   **11.** January, July
12. $900   **13.** 40 yr   **14.** $1300
15.

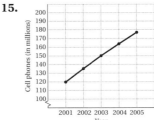

Exercise Set 7.3, p. 468

1. 190 calories   **3.** 1 slice of chocolate cake with fudge frosting   **5.** 1 cup of premium chocolate ice cream
7. About 120 calories   **9.** About 920 calories
11. About 28 lb   **13.** 1970: $11,000; 1997: $46,000; $35,000; 318%   **15.** 1970: $6000; 1997: $22,000; $16,000; 267%   **17.** $4000   **19.** $9000

21.

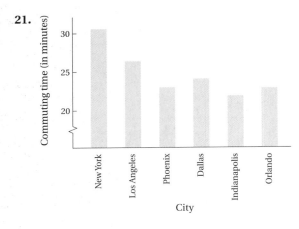

Commuting time (in minutes) / City

23. Indianapolis   **25.** 24.8 min   **27.** 1995 and 1996, 1996 and 1997, 1998 and 1999   **29.** 51%   **31.** $3.9 billion
33. 15.8 yd   **35.** 1988 and 1995
37.

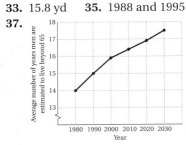

Average number of years men are estimated to live beyond 65 / Year

39. 25%   **41.** 10.1%   **43.** 1995 and 1996
45. About 308 murders   **47.** 14.1%   **49.** D_W
51. 18 min   **52.** 18%   **53.** 82.5   **54.** $66\frac{2}{3}$%
55. 53.125%   **56.** 68.75%   **57.** 67.3%   **58.** 97.81%
59. 118.75%   **60.** 103.125%   **61.** 51.2%   **62.** 99.6%
63.

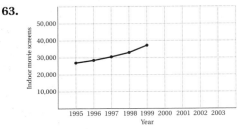

Indoor movie screens / Year

Using the line through the points for 1996 and 1998 (instead of a line through the point for 1999) probably gives more realistic estimates of the number of movie screens for the years 2000, 2001, and 2003.   2000: 37,500; 2001: 40,000; 2003: 44,500. Answers may vary.

Margin Exercises, Section 7.4, pp. 474–475

1. Spaying or neutering   **2.** 87%   **3.** $1122
4. 8% + 3%, or 11%

5.   Times of Engagement of Married Couples

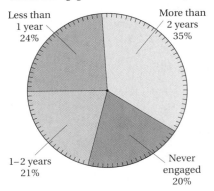

Exercise Set 7.4, p. 476

1. 3.7%   **3.** 270 recordings   **5.** 84.3%   **7.** Food
9. 14%
11.

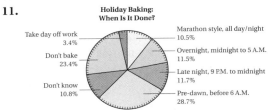

Holiday Baking: When Is It Done?

13.

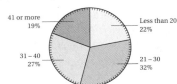

Weight Gain During Pregnancy

15.

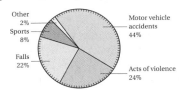

Causes of Spinal Injuries

17. D_W   **19.** $\frac{11}{25}$ **20.** $\frac{6}{25}$ **21.** $\frac{11}{50}$ **22.** $\frac{2}{25}$ **23.** $\frac{1}{50}$
24. $\frac{7}{25}$ **25.** $\frac{1}{5}$ **26.** $\frac{3}{25}$ **27.** $\frac{9}{100}$ **28.** $\frac{31}{100}$

Summary and Review: Chapter 7, p. 479

1. $29.50   **2.** $41.00   **3.** $17.50   **4.** $16.00   **5.** No
6. $18.75   **7.** 14,000 officers   **8.** Los Angeles
9. Houston   **10.** 12,500   **11.** 26   **12.** 11 and 17
13. 0.2   **14.** 700 and 800   **15.** $17   **16.** 20
17. $110.50; $107   **18.** 33 mpg   **19.** 96
20. About 420 calories   **21.** About 440 calories
22. Big Bacon Classic   **23.** Plain Single   **24.** Plain Single
25. Chicken Club   **26.** About 50 calories

27. About 220 calories   **28.** Under 20   **29.** 12 accidents
30. 13 accidents per 100 drivers   **31.** 45–74
32. 11 accidents per 100 drivers   **33.** Under 20
34. 22%   **35.** 11%   **36.** 1600 travelers   **37.** 25%
38.

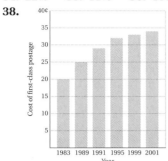

39.

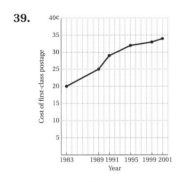

40. Battery A: average $\approx$ 43.04 hr; battery B:
average = 41.55 hr; battery A is better.   **41.** 38.5
42. 13.4   **43.** 1.55   **44.** 1840   **45.** $16.\overline{6}$ **46.** 321.$\overline{6}$
47. 38.5   **48.** 14   **49.** 1.8   **50.** 1900   **51.** $17
52. 375   **53.** 3.1
54. DW The average, the median, and the mode are
"center points" that characterize a set of data. You might use
the average to find a center point that is midway between
the extreme values of the data. The median is a center point
that is in the middle of all the data. That is, there are as
many values less than the median than there are values
greater than the median. The mode is a center point that
represents the value or values that occur most frequently.
55. DW The equation could represent a person's average
income during a 4-yr period. Answers may vary.
56. 12,600 mi   **57.** 5.7592 billion   **58.** 222.$\overline{2}$%
59. 50%   **60.** $\frac{9}{10}$ **61.** $\frac{5}{12}$ **62.** $a = 316$, $b = 349$

Test: Chapter 7, p. 483

1. [7.2a] 179 lb   **2.** [7.2a] 111 lb   **3.** [7.2a] 5 ft, 3 in.;
medium frame   **4.** [7.2a] 5 ft, 11 in.; medium frame
5. [7.2b] 1995   **6.** [7.2b] 1992   **7.** [7.2b] 8 hr
8. [7.2b] 10 hr   **9.** [7.1a] 49.5   **10.** [7.1a] 2.6
11. [7.1a] 15.5   **12.** [7.1b, c] 50.5; 52
13. [7.1b, c] 3; 1 and 3   **14.** [7.1b, c] 17.5; 17 and 18
15. [7.1a] 38 mpg   **16.** [7.1a] 76   **17.** [7.3c] 53%
18. [7.3c] 41%   **19.** [7.3c] 1967   **20.** [7.3c] 2006

21. [7.3b]

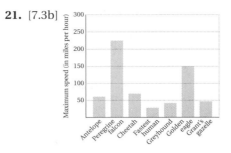

22. [7.2a], [7.3a] 197 mph   **23.** [7.2a], [7.3a] No; the
greyhound can run 14 mph faster than a human.
24. [7.2a], [7.3a] 89 mph   **25.** [7.2a], [7.3a] 61 mph
26. [7.4b]

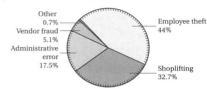

27. [7.2a], [7.4a] Employee theft: $10.12 billion; shoplifting:
$7.521 billion; administrative error: $4.025 billion; vendor
fraud: $1.173 billion; other: $0.161 billion
28. [7.3b]

29. [7.3d]

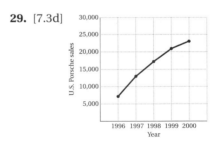

30. [7.1d] Bar A: average $\approx$ 8.417; bar B: average $\approx$ 8.417;
equal quality   **31.** [7.1a] 2.9   **32.** [2.7b] $\frac{25}{4}$, or $6\frac{1}{4}$
33. [6.3b], [6.4b] 68   **34.** [6.5a] 15,600 TV sets
35. [5.4a] 340 at-bats   **36.** [7.1a, b] $a = 74$, $b = 111$

Cumulative Review: Chapters 1–7, p. 487

1. [4.3b] $51,000,000,000   **2.** [7.1a] 27 mpg
3. [1.1a] 5 hundreds   **4.** [1.9c] 128   **5.** [2.1a] 1, 2, 3, 4, 5,
6, 10, 12, 15, 20, 30, 60   **6.** [4.1d] 52.0   **7.** [3.4a] $\frac{33}{10}$
8. [4.3b] $2.10   **9.** [4.3b] $3,250,000,000,000
10. [5.3a] No   **11.** [3.5a] $6\frac{7}{10}$ **12.** [4.2a] 44.6351
13. [3.3a] $\frac{1}{3}$ **14.** [4.2b] 325.43   **15.** [3.6a] 15
16. [1.5a] 2,740,320   **17.** [2.7b] $\frac{9}{10}$ **18.** [3.4b] $4361\frac{1}{2}$

19. [5.3b] $9\frac{3}{5}$ **20.** [2.7c] $\frac{3}{4}$ **21.** [4.4b] 6.8
22. [1.7b] 15,312 **23.** [5.2b] 20.6 ¢/oz
24. [6.5a] 3324 students **25.** [3.6c] $\frac{1}{4}$ yd
26. [2.6b] $\frac{3}{8}$ cup **27.** [4.7a] 6.2 lb **28.** [1.8a]
2572 billion kWh **29.** [6.5a] 2.5%; 0.2% **30.** [6.5b] 172%
31. (a) [3.6c] 1870 sq yd; **(b)** [3.6c] 6400 sq yd; **(c)** [1.8a]
4530 sq yd **32.** [7.1a, b] $37.84; $38.00
33. [7.3b]

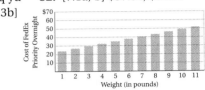

34. [7.3d]

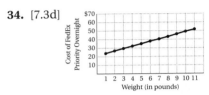

35. [3.2b], [3.3d] $\frac{1}{4}$ **36.** [5.4a] 1122 defective valves
37. [4.7a] $9.55 **38.** [6.6b] 7% **39.** [7.2a], [6.5a]
Clothing: $129.54; entertainment: $83.82; food: $83.82;
other: $83.82 **40.** [2.6b] $254; $127 **41.** [7.2a], [6.5a]
Clothing: $44.064 billion; entertainment: $28.512 billion;
food: $28.512 billion; other: $28.512 billion
42. [6.5b] 73% **43.** [6.5b], [7.1a] 12% decrease

Chapter 8

Pretest, Chapter 8, p. 492

1. 96 **2.** $\frac{5}{12}$ **3.** 8460 **4.** 0.92 **5.** 190; 0.19
6. 457,000; 45,700 **7.** 2.638 **8.** 149,637,000
9. 0.1160; 4.173; 0.106; 106 **10.** 9140; 3600; 91.4; 91,400
11. 2.304 **12.** 2400 **13.** 80 **14.** 8800 **15.** 4800
16. 0.62 **17.** 3.4 **18.** 420 **19.** 384 **20.** 64
21. 2560 **22.** 24 **23.** 25°C **24.** 98.6°F **25.** 400 mcg
26. 88.71 mL **27.** 1000 g **28.** 144 **29.** 2,000,000

Margin Exercises, Section 8.1, pp. 493–496

1. 2 **2.** 3 **3.** $1\frac{1}{2}$ **4.** $2\frac{1}{2}$ **5.** 288 **6.** $8\frac{1}{2}$
7. 240,768 **8.** 6 **9.** $1\frac{5}{12}$ **10.** 8 **11.** $11\frac{2}{3}$, or $11.\overline{6}$
12. 5 **13.** 0.5 **14.** 288

Exercise Set 8.1, p. 497

1. 12 **3.** $\frac{1}{12}$ **5.** 5280 **7.** 108 **9.** 7 **11.** $1\frac{1}{2}$
13. 26,400 **15.** $5\frac{1}{4}$, or 5.25 **17.** $3\frac{1}{3}$ **19.** 37,488
21. $1\frac{1}{2}$ **23.** $1\frac{1}{4}$, or 1.25 **25.** 110 **27.** 2 **29.** 300
31. 30 **33.** $\frac{1}{36}$ **35.** 126,720 **37.** $\mathbf{D_W}$ **39.** $\frac{37}{400}$
40. $\frac{7}{8}$ **41.** $\frac{11}{40}$ **42.** $\frac{57}{100}$ **43.** 137.5%
44. $66.\overline{6}$%, or $66\frac{2}{3}$% **45.** 25% **46.** 43.75%

47. $\frac{373}{100,000}$; $\frac{100,000}{373}$ **48.** $\frac{73}{6400}$; $\frac{6400}{73}$ **49.** Length: 5400 in., or
450 ft; breadth: 900 in., or 75 ft; height: 540 in., or 45 ft

Margin Exercises, Section 8.2, pp. 500–504

1. 2 cm, or 20 mm **2.** 2.3 cm, or 23 mm **3.** 3.8 cm, or
38 mm **4.** cm **5.** km **6.** mm **7.** m **8.** cm
9. m **10.** 23,000 **11.** 400 **12.** 178 **13.** 9040
14. 7.814 **15.** 781.4 **16.** 0.967 **17.** 8,900,000
18. 6.78 **19.** 97.4 **20.** 0.1 **21.** 8.451

Exercise Set 8.2, p. 505

1. (a) 1000; **(b)** 0.001 **3. (a)** 10; **(b)** 0.1 **5. (a)** 0.01;
(b) 100 **7. (a)** 6700; **(b)** Substitute **9. (a)** 0.98;
(b) Multiply by 1 **11.** 8.921 **13.** 0.05666 **15.** 566,600
17. 4.77 **19.** 688 **21.** 0.1 **23.** 100,000 **25.** 142
27. 0.82 **29.** 450 **31.** 0.000024 **33.** 0.688 **35.** 230
37. 3.92 **39.** 180; 0.18 **41.** 278; 27.8 **43.** 48,440;
48.44 **45.** 4000; 400 **47.** 0.027; 0.00027 **49.** 442,000;
44,200 **51.** $\mathbf{D_W}$ **53.** 0.234 **54.** 0.0234 **55.** 13.8474
56. $80\frac{1}{2}$ **57.** 0.9 **58.** 0.108 **59.** 1.0 m **61.** 1.4 cm

Margin Exercises, Section 8.3, pp. 507–508

1. 91.4 **2.** 804.5 **3.** 1479.843 **4.** 913.912
5. 343.735 m **6.** 3.175

Exercise Set 8.3, p. 509

1. 100.65 **3.** 727.409592 **5.** 104.585 **7.** 289.62
9. 112.63 **11.** 9.14 **13.** 83.8581 **15.** 1250.061
17. 19.05 **19.** 1376.136 **21.** 0.51181
23.–33. Answers may vary, depending on the conversion
factor used.

	yd	cm	in.	m	mm
23.	0.2604	23.8	9.37006	0.238	238
25.	0.2361	21.59	$8\frac{1}{2}$	0.2159	215.9
27.	52.9934	4844	1907.0828	48.44	48,440
29.	4	365.6	144	3.656	3656
31.	0.000295	0.027	0.0106299	0.00027	0.27
33.	483.548	44,200	17,401.54	442	442,000

35. $\mathbf{D_W}$ **37.** 0.561 **38.** 67.34% **39.** 112.5%, or $112\frac{1}{2}$%
40. 25 **41.** 100 **42.** 961 **43.** 113,000,000
44. $1,300,000,000 **45.** $1,600,000,000,000
46. 45,000,000 **47.** 1 in. = 25.4 mm **49.** 85.725 m

Margin Exercises, Section 8.4, pp. 511–515

1. 80 **2.** 4.32 **3.** 32,000 **4.** kg **5.** kg **6.** mg
7. g **8.** t **9.** 6200 **10.** 3.048 **11.** 77 **12.** 234.4
13. 6700 **14.** 0.001 **15.** 0.5 mg

Exercise Set 8.4, p. 516

1. 2000 **3.** 3 **5.** 64 **7.** 12,640 **9.** 0.1 **11.** 5
13. 26,000,000,000 lb **15.** 1000 **17.** 10
19. $\frac{1}{100}$, or 0.01 **21.** 1000 **23.** 10 **25.** 234,000
27. 5.2 **29.** 6.7 **31.** 0.0502 **33.** 8.492 **35.** 58.5
37. 800,000 **39.** 1000 **41.** 0.0034 **43.** 1000
45. 0.325 **47.** 125 mcg **49.** 0.875 mg; 875 mcg
51. 4 tablets **53.** 8 cc **55.** $\mathbf{D_W}$ **57.** $\frac{7}{20}$ **58.** $\frac{99}{100}$
59. $\frac{171}{200}$ **60.** $\frac{171}{500}$ **61.** $\frac{3}{8}$ **62.** $\frac{2}{3}$ **63.** $\frac{5}{6}$ **64.** $\frac{1}{6}$
65. 7500 sheets **66.** 46 test tubes, 2 cc left over
67. 144 packages **69. (a)** 109.134 g; **(b)** 9.104 g;
(c) Golden Jubilee: 3.85 oz; Hope: 0.321 oz

Margin Exercises, Section 8.5, pp. 520–522

1. 40 **2.** 20 **3.** mL **4.** mL **5.** L **6.** L **7.** 970
8. 8.99 **9.** 2.4 L **10. (a)** About 59.14 mL;
(b) about 0.059 L

Exercise Set 8.5, p. 523

1. 1000; 1000 **3.** 87,000 **5.** 0.049 **7.** 0.000401
9. 78,100 **11.** 320 **13.** 10 **15.** 32 **17.** 20
19. 14 **21.** 88

	gal	qt	pt	cups	oz
23.	1.125	4.5	9	18	144
25.	16	64	128	256	2048
27.	0.25	1	2	4	32
29.	0.1171875	0.46875	0.9375	1.875	15

	L	mL	cc	cm³
31.	2	2000	2000	2000
33.	64	64,000	64,000	64,000
35.	0.443	443	443	443

37. 2000 mL **39.** 0.32 L **41.** 59.14 mL **43.** 500 mL
45. 125 mL/hr **47.** 9 **49.** $\frac{1}{5}$ **51.** 6 **53.** 15
55. $\mathbf{D_W}$ **57.** 45.2% **58.** 99.9% **59.** 33.$\overline{3}$%, or $33\frac{1}{3}$%
60. 66.$\overline{6}$%, or $66\frac{2}{3}$% **61.** 55% **62.** 105% **63.** 88%
64. 8% **65.** $\frac{2.7}{13.1}$; $\frac{13.1}{2.7}$ **66.** $\frac{1}{4}$; $\frac{3}{4}$
67. 1.75 gal/week; 7.5 gal/month; 91.25 gal/year;
70,250,000 gal/day; 25,641,250,000 gal/year **69.** 0.21$\overline{3}$ oz

Margin Exercises, Section 8.6, pp. 526–528

1. 120 **2.** 1461 **3.** 1440 **4.** 1 **5.** 80°C **6.** 0°C
7. −20°C **8.** 80°F **9.** 100°F **10.** 50°F **11.** 176°F
12. 95°F **13.** 35°C **14.** 45°C

Calculator Corner, p. 528

1. 41°F **2.** 122°F **3.** 20°C **4.** 45°C

Exercise Set 8.6, p. 529

1. 24 **3.** 60 **5.** $365\frac{1}{4}$ **7.** 0.05 **9.** 8.2 **11.** 6.5
13. 10.75 **15.** 336 **17.** 4.5 **19.** 56 **21.** 86,164.2 sec
23. 77°F **25.** 104°F **27.** 186.8°F **29.** 136.4°F
31. 35.6°F **33.** 41°F **35.** 5432°F **37.** 30°C
39. 55°C **41.** 81.$\overline{1}$°C **43.** 60°C **45.** 20°C **47.** 6.$\overline{6}$°C
49. 37°C **51. (a)** 136°F = 57.$\overline{7}$°C, $56\frac{2}{3}$°C = 134°F;
(b) 2°F **53.** $\mathbf{D_W}$ **55.** 87.5% **56.** 58%
57. 66.$\overline{6}$%, or $66\frac{2}{3}$% **58.** 43.61% **59.** 37.5%
60. 62.5% **61.** 66.$\overline{6}$%, or $66\frac{2}{3}$% **62.** 20% **63.** $24.30
64. 12% **65.** 187,200 **66.** About 324
67. About 0.03 yr **69.** About 31,688 yr **71.** 0.25

Margin Exercises, Section 8.7, pp. 532–533

1. 9 **2.** 45 **3.** 2880 **4.** 2.5 **5.** 3200 **6.** 1,000,000
7. 1 **8.** 28,800 **9.** 0.043 **10.** 0.678

Exercise Set 8.7, p. 534

1. 144 **3.** 640 **5.** $\frac{1}{144}$ **7.** 198 **9.** 396 **11.** 12,800
13. 27,878,400 **15.** 5 **17.** 1 **19.** $\frac{1}{640}$, or 0.0015625
21. 5,210,000 **23.** 140 **25.** 23.456 **27.** 0.085214
29. 2500 **31.** 0.4728 **33.** $\mathbf{D_W}$ **35.** $240 **36.** $212
37. (a) $484.11; **(b)** $15,984.11 **38. (a)** $209.59;
(b) $8709.59 **39. (a)** $220.93; **(b)** $6620.93
40. (a) $37.97; **(b)** $4237.97 **41.** 10.76 **43.** 1.67
45. 1876.8 m²

Summary and Review: Chapter 8, p. 537

1. $2\frac{2}{3}$ **2.** 30 **3.** 0.03 **4.** 0.004 **5.** 72 **6.** 400,000
7. $1\frac{1}{6}$ **8.** 0.15 **9.** 218.8 **10.** 32.18 **11.** 10; 0.01
12. 305,000; 30,500 **13.** 112 **14.** 0.004 **15.** $\frac{4}{15}$, or 0.2$\overline{6}$
16. 0.464 **17.** 180 **18.** 4700 **19.** 16,140 **20.** 830
21. $\frac{1}{4}$, or 0.25 **22.** 0.04 **23.** 200 **24.** 30 **25.** 0.0007
26. 0.06 **27.** 1600 **28.** 400 **29.** 1.25 **30.** 50
31. 160 **32.** 7.5 **33.** 13.5 **34.** 6 mL **35.** 3000 mL
36. 250 mcg **37.** 80.6°F **38.** 20°C **39.** 36
40. 300,000 **41.** 14.375 **42.** 0.06 **43.** $\mathbf{D_W}$ The
metric system was adopted by law in France in about 1790,
during the rule of Napoleon I. **44.** $\mathbf{D_W}$ 1 gal = 128 oz, so
1 oz of water (as capacity) weighs $\frac{8.3453}{128}$ lb, or about 0.0652 lb.
An ounce of pennies weighs $\frac{1}{16}$ lb, or 0.0625 lb. Thus an
ounce of water (as capacity) weighs more than an ounce of
pennies. **45.** 47% **46.** 92% **47.** 0.567 **48.** $\frac{73}{100}$
49. 17.66 sec

Test: Chapter 8, p. 539

1. [8.1a] 48 **2.** [8.1a] $\frac{1}{3}$ **3.** [8.2a] 6000 **4.** [8.2a] 0.87
5. [8.3a] 182.8 **6.** [8.3a] 1490.4 **7.** [8.2a] 5; 0.005
8. [8.2a] 1854.2; 185.42 **9.** [8.5a] 3.08 **10.** [8.5a] 240
11. [8.4a] 64 **12.** [8.4a] 8220 **13.** [8.4b] 3800
14. [8.4b] 0.4325 **15.** [8.4b] 2.2 **16.** [8.6a] 300
17. [8.6a] 360 **18.** [8.5a] 32 **19.** [8.5a] 1280
20. [8.5a] 40 **21.** [8.4c] 370 **22.** [8.6b] 35°C

23. [8.6b] 138.2°F   **24.** [8.3a] 1.094; 100; 39.370; 1000
25. [8.3a] 36377.2; 14,328; 363.772; 363,772
26. [8.5b] 2500 mL   **27.** [8.4c] 1500 mcg
28. [8.5b] About 118.28 mL   **29.** [8.7a] 1728
30. [8.7b] 0.0003   **31.** [6.1b] 93%   **32.** [6.2a] 81.25%
33. [6.1b] 0.932   **34.** [6.2b] $\frac{1}{3}$ **35.** [8.3a] About 39.47 sec

Cumulative Review: Chapters 1–8, p. 541

1. [4.3b] 24,500,000   **2.** [5.2a] 26 mpg   **3.** [6.5b] 30; 30.3%
4. [6.5b] 801; 18.3%   **5.** [6.5b] 146; 57.9%
6. [6.5b] 179; 11.7%   **7.** [6.5b] 63; 13
8. [1.5a] 50,854,100   **9.** [2.6a] $\frac{1}{12}$ **10.** [4.5c] 15.2
11. [3.6b] $\frac{2}{3}$ **12.** [4.4a] 35.6   **13.** [3.5b] $\frac{5}{6}$ **14.** [2.2a] Yes
15. [2.2a] No   **16.** [2.1d] $3 \cdot 3 \cdot 11$ **17.** [3.1a] 245
18. [4.5b] 35.8   **19.** [4.1a] One hundred three and
sixty-four thousandths
20. [7.1a, b] 17.8$\overline{3}$, 17.5   **21.** [6.1b] 8%
22. [6.2a] 60%   **23.** [4.4b] 150.5
24. [1.7b] 19,248   **25.** [2.7c] $\frac{15}{2}$ **26.** [3.3c] $\frac{2}{35}$
27. [8.1a] 6   **28.** [8.4a] 0.375   **29.** [8.6b] 59°F
30. [8.5a] 87   **31.** [8.6a] 0.15   **32.** [8.2a] 0.17
33. [8.3a] 3539.8   **34.** [8.5a] About 5.7
35. [8.4c] 230
36. [7.3b]

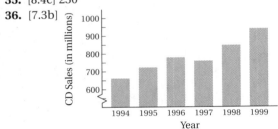

37. [7.3d]

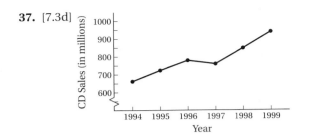

38. [7.4b]

Total CD Sales from 1994 – 1999: 4709 million

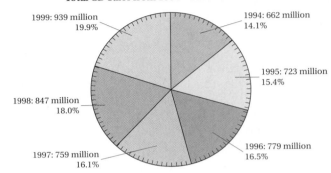

39. [6.5a] Yes   **40.** [1.8a] 165,000,000,000 lb
41. [4.7a], [5.2a] 255.8 mi; 15.9875 mpg
42. [5.4a] $8\frac{3}{4}$ lb   **43.** [2.6b] $13,200
44. [3.2b] $1\frac{1}{4}$ hr   **45.** [8.4a] 240 oz
46. [1.8a] 60 dips
47. [3.5c], [3.6c] About 25 or 26 dips
48. [4.7a] $119.40   **49.** [1.8a], [4.7a] $89.70
50. [1.8a], [8.4a] 480 lb; 7680 oz

51. [1.8a], [4.7a]

Day	Total Number of Pounds in the Tubs at the Start of the Day	Number of Pounds Sold	Number of Ounces Sold	Number of Dips Sold	Amount of Revenue at $1.99/cone	Amount of Revenue at $2.99/cone
Monday	480	93	1488	372	$740.28	$556.14
Tuesday	387	63	1008	252	501.48	376.74
Wednesday	324	67	1072	268	533.32	400.66
Thursday	257	105	1680	420	835.80	627.90
Friday	152	78	1248	312	620.88	466.44
Saturday	224	86	1376	344	684.56	514.28
Sunday	138	53	848	212	421.88	316.94
Monday	85					

Source: Sundae's Homemade Ice Cream & Coffee Co.

$4338.20; $3259.10

52. [4.7a] At least $3259.10, probably over $3500; answers may vary
53. [1.8a], [4.7a] 1-dip cones: $2122.67; 2-dip cones: $2068.17
54. [1.8a], [4.7a] 1-dip cones: $2215.53; 2-dip cones: $1190.93

Chapter 9

Pretest: Chapter 9, p. 546

1. 131 mm **2.** 100 ft^2 **3.** 92 in^2 **4.** 22 cm^2
5. $32\frac{1}{2}$ ft^2 **6.** 4 m^2 **7.** 9.6 m **8.** 30.144 m
9. 72.3456 m^2 **10.** 160 cm^3 **11.** 1256 ft^3
12. 33,493.$\overline{3}$ yd^3 **13.** 150.72 cm^3 **14.** 60°
15. (a) Scalene; (b) right **16.** 9 **17.** 9.849
18. $b = \sqrt{45}$; $b \approx 6.708$

Margin Exercises, Section 9.1, pp. 547–549

1. 26 cm **2.** 46 in. **3.** 12 cm **4.** 17.5 yd **5.** $27\frac{5}{6}$ in.
6. 40 km **7.** 21 yd **8.** 31.2 km **9.** 70 ft; $346.50

Exercise Set 9.1, p. 550

1. 17 mm **3.** 15.25 in. **5.** 18 km **7.** 30 ft
9. 79.14 cm **11.** 88 ft **13.** 182 mm
15. 826 m; $1197.70 **17.** 122 cm **19.** (a) 228 ft;
(b) $1046.52 **21.** ^{D}**W** **23.** $19.20 **24.** $96
25. 1000 **26.** 1331 **27.** 225 **28.** 484 **29.** 49
30. 64 **31.** 5% **32.** 11% **33.** 9 ft

Margin Exercises, Section 9.2, pp. 552–557

1. 8 cm^2 **2.** 56 km^2 **3.** $18\frac{3}{8}$ yd^2 **4.** 144 km^2
5. 118.81 m^2 **6.** $12\frac{1}{4}$ yd^2 **7.** 43.8 cm^2 **8.** 12.375 km^2
9. 96 m^2 **10.** 18.7 cm^2 **11.** 100 m^2 **12.** 88 cm^2
13. 228 in^2

Exercise Set 9.2, p. 558

1. 15 km^2 **3.** 1.4 in^2 **5.** $6\frac{1}{4}$ yd^2 **7.** 8100 ft^2 **9.** 50 ft^2
11. 169.883 cm^2 **13.** $41\frac{2}{9}$ in^2 **15.** 484 ft^2
17. 3237.61 km^2 **19.** $28\frac{57}{64}$ yd^2 **21.** 32 cm^2 **23.** 60 in^2
25. 104 ft^2 **27.** 45.5 in^2 **29.** 8.05 cm^2 **31.** 297 cm^2
33. 7 m^2 **35.** 1197 m^2 **37.** (a) 5599.75 ft^2;
(b) about $45 **39.** 630.36 m^2 **41.** (a) 819.75 ft^2;
(b) 10 gal; (c) $179.50 **43.** 80 cm^2 **45.** 675 cm^2
47. 21 cm^2 **49.** 852.04 ft^2 **51.** ^{D}**W** **53.** 234
54. 230 **55.** 336 **56.** 24 **57.** 0.724 **58.** 0.0724
59. 2520 **60.** 6 **61.** 28 **62.** 39,600 **63.** 12
64. 46,252.8 **65.** 16,914 in^2

Margin Exercises, Section 9.3, pp. 563–567

1. 9 in. **2.** 5 ft **3.** 62.8 m **4.** 88 m **5.** 34.296 yd
6. $78\frac{4}{7}$ km^2 **7.** 339.62 cm^2 **8.** 12-ft diameter flower
bed, by about 13.04 ft^2

Exercise Set 9.3, p. 568

1. 14 cm; 44 cm; 154 cm^2 **3.** $1\frac{1}{2}$ in.; $4\frac{5}{7}$ in.; $1\frac{43}{56}$ in^2
5. 16 ft; 100.48 ft; 803.84 ft^2 **7.** 0.7 cm; 4.396 cm;
1.5386 cm^2 **9.** 3 cm; 18.84 cm; 28.26 cm^2 **11.** 153.86 ft^2
13. 2.5 cm; 1.25 cm; 4.90625 cm^2 **15.** 3.454 ft
17. 65.94 yd^2 **19.** 45.68 ft **21.** 26.84 yd **23.** 45.7 yd
25. 100.48 m^2 **27.** 6.9972 cm^2 **29.** 64.4214 in^2
31. ^{D}**W** **33.** 16 **34.** 289 **35.** 125 **36.** 64
37. 543 **38.** 0.00543 **39.** 50.1% **40.** $730
41. 10 cans **42.** 5 lb **43.** 3.1416
45. $3d$; πd; circumference of one ball, since $\pi > 3$

Margin Exercises, Section 9.4, pp. 572–576

1. 12 cm^3 **2.** 20 ft^3 **3.** 128 ft^3 **4.** 785 ft^3
5. 67,914 m^3 **6.** $91,989\frac{1}{3}$ ft^3 **7.** 38.77272 cm^3
8. 1695.6 m^3 **9.** 528 in^3 **10.** 83.7$\overline{3}$ mm^3

Calculator Corner, p. 575

1. Left to the student **2.** Left to the student

Exercise Set 9.4, p. 577

1. 768 cm^3 **3.** 45 in^3 **5.** 75 m^3 **7.** $357\frac{1}{2}$ yd^3
9. 803.84 in^3 **11.** 353.25 cm^3 **13.** 41,580,000 yd^3
15. $4,186,666\frac{2}{3}$ in^3 **17.** 124.72 m^3 **19.** $1950\frac{101}{168}$ ft^3
21. 113,982 ft^3 **23.** 24.64 cm^3 **25.** 0.423115 yd^3
27. 367.38 m^3 **29.** 143.72 cm^3 **31.** 32,993,440,000 mi^3
33. 646.74 cm^3 **35.** 61,600 cm^3 **37.** 5832 yd^3
39. ^{D}**W** **41.** 396 **42.** 3 **43.** 14 **44.** 253,440
45. 12 **46.** 335,808 **47.** 24 **48.** 14 **49.** 0.566
50. 13,400 **51.** About 57,480 in^3 **53.** 9.425 L
55. The diameter of the earth at the equator is about
7930 mi. The diameter of the earth between the north and
south poles is about 7917 mi. If we use the average of these
two diameters (7923.5 mi), the volume of the earth is
$\frac{4}{3} \cdot \pi \cdot \left(\frac{7923.5}{2}\right)^3$, or about 260,000,000,000 mi^3. **57.** 0.331 m^3

Margin Exercises, Section 9.5, pp. 582–587

1. Angle DEF, angle FED, $\angle DEF$, $\angle FED$, or $\angle E$
2. Angle PQR, angle RQP, $\angle PQR$, $\angle RQP$, or $\angle Q$ **3.** 127°
4. 33° **5.** Right **6.** Acute **7.** Obtuse **8.** Straight
9. $\angle 1$ and $\angle 2$; $\angle 1$ and $\angle 4$; $\angle 2$ and $\angle 3$; $\angle 3$ and $\angle 4$
10. 45° **11.** 72° **12.** 5°
13. $\angle 1$ and $\angle 2$; $\angle 1$ and $\angle 4$; $\angle 2$ and $\angle 3$; $\angle 3$ and $\angle 4$
14. 142° **15.** 23° **16.** 90° **17.** (a) $\triangle ABC$;
(b) $\triangle ABC$, $\triangle MPN$; (c) $\triangle DEF$, $\triangle GHI$, $\triangle JKL$, $\triangle QRS$
18. Yes **19.** No **20.** (a) $\triangle DEF$; (b) $\triangle GHI$, $\triangle QRS$;
(c) $\triangle ABC$, $\triangle MPN$, $\triangle JKL$ **21.** 180° **22.** 64°

Exercise Set 9.5, p. 588

1. Angle GHI, angle IHG, $\angle GHI$, $\angle IHG$, or $\angle H$ **3.** 10°
5. 180° **7.** 130° **9.** Obtuse **11.** Acute **13.** Straight
15. Obtuse **17.** Acute **19.** Obtuse **21.** 79°

23. 23° **25.** 32° **27.** 61° **29.** 177° **31.** 41°
33. 95° **35.** 78° **37.** Scalene; obtuse **39.** Scalene;
right **41.** Equilateral; acute **43.** Scalene; obtuse
45. 46° **47.** 120° **49.** $\mathbf{D_W}$ **51.** $160 **52.** $22.50
53. $148 **54.** $1116.67 **55.** $33,597.91
56. $413,458.31 **57.** $641,566.26 **58.** $684,337.34
59. $m\angle 2 = 67.13°$; $m\angle 3 = 33.07°$; $m\angle 4 = 79.8°$;
$m\angle 5 = 67.13°$ **61.** $m\angle ACB = 50°$; $m\angle CAB = 40°$;
$m\angle EBC = 50°$; $m\angle EBA = 40°$; $m\angle AEB = 100°$;
$m\angle ADB = 50°$

Margin Exercises, Section 9.6, pp. 591–595

1. 81 **2.** 100 **3.** 121 **4.** 144 **5.** 169 **6.** 196
7. 225 **8.** 256 **9.** 289 **10.** 324 **11.** 400 **12.** 625
13. 3 **14.** 4 **15.** 11 **16.** 10 **17.** 9 **18.** 8
19. 18 **20.** 20 **21.** 15 **22.** 13 **23.** 1 **24.** 0
25. 2.236 **26.** 8.832 **27.** 12.961 **28.** $c = 13$
29. $a = \sqrt{75}$; $a \approx 8.660$ **30.** $b = \sqrt{120}$; $b \approx 10.954$
31. $a = \sqrt{175}$; $a \approx 13.229$ **32.** $\sqrt{424}$ ft ≈ 20.6 ft

Calculator Corner, p. 592

1. 6.6 **2.** 9.7 **3.** 19.8 **4.** 17.3 **5.** 24.9 **6.** 24.5
7. 121.2 **8.** 115.6 **9.** 16.2 **10.** 85.4

Exercise Set 9.6, p. 596

1. 10 **3.** 21 **5.** 25 **7.** 19 **9.** 23 **11.** 100
13. 6.928 **15.** 2.828 **17.** 4.243 **19.** 2.449
21. 3.162 **23.** 8.660 **25.** 14 **27.** 13.528
29. $c = \sqrt{34}$; $c \approx 5.831$ **31.** $c = \sqrt{98}$; $c \approx 9.899$
33. $a = 5$ **35.** $b = 8$ **37.** $c = 26$ **39.** $b = 12$
41. $b = \sqrt{1023}$; $b \approx 31.984$ **43.** $c = 5$
45. $\sqrt{250}$ m ≈ 15.8 m **47.** $\sqrt{8450}$ ft ≈ 91.9 ft
49. $h = \sqrt{500}$ ft ≈ 22.4 ft
51. $\sqrt{211,200,000}$ ft $\approx 14,532.7$ ft **53.** $\sqrt{832}$ ft ≈ 28.8 ft
55. $\mathbf{D_W}$ **57.** 1000 **58.** 100 **59.** 100,000 **60.** 10,000
61. $208 **62.** $11.25 **63.** $115.02 **64.** $994,274.04
65. $1,686,435.46 **67.** The areas are the same.
69. Length: 36.6 in.; width: 20.6 in.

Summary and Review: Chapter 9, p. 601

1. 23 m **2.** 4.4 m **3.** 228 ft; 2808 ft^2 **4.** 36 ft; 81 ft^2
5. 17.6 cm; 12.6 cm^2 **6.** 60 cm^2 **7.** 35 mm^2
8. 22.5 m^2 **9.** 29.64 cm^2 **10.** 88 m^2 **11.** $145\frac{5}{9}$ in^2
12. 840 ft^2 **13.** 8 m **14.** $\frac{14}{11}$ in., or $1\frac{3}{11}$ in. **15.** 14 ft
16. 20 cm **17.** 50.24 m **18.** 8 in. **19.** 200.96 m^2
20. $5\frac{1}{11}$ in^2 **21.** 1038.555 ft^2 **22.** 93.6 m^3
23. 193.2 cm^3 **24.** 31,400 ft^3 **25.** $33.49\overline{3}$ cm^3
26. 4.71 in^3 **27.** 942 cm^3 **28.** 26.28 ft^2; 20.28 ft
29. 54° **30.** 180° **31.** 140° **32.** 90° **33.** Acute
34. Straight **35.** Obtuse **36.** Right **37.** 49°
38. 136° **39.** 60° **40.** Scalene **41.** Right **42.** 8
43. 9.110 **44.** $c = \sqrt{850}$; $c \approx 29.155$
45. $b = \sqrt{51}$; $b \approx 7.141$ **46.** $c = \sqrt{89}$ ft; $c \approx 9.434$ ft
47. $a = \sqrt{76}$ cm; $a \approx 8.718$ cm **48.** About 28.8 ft

49. About 44.7 ft **50.** About 85.9 ft
51. $\mathbf{D_W}$ See the volume formulas listed at the beginning of
the Summary and Review Exercises for Chapter 9.
52. $\mathbf{D_W}$ Volume of two spheres, each with radius r:
$2\left(\frac{4}{3}\pi r^3\right) = \frac{8}{3}\pi r^3$; volume of one sphere with radius $2r$:
$\frac{4}{3}\pi(2r)^3 = \frac{32}{3}\pi r^3$. The volume of the sphere with radius $2r$
is four times the volume of the two spheres, each with
radius r: $\frac{32}{3}\pi r^3 = 4 \cdot \frac{8}{3}\pi r^3$. **53.** $39.04 **54.** 27
55. 22.09 **56.** 103.823 **57.** $\frac{1}{16}$ **58.** 13,200 **59.** 4
60. 45.68 **61.** 45,680 **62.** 100 ft^2 **63.** 7.83998704 m^2
64. 42.05915 cm^2

Test: Chapter 9, p. 605

1. [9.1a], [9.2a] 32.82 cm; 65.894 cm^2
2. [9.1a], [9.2a] $19\frac{1}{2}$ in.; $23\frac{49}{64}$ in^2 **3.** [9.2b] 25 cm^2
4. [9.2b] 12 m^2 **5.** [9.2b] 18 ft^2 **6.** [9.3a] $\frac{1}{4}$ in.
7. [9.3a] 9 cm **8.** [9.3b] $\frac{11}{14}$ in. **9.** [9.3c] 254.34 cm^2
10. [9.3d] 65.46 km; 103.815 km^2 **11.** [9.4a] 84 cm^3
12. [9.4e] 420 in^3 **13.** [9.4b] 1177.5 ft^3
14. [9.4c] $4186.\overline{6}$ yd^3 **15.** [9.4d] 113.04 cm^3
16. [9.5a] 90° **17.** [9.5a] 35° **18.** [9.5a] 180°
19. [9.5a] 113° **20.** [9.5b] Right **21.** [9.5b] Acute
22. [9.5b] Straight **23.** [9.5b] Obtuse **24.** [9.5e] 35°
25. [9.5d] Isosceles **26.** [9.5d] Obtuse
27. [9.5c] Complement: 25°; supplement: 115°
28. [9.6a] 15 **29.** [9.6b] 9.327 **30.** [9.6c] $c = 40$
31. [9.6c] $b = \sqrt{60}$; $b \approx 7.746$ **32.** [9.6c] $c = \sqrt{2}$;
$c \approx 1.414$ **33.** [9.6c] $b = \sqrt{51}$; $b \approx 7.141$
34. [9.6d] About 15.8 m **35.** [6.7a] $680
36. [1.9b] 1000 **37.** [1.9b] $\frac{1}{16}$ **38.** [1.9b] 9.8596
39. [1.9b] 0.00001 **40.** [8.1a] 42 **41.** [8.1a] 250
42. [8.2a] 2300 **43.** [8.2a] 3400 **44.** [8.1a], [9.2a] 2 ft^2
45. [8.1a], [9.2b] 1.875 ft^2 **46.** [8.1a], [9.4a] 0.65 ft^3
47. [8.1a], [9.4d] 0.033 ft^3 **48.** [9.4b] 0.055 ft^3

Cumulative Review: Chapters 1–9, p. 608

1. [4.3b] 1,500,000 **2.** [8.1a], [8.3a] $437\frac{1}{3}$ yd; about 400 m
3. [7.3c] 236 in 1993 and 1995 **4.** [7.3c] 258 in 2000
5. [7.1a, b, c] Average: 243.375; median: 239.5; mode: 236
6. [7.1a] 249.5 **7.** [7.1a] 237.25; the average over the
years 1993–1996 is 12.25 lower than the average over the
years 1997–2000. **8.** [6.5b] About 8.4%
9. [8.5b] 350 mL **10.** [9.4e] 93,750 lb **11.** [3.5a] $4\frac{1}{6}$
12. [4.5c] 49.2 **13.** [4.2b] 87.52 **14.** [1.6b] 1234
15. [1.9d] 2 **16.** [1.9c] 1565 **17.** [4.1b] $\frac{1209}{1000}$
18. [6.2b] $\frac{17}{100}$ **19.** [3.3b] < **20.** [2.5c] = **21.** [8.4a] $\frac{3}{8}$
22. [8.6b] 59 **23.** [8.5a] 87 **24.** [8.6a] $\frac{3}{20}$ **25.** [8.7a] 27
26. [8.2a] 0.17 **27.** [8.4c] 2.437 **28.** [8.3a] 8811.846
29. [8.5a] 9.54 **30.** [5.3b] $14\frac{2}{5}$ **31.** [5.3b] $4\frac{2}{7}$
32. [4.4b] 113.4 **33.** [3.3c] $\frac{1}{8}$ **34.** [9.1a], [9.2b] 380 cm;
5500 cm^2 **35.** [9.1a], [9.2b] 32.3 ft; 56.55 ft^2
36. [9.3a, c] 70 in.; 3850 in^2
37. [9.4c] $179,666\frac{2}{3}$ in^3 **38.** [7.1a] 99
39. [1.8a] 528 million pets **40.** [6.7a] $164

41. [6.7b] $57,381.45   **42.** [9.6d] 17 m   **43.** [6.6a] 6%
44. [3.5c] $2\frac{1}{8}$ yd   **45.** [4.7a] $21.82
46. [5.2b] The 8-qt box   **47.** [2.6b] $\frac{7}{20}$ km   **48.** [9.5e] 30°
49. [9.5d] Isosceles   **50.** [9.5d] Obtuse
51. [9.3c] Area of a circle of radius 4 ft: $16 \cdot \pi$ ft^2;
[9.2a] Area of a square of side 4 ft: 16 ft^2;
[9.3b] Circumference of a circle of radius 4 ft: $8 \cdot \pi$ ft;
[9.4d] Volume of a cone with radius of base 4 ft and height
8 ft: $\frac{128}{3} \cdot \pi$ ft^3; [9.2a] Area of a rectangle of length 8 ft and
width 4 ft: 32 ft^2; [9.2b] Area of a triangle with base 4 ft and
height 8 ft: 16 ft^2; [9.4c] Volume of a sphere of radius 4 ft:
$\frac{4^4}{3} \cdot \pi$ ft^3; [9.4b] Volume of a right circular cylinder with
radius of base 4 ft and height 8 ft: $128 \cdot \pi$ ft^3; [9.1a] Perimeter
of a square of side 4 ft: 16 ft; [9.1a] Perimeter of a rectangle
of length 8 ft and width 4 ft: 24 ft   **52.** [8.1a], [9.4e] 272 ft^3
53. [8.1a], [9.4b] 94,200 ft^3 **54.** [8.1a], [9.4a] 1.342 ft^3
55. [9.4b, c] 1204.260429 ft^3; about 22,079,692.8 ft, or
4181.76 mi

Chapter 10

Pretest: Chapter 10, p. 614

1. >   **2.** >   **3.** >   **4.** <   **5.** −0.625   **6.** −0.$\overline{6}$
7. −0.$\overline{90}$ **8.** 12   **9.** 2.3   **10.** 0   **11.** −5.4   **12.** $\frac{2}{3}$
13. −17   **14.** 38.6   **15.** −$\frac{17}{15}$ **16.** −5   **17.** 63
18. −$\frac{5}{12}$ **19.** −98   **20.** 8   **21.** 24   **22.** 26
23. 50°C higher

Margin Exercises, Section 10.1, pp. 616–620

1. 8; −5   **2.** 950,000,000   **3.** The integer −6
corresponds to the decrease in the stock value.
4. −10; 148   **5.** −137; 289
6.
7.
8.
9. −0.375   **10.** −0.$\overline{54}$ **11.** 1.$\overline{3}$ **12.** <   **13.** <
14. >   **15.** >   **16.** >   **17.** <   **18.** <   **19.** >
20. 8   **21.** 0   **22.** 9   **23.** $\frac{2}{3}$ **24.** 5.6

Exercise Set 10.1, p. 621

1. −1286; 14,410   **3.** 24; −2   **5.** −5,600,000,000,000
7. −34; 15
9.
11.

13. −0.875   **15.** 0.8$\overline{3}$ **17.** 1.1$\overline{6}$ **19.** 0.$\overline{6}$ **21.** −0.5
23. −8.28   **25.** >   **27.** <   **29.** <   **31.** <   **33.** >
35. <   **37.** >   **39.** <   **41.** <   **43.** <   **45.** 3
47. 18   **49.** 325   **51.** 3.625   **53.** $\frac{2}{3}$ **55.** $\frac{0}{4}$, or 0
57. D$_W$ **59.** $2 \cdot 3 \cdot 3 \cdot 3$, or $2 \cdot 3^3$
60. $2 \cdot 2 \cdot 2 \cdot 2 \cdot 2 \cdot 2 \cdot 3$, or $2^6 \cdot 3$ **61.** $2 \cdot 3 \cdot 17$
62. $2 \cdot 2 \cdot 5 \cdot 13$, or $2^2 \cdot 5 \cdot 13$
63. $2 \cdot 2 \cdot 2 \cdot 2 \cdot 2 \cdot 3 \cdot 3 \cdot 3$, or $2^5 \cdot 3^3$
64. $2 \cdot 2 \cdot 3 \cdot 3 \cdot 13$, or $2^2 \cdot 3^2 \cdot 13$ **65.** 18   **66.** 72
67. 96   **68.** 72   **69.** 1344   **70.** 252   **71.** >   **73.** =
75. $-100, -8\frac{7}{8}, -8\frac{5}{8}, -\frac{67}{8}, -5, 0, 1^7, |3|, \frac{14}{4}, 4, |-6|, 7$

Margin Exercises, Section 10.2, pp. 623–626

1. −3   **2.** −3   **3.** −5   **4.** 4   **5.** 0   **6.** −2   **7.** −11
8. −12   **9.** 2   **10.** −4   **11.** −2   **12.** 0   **13.** −22
14. 3   **15.** 0.53   **16.** 2.3   **17.** −7.7   **18.** −6.2
19. −$\frac{2}{9}$ **20.** −$\frac{19}{20}$ **21.** −58   **22.** −56   **23.** −14
24. −12   **25.** 4   **26.** −8.7   **27.** 7.74   **28.** $\frac{8}{9}$ **29.** 0
30. −12   **31.** −14; 14   **32.** −1; 1   **33.** 19; −19
34. 1.6; −1.6   **35.** −$\frac{2}{3}$; $\frac{2}{3}$ **36.** $\frac{9}{8}$; −$\frac{9}{8}$ **37.** 4   **38.** 13.4
39. 0   **40.** −$\frac{1}{4}$

Calculator Corner, p. 626

1. 2   **2.** 13   **3.** −15   **4.** −8   **5.** −3.3   **6.** −13.4

Exercise Set 10.2, p. 627

1. −7   **3.** −4   **5.** 0   **7.** −8   **9.** −7   **11.** −27
13. 0   **15.** −42   **17.** 0   **19.** 0   **21.** 3   **23.** −9
25. 7   **27.** 0   **29.** 45   **31.** −1.8   **33.** −8.1   **35.** −$\frac{1}{5}$
37. −$\frac{8}{7}$ **39.** −$\frac{3}{8}$ **41.** −$\frac{29}{35}$ **43.** −$\frac{11}{15}$ **45.** −6.3
47. $\frac{7}{16}$ **49.** 39   **51.** 50   **53.** −1093   **55.** −24
57. 26.9   **59.** −9   **61.** $\frac{14}{3}$ **63.** −65   **65.** $\frac{5}{3}$ **67.** 14
69. −10   **71.** D$_W$ **73.** 357.5 ft^2 **74.** 45.51 ft^2
75. 54,756 mi^2 **76.** 1.4196 mm^2 **77.** 66.97 ft^2
78. 7976.1 m^2 **79.** All positive   **81.** −640,979
83. Negative

Margin Exercises, Section 10.3, pp. 629–631

1. −10   **2.** 3   **3.** −5   **4.** −1   **5.** 2   **6.** −4   **7.** −2
8. −11   **9.** 4   **10.** −2   **11.** −6   **12.** −16   **13.** 7.1
14. 3   **15.** 0   **16.** $\frac{3}{2}$ **17.** Three minus eleven; −8
18. Twelve minus five; 7   **19.** Negative twelve minus
negative nine; −3   **20.** Negative twelve point four minus
ten point nine; −23.3   **21.** Negative four-fifths minus
negative four-fifths; 0   **22.** −9   **23.** 17   **24.** 12.7
25. 214°F

Exercise Set 10.3, p. 632

1. −4   **3.** −7   **5.** −6   **7.** 0   **9.** −4   **11.** −7
13. −6   **15.** 0   **17.** 11   **19.** −14   **21.** 5   **23.** −7
25. −5   **27.** −3   **29.** −23   **31.** −68   **33.** −73

35. 116 **37.** −2.8 **39.** −$\frac{1}{4}$ **41.** $\frac{1}{12}$ **43.** −$\frac{17}{12}$ **45.** $\frac{1}{8}$
47. 19.9 **49.** −9 **51.** −0.01 **53.** −2.7 **55.** −3.53
57. −$\frac{1}{2}$ **59.** $\frac{6}{7}$ **61.** −$\frac{41}{30}$ **63.** −$\frac{1}{156}$ **65.** 37
67. −62 **69.** 6 **71.** 107 **73.** 219 **75.** Down 3 ft
77. −3° **79.** 30,314 ft **81.** −$85 **83.** 100°F
85. $\mathbf{D_W}$ **87.** 96.6 cm^2 **88.** 2 · 3 · 5 · 5 · 5, or 2 · 3 · 5^3
89. 108 **90.** 125.44 km^2 **91.** 64 **92.** 125
93. 8 cans **94.** 288 oz **95.** True **97.** True
99. True **101.** True

Margin Exercises, Section 10.4, pp. 635–636

1. 20; 10; 0; −10; −20; −30 **2.** −18 **3.** −100 **4.** −80
5. −$\frac{5}{9}$ **6.** −30.033 **7.** −$\frac{7}{10}$ **8.** −10; 0; 10; 20; 30
9. 12 **10.** 32 **11.** 35 **12.** $\frac{20}{63}$ **13.** $\frac{2}{3}$ **14.** 13.455
15. −30 **16.** 30 **17.** −32 **18.** −$\frac{8}{3}$ **19.** −30
20. −30.75 **21.** −$\frac{5}{3}$ **22.** 120 **23.** −120 **24.** 6

Exercise Set 10.4, p. 637

1. −16 **3.** −24 **5.** −72 **7.** 16 **9.** 42 **11.** −120
13. −238 **15.** 1200 **17.** 98 **19.** −12.4 **21.** 24
23. 21.7 **25.** −$\frac{2}{5}$ **27.** $\frac{1}{12}$ **29.** −17.01 **31.** −$\frac{5}{12}$
33. 420 **35.** $\frac{2}{7}$ **37.** −60 **39.** 150 **41.** −$\frac{2}{45}$
43. 1911 **45.** 50.4 **47.** $\frac{10}{189}$ **49.** −960 **51.** 17.64
53. −$\frac{5}{784}$ **55.** −756 **57.** −720 **59.** −30,240
61. $\mathbf{D_W}$ **63.** 2 · 2 · 2 · 2 · 2 · 2 · 2 · 2 · 2 · 3 · 3, or 2^9 · 3^2
64. 180 **65.** 33$\frac{1}{3}$%, or 33.$\overline{3}$% **66.** 262.44 **67.** 72
68. 40% **69.** 40 ft^2 **70.** 240 cartons **71.** −$23
73. **(a)** One must be negative and one must be positive.
(b) Either or both must be zero. **(c)** Both must be negative
or both must be positive.

Margin Exercises, Section 10.5, pp. 639–644

1. −2 **2.** 5 **3.** −3 **4.** 9 **5.** −6 **6.** −$\frac{30}{7}$ **7.** Not
defined **8.** 0 **9.** $\frac{3}{2}$ **10.** −$\frac{4}{5}$ **11.** −$\frac{1}{3}$ **12.** −5
13. $\frac{1}{5.78}$ **14.** −$\frac{7}{2}$
15.

Number	Opposite	Reciprocal
$\frac{2}{3}$	−$\frac{2}{3}$	$\frac{3}{2}$
−$\frac{5}{4}$	$\frac{5}{4}$	−$\frac{4}{5}$
0	0	Undefined
1	−1	1
−4.5	4.5	−$\frac{1}{4.5}$

16. $\frac{4}{5}$ · $\left(-\frac{5}{3}\right)$ **17.** 5 · $\left(-\frac{1}{8}\right)$ **18.** −10 · $\left(\frac{1}{7}\right)$ **19.** −$\frac{2}{3}$ · $\frac{7}{4}$
20. −5 · $\left(\frac{1}{7}\right)$ **21.** −$\frac{20}{21}$ **22.** −$\frac{12}{5}$ **23.** $\frac{16}{7}$ **24.** −7
25. −3.4°F per minute **26.** −1237 **27.** 8 **28.** 381
29. −12

Exercise Set 10.5, p. 645

1. −6 **3.** −13 **5.** −2 **7.** 4 **9.** −8 **11.** 2
13. −12 **15.** −8 **17.** Not defined **19.** −9 **21.** −$\frac{7}{15}$
23. $\frac{1}{13}$ **25.** −$\frac{9}{8}$ **27.** $\frac{5}{3}$ **29.** $\frac{9}{14}$ **31.** $\frac{9}{64}$ **33.** −2
35. $\frac{11}{13}$ **37.** −16.2 **39.** Not defined **41.** −54°C
43. $12.71 **45.** 32 m below sea level **47.** −7.4%
49. 42.4% **51.** −7 **53.** −7 **55.** −334 **57.** 14
59. 1880 **61.** 12 **63.** 8 **65.** −86 **67.** 37 **69.** −1
71. −10 **73.** −67 **75.** −7988 **77.** −3000 **79.** 60
81. 1 **83.** 10 **85.** −$\frac{13}{45}$ **87.** −$\frac{4}{3}$ **89.** $\mathbf{D_W}$
91. 2 · 2 · 2 · 2 · 2 · 2 · 3 · 5, or 2^6 · 3 · 5
92. 5 · 5 · 41, or 5^2 · 41 **93.** 665 **94.** 320 **95.** 1710
96. 4%
97.

bar graph: Sound (y-axis) vs Loudness (in decibels) (x-axis: 0 to 140). Categories: Light whisper, Normal conversation, Noisy office, Thunder, Moving car, Chain saw, Jet takeoff

99. −159 **101.** Negative **103.** Negative

Summary and Review: Chapter 10, p. 648

1. −45; 72 **2.** 38 **3.** 7.3 **4.** $\frac{5}{2}$ **5.** −0.2 **6.** −1.25
7. −0.8$\overline{3}$ **8.** −0.41$\overline{6}$ **9.** −0.$\overline{27}$
10.

number line with −2.5 marked, from −6 to 6

11.

number line with $\frac{8}{9}$ marked, from −6 to 6

12. < **13.** > **14.** > **15.** < **16.** −3.8 **17.** $\frac{3}{4}$
18. 34 **19.** 5 **20.** $\frac{8}{3}$ **21.** −$\frac{1}{7}$ **22.** −10 **23.** −3
24. −$\frac{7}{12}$ **25.** −4 **26.** −5 **27.** 4 **28.** −$\frac{7}{5}$
29. −7.9 **30.** 54 **31.** −9.18 **32.** −$\frac{2}{7}$ **33.** −210
34. −7 **35.** −3 **36.** $\frac{3}{4}$ **37.** 40.4 **38.** −62 **39.** −5
40. 2 **41.** −180 **42.** 8-yd gain **43.** −$130
44. $4.64 per share **45.** $18.95
46. $\mathbf{D_W}$ Yes; the numbers 1 and −1 are their own
reciprocals: 1 · 1 = 1 and −1(−1) = 1.
47. $\mathbf{D_W}$ We know that $a + (-a) = 0$, so the opposite of $-a$
is a. That is, $-(-a) = a$.
48. 210 cm^2 **49.** 270
50. 2 · 2 · 2 · 3 · 3 · 3 · 3, or 2^3 · 3^4 **51.** 36%
52. 403 and 397 **53.** 12 and −7
54. **(a)** −7 + (−6) + (−5) + (−4) + (−3) + (−2) +
(−1) + 0 + 1 + 2 + 3 + 4 + 5 + 6 + 7 + 8 = 8; **(b)** 0
55. −1; Consider reciprocals and pairs of products of
negative numbers.
56. −$\frac{5}{8}$ **57.** −2.1

Test: Chapter 10, p. 651

1. [10.1d] $<$ **2.** [10.1d] $>$ **3.** [10.1d] $>$ **4.** [10.1d] $<$
5. [10.1c] -0.125 **6.** [10.1c] $-0.\overline{4}$ **7.** [10.1c] $-0.\overline{18}$
8. [10.1e] 7 **9.** [10.1e] $\frac{9}{4}$ **10.** [10.1e] -2.7
11. [10.2b] $-\frac{2}{3}$ **12.** [10.2b] 1.4 **13.** [10.2b] 8
14. [10.5b] $-\frac{1}{2}$ **15.** [10.5b] $\frac{7}{4}$ **16.** [10.3a] 7.8
17. [10.2a] -8 **18.** [10.2a] $\frac{7}{40}$ **19.** [10.3a] 10
20. [10.3a] -2.5 **21.** [10.3a] $\frac{7}{8}$ **22.** [10.4a] -48
23. [10.4a] $\frac{3}{16}$ **24.** [10.5a] -9 **25.** [10.5c] $\frac{3}{4}$
26. [10.5c] -9.728 **27.** [10.5e] 109 **28.** [10.3b] 14°F
higher **29.** [10.3b] Up 15 pts **30.** [10.5d] 16,080
31. [10.5d] About 0.9429°C each minute
32. [9.2a] 55.8 ft^2 **33.** [6.3b], [6.4b] 48%
34. [2.1d] $2 \cdot 2 \cdot 2 \cdot 5 \cdot 7$, or $2^3 \cdot 5 \cdot 7$ **35.** [3.1a] 240
36. [10.1e], [10.3a] 15 **37.** [10.3b] 2385 m
38. (a) [10.3a] $-4, -9, -15$; **(b)** [10.3a] $-2, -6, -10$;
(c) [10.3a] $-18, -24, -31$; **(d)** [10.5c] $-0.25, 0.125, -0.0625$

Cumulative Review: Chapters 1–10, p. 653

1. [6.5a] 3.52 oz **2.** [4.3b] 2,600,000,000
3. [6.5b] 19.2% **4.** [9.4e] 64 ft^3 **5.** [6.1b] 0.263
6. [10.1c] $-0.\overline{45}$ **7.** [8.4b] 834 **8.** [8.7b] 0.0275
9. [10.1e] 4.5 **10.** [10.3a] -11 **11.** [5.2a] 12.5 m/sec
12. [9.3a, b, c] 35 mi; 220 mi; 3850 mi^2
13. [7.3b]

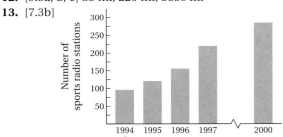

14. [7.3d]

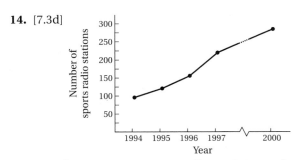

15. [7.1a, b, c] Average: 175.6; median 156; no mode
16. [6.5b] 82.7% **17.** [6.5b] 29.5% **18.** [9.6a] 15
19. [9.6b] 8.31 **20.** [10.4a] -10 **21.** [10.5a] 3
22. [10.2a] 8 **23.** [10.3a] 8 **24.** [4.3a] 0.01485
25. [10.3a] -3 **26.** [10.3a] $-\frac{3}{22}$ **27.** [10.4a] $-\frac{1}{24}$
28. [3.5b] $1\frac{5}{6}$ **29.** [10.5c] $-\frac{1}{4}$ **30.** [10.5e] -57.5
31. [1.5a] 9,640,500,000 **32.** [10.2a] 32.49
33. [10.5e] Not defined **34.** [6.3b], [6.4b] 87.5%
35. [6.3b], [6.4b] 32 **36.** [9.4e] 307.72 cm^3
37. [7.1a] 44.8$\overline{3}$° **38.** [6.5a] 78 students

39. [9.2c] 6902.5 m^2 **40.** [3.7b] $10\frac{1}{2}$ **41.** [3.7b] $11\frac{1}{2}$
42. [3.7b] 130 **43.** [10.5b] (c) **44.** [10.5e] (e)
45. [8.4b] (c) **46.** [9.3c] (d) **47.** [3.5a] Off duty: 7 hr;
sleeper berth: $2\frac{3}{4}$ hr; driving: $7\frac{3}{4}$ hr; on duty (not driving):
$6\frac{1}{2}$ hr **48.** [9.4e] 60 cm^3

Chapter 11

Pretest: Chapter 11, p. 658

1. $\frac{5}{16}$ **2.** 78% x, or $0.78x$ **3.** $9z - 18$
4. $-4a - 2b + 10c$ **5.** $4(x - 3)$ **6.** $3(2y - 3z - 6)$
7. $-3x$ **8.** $2x + 2y + 18$ **9.** -7 **10.** -1 **11.** 2
12. 3 **13.** Width: 34 m; length: 39 m **14.** \$780

Margin Exercises, Section 11.1, pp. 659–665

1. 64 **2.** 28 **3.** 60; -27.2 **4.** 192 ft^2 **5.** -25
6. 16 **7.**

Value	$1 \cdot x$	x
$x = 3$	3	3
$x = -6$	-6	-6
$x = 4.8$	4.8	4.8

8.

Value	$2x$	$5x$
$x = 2$	4	10
$x = -6$	-12	-30
$x = 4.8$	9.6	24

9.

Value	$5(x + y)$	$5x + 5y$
$x = 2, y = 8$	50	50
$x = -3, y = 4$	5	5
$x = -10, y = 5$	-25	-25

10. 90; 90 **11.** 64; 64 **12.** 14; 14 **13.** 30; 30
14. 12; 12 **15.** $5x$; $-4y$; 3 **16.** $-4y$; $-2x$; $3z$
17. $3x - 15$ **18.** $5x + 5$ **19.** $\frac{5}{4}x - \frac{5}{4}y + 5$
20. $-2x + 6$ **21.** $-5x + 10y - 20z$ **22.** $6(z - 2)$
23. $3(x - 2y + 3)$ **24.** $2(8a - 18b + 21)$
25. $-4(3x - 8y + 4z)$ **26.** $3x$ **27.** $6x$ **28.** $-8x$
29. $0.59x$ **30.** $3x + 3y$ **31.** $-4x - 5y - 7$
32. $\frac{1}{10}x + \frac{7}{9}y - \frac{2}{3}$

Calculator Corner, p. 660

1. 51 **2.** 98 **3.** -6 **4.** -19 **5.** 9 **6.** 12
7. 52; 52 **8.** 156; 156 **9.** 228; 228 **10.** 84; 84

Exercise Set 11.1, p. 666

1. 42 **3.** 3 **5.** −1 **7.** 6 **9.** −20 **11.** 240; 240
13. 160; 160 **15.** $2b + 10$ **17.** $7 − 7t$ **19.** $30x + 12$
21. $7x + 28 + 42y$ **23.** $−7y + 14$ **25.** $45x + 54y − 72$
27. $\frac{3}{4}x − \frac{9}{4}y − \frac{3}{2}z$ **29.** $−3.72x + 9.92y − 3.41$
31. $2(x + 2)$ **33.** $5(6 + y)$ **35.** $7(2x + 3y)$
37. $5(x + 2 + 3y)$ **39.** $8(x − 3)$ **41.** $4(8 − y)$
43. $2(4x + 5y − 11)$ **45.** $6(−3x − 2y + 1)$, or
$−6(3x + 2y − 1)$ **47.** $19a$ **49.** $9a$ **51.** $8x + 9z$
53. $−19a + 88$ **55.** $4t + 6y − 4$ **57.** $8x$ **59.** $5n$
61. $−16y$ **63.** $17a − 12b − 1$ **65.** $4x + 2y$
67. $\frac{39}{20}x + \frac{1}{2}y + 12$ **69.** $0.8x + 0.5y$ **71.** $\mathbf{D_W}$
73. 30 yd; 94.2 yd; 706.5 yd^2 **74.** 16.4 m; 51.496 m;
211.1336 m^2 **75.** 19 mi; 59.66 mi; 283.385 mi^2
76. 4800 cm; 15,072 cm; 18,086,400 cm^2 **77.** 10 mm;
62.8 mm; 314 mm^2 **78.** 132 km; 828.96 km; 54,711.36 km^2
79. 2.3 ft; 14.444 ft; 16.6106 ft^2 **80.** 5.15 m; 32.342 m;
83.28065 m^2 **81.** $q(1 + r + rs + rst)$

Margin Exercises, Section 11.2, pp. 668–669

1. (a) $2 + \boxed{−2} = 0$; (b) 9 **2.** −5 **3.** 13.2 **4.** −6.5

Exercise Set 11.2, p. 670

1. 4 **3.** −20 **5.** −14 **7.** 7 **9.** −26 **11.** −18
13. 15 **15.** −14 **17.** 2 **19.** 20 **21.** −6 **23.** $\frac{7}{3}$
25. $−\frac{7}{4}$ **27.** $\frac{41}{24}$ **29.** $−\frac{1}{20}$ **31.** 5.1 **33.** 12.4 **35.** −5
37. $1\frac{5}{6}$ **39.** $−\frac{10}{21}$ **41.** $\mathbf{D_W}$ **43.** −11 **44.** $−\frac{1}{24}$
45. −34.1 **46.** −1.7 **47.** 5 **48.** $−\frac{31}{24}$ **49.** 5.5
50. 8.1 **51.** 24 **52.** $−\frac{5}{12}$ **53.** 283.14 **54.** −15.68
55. 8 **56.** $−\frac{16}{15}$ **57.** −14.3 **58.** −4.9 **59.** 342.246
61. $−\frac{26}{15}$ **63.** −10 **65.** $−\frac{5}{17}$

Margin Exercises, Section 11.3, pp. 672–674

1. 15 **2.** $−\frac{7}{4}$ **3.** −18 **4.** 10 **5.** $−\frac{4}{5}$ **6.** 7800
7. −3 **8.** 28

Exercise Set 11.3, p. 675

1. 6 **3.** 9 **5.** 12 **7.** −40 **9.** 5 **11.** −7 **13.** −6
15. 6 **17.** −63 **19.** 36 **21.** −21 **23.** $−\frac{3}{5}$ **25.** $−\frac{3}{2}$
27. $\frac{9}{2}$ **29.** 7 **31.** −7 **33.** 8 **35.** 15.9 **37.** −50
39. −14 **41.** $\mathbf{D_W}$ **43.** 62.8 ft; 20 ft; 314 ft^2
44. 75.36 cm; 12 cm; 452.16 cm^2 **45.** 8000 ft^3
46. 31.2 cm^3 **47.** 53.55 cm^2 **48.** 64 mm^2 **49.** 68 in^2
50. 38.25 m^2 **51.** −8655 **53.** No solution
55. No solution

Margin Exercises, Section 11.4, pp. 677–683

1. 5 **2.** 4 **3.** 4 **4.** 39 **5.** $−\frac{3}{2}$ **6.** −4.3 **7.** −3
8. 800 **9.** 1 **10.** 2 **11.** 2 **12.** $\frac{17}{2}$ **13.** $\frac{8}{3}$
14. $−\frac{43}{10}$, or −4.3 **15.** 2 **16.** 3 **17.** −2 **18.** $−\frac{1}{2}$

Calculator Corner, p. 683

1. Left to the student **2.** Left to the student

Exercise Set 11.4, p. 684

1. 5 **3.** 8 **5.** 10 **7.** 14 **9.** −8 **11.** −8 **13.** −7
15. 15 **17.** 6 **19.** 4 **21.** 6 **23.** −3 **25.** 1
27. −20 **29.** 6 **31.** 7 **33.** 2 **35.** 5 **37.** 2
39. 10 **41.** 4 **43.** 0 **45.** −1 **47.** $−\frac{4}{3}$ **49.** $\frac{2}{5}$
51. −2 **53.** −4 **55.** $\frac{4}{5}$ **57.** $−\frac{28}{27}$ **59.** 6 **61.** 2
63. 6 **65.** 8 **67.** 1 **69.** 17 **71.** $−\frac{5}{3}$ **73.** −3
75. 2 **77.** $−\frac{51}{31}$ **79.** 2 **81.** $\mathbf{D_W}$ **83.** 6.5 **84.** 6.5
85. −6.5 **86.** −6.5 **87.** $7(x − 3 − 2y)$
88. $25(a − 25b + 3)$ **89.** $14(3t + m − 4)$
90. $16(a − 4b + 14 − 2q)$ **91.** −8 **92.** < **93.** 2
95. −2 **97.** 8 **99.** 2 cm

Margin Exercises, Section 11.5, pp. 689–696

1. $x − 12$ **2.** $y + 12$, or $12 + y$ **3.** $m − 4$
4. $\frac{1}{2} \cdot p$ or $\frac{p}{2}$ **5.** $6 + 8x$, or $8x + 6$ **6.** $a − b$
7. $59\%x$, or $0.59x$ **8.** $xy − 200$ **9.** $p + q$ **10.** $62\frac{2}{3}$ mi
11. 16 in.; 4 in.; 32 in. **12.** 587.8 mi **13.** 30°, 90°, 60°
14. Width: 9 ft; length: 12 ft **15.** $8400

Exercise Set 11.5, p. 697

1. $2x − 3$ **3.** $97\%y$, or $0.97y$ **5.** $5x + 4$, or $4 + 5x$
7. $\frac{1}{3}x$, or $240 − x$ **9.** 32 **11.** 305 ft **13.** $3.67
15. 57 **17.** −12 **19.** $699\frac{1}{3}$ mi **21.** First: 30 m;
second: 90 m; third: 360 m **23.** Length: 94 ft; width: 50 ft
25. Length: 265 ft; width: 165 ft; area: 43,725 ft^2
27. About 412.6 mi **29.** First: 22.5°; second: 90°;
third: 67.5° **31.** $350 **33.** $852.94 **35.** 12 mi
37. $36 **39.** Length: 48 ft; width: 14 ft **41.** $\mathbf{D_W}$
43. $−\frac{47}{40}$ **44.** $−\frac{17}{40}$ **45.** $−\frac{3}{10}$ **46.** $−\frac{32}{15}$ **47.** 1.6
48. 409.6 **49.** −9.6 **50.** −41.6
51. $1863 = 1776 + (4s + 7)$ **53.** Length: 12 cm;
width: 9 cm **55.** Half dollars: 5; quarters: 10; dimes: 20;
nickels: 60

Summary and Review: Chapter 11, p. 702

1. 4 **2.** $15x − 35$ **3.** $−8x + 10$ **4.** $4x + 15$
5. $−24 + 48x$ **6.** $2(x − 7)$ **7.** $6(x − 1)$ **8.** $5(x + 2)$
9. $3(4 − x)$ **10.** $7a − 3b$ **11.** $−2x + 5y$ **12.** $5x − y$
13. $−a + 8b$ **14.** −22 **15.** 7 **16.** −192 **17.** 1
18. $−\frac{7}{3}$ **19.** 25 **20.** $\frac{1}{2}$ **21.** $−\frac{15}{64}$ **22.** 9.99 **23.** −8
24. −5 **25.** $−\frac{1}{3}$ **26.** 4 **27.** 3 **28.** 4 **29.** 16
30. 6 **31.** −3 **32.** −2 **33.** 4 **34.** $19\%x$, or $0.19x$
35. Length: 365 mi; width: 275 mi **36.** $2117
37. 27 appliances **38.** 35°, 85°, 60° **39.** $220
40. $26,087 **41.** $138.95 **42.** Amazon: 6437 km;
Nile: 6671 km **43.** $791 **44.** Width: 11 cm;
length: 17 cm **45.** $\mathbf{D_W}$ The distributive laws are used to
multiply, factor, and collect like terms in this chapter.

46. DW **(a)** $4 - 3x = 9$
$$3x = 9 \quad (1)$$
$$x = 3 \quad (2)$$

1. 4 was subtracted on the left side but not on the right side. Also, the minus sign preceding $3x$ has been dropped.
2. This step would give the correct result if the preceding step were correct.

The correct steps are
$$4 - 3x = 9$$
$$-3x = 5 \quad (1)$$
$$x = -\tfrac{5}{3}. \quad (2)$$

(b) $2(x - 5) = 7$
$$2x - 5 = 7 \quad (1)$$
$$2x = 12 \quad (2)$$
$$x = 6 \quad (3)$$

1. When a distributive law was used to remove parentheses, x was multiplied by 2 but 5 was not.
2. and **3.** These steps would give the correct result if the preceding step were correct.

The correct steps are
$$2(x - 5) = 7$$
$$2x - 10 = 7 \quad (1)$$
$$2x = 17 \quad (2)$$
$$x = \tfrac{17}{2}. \quad (3)$$

47. 40 ft; 125.6 ft; 1256 ft² **48.** 1702 cm³ **49.** −45
50. −60 **51.** 23, −23 **52.** 20, −20

Test: Chapter 11, p. 705

1. [11.1a] 6 **2.** [11.1b] $18 - 3x$ **3.** [11.1b] $-5y + 5$
4. [11.1c] $2(6 - 11x)$ **5.** [11.1c] $7(x + 3 + 2y)$
6. [11.1d] $-5x - y$ **7.** [11.1d] $4a + 5b$ **8.** [11.2a] 8
9. [11.2a] 26 **10.** [11.3a] −6 **11.** [11.3a] 49
12. [11.4b] −12 **13.** [11.4a] 2 **14.** [11.4a] −8
15. [11.2a] $-\tfrac{7}{20}$ **16.** [11.4b] 2.5 **17.** [11.4c] 7
18. [11.4c] $\tfrac{5}{3}$ **19.** [11.5a] $x - 9$ **20.** [11.5b] Width: 7 cm; length: 11 cm **21.** [11.5b] About $141.5 billion
22. [11.5b] 3 m, 5 m **23.** [11.5b] $880 **24.** [11.5b] 6
25. [11.5b] 25.625° **26.** [10.4a] 180 **27.** [10.2a] $-\tfrac{2}{9}$
28. [9.3a, b, c] 140 yd; 439.6 yd; 15,386 yd²
29. [9.4a] 1320 ft³ **30.** [10.1e], [11.4b] 15, −15
31. [11.5b] 60 tickets

Cumulative Review/Final Examination: Chapters 1–11, p. 707

1. [5.2b] Unit prices: 7.647 ¢/oz, 7.133 ¢/oz, 7.016 ¢/oz, 5.828 ¢/oz, 4.552 ¢/oz; brand E has the lowest unit price.
2. [4.1a] Seven and four hundred sixty-three thousandths
3. [9.3b], [9.4c] **(a)** 4400 mi; **(b)** about 1,437,333,333 mi³
4. [6.5a] $13,444 **5.** [6.5a] 15% **6.** [6.5a] $2151.04
7. [6.5a] 30% **8.** [6.5a] $537.76 **9.** [1.1a] 7
10. [1.1b] 7 thousands + 4 hundreds + 5 ones

11. [1.2a] 1012 **12.** [1.2a] 21,085 **13.** [3.2a] $\tfrac{5}{26}$
14. [3.5a] $5\tfrac{7}{9}$ **15.** [4.2a] 493.971 **16.** [4.2a] 802.876
17. [1.3b] 152 **18.** [1.3b] 674 **19.** [3.3a] $\tfrac{5}{24}$
20. [3.5b] $2\tfrac{17}{24}$ **21.** [4.2b] 19.9973 **22.** [4.2b] 34.241
23. [1.5a] 4752 **24.** [1.5a] 266,287 **25.** [3.6a] $4\tfrac{1}{12}$
26. [2.6a] $\tfrac{6}{5}$ **27.** [2.4a] 10 **28.** [4.3a] 259.084
29. [1.6b] 573 **30.** [1.6b] 56 R 10 **31.** [3.4b] $56\tfrac{5}{17}$
32. [2.7b] $\tfrac{3}{2}$ **33.** [3.6b] $\tfrac{7}{90}$ **34.** [4.4a] 39 **35.** [1.4a] 68,000
36. [4.1d] 0.428 **37.** [4.5b] 21.84 **38.** [2.2a] Yes
39. [2.1a] 1, 3, 5, 15 **40.** [3.1a] 800 **41.** [2.5b] $\tfrac{7}{10}$
42. [2.5b] 55 **43.** [3.4a] $3\tfrac{3}{5}$ **44.** [2.5c] ≠ **45.** [3.3b] <
46. [4.1c] 1.001 **47.** [1.4c] > **48.** [2.3a] $\tfrac{3}{5}$
49. [4.1b] 0.037 **50.** [4.5a] 0.52 **51.** [4.5a] $0.\overline{8}$
52. [6.1b] 0.07 **53.** [4.1b] $\tfrac{463}{100}$ **54.** [3.4a] $\tfrac{29}{4}$ **55.** [6.2b] $\tfrac{2}{5}$
56. [6.2a] 85% **57.** [6.1b] 150% **58.** [1.7b] 555
59. [4.4b] 64 **60.** [2.7c] $\tfrac{5}{4}$ **61.** [5.3b] $76\tfrac{1}{2}$, or 76.5
62. [7.4b]

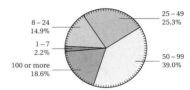

Number of Rounds of Golf Per Year

63. [7.3b]

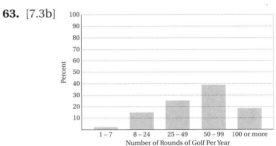

64. [9.5e] 118° **65.** [9.5d] Obtuse **66.** [1.8a] $675
67. [1.8a] 65 min **68.** [4.7a] $25.75 **69.** [4.7a] 485.9 mi
70. [1.8a] $8100 **71.** [1.8a] $423 **72.** [2.4c] $\tfrac{3}{10}$ km
73. [4.7a] $84.96 **74.** [5.4a] 13 gal **75.** [6.7a] $150
76. [6.6b] 7% **77.** [6.5b] 30,160 **78.** [7.1a, b, c] 28; 26; 18 **79.** [1.9b] 324 **80.** [1.9b] 400 **81.** [9.6a] 3
82. [9.6a] 11 **83.** [9.6b] 4.472 **84.** [8.1a] 12
85. [8.2a] 428 **86.** [8.6a] 72 **87.** [8.4b] 20
88. [8.4a] 80 **89.** [8.4b] 0.08 **90.** [8.5a] 8.19
91. [8.5a] 5 **92.** [9.6c] $c = \sqrt{50}$ ft; $c \approx 7.071$ ft
93. [9.1a], [9.2a] 25.6 m; 25.75 m² **94.** [9.2b] 25 in²
95. [9.2b] 128.65 yd² **96.** [9.2b] 61.6 cm²
97. [9.3a, b, c] 20.8 in.; 65.312 in.; 339.6224 in²
98. [9.4a] 52.9 m³ **99.** [9.4b] 803.84 ft³
100. [9.4d] $267.94\overline{6}$ cm³ **101.** [1.9c] 238 **102.** [1.9c] 172
103. [10.1e], [10.4a] 3 **104.** [10.2a] 14 **105.** [10.3a] $\tfrac{1}{3}$
106. [10.4a] 30 **107.** [10.4a] $-\tfrac{2}{7}$ **108.** [10.5a] −8
109. [11.4a] −5 **110.** [11.3a] 4 **111.** [11.4b] −8
112. [11.4c] $\tfrac{2}{3}$ **113.** [11.5a] $y + 17$, or $17 + y$
114. [11.5a] $38\%x$, or $0.38x$ **115.** [11.5b] $45
116. [11.5b] $1050 **117.** [11.5a] 50 m; 53 m; 40 m
118. [11.4b] 0 **119.** [11.4b] 2 **120.** [11.1d] (c)
121. [11.1c] (e) **122.** [10.5c] (d) **123.** [10.2a] (a)

Index

Photo Credits

8, The Image Bank 9, 10, Tony Stone Images 37, courtesy of Cytec Fiberite, Inc. 44, Camel 68, UPI Newspictures 78, EyeWire 93, Tony Stone Images 96, Dick Morton 115 (left), © 2001 Corbis; (right top), PhotoDisc; (right bottom), Karen Bittinger 117, Corbis 117 (Aguilera), Michael Germana, UPI Photo 123, © Ariel Skelley/Corbisstockmarket 130, John Coletti/Tony Stone Images 139, The Image Bank 149, AP/Wide World Photos 153, Greg Probst/Tony Stone Images 160, 178, PhotoDisc 185, FoodPix 191, PhotoDisc 196, Jim Lawler 200, Tony Stone Images 206, Karen Bittinger 210, Tony Stone Images 212, PhotoDisc 215, David Young-Wolff/PhotoEdit 223, Corbis 227, Karen Bittinger 234, © AFP/Robert Sullivan/CORBIS 240, Creative Labs 255, © Bettmann/Corbis 273, Daniel J. Cox, Tony Stone Images 277, © Chuck Savage/Corbis Stockmarket 278, Chris Thomaidis, Tony Stone Images 281, Gary Gay, The Image Bank 294 (left), Russ Einhorn, UPI Photo; (right), © Lee Celano, AFB/Corbis 297, © Janette Beckman/CORBIS 301, © Tom Nebbia/Corbis Stockmarket 302, Bob Gage, FPG International 308, © Todd Haimann/Corbis 312, Elis Years, FPG International 313, David Muench, Corbis 314, © AFP Photo, Bruce Weaver/Corbis 316, Chris Trotman, Corbis 317, Albert J. Copley, PhotoDisc 320 (top), EyeWire Collection; (right), AFP Photo/John G. Mabanglo, Corbis 321, Dick Morton 323 (top), Michael J. Miller, UPI Photo Service; (bottom), Jeremy Woodhouse, PhotoDisc 325, Annie Griffiths Belt, Corbis 326 (top), AFP Photo/Henry Ray Abrams © AFP Corbis; (bottom), © Reuters NewMedia, Inc./Corbis 334, Brian Spurlock 335, Buccina Studios, PhotoDisc 338, Nicholas Devore III/Photographers Aspen 340, Barry Rosenthal, FPG International 341, © Kevin Fleming, Corbis 342, Jim Zuckerman/Westlight 352, Wendy Kaveney 353 (left), Dick Morton; (right), Patrick Clark, PhotoDisc 354, Phil Schermeister, Corbis 358, Keith Brofsky, Photodisc 363, FPG International 364, Kenneth Chen/Envision 368 (top), Photodisc; (bottom), Hayden Roger Celestin, UPI Photo Service 371, 380, PhotoDisc 381, Corbis 382, 386, PhotoDisc 396, Charles Gupton, Stone/Getty Images 399, Jessie Cohen, National Zoo, UPI Photo Service 400, Christine Chew, UPI Photo Service 410, Ezio Petersen, UPI Photo Service 411, Christopher Bissell, Stone/Getty Images 421, PhotoLink 423, Walter Hodges/Tony Stone Images 424, PhotoDisc 431, Chad Baker/Ryan McVay, PhotoDisc 435, EyeWire Collection 443, Hulton Archive 444 (top), American Honda; (bottom), Penny Tweedie/Tony Stone Images 448, Tom Stratman 449, PBA 450, © Corbis Images 455, Bob Handelman, Tony Stone Images 473, G. K. and Vikki Hart, The Image Bank 487, Corbis 489, FPG International 507, Owen Franken/Stock, Boston 515, © Tom and Dee Ann McCarthy/Corbis Stock Market 518, David Young-Wolff/PhotoEdit 519, © Richard T. Nowitz/ CORBIS 525, © Catherine Karnow/CORBIS 530, © Galen Rowell/Corbis 535, © Jeremy Woodhouse/Photodisc 538, © Peter Beck/Corbis Stock Market 543, Debby Winchester/Karen Bittinger 569, © Nancy Dudley/Stock Boston 581, © 1988 Al Satterwhite 608, © National Geographic Image Collection 621, © AP/Wide World Photos 633, © Ruth Dixon/Stock Boston 650, © Comstock Images 698 (top), EyeWire Collection; (bottom), © Kevin Horan, Tony Stone Images 712, © ML Sinibaldi/The Stock Market

Index of Applications

Index of Study Tips